AF499990

System Modelling and Optimization

IFIP – The International Federation for Information Processing

IFIP was founded in 1960 under the auspices of UNESCO, following the First World Computer Congress held in Paris the previous year. An umbrella organization for societies working in information processing, IFIP's aim is two-fold: to support information processing within its member countries and to encourage technology transfer to developing nations. As its mission statement clearly states,

> IFIP's mission is to be the leading, truly international, apolitical organization which encourages and assists in the development, exploitation and application of information technology for the benefit of all people.

IFIP is a non-profitmaking organization, run almost solely by 2500 volunteers. It operates through a number of technical committees, which organize events and publications. IFIP's events range from an international congress to local seminars, but the most important are:

- the IFIP World Computer Congress, held every second year;
- open conferences;
- working conferences.

The flagship event is the IFIP World Computer Congress, at which both invited and contributed papers are presented. Contributed papers are rigorously refereed and the rejection rate is high.

As with the Congress, participation in the open conferences is open to all and papers may be invited or submitted. Again, submitted papers are stringently refereed.

The working conferences are structured differently. They are usually run by a working group and attendance is small and by invitation only. Their purpose is to create an atmosphere conducive to innovation and development. Refereeing is less rigorous and papers are subjected to extensive group discussion.

Publications arising from IFIP events vary. The papers presented at the IFIP World Computer Congress and at open conferences are published as conference proceedings, while the results of the working conferences are often published as collections of selected and edited papers.

Any national society whose primary activity is in information may apply to become a full member of IFIP, although full membership is restricted to one society per country. Full members are entitled to vote at the annual General Assembly, National societies preferring a less committed involvement may apply for associate or corresponding membership. Associate members enjoy the same benefits as full members, but without voting rights. Corresponding members are not represented in IFIP bodies. Affiliated membership is open to non-national societies, and individual and honorary membership schemes are also offered.

System Modelling and Optimization

Proceedings of the Seventeenth IFIP TC7 Conference on System Modelling and Optimization, 1995

Edited by

Jaroslav Doležal

Honeywell Technology Center
Prague
Czech Republic

and

Jiří Fidler

Institute of Information Theory and Automation
Academy of Sciences of the Czech Republic
Prague
Czech Republic

SPRINGER-SCIENCE+BUSINESS MEDIA, B.V.

First edition 1996

Originally published by Chapman & Hall in 1996
MyCopy version of the original edition 1996

DOI 10.1007/978-0-387-34897-1

A catalogue record for this book is available from the British Library

Printed on permanent acid-free text paper, manufactured in accordance with ANSI/NISO Z39.48-1992 and ANSI/NISO Z39.48-1984 (Permanence of Paper).
www.springer.com/mycopy

CONTENTS

Preface xi

International Program Committee
Local Organizing Committee xiii

Supporters
Co-sponsors xiv

Part One Invited Papers

1 On the convergence of a trust region SQP algorithm for nonlinearly constrained optimization problems
P.T. Boggs, J.W. Tolle and A.J. Kearsley 3

2 Decomposition and suboptimal control in dynamical systems
F.L. Chernousko 13

3 Network flow – theory and applications with practical impact
M. Iri 24

4 The mathematical theory of evidence – a short introduction
J. Kohlas 37

5 Algebraic methods in control, theory and applications
V. Kučera 54

6 One method for robust control of uncertain systems – theory and practice
G. Leitmann 64

7 Stochastic optimization methods in engineering
K. Marti 75

Part Two Contributed Papers

Automatic Control

8 Robust stabilization of nonlinear systems by optimal controllers
J. Kabziński 91

9 Weighted H^2 approximation of transfer functions
J. Leblond and M. Olivi 99

10 On design of H_∞ optimal controls for uncertain nonlinear systems
S. Tong, Z. Zhang and Z. Han 107

Biomedical Systems

11 Constrained optimization algorithms and automatic differentiation for parameter estimation with application to granulocytics models
B. Tibken and E.P. Hofer 115

12 Expert system for diagnosis of womens' menstrual cycle using natural family planning method
A. Urbaniak 120

13 Metabolic flux determination by 13-C tracer experiments: analysis of sensitivity, identifiability and redundancy
W. Wiechert 128

Discrete Event Systems

14 Binding-time analysis applied to mathematical algorithms
R. Glück, R. Nakashige and R. Zöchling 137

15 Invariant state progress and relation modelling of DEDS
G. Juhás and M. Kocian 147

Discrete-Time Systems

16 Remarks on the observability of nonlinear discrete time systems
F. Albertini and D. D'Alessandro 155

17 Risk-sensitive control and dynamic games: the discrete-time case
P. Dai Pra and C. Rudari 163

18 Dynamic portfolio optimization based on reference trajectories
A.M.J. Skulimowski 171

19 Stability analysis of time-varying discrete interval systems
K. Sladký 179

Distributed Parameter Systems

20 The relaxation theory applied to optimal control problems of semilinear elliptic equations
E. Casas 187

21 On the use of space invariant imbedding to solve optimal control problems for second order elliptic equations
J. Henry and J.P. Yvon 195

22 Semismoothness in parametrized quasi-variational inequalities
J.V. Outrata 203

23 Optimal control problem governed by a semilinear parabolic equation
J.P. Raymond and H. Zidani 211

24 Shape optimization of hyperelastic rod
B. Rousselet, J. Piekarski and A. Myslinksi 218

Engineering Applications

25 Dynamic modelling and optimal hierarchical control of a multiple-effect evaporator – superconcentrator plant
P. Gil, H. Duarte-Ramos and A. Dourado Correia 227

26 On the use of consistent approximations for the optimal design of beams
C. Kirjner-Neto and E. Polak 235

27 A game-theoretical model for a controlled process of heat transfer
O.A. Malafeev and M.S. Troeva 243

28 Constrained predictive control of a counter-current extractor
M. Mulholland and N.K. Narotam 251

29 Optimal policies under different assumptions about target values: an optimal control analysis for Austria
R. Neck and S. Karbuz 259

30 Optimal usage of saline and non saline irrigation water; a policy tool
A. Sadeh 267

Fuzzy Systems

31 Fuzzy integer sharing problem with fuzzy capacity constraints
H. Ishii and T. Itoh 273

32 A fuzzy-PID-concept with minimal rule set
F. Ross and U. Döring 279

Game Problems

33 A numerical procedure for minimizing the maximum cost
S.C. Di Marco and R.L.V. Gonzáles 285

34 Game of pursuit with zero stop probability
H.S. Kang 292

35 Solution concepts in multicriteria cooperative games without side payments
L. Kruś and P. Bronisz 300

Immunology

36 Computer models for maximising tumor cell kill and for minimizing side effects in radiation therapy
W. Düchting, T. Ginsberg and W. Ulmer 309

37 Decision making problems: AIDS prevention and energy development
G. Dzemyda, V. Šaltenis and V. Tiešis 317

38 A mathematical model of HIV infection: the role of $CD8^+$ lymphocytes
T. Hraba and J. Doležal 325

39 Mathematical modelling of conjugate formation by cytotoxic lymphocytes and tumour cells
J. Waniewski, K. Palucka and A. Porwit 331

Information Systems

40 Reliability optimization of complex systems using sharp lower bounds
M. Souissi and Y. Smeers 339

41 Knowledge retrieval for autonomous agents
E. Szczerbicki 347

42 Simulation and optimization of complex systems reliability characteristics in grouped data structure
Ye.B. Tsoi and S.V. Tishkovskaya 355

Multicriterial Problems

43 A modular system of software tools for multicriteria model analysis
J. Granat, T. Kręglewski, J. Paczyński, A. Stachurski and A.P. Wierzbicki 363

44 Methodology and modular tool for aspiration-led analysis of LP models
M. Makowski 371

45 Interactive multiobjective optimization system NIMBUS applied to nonsmooth structural design problems
K. Miettinen, M.M. Mäkelä and R.A.E. Mäkinen 379

Nondifferentiable Optimization

46 Preliminary computational experience with a descent level method for convex nondifferentiable optimization
U. Brännlund, K.C. Kiwiel and P.O. Lindberg 387

47 Bundle methods applied to the unit-commitment problem
C. Lemaréchal, C. Sagastizábal, F. Pellegrino and A. Renaud 395

48 Nondifferentiable optimization solver: basic theoretical assumptions
A. Stachurski 403

Optimal Control

49 Discrete approximation of nonlinear control problems
R. Lepp 411

50 Convergence of Lagrange–Newton method for control-state and pure state constrained optimal control problems
K. Malanowski 419

51 Descent methods for optimal periodic hereditary control problems
K. Nitka-Styczeń 427

52 Aircraft trajectory optimization using nonlinear programming
T. Raivio, H. Ehtamo and R.P. Hämäläinen 435

53 Feedback control of state constrained optimal control problems
D.A. Redfern and C.J. Goh 442

Optimization Algorithms and Methods

54 Primal-dual interior point method for multicommodity network flows with side constraints and comparison with alternative methods
J. Castro and N. Nabona 451

55 Dual Bregman proximal methods for large-scale 0–1 problems
K.C. Kiwiel, P.O. Lindberg and A. Nõu 459

56 On long-step surrogate projection methods for solving convex feasibility problems
K.C. Kiwiel and B. Lopuch 466

57 Theoretical and experimental analyis of random linkage algorithms for global optimization
M. Locatelli and F. Schoen 473

58 A dynamic list heuristic for 2D-cutting
L.A.N. Lorena and F.B. Lopes 481

59 About solving linear integer programs through hermite normal form decomposition
J. Maublanc and A. Quilliot 489

60 Software system for solving multi-scale optimization problems
E. Semenkin and K. Abramovich 497

61 Dual barrier-projection and barrier-Newton methods in linear programming
V.G. Zhadan 502

Production Systems

62 Flow and release optimization in manufacturing systems represented as timed event graphs
A. Di Febbraro, R. Minciardi, M. Profumo and S. Sacone 511

63 A control model for assembly manufacturing systems
A. Dolgui, M.C. Portmann and J.M. Proth 519

64 Numerical experiment on the 2D cutting-stock algorithms based on local optimization
T. Sakamoto 527

Scheduling Problems

65 An algorithm for the transportation problem with given frequencies
L. Bertazzi, M.G. Speranza and W. Ukovich 535

66 The traveling salesman problem with precedence constraints and binary costs
L. Bianco, P. Dell'Olmo and S. Giordani 543

67 Cost oriented competing processes – a new handling of assignment problems
J. Starke 551

68 Modelling and solving of the allocation problem of non-convex polygons with rotations
Yu.G. Stoyan and M.V. Novozhilova 559

Stochastic Problems

69 Parameters identification of a time-varying stochastic dynamic systems using Viterbi algorithm
T. Al Ani and Y. Hamam 567

70 Management of bond portfolios via stochastic programming – postoptimality and sensitivity analysis
J. Dupačová and M. Bertocchi 574

71 A note on objective functions in multistage stochastic nonlinear programming problems
V. Kaňková 582

Transportation Systems

72 Dynamic search for shortest multimodal paths in a transportation network
A. Di Febbraro and S. Sacone 591

73 Arc routing for rural Irish networks
P. Keenan and M. Naughton 599

74 Arc routing vehicle routing problems with vehicle/site dependencies
J. Sniezek and L. Bodin 607

Index of contributors 615

Keyword index 616

Preface

The 17th IFIP TC7 Conference on **System Modelling and Optimization** took place in Prague, July 10–14, 1995. It was organized by the Institute of Information Theory and Automation of the Academy of Sciences of the Czech Republic, one of the leading institutions in the field of information theory, automatic control and computer science in the Czech Republic. Among co-sponsors acted also the Czech Society for Cybernetics and Informatics, the new representative of the Czech Republic in the International Federation for Information Processing.

IFIP general sponsor was represented by TC7 on System Modelling and Optimization. TC7 aims to promote theoretical and applied research in areas of system modelling and optimization. Biannually organized general conferences bring together TC7 working groups and wide scientific and engineering community interested in latest achievements, exchange of information and practical impacts of the scheduled conference topics which included but were not limited to

Optimization Theory – duality and optimality, stability and sensitivity, robustness, system analysis, decomposition, identification, multicriterial optimization

Optimal Control – linear and nonlinear systems, game theory, decision making, control and estimation, distributed parameter systems, inverse problems, singularly perturbed systems, algebraic methods in optimal control

Mathematical Programming – theory and algorithms, linear and nonlinear programming, global optimization, nonsmooth optimization, numerical optimization and optimization packages, symbolic and automatic differentiation in optimization, interior point methods, parallel computing, computational geometry, large-scale mathematical programming

Discrete Systems – combinatorial optimization, theory and algorithms, integer programming; discrete event systems, reliability and structural optimization, Petri nets, knowledge-based systems, decision support systems, intelligent systems, planning and scheduling

Stochastic Optimization – stochastic programming, stochastic control theory, stochastic modelling, fuzzy systems, fuzzy control, neural networks

Applied Modelling and Optimization – biological and medical systems, distribution and logistic systems, flexible production systems, structural systems, optimization-based computer-aided modelling and design, applications in engineering (electrical, mechanical, civil), energy, communications, finance, economics, ecology, environment etc.

After having last two conferences in Western Europe, this time Prague replaced in the very last moment originally negotiated organizer from the U.S.A. As Academy of Sciences in the Czech Republic is primarily represented by fundamental research institutions and lacks on university type facilities for efficient organization of large conferences, such facilities had to be rented. And practical lack on any remarkable support from local sourced forced the organizers to require the higher conference fee than at previous conferences. There was also only a little opportunity to subsidize participation from other economically poor areas.

In spite of the effort to avoid as much a possible *no-show* participants, a not negligible number of mainly Russian scientists cancelled their preliminary confirmed participation at the very last moment or not at all. Also several *incognito* participants not only from soft currency areas were experienced. Such circumstances also do not contribute to balanced conference budget. Anyhow, according to the conference statistics there were 185 registered participants from 37 countries. This fact documents the world-wide impact of TC7 general conferences.

This collection contains invited papers and contributions selected during the second review phase taking place during the Conference. Originally over 400 submission were received. As TC7's traditional publisher Springer-Verlag is no more interested in conference proceedings publication, a new IFIP contractual publisher Chapman & Hall was contacted, which showed enough flexibility to handle our case. Such change caused necessarily some modifications in the used publication process, mainly with respect to the shorter deadline of the final manuscript submission and the imposed overall page limit. Complying with such requirements needed really effective cooperation with all potential authors and editors express their sincere gratitude for such effort. Final competition was a keen one, practically only one third of the presented papers could be included, however it is believed that the solicited contributions reflect considerably well outcome of the conference and the diversity of addressed topics.

Finally also cordial thanks are expressed to all active members of the International Program Committee for not an easy task to solicit the best contributions for this volume and to members of the Local Organizing Committee without their efficient help the organization of such event would be hardly possible. Special thanks are due to I. Pauknerová, who managed the registration and other financial and administrative matters. We do hope that this meeting in Prague was successful not only from scientific, but also from social and cultural points of view as Prague nowadays offers numerous opportunities in this respect.

Prague, August 17, 1995 J. Doležal, J. Fidler

International Program Committee

A. V.	Balakrishan	University of California, USA
A.	Ben-Tal	Technion, IL
A. J. M.	Beulens	Haagse Hogeschool, NL
R.	Burkard	Technical University of Graz, A
J.	Doležal	Academy of Sciences, CZ
I. V.	Evstigneev	Academy of Sciences, RUS
E. G.	Evtushenko	Academy of Sciences, RUS
G.	Feichtinger	Technical University of Vienna, A
Y. M.	Hamam	Ecole Superieure Noisy-Le-Grand, F
J.	Henry	INRIA, F
K. S.	Hindi	University of Manchester, GB
M.	Iri	Chuo University, Tokyo, J
P.	Kall	University of Zurich, CH
A.	Kalliauer	Österr. Elektr. Wirtsch., A
V.	Kučera	IFAC, Academy of Sciences, CZ
I.	Lasiecka	University of Virginia, USA
M.	Lucertini	University of Roma, I
K.	Malanowski	Polish Academy of Sciences, PL
K.	Marti	Military Univ. of Munich, D
M. J. D.	Powell	University of Cambridge, GB
R.	Rackwitz	Technical University of Munich, D
W.J.	Runggaldier	University of Padova, I
H.-J.	Sebastian	Technical University of Aachen, D
J.	Stoer	University of Würzburg, D
K. L.	Teo	University of Western Australia, AUS
P.	Thoft-Christensen	University of Aalborg, DK (chairman)
A. L.	Tits	University of Maryland, USA
Ph. L.	Toint	University of Namur, B
J.-P.	Vial	Universtiy of Geneva, CH
J.-P.	Yvon	University of Technology Compiègne, F
J.	Zowe	University of Bayreuth, D

Local Organizing Committee

Z. Beran
P. Buráň
J. Doležal (Chairman)
J. Fidler
V. Kaňková
I. Pauknerová
M. Součková

SUPPORTERS

- ÚTIA (Institute of Information Theory and Automation)
- HTC-MO (Honeywell Technology Center Minneapolis)
- HTC-P (Honeywell Technology Center Prague)

CO-SPONSORS

- IFAC (International Federation of Automatic Control)
- CSCI (Czech Society for Cybernetics and Informatics)
- ÚTIA (Institute of Information Theory and Automation)

PART ONE

Invited Papers

1

On the convergence of a trust region SQP algorithm for nonlinearly constrained optimization problems

Paul T. Boggs
Applied and Computational Mathematics Division
National Institute of Standards and Technology, Gaithersburg, MD 20899, U.S.A.

Jon W. Tolle
Mathematics Department
University of North Carolina, Chapel Hill, NC 27599, U.S.A.

Anthony J. Kearsley
Department of Computational and Applied Mathematics
Rice University, Houston, Texas 77251-1892, U.S.A.

Abstract

In (Boggs, Tolle and Kearsley, 1994) the authors introduced an effective algorithm for general large scale nonlinear programming problems. In this paper we describe the theoretical foundation for this method. The algorithm is based on a trust region, sequential quadratic programming (SQP) technique and uses a special auxiliary function, called a merit function or line-search function, for assessing the steps that are generated. A global convergence theorem for a basic version of the algorithm is stated and its proof is outlined.

Keywords

Sequential Quadratic Programming, Merit Functions, Global Convergence, Trust Region

1 INTRODUCTION

We consider the inequality-constrained minimization problem,

$$\begin{aligned} &\min_x \; f(x) \\ &\text{subject to:} \;\; g(x) \leq 0 \end{aligned} \tag{1}$$

where $x \in \mathcal{R}^n$, and $f : \mathcal{R}^n \rightarrow \mathcal{R}$, and $g : \mathcal{R}^n \rightarrow \mathcal{R}^m$ are smooth functions. One of the most successful methods for solving (1) is the sequential quadratic programming (SQP) method in which at each iteration a quadratic program is solved to obtain the step direction toward the next iterate. In particular, given a current approximation x^k to the solution x^*, one forms the quadratic program

$$\begin{aligned}&\min_{\delta} \nabla f(x^k)^t\delta + \tfrac{1}{2}\delta^t B^k \delta \\ &\text{subject to: } \nabla g(x^k)^t\delta + g(x^k) \le 0\end{aligned} \tag{2}$$

where B^k is usually taken to be a current approximation to the Hessian of the Lagrangian of (1). Let δ^k be the solution of (2). Then the next approximation, x^{k+1}, is calculated by

$$x^{k+1} = x^k + \alpha\delta^k \tag{3}$$

where α is a scalar *steplength.* This procedure is repeated until convergence. An enormous amount of research has been published on the theory of SQP methods and algorithms based on this method are among the most effective in solving general constrained nonlinear problems. For a survey of this topic the reader is referred to (Boggs and Tolle, 1995).

In (Boggs, Tolle and Kearsley, 1994) the authors have developed an SQP method that is designed specifically for large scale problems of the form (1). An implementation of this algorithm has performed quite well on a wide variety of problems, including examples from optimal control, molecular chemistry, statistics, and engineering. The purpose of this paper is to describe the theoretical underpinnings for this method.

A key element in our algorithm is the use of a special merit function. By a merit function we mean a scalar-valued function that can be used as a test for ensuring that the potential step given by (3) will make x^{k+1} a better approximation to the solution than x^k. A merit function for testing steps is an essential ingredient of a globally convergent algorithm. For unconstrained optimization problems, the objective function serves this purpose; one simply chooses α such that $f(x^{k+1}) < f(x^k)$ (with appropriate restrictions on α to assure that a sufficient decrease is achieved). For constrained optimization, the possible decrease in the objective value must be weighed against the requirement that feasibility must also be achieved — at least in the limit. Therefore, if one is using a method, such as an SQP algorithm, that does not maintain feasibility at each step, the objective function by itself is not an acceptable merit function. Most merit functions for constrained optimization algorithms are specially designed functions that have an unconstrained minimum at the solutions of (1). (See (Boggs and Tolle, 1995) for examples.) The merit function proposed here is different in that it does not have an unconstrained local minimum at x^* but, rather, a constrained minimum at (x^*, z^*) where the variables z are *nonnegatively constrained* slack variables. This merit function has excellent theoretical properties relative to the SQP method but needs to be augmented by a sequence of approximate merit functions in order to produce a globally convergent algorithm.

An outline of the paper is as follows. In Section 2 we define our proposed merit function and its approximations. We then set forth the conditions that are assumed to hold throughout and state the basic properties of the merit function and its approximations. In Section 3 we incorporate these ideas in a somewhat detailed algorithm. We then state a global convergence theorem. In Section 4 we describe enhancements to the algorithm that improve the performance in the large scale case. These include the automatic adjustment of the penalty parameter that occurs in the merit function and a trust region strategy for use in approximately solving the quadratic subproblems. This leads us to the discussion of an interior point quadratic program solver and an approach to dealing with inconsistent quadratic subproblems.

2 THE MERIT FUNCTIONS AND THEIR PROPERTIES

In order to obtain a useful merit function for the inequality-constrained problem we maintain, along with the iterates x^k, a sequence of nonnegatively constrained slack variables $\{z^k\}$. That is, if x^k is a current approximation to the solution of problem (1.1) we consider z^k to be the corresponding approximation to the optimal slack vector. Accordingly, if δ^k is the step computed at x^k using the quadratic program (1.2), then the corresponding step for z^k is taken to be

$$q^k = -(\nabla g(x^k)^t \delta^k + g(x^k) + z^k), \tag{4}$$

i.e., $z^k + q^k$ is the slack vector for (1.2). (This choice for q^k is further motivated in (Boggs, Tolle and Kearsley, 1995).) We then choose the new iterates as

$$(x^{k+1}, z^{k+1}) = (x^k, z^k) + \alpha\,(\delta^k, q^k)$$

for some steplength α.

Two important points need to be emphasized here. First, if z^k is nonnegative and $\alpha \in [0,1]$ then $z^{k+1} \geq 0$. Thus the nonnegativity of the slack variables is easily maintained. Second, the slack variable updates are not obtained from an optimization procedure, they are computed algebraically. Consequently, we are *not* introducing any complexity into the computational process.

The purpose for the introduction of the slack variables is to make it possible to derive a useful merit function The following merit function is derived from the equality-constrained problem which results from the addition of slack variables; details are contained in (Boggs, Tolle and Kearsley, 1995).

$$\psi_d(x,z) = f(x) + \bar{\lambda}(x,z)^t \bar{c}(x,z) + \frac{1}{d}\bar{c}(x,z)^t \bar{A}(x,z)^{-1}\bar{c}(x,z)$$

where

$$\begin{aligned} \bar{c}(x,z) &= g(x) + Ze \\ \bar{A}(x,z) &= \nabla g(x)^t \nabla g(x) + Z \\ \bar{\lambda}(x,z) &= -\bar{A}(x,z)^{-1}\nabla g(x)^t \nabla f(x), \end{aligned}$$

d is a small parameter, and

$$Z = \text{diag}\,\{z_1, \ldots, z_m\}.$$

The function $\bar{\lambda}(x,z)$ can be interpreted as a least squares approximation to the Lagrange multiplier vector for (1.1).

Since the z_i are to be interpreted as nonnegative slack variables for (1), the minimization of ψ_d has the form:

$$\begin{aligned} &\min_{x,z} \ \psi_d(x,z) \\ &\text{subject to: } \ z \geq 0. \end{aligned} \tag{5}$$

A fundamental property of this merit function (under the assumptions to be discussed shortly) is that the local solutions to (5) correspond to local solutions of (1.1).

Since $\psi_d(x,z)$ involves the gradients of the objective function and the constraints, carrying out line searches for this function is expensive. To ameliorate this difficulty, we define a sequence of (local) approximate merit functions by keeping the gradient terms fixed at the kth iteration. This produces functions that only require the evaluation of the function and the constraints, with no extra gradient evaluations, to check a prospective point. Our approximate merit function at the kth iterate is

$$\psi_d^k(x,z) = f(x) + \bar{c}(x,z)^t \bar{\lambda}^k + \frac{1}{d}\bar{c}(x,z)^t \bar{A}_k^{-1}\bar{c}(x,z)$$

where

$$\bar{A}_k = \nabla g(x^k)^t \nabla g(x^k) + Z^k$$

and

$$\bar{\lambda}^k = -\bar{A}_k^{-1}\nabla g(x^k)^t \nabla f(x^k).$$

While the $\psi_d^k(x,z)$ are cheap to evaluate, they cannot be used as a single merit function for measuring the progress towards a solution. However, as will be shown in Section 3, they can be employed as surrogates for $\psi_d(x,z)$ in a globally convergent algorithm.

To describe the basic theoretical properties of the merit functions and its local approximations, we first introduce appropriate notation and formulate some basic assumptions. We note that the feasible region for (1.1) can be thought of as the set of x such that $g(x) + z = 0$ for some $z \geq 0$. Consequently, using the notation given above, we interpret

$$\mathcal{C}_0 = \{(x,z) : \bar{c}(x,z) = 0, z \geq 0\}$$

as the *feasible set* and call

$$\mathcal{C}_\eta = \{(x,z) : ||\bar{c}(x,z)||^2 \leq \eta, z \geq 0\}$$

the η*-tube* surrounding the feasible set. We will denote by $\mathcal{G}$ a compact set of $\mathcal{R}^n \times \mathcal{R}^m_+$ and by $\mathcal{S}$ the set of points of $\mathcal{G}$ satisfying the first order conditions for (1.1), i.e.,

$$\mathcal{S} = \{(x,z) \in \mathcal{G} : \nabla f(x) + \lambda^t \nabla g(x) = 0 \text{ and } \lambda^t g(x) = 0 \text{ for some } \lambda \geq 0\}.$$

Also we will assume that the matrices used in (1.2) are chosen from $\mathcal{B}$, a compact set of positive definite $n \times n$ matrices.

The basic assumptions are the following:

A1: For each $(x, z) \in \mathcal{G}$ the matrix $\bar{A}(x, z)$ is positive definite.
A2: For each $(x, z) \in \mathcal{G}$ and $B^k \in \mathcal{B}$ the quadratic program (1.2) is feasible and its solution satisfies the strong second order sufficient conditions.
A3: $\mathcal{S}$ is not empty.

Note that the second assumption implies that the step (δ^k, q^k) will be a continuous function of (x^k, z^k) and B^k.

The merit functions can be shown to have the following basic properties.

Proposition 1: There exists a $\bar{d} > 0$ and a constant $\theta_1 > 0$ such that if $d \leq \bar{d}$ then for each $(x^k, z^k) \in \mathcal{G}$ and $B^k \in \mathcal{B}$

$$\nabla \bar{c}(x^k, z^k)^t (\delta^k, q^k) = -||\bar{c}(x^k, z^k)||^2 \tag{6}$$

and

$$\nabla \psi_d^k(x^k, z^k)^t (\delta^k, q^k) \leq -\theta_1 ||(\delta^k, q^k)||^2. \tag{7}$$

Proposition 2: For each d sufficiently small there exist an $\eta(d) > 0$ and a $\theta(d) > 0$ so that $(x^k, z^k) \in C_{\eta(d)}$ and $B^k \in \mathcal{B}$ imply

$$\nabla \psi_d(x^k, z^k)^t (\delta^k, q^k) \leq -\theta(d) ||(\delta^k, q^k)||^2. \tag{8}$$

Proposition 2 shows that at (x^k, z^k) sufficiently close to feasibility the step (δ^k, q^k) is a descent direction for the merit function while Proposition 1 gives the stronger results that the step is a descent direction everywhere for both the corresponding approximate merit function and the measure of infeasibility, $\bar{c}$. The rates of descent in the direction of the step can be used to guarantee that the *Wolfe conditions* (see (Nocedal, 1992)) hold for each of the merit functions and the infeasibility function in the sets indicated in the propositions. For a given function ϕ and descent direction v at a point w the Wolfe conditions require the steplength α to satisfy

$$\phi(w + \alpha v) \leq \phi(w) + \sigma_1 \alpha \nabla \phi(w)^t v$$

and

$$\nabla \phi(w + \alpha v)^t v \geq \sigma_2 \nabla \phi(w)^t v$$

for constants σ_1 and σ_2. The following result, which is a consequence of (7) and (8), relates the decreases in $\psi_d^k(x, z)$ and $\psi_d(x, z)$ and is important in establishing the global convergence of the algorithm. It shows that, under certain conditions, if α satisfies the first Wolfe condition for ψ_d^k (with constant σ) it also satisfies the Wolfe condition for ψ_d (with constant $\gamma \sigma$).

Proposition 3: Let $d > 0$ be sufficiently small and let $\gamma \in (0,1)$ and $\sigma \in (0,1/2)$ be specified. Let ϵ be a sufficiently small positive number and set

$$\mathcal{S}_\epsilon = \{(x,z) : ||(x,z) - (x^*,z^*)|| < \epsilon \text{ for some } (x^*,z^*) \in \mathcal{S}\}.$$

Then there exists an $\eta(\epsilon)$ such that if $(x^k, z^k) \in \mathcal{C}_{\eta(\epsilon)} - \mathcal{S}_\epsilon$ and

$$\psi_d^k(x^k + \alpha\,\delta^k, z^k + \alpha\, q^k) \leq \psi_d^k(x^k, z^k) + \sigma\,\alpha\,\nabla\psi_d^k(x^k, z^k)^{\mathrm{t}}(\delta^k, q^k)$$

then

$$\psi_d(x^k + \alpha\,\delta^k, z^k + \alpha\, q^k) \leq \psi_d(x^k, z^k) + \gamma\,\sigma\,\alpha\nabla\psi_d^k(x^k, z^k)^{\mathrm{t}}(\delta^k, q^k).$$

It is worth noting, although not emphasized here, that if the iterates are converging to a solution q-superlinearly, neither the merit function nor its approximations will interfere with this process, i.e., a steplength of $\alpha = 1$ is eventually acceptable. For details, see (Boggs, Tolle and Kearsley, 1995).

3 A GLOBALLY CONVERGENT ALGORITHM

We first give a description of the algorithm and then proceed to state the basic global convergence theorem. The underlying idea of the algorithm is a standard one; the iterates are forced towards feasibility and then towards a solution. What distinguishes this algorithm is the use of the approximate merit functions that, far from feasibility, determine efficient steplengths that are likely to force the iterates toward optimality as well as feasibility and, near feasibility, provide relatively simple surrogates for the true merit function.

In the description of the algorithm, we drop the superscripts k on the iterates and the steps, denoting the current iterate by (x,z) and the step direction at this iterate by (δ, q). It is assumed that constants $d > 0$, $\eta > 0$, and $\sigma \in (0,1/2)$ have been specified.

Algorithm

1. Set FLAG = FALSE.
2. If $||\bar{c}(x,z)||^2 < \eta$, set FLAG = TRUE.
3. Compute (δ, q) and set $\alpha = 2$.
4. Set $\alpha = \alpha/2$ and $(x_\alpha, z_\alpha) = (x,z) + \alpha\,(\delta, q)$.
5. If

   $$\psi_d^k(x_\alpha, z_\alpha) \geq \psi_d^k(x,z) + \alpha\,\sigma\,\nabla\psi_d^k(x,z)^{\mathrm{t}}(\delta, q),$$

   return to Step 4.
6. If FLAG = FALSE, then

a. if

$$||\bar{c}(x_\alpha, z_\alpha)||^2 \geq (1 - 2\,\sigma\,\alpha)\,||\bar{c}(x,z)||^2,$$

return to Step 4;
b. else set $(x,z) = (x_\alpha, z_\alpha)$ and return to Step 2.

7. If FLAG = TRUE, then

a. if

$$||\bar{c}(x_\alpha, z_\alpha)||^2 \geq \eta,$$

return to Step 4;
b. elseif

$$\psi_d(x_\alpha, z_\alpha) \geq \psi_d(x,z)$$

set $\eta = \frac{1}{2}||\bar{c}(x,z)||^2$ and return to Step 1;
c. elseif

$$\psi_d(x_\alpha, z_\alpha) \geq \psi_d(x,z) + \tfrac{1}{2}\alpha\,\sigma_1\,\nabla\psi_d(x,z)^t(\delta, q),$$

set $\eta = \frac{1}{2}||\bar{c}(x,z)||^2$, set $(x,z) = (x_\alpha, z_\alpha)$, and return to Step 1;
d. else set $(x,z) = (x_\alpha, z_\alpha)$ and return to Step 3.

We first note that the steplength parameter α is always chosen from (0,1) and hence the variable z will remain nonnegative. The determination of the α is, of course, the crucial factor in establishing that the resulting sequence of iterates converges to a solution of the problem. In our algorithm, the criteria for choosing the parameter depend upon whether or not the current iterate is sufficiently close to feasibility; i.e., on whether $(x,z) \in \mathcal{C}_\eta$ for some sufficiently small η . Unfortunately, an appropriate value of η cannot be determined *a priori* and hence the value must be adjusted as the algorithm proceeds.

In the algorithm the logical variable FLAG is set to TRUE when the current iterate (x,z) is in $\mathcal{C}_\eta$ for the current value of η and is FALSE otherwise. Thus given a value of η, if (x,z) is not in $\mathcal{C}_\eta$ the algorithm forces the iterates toward feasibility. In particular, the inequality in Step 6 forces α to be chosen so that the first Wolfe condition for the function $||\bar{c}(x,z)||^2$ is satisfied for constant σ. It follows from (6) that eventually the iterates will enter the current η-tube. By Step 5 the α is also chosen to ensure that the Wolfe condition for the approximate merit function is also satisfied at *every* step. That this is possible is a consequence of (7). Once the iterate is inside the given η-tube, it is not allowed thereafter to leave it; α is reduced if necessary to guarantee this (Step 7a). Since the step is always in the direction of feasibility in (x,z)-space this restriction is not severe. Now the computed step is tested to see if the true merit function $\psi_d(x,z)$ satisfies the Wolfe condition for the constant $\sigma/2$ (Steps 7b–7d). If η is small enough then Proposition 3 ensures that such a decrease will occur for any α that yields a decrease in the approximate merit function $\psi_d^k(x,z)$ for the constant σ. If the Wolfe condition for $\psi_d(x,z)$ is not satisfied for the value of α then we take this as a signal that the current value of η is too large and we decrease η to one-half the current value of $\bar{c}(x,z)$. Thus when the value of η is decreased, it is decreased by at least a factor of one-half, so that the sequence of η values either tends to zero or else the the Wolfe condition is satisfied for $\psi_d(x,z)$ for all (x^k,z^k) for k sufficiently large.

There is another parameter central to our algorithm, namely the penalty parameter d. Here, we make the assumption that this parameter is initially small enough so that the basic propositions of the preceding section are satisfied. In the implementation of our algorithm we do have a heuristic procedure for adjusting d, but for the theoretical convergence we do not include this modification (but see Section 4).

Theorem: Led $d > 0$ be sufficiently small and let an initial $\eta > 0$ and an initial point (x^0, z^0) be given. Assume that the iterates lie in a compact set in which assumptions **A1–A3** hold. If (x^*,z^*) is a limit point of the sequence of iterations then $(x^*, z^*) \in \mathcal{C}_0$ and there exists a $\lambda^* \geq 0$ such that

$$\nabla f(x^*) + \nabla g(x^*)^t \lambda^* = 0 \tag{9}$$

and

$$g(x^*)^t \lambda^* = 0. \tag{10}$$

Outline of Proof: Let $\{\eta_j\}$ be the set of *assigned* values of η in the algorithm. If there are only a finite number, say K, of values then at some point all of the iterates are in the tube $\mathcal{C}_{\eta_K}$ and satisfy the first Wolfe condition for ψ_d. By an argument similar to that employed in proving global convergence of descent methods for unconstrained optimization algorithms, it can then be shown that the steps $(\delta, q) \to 0$. In this case the assumptions guarantee that $\delta = 0$ is a solution of (2) for some $B \in \mathcal{B}$ and hence (9) and (10) are satisfied at any limit point. If, on the other hand, $\eta_j \to 0$ then it is possible that ψ_d does not satisfy the desired Wolfe condition at a sequence of iterations. Nevertheless, any limit point must be in $\mathcal{C}_0$ and it follows from the assumptions and Proposition 3 that some subsequence of the steps must converge to zero. Then a similar argument to that above yields the desired conditions.

It is observed that the standard global convergence proofs for unconstrained optimization require that the second Wolfe condition be satisfied. This condition is imposed to keep the steps from becoming too small. Here, however, we have followed the common practice of using a backtracking line search to avoid overly small steps.

4 ALGORITHMIC ENHANCEMENTS FOR LARGE SCALE PROBLEMS

Almost all algorithms need to have certain enhancements to perform effectively in practice. In this section we describe the most important enhancements that we have used in adapting the algorithm to solve large scale problems. The numerical experiments for this implementation are reported in (Boggs, Tolle and Kearsley, 1994).

4.1 The dynamic adjustment of d

It follows from Proposition 1 that if (x^k, z^k) isn't feasible, then the penalty parameter d can be made small enough to ensure that the step is a direction of descent for ψ_d^k. In our implementation if, at some non feasible point, a good decrease in ψ_d^k is not achieved, we decrease d by an appropriate factor. This can be done without fundamentally altering the convergence theory. However, using too small of a value of d often causes the algorithm to slow dramatically, since it tends to force the iterates to stay close to the constraints. To avoid this, it is desirable to incorporate a means for allowing an increase in d when the steps are good descent steps for ψ_d^k. Since increasing d can theoretically cause the iterates to cycle, care has to be taken in implementing such a procedure. We have used a heuristic strategy for increasing d while avoiding cycling that has been effective in our numerical experiments. However, a theoretical proof of global convergence when this process is implemented is lacking.

4.2 A trust region approach for solving the quadratic subproblems

In applying the SQP method to large scale problems, it is necessary to obtain a solution (or else an approximate solution) to the quadratic subproblem quickly. Since the data for (2) are likely to be sparse this suggests the use of an iterative scheme for the solution. We have used such a procedure, an interior point algorithm, called the O3D method, that has special properties that make it particularly compatible with our SQP algorithm. The defining characteristic of this quadratic program solver is that it proceeds by solving a sequence of three-dimensional subspace approximations to the quadratic program. At a current iterate (of the quadratic program) three independent directions are generated and the quadratic program restricted to the subspace determined by these directions is solved. The next iterate is taken to be 99% of the distance to the boundary in the direction thus generated. A detailed description of this method is found in (Boggs, Domich and Rogers, 1994).

The O3D solver is employed in the following way for the SQP algorithm. At iteration k in the main algorithm, the quadratic program (2) is formed and the iteration procedure is begun. The algorithm is stopped when a prescribed tolerance for an optimality condition is satisfied *or* when an implicit *trust region constraint* of the form

$$||\delta|| \leq \tau_k$$

is violated. If the final iterate, δ_j^k, is optimal (or very nearly optimal) for (2) then, of course, the theory described in the preceding section applies. The significance of this trust region approach lies in the case where the jth iterate is not optimal. The remarkable fact is that if q^k is set equal to $-(\nabla g(x^k)^t\delta_j^k + g(x^k) + z^k)$ then the step (δ_j^k, q^k) satisfies the strong descent conditions given in Propositions 1 and 2 despite the fact that it is not optimal. This implies that the algorithm described in the preceding section can be applied using approximate solutions to the quadratic subproblems rather than the true optimal solutions. Moreover, together with the fact that the trust region constraint will become irrelevant near the solution, it means that a global convergence analysis of this trust region algorithm modeled after that of Section 3 is feasible.

The trust region parameter, τ_k, is adjusted in a manner similar in spirit to that used in most trust region methods; that is, the decision to increase or decrease τ_k is based on a comparison of the predicted and actual reductions of the merit function. In our implementation we use either ψ_d or ψ_d^k depending on the current status of the point (x^k, z^k). For details of this procedure, see (Boggs, Tolle and Kearsley, 1994).

4.3 Inconsistent quadratic subproblems

One of the major theoretical weaknesses of SQP methods is the assumption that the quadratic subproblems, (2), are feasible. It is a troublesome practical problem as well since it is a reasonably common occurrence to have the linearized constraints be inconsistent. Therefore, any implementation must take into account this possibility.

The use of the O3D quadratic program solver gives a natural way to handle this difficulty in our algorithm. The O3D solver uses an artificial variable together with an adaptive "big M" method for generating a starting point, so if the quadratic problem is infeasible then the algorithm will terminate with a step in the x variables, δ_j^k, that (together with a value of the artificial variable) is either optimal in the modified problem or exceeds the trust region constraint. In either case it can be shown that, under an additional assumption on the "optimal" artificial variable, the step (δ_j^k, q^k) is still a descent step for $\bar{c}, \psi_d^k$, and ψ_d in the sets prescribed in Propositions 1 and 2. Therefore using this step allows us to move away from a point where the quadratic subproblem is inconsistent in a direction that is compatible with the goals of the algorithm. A theoretical difficulty that arises from this approach is that the resulting point could be such that z^{k+1} is negative (since $\nabla g(x^k)^t \delta_j^k + g(x^k)$ will have positive components). Our procedure has been to accept such steps and simply reset any negative components of z to be positive. Although this procedure has been quite successful in practice, it has not been possible to prove global convergence in this case.

REFERENCES

Boggs, P.T., Domich, P.D. and Rogers, J.E. (1994) *An Interior-Point Method for General Large Scale Quadratic Programming Problems.* Internal Report 5406, National Institute of Standards and Technology (accepted in *Annals of Operations Research*).

Boggs, P.T. and Tolle, J.W. (1995) Sequential Quadratic Programming. *Acta Numerica,* **1995**, 1–52.

Boggs, P.T., Tolle, J.W. and Kearsley, A.J. (1994) *A Practical Algorithm for General Large Scale Nonlinear Optimization Problems.* Internal Report, National Institute of Standards and Technology.

Boggs, P.T., Tolle, J.W. and Kearsley, A.J. (1995) *A Basic Global Convergence Analysis of a Large Scale Algorithm for General Nonlinear Programming Problems.* Internal Report, National Institute of Standards and Technology.

Nocedal, J. (1992) Theory of Algorithms for Unconstrained Optimization. *Acta Numerica,* **1991**, 199–242.

2

Decomposition and suboptimal control in dynamical systems

Felix L. Chernousko
Institute for Problems in Mechanics of the Russian Academy of Sciences
pr. Vernadskogo 101, Moscow 117526, Russia.
Tel: 007-095-4340207. Fax: 007-095-9382048.
e-mail: chern@ipm.msk.su

Abstract
Constructive methods of control for dynamical systems subject to bounded controls are proposed. Lagrangian systems governed by nonlinear ordinary differential equations and linear systems governed by partial differential equations are considered. The approach is based on the decomposition of the original system with a finite or infinite number of degrees of freedom into simple subsystems with one degree of freedom each. Methods of optimal control and differential games are applied to the subsystems. Explicit feedback control laws are obtained which bring the original dynamical systems to the prescribed terminal states in finite time. Upper estimates on the control time are obtained. Controllability of systems subject to bounded controls is discussed. The proposed methods are applicable to robots and elastic systems.

Keywords
Decomposition, nonlinear dynamical systems, optimal control, differential games, controllability

1 INTRODUCTION

The well known classical methods of automatic control present the control u as a linear operator of the state x: $u = Lx$. Thus, in the vicinity of the terminal state $x = 0$, the control u becomes small: $u \to 0$. Here, not all control possibilities are used, and the control time is infinite: $x \to 0$ as $t \to \infty$. On the other hand, far away from the terminal state, the control u becomes large enough and may violate constraints usually imposed on control.

Methods of optimal control (Pontryagin et al, 1972) are free from these drawbacks, take account of control constraints and can bring the system to the terminal state in minimal time. However, it is very difficult to obtain optimal feedback controls for nonlinear systems.

In this paper, we propose a constructive approach for obtaining bounded controls in dynamical systems. We consider systems of two types: nonlinear systems described by the Lagrangian ordinary differential equations and linear systems described by partial differential equations. To obtain the control laws, we use the decomposition of our systems with a finite or infinite number of degrees of freedom into simple subsystems with one degree of freedom each. After that, methods of optimal control and differential games are applied to the subsystems. Explicit feedback control laws are given. The results presented here were obtained in (Chernousko, 1990, 1992a, 1992b, 1993a, 1993b, Ananyevskii et al, 1995) where full proofs, upper estimates on the total time of motion, more details and examples can be found.

2 LAGRANGIAN SYSTEMS

Consider a nonlinear dynamical system governed by Lagrange's equations

$$\frac{d}{dt}\frac{\partial T}{\partial \dot{q}_k} - \frac{\partial T}{\partial q_k} = Q_k + F_k(q, \dot{q}, t), \qquad i = 1, \ldots, n. \tag{1}$$

Here t is time, $\boldsymbol{q} = (q_1, \ldots, q_n)$ is an n - vector of generalized coordinates, T is the kinetic energy, Q_k are control forces subject to the constraints

$$|Q_k| \leq Q_k^0, \qquad k = 1, \ldots, n \tag{2}$$

where Q_k^0 are given constants, F_k are all other forces including uncertain disturbances. The kinetic energy of the system T is given by

$$T = \frac{1}{2}(\boldsymbol{A}(\boldsymbol{q})\dot{\boldsymbol{q}}, \dot{\boldsymbol{q}}) = \frac{1}{2}\sum_{i,j=1}^{n} A_{ij}(\boldsymbol{q})\dot{q}_i\dot{q}_j \tag{3}$$

where $\boldsymbol{A}(\boldsymbol{q})$ is a symmetrical positive definite $n \times n$ - matrix, and $(\cdot\,,\,\cdot)$ denotes a scalar product of vectors. Substituting (3) into (1) we obtain

$$\boldsymbol{A}(\boldsymbol{q})\ddot{\boldsymbol{q}} = \boldsymbol{Q} + \boldsymbol{F}(\boldsymbol{q}, \dot{\boldsymbol{q}}, t) + \boldsymbol{S}(\boldsymbol{q}, \dot{\boldsymbol{q}}) \tag{4}$$

where the following notation is introduced

$$\boldsymbol{Q} = (Q_1, \ldots, Q_n), \quad \boldsymbol{F} = (F_1, \ldots, F_n), \quad \boldsymbol{S} = (S_1, \ldots, S_n),$$

$$S_k(\boldsymbol{q}, \dot{\boldsymbol{q}}) = \sum_{i,j=1}^{n} \Gamma_{kij}(\boldsymbol{q})\dot{q}_i\dot{q}_j, \qquad k = 1, \ldots, n, \tag{5}$$

$$\Gamma_{kij}(\boldsymbol{q}) = \frac{1}{2}\left(\frac{\partial A_{ij}}{\partial q_k} - \frac{\partial A_{kj}}{\partial q_i} - \frac{\partial A_{ki}}{\partial q_j}\right).$$

The problem is to find a feedback control $\boldsymbol{Q}(\boldsymbol{q}, \dot{\boldsymbol{q}})$ satisfying (2) and bringing the system (1) (or (4)) from any given initial state

$$\boldsymbol{q}(t_0) = \mathbf{q}^0, \qquad \dot{\boldsymbol{q}}(t_0) = \dot{\mathbf{q}}^0 \tag{6}$$

to the prescribed terminal state

$$\boldsymbol{q}(t_1) = \mathbf{q}^1, \qquad \dot{\boldsymbol{q}}(t_1) = 0 \tag{7}$$

in finite time. Here, the instant t_0 and the vectors $\mathbf{q}^0$, $\dot{\mathbf{q}}^0$ and $\mathbf{q}^1$ are fixed whereas t_1 is free. The case of nonzero terminal velocity $\dot{\mathbf{q}}(t_1) \neq 0$ was considered in (Ananyevskii et al, 1995). We shall discuss two approaches to the control problem stated above.

3 THE FIRST APPROACH

Multiplying (4) by the matrix $\boldsymbol{A}(\mathbf{q}_1)\boldsymbol{A}^{-1}(\boldsymbol{q})$, we get

$$\mathbf{A}_1\ddot{\boldsymbol{q}} = \boldsymbol{Q} + \boldsymbol{R}(\boldsymbol{q}, \dot{\boldsymbol{q}}, t, \boldsymbol{Q}), \qquad \mathbf{A}_1 = \boldsymbol{A}(\boldsymbol{q}_1)$$

$$\boldsymbol{R} = \mathbf{A}_1\boldsymbol{A}^{-1}(\boldsymbol{q})(\boldsymbol{F} + \boldsymbol{S}) + \mathbf{A}_1\boldsymbol{A}^{-1}(\boldsymbol{q})\boldsymbol{Q} - \boldsymbol{Q} \tag{8}$$

Suppose all motions under consideration as well as the initial and terminal states (6) and (7) lie in some domain W of the $2n$-space $(\boldsymbol{q}, \dot{\boldsymbol{q}})$ where

$$|R_k| \leq R_k^0 < Q_k^0, \qquad k = 1, \ldots, n \tag{9}$$

for all $t \geq t_0$, $(\boldsymbol{q}, \dot{\boldsymbol{q}}) \in W$ and Q_k satisfying (2). Here R_k^0 are constants. The inequalities (9) can be verified by means of the following Lemma given in (Chernousko, 1990, 1993a, 1993b).

Lemma *Suppose the following inequalities*

$$\|A_1 z\| \geq \mu_1 \|z\|, \qquad \|[A(q) - A_1]z\| \leq \mu\|z\|,$$

$$|F_k + S_k| \leq \nu Q_k^0, \quad k = 1, \ldots, n, \quad 0 < \mu < \mu_1, \ \nu > 0$$

hold for any $z \in R^n$, all $t \geq t_0$, and all $(q, \dot{q}) \in W$ where μ_1, μ, and ν are constants. Then for all $t \geq t_0$, all $(q, \dot{q}) \in W$, and all Q satisfying (2) we have

$$\begin{aligned} |R_k| &\leq \nu Q_k^0 + \mu(\mu_1 - \mu)^{-1}(1 + \nu)Q^0, \\ Q^0 &= \left[\sum_{k=1}^{n}(Q_k^0)^2\right]^{1/2}, \qquad k = 1, \ldots, n. \end{aligned}$$

It follows from the Lemma that the inequalities (9) are satisfied, if $\boldsymbol{A}(\boldsymbol{q})$ is close to $\mathbf{A}_1$ and Q_k^0 are sufficiently large. By the change of variables $\mathbf{A}_1(\boldsymbol{q} - \mathbf{q}_1) = \boldsymbol{y}$ we reduce the system (8), constraints (2) and (9), and terminal conditions (7) to the form

$$\ddot{y}_k = Q_k + R_k, \quad |Q_k| \leq Q_k^0, \quad |R_k| \leq R_k^0 < Q_k^0,$$

$$y_k(t_1) = \dot{y}_k(t_1) = 0 \tag{10}$$

We regard the terms R_k in (10) as independent and arbitrary uncertain disturbances subject to the constraints (9). Applying the approach of differential games, we can treat (10) as a set of n separate independent subsystems with one degree of freedom each. Here, Q_k and R_k are the controls of the two players acting on the kth subsystem. We shall find the feedback control Q_k of the first player and thus obtain the solution of our control problem.

Using the change of variables

$$y_k = Q_k^0 x\,,\ Q_k = Q_k^0 u\,,\ R_k = Q_k^0 v\,,\ \rho = R_k^0/Q_k^0 < 1\,, \tag{11}$$

we reduce the kth subsystem (10) to the normalized form

$$\ddot{x} = u + v\,,\ |u| \leq 1\,,\ |v| \leq \rho < 1\,,\ x(\tau) = \dot{x}(\tau) = 0 \tag{12}$$

where τ is the terminal time.

Let us find the feedback control $u(x, \dot{x})$ as a time-optimal control of the first player in the differential game (12). This linear differential game is reduced (Krasovskii, 1970) to the time-optimal control problem for the system

$$\ddot{x} = (1-\rho)u\,,\ |u| \leq 1\,,\ x(\tau) = \dot{x}(\tau) = 0\,,\ \tau \to \min \tag{13}$$

obtained from (12) through replacing v by $(-\rho u)$. The optimal feedback control for the system (13) is (Pontryagin et al, 1972)

$$\begin{aligned} u(x,\dot{x}) &= \operatorname{sign}\psi_\rho(x,\dot{x}) \quad \text{if} \quad \psi_\rho \neq 0\,, \\ u(x,\dot{x}) &= \operatorname{sign} x = -\operatorname{sign}\dot{x} \quad \text{if} \quad \psi_\rho = 0\,, \\ \psi_\rho(x,\dot{x}) &= -x - \dot{x}\,|\dot{x}|\,[2(1-\rho)]^{-1} \end{aligned} \tag{14}$$

where ψ_ρ is the switching function. Returning to the original variables using $\boldsymbol{y} = \mathbf{A}_1(\boldsymbol{q} - \mathbf{q}_1)$ and (13), we obtain the desired feedback control as follows

$$Q_k(\boldsymbol{q},\dot{\boldsymbol{q}}) = Q_k^0 u(x,\dot{x})\,, \quad x = (Q_k^0)^{-1}[\mathbf{A}_1(\boldsymbol{q} - \mathbf{q}_1)]_k\,,$$

$$\dot{x} = (Q_k^0)^{-1}(\mathbf{A}_1\dot{\boldsymbol{q}})_k \tag{15}$$

where $u(x,\dot{x})$ is defined in (14) and ρ is given by (11) for each k. The total time of control is finite and does not exceed the maximum (over all k) of the control times for the subsystems (13). We have the following upper estimate for the terminal instant t_1:

$$t_1 - t_0 \leq \max_k \tau(x_k^0, \dot{x}_k^0)\,, \qquad k = 1, \ldots, n\,,$$

$$x_k^0 = (Q_k^0)^{-1}\Big[\mathbf{A}_1(\mathbf{q}^0 - \mathbf{q}^1)\Big]_k\,, \quad \dot{x}_k^0 = (Q_k^0)^{-1}(\mathbf{A}_1\mathbf{q}^0)_k\,,$$

$$\tau(\xi,\eta) = (1-\rho)^{-1}\left\{[2\eta^2 - 4(1-\rho)\xi\gamma]^{1/2} - \eta\gamma\right\},$$

$$\gamma = \operatorname{sign}\psi_\rho(\xi,\eta)\,, \qquad \rho = R_k^0/Q_k^0 < 1\,.$$

4 SOME GENERALIZATIONS

The approach described above consists of two stages: 1) decomposition of the original system into simple subsystems; 2) design of the feedback control for the subsystem. Both stages can be modified.

At the first stage, the system (4) can be reduced to the set of subsystems other than (13). Namely, the following subsystems with linear or nonlinear damping can be used

$$\ddot{x} + \lambda \dot{x} = u + v\,,\ \lambda = \text{const} > 0; \qquad \ddot{x} + f(\dot{x}) = u + v \tag{16}$$

where u and v are again the control and disturbance, respectively. Here the smooth function $f(z)$ has the following properties:

$$zf(z) > 0\,, \qquad f'(z) > 0 \quad \text{if} \quad z \neq 0\,, \qquad f(0) = 0\,.$$

Some results for equations (16) are presented in (Chernousko, 1993a, 1993b).

At the second stage, the game approach can be replaced by other methods for control design. The simplest possibility is to ignore the disturbance v while designing the control u and then to use thus obtained control for the system subject to the disturbance. Applying this simplified approach to the system (12), we should take the control $u(x, \dot{x})$ in the form (14) with $\rho = 0$. The analysis presented in (Chernousko 1990, 1993a, 1993b) shows that the dynamics of the system (14) under this simplified control depends essentially on the parameter ρ, the critical value of ρ being equal to the golden section ratio

$$\rho^* = (5^{1/2} - 1)/2 = 0.618\ldots\,.$$

If $\rho < \rho^*$, then the system (12) under the simplified control and arbitrary disturbance $v, |v| \leq \rho$, reaches the terminal state $x = \dot{x} = 0$ in finite time. If $\rho = \rho^*$, then the system stays in the bounded domain of the phase plane, and there exists an admissible disturbance such that the system has a periodic solution and never reaches the terminal state. If $\rho > \rho^*$, then there exist admissible disturbances such that the system (12) has unbounded solutions: $x \to \infty$ as $t \to \infty$. Hence, our system (12) under the simplified control is controllable, if and only if $\rho < \rho^*$, whereas under the game control it is controllable, if and only if $\rho < 1$. The same conclusion is true also for the systems (16). Hence, the bounded disturbance can be ignored while designing the control, if the ratio of the amplitudes of the disturbance and control is less than the golden section ratio ρ^*. Obviously, the game approach is preferable to the simplified one.

5 THE SECOND APPROACH

We consider again the system (1) (or (4)) and assume that all its motions under investigation belong to the set D of the $\boldsymbol{q}$-space. The set D may coincide with the whole space R^n. Suppose that

$$m\|\mathbf{z}\|^2 \leq (\boldsymbol{A}(\boldsymbol{q})\mathbf{z}, \mathbf{z}) \leq M\|\mathbf{z}\|^2\,, \quad 0 < m < M\,,$$

$$|\partial A_{ij}/\partial q_k| \leq C\,, \qquad i, j, k = 1, \ldots, n\,. \tag{17}$$

for any $\mathbf{z} \in R^n$ and all $\boldsymbol{q} \in D$. Here m, M, and C are positive constants. Let the uncontrollable forces $\boldsymbol{F}$ in (1) and (4) consist of two terms

$$\boldsymbol{F} = \boldsymbol{G} + \boldsymbol{\Phi}, \quad \boldsymbol{G} = (G_1, \ldots, G_n), \quad \boldsymbol{\Phi} = (\lambda_1, \ldots, \lambda_n) \tag{18}$$

subjected to different constraints.

The dissipative forces $\boldsymbol{G}(\boldsymbol{q}, \dot{\boldsymbol{q}}, t)$ satisfy two conditions:

$$1)\, (\boldsymbol{G}, \dot{\boldsymbol{q}}) \leq 0 \quad \text{for all } \boldsymbol{q} \in D\,, \text{ and all } t \geq t_0\,;$$
$$2)\, |G_k| \leq G_k^0(\alpha)\,, \quad k = 1, \ldots, n\,, \text{ if } |\dot{q}_i| \leq \alpha < \varepsilon_0 \text{ for all } i = 1, \ldots, n\,.$$

Here $\varepsilon_0 > 0$ is a positive number, $G_k^0(\alpha)$ are continuous monotone increasing functions for $\alpha \in [0, \varepsilon_0]$, and $G_k^0(\alpha) = 0$, $k = 1, \ldots, n$.

The forces $\boldsymbol{\Phi}$ in (18) are arbitrary but bounded disturbances such that

$$|\lambda_k| \leq \lambda_k^0 < Q_k^0\,, \qquad i = 1, \ldots, n \tag{19}$$

where $\lambda_k^0 > 0$ are given constants.

Let us introduce the sets in the $2n$-space $(q, \dot{q})$

$$\Omega_1 = \{(\boldsymbol{q}, \dot{\boldsymbol{q}}) \;:\; \boldsymbol{q} \in D\,;\; \exists k\,, |\dot{q}_k| > \varepsilon\}\,, \tag{20}$$

$$\Omega_2 = \{(\boldsymbol{q}, \dot{\boldsymbol{q}}) \;:\; \boldsymbol{q} \in D\,;\; \forall k\,, |\dot{q}_k| \leq \varepsilon\}\,.$$

The number $\varepsilon > 0$ will be chosen later. We define our feedback control in Ω_1 as follows

$$\begin{aligned} Q_k(\boldsymbol{q}, \dot{\boldsymbol{q}}) &= -Q_k^0 \operatorname{sign} \dot{q}_k \qquad \text{if } \dot{q}_k \neq 0\,, \\ Q_k(\boldsymbol{q}, \dot{\boldsymbol{q}}) &= 0 \qquad \text{if } \dot{q}_k = 0\,,\; k = 1, \ldots, n\,. \end{aligned} \tag{21}$$

By taking into account equations (18), (21), and the assumptions made, we obtain from (4)

$$dT/dt \leq (\boldsymbol{Q} + \boldsymbol{\Phi}, \dot{\boldsymbol{q}}) \leq -\sum_{k=1}^{n} (Q_k^0 - \lambda_k^0)\, |\dot{q}_k| \;\leq -r\, |\dot{\boldsymbol{q}}|\,,$$

$$r = \min_k (Q_k^0 - \lambda_k^0) > 0\,, \qquad k = 1, \ldots, n\,.$$

The upper estimate (17) for T implies that $|\dot{\boldsymbol{q}}| \geq (2T/M)^{1/2}$. Hence, we have in Ω_1:

$$2T^{1/2} dT^{1/2}/dt \leq -r(2T/M)^{1/2}\,, \quad \text{or} \quad dT^{1/2}/dt \leq -r(2M)^{-1/2}\,.$$

Integrating the latter inequality, we obtain that,if the motion starts in Ω_1, then the control (21) steers our system (4) to the boundary between Ω_1 and Ω_2 in finite time. The system reaches this boundary at the instant t_* such that

$$t_* - t_0 \leq (2M)^{1/2} r^{-1} \left[T_0^{1/2} - (m/2)^{1/2} \varepsilon\right]\,, \quad T_0 = (A(\mathbf{q}^0)\dot{\mathbf{q}}^0, \dot{\mathbf{q}}^0)/2\,.$$

In Ω_2, we rewrite equation (4) as follows

$$\ddot{q} = \boldsymbol{U} + \boldsymbol{V},\ \boldsymbol{U} = \boldsymbol{A}^{-1}\boldsymbol{Q},\ \boldsymbol{V} = \boldsymbol{A}^{-1}(\boldsymbol{G} + \boldsymbol{\Phi} + \boldsymbol{S}). \tag{22}$$

Let us regard $\boldsymbol{U}$ and $\boldsymbol{V}$ as a control and disturbance, respectively. It follows from (17)-(20) that

$$|V_k| \leq V^0,\ V^0 = m^{-1}[G^0(\varepsilon) + \lambda^0 + 1.5Cn^{5/2}\varepsilon^2], \tag{23}$$

$$G_k^0(\varepsilon) = \left\{ \sum_{k=1}^{n} \left[G_k^0(\varepsilon)\right]^2 \right\}^{1/2},\quad \lambda^0 = \left[\sum_{k=1}^{n} (\lambda_k^0)^2 \right]^{1/2}.$$

We impose the following constraints on $\boldsymbol{U}$

$$|U_k| \leq U^0,\ U^0 = rM^{-1}n^{-1/2},\ r = \min_k Q_k,\ k = 1,\ldots,n. \tag{24}$$

It follows from (17), (22), and (24) that

$$\boldsymbol{Q} = \boldsymbol{A}\boldsymbol{U},\ |Q_k| \leq \|\boldsymbol{A}\boldsymbol{U}\| \leq M\|\boldsymbol{U}\| \leq Mn^{1/2}U^0 = r \leq Q_k^0.$$

Hence, if the inequalities (24) hold, our original constraints (2) are satisfied. Equations (22) and constraints (23) and (24) can be rewritten in the form

$$\ddot{q}_k = U_k + V_k,\ |U_k| \leq U^0,\ |V_k| \leq V^0,\ k = 1,\ldots,n \tag{25}$$

quite similar to (12). The only difference is that the solution of the system (25) must satisfy the state constraint

$$|\dot{q}_k(t)| \leq \varepsilon, \qquad k = 1,\ldots,n \tag{26}$$

following from the requirement that our system should not leave the set Ω_2 from (20). The feedback control satisfying this additional condition can be obtained as a modification of (14), if $\rho = V^0/U^0 < 1$. We take

$$\begin{aligned}
U_k(q_k, \dot{q}_k) &= U^0 \operatorname{sign}\left[\psi(x_k) - \dot{q}_k\right] \quad \text{if } \dot{q}_k \neq \psi(x_k), \\
U_k(q_k, \dot{q}_k) &= U^0 \operatorname{sign} x_k = -U^0 \operatorname{sign} \dot{q}_k \quad \text{if } \dot{q}_k = \psi(x_k); \\
\psi(x) &= -\left[2U^0(1-\rho)|x|\right]^{1/2} \operatorname{sign} x \quad \text{if } |x| \leq x^*, \\
\psi(x) &= -\delta \operatorname{sign} x \quad \text{if } |x| \geq x^*;\ x_k = q_k - q_k^1,\ x^* = \delta^2\left[2U^0(1-\rho)\right]^{-1}; \\
& \delta \in (0,\varepsilon),\ \rho = V^0/U^0 < 1,\ k = 1,\ldots,n.
\end{aligned} \tag{27}$$

Here, δ is an arbitrary number from the interval $(0, \varepsilon)$, and x^* is chosen so that the function $\psi(x)$ is continuous. The switching curve $\dot{q}_k = \psi(x_k)$ defined by (27) is symmetrical with respect to the point $x_k = \dot{q}_k = 0$, lies in the band (26), and consists of two arcs of parabolas for $|x_k| \leq x^*$ and two rays $\dot{q}_k = \pm\delta$ for $|x_k| \geq x^*$. As shown in (Chernousko,

1992a, 1993b), if $\rho < 1$, the feedback control (27) brings the system (25) to the terminal state (7) in finite time under any admissible V_k, $|V_k| \leq V^0$, and the constraint (26) is satisfied. To ensure the condition $\rho < 1$, we should choose ε so that

$$\rho = V^0/U^0 = (M/m)n^{1/2}r^{-1}[G^0(\varepsilon) + \lambda^0 + 1.5Cn^{5/2}\varepsilon^2] < 1. \tag{28}$$

Since $G^0(\varepsilon) \to 0$ as $\varepsilon \to 0$, the condition (28) can be satisfied, if and only if

$$\lambda^0 < (m/M)rn^{-1/2}. \tag{29}$$

Hence, if (29) holds, our control problem is solvable by the second approach described in this section. The condition (29) is always satisfied, if the disturbances are absent ($\boldsymbol{\Phi} = 0$).

Let (29) hold. Then the parameter ε in (20) should be chosen so that $\varepsilon \leq \varepsilon_0$ and (28) holds. Then the feedback control $\boldsymbol{Q}(\boldsymbol{q}, \dot{\boldsymbol{q}})$, defined by (21) in Ω_1 and by $\boldsymbol{Q} = \boldsymbol{A}(\boldsymbol{q})\boldsymbol{U}$ and (27) in Ω_2, brings our system (4) from any initial state (6) to the prescribed terminal state (7) in finite time under any admissible disturbances. The upper estimate on the total time of motion is given by

$$\begin{aligned} t_1 - t_0 \;\leq\; & \delta^{-1}\max_k |q_k^0 - q_k^1| + (2M)^{1/2}r^{-1}\left[T_0^{1/2} - (m/2)^{1/2}\varepsilon\right] \\ & +(M/m)^{1/2}r^{-1}\delta^{-1}(T_0 - m\varepsilon^2/2) \\ & +(2\varepsilon^2 + 4\varepsilon\delta + 3\delta^2)\delta^{-1}\left[2U^0(1-\rho)\right]^{-1}, \qquad k = 1, \ldots, n, \\ \rho = V^0/U^0, \qquad & \delta \in (0, \varepsilon). \end{aligned}$$

Note that the parameter δ should be taken close to ε, because the greater is δ, the smaller is the time of motion.

6 DISCUSSION

Let us compare the two approaches described in Sections 3 and 5. The first approach is based on the assumptions imposed in the $2n$-dimensional $(\boldsymbol{q}, \dot{\boldsymbol{q}})$-space, whereas the second one deals with assumptions in the n-dimensional $\boldsymbol{q}$-space. Hence, the assumptions of the second approach are less restrictive. On the other hand, the second approach incorporates motions with low velocities in the set Ω_2 and, hence, the total time of motion may be greater for the second approach.

Both approaches do not presume that the external forces are known. These forces may be uncertain, with only the bounds on them being essential.

Equations (1) or (4) are typical for robotic manipulators. Here, q_k are generalized coordinates of a robot (angles of rotation and linear displacements for revolute and prismatic joints, respectively), whereas Q_k are control torques, forces, or electric voltages. Computer simulation of the proposed control laws for robots with different kinematics showed quite satisfactory results.

7 DISTRIBUTED PARAMETERS SYSTEMS

Let us consider briefly the decomposition of controlled distributed parameters systems described by linear partial differential equations of one of the two following types

$$w_t = Aw + v \tag{30}$$

$$w_{tt} = Aw + v \tag{31}$$

Here $w(\boldsymbol{x}, t)$ is a scalar state depending on time t and n-vector $\boldsymbol{x}$, v is a scalar control, and A is a linear selfadjoint differential operator of the even order $2m$ (in $\boldsymbol{x}$) independent of t. For example, if $m = 1$, $A = \Delta$, then (30) is a parabolic equation, whereas (31) is a wave equation; if $m = 2$, $A = -\Delta^2$, then (31) describes elastic vibrations of a beam ($n = 1$) or a plate ($n = 2$). Equations (30), (31) are considered in the domain $\boldsymbol{x} \in \Omega$, $t \geq 0$ under the homogeneous boundary conditions

$$Mw = 0\,, \qquad M = (M_1, \ldots, M_m)\,, \qquad \boldsymbol{x} \in \Gamma = \partial\Omega \tag{32}$$

where M_j is a linear differential operator of the order less than $2m$ with coefficients independent of t. The initial conditions for equation (31) are

$$w(\boldsymbol{x}, 0) = w_0(\boldsymbol{x})\,, \qquad w_t(\boldsymbol{x}, 0) = w_{t0}(\boldsymbol{x})\,, \qquad \boldsymbol{x} \in \Omega\,, \tag{33}$$

whereas in case of (30) the second condition (33) should be omitted. The control v in (30) and (31) is subject to the constraint

$$|v(\boldsymbol{x}, t)| \leq v^0\,, \qquad \boldsymbol{x} \in \Omega\,, \qquad t \geq 0 \tag{34}$$

where $v^0 > 0$ is a given constant.

We are to find the control v satisfying (34) and such that the corresponding solution of the initial boundary value problem described by (31)-(33) becomes zero at $t = T$: $w(\boldsymbol{x}, T) = w_t(\boldsymbol{x}, T) = 0$. For equation (30), one should omit the latter equality and the second condition (33). The time T is not fixed a priori.

Let us consider the eigenvalue problem

$$A\varphi = -\lambda\varphi\,, \qquad \boldsymbol{x} \in \Omega\,; \qquad M\varphi = 0\,, \qquad \boldsymbol{x} \in \Gamma \tag{35}$$

and assume that all eigenvalues $\lambda_k > 0$, $k = 1, 2, \ldots$, and $\lambda_1 \leq \lambda_2 \leq \ldots$. If there is also a zero eigenvalue, we put $\lambda_0 = 0$. We present the state and control as series in orthonormal eigenfunctions φ_k

$$w(\boldsymbol{x}, t) = \sum q_k(t)\varphi_k(\boldsymbol{x})\,, \qquad v(\boldsymbol{x}, t) = \sum u_k(t)\varphi_k(\boldsymbol{x}) \tag{36}$$

Substituting (36) into (30) and (31) and using (35), we obtain the respective equations for q_k

$$\dot{q}_k + \lambda_k q_k = u_k\,, \qquad \ddot{q}_k + \lambda_k q_k = u_k\,, \qquad \lambda_k \geq 0\,. \tag{37}$$

The initial conditions (33) are replaced by

$$q_k(0) = \int_\Omega w_0 \varphi_k dx\,, \qquad \dot{q}_k(0) = \int_\Omega w_{t0} \varphi_k dx\,. \tag{38}$$

Let us impose the constraints on controls u_k:

$$|u_k| \le U_k\,, \qquad t \ge 0\,. \tag{39}$$

The bounds $U_k > 0$ should be chosen so that the constraint (34) holds. It is sufficient to require that

$$\sum U_k N_k \le v^0\,, \qquad N_k = \max_{\boldsymbol{x}\in\Omega} |\varphi_k(\boldsymbol{x})|\,. \tag{40}$$

Since each equation (37) is controllable, we can choose the time-optimal control u_k bringing the kth mode to the zero terminal state under the constraint (39). For the first-order equations (37), we have

$$\begin{aligned} u_k(q_k) &= -U_k \operatorname{sign} q_k\,, \qquad q_k \ne 0\,, \qquad u_k(0) = 0\,, \\ T_k &= \lambda_k^{-1} \ln\left[1 + \lambda_k\, |q_k(0)|\, U_k^{-1}\right], \qquad \lambda_k > 0\,, \\ T_0 &= |q_0(0)|\, U_0^{-1}\,, \qquad \lambda_0 = 0 \end{aligned} \tag{41}$$

where T_k is the minimal time for the kth mode. To minimize the total time of motion $T = \max_k T_k$, we should choose U_k in such a way that all T_k are equal: $T_k = T$, $k \ge 0$. We obtain the following condition on T

$$\sum \lambda_k \left[\exp(\lambda_k T) - 1\right]^{-1} |q_k(0)|\, N_k \le v^0\,. \tag{42}$$

It can be shown that the series (42) converges for all $T > 0$, and its sum decreases monotonically from ∞ to 0 as T changes from 0 to ∞. Hence, there exists T such that (42) holds, and our control problem is solved by the proposed method. The control for each kth mode is given in an explicit form.

For the second-order equations (37), the situation is more complicated. The feedback time-optimal control $u_k(q_k, \dot{q}_k)$ for each subsystem (37) under the constraints (39) is again given in an explicit form (Pontryagin et al, 1972), but the corresponding minimal times T_k are not expressed explicitly. We have obtained upper estimates on these times and used them to derive sufficient controllability conditions for the system (31). These conditions can be presented as convergence conditions for certain series (Chernousko, 1992b). For example, if the following two series

$$\sum \lambda_k^{1/2} |q_k(0)|\, N_k < \infty\,, \qquad \sum |\dot{q}_k(0)|\, N_k < \infty \tag{43}$$

converge, then the system (31) is controllable by the proposed method.

Using (43), sufficient controllability conditions for various specific cases were obtained, including control of elastic systems (membranes, beams, and plates of various shapes under different boundary conditions) by means of bounded distributed forces.

For example, let us consider vibrations of an elastic beam controlled by distributed forces. Equation (31) and the control constraint (34) can be presented as follows

$$w_{tt} = -w_{xxxx} + v\,, \qquad |v| \leq v^0\,, \qquad a \leq x \leq b\,.$$

The boundary conditions for the case of hinged support at the both ends of the beam are $w = w_{xx} = 0$ for $x = a$, $x = b$. It can be shown that the both series (43) converge, if the initial distributions of elastic displacements $w_0(x) = w(x, 0)$ and velocities $w_{t0}(x) = w_t(x, 0)$ of the beam satisfy the following smoothness conditions:

$$w_0 \in C^4\,, \qquad w_{t0} \in C^2\,, \qquad x \in [a, b]\,.$$

8 CONCLUSION

The results presented above are based on the decomposition of the original system into simple subsystems and use optimal control methods. Since the time of motion is minimized for each subsystem, these methods may be regarded as suboptimal. The total time of motion is finite: the motion stops when the last degree of freedom reaches its terminal state. Upper estimates on terminal times are given.

9 ACKNOWLEDGEMENTS

The paper is based on research sponsored by the International Science Foundation, Grant M4F000, and by the Russian Foundation of Basic Research, Project 93-01-16286.

REFERENCES

Ananyevskii, I.M., Dobrynina, I.S. and Chernousko, F.L. (1995) Method of decomposition in the control problem for a mechanical system. *Izvestiya of the Russian Academy of Sciences, Control Theory and Systems*, **2**, 3-14.

Chernousko, F.L. (1990) Decomposition and suboptimal control in dynamical systems. *Journal of Applied Mathematics and Mechanics (PMM)*, **54**, 6, 727-34.

Chernousko, F.L. (1992a) Control synthesis in a non-linear dynamical system. *Journal of Applied Mathematics and Mechanics (PMM)*, **56**, 2, 157-66.

Chernousko, F.L. (1992b) Bounded controls in distributed-parameter systems. *Journal of Applied Mathematics and Mechanics (PMM)*, **56**, 5, 707-23.

Chernousko, F.L. (1993a) Decomposition and suboptimal control in dynamic systems. *Optimal Control Applications and Methods*, **14**, 2, 125-43.

Chernousko, F.L. (1993b) The Decomposition of Controlled Dynamic Systems, in *Advances in Nonlinear Dynamics and Control : A Report from Russia* (ed. A.B. Kurzhanski), Birkhäuser, Boston.

Krasovskii, N.N. (1970) *Game Problems of the Encounter of Motions.* Nauka, Moscow.

Pontryagin, L.S., Boltyansky, V.G., Gamkrelidze, R.V. and Mishchenko, E.F. (1972) *Mathematical Theory of Optimal Processes.* Wiley-Interscience, New York.

3

Network flow — Theory and applications with practical impact

Masao Iri
Department of Information and System Engineering
Faculty of Science and Engineering, Chuo University
1-13-27 Kasuga, Bunkyo-ku, Tokyo 112, Japan.
Tel: +81-3-3817-1690. Fax: +81-3-3817-1681.
e-mail: iri@ise.chuo-u.ac.jp

Keywords
Network flow, history, applications, duality, future

1 INTRODUCTION

The research in this subject has now a long history and a large variety of related fields in science, engineering and operations research, and is still developing fast, especially in its algorithmic aspects, attracting many excellent senior as well as young researchers. And the space allotted to this paper is limited. So, in this paper I will not lay so much stress on particular frontiers of research in the subject as on the wider and historical viewpoint.

Network-flow theory is one of the best studied and developed fields of optimization, and has important relations to quite different fields of science and technology such as combinatorial mathematics, algebraic topology, electric circuit theory, nonlinear continuum theory including plasticity theory, geographic information systems, VLSI design, *etc.*, *etc.*, besides standard applications to transportation, scheduling, *etc.* in operations research.

An important point of view from which we should look at network flow theory is the "duality" viewpoint, to which due attention was paid at the earlier stages of development of the theory but which tends to be gradually ignored as time goes.

Since the papers and books published on network flow are too many to cite here, I do not intend to compile even an incomplete list of references but only to cite a few which I directly mention and which I think tend to become obscured during the long history in spite of their ever lasting importance. Happily enough, we saw a great book (Ahuja, Magnanti and Orlin, 1993) recently published on network flows containing most up-to-date informations, as well as an extensive bibliography, on related mathematical theories, algorithms and applications. So, I will try to make the overlap of this survey paper of mine with this book as little as possible, i.e., to make the paper complementary to this book as much as possible. As may be natural, the standpoint of my survey is substantially influenced by the rather old book of mine (Iri, 1969), and details of a number of topics which are only briefly mentioned here will be found in this book.

2 WHAT IS NETWORK FLOW?

A flow network $N = (G; \xi, \eta, \zeta, \cdots)$ is a graph $G = (V, E, \partial^+, \partial^-)$ on which real quantities called flows ξ, tensions η and potentials ζ are attached to its edges E and vertices V with the Kirchhoff laws satisfied, where some additional constraints are imposed on flows and tensions, and an appropriate objective (or objectives) is usually assumed. (See Figure 1.)

$G = (V, E, \partial^+, \partial^-)$ $\quad V$: vertex set, $\quad E$: edge set

$\qquad \partial^+ : E \to V, \quad \partial^- : E \to V \quad$ (incidence functions)

$N = (G; \xi, \eta, \zeta, \cdots) \quad \xi : E \to \mathbf{R}, \quad \eta : E \to \mathbf{R}, \quad \zeta : V \to \mathbf{R}$

Kirchhoff's laws:

$$\sum\{\xi(e) \mid e \in E,\ \partial^+ e = v\} = \sum\{\xi(e) \mid e \in E,\ \partial^- e = v\} \quad \text{for } \forall v \in V;$$

$$\exists\zeta : \ \eta(e) = \zeta(\partial^+ e) - \zeta(\partial^- e) \quad \text{for } \forall e \in E.$$

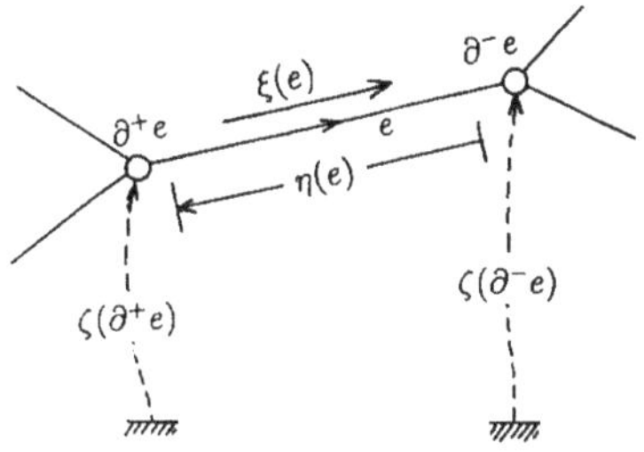

Figure 1 Basic elements of a network.

The most typical three network flow problems are

1° Maximum Flow Problem,

2° Shortest Path Problem,

3° Minimum-Cost Flow/Tension Problem,

each of which admits a natural interpretation in applications.

3 DAWN OF THE THEORY

Although there had been a number of earlier pioneering works, we owe the present-day formalism of network flow theory essentially to works by Minty, Fulkerson, Berge, et al. in the 1950–1960's, which are found in (Minty, 1960), (Berge et Ghouila-Houri, 1962), (Ford and Fulkerson, 1962), (Minty, 1966). Personally I should be glad if I were allowed to count the already-mentioned book of mine (Iri, 1969) among them.

4 EXPANDING HORIZONS OF APPLICATION

Once the basic formalism of the theory was established, the horizons of application rapidly expanded to various fields of operations research such as assignment, transportation and scheduling, and even to branches of mathematics itself such as combinatorial mathematics. However, from the standpoint of more application-oriented people including myself, this type of statement may seem to be inappropriate. It might better be stated that accumulation of a pile of individual practically solved examples and unsolved needs motivated people to establish a novel theory. Among those examples we may count the now "prehistoric" works such as (Monge, 1781) and (Kantorovitch, 1942).

5 RELATED FIELDS

There are a number of classical as well as modern fields of physical sciences and engineering where the same or similar logical structure is found as network-flow theory. In particular, the theory of nonlinear monotone resistive electric networks is *exactly* the same as (single-commodity) network-flow theory, although, regrettably, there is sometimes allergy to "electricity" among network-flow people — at least around me in Japan.

It is interesting to note that, in the earlier age of development of network-flow theory, "electrical" viewpoint was emphasized by a few people almost simultaneously, *e.g.* , in (Dennis, 1959), (Iri, 1959), (Minty, 1960).

In an electric network whose topology is determined by a graph $G = (V, E, \partial^+, \partial^-)$,

electric current $\xi(e)$ in edge $e \in E$ is interpreted as the flow in it,

electric voltage $\eta(e)$ across edge $e \in E$ as the tension across it,

and

electric potential $\zeta(v)$ at vertex $v \in V$ as the potential there.

For each edge $e \in E$ the current/flow and the voltage/tension are related to each other in terms of a monotone nondecreasing function, of which noteworthy special cases are shown in Figure 2.

An electric network to represent a (two-terminal) *maximum flow problem* can be composed of a single edge of type (ii) (whose two end vertices represent the two terminals of the network) and all the other edges of types (i) and (iii). An electrical network to represent a (two-terminal) *shortest path problem* can be composed of a single edge of type (i) and all the other edges of types (ii) and (iii). An electric network to represent a *minimum-cost flow problem* (including transportation, assignment, *etc.*) as well as a minimum-cost tension problem (including PERT/CPM-type scheduling problems, *etc.*) can be composed of edges of types (i), (ii) and (iii).

All these problems are "linear" network flow problems, but the electrical networks for them contain "highly nonlinear" elements, namely, diodes. If edges of type (iv) (which are most popular "linear" elements in electric networks) are contained in the network, then we shall have a "quadratic-cost" network flow problem.

(i) currrent source for which $\xi(e)$ =given constant irrespective of $\eta(e)$:

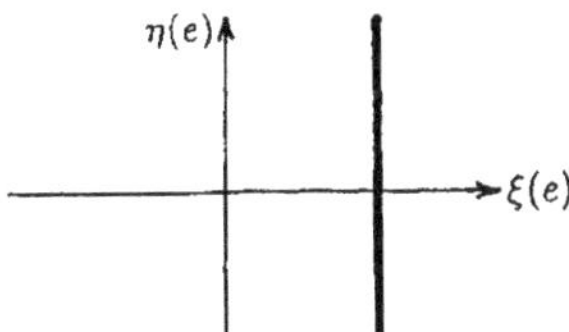

(ii) voltage source for which $\eta(e)$ =given constant irrespective of $\xi(e)$:

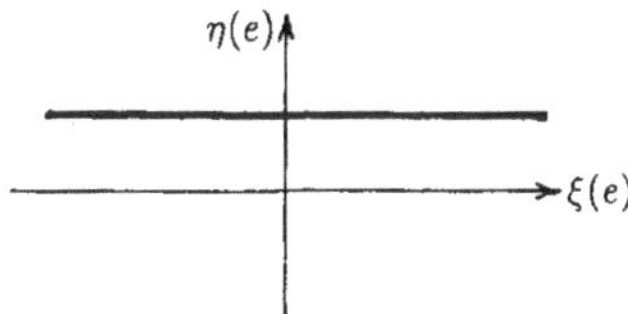

(iii) diode (inequality) for which $\xi(e) \geq 0$, $\eta(e) \geq 0$, $\xi(e) \cdot \eta(e) = 0$:

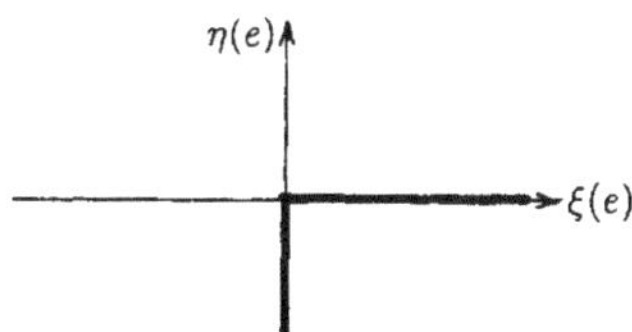

(iv) linear resistor for which $\eta(e) = r(e)\xi(e)$
where $r(e)$ is the resistance (a positive constant):

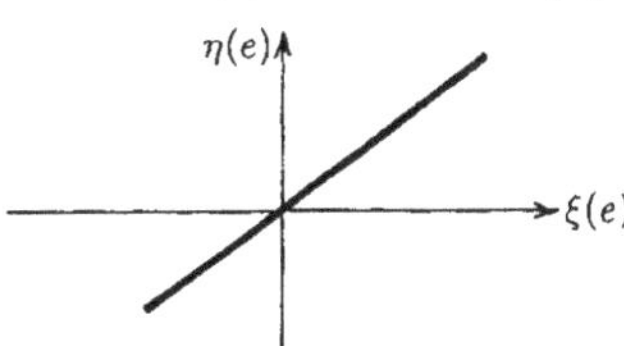

Figure 2 Typical flow-tension (or current-voltage) relations.

6 ALGORITHMS DEVELOPED

Around 1970 there was a discontinuous progress in graph and network-flow algorithms owing to Hopcroft and Tarjan's epoch-making works. Since then, a tremendously large number of works have been done to make the related algorithms faster and faster, and now we have a big family of network algorithms some of which make use of highly sophisticated data-structures sometimes adversely called "acrobatic".

Since linear network-flow problems can be solved much faster by means of proper algorithms than the corresponding linear-programming problems are solved by means of the simplex or interior-point algorithms, and that the tractable problem size is far larger, the reducibility of a linear-programming problem to a network-flow problem naturally attracts the researchers' interests. Here again, it is interesting to note that this reducibility problem is equivalent to the so-called "graph-realization problem" which has been studied for the purpose of application to the topological synthesis of electrical networks.

The history of research on this topic is long, starting from those in early days, such as (Iri, 1962), (Iri, 1966), (Iri, 1968), through a number of researches, *e.g.* , (Bixby and Cunningham, 1980), until the latest, *e.g.* , (Gautier and Granot, 1994). Probably, (Fujishige, 1980) gives a best result for the graph realization algorithm.

7 SINGLE-COMMODITY VS. MULTICOMMODITY

There are more practical problems which are modelled as a multicommodity flow problem than those modelled as a single-commodity flow problem, but from the point of view of solution algorithm, the former shares the advantage of the latter only partially. This is mainly because the nice property of "total unimodularity" possessed by the single-commodity flow is not valid for the multicommodity any longer.

Although there have been a lot of progress in algorithmic and technical aspects of solutions of multicommodity flow problems, the old survey (Jewell, 1966) still remains to be an excellent general introduction to multicommodity flow problems.

Traditional empirical "incremental assignment" algorithms developed and used by rather practical traffic assignment people seemingly motivated a number of decomposition principles and column-generation techniques, and the incremental-type algorithms have become theoretically better grounded. Efficient solution algorithms have been found for special classes of problems including various VLSI design problems.

There are the dual counterparts (see §8) of multicommodity flow problems which may be called "multi-commodity tension" problems and which represent some characteristics of multi-job scheduling problems, although they are less popular.

8 DUALITY

The *duality* is one of the most conspicuous features of network-flow theory. One can better understand the theory with the concept of duality and apply the theory to wider practical fields than without it. It seems that more stress was laid on the duality in the earlier stages of development of the theory than nowadays. (Berge et Ghouila-Houri, 1962) was duality-conscious, whereas (Ford and Fulkerson, 1962) was less conscious. Viewed from matroid theory (cf. (Minty, 1966)) the duality is a very natural structure of network-flow theory, i.e., that "part" of network flow theory which can be stated in terms of concepts related to *edges only* without reference to vertices possesses a "dual" logical structure. This duality of the theory comes from the duality of graph theory, and every proposition for flows has its dual counterpart for tensions, and vice versa.

Sometimes the concept of the "duality of the theory" (i.e., the dualistic structure of the theory itself, which is a "meta-theoretical" concept as it were) is confused with the "duality theorem" on optimization problems or their "dual problems", i.e., with the "mathematical-programming duality". But the two must be clearly distinguished from each other although they may coincidentally lead us to one and the same pair of propositions for some problems. Thus, in some cases, for a pair of dual problems or duality theorems in the sense of mathematical programming, we may have the meta-theoretically dual pair, i.e., we may have a quadruple of problems or theorems. Figure 3 gives an example.

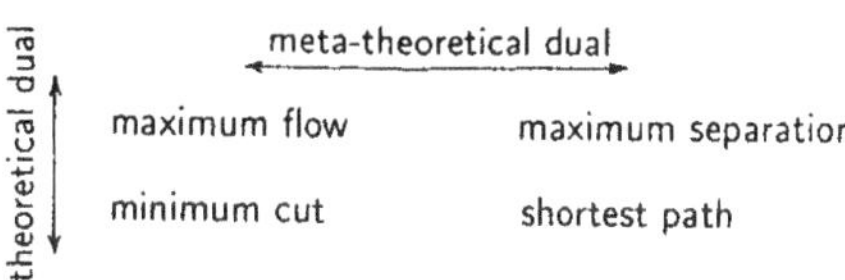

Figure 3 Two kinds of duality.

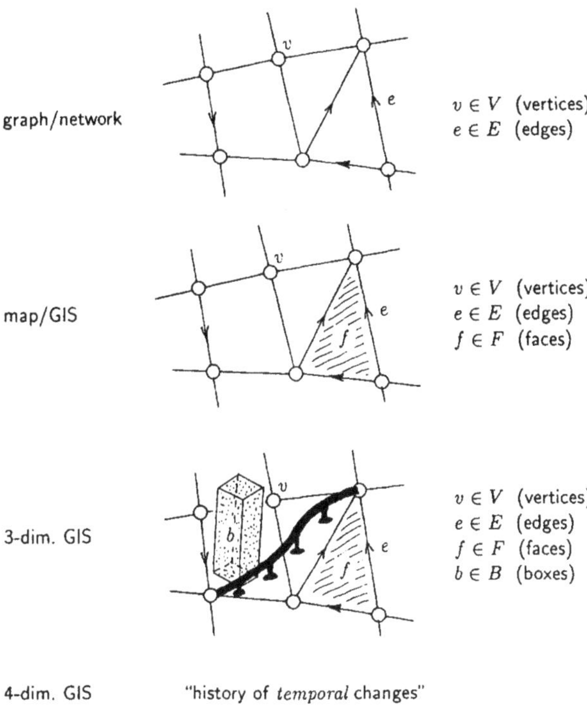

Figure 4 From network to 4-dimensional GIS.

In addition, it is worthwhile to note that traditionally well-known "triangular inequalities" for *distances* $d_{\alpha\beta}$'s (between pairs of vertices in a network) and for *terminal-pair capacities* $c_{\alpha\beta}$'s (in a capacitated network) ($v_\alpha, v_\beta, v_\gamma \in V$):

$$d_{\alpha\gamma} \leq d_{\alpha\beta} + d_{\beta\gamma}, \qquad c_{\alpha\gamma} \geq \min(c_{\alpha\beta}, c_{\beta\gamma})$$

are not the meta-theoretical dual of each other, so that we do have two dual counterparts of theirs. To find them will be an enjoyable exercise. See, for details, (Iri, 1981).

9 EXTENSIONS

Techniques and, what is more important, *spirits*, fostered for network flows are extended in various directions.

9.1 Dimension

Topologically, a graph may be regarded as a one-dimensional topological complex. Extension in dimension will lead us to 2-dimensional complexes, of which the simplest but practically most useful examples are map data systems or GIS (geographic information systems, now becoming a big industry). The latter may be extended to 3-dimensional GIS to deal with underground and sky information, and further to 4-dimensional GIS to deal also with time sequences or temporal changes. (See Figure 4.)

9.2 Underlying algebraic structure

With respect to the shortest-path problem which makes a simplest class of network-flow problems, the so-called "min-addition algebra" plays an interesting role. It is an algebraic structure on the set of real numbers and ∞ in which ordinary "addition" and "multiplication" are replaced, respectively, by "min" and "addition". Earlier references may be (Carré, 1971a), (Carré, 1971b). Extensions and applications of this kind of algebraic structure have been intensively studied by Burkard's school in Graz. (See also §9.3.)

Mechanical structures such as trusses and frames may be regarded as an extension in this direction where vector-tensor flows and tensions are attached to the underlying graph, and the graph is not merely a "topological" existence but "geometric" in the sense that the continuity of forces and moments as well as the compatibility of deformations (corresponding to Kirchhoff's laws for ordinary flows and tensions) is expressed in terms not only of the topology of the structure but also of its geometry, i.e., lengths and angles.

9.3 Leaky Kirchhoff

Network flows with gains and losses come out by making Kirchhoff's laws "leaky" (see Figure 5). An earlier but rather complete presentation is found in (Jewell, 1962).

It is interesting to see the "min-addition algebra" naturally extended to this kind of problems to produce a new class of algebras which, probably, has been given no proper name yet (Iri, Amari and Takata, 1961), (Iri, Amari and Takata, 1968).

This extension was obviously motivated by application to machine loading problems, which we may find in the monumental book on operations research, i.e., (Churchman,

Ackoff and Arnoff, 1957), and rather extensive studies by (then) Roumanian group (K. Eisemann, J. R. Lourie, E. Balas and P. L. Ivanescu) are found in three papers (pp. 154–202) of Vol. 11, No. 1 (1964) of *Management Science.*

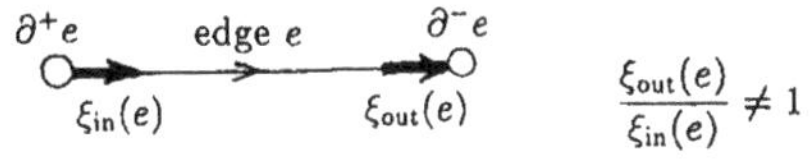

$\partial^+ e$ edge e $\partial^- e$

$\xi_{in}(e)$ $\xi_{out}(e)$ $\frac{\xi_{out}(e)}{\xi_{in}(e)} \neq 1$

Figure 5 Edge with gain/loss.

9.4 Extra constraints

A small number of extra constraints imposed on an ordinary network flow problem can be treated without seriously deteriorating the efficiency of the solution algorithms. Thus flows with bundle constraints as well as tensions with bundle constraints were studied in earlier days. The idea of flow bundle is found in (Berge et Ghouila-Houri, 1962) and that of tension bundle (although not so named) in (Jewell, 1965). This type of problems seems still attracting researchers' interest (see, *e.g.* , (Gautier and Granot, 1994)). Further extension is to consider "submodular" constraints on flows in edges (Fujishige, 1991).

However, if we admit to introduce a large number of bundle constraints and to make Kirchhoff's laws leaky, we shall be led to fairly general optimization problems (containing linear programming problems), Rockafellar's "monotropic optimization" being essentially of this type (Rockafellar, 1984).

9.5 Multiple objectives

Compromise between two (or more) conflicting objectives is an interesting research subject which is important in application. For example, the problem of choice between "time" and "cost" when we use a high-way network is of this kind. We can trace the origin of the idea back to (Lawler, 1967), at least, and the extension in this direction will add a flavour of "fractional programming" to network flows. See, *e.g.* , (Megiddo, 1979).

9.6 Discretum to continuum

One of the earliest problems in the history of optimization was the *continuous* transportation problem (Monge, 1781), (Kantorovitch, 1942). It was a problem on flows in a field, not in a network. But recent trend of research consists in "generalizing" results in network flow theory to flows (and tensions) in continua. Also, there are interesting and challenging problems of *approximating a continuum (i.e., field) by a discretum (i.e., network) and vice versa.* (See (Iri, 1980), (Taguchi and Iri, 1982).)

It seems that there are several different approaches to model, or define, a problem of flows in continua, *e.g.* , (Hu, 1969), (Strang, 1983), (Blum, 1990), (Nozawa, 1994).

10 MONOTONICITY, NON-AMPLIFICATION, SENSITIVITY AND RECIPROCITY

These concepts are familiar in electric network theory but not so much in network flow theory. However, in principle, the concepts themselves as well as the related results in the former theory can be transferred to the latter. They may be summarized as follows.

1° Monotonicity:— If the flow in and the tension across every edge of a network are in monotone nondecreasing relation to each other, then, between any two vertices in the network, the flow and the tension are also in a monotone nondecreasing relation, i.e., the two-terminal characteristic is monotone.

2° Non-amplification:— If a current source of value ε is attached in parallel to an edge (*e.g.* , the capacity of the edge is perturbed by ε), then the corresponding perturbation of the flow in any edge does not exceed ε in absolute value (*e.g.* , the value of the maximum flow does not change by more than ε). Dually, if a voltage source of value ε is inserted in series to an edge (*e.g.* , the length of the edge is perturbed by ε), then the corresponding perturbation of the tension across any edge does not exceed ε (*e.g.* , the shortest distance does not change by more than ε).

3° Sensitivity:— The sensitivity is the variation of flows and tensions against some perturbation in the flow-tension relation of an edge. The "out-of-kilter" method (called so in (Ford and Fulkerson, 1962) but the idea had been presented in (Minty, 1960) in a more general form) affords a method of calculating the variation concretely.

4° Reciprocity:— We assume that the pair of flow and tension of every edge of the network is at a point where the relation between them has a well-defined tangent (i.e., the relation is "differentiable" at that point). Let, furthermore, the variation in the tension across an edge e_j when a current source of value $\Delta\xi(e_i)$ is attached in parallel to another edge e_i be denoted by $\Delta\eta(e_j)$. Then we have

$$\frac{\Delta\eta(e_2)}{\Delta\xi(e_1)} = \frac{\Delta\eta(e_1)}{\Delta\xi(e_2)}.$$

The dual statement is valid, too.

The common-sense "monotonicity" conjecture that *any additional investment on a system will result in some improvement on the system function/performance* sometimes fails, in particular for multicommodity flow networks. The famous Braess paradox (Braess, 1968) says that there is the case where construction of a new road will result in increase of each user's travel time, thus deteriorating the system performance in any sense. A few other paradoxes of similar kind are known in traffic assignment theory. Quite recently, a popular-science article (Arnott and Small, 1994) was published, which indicates that the problem is still of hot interest in the field of urban design.

The basic principles underlying traffic assignment are so-called "Wardrop's laws" (Wardrop, 1952), where two different concepts of optimality, i.e., "system optimization" and "user optimization", are enunciated and their mutual relations discussed.

11 RESTRUCTURING THE THEORY NEEDED?

In ordinary network-flow theory, cancellation of flows in an edge is allowed — or even necessary in many algorithms of augmenting-path type — when two or more flows are superposed (or added), which is a barrier hard for the beginner to jump over. By making use of it, we can get an optimum plan. And we anticipate users of a system to obey the plan afforded by the planner. However, in the real world of the new century, which will certainly be a century of multiple values, we cannot expect users of a system to be obedient to the planner, i.e., we must be prepared to assume that users may use the system in whatever way physically possible. In such circumstances, cancellation of flows will not be a realistic operation. So, there arises an idea of restructuring, from scratch, network flow theory without allowing cancellation of flows. Then what will happen as the result of such restructuring? A lot remains to be investigated, but there is an essay in that direction (Iri 1994). Among a number of examples there, I will take up one.

In the two-terminal capacitated network of Figure 6 (left), where all the edges are one-way and the numbers beside edges indicate the capacities of edges, the value of a maximum flow in the ordinary sense is obviously equal to 2. Let us then consider the unit flow along the path shown in Figure 6 (right). Since the flow saturates all the three edges on the path which together make a cut between the entrance vertex and the exit vertex, we cannot augment the flow any more without cancelling the flow in the vertical edge. Thus, the flow of Figure 6 (right) is a "maximal" (or "blocking", as is sometimes called) flow from the entrance to the exit — of the minimum value in this case.

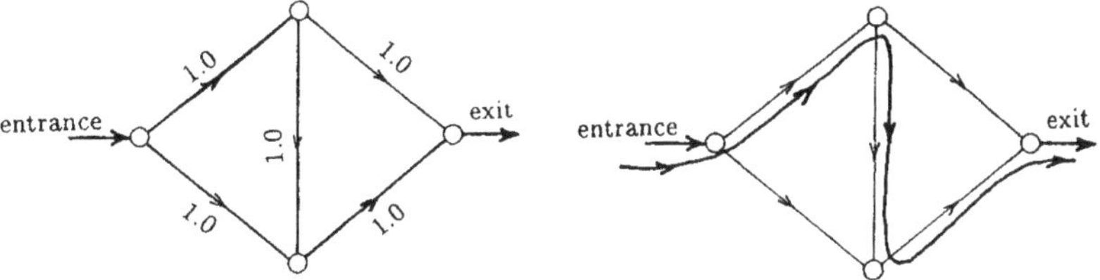

Figure 6 A capacitated network (left) and a unit flow along a path (right).

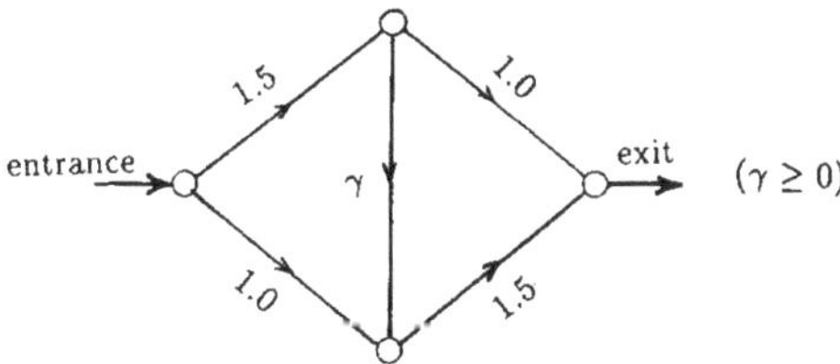

Figure 7 Another capacitated network with a parametric capacity γ.

It may be interesting to note that the dependence of the minimum value of maximal two-terminal flows on edge capacities is far from being monotone. For example, look at the network of Figure 7, where the numbers indicate edge capacities as before and γ is the parametric capacity of the vertical edge. Then the dependence of the minimum value of maximal flows on γ will be seen to be as shown in Figure 8.

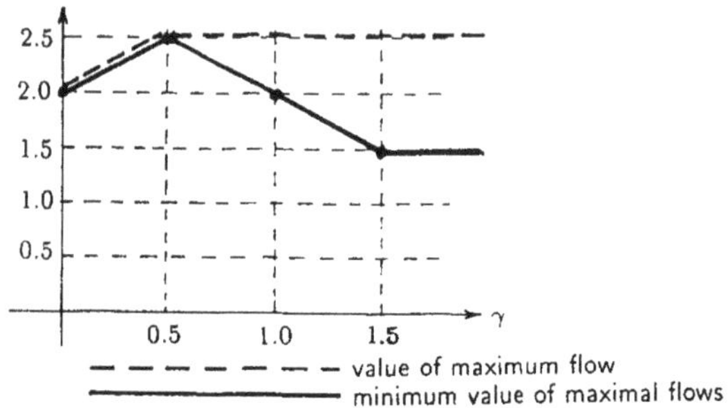

Figure 8 Minimum value of maximal flows against γ.

REFERENCES

Ahuja, R.K., Magnanti, T.L. and Orlin, J.B. (1993) *Network Flows — Theory, Algorithms, and Applications.* Prentice Hall, Englewood Cliffs, New Jersey.

Arnott, R. and Small, K. (1994) The economics of traffic congestion. *American Scientists*, **82**, 446–55.

Berge, C. et Ghouila-Houri, A. (1962) *Programmes, Jeux et Réseaux de Transport.* Dunod, Paris.

Bixby, R.E. and Cunningham, W.H. (1980) Converting linear programs to network programs. *Mathematics of Operations Research*, **5**, 321–57.

Blum, W. (1990) A measure-theoretical max-flow-min-cut problem. *Mathematische Zeitschrift*, **205**, 451–70.

Braess, D. (1968) Über ein Paradoxon aus der Verkehrsplanung. *Unternehmensforschung*, **12**, 258–684.

Carré, B.A. (1971a) An algebra for network routing problems. *Journal of the Institute of Mathematics and Its Applications*, **7**, 273–94.

Carré, B.A. (1971b) An elimination method for minimal cost network flow problem, in *Large Sparse Sets of Linear Equations* (ed. J.K. Reid), Academic Press.

Churchman, C.W., Ackoff, R.L. and Arnoff, E.L. (1957) *Introduction to Operations Research.* John Wiley & Sons, New York, 377–9.

Dennis, J.B. (1959) *Mathematical Programming and Electrical Networks.* MIT Technology Press, Cambridge, Massachusetts, and John Wiley & Sons, New York.

Ford,L.R., Jr., and Fulkerson, D.R. (1962) *Flows in Networks.* Princeton University Press, Princeton, New Jersey.

Fujishige, S. (1980) An efficient PQ-graph algorithm for solving the graph-realization problem. *Journal of Computers and System Sciences*, **21**, 63–86.

Fujishige, S. (1991) *Submodular Functions and Optimization.* Annals of Discrete Mathematics **47**, North-Holland, Amsterdam-New York-Oxford-Tokyo.

Gautier, A. and Granot, F. (1994) On the equivalence of constrained and unconstrained flows. *Discrete Applied Mathematics*, **55**, 113–32.

Hu, T.C. (1969) *Integer Programming and Network Flows.* Addison-Wesley, Massachusetts.

Iri, M. (1959) Algebraic and topological theory of the problems of transportation networks with the help of electric circuit models. *RAAG Research Notes*, Third Series, No. 13.

Iri, M., Amari, S. and Takata, M. (1961) Linear programming with weak graphical representation. *RAAG Research Notes*, Third Series, No. 47.

Iri, M. (1962) A necessary and sufficient condition for a matrix to be a loop or cut-set matrix of a graph and a practical method for the topological synthesis of networks. *RAAG Research Notes*, Third Series, No. 50.

Iri, M. (1966) A criterion for the reducibility of a linear programming problem to a linear network-flow problem. *RAAG Research Notes*, Third Series, No. 98.

Iri, M. (1968) On the synthesis of loop and cutset matrices and the related problems. *RAAG Memoirs*, **4**, A-XIII, 4–38.

Iri, M., Amari, S. and Takata, M. (1968) Algebraical and topological theory and methods of linear programming with weak graphical representation. *RAAG Memoirs*, **4**, Misc-VII, 539–620.

Iri, M. (1969) *Network Flow, Transportation and Scheduling — Theory and Algorithms.* Academic Press, New York and London.

Iri, M. (1980) Theory of flows in continua as approximation to flows in networks, in *Survey of Mathematical Programming* (ed. A. Prékopa), (Proceedings of the IXth International Symposium on Mathematical Programming, Budapest, 1976), North-Holland, **2**.

Iri, M. (1981) "Dualities" in graph theory and in the related fields viewed from the metatheoretical standpoint, in *Graph Theory and Algorithms*, (eds. N. Saito and T. Nishizeki), Lecture Notes in Computer Science **108**, Springer-Verlag, Berlin-Heidelberg-New York.

Iri, M. (1994) *An Essay in the Theory of Uncontrollable Flows and Congestion.* TRISE (Technical Report on Information and System Engineering) 94-03, Faculty of Science and Engineering, Chuo University. (Also presented at the XVth International Symposium on Mathematical Programming, August 1994, University of Michigan, Ann Arbor, Michigan.)

Jewell, W.S. (1962) Optimal flows with gains. *Operations Research*, **10**, 476–99.

Jewell, W.S. (1965) Divisible activities in critical path analysis. *Operations Research*, **13**, 747–60.

Jewell, W.S. (1966) *A Primal-Dual Multi-commodity Flow Algorithm.* Report ORC 66-24, OR Center, University of California, Berkeley, California.

Kantorovitch, L. (1942) On the translocation of masses. *Comptes Rendus (Doklady) de l'Academie des Sciences de l'URSS*, **XXXVII**, 199–201 (Mathematics).

Lawler, E.L. (1967) Optimal cycles in doubly weighted linear graphs, in *Theory of Graphs* (ed. P. Rosenstiehl), Dunod, Paris.

Megiddo, N (1979) Combinatorial optimization with rational objective functions. *Mathematics of Operations Research*, **4**, 414–24.

Minty, G.J. (1960) Monotone networks. *Proceedings of the Royal Society*, **A251**, 194–212.

Minty, G.J. (1966) On the axiomatic foundation on the theories of directed linear graphs, electrical networks and network-programming. *Journal of Mathematics and Mechanics*, **15**, 485–520.

Monge, G. (1781) Déblai et remblai. *Mémoires de l'Academie des Sciences.*

Nozawa, R. (1994) Examples of max-flow and min-cut problems with duality gaps in continuous networks. *Mathematical Programming*, **63**, 213–34.
Rockafellar, R.T. (1984) *Network Flows and Monotropic Optimization.* John Wiley & Sons, New York.
Strang, G. (1983) Maximal flow through a domain. *Mathematical Programming*, **26**, 123–43.
Taguchi, A. and Iri, M. (1982) Continuum approximation to dense networks and its application to the analysis of urban road networks. *Mathematical Programming Study*, **20**, 178–217.
Wardrop, J.G. (1952) Some theoretical aspects of road traffic research. *Proceedings of the Institute of Civil Engineers*, **1**, 325–78.

4

The mathematical theory of evidence – A short introduction

Jürg Kohlas
Institute of Informatics
University of Fribourg, CH 1700 Fribourg, Switzerland.
e-mail: Juerg.Kohlas@unifr.ch

Abstract
Evidence or Dempster-Shafer theory is used to model information which is both uncertain and imprecise. Such a piece of information can be captured by the mathematical model of a hint. It is shown how hints can be combined and used to judge hypotheses by degrees of support and plausibility. Applications of this theory to statistical inference, diagnostics and risk analysis, and to decision analysis are discussed. The practical implementation of Dempster-Shafer theory depends on appropriate computational architectures both for modeling and for the inference mechanisms. A fundamental scheme for local computation in hypertrees is presented.

1 ORIGINS

The term "Evidence Theory" was coined by Glenn Shafer in his book "A Mathematical Theory of Evidence", published by Princeton University Press in 1976. This work was initiated by a course on statistical inference taught by Arthur Dempster at Harvard University. Dempster developed there a theory of lower and upper probabilities in an attempt to reconcile Bayesian statistics with Fisher's fiducial argument (Dempster 1967, 1968). As Shafer states in the preface of his book "... It offers a reinterpretation of Dempster's work, a reinterpretation that identifies his "lower probabilities" as epistemic probabilities or degrees of belief, takes the rule for combining such degrees of belief as fundamental, and abandons the idea that they arise as lower bounds over classes of Bayesian probabilities." The rule mentioned in this statement became known as Dempster's rule.

"Evidence" is an appealing term and it is no surprise that the "Theory of Evidence" found much interest among knowledge engineers who have to model uncertain information. After a lot of ad hoc attempts to model uncertainty in expert systems, a serious theory seemed finally available to treat one of the basic problems of the field. However, as the theory is not really compatible with the simple paradigms of early rule based expert systems, misinterpretations and oversimplistic misuses of the theory caused much

Research partly supported by Grants No. 21-30186.90 and 21-32660.91 of the Swiss National Foundation for Scientific Research and the Swiss Federal Office for Science and Education, Esprit Basis Research Activity Project DRUMS II (Defeasible Reasoning ans Uncertainty Management Systems).

deception. Only slowly begins the true nature and the real meaning of evidence theory to emerge and to be understood. And this will hopefully be the starting point for fruitful and appropriate applications to the modeling of uncertainty, decision analysis and knowledge engineering in general.

Evidence is a notion which probably can never be fully captured by a single formal theory. In this introduction however "Theory of Evidence" will be understood in a narrow sense as the theory introduced by Dempster and Shafer and variants thereof. This particular theory is also often called Dempster-Shafer theory. It is clear today that this theory can be given various different, but essentially equivalent mathematical forms. Some of them are based on probability theory, others are axiomatic theories, a priori without a reference to probability theory. According to this distinction we classify the approaches broadly into *probabilistic* approaches and *non probabilistic* ones. The former try to integrate evidence theory into the framework of classical probability theory, whereas the latter deliberately go beyond classical probability. May be it would be more correct to call them rather non-standard probability theories than non-probabilistic ones. Despite the differences in approach and interpretation, all of them lead essentially to mathematically equivalent theories – at least in the finite case. That is they share the same basic theorems and the same computational procedures apply. In this introduction a probabilistic approach to evidence theory will be given, which is based upon Dempster's original work.

An important pragmatic consideration regarding the usefulness of evidence theory is the question, whether there are indeed important practical applications which can be described in a natural and useful way by the structures provided by the theory of evidence. This question is addressed in section 3. From the point of view of computation, evidence theory is complex in the usual technical sense. Nevertheless there exist methods which greatly improve the efficiency of computations. Also, based on these methods, software packages begin to appear which permit to apply evidence theory to practical problems. This subject is discussed in section 4.

Today there are not many comprehensive and introductory presentations and surveys of evidence theory. Beside the book of Shafer (1976 a), which is still a recommended reading, there are recent surveys by Smets, 1988, Shafer, 1990 and 1992, Kohlas, Monney, 1994 a. Recent books on uncertainty including chapters on evidence theory are by Kruse et al. (1991) and Hajek et al., 1992. Finally, let's mention the recent book by Fedrizzi et al. (1993) and Kohlas, Monney (1995) devoted entirely to Dempster-Shafer theory.

2 A MATHEMATICAL MODEL OF HINTS

Suppose that a certain precise question,whose answer is unknown, has to be studied and the elements θ of Θ represent the possible answers to the question. This means that exactly one of the $\theta \in \Theta$ is the correct answer, but it is unknown which one. However there is some information or evidence available relative to this question. This information allows for several, distinct interpretations, depending on some unknown circumstances and these interpretations are represented by the elements ω of Ω. This means that there is exactly one correct interpretation ω in Ω, but again it is unknown which one. Not all interpretations are equally likely and the probabilities $p(\omega)$ describe these different likelihoods. If $\omega \in \Omega$ is the correct interpretation, then the unknown answer θ is known to be in the set $\Gamma(\omega)$. Such a piece of information is called a **hint** (Kohlas, 1993 b and

1995). This model allows to give a specific and clear sense to the further important notions introduced by Dempster (1967).

If H is a subset of Θ, the hypothesis that the unknown answer θ is in H can be considered. In order to judge the hypothesis H in the light of the hint $(\Omega, P, \Gamma, \Theta)$, we may ask, which of the possible interpretations make H necessarily true. These are all interpretations ω having a non empty set $\Gamma(\omega)$ which is contained in H because if such an interpretation is the correct one, then $\theta \in \Gamma(\omega)$ and thus necessarily $\theta \in H$. Define

$$u(H) = \{\omega \in \Omega : \emptyset \neq \Gamma(\omega) \subseteq H\}. \tag{1}$$

In the same spirit, we may ask what are the interpretations which would make H not necessarily true, but at least possible. These interpretations are those in

$$v(H) = \{\omega \in \Omega : \Gamma(\omega) \cap H \neq \emptyset\} \tag{2}$$

because if $\omega \in v(H)$ is the correct interpretation, then $\theta \in \Gamma(\omega)$ and it is at least possible that θ is in H, although not sure.

Note that $v(\Theta)$ contains all interpretations with $\Gamma(\omega) \neq \emptyset$. Now, because Θ is supposed to contain the true answer, an interpretation ω which is not in $v(\Theta)$ is not really a possible interpretation. Thus, the correct interpretation must be in $v(\Theta)$ and this supplementary information allows to pass to the conditional probability $p(\omega|v(\Theta)) = p(\omega)/P(v(\Theta))$ for the possible interpretations in $v(\Theta)$.

As the correct interpretation is unknown, it is not possible to confirm whether it is in $u(H)$ or $v(H)$, that is whether H is true or only possible. But it is at least possible to compute the probabilities that the correct interpretation is in $u(H)$ or $v(H)$:

$$sp(H) = P(u(H)|v(\Theta)) = P(u(H))/P(v(\Theta)) \tag{3}$$
$$pl(H) = P(v(H)|v(\Theta)) = P(v(H))/P(v(\Theta)). \tag{4}$$

Because the ω in $u(H)$ can all be regarded as "arguments" in favour of the hypothesis H, arguments however, which are not sure to hold, but only more or less probable, $sp(H)$ can be regarded as the probability with which H can be infered or proved from the available information $(\Omega, P, \Gamma, \Theta)$. From this point of view $sp(H)$ can be regarded as the *degree of support* of the hypothesis H induced by the hint $(\Omega, P, \Gamma, \Theta)$. Similarly, all the interpretations ω in the set $v(H)$ can be regarded as "arguments" for the possibility of H. Then $pl(H)$ can be considered as the *degree of possibility* or *plausibility* of H .

Considering all possible hypotheses $H \subseteq \Theta$, sp and pl become functions from the power set of 2^Θ to $[0, 1]$ and as such they are called support (also belief) and plausibility functions. The following theorem collects a number of fundamental properties of these functions.

Theorem 2.1 *If sp and pl are defined by (3) and (4) relative to a hint* $(\Omega, P, \Gamma, \Theta)$, *then*

$$sp(\emptyset) = pl(\emptyset) = 0; \;\; sp(\Theta) = pl(\Theta) = 1 \tag{5}$$

$$sp(H) \leq pl(H) \text{ for all } H \subseteq \Theta \quad (6)$$
$$sp(H) = 1 - pl(H^c), \; pl(H) = 1 - sp(H^c), \quad (7)$$

and

$$sp(H) \geq \sum \{(-1)^{|I|+1} sp(\cap_{i \in I} H_i) : \emptyset \neq I \subseteq \{1, \ldots, n\}\} \quad (8)$$

for all $n \geq 1$ and sets H, H_i in Θ such that $H \supseteq H_i$ and

$$pl(H) \leq \sum \{(-1)^{|I|+1} pl(\cup_{i \in I} H_i) : \emptyset \neq I \subseteq \{1, \ldots, n\}\} \quad (9)$$

for all $n \geq 1$ and sets H, H_i in Θ such that $H \subseteq H_i$.

A proof of these properties may be found for example in Shafer (1976 a) or Kohlas, Monney (1995). The inequality (8) says that sp is monotone of order ∞ and (9) states that pl is alternating of order ∞.

It is possible that one has two or more hints or sources of information $\mathcal{H}_i = (\Omega_i, P_i, \Gamma_i, \Theta)$, $i = 1, \ldots, m$ relative to the same question. Then this information has to be combined in order to integrate all the available information. As exactly one interpretation $\omega_i \in \Omega_i$ is the correct one for every hint, there is exactly one correct combined interpretation $(\omega_1, \ldots, \omega_m)$. This combined interpretation restricts the correct answer θ to the question considered into the subset $\Gamma(\omega_1, \ldots, \omega_m) = \Gamma_1(\omega_1) \cap \ldots \cap \Gamma_m(\omega_m)$ of Θ. Thus $\Omega_1 \times \ldots \times \Omega_m$ is the set of all combined interpretations to be considered. They have a joint probability $p_{1 \ldots m}(\omega_1, \ldots, \omega_m)$, which has as marginal probabilities $p_i(\omega_i), i = 1, \ldots, m$. This consideration leads then to the combined hint

$$\mathcal{H}_1 \oplus \ldots \oplus \mathcal{H}_m = (\Omega_1 \times \ldots \times \Omega_m, P_{1 \ldots m}, \Gamma, \Theta). \quad (10)$$

In many cases one may assume that the interpretations of the distinct hints are stochastically independent, such that $P_{1 \ldots m}$ is the product measure on $\Omega_1 \times \ldots \times \Omega_m$. For this case of independent hints or sources of information, (10) is called *Dempster's rule of combination*. The combined interpretations $(\omega_1, \ldots, \omega_m)$ for which $\Gamma_1(\omega_1) \cap \ldots \cap \Gamma_m(\omega_m)$ is empty are called contradictory. If all combined interpretations are contradictory, the hints are not combinable, they are in complete contradiction.

It is possible that the subsets $\Gamma(\omega)$ are all identical and equal to a subset B of Θ for all interpretations ω. This says simply that the correct answer is for sure in the subset B. Such a hint is called *deterministic* and is simply denoted by B. If $\mathcal{H}$ is any other hint relative to the same question, then $\mathcal{H} \oplus B$ can be formed. This is called conditioning of the hint $\mathcal{H}$. The following theorem holds:

Theorem 2.2 *(see for example Shafer, 1976 a, Kohlas, Monney, 1995). Let $sp, sp(.|B)$ and $pl, pl(.|B)$ denote the support and plausibility functions of $\mathcal{H}$ and $\mathcal{H} \oplus B$ respectively. Then*

$$sp(H|B) = (sp(H \cup B^c) - sp(B^c))/(1 - sp(B^c)), \quad (11)$$
$$pl(H|B) = pl(H \cap B)/pl(B). \quad (12)$$

This is called Dempster's rule of conditioning. In particular, it is possible that $B = \Theta$. Then the deterministic hint is called *vacuous*, it carries no information whatsoever concerning the question considered. For a vacuous hint $sp(H) = 0$ for all $H \neq \Theta$ and $pl(H) = 1$ for all $H \neq \emptyset$. This is a perfect representation of complete ignorance. Clearly we have $\mathcal{H} \oplus \Theta = \mathcal{H}$.

If for all ω the sets $\Gamma(\omega)$ are singletons, then the hint is called *precise* or *Bayesian* (by Shafer, 1976 a). It is then essentially a random variable. In this case $sp = pl$ and sp becomes in fact a probability measure on Θ and Dempster's rule of conditioning reduces to the usual definition of conditional probability. This is generalized in the following theorem.

Theorem 2.3 *(see for example Shafer, 1976 a, Kohlas, Monney, 1995). Let $\mathcal{H}_1$ be a precise hint, $\mathcal{H}_2$ an arbitrary hint. Then $\mathcal{H} = \mathcal{H}_1 \oplus \mathcal{H}_2$ is also a precise hint. Denote by p, p_1 the probabilities induced by $\mathcal{H}$ and $\mathcal{H}_1$, and by pl_2 the plausibility function relative to $\mathcal{H}_2$. Then, for all $\theta \in \Theta$,*

$$p(\{\theta\}) = k\ p_1(\{\theta\})pl(\{\theta\}), \quad k^{-1} = \sum_{\theta \in \Theta} p_1(\{\theta\})pl(\{\theta\}). \tag{13}$$

This shows that little information is needed in this case from the second hint $\mathcal{H}_2$. The plausibilities of the singletons are sufficient, in fact it is even sufficient to know only the relative values of these plausibilities. This is a quite remarkable result, which permits to link evidence theory with the usual Bayesian analysis, a subject which cannot be pursued here.

3 APPLICATIONS OF EVIDENCE THEORY

Do the structures introduced above correspond to anything interesting in our surrounding world? Does evidence theory have interesting, nontrivial and convincing applications? Misunderstandings of evidence theory led to disappointing results and as a consequence to a rejection of the theory by some authors. Nevertheless, there are today some domains where evidence theory begins to emerge as a natural approach which gives convincing results. Among these fields are statistical inference, diagnostics, risk analysis and decision analysis. These applications will be surveyed briefly in this section. But in addition to these applications a few others have to be mentioned. At the Stanford Research Institute International, evidence theory was studied very early from an application point of view, see Lowrance et al. 1986, Lowrance, 1988. This institute developed also a Lisp based general system (a shell) for evidential reasoning. Shafer himself applied evidence theory to auditing (Shafer, Srivastava, 1990). Other applications concern image processing (Wesley, 1986; Provan, 1990 b; Lohmann, 1991), geometric and temporal reasoning (Kohlas, Monney, 1991; Monney, 1991 a and b), including project scheduling and financial modelling (Kohlas, 1989). Evidence theory was of course considered as a general approach to uncertainty management in expert systems or knowledge bases. A recent architecture of knowledge bases incorporating evidence theory has been developed by Saffiotti (1991 a and b); he addresses in particular the important problem of using an appropriate language like first order logic as a base for the automatic construction of a propositional or extended mathematical model of evidence.

3.1 Statistical Inference from an Evidential Point of View

Dempster's motivation to introduce multivalued mappings and the related lower and upper probabilities was statistical inference (Dempster, 1966, 1967, 1968, 1990). The goal of this new approach was to generalize Bayesian inference and also to reconcile the latter with Fisher's fiducial probabilities. Dempster was able to derive lower and upper probabilities on the unknown parameters to be studied by statistical inference by specifying a multivalued mapping and without using an a priori distribution on the parameter space. If an a priori distribution is known, then Dempster's method can incorporate it and leads to the classical Bayesian approach, but his method works also, if the a priori information about the parameter is less specific than a distribution, if it is only described by a belief function. It unifies and generalizes both Bayesian and Fisher statistics.

This approach will be sketched here in a way proposed by Kohlas (1992), which shows that the lower and upper probabilities of Dempster can in fact be seen as degrees of support and plausibility of hypotheses about the unknown parameter derived from observations and a process model specifying how observations are generated. Let Π be the set of possible values of an unknown parameter π. Inference about this parameter is made by means of an observation process which yields an observation x in some obervation space X. It is assumed that the actual observation x depends only on the unknown parameter π and a further unknown chance element ω which belongs to a known probability space $(\Omega, \mathcal{A}, P)$. This is expressed by a known function

$$x = f(\pi, \omega) \tag{14}$$

Equation (14) is called a process model of observation. Now, if the experiment is carried out, then an observation $x \in X$ becomes available. Given such an observation x and the process model (14), what can be said about the unknown parameter π?

Define $\Gamma_x(\omega) = \{\pi \in \Pi : x = f(\pi, \omega)\}$, the set of all possible parameter values if x is the given observation, and ω the random element which generated the observation. Then $\mathcal{H}_x = (\Omega, \mathcal{A}, P, \Gamma_x, \Pi)$ is a hint about the unknown parameter value π. Hence, if $H \subseteq \Pi$ is any hypothesis about the unknown parameter, then $u_x(H) = \{\omega \in \Omega : \emptyset \neq \Gamma_x(\omega) \subseteq H\}$ represents all random elements ω for which the hypothesis is necessarily true and $v_x(H) = \{\omega \in \Omega : \Gamma_x(\omega) \cap H \neq \emptyset\}$ the random elements for which the hypothesis is possibly true. Thus, $sp_x(H) = P(u_x(H))/P(v_x(\Pi))$ and $pl_x(H) = P(v_x(H))/P(v_x(\Pi))$ are the degrees of support and plausibility of H given the observation x (assuming that $u_x(H), v_x(H)$ are measurable). Note that $\Gamma_x(\omega)$ may well be empty. But such an ω cannot possibly have generated the observation x, because there is no parameter π which in conjunction with ω generates x. Thus, the conditioning in sp_x and pl_x is well founded. Application of this approach is possible for so called measurement or also structural models (Frazer, 1968; Kohlas, 1992). But also sampling problems may be formulated and analyzed this way (Dempster, 1967 b, 1968 a and b, 1990; Shafer, 1982 a; Kohlas, 1992). Finally, the problem of filtering stochastic dynamic systems may be approached from this point of view (Kohlas, 1991).

A further reference for the application of belief functions in statistical inference is given by Wasserman (1990); this paper takes however not the evidential point of view, but uses belief functions as a means to describe convex sets of a priori probability measures for a

Bayesian analysis. In the cited paper, other references to similar uses of lower and upper probability in statistical inference may be found.

3.2 Diagnostics and Risk Analysis

Diagnostics is the task to infer explanations for a set of observations. In many cases these observations are symptoms of diseases or faults of a medical or technical system. The explanations furnish then the possible diseases causing the observed facts. Symptoms can be considered as evidence for certain possible diseases and evidence theory is therefore an obvious approach to diagnostics.

An early example of a diagnostic expert system was MYCIN. The authors of this system already noted that symptoms are not infallible evidence but may have exceptions. Also they stated that symptoms are evidences which bear on sets of hypotheses rather than on individual hypotheses. They concluded that probability theory was not an appropriate tool for their purposes and they developed their own ad hoc uncertainty calculus. Later, it became clear that evidence theory would have been well adapted to the structure of MYCIN (Gordon, Shortliffe, 1985, Shafer, Logan, 1987). Applications of evidence theory to diagnosis have been discussed by Smets (1979, 1981). GERTIS is a newer expert system which is of a similar type (Yen, 1989), although it mixes evidential theory and Bayesian analysis.

In the mean time, a quite different architecture of diagnostic systems emerged, the model based inference systems. Here a model of the properly functioning system serves as a basis to predict output values from given input values. Differences between these predicted values and the observed ones are then used to derive possible diagnoses (a fundamental paper in this domain of logic based diagnosis is Reiter, 1987). These approaches are essentially symbolic and provide no means to rank the possible diagnoses according to their likelihood or to judge the likelihood that a particular component of a system is faulty. It will be explained how these models can be extended in a very natural way by evidence theory to include likelihood measures. Some of these models may not only be used for diagnostic purposes, but for predictive risk analysis as well.

(i) *Model Based Diagnostics.* A model of a properly functioning system can often be described by a set of variables and relations between them. Let $M = \{1, \ldots, m\}$ be an index set, X_i with $i \in M$ a set of variables taking values in the frames Θ_i. Let R be a family of subsets of M and for every $J \in R$ let $R_J \subseteq \prod_{i \in J} \Theta_i = \Theta_J$ be a given relation between the variables X_i for $i \in J$. These relations correspond to physical devices which realize those relations as input-output relations (where some variables are considered as inputs and the others as outputs). As an example, the X_i may be integer variables, taking values in $\mathcal{N}$, and the relations are arithmetic relations like additions $X_i + X_j = X_k$ or multiplications $X_i \times X_j = X_k$. Or else the X_i are Boolean variables with values $\{0, 1\}$ and the relations are Boolean relations as they arise for example in digital circuits.

If the values of some variables are observed, then such a model permits to test whether the observations are consistent with the assumed relations between the variables. If this is not the case, then some relations must be violated and the corresponding devices must be faulty.

Now fault models of the components can be introduced. Let $\Omega_J = \{\omega_0, \omega_1, \ldots, \omega_s\}$ denote a set of different mutually exclusive working modes of the component associated to the relation $J \subseteq M$. It is supposed that the probabilities $p_J(\omega_i)$ of these different modes are known. ω_0 is assumed to correspond to a properly functioning component, whereas ω_1 to ω_s represent different faulty modes. To any mode ω_i corresponds a relation $\Gamma_J(\omega_i) \subseteq \Theta_J$; $\Gamma_J(\omega_0) = R_J$ is the above relation of a properly functioning device. An and-gate as used in a digital circuit may for example have two fault modes, one where the output is 0 for every input configuration and one where the output is entirely unpredictable for each input. The later mode corresponds to the vacuous relation $\{0,1\} \times \{0,1\} \times \{0,1\}$, all combinations of input and output are possible. This associates clearly a hint $(\Omega_J, p_J, \Gamma_J, \Theta_J)$ to every component in the system. The observation of certain variables X_i provides further deterministic and precise hints relative to Θ_i.

These hints refer to different but related frames. This situation differs from those considered so far and especially in section 2, where hints refered always to the same frame of discernment. The situation encountered here is however typical for most applications of evidence theory. Dempster's rule for the combination of hints can be extended to such situations as will be explained in section 4. Hence these hints may be combined to obtain degrees of support and plausibility for different diagnoses, that is for hypotheses about the possible faulty configurations of modes which may explain the observed values of variables. The computational methods for this analysis are described in section 4. This shows that a well known model in diagnosis theory can very naturally be extended using evidence theory so as to incorporate information about the likelihood of possible diagnoses. This approach has been described in Kohlas, Monney, Haenni (1995).

Such systems may also often be conveniently formulated in the framework of logic with uncertain arguments and related to truth maintenance systems (Provan, 1988). A further application of the model introduced above is to the detection of bad data in measurement models (Kohlas, 1992).

(ii) *Risk Analysis.* Boolean systems as described above arise not only in modeling digital circuits, but also in fault or event trees as used for reliability or risk analysis. These structures can be used for diagnostic purposes, if some events are observed exactly in the way described above. They are however also used for predictive studies of risks. Almond (1991, 1992 a) discusses the statistical analysis of failure rates of components, using methods similar as those described in subsection 3.1. This leads to only partially known component models, in the sense that only belief functions for the unknown parameters can be given. This in turn leads to belief functions for the failure of components and thus for the events in the fault tree. This is an example of how evidence theory may be applied in risk analysis.

3.3 Decision Analysis

(i) *Generalized Decision Principles.* Analysis of uncertain perspectives and data is an important activity in decision support. One aspect is the problem of choice. In the standard choice model, a set $A = \{a_1, \ldots, a_m\}$ of possible activities and a set $S = \{s_1, \ldots, s_n\}$ of possible states of the world is given, and it is assumed that for every activity a_i and state s_j a utility index u_{ij} is defined which expresses the preferences of the decision maker. Classical decision theory considers two extreme situations relative to the knowledge of the state of the world. Either there is full ignorance about the state; in this case decision principles like the Hurwicz principle, Laplace's principle of insufficient reason or the principle of minimal regret are proposed to select an activity. Or else a probability distribution $p(s_j)$ over the set S is assumed to be given. Then the activity is selected by maximizing the expected value of the utility of an action a_i

$$E_i(u_{ij}) = \sum_{j=1}^{n} u_{ij} p(s_j). \tag{15}$$

In our terminology, the first case corresponds to a vacuous hint about the state of the world and the second to a precise hint about the state. So it is natural to look at intermediate cases, where an arbitrary hint $\mathcal{H} = (\Omega, p, \Gamma, S)$ about the state is given. If a fixed interpretation ω is known to be the correct one, then the true state is known to be in the set $\Gamma(\omega)$. With respect to this set of states we have a decision problem with full ignorance and we may apply any one of the decision principles to evaluate the actions. We may then take the expected value over these evaluations for every interpretation ω in Ω. Each decision principle leads in this way to a *generalized decision principle.*

First of all, note that an action a_i is dominated by an action a_k, if for all ω max $\{u_{ij} : s_j \in \Gamma(\omega)\} \leq$ min $\{u_{kj} : s_j \in \Gamma(\omega)\}$. Dominated actions may be eliminated from further considerations.

Laplace' principle of insufficient reason amounts to assume a uniform distribution over the states in $\Gamma(\omega)$ for every ω, taking the expected value of the utilities relative to these uniform distributions and then taking the expected value over Ω:

$$\sum_{\omega \in \Omega} p(\omega) \sum \{u_{ij}/|\Gamma(\omega)| : s_j \in \Gamma(\omega)\}, \text{ for all } a_i \in A. \tag{16}$$

This generalized principle of insufficient reason has been proposed by Dubois, Prade (1982), Williams (1982), Smets (1989), and Smets, Kennes (1990).

Hurwicz's principle uses an index of optimism α between 0 and 1 to weigh the pessimistic minimal utility value and the optimistic maximal utility value over the possible states. The generalized Hurwicz' principle considers thus the following index for every action a_i:

$$\sum_{\omega \in \Omega} p(\omega)(\alpha \max \{u_{ij} : s_j \in \Gamma(\omega)\} + (1 - \alpha) \min \{u_{ij} : s_j \in \Gamma(\omega)\}). \tag{17}$$

This approach has been proposed by Strat (1990). It can also be seen as a linear interpolation between lower and upper bounds of expectation intervals defined by the hint. Strat shows also how to use this principle in the context of generalized decision trees. For $\alpha = 1$ the principle reduces to the generalized maximax principle and for $\alpha = 0$ to the generalized maximin principle.

The regret principle computes for every state s_j and action a_i the regret $r_{ij} = \max \{u_{ij} : i = 1, \ldots, m\} - u_{ij}$. It applies then the generalized minimax principle to the matrix (r_{ij}). This procedure has not been yet proposed in the literature, but it is in the line of the above principles.

(ii) *Expected Utility Theory.* Let us generalize the basic decision model introduced above. Instead of describing the outcome of an action a_i and a state s_j directly by an numerical utility index u_{ij}, suppose that the outcome is described by some more basic, possibly nonnumeric element $\theta_{ij} = \theta(a_i, s_j) \in \Theta$. The hint $\mathcal{H}$ on the set of possible states s is transformed for each action a_i by the function $\theta(a_i, \cdot) : S \to \Theta$ into a hint $\mathcal{H}_i$ on Θ. The problem of choice of an action a_i reduces then to the choice of a hint on Θ or to a comparison of hints or belief functions on Θ. Now, a hint on Θ can be seen as probability distribution over the subsets of Θ. Let $\mathcal{P}_\Theta$ denote the set of probability distributions over 2^Θ, where Θ is assumed to be finite. Then the following theorem holds:

Theorem 3.1 *(see for example Fishburn, 1970, theorem 8.2) Let $\prec$ denote a binary relation on $\mathcal{P}_\Theta$. Then there is a real-valued function u (called a utility function) on 2^Θ that satisfies*

$$P \prec Q \text{ iff } \sum_{B \subseteq \Theta} P(B)u(B) < \sum_{B \subseteq \Theta} Q(B)u(B) \tag{18}$$

for all P, Q in $\mathcal{P}_\Theta$
if and only if, for all P, Q, R in $\mathcal{P}_\Theta$

1. *$\prec$ on $\mathcal{P}_\Theta$ is a weak order,*
2. *(Independence). $P \prec Q$ implies $\alpha P + (1-\alpha)R < \alpha Q + (1-\alpha)R$ for all $0 < \alpha < 1$,*
3. *(Contuinity). $P \prec Q, Q \prec R$ implies $\alpha P + (1-\alpha)R < Q$ and $Q < \beta P + (1-\beta)R$ for some $\alpha, \beta \in (0,1)$.*

Moreover, u is unique up to a positive linear transformation $au + b$.

Inequality (18) states that under the conditions of the theorem the action can be selected by maximizing an expected utility. This has been noted by Jaffray (1989). (18) may however not be very practical given the large numbers of subsets of Θ for which utilities $u(B)$ must be determined. Jaffray (1989) adds a further condition to (1) to (3) above in theorem 3.1, which reduces (18) essentially to the generalized Hurwicz principle. Jaffray, Wakker (1992) relate the above result to Savage's sure thing principle and show how a weaker version of this principle leads to the above result. Jaffray (1992 a) then considers this expected utility approach also in a dynamic setting.

(iii) *Support for Actions.* Another approach, but very much in the spirit of evidence theory, is to look for arguments for or against the hypothesis that a given action is satisfactory (achieves the specified goals), or that it is better than another one, or even that it is the best among a given set of actions. Hints and belief functions about the consequences of an action may be used for this purpose. According to Strat (1990) such a "constructive" approach to decision theory has been proposed by Shafer (1982 b) in an unpublished paper. This approach seems not to have been pursued in the literature (see however Schaller, 1991). It would be close in spirit to the outranking methods of multicriteria analysis (see for example Roy, 1985). It should be emphasized that evidential decision analysis as understood here is different from decision analysis with partially known probabilities.

4 COMPUTATIONAL ASPECTS OF EVIDENCE THEORY

The practical implementation of evidence theory leads to several difficult computational problems. How can the available evidence be best encoded for representation on a computer and what are the corresponding memory requirements? Which are the best algorithms to compute degrees of support and plausibility or to combine pieces of evidence by Dempster's rule and what are their complexities? Are there special structures of evidence which facilitate representation and computation? These are the questions which are briefly addressed in this section. References to computer packages which implement evidence theory will also be given.

In practical applications of evidence theory the reasoning concerns a group of distinct variables $X_1, \ldots, X_m$, each one with its own domain (or frame of discernment) Θ_i. Then arbitrary groupings of variables $\{X_j : j \in J \subseteq M$ (with $M = \{1, \ldots, m\}$) can be considered and such a group of variables has the frame $\Theta_J = \prod_{j \in J} \Theta_j$, the product space of the frames of the variables in the group. Denote Θ_M by Θ; this is the overall frame of the model to be considered. The available pieces of evidence very often do not concern all variables together, but only some groups of them. Such pieces of evidence concerning a group J of variables may be described by the hints $\mathcal{H}_i, i = 1, \ldots, r$, relative to the frame Θ_J. More precisely, let J_i denote the group of variables of the hint $\mathcal{H}_i$. As an example, consider model based diagnostics described in section 3.2 (ii) above.

As these different hints refer to different frames, they can not be directly combined by Dempster's rule. A hint $\mathcal{H} = (\Omega, p, \Gamma, \Theta_J)$ relative to the group of variables J can be extended to a group $K \supset J$ simply by replacing the focal sets $\Gamma(\omega)$ by their cylindrical extension $\Gamma(\omega) \uparrow K = \Gamma(\omega) \times \Theta_{K-J}$. The new hint $\mathcal{H} \uparrow K = (\Omega, p, \Gamma \uparrow K, \Theta_K)$ is called the *vacuous extension* of $\mathcal{H}$ from J to K. Then the hints relative to different groups of variables may all be vacuously extended to M and then combined on the common frame Θ,

$$\mathcal{H} = (\mathcal{H}_1 \uparrow M) \oplus \ldots \oplus (\mathcal{H}_r \uparrow M) \tag{19}$$

In general, let us define Dempster's rule by $\mathcal{H}_1 \oplus \mathcal{H}_2 = (\mathcal{H}_1 \uparrow M) \oplus (\mathcal{H}_2 \uparrow M)$. A representation of $\mathcal{H}$ by its pieces $\mathcal{H}_i$ is clearly advantageous from a memory point of view. Not only has each $\mathcal{H}_i$ in general much less focal sets than $\mathcal{H}$, but also their focal sets are generally in a much smaller dimension $|J_i|$ than $|M|$.

It seems however at first sight that this last advantage is lost, when the hints are combined, because all of them must be extended to M. This however can be avoided in many cases by the ability to combine hints locally. Trivially, if two hints refer to the same group J of variables, then they may be combined on Θ_J beforehand rather than on Θ. Also if one hint refers to a group J, the other to a group K and $J \subseteq K$, then the two hints can be combined on Θ_K. If all these combinations are executed beforehand, then a family $\mathcal{J} = \{J_1, \ldots, J_r\}$ of incomparable groups of variables remain (i.e. all groups are distinct and none is contained in another one). $\mathcal{J}$ is called a scheme (Thoma, 1991) and it defines a *hypergraph* (with hyperedges J_i).

Now, even beyond the trivial cases above, *local combinations* are possible, due to a basic theorem given below. If $\mathcal{H} = (\Omega, p, \Gamma, \Theta_J)$ is a hint relative to a group of variables J, and K is a subset of J, then $\mathcal{H} \downarrow K = (\Omega, p, \Gamma \downarrow K, \Theta_K)$ is the restriction or projection of $\mathcal{H}$ from J to K, if $\Gamma(\omega) \downarrow K$ is the projection of $\Gamma(\omega)$ from Θ_J to Θ_K. Contrary to the vacuous extension, where information is neither added nor lost (i.e. $(\mathcal{H} \uparrow K) \downarrow J = \mathcal{H}$ if $J \subseteq K$), projecting hints causes in general a loss of information, that is $(\mathcal{H} \downarrow K) \uparrow J \neq \mathcal{H}$, if $K \subseteq J$. Vacuous extension, followed by projection permit to transport a hint relative to a group of variables J to any other group K, $\mathcal{H}|K = (\mathcal{H} \uparrow J \cup K) \downarrow K$. Then the following theorems of local computations hold:

Theorem 4.1 *(see for example Shafer et al. 1987). If $\mathcal{H}_1$ and $\mathcal{H}_2$ are hints relative to the groups of variables H and K respectively and $H \cap K \subseteq J$, then*

$$(\mathcal{H}_1 \oplus \mathcal{H}_2)|J = (\mathcal{H}_1|J) \oplus (\mathcal{H}_2|J). \tag{20}$$

Theorem 4.2 *(Shafer et al. 1987). If $\mathcal{H}$ is a hint relative to a group of variables H, and K, J denote groups of variables such that $H \cap K \subseteq J$, then*

$$\mathcal{H}|K = (\mathcal{H}|J)|K. \tag{21}$$

This last theorem shows that transporting a hint from a group of variables H to another group K can be done simpler by projecting the hint first to $H \cap K$ and then extending it to K, rather than first extending it to $H \cup K$ and then projecting it to K. Theorem 4.1 shows that in some cases hints may be combined locally on certain smaller frames Θ_J relative to a group of variables J rather than on the overall frame Θ.

If $\mathcal{J}$ is a hypergraph, then it may be possible to arrange its hyperedges J_i into a family $\mathcal{E}$ of pairs $\{J_i, J_k)\}$such that

1. $(\mathcal{J}, \mathcal{E})$ is a tree,
2. if J_i and J_k are distinct vertices of the tree $(\mathcal{J}, \mathcal{E})$, then $J_i \cap J_k$ is contained in every vertex on the unique path from J_i to J_k in the tree $(\mathcal{J}, \mathcal{E})$.

Such a tree is called a *Markov tree* (also called a join tree in data base theory, see Maier, 1983). If such a tree exists, then $\mathcal{J}$ is called a hypertree or an acyclic hypergraph. If one now considers any vertex J of a Markov tree $T = (\mathcal{J}, \mathcal{E})$, then it has a set $N(J)$ of neighboring vertices and if node J is eliminated then T decomposes into a set of subtrees $T(J_k)$, each one associated with a vertex $J_k \in N(J)$. It can be shown that every subtree of a Markov tree is still a Markov tree. If $T = (\mathcal{J}, \mathcal{E})$ is a Markov tree and $\mathcal{J}$ is a scheme, then there is a hint $\mathcal{H}_J$ associated with every vertex J of T. Let $\mathcal{H}(T)$ denote $\oplus \{\mathcal{H}_J : J \in \mathcal{H}\}$. Property (2) of a Markov tree permits the application of both theorems 4.1 and 4.2 to give the following result:

Theorem 4.3 *(Shafer et al. 1987). If T is a Markov tree and J is a vertex of T, then*

$$\mathcal{H}(T)|J = \mathcal{H}_J \oplus [\oplus \{(\mathcal{H}(T(J_k))|J_k)|J : J_k \in N(J)\}]. \tag{22}$$

This is a recursive formula for combining hints on Markov trees, which can be used to compute projections of the combined hint to groups of variables in the hypergraph $\mathcal{J}$, using local combinations only.

Theorem 4.3 is a basic result for the computational aspects of evidence theory. In fact it is a special case of a much more general abstract framework (Shenoy, Shafer, 1990; Shafer, 1991) which covers such diverse models as nonserial dynamic programming (Bertele, Brioschi, 1972), probabilistic (Bayes or Markov) networks (Lauritzen, Spiegelhalter, 1988), constraint propagation, influence diagrams (Oliver, Smith, 1990) and others.

Sometimes the hypergraph has an evident Markov tree structure. Such is the case for example for models of dynamical processes (Kohlas, 1991). However, it is not always possible to construct a Markov tree for a given hypergraph $\mathcal{J}$. In such a case, the hypergraph must be enlarged by augmenting the hyperedges until a Markov tree can be obtained; this is called constructing a hypertree cover of the hypergraph. Although it is not difficult to find a hypertree cover of a hypergraph, it is difficult to find a good one, that is one with hyperedges of small cardinality. There is a growing literature on this subject; see for example Rose, 1970, Bertele and Brioschi, 1972, Tarjan and Yannakakis, 1984, Kong, 1986, Arnborg et al., 1987, Mellouli, 1987, Zhang, 1988 and Almond, Kong, 1991.

Meanwhile, several software packages based on this framework have been developed. Examples are DELIEF (Zarley et al. 1988), MacEvidence (Hsia, Shenoy, 1989), PULCINELLA (Safiotti, Umkehrer, 1991 and 1992), BELIEF (Almond, 1992 b, c) and TRESBEL (Xu, 1992).

5 CONCLUSION AND OUTLOOK

The theory of evidence presented so far may be put into a larger perspection as a theory of probabilistic argumentation systems (Kohlas, 1995). Argumentation systems are in their most general framework algebraic structures which link hypotheses to supporting arguments. This leads then first to an algebraic or symbolic version of evidence theory. Both theoretically and practically important models of such argumentation systems can be derived from assumption-based reasoning formed in general logics (Besnard, Kohlas, 1994) or in particular in propositional logic (Kohlas, 1993 a, Kohlas, Monney, 1995).

The arguments may in a second step be weighted by their probabilities. Thie permits then to measure by what probabilities hypotheses are supported by the argumentation system. In these extended probabilistic argumentation systems the hypotheses are numerically evaluated by degrees of support. This leads then to a very general theory of evidence. The theory of hints sketched above is but one, although important model of this general theory of evidence.

REFERENCES

Arnborg, S., Corneil, D.G., Proskurowski, A. (1987) Complexity of Finding Embeddings in a k-Tree. *SIAM J. Algebraic and Discrete Methods*, **8**, 277-84.

Almond, R. (1991) Belief Function Models For Simple Series and Parallel Systems. Tech. Report 207, Department of Statistics, Univ. of Washington.

Almond, R. (1992 a) Models for Incomplete Failure Data. Statistical Sciences. Inc., Seattle, WA.

Almond, R. (1992 b) Using the Belief Package. Harvard Univ., Dept. of Statatistics.

Almond, R. (1992 c) Graphical-Belief: Project Overview. Statistical Sciences, Inc., Seattle, WA.

Almond, R., Kong, A. (1991) Optimality Issues in Constructing Markov Tree from Graphical Models. Res. Rep. A-3, Harvard Univ., Dept. of Stat.

Bertele, U., Brioschi, F. (1972) *Nonserial Dynamic Programming.* Academic Press.

Besnard, P., Kohlas, J. (1994) *Evidence Theory Based on General Consequence Relations.* Tech. Rep. Institute of Informatics, University of Fribourg. To be published in *Int. J. of Foundations of Computer Science,* 1995.

Bonissone, P.P., Henrion, M., Kanal, L.N., Lemmer, J.F. (eds.) (1991) *Uncertainty in Artificial Intelligence 6.* North Holland, Amsterdam.

Bouchon-Meunier, B., Yager, R.R., Zadeh, L.A. (eds.) (1991) *Uncertainty in Knowledge Bases.* Springer Lecture Notes in Computer Science, No. 521.

Dempster, A. (1967 a) Upper and Lower Probabilities Induced by a Multivalued Mapping. *Annals Math. Stat.*, **38**, 325-39.

Dempster, A. (1967 b) Upper and Lower Probability Inferences Based on a Sample from a Finite Univariate Population. *Biometrika*, **54**, 515-28.

Dempster, A. (1968 a) A Generalization of Bayesian Inference. *J. Royal Stat. Soc., Series B*, **30**, 205-47.

Dempster, A. (1968 b) Upper and Lower Probabilities Generated by a Random Closed Interval. *Annals Math. Stat.*, **39**, 957-66.

Dempster, A. (1990) Bayes, Fisher, and Belief Functions, in *Bayesian and Likelihood Methods in Statistics and Econometrics.* (eds. S. Geisser, J.S. Hodges, S.J. Press, A. Zellner)

Dubois, D., Prade, H. (1982) On Several Representations of an Uncertain Body of Evidence in *Fuzzy Information and Decision Processes* (eds. M.M. Gupta, E. Sanchez) North Holland, Amsterdam, 167-81.

Fishburn, P.C. (1970) *Utility Theory for Decision Making.* Wiley, New York.

Frazer, D.A.S. (1968) *The Structure of Inference.* Wiley, New York.

Fedrizzi, M., Kacprzyk, J., Yager, R.R. (eds.) (1993) *Advances in the Dempster-Shafer Theory of Evidence.* Wiley.

Gordon, J., Shortliffe, E.H. (1985) A Method for Managing Evidential Reasoning in a Hierarchical Hypothesis Space. *Artificial Intelligence*, **26**, 323-57.

Hajek, P., Havranek, T., Jirousek, R. (1992) *Uncertain Information Processing in Expert Systems.* CRC Press, London.

Hsia, Y., Shenoy, P.P. (1989) Mac Evidence: A Visual Evidential Language for Knowledge Based Systems. Working Paper No. 211, Univ. of Kansas, School of Business.

Jaffray, J.Y. (1989) Linear Utility Theory for Belief Functions. *Operations Research Letter*, **8**, 107-12.

Jaffray, J.Y. (1992) Dynamic Decision Making with Belief Functions. *Cahiers de Recherche en Economie, Mathématiques et Applications*, Univ. de Paris I, No. 92-15.

Jaffray, J.Y., Wakker, P. (1992) Decision Making with Belief Functions: Compatibility and Incompatibility with the Sure-Thing Principle. Unpubl. Paper, Univ. de Paris VI, LID Paris, France, University of Nijmegen, NICI Nijmegen, The Netherlands.

Kohlas, J. (1989) Modeling Uncertainty with Belief Functions in Numerical Models. *European Journal of Operational Research*, **40**, 377-88.

Kohlas, J. (1991) Describing Uncertainty in Dynamical Systems by Uncertain Restrictions, in *Modeling, Estimation and Control of Systems with Uncertainty* (eds. G.B. Di Masi et al.), Birkhäuser.

Kohlas, J. (1992) Evidential Reasoning About Parametric Models. Inst. for Automation and Op. Res., University of Fribourg, Switzerland, Tech. Rep. No. 194.

Kohlas, J. (1993 a) Symbolic Evidence, Arguments, Supports and Valuation Networks, in *Symbolic an Quantitative Approaches to Reasoning and Uncertainty* (eds. M. CLARKE et al.), Springer.

Kohlas, J. (1993 b) Support and Plausibility Functions Induced by Filter-Valued Mappings. *Int. J. Gen. Sys.*, **21**, 343-63.

Kohlas, J. (1995) Mathematical Foundations of Evidence Theory, in *Mathematical Models for Handling Partial Knowledge in Artificial Intelligence* (eds. G. Coletti et al.) Plenum Publ. Corp.

Kohlas, J., Monney, P.A. (1991) Propagating Belief Functions Through Constraint Systems. *Int. J. Approximate Reasoning*, **5**, 433-61. (a shorter version is published in Bouchon et al., 1991, 50-57).

Kohlas, J., Monney, P.A. (1993) Probabilistic Assumption-Based Reasoning. Inst. for Automation and Op. Res., University of Fribourg, Switzerland, Tech. Rep. No. 208.

Kohlas, J., Monney, P.A. (1994 a) Theory of Evidence - A Survey of is Mathematical Foundations, Applications and Computational Aspects. *ZOR, Math. Methods of O.R.*, **39**, 35-68.

Kohlas, J., Monney, P.A. (1994 b) Representation of Evidence by Hints, in *Advances in the Dempster-Shafer Theory of Evidence* (eds. J. Yager et al.), John Wiley, New York.

Kohlas, J., Monney, P.A. (1995) *A Mathematical Theory of Hints.* Lecture Notes in Economics and Mathematical Systems, Volume 425, Springer-Verlag.

Kohlas, J., Monney, P.A., Haenni, R. (1995) Model-Based Diagnostics Using Hints. To appear in: *Springer Lecture Notes in Computer Science.*

Kong, A. (1986) Multivariate Belief Functions and Graphical Models. Doct. diss. Harvard Univ., Dept. of Stat.

Kruse, R., Schwecke E., Heinsohn J. (1991) *Uncertainty and Vagueness in Knowledge-Based Systems.* Springer.

Kruse, R.,; Siegel, P. (eds.) (1991) Symbolic and Quantitative Approaches to Uncertainty. Springer Lecture Notes in Computer Science No. 548.

Lauritzen, S.L., Spiegelhalter, D.J. (1988) Local Computations with Probabilities on Graphical Structures and their Application to Expert Systems. *J. Royal Stat. Soc.*, Series B, **50**, 157-224.

Lohmann, G. (1991) An Evidential Reasoning Approach to the Classification of Satellite Images. In Kruse, R., Siegel, P. (1991): 227-31.

Lowrance, J.D. (1988) Automated Argument Construction. *J. Stat. Planning Inference*, **20**, 369-87.

Lowrance, J.D., Garvey, T.D., Strat, T.M. (1986) A Framework for Evidential Reasoning Systems, AAAI-86, 869-903.

Maier, D. (1983) *The Theory of Relational Data Bases.* Computer Science Press.

Matheron, G. (1975) *Random Sets and Integral Geometry.* Wiley, New York.

Mellouli, K. (1987) On the Propagation of Beliefs in Networks Using the Dempster-Shafer Theory of Evidence. Doct. Diss., Univ. of Kansas, School of Business.

Monney, P.A. (1991 a) *Le raisonnement temporel et géométrique sous incertitude.* Doctoral Thesis, Math. Institute, Univ. Fribourg, Switzerland.

Monney, P.A. (1991 b) Planar Geometric Reasoning With the Theory of Hints, in *Computational Geometry. Methods, Algorithms and Applications.* (eds. H. Bieri, H. Noltemeier), Lecture Notes in Computer Science, vol. 553, p. 141-59.

Oliver, R.M., Smith, J.Q. (eds.) (1990) *Influence Diagrams, Belief Nets and Decision Analysis.* Wiley

Provan, G.M. (1988) Solving Diagnostic Problems Using Extended Truth Maintenance Systems. Proc. European Conf on Art. Int., 547-52.

Provan, G.M. (1990 a) A Logic-Based Analysis of Dempster-Shafer Theory. *Int. J. Approximate Reasoning*, **4**, 451-95.

Provan, G.M. (1990 b) The Application of Dempster-Shafer Theory to a Logic-Based Visual Recognition System, in M. Henrion et al. (1990).

Reiter, R. (1987) A Theory of Diagnosis from First Principles. *Artificial Intelligence*, **32**, 57-95.

Rose, D.J. (1970) Triangulated Graphs and the Elimination Process. *J. Math. Anal. and Appl.*, **32**, 597-609.

Roy, B. (1985) *Méthodologie Multicritère d'Aide à la Décision.* Editions Economica, Paris.

Saffiotti, A. (1991 a) Using Dempster-Shafer Theory in Knowledge Representation, in Bonissone et al. (1991).

Saffiotti, A. (1991 b) A Hybrid Belief System for Doubtful Agents, in Bouchon et. al. (1991).

Saffiotti, A., Umkehrer, E. (1991) PULCINELLA: A General Tool for Propagating Uncertainty in Valuation Networks. Proc. 7-th Conf on Uncertainty in AI, Los Angeles Ca.

Schaller, J.P. (1991) *Multiple Criteria Decision Aid Under Incomplete Information.* Doctoral Thesis, Economic Sciences Faculty, Univ. of Fribourg, Switzerland.

Shafer, G. (1976 a) *A Mathematical Theory of Evidence.* Princeton University Press.

Shafer, G. (1976 b) A Theory of Statistical Evidence, in *Foundations of Probability Theory, Statistical Inference, and Statistical Theories of Science.* (eds. Harper, Hooker), Vol. II, Reidel, Dortrecht.

Shafer, G. (1979) Allocations of Probability. *Annals of Prob.*, **7**, 827-39.

Shafer, G. (1982 a) Belief Functions and Parametric Models. *J. Royal Statist. Soc. B*, **44**, 322-52.

Shafer, G. (1982 b) Constructive Decision Theory. Univ. of Kansas Dept. of Math. Working Paper.

Shafer, G. (1991) An Axiomatic Study of Computations in Hypertrees. Working Paper 232, Business School, Univ. of Kansas.

Shafer, G., Logan, R. (1987) Implementing Dempster's Rule for Hierarchical Evidence. *Artificial Intelligence*, **33**, 271-89.

Shafer, G., Pearl, J. (eds.) (1990) *Readings in Uncertain Reasoning.* Morgan-Kaufman Publ. San Mateo, Cal.

Shafer, G., Shenoy, P.P., Mellouli, K. (1987) Propagating Belief Functions in Qualitative Markov Trees. *Int. J. Approximate Reasoning*, **1**, 349-400.

Shafer, G., Srivastava, R.P. (1990) The Bayesian and Belief Function Formalisms: A General Perspective for Auditing, Auditing. *J. Pract. Theory,* **9** (Suppl.).

Shenoy, P.P., Shafer, G. (1990) Axioms for Probability and Belief-Function Propagation, in Shafer, Pearl, 1990.

Smets, P. (1979) Modèle Quantitatif du Diagnostic Médical. *Bulletin de l'Académie de Médecine de Belgique*, **134**, 330-343.
Smets, P. (1981) Medical Diagnosis: Fuzzy Sets and Degrees of Belief. *Fuzzy Sets and Systems*, **5**, 259-66.
Smets, P. (1988) Belief Functions, in *Non-Standard Logics for Automated Reasoning* (eds. P. Smets), Academic Press, London.
Smets, P. (1989) Constructing the Pignistic Probability Function in a Context of Uncertainty. Proc. of the Fifth Workshop on Uncertainty in AI, Windsor, Canada, 319-326. Also in Henrion et al. (1990).
Smets, P., Kennes, R. (1990) The Transferable Belief Model. IRIDIA Technical Report TR/IRIDIA/90-14.2, Bruxelles.
Strat, T.M. (1990) Decision Analysis Using Belief Functions. *Int. J. Approximate Reasoning*, **4**, 391-418.
Tarjan, R.E., Yannakakis, M. (1984) Simple Linear Time Algorithms to Test Chordality of Graphs, Test Acyclicity of Hypergraphs, and Selectively Reduce Acyclic Hypergraphs. *SIAM J. Computing*, **13**, 566-79.
Thoma, H.M. (1991) Belief Function Computations, in I.R. Goodman et al. (eds.) (1991).
Wasserman, L.A. (1990 a) Prior Envelopes Based on Belief Functions. *Annals of Stat.*, **18**, 454-64.
Wasserman, L.A. (1990 b) Belief Functions and Statistical Inference. *Can. J. Stat.*, **18**, 183-96.
Williams, P.M. (1982) Discussion of Shafer (1982 a), *J. Royal Statist. Soc. B*, **44**, 342.
Wesley, L.P. (1986) Evidential Knowledge-Based Computer Vision. *Optical Engineering*, **25**, 363-79.
Xu, H. (1992) An Efficient Tool for Reasoning with Belief Functions. Proc. IPMU'92 , 65-8.
Yen, J. (1989) GERTIS: A Dempster-Shafer Approach to Diagnosing Hierarchical Hypotheses. *Communications ACM*, **32**, 573-85.
Zarley, D., Hsia, Y.T., Shafer, G. (1988): Evidential Reasoning Using DELIEF, Proc. Nat. Conf, A.I., 1988; Reprinted in Shafer, Pearl, 1990, 619-23.
Zhang, L. (1988): Studies on Finding Hypertree Covers for Hypergraphs. Working Paper No. 198, Univ. of Kansas, School of Business.

5

Algebraic methods in control, theory and applications

Vladimír Kučera
Institute of Information Theory and Automation
Academy of Sciences of the Czech Republic
Pod vodárenskou věží 4, P. O. Box 18, 182 08 Praha, CZ.
Tel: (42-2) 6884669. Fax: (42-2) 6884903.
e-mail: kucera@utia.cas.cz

Abstract
Fractional representations are a useful algebraic tool for control system design. They result in a simple parametrization of all controllers that stabilize a given plant. One could then, in principle, choose the best controller for various applications.

Keywords
Algebraic system theory; control system synthesis; feedback control; linear systems; transfer functions.

1 INTRODUCTION

This paper is a tutorial whose aim is to explain a useful algebraic tool for control system design, the so-called *fractional representation* approach. This approach is based on the input-output properties of linear systems. The central idea is that of representing the transfer function of a (not necessarily stable) system as the ratio of two *stable* transfer functions. This is a natural step for the linear systems whose transfer functions are rational, i.e., for the lumped-parameter systems. Under certain conditions, however, this approach is productive also for the distributed-parameter systems.

The starting point of the design is to obtain a simple parametrization of *all* controllers that stabilize a given plant. One could then, in principle, choose the best controller for various applications. The key point is that the parameter appears in the closed-loop system transfer function in a *linear* manner, thus making it easier to meet additional design specifications.

The actual design of control systems is an engineering task that cannot be reduced to algebra. Design contains many additional aspects which have to be taken into account: sensor placement,computational constraints, actuator constraints, redundancy, performance robustness, among many others. There is a need for understanding of the control process, a feeling for what kinds of performance objectives are unrealistic, or even dangerous, to ask for. The algebraic approach to be presented, nevertheless, is an elegant and useful tool for the mathematical part of the controller design (Kučera, 1993).

2 SYSTEMS AND SIGNALS

The fundamentals of the fractional representation approach will be explained for linear systems with rational transfer functions whose input u and output y are scalar quantities. We suppose that u and y live in a space of functions mapping a time set into a value set. The time set is the set of non-negative reals, R_+, and the value set is taken to be the set of real numbers R.

Let the input and output spaces of a system be the spaces of locally (Lebesgue) integrable functions f from R_+ into R and define a p-norm

$$||f||_{L_p} = \left[\int_0^\infty |f(t)|^p dt\right]^{1/p} \quad \text{if } 1 \leq p < \infty$$

$$||f||_{L_\infty} = \operatorname{ess}\sup_{t\geq 0} |f(t)| \quad \text{if } p = \infty.$$

The corresponding normed space is denoted by L_p.

Systems having the desirable property of preserving the functional space are called *stable.* More precisely, a system is said to be L_p-stable if any input $u \in L_p$ gives rise to an output $y \in L_p$. The systems that are L_∞-stable are also termed to be *bounded-input bounded-output* (BIBO) *stable.*

The transfer function of a system is the Laplace transform of its impulse response $g(t)$,

$$G(s) = \int_0^\infty g(t)e^{-st}dt.$$

It is well known that a system with a rational transfer function $G(s)$ is BIBO stable if and only if $G(s)$ is proper and stable, i.e., bounded at infinity with all poles having negative real parts (Desoer and Vidyasagar, 1975).

3 FRACTIONAL REPRESENTATIONS

Consider a rational function $G(s)$. By definition, it can be expressed as the ratio

$$G(s) = \frac{B(s)}{A(s)}$$

of two qualified rational functions A and B.

A well known example is the polynomial description, in which case A and B are coprime polynomials, i.e., polynomials having no roots in common.

Another example is to take for A and B two coprime, *proper and stable* rational functions. When $G(s)$ is, say,

$$G(s) = \frac{s+1}{s^2+1}$$

then one can take

$$A(s) = \frac{s^2+1}{(s+\lambda)^2}, \quad B(s) = \frac{s+1}{(s+\lambda)^2}$$

where λ is a positive real number. We recall that two proper and stable rational functions are coprime if they have no infinite nor unstable zeros in common. Therefore, in the example above, the denominator of A and B can be any strictly Hurwitz polynomial of degree exactly two; if the degree is lower, then A would not be proper and if it is higher, then A and B would have a common zero at infinity. The set of proper and stable rational functions will be denoted by $R_{ps}(s)$. It has the algebraic structure of a ring, in particular a euclidean domain (Vidyasagar, 1985).

4 FEEDBACK SYSTEMS

To control a system means altering its dynamics so that a desired behaviour is obtained. This can be done by feedback. A typical feedback system consists of two subsystems, S_1 and S_2, connected as shown in Fig. 1

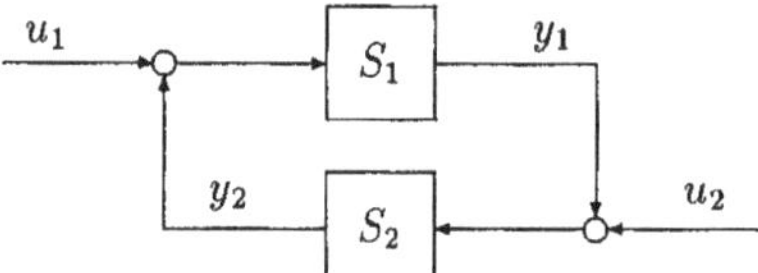

Figure 1 Feedback system

In most applications, it is desirable that the feedback system be BIBO stable in the sense that whenever the inputs u_1 and u_2 are bounded in magnitude so too are the outputs y_1 and y_2.

In order to study this property, we express the transfer functions of S_1 and S_2 as ratios of proper stable rational functions and seek for conditions under which the transfer function of the feedback system is proper and stable.

Thus we write

$$S_1(s) = \frac{B(s)}{A(s)}, \quad S_2(s) = -\frac{Y(s)}{X(s)}$$

where A, B and X, Y are two couples of coprime rational functions from $R_{ps}(s)$. The transfer matrix of the feedback system

$$\begin{bmatrix} y_1 \\ y_2 \end{bmatrix} = \begin{bmatrix} \frac{S_1}{1-S_1S_2} & \frac{S_1S_2}{1-S_1S_2} \\ \frac{S_1S_2}{1-S_1S_2} & \frac{S_2}{1-S_1S_2} \end{bmatrix} \begin{bmatrix} u_1 \\ u_2 \end{bmatrix}$$

is then given by

$$\begin{bmatrix} y_1 \\ y_2 \end{bmatrix} = \frac{1}{AX+BY} \begin{bmatrix} BX & -BY \\ -BY & -AY \end{bmatrix} \begin{bmatrix} u_1 \\ u_2 \end{bmatrix}.$$

We observe that the numerator matrix has all its entries in $R_{ps}(s)$ and that no infinite nor unstable zeros of the denominator can be absorbed in all these entries. We therefore conclude that the four transfer functions belong to $R_{ps}(s)$ if and only if $AX + BY$ is a *unit* of $R_{ps}(s)$, i.e., the inverse of $AX + BY$ belongs to $R_{ps}(s)$.

We illustrate with the example where S_1 is a differentiator and S_2 is an invertor such that

$$S_1(s) = s, \quad S_2(s) = -1.$$

We take

$$A(s) = \frac{1}{s+\lambda}, \quad B(s) = \frac{s}{s+\lambda}$$

for any real $\lambda > 0$ and

$$X(s) = 1, \quad Y(s) = 1.$$

Then

$$(AX + BY)^{-1}(s) = \frac{s+\lambda}{s+1}$$

resides in $R_{ps}(s)$ and hence the closed-loop system is BIBO stable.

To summarize, the fractional representation used should be matched with the goal of the analysis. The analysis is then more transparent and leads to a simple algebraic condition: $AX + BY$ is a unit of the underlying ring (Desoer et al., 1980).

5 PARAMETRIZATION OF STABILIZING CONTROLLERS

The design of feedback control systems consists of the following: given one subsystem, say S_1, we seek to determine the other subsystem, S_2, so that the resulting feedback system shown in Fig. 1 meets the design specifications. We call S_1 the *plant* and S_2 the *controller.*

The primary design specification is some sort of stability. We shall focus on achieving BIBO stability. Any controlller that BIBO stabilizes the plant will be called a *stabilizing* controller for that plant.

Suppose the plant gives rise to the transfer function

$$S_1(s) = \frac{B(s)}{A(s)}$$

for some coprime elements A and B of $R_{ps}(s)$. It follows from the foregoing analysis (Kučera, 1979) that a stabilizing controller exists and that all controllers that stabilize the given plant are generated by all solution pairs X, Y with $X \neq 0$ of the Bézout equation

$$AX + BY = 1$$

over $R_{ps}(s)$. There is no loss of generality in setting $AX + BY$ to the identity rather than to an arbitrary unit of $R_{ps}(s)$: this unit is absorbed by X and Y and therefore cancels in forming

$$S_2(s) = -\frac{Y(s)}{X(s)}.$$

The solution set of the Bézout equation can be parametrized (Kučera, 1979) as

$$X = X' + BW, \quad Y = Y' - AW$$

where X', Y' is a particular solution of the equation and W is a free parameter, which is an arbitrary function in $R_{ps}(s)$.

The parametrization of the family of all stabilizing controlllers for the plant now falls out (Kučera, 1974; Youla et al., 1976) as

$$S_2(s) = -\frac{Y'(s) - A(s)W(s)}{X'(s) + B(s)W(s)}$$

where the parameter W varies over $R_{ps}(s)$ while satisfying $X' + BW \neq 0$. This condition is not very restrictive, as $X' + BW$ can identically vanish for at most one choice of W.

As an example, we shall stabilize an integrator plant S_1. Its transfer function can be expressed as

$$S_1(s) = \frac{\frac{1}{s+1}}{\frac{s}{s+1}}$$

where $s + 1$ is an arbitrarily chosen Hurwitz polynomial of degree one. One stabilizing controller for S_1 is easy to find, namely

$$S_2(s) = -1.$$

It corresponds to a particular solution $X' = 1,\ Y' = 1$ of the Bézout equation

$$\frac{s}{s+1}X + \frac{1}{s+1}Y = 1.$$

The solution set in $R_{ps}(s)$ of this equation is

$$X(s) = 1 + \frac{1}{s+1}W(s), \quad Y(s) = 1 - \frac{1}{s+1}W(s).$$

Hence all controllers S_2 that BIBO stabilize S_1 have the transfer function

$$S_2(s) = -\frac{1 - \frac{s}{s+1}W(s)}{1 + \frac{1}{s+1}W(s)}$$

where W is any function in $R_{ps}(s)$.

It is clear that the result is independent of the particular fraction taken to represent S_1. Indeed, if $s + 1$ is replaced by another Hurwitz polynomial $s + \lambda$ in the above example, one obtains

$$S_2(s) = -\frac{\lambda - \frac{s}{s+\lambda}W'(s)}{1 + \frac{1}{s+\lambda}W'(s)},$$

which is the same set when

$$W'(s) = \left(\frac{s+\lambda}{s+1}\right)^2 W(s) + \frac{s+\lambda}{s+1}(\lambda - 1).$$

6 PARAMETRIZATION OF CLOSED–LOOP TRANSFER FUNCTIONS

The utility of the fractional approach derives not merely from the fact that it provides a parametrization for all controllers that stabilize a given plant in terms of a free parameter

W, but also from the simple manner in which this parameter enters the resulting (proper and stable) closed-loop transfer functions.

In fact,

$$\begin{bmatrix} y_1 \\ y_2 \end{bmatrix} = \begin{bmatrix} B(X' + BW) & -B(Y' - AW) \\ -B(Y' - AW) & -A(Y' - AW) \end{bmatrix} \begin{bmatrix} u_1 \\ u_2 \end{bmatrix}$$

and we observe that all four transfer functions are *linear* in the free parameter W.

This result serves to parameterize the performance specifications and it is the starting point for the selection of the best controller for the application at hand. The search for S_2 is thus replaced by a search for W. The crucial point is that the resulting selection/optimization problem is linear in W while it is non-linear in S_2.

7 OPTIMAL PERFORMANCE

The performance specifications often involve a norm minimization.

Let us consider the problem of *disturbance attenuation*. We are given a plant S_1 having two inputs: the control input u and an unmeasurable disturbance d, see Fig. 2.

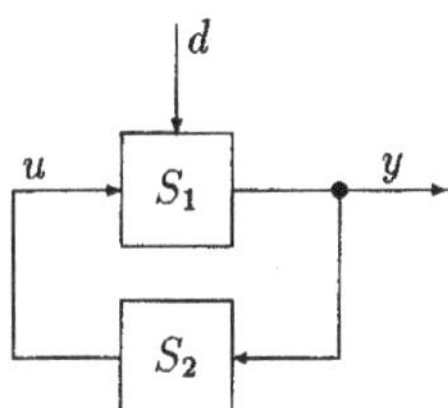

Figure 2 Disturbance attenuation

The objective is to dertermine a BIBO stabilizing controller S_2 for the plant S_1 such that the effect of d on the plant output y is minimized in some sense.

We describe the plant by two transfer functions

$$S_{1u}(s) = \frac{B(s)}{A(s)}, \quad S_{1d}(s) = \frac{C(s)}{A(s)}$$

where A, B and C is a triple of coprime functions from $R_{ps}(s)$. The set of stabilizing controllers for S_1 is given by the transfer function

$$S_2(s) = -\frac{Y'(s) - A'(s)\,W(s)}{X'(s) + B'(s)\,W(s)}$$

where A', B' is a *coprime* fraction over $R_{ps}(s)$ for S_{1u},

$$\frac{B(s)}{A(s)} = \frac{B'(s)}{A'(s)}$$

and X', Y' is a particular solution over $R_{ps}(s)$ of the Bézout equation

$$A'X + B'Y = 1$$

such that $X' + B'W \neq 0$.

The transfer function, $G(s)$, between d and y in a stable feedback system is

$$G = \frac{S_{1d}}{1 - S_{1u}S_2} = C(X' + B'W)$$

and it is indeed linear in the proper stable rational parameter W.

Now suppose that the disturbance d is any function from L_∞, i.e., any essentially bounded real function on R_+. Then (Doyle et al., 1992)

$$||y||_{L_\infty} \leq ||G||_1 \; ||d||_{L_\infty},$$

where

$$||G||_1 = \int_0^\infty |g(t)| \, dt$$

and $g(t)$ is the impulse response corresponding to $G(s)$. The parameter W can be used to minimize the norm $||G||_1$ and hence the maximum output amplitude.

If d is stationary white noise, the steady–state output variance equals (Kučera, 1979)

$$Ey^2 = ||G||_2^2 \, Ed^2,$$

where

$$||G||_2^2 = \int_0^\infty |g(t)|^2 \, dt = \frac{1}{2\pi j} \oint G(-s)G(s) \, ds.$$

The last integral is a contour integral up the imaginary axis and then around an infinite semicircle in the left half–plane. Again, W can be selected so as to minimize the norm $||G||_2$, thus minimizing the steady–state output variance.

Finally suppose that d is any function from L_2, i.e., any finite–energy real function on R_+. Then one obtains (Doyle et al., 1992)

$$||y||_{L_2} \leq ||G||_\infty \; ||d||_{L_2},$$

where

$$||G||_\infty = \sup_{Re\, s > 0} |G(s)| \, .$$

Therefore choosing W so as to make the norm $||G||_\infty$ minimal, one minimizes the maximum output energy.

Here is an illustrative example. The plant is given by

$$S_{1u}(s) = \frac{1}{s}, \quad S_{1d}(s) = 1 + \frac{1}{s}$$

and we seek to find a stabilizing controller S_2 such that $G(s)$ has minimum ∞–norm.

We write

$$A(s) = \frac{s}{s+1}, \quad B(s) = \frac{1}{s+1}, \quad C(s) = 1$$

and recall that the set of stabilizing controllers is given by

$$S_2(s) = -\frac{1 - \frac{s}{s+1} W(s)}{1 + \frac{1}{s+1} W(s)}$$

where W is a free parameter in $R_{ps}(s)$. Then

$$G(s) = \frac{S_{1d}(s)}{1 - S_{1u}(s)S_2(s)} = 1 + \frac{1}{s+1} W(s)$$

so that the least norm $||G||_\infty = 0$ is achieved by $W(s) = -(s+1)$. This parameter, however, does not belong to $R_{ps}(s)$. So we approximate it by

$$W_\varepsilon(s) = -\frac{s+1}{\varepsilon s + 1}$$

for any real $\varepsilon > 0$. Then

$$G_\varepsilon(s) = \frac{\varepsilon s}{\varepsilon s + 1}$$

and the least ∞–norm attainable by a proper and stable rational parameter W_ε is $||G_\varepsilon||_\infty = 1$. The optimal stabilizing controller guarantees $||y||_{L_2} \leq ||d||_{L_2}$ for any $d \in L_2$.

8 ROBUST STABILIZATION

The actual plant can differ from its nominal model. We suppose that a nominal plant description is available together with a description of the plant uncertainty. The objective is to design a controller that stabilizes the nominal plant as well as all plants lying within the specified domain of uncertainty. Such a controlller is said to *robustly* stabilize the plant.

The plant uncertainty can be modelled conveniently in terms of the fractional representation over $R_{ps}(s)$. We endow $R_{ps}(s)$ with the ∞–norm: for any function $F(s)$ from $R_{ps}(s)$,

$$||F||_\infty = \sup_{Re\, s>0} |F(s)|.$$

For any two such function, $F_1(s)$ and $F_2(s)$, we define

$$||\,[F_1 \quad F_2]\,||_\infty = ||\begin{bmatrix} F_1 \\ F_2 \end{bmatrix}||_\infty = \sup_{Re\, s>0} (|F_1(s)|^2 + |F_2(s)|^2)^{1/2}.$$

Let S_{10} be a nominal plant giving rise to a transfer function

$$S_{10}(s) = \frac{B(s)}{A(s)}$$

where A and B are coprime functions from $R_{ps}(s)$. We denote $S_1(A, B, \mu)$ the family of plants having transfer functions

$$S_1(s) = \frac{B(s) + \Delta B(s)}{A(s) + \Delta A(s)}$$

where ΔA and ΔB are functions from $R_{ps}(s)$ such that

$$||\,[\Delta A \quad \Delta B]\,||_\infty < \mu$$

for some non–negative real number μ.

Now, let S_2 be a BIBO stabilizing controller for S_{10}. Therefore

$$S_2(s) = -\frac{Y'(s) - A(s)\, W(s)}{X'(s) + B(s)\, W(s)}$$

where $AX' + BY' = 1$ and W is an element of $R_{ps}(s)$. Then S_2 will BIBO stabilize *all* plants from $S_1(A, B, \mu)$ if and only if

$$(A + \Delta A)(X' + BW) + (B + \Delta B)(Y' - AW) = 1 + [\Delta A \quad \Delta B] \begin{bmatrix} X' + BW \\ Y' - AW \end{bmatrix}$$

is a unit of $R_{ps}(s)$. This is the case whenever

$$|| [\Delta A \quad \Delta B] \begin{bmatrix} X' + BW \\ Y' - AW \end{bmatrix} ||_\infty < 1$$

so we have the following condition of robust stability (Vidyasagar, 1985):

$$\mu || \begin{bmatrix} X' + BW \\ Y' - AW \end{bmatrix} ||_\infty \leq 1 .$$

The best controller that robustly stabilizes the plant corresponds to the parameter W that minimizes the ∞–norm above.

To illustrate, consider the family of plants

$$S_1 = (\frac{s}{s+1}, \frac{1}{s+1}, \mu)$$

around the nominal model

$$S_{10}(s) = \frac{1}{s}.$$

This family accommodates, among other things, gain perturbations and stable second-order dynamics of the form

$$S_1(s) = \frac{1 + \mu}{s(1 + \varepsilon s)} .$$

All controllers that BIBO stabilize the nominal plant S_{10} are given by

$$S_2(s) = -\frac{1 - \frac{s}{s+1} W(s)}{1 + \frac{1}{s+1} W(s)}$$

where W is a free parameter in $R_{ps}(s)$.

Suppose we wish to obtain a robust *proportional* controller, $S_2(s) = -\lambda$. Then

$$W(s) = (1 - \lambda)\frac{s+1}{s+\lambda}$$

and $\lambda > 0$. The norm

$$|| \begin{bmatrix} 1 + \frac{1}{s+1} W \\ 1 - \frac{s}{s+1} W \end{bmatrix} ||_\infty = \sqrt{1 + \lambda^2} \, || \frac{s+1}{s+\lambda} ||_\infty$$

is calculated to be

$$N(\lambda) = \begin{cases} \frac{\sqrt{1+\lambda^2}}{\lambda}, & \text{if} \quad 0 < \lambda < 1 \\ \sqrt{1 + \lambda^2}, & \text{if} \quad \lambda \geq 1 . \end{cases}$$

Thus our proportional controller is robust if and only if

$$N(\lambda) \leq \frac{1}{\mu}.$$

Which proportional controller will maximize the stability margin μ? The one which minimizes $N(\lambda)$. Since

$$\min_{\lambda>0} N(\lambda) = \sqrt{2}$$

is attained by $\lambda = 1$, which corresponds to $W(s) = 0$, the consequent controller $S_2(s) = -1$ achieves the stability margin as large as

$$\mu = \frac{1}{\sqrt{2}}.$$

REFERENCES

Desoer, C. A. and M. Vidyasagar (1975) *Feedback Systems: Input–Output Properties.* Academic Press, New York.

Desoer, C. A., R. W. Liu, J. Murray and R. Saeks (1980) Feedback systems design: The fractional representation approach to analysis and synthesis. *IEEE Trans. Aut. Control,* AC–**25**, 399–412.

Doyle, J. C., B. A. Francis and A. R. Tannenbaum (1992) *Feedback Control Theory.* Macmillan, New York.

Kučera, V. (1974) Closed–loop stability of discrete linear single–variable systems. *Kybernetika,* **10**, 146–171.

Kučera, V. (1979) *Discrete Linear Control: The Polynomial Equation Approach.* Wiley, Chichester.

Kučera, V. (1993) Diophantine equations in control – A survey. *Automatica,* **29**, 1361–1375.

Vidyasagar, M. (1985) *Control System Synthesis: A Factorization Approach.* MIT Press, Cambridge, MA.

Youla, D. C., J. J. Bongiorno and H. A. Jabr (1976) Modern Wiener–Hopf design of optimal controllers, Part I: The single–input case. *IEEE Trans. Aut. Control,* AC–**21**, 3–14.

6

One method for robust control of uncertain systems – Theory and practice

George Leitmann
College of Engineering, University of California
Berkeley, CA 94720, U.S.A.

Abstract
We present a controller design methodology for uncertain systems which is based on the constructive use of Lyapunov stability theory. The uncertainties, which are deterministic, are characterized by certain structural conditions and known as well as unknown bounds. As a consequence of the Lyapunov approach, the methodology is not restricted to linear or time-invariant systems. The robustness of these controllers in the presence of singular perturbations is considered. The situation in which the full state of the system is not available for measurement is also considered as are other generalizations. Applications of the proposed discussed in the complete version of the paper.

Keywords
uncertain systems; deterministic control; robust control; resource management

1 INTRODUCTION

A fundamental feedback control problem* is that of obtaining some specified desired behavior from a system about which there is incomplete or uncertain information. Here we consider systems whose uncertainties are characterized deterministically rather that stochastically or fuzzily; for a stochastic approach see (Bass, 1957), and for fuzzy one see (Klir and Folger, 1988).

Our model of an uncertain system is of the form

$$\dot{x}(t) = F(t, x(t), u(t), \omega) \tag{1}$$

where $t \in I\!R$ is the "time" variable, $x(t) \in I\!R^n$ is the state and $u(t) \in I\!R^m$ is the control input. All the uncertainty in the system is represented by the lumped uncertain element $\omega \in \Omega$. It could be an element of $I\!R^q$ representing constant unknown parameters and inputs; it could also be a function from $I\!R$ into $I\!R^q$ representing unknown time varying

*Throughout this paper, references are intended to be representative rather than exhaustive. For a more complete bibliography see (Leitmann, 1990 and 1993).

parameters and inputs; it could also be a function from $IR \times IR^n \times IR^m$ into IR^q representing nonlinear elements which are difficult to characterize exactly; it could be merely an index. $F : IR \times IR^n \times IR^m \times \Omega \rightarrow IR^n$ is known. The only information assumed about ω is the knowledge of a nonempty set Ω to which it belongs. A related characterization of uncertainties is via inclusions see (Kurzhanskii, 1983).

Discrete systems are usually modelled by a difference equation

$$x(k+1) = F(k, x(k), u(k)) \tag{2}$$

where $k \in \mathbb{Z}$ is the "time", $x(k) \in IR^n$ is the state, $u(k) \in IR^m$ is the control, and F is not known but rather belongs to a set F, with F known.

2 CONTINUOUS SYSTEM CONTROL

For continuous systems modelled by ordinary differential equations of the form (1) we consider control to be given by a memoryless state feedback controller

$$u(t) = p(t, x(t)) \ . \tag{3}$$

Ideally we wish to choose $p : IR \times IR^n \rightarrow IR^m$ so that the feedback controlled system

$$\dot{x}(t) = f(t, x(t), \omega) \ , \tag{4}$$

where

$$f(t, x, \omega) := F(t, x, p(t, x), \omega) \ , \tag{5}$$

has the property of g.u.a.s. (global uniform asymptotic stability) about the zero state for all $\omega \in \Omega$ and for all initial states in IR^n. However to assure g.u.a.s. of an uncertain system one sometimes has to resort to controllers which are discontinuous in the state; see (Gutman and Leitmann, 1976). To avoid such discontinuous controllers, we relax the problem to that of obtaining a family of controllers which assure that the behavior of (1) can be made arbitrarily close to g.u.a.s.; such a family is called a practically stabilizing family see (Corless and Leitmann, 1988, Corless et al., 1990).

2.1 A Specific Class of Uncertain Continuous Systems

An uncertain continuous system under consideration here is described by (1) and satisfies the following assumption.

Assumption C.1.[†] There exist a continuous function $B : IR \times IR^n \rightarrow IR^{n,m}$, a candidate Lyapunov function $V : IR \times IR^n \rightarrow IR_+$, a class K function $\gamma : IR_+ \rightarrow IR_+$, functions $\beta_1, \beta_2 : IR \times IR^n \times \Omega \rightarrow IR_+$ and continuous functions $\kappa, \rho : IR \times IR^n \rightarrow IR_+$ such that

$$F(t, x, u, \omega) = f_s(t, x, \omega) + B(t, x) g(t, x, u, \omega) \tag{6}$$

[†]For definition see (Corless et al., 1988; Corless et al., 1990.)

for some functions f_s and g which satisfy:
1) For each $\omega \in \Omega$, $f_s(\cdot, \omega)$ is continuous and

$$\frac{\partial V}{\partial t}(t, x) + \frac{\partial V}{\partial x}(t, x) f_s(t, x, \omega) \leq -\gamma(\|x\|) \tag{7}$$

for all $t \in IR$, $x \in IR^n$.
2) For each $\omega \in \Omega$, $g(\cdot\, . \,\omega)$ is continuous and

$$u^T g(t, x, u, \omega) \geq -\beta_1(t, x, \omega)\|u\| + \beta_2(t, x, \omega)\|u\|^2 \tag{8}$$

where

$$\beta_1(t, x, \omega) \leq \beta_2(t, x, \omega)\rho(t, x) \tag{9}$$

$$\beta_1(t, x, \omega) \leq \kappa(t, x) \tag{10}$$

for all $t \in IR$, $x \in IR^n$, $u \in IR^m$.

2.2 Proposed Controllers

Here we present some practically stabilizing controller sets for the system considered in the previous section. These controllers can be regarded as continuous approximations of those presented in (Gutman and Leitmann, 1976).

Consider any uncertain system described above and let $(B, V, \gamma, \rho, \kappa)$ be a quintuple which assures the satisfaction of assumption $C.1$. Choose any continuous functions ρ_c, κ_c which satisfy

$$\rho_c(t, x) \geq \rho(t, x), \ \kappa_c(t, x) \geq \kappa(t, x) \tag{11}$$

and define

$$\alpha(t, x) := B(t, x)^T \frac{\partial V}{\partial x}(t, x)^T, \tag{12}$$

$$\eta(t, x) := \kappa_c(t, x)\alpha(t, x). \tag{13}$$

A practically stabilizing family of controllers is the set

$$\mathsf{P} := \{p_\epsilon \mid \epsilon > 0\} \tag{14}$$

where p_ϵ is any continuous function which satisfies

$$\|\alpha(t, x)\| p_\epsilon(t, x) = -\|p_\epsilon(t, x)\| \alpha(t, x) \tag{15}$$

i.e., $p_\epsilon(t, x)$ is opposite in direction to $\alpha(t, x)$, and

$$\|\eta(t, x)\| \neq 0 \implies \|p_c(t, x)\| \geq \rho_c(t, x)[1 - \|\eta(t, x)\|^{-1}\epsilon]. \tag{16}$$

As an example of a function satisfying the above requirements on p_ϵ, consider

$$p_\epsilon(t,x) := \begin{cases} -\frac{\eta(t,x)}{\epsilon}\rho_c(t,x) & if\|\eta(t,x)\| \leq \epsilon \\ -\frac{\eta(t,x)}{\|\eta(t,x)\|}\rho_c(t,x) & if\|\eta(t,x)\| > \epsilon; \end{cases} \tag{17}$$

see (Corless and Leitmann, 1981).

As another example, consider

$$p_\epsilon(t,\ x) \ := \ - \frac{\eta(t,\ x)}{\|\eta(t,\ x)\| \ + \ \epsilon} \ \rho_c(t,\ x) \ ; \tag{18}$$

see (Ambrosino et al., 1985).

Controllers of a discontinuous type as well as their continuous approximations, related to those proposed here, have been deduced by employing the theory of variable structure control; see (Bartolini and Zolezzi, 1985). Some early treatments of controller design for uncertain systems were based on "games against nature"; see (Gutman and Leitmann, 1975). Another class of controllers for systems of type (1) are deduced in (Barmish et al., 1983).

2.3 Matching Conditions

Given a system described by (1) the choice of $B,\ f_s,\ g$ to assure satisfaction of Assumption $C.1$ (if possible) may not be obvious. This choice is usually easier if the uncertainties are *matched* in the sense that there exist functions $f_0,\ B,\ g$ with $B(t,\ x)\ \in\ I\!R^{n,m}$ such that

$$F(t,\ x,\ u,\ \omega) \ = \ f_0(t,\ x) \ + \ B(t,\ x)g(t,\ x,\ u,\ \omega) \ ; \tag{19}$$

that is, the uncertainty ω and the control enter the system description via the same matrix $B(t,\ x)$.

Much of the literature concerns systems in which the uncertainties are matched. (Barmish and Leitmann, 1982) and (Chen and Leitmann, 1987) consider systems with unmatched uncertainties; there the norm of the unmatched portion of the uncertain term must be smaller than a certain threshold value. In (Stalford, 1987a) linear systems are considered in which the uncertainty satisfies generalized matching conditions, that is, structural conditions which are less restrictive than the matching condition. In these cases, as in the matched case, the norm bounds of the uncertain terms can be arbitrarily large. Linear time-invariant systems with scalar control input are treated in (Stalford, 1987b), while (Schmitendorf, 1988) requires the existence of a positive definite solution of a certain Riccati equation.

2.4 Other Problems

While *global* uniform asymptotic stability or at least practical stability can be guaranteed provided the control is not constrained, only *local* stability can be assured if the available control is subject to constraints. One class of stabilization problems with control constraints is considered by (Soldatos et al., 1991, Corless and Leitmann, 1993). Controllers which assure not only practical stability but also exponential convergence at a prescribed rate are treated in (Corless et al., 1988, Corless, 1993). (Corless and Leitmann, 1983) deal with systems in which the uncertainty bounds are not known exactly but depend on unknown constants; the controllers presented there are parameter adaptive controllers. Problems in which one wishes to keep the system state within or outside a prescribed region of the state space are considered in (Corless et al., 1987). Systems with delay are considered in (Thowsen, 1983) and (Lee and Leitmann, 1988). (Ha and Gilbert, 1987) treat controllers which linearize a nominal system in addition to assuring stability of the actual one. Large scale uncertain systems with decentralized control are discussed in (Chen, 1987a) and (Siljak, 1991).

3 DISCRETE SYSTEMS CONTROL

The control of uncertain discrete systems modelled by difference equations of the form (2) has been treated in (Corless and Manela, 1986, Magana and Zak, 1988 and Sezer and Siljak, 1988). Unlike in the continuous case reviewed in the previous section, arbitrarily large uncertainties cannot be tolerated, in general, and the region of ultimate attraction cannot be made arbitrarily small. (Corless and Manela, 1986) consider the matched case, namely

$$x(k+1) = f(k, x(k)) + B(k, x(k))\,[u(k) + e(k, x(k), u(k))] \tag{20}$$

where $k \in \mathbb{Z}$, $x(k) \in \mathbb{R}^n$ *and* $u(k) \in \mathbb{R}^m$. The functions $f : \mathbb{Z} \times \mathbb{R}^n \rightarrow \mathbb{R}^n$ and $B : \mathbb{Z} \times \mathbb{R}^n \rightarrow \mathbb{R}^{n,m}$ are assumed known, with

$$rank[B(k, x(k))] = m\ . \tag{21}$$

The function $e : \mathbb{Z} \times \mathbb{R}^n \times \mathbb{R}^m \rightarrow \mathbb{R}^m$ is not known; however, it is assumed that the class of functions E to which it belongs is known. We make the following two assumptions before stating a stabilization theorem.

Assumption D.1. [‡] Given a positive definite $P \in \mathbb{R}^{n,n}$ there exist non-negative scalars ρ_0, ρ_1 *and* ρ_2 such that for all $e \in \mathsf{E}$

$$\|B(k, x)e(k, x, u)\|_P \leq \rho_0 + \rho_1\|x\|_P + \rho_2\|u\|_{R(k,\, x)} \tag{22}$$

for all $(k, x, u) \in \mathbb{Z} \times \mathbb{R}^n \times \mathbb{R}^m$, where $R(k, x) := B(k, x)^T P B(k, x)$. Next we define

$$\psi(k, x) := [B(k, x)^T P B(k, x)]^{-1} B(k, x)^T P\ , \tag{23}$$

$$\phi(k, x) := B(k, x)\psi(k, x)\ , \tag{24}$$

$$\bar{f}(k, x) := \phi\,(k, x) f(k, x)\ , \tag{25}$$

[‡]Let $P \in \mathbb{R}^{n,n}$ be a positive definite matrix. We define the norm

$\|\cdot\|_P : \mathbb{R}^n \rightarrow \mathbb{R}_+$ *by* $\|r\|_P := \sqrt{r^T P r}$.

and

$$\tilde{f}(k,\ x)\ :=\ f(k,\ x)\ -\ \bar{f}(k,\ x)\ . \tag{26}$$

Assumption D.2. There exist a positive definite matrix $P \in I\!R^{n,n}$ and a non-negative scalar $\tilde{c} < 1$ such that

$$\|\tilde{f}(k,\ x)\|_P\ \leq\ \tilde{c}\,\|x\|_P \tag{27}$$

for all $(k,\ x)\ \in\ Z\!\!Z\ \times\ I\!R^n$. If $\rho_2 \neq 0$, then there also exist non-negative scalars c_0 *and* c_1 such that

$$\|\bar{f}(k,\ x)\|_P\ \leq\ c_0\ + c_1\|x\|_P \tag{28}$$

for all $(k,\ x)\ \in\ Z\!\!Z\ \times\ I\!R^n$.

3.1 Proposed Controllers

Consider an uncertain discrete system (20) satisfying assumptions $D.1. - D.2$ and subject to the control $u(k)\ =\ p(k,\ x(k))$ where $p(k,\ x(k))$ is defined as follows:

$$p(k, x(k)) := \begin{cases} 0 & if\ \rho_2 \geq 1 \\ \\ -\Psi(k, x(k))f(k, x(k)) & if\ \rho_2 < 1. \end{cases} \tag{29}$$

Suppose that

$$\tilde{c}^2 + (\rho_1 + c_1^*)^2 < 1 \tag{30}$$

where

$$c_1^* := \begin{cases} 0 & if\ \rho_2 = 0 \\ \rho_2^* c_1 & if\ \rho_2 \neq 0 \end{cases} \tag{31}$$

and

$$\rho_2^*\ :=\ min\ \{\rho_2,\ 1\}. \tag{32}$$

Then for all $e\ \in\ \mathsf{E}$, the feedback controlled system (20) is g.u.a.s. about the set

$$B_P(d)\ := \{x\ \in\ I\!R^n|\ \ \|x\|_P\ \leq\ d\} \tag{33}$$

where

$$d\ :=\ \frac{\rho_0\ +\ c_0^*}{\sqrt{1\ -\ \tilde{c}^2\ -\ (\rho_1\ +\ c_1^*)}} \tag{34}$$

and

$$c_0^* := \begin{cases} 0 & if\ \rho_2 = 0 \\ \rho_2^* c_0 & if\ \rho_2 \neq 0. \end{cases} \tag{35}$$

4 ROBUSTNESS IN THE PRESENCE OF SINGULAR PERTURBATIONS

Consider an uncertain singularly perturbed system described by

$$\begin{aligned} \dot{x} &= F(t,\ x,\ y,\ u,\ \mu,\ \omega) \\ \mu\dot{y} &= G(t,\ x,\ y,\ u,\ \mu,\ \omega) \end{aligned} \tag{36}$$

where $(x,\ y) \in I\!R^n \times I\!R^l$ describe the state of the system. $\mu \in (0,\ \infty)$ is the singular perturbation parameter, and all the other variables are as described above. Here one wants to obtain memoryless feedback controllers (generating u) which assure that, for all $\omega \in \Omega$ and for all sufficiently small μ, the behavior of the feedback controlled system is close to that of g.u.a.s.

Assuming that, for each $x,\ u,\ \omega$ there exists a unique vector $H(x,\ u,\ \omega) \in I\!R^l$ such that

$$G(t,\ x,\ H(x,\ u,\ \omega),\ u,\ 0,\ \omega) = 0 \tag{37}$$

for all t, the reduced order system associated with (36) (let $\mu = 0$ in (36)) is given by

$$\dot{x} = \bar{F}(t,\ x,\ u,\ \omega) \tag{38}$$

where

$$\bar{F}(t,\ x,\ u,\ \omega) := F(t,\ x,\ H(x,\ u,\ \omega),\ u,\ 0,\ \omega). \tag{39}$$

For each $t,\ x,\ u,\ \omega$ the boundary layer system associated with (36) is given by

$$\frac{dy}{d\tau}(\tau) = G(t,\ x,\ y(\tau),\ u,\ 0,\ \omega). \tag{40}$$

(Corless et al., 1990) require that the boundary layer system satisfies g.u.a.s. about its equilibrium state $H(x,\ u,\ \omega)$ and present stabilizing controllers whose designs are based on the reduced order system. This situation occurs for systems with stable "neglected dynamics." In (Garofalo and Leitmann, 1990) the boundary layer system is not required to be stable. The "stabilizing" controllers presented there are composite controllers in the sense that they consist of two parts; one part is utilized to stabilize the boundary layer system and the other part is based on a nominal reduced order system.

5 OUTPUT FEEDBACK

Heretofore it was assumed that the complete state is available for feedback. Consider now the more general situation in which the output $y(t) \in I\!R^s$ available for feedback is related to the state by

$$y(t) = c(t, x(t), \omega) \tag{41}$$

for some function $c : I\!R \times I\!R^n \times \Omega \to I\!R^s$.

Memoryless output feedback controllers are treated in (Galimidi and Barmish, 1986, Steinberg and Ryan, 1986 and Chen 1987b). The dynamic output feedback controllers in the literature utilize state estimators. The state is fed to a memoryless controller whose design is based on having the complete state available for feedback. Full order observers are utilized in (Barmish and Galimidi, 1986 and Walcott and Zak, 1987). (Breinl and Leitmann, 1983 and Breinl and Leitmann, 1987) utilize reduced order observers. There the uncertain terms must satisfy certain structural conditions and the differential equation describing the evolution of the state estimation error is decoupled from the state equation.

6 APPLICATIONS

Controller designs based on a constructive use of Lyapunov stability theory or closely related methods have been applied to a variety of uncertain systems. In the realm of engineering these applications include tracking control for robotic manipulators including hybrid tracking and force control (Reithmeier and Leitmann, 1991), suspension control for magnetically levitated vehicles (Breinl and Leitmann, 1983 and Breinl and Leitmann, 1987), control of seismically excited structures (Kelly et al., 1987), of high speed rotors (Weltin, 1988), and of nuclear power plants (Parlos et al., 1988), as well as various aircraft and aerospace systems (Singh, 1987, Stalford, 1987b, Leitmann and Pandey, 1991). Experimental results may be found in (Horowitz et al., 1989 and Kang et al., 1991). (Deissenberg, 1986 and Leitmann and Wan, 1978) concern applications in economics. Resource allocation in fisheries is discussed in (Kaitala and Leitmann, 1990, Kaitala and Leitmann, 1992, Hilden et al., 1993). Harvesting problems are treated in (Lee and Leitmann, 1983 and Corless and Leitmann, 1985). (Lee and Leitmann, 1987 and Lee and Leitmann, 1988) deal with pollution control in rivers. (Lee and Leitmann, 1991) treat a problem in pedagogy. In the complete version of the paper presented here, two examples of resource management based on uncertain models, one continous and the other discrete, are discussed; see (Leitmann, 1995).

REFERENCES

Ambrosino, G., Celentano, G. and Garofalo, F. (1985) Robust Model Tracking Control for a Class of Nonlinear Plants. *IEEE Trans. Automatic Control*, **AC-30**, 275.

Barmish, B.R., Corless, M., Leitmann, G. (1983) A new class of stabilizing controllers for uncertain dynamical systems. *SIAM J. on Control and Optimization*, **21**, No. 2, 246.

Barmish, B.R., Galimidi, A.R. (1986) Robustness of Luenberger Observers: Linear System Stabilized via Nonlinear Control. *Automatica* **22**, 246.

Barmish, B.R., Leitmann, G. (1982) On Ultimate Boundedness Control of Uncertain Systems in Absence of Matching Conditions. *IEEE Trans. Automatic Control*, **AC-27**, 1253.

Bartolini G., Zolezzi, T. (1985) Variable Structure Systems Nonlinear in the Control Law. *IEEE Trans. Automatic Control*, **AC-30**, 681.

Bass, R.W. (1985) Discussion of: "Die Stabilität von Regelsystemen mit nachgebender Rückführung" by A.M. Letov. in *Proc. Heidelberg Conf. Automatic Control.*

Breinl, W., Leitmann, G. (1983) Zustandsrückführung für dynamische Systeme mit Parameterunsicherheiten. *Zeitschrift für Regelungstechnik*, **31**, 95.

Breinl, W., Leitmann, G. (1987) State Feedback for Uncertain Dynamical Systems. *Applied Mathematics and Computation*, **22**, 65.

Chen, Y.H. (1987a) Deterministic Control of Large Scale Uncertain Dynamical Systems. *J. Franklin Inst.*, **323**, 125.

Chen, Y.H. (1987b) Robust Output Feedback Controller: Direct Design. *Int. J. Control*, **46**, 1083.

Chen, Y.H., Leitmann, G. (1987) Robustness of Uncertain Systems in the Absence of Matching Assumptions. *Int. Journal of Control*, **45**, 1527.

Corless, M. (1993) Control of Uncertain Nonlinear Systems. *ASME J. of Dynamical Systems, Measurement, and Control*, **115**, 362.

Corless, M., Garofalo, F., Leitmann, G. (1988) Guaranteeing Exponential Convergence for Uncertain Systems, *Proc. Int. Workshop on Robustness in Identification and Control*, Torino, Italy.

Corless, M., Leitmann, G. (1981) Continuous State Feedback Guaranteeing Uniform Ultimate Boundedness for Uncertain Dynamic Systems. *IEEE Trans. Automatic Control*, **AC-26**, 1139.

Corless, M., Leitmann G. (1983) Adaptive Control of Systems Containing Uncertain Functions and Unknown Functions with Uncertain Bounds. *J. Optimiz. Theory Applic.*, **41**, 155.

Corless, M., Leitmann, G. (1985) Adaptive Long-Term Management of Some Ecological Systems Subject to Uncertain Disturbances, in *Optimal Control Theory and Economic Analysis 3* (ed. G. Feichtinger), Elsvier Science Publishers, Amsterdam, Holland.

Corless, M., Leitmann, G. (1988) Deterministic Control of Uncertain Systems, in *Proc. Conf. on Modeling and Adaptive Control*, Sopron, Hungary, Lecture Notes in Control and Information Sciences, Springer Verlag, IIASA 105.

Corless, M., Leitmann, G. (1990) Deterministic Control of Uncertain Systems: A Lyapunov Theory Approach, in *Deterministic Control of Uncertain Systems*, Chapter 11, (ed. A.S.I. Zinober), Peter Peregrinus, London.

Corless, M., Leitmann, G. (1993) Bounded Controllers for Robust Exponential Convergence. *Optimiz. Theory Appl.*, **76**, 1.

Corless, M., Leitmann, G., Ryan, E.P. (1990) Uncertain Systems With Neglected Dynamics, in *Deterministic Control of Uncertain Systems*, Chapter 12, (ed. A.S.I. Zinober), Peter Perigrinus, London.

Corless, M., Leitmann, G., Skowronski, J.M. (1987) Adaptive Control for Avoidance or Evasion in an Uncertain Environment. *Computers and Mathematics with Applicaitons*, **13**, 1.

Corless, M., Manela, J. (1986) Control of Uncertain Discrete-Time Systems, in *Proc. American Control. Conf.*, Seattle, Washington.

Galimidi, A.R., Barmish, B.R. (1986) The Constrained Lyapunov Problem and its Application to Robust Output Feedback Stabilization, *IEEE Trans. Automatic Control*, **AC-31**, 410.

Garofalo, F., Leitmann, G. (1990) Nonlinear Composite Control of a Class of Nominally Linear Singularly Perturbed Uncertain Systems, in *Deterministic Control of Uncertain Systems*, Chapter 13, (ed. A.S.I. Zinober), Peter Peregrinus, London.

Gutman, S., Leitmann, G. (1975) On a Class of Linear Differential Games. *J. Optimiz. Theory Appl.*, **17**, 511.

Gutman, S., Leitmann, G. (1976) Stabilizing Feedback Control for Dynamical Systems with Bounded Uncertainty, in *Proc. IEEE Conf. Decision Control*, Clearwater, Florida.

Ha, I.-J., Gilbert, E.G. (1987) Robust Tracking in Nonlinear Systems. *IEEE Trans. Automatic Control*, **AC-32**, 763.

Hilden, M., Kaitala, V., Leitmann, G. (1994) Stabilizing Management and Structural Development of Open Access Fisheries, in *Advances in Dynamic Games and Applications*, (eds. T. Basar and A. Haurie), Birkhäuser Verlag, Basel, Switzerland.

Horowitz, R., Stephens, H.I., Leitmann, G. (1989) Experimental Implementation of a Deterministic Controller for a D.C. Motor with Uncertain Dynamics. *J. Dynam. Syst. Meas. Control*, **111**, 244.

Kaitala, V., Leitmann, G. (1990) Stabilizing Employment in a Fluctuating Recource Economy, *J. Optimiz. Theory Appl.*, **67**, 1.

Kaitala, V., Leitmann, G. (1992) Income Subsidizing and Fisheries Development - An Analysis of Stabilizing Management, in *Dynamic Economic Models and Optimal Control*, (ed. G. Feichtinger), Elsevier Science Publishers, Amsterdam, Holland.

Kang, C.G., Horowitz, R., Leitmann, G. (1991) Robust Deterministic Control for Robotic Manipulators, in *Proceed. ASME, Annual Meeting.*

Klir, G.J., Folger, T.A. (1988) *Fuzzy Sets, Uncertainty and Information.* Prentice Hall, Englewood Cliffs, N.J.

Kurzhanskii, A.B. (1983) Evolution Equations for Problems of Control and Estimation of Uncertain Systems. *Proc. Int. Cong. Math.*, Warsaw, Poland.

Lee, C.S., Leitmann, G. (1983) On Optimal Long-Term Management of Some Ecological Systems Subject to Uncertain Disturbances. *Int. J. Syst. Sci.*, **14**, 979.

Lee, C.S. Leitmann, G. (1987) Uncertain Dynamical Systems: An Application to River Polution Control, in *Proc. Modeling and Management of Resources Under Uncertainty*, Honolulu, Lecture Notes in Biomathematics, **72**, Springer Verlag, Berlin.

Lee, C.S., Leitmann, G. (1988) Continuous Feedback Guaranteeing Uniform Ultimate Boundedness for Uncertain Linear Delay Systems: An Application to River Pollution Control. *Comput. Math. Appl.*, **16**, 929.

Lee, C.S., Leitmann, G. (1991) Some stabilizing study strategies for a student-related problem under uncertainty. *Dynamics and Stability of Systems*, **6**, No. 1, 63.

Leitmann, G. (1990) Deterministic Control of Uncertain Systems Via a Constructive Use of Lyapunov Stability Theory, in *Proceed. 14th IFIP Conf.*, Leipzig, 1989, Lecture Notes in Control and Information Sciences, **143**, Springer Verlag, Berlin.

Leitmann, G. (1993) On One Approach to the Control of Uncertain Systems. *ASME J. of Dynamical Systems, Measurement, and Control*, **115**, 373.

Leitmann, G. (1995) One Method for Robust Control of Uncertain Systems – Theory and Practice. *Kybernetika*, to appear.
Leitmann, G., Pandey, S. (1991) Aircraft Control for Flight in an Uncertain Environment: Takeoff in Windshear. *J. Optimiz. Theory Appl.*, **70**, (1).
Leitmann, G., Wan, H.Y. Jr. (1978) A Stabilization Policy for an Economy with Some Unknown Characteristics. *J. Franklin Inst.*, **306**, 23.
Magana, M.E., Zak, S.H. (1988) Robust Output Feedback Stabilization of Discrete-Time Uncertain Dynamical Systems, *IEEE Trans. Automatic Control*, **AC-33**, 1082.
Parlos, A.G., Henry, A.F. Schweppe, F.C., Gould, L.A., Lanning, D.D. (1988) Nonlinear Multivariable Control of Nuclear Power Plants Based on the Unknown-but-bounded Disturbance Model. *IEEE Trans. Automatic Control*, **AC-33**, 130.
Reithmeier, E., Leitmann, G. (1991) Tracking and Force Control for a Class of Robotic Manipulators. *Dynamics and Control*, 1(2).
Schmitendorf, W.E. (1988) Stabilizing Controllers for Uncertain Linear Systems with Additive Disturbances. *Int. J. Control***47**, 85.
Sezer, M.E., Siljak, D.D. (1988) Robust Stability of Discrete Systems. *Int. J. Control*, **48**, 2055.
Siljak, D.D. (1991) *Decentralized Control of Complex Systems.* Academic Press, New York.
Singh, S.N. (1987) Attitude Control of a Three Rotor Gyrostat in the Presence of Uncertainty. *J. Astronautical Sciences*,**35**.
Soldatos, A.G., Corless, M., Leitmann, G. (1991) Stabilizing Uncertain Systems with Bounded Control, in *Third Workshop on Control Mechanics*, Lecture Notes in Control and Information Sciences, 409, Springer Verlag, Berlin.
Stalford, H.L. (1987a) Robust Control of Uncertain Systems in the Absence of Matching Conditions: Scalar Input, in *Proc. IEEE Conf. Decision and Control.*
Stalford, H.L. (1987b) On Robust Control of Wing Rock Using Nonlinear Control, in *Proc. American Control Conf.*, Minneapolis, Minn.
Steinberg, A., Ryan, E.P. (1986) Dynamic Output Feedback Control of a Class of Uncertain Systems. *IEEE Trans. Control*, **AC-31**, 1163.
Thowsen, A. (1983) Uniform Ultimate Boundedness of the Solutions of Uncertain Dynamic Delay Systems with State-Dependent and Memoryless Feedback. *Int. J. Control*, **37**, 1135.
Walcott, B.L., Zak, S.H. (1987) State Observation of Nonlinear Uncertain Dynamical Systems. *IEEE Trans.Automatic Control*, **AC-32**, 166.
Weltin, U. (1988) Aktive Schwingungsdämpfung von Rotoren mit Parameterunsicherheiten. *Dr. - Ing. Dissertation*, TU Darmstadt, Germany.

7

Stochastic optimization methods in engineering

Kurt Marti
Federal Armed Forces University Munich
Aero-Space Engineering and Technology
D-85577 Neubiberg/Munich, Germany.
Tel: +49-89-6004-2541/2109.
Fax: +49-89-6004-3560.
e-mail: Kurt.Marti@rz.unibw-muenchen.de

Abstract

Yield stresses, allowable stresses, moment capacities (plastic moments), external loadings, manufacturing errors are not given fixed quantities in practice, but have to be modelled as random variables with a certain joint probability distribution. Hence, problems from limit (collapse) load analysis or plastic analysis and from plastic and elastic design of structures are treated in the framework of stochastic optimization. Using especially reliability-oriented optimization methods, the behavioral constraints are quantified by means of the corresponding probability p_s of survival. Lower bounds for p_s are obtained by selecting certain redundants in the vector of internal forces; moreover, upper bounds for p_s are constructed by considering a pair of dual linear programs for the optimizational representation of the yield or safety conditions. Whereas p_s can be computed *e.g.* by sampling methods or by asymptotic expansion techniques based on Laplace integral representations of certain multiple integrals, efficient techniques for the computation of the sensitivities (of various orders) of p_s with respect to input or design variables have yet to be developed. Hence several new techniques are suggested for the numerical computation of derivatives of p_s .

Keywords

Stochastic optimization, structural design, structural analysis, probability functions

1 LIMIT (COLLAPSE) LOAD ANALYSIS OF STRUCTURES AS A LINEAR PROGRAMMING PROBLEM

The **collapse load** can be defined [4] "as the load required to generate enough number of local plastic yield points (referred as plastic hinges for bending type members) to cause the structure to become a mechanism and develop excessive deflections". Assuming that the material behaves as an elastic-perfectly plastic material [5], a conservative estimate of the collapse load factor λ_T is based on the following formulation as a linear program (LP):

$$\text{maximize } \lambda \tag{1}$$

s.t.

$$F^L \le F \le F^U \tag{1.1}$$

$$CF = \lambda R_o . \tag{1.2}$$

Here, (1.2) is the equilibrium equation of a statically indeterminate loaded structure involving an $m \times n$ matrix $C = (c_{ij})$, $m<n$, of given coefficients c_{ij}, $1 \le i \le m$, $1 \le j \le n$, depending on the undeformed geometry of the structure having n_o members (elements); we suppose that $\text{rank} C = m$.

In case of a truss [5] we have $n = n_o$, and the matrix C contains the direction cosines of the members. Furthermore, R_o is an external load m-vector, and F denotes the n-vector of internal forces and bending-moments in the relevant points (sections, nodes or elements) of the structure; in case of a truss F involves only the member forces. Finally, (1.1) are the yield conditions with the vector of lower and upper bounds F^L, F^U. For a truss we have that

$$F_j^L = \sigma_{yj}^L A_j, \quad F_j^U = \sigma_{yj}^U A_j, \quad j=1,\dots,n(=n_o), \tag{1.3}$$

where A_j is the (given) cross-sectional area, and σ_{yj}^L, σ_{yj}^U, resp., denotes the yield stress in compression (negative values) and tension of the j-th member of the truss. In case of a plane frame F is composed of subvectors

$$F^{(k)} = \begin{pmatrix} F_1^{(k)} \\ F_2^{(k)} \\ F_3^{(k)} \end{pmatrix}, \tag{1.4}$$

where $F_1^{(k)}$ denotes the normal (axial) force, and $F_2^{(k)}$, $F_3^{(k)}$ are the bending-moments at the two ends of the k-th member. In this case F^L, F^U contain - for each member k - the subvectors

$$F^{(k)L} = \begin{pmatrix} \sigma_{yk}^L A_k \\ -M_{p\ell,k} \\ -M_{p\ell,k} \end{pmatrix} \quad F^{(k)U} = \begin{pmatrix} \sigma_{y,k}^U A_k \\ M_{p\ell,k} \\ M_{p\ell,k} \end{pmatrix}, \tag{1.4.1}$$

resp., where $M_{p\ell,k}$, $k=1,\dots,n_o$, denote the plastic moments (moment capacities) given [4],[5] by

$$M_{p\ell,k} = \sigma^U_{yk} W_{p\ell,k}, \tag{1.4.2}$$

and $W_{p\ell,k} = W_{p\ell,k}(A_k)$ is the plastic modulus of the cross-section of the k-th member (beam).

2 PLASTIC AND ELASTIC DESIGN OF STRUCTURES

In the plastic design of trusses, frames [5] having n_o members, the n-vectors F^L, F^U of lower and upper bounds

$$F^L = F^L(\sigma^L_y, \sigma^U_y, X), \quad F^U = F^U(\sigma^L_y, \sigma^U_y, X) \tag{2}$$

for the n-vector F of internal member forces and bending-moments F_j, $j=1,\dots,n$, are determined [3],[5] by the yield stresses, i.e. compressive limiting stresses (negative values) $\sigma^L_y = (\sigma^L_{y1}, \dots, \sigma^L_{yn_o})'$, the tensile yield stresses $\sigma^U_y = (\sigma^U_{y1}, \dots, \sigma^U_{yn_o})'$, and the r-vector

$$X = (X_1, X_2, \dots, X_r)' \tag{2.1}$$

of design variables of the structure. In case of trusses we simply have that, cf. (1.3),

$$F^L = \sigma^L_{yd} A(X), \quad F^U = \sigma^U_{yd} A(X), \tag{2.2}$$

where $n=n_o$, and

$\sigma^L_{yd}, \sigma^U_{yd}$ denote the n×n diagonal matrices having the diagonal elements σ^L_{yj}, σ^U_{yj}, resp., $j=1,\dots,n$, moreover,

$$A(X) = (A_1(X), \dots, A_n(X))'$$

is the n-vector of cross-sectional areas $A_j = A_j(X)$, $j=1,\dots,n$, depending on the r-vector X of design variables X_k, $k=1,\dots,r$.

Corresponding to (1.2), here we have again the equilibrium equation

$$CF = R_u, \tag{2.3}$$

where R_u describes [5] the ultimate load (representing constant external loads or self-weight expressed in linear terms of A(X)).

The **plastic design** of structures can be represented then [1],[2],[5] by the following optimization problem

$$\min G(X) \tag{3}$$

s.t.

$$F^L(\sigma^L_y, \sigma^U_y, X) \le F \le F^U(\sigma^L_y, \sigma^U_y, X) \tag{3.1}$$

$$CF = R_u, \tag{3.2}$$

where $G=G(X)$ is a certain objective function, e.g. the volume or weight of the structure.

For the **elastic design** we have to replace the yield stresses σ_y^L, σ_y^U by the allowable stresses σ_y^L, σ_y^U, and instead of ultimate loads we consider service loads R_s. Hence, instead of (3) we get the related program

$$\min G(X) \tag{4}$$
s.t.
$$F^L(\sigma_a^L, \sigma_a^U, X) \le F \le F^U(\sigma_a^L, \sigma_a^U, X) \tag{4.1}$$
$$CF = R_s \tag{4.2}$$
$$X^L \le X \le X^U, \tag{4.3}$$

where X^L, X^U still denote lower and upper bounds for X.

3 ANALYSIS AND DESIGN OF STRUCTURES IN CASE OF RANDOM DATA

In practice, yield stresses, allowable stresses, the loads applied to the structure, other material properties and the manufacturing errors are not given fixed quantities, but must be treated as random variables on a certain probability space $(\Omega, \mathcal{A}, P)$. Hence, (1),(3),(4) are stochastic linear/nonlinear programs (SLP/SNLP) which obviously have the same basic structure represented by a random objective function

$$Z(\omega) := G(\omega, X), \quad \omega \in \Omega, \tag{5}$$

and by stochastic constraints of the type

$$CF = R(\omega) \tag{6}$$
$$F^L(\omega) \le F \le F^U(\omega), \tag{7}$$

where $R=R(\omega)$, $\omega \in \Omega$, is a random load m-vector given by

$$R(\omega) = \lambda R_o(\omega), \; R(\omega) = R_u(\omega), \; R(\omega) = R_s(\omega), \tag{6.1}$$

resp., and for the n-vector $F=(F_j)$ of internal member forces and bending-moments we have the n-vectors of random bounds

$$F^L(\omega) = F^L(\omega, X), \; F^U(\omega) = F^U(\omega, X), \; \omega \in \Omega, \tag{7.1}$$

depending on an r-vector X of design variables X_k, $k=1,\ldots,r$.

Obviously, each realization of the random element $\omega \in \Omega$ yields new loading conditions, represented by the vector $R=R(\omega)$, and therefore, cf. (6), new arrangements $F=F(\omega)$ of internal member forces and bending-moments. Hence, in the present case of "multiple loadings" caused by the random variations of $R=R(\omega)$, the survival of the structure, i.e. the existence of certain arrangements $F(\omega)$ of internal member forces/bending-moments not overwhelming the strength of the structure, can be evaluated by the probability of survival

$$p_s := P(\text{There is } F=F(\omega) \text{ such that } CF(\omega) = R(\omega) \text{ and } F^L(\omega) \le F(\omega) \le F^U(\omega)), \tag{8}$$

see [2],[10], assuming that, cf. (26),(33),

$$S(F^L(.), F^U(.)) := \{\omega \in \Omega: \text{there is a vector } F=F(\omega) \text{ fulfilling (6) and (7)}\} \tag{8.1}$$

is a measurable set. Denoting by $[F,^L,F^U]$ the n-dimensional interval

$$[F^L,F^U]:=\{F\in\mathbb{R}^n:\ F^L\le F\le F^U\}, \tag{8.2}$$

we find that

$$S(F^L(.),F^U(.)) = \{\omega\in\Omega:\ R(\omega)\in C[F^L(\omega),F^U(\omega)]\} \tag{8.3}$$

with $C[F^L,F^U] = \{CF:\ F^L\le F\le F^U\}$, and therefore

$$\begin{aligned} p_s &= P(S(F^L(.),F^U(.))) = P(R(\omega)\in C[F^L(\omega),F^U(\omega)]) \\ &= \int P(R(\omega)\in C[F^L,F^U]\,|\,F^L(\omega)=F^L,F^U(\omega)=F^U)P_{(F^L(.),F^U(.))}(dF^L,dF^U), \end{aligned} \tag{9}$$

where $P_{(F^L(.),F^U(.))}$ denotes the distribution of the bounds $(F^L(\omega),F^U(\omega))$.

Since the bounds F^L,F^U in (7) depend also on the vector X of design variables X_k, $k=1,\dots,r$, cf. (7.1), we have $p_s=P(X)$ with the probability function

$$P(X) = P(R(\omega)\in C[F^L(\omega,X),F^U(\omega,X)]); \tag{10}$$

furthermore, if the external load $R=R(\omega)$ is given by

$$R(\omega) = R(\omega,\lambda):= \sum_{i=1}^{m_R}\lambda_i R^{(i)}(\omega) \tag{10.1}$$

with random m-vectors $R^{(i)}= R^{(i)}(\omega)$, $i=1,\dots,m_R$, and deterministic coefficients λ_i, $i=1,\dots,m_R$, then $p_s=P(X,\lambda)$ with

$$P(\lambda,X):= P(\sum_{i=1}^{m_R}\lambda_i R^{(i)}(\omega)\in C[F^L(\omega,X),F^U(\omega,X)]). \tag{10.2}$$

Especially, in case of trusses, see (1),(3),(4) and (2.2), for dealing with the probability of survival p_s we have the following probability functions:

$$P_A(\lambda,X):= P(\lambda R_o(\omega)\in C[A(X)_d\sigma_y^L(\omega),A(X)_d\sigma_y^U(\omega)]),\ \lambda\in\mathbb{R} \tag{10.3}$$

$$P_u(X):= P(R_u(\omega)\in C[A(X)_d\sigma_y^L(\omega),A(X)_d\sigma_y^U(\omega)]) \tag{10.4}$$

$$P_s(X):= P(R_s(\omega)\in C[A(X)_d\sigma_y^L(\omega),A(X)_d\sigma_y^U(\omega)]), \tag{10.5}$$

where $A_d=A(X)_d$ denotes the n×n diagonal matrix having the diagonal elements $A_j=A_j(X)$, $j=1,\dots,n(=n_o)$.

Since $p_s=P(X)$ are very complicated expressions in general, in the following we are looking for approximations (lower, upper bounds) of $p_s=P(X)$ by simpler probability functions.

4 LOWER AND UPPER BOUNDS FOR $p_s = P(X)$

According to (8.3) we have that

$$S(F^L(.),F^U(.)) \subset \bigcap_{i=1}^{m} S_i(F^L(.),F^U(.)) \tag{11}$$

where

$$S_i(F^L(.),F^U(.)) := \{\omega\in\Omega: R_i(\omega) \in C_i[F^L(\omega),F^U(\omega)]\}, \; i=1,\dots,m, \tag{11.1}$$

and C_i denotes the i-th row of C. Hence, we get the inequality

$$P(X) \le \min_{1\le i\le m} P_i(X), \tag{12}$$

where

$$P_i(X) := P(R_i(\omega) \in C_i[F^L(\omega,X),F^U(\omega,X)]). \tag{12.1}$$

Note that related inequalities (with more exact bounds) follow from more general Bonferroni-type inequalities [6]. We find that

$$C_i[F^L(\omega,X),F^U(\omega,X)] = [\gamma_i^L(\omega,X),\gamma_i^U(\omega,X)] \tag{12.2}$$

is an interval in $\mathbb{R}^1$ having the bounds

$$\begin{aligned} \gamma_i^L(\omega,X) &= \min_{1\le\iota\le J} C_i G^\iota(\omega,X) \\ \gamma_i^U(\omega,X) &= \max_{1\le\iota\le J} C_i G^\iota(\omega,X), \end{aligned} \tag{12.3}$$

where $G^\iota = G^\iota(\omega,X)$, $\iota=1,\dots,J$, are the extreme points of the interval $[F^L(\omega,X), F^U(\omega,X)]$. Since the components $G_j^\iota(\omega,X)$, $j=1,2,\dots,n$, of $G^\iota(\omega,X)$, $\iota=1,\dots,J$, are certain elements of $(F^L(\omega,X),F^U(\omega,X))$, the measurability of the bounds F^L,F^U with respect to $\omega\in(\Omega, ,P)$ yields the measurability of γ_i^L and γ_i^U, $i=1,\dots,m$, with respect to ω. Hence, $S_i(F^L(.),F^U(.))$, $i=1,\dots,m$, are measurable sets, and we may write

$$P_i(X) = P(\gamma_i^L(\omega,X) \le R_i(\omega) \le \gamma_i^U(\omega,X)), \; i=1,\dots,m. \tag{12.4}$$

4.1. Lower bounds by selection of redundants

For the construction of lower bounds for P(X), the vector $F=F(\omega)$ is partitioned

$$F(\omega) = \begin{pmatrix} F_I \\ N \end{pmatrix} \tag{13}$$

into a certain (n-m)-vector $N=(F_{j_\ell})_{1\le\ell\le n-m}$ of **redundants** F_{j_ℓ}, $\ell=1,\dots,n-m$, and an m-vector F_I of statically determined member forces/bending-moments. Hence, with a corresponding partition of the m×n matrix C into m×m, m×(n-m) submatrices C_I, C_{II}, resp., where

$$\text{rank}C_I = \text{rank}C = m, \tag{13.1}$$

the equilibrium equation (6) yields for $F(\omega)$ the representation

$$F(\omega) = \begin{pmatrix} F_I \\ N \end{pmatrix} = \begin{pmatrix} C_I^{-1}R(\omega) \\ 0 \end{pmatrix} + \begin{pmatrix} C_I^{-1}C_{II} \\ I \end{pmatrix} N. \tag{14}$$

Consequently, selecting for each $\omega \in \Omega$ a vector of redundants $N=N(\omega)=(F_{j_\ell}(\omega))_{1\le\ell\le n-m}$ such that $N(.)$ is a measurable function on $(\Omega,\mathfrak{A},P)$, we get

$$S(F^L(.,X),F^U(.,X)) \supset \tilde{S}(X,N(.)), \tag{15}$$

where $\tilde{S}(X,N(.))$ is the measurable set given by

$$\tilde{S}(X,N(.)) := \{\omega\in\Omega: F^L(\omega,X) \le \begin{pmatrix} C_I^{-1}R(\omega)-C_I^{-1}C_{II}N(\omega) \\ N(\omega) \end{pmatrix} \le F^U(\omega,X)\}. \tag{15.1}$$

Thus, for $P(X)$ we find the lower bound

$$P(X) \ge \tilde{P}(X,N(.)), \tag{16}$$

where

$$\tilde{P}(X,N(.)) := P\begin{pmatrix} F_I^L(\omega,X) \le C_I^{-1}R(\omega) - C_I^{-1}C_{II}N(\omega) \le F_I^U(\omega,X) \\ F_{II}^L(\omega,X) \le N(\omega) \le F_{II}^U(\omega,X) \end{pmatrix}, \tag{16.1}$$

and F_I^L,F_{II}^L and F_I^U,F_{II}^U denotes the partition of F^L,F^U, resp., corresponding to (13). Note that the inequality (16) holds for any choice $N=(F_{j_\ell})_{1\le\ell\le n-m}$ of an (n-m)-subvector of redundants such that (13.1) holds and any representation of N as a random vector $N=N(\omega)$ on $(\Omega, ,P)$; especially, N can be selected as a **deterministic** vector of redundants:

$$N(\omega) = z \text{ a.s. (almost sure)}, \tag{17}$$

where $z\in\mathbb{R}^{n-m}$ is a deterministic vector; in this case we set $\tilde{P}(X,N(.)) = \tilde{P}(X,z)$.

4.1.1. Special cases

a) In case of trusses, cf. (1.3),(2.2), we have that

$$\begin{aligned} F_I^L(\omega,X) &= A_I(X)_d\sigma_I^L(\omega), \quad F_I^U(\omega,X) = A_I(X)_d\sigma_I^U(\omega) \\ F_{II}^L(\omega,X) &= A_{II}(X)_d\sigma_{II}^L(\omega), \quad F_{II}^U(\omega,X) = A_{II}(\omega)_d\sigma_{II}^U(\omega), \end{aligned} \tag{18}$$

where $A_I,A_{II},\sigma_I^L,\sigma_{II}^L,\sigma_I^U,\sigma_{II}^U$ are the partitions of A,σ^L,σ^U, resp., corresponding to the partition F_I,F_{II} of F, and $A_I(X)_d$ denotes the diagonal matrix which has the components of $A_I(X)$ as its diagonal elements. Thus, we get

$$\tilde{P}(X,z) = P\begin{pmatrix} A_I(X)_d\, \sigma_I^L(\omega) \le C_I^{-1}R(\omega) - C_{II}^{-1}z \le A_I(X)_d\sigma_I^U(\omega) \\ A_{II}(X)_d\sigma_{II}^L(\omega) \le z \le A_{II}(X)_d\sigma_I^U(\omega) \end{pmatrix}$$

$$= P\begin{pmatrix} \sigma_I^L(\omega) \le A_I(X)_d^{-1}(C_I^{-1}R(\omega) - C_I^{-1}C_{II}z) \le \sigma_I^U(\omega) \\ A_{II}(X)_d\sigma_{II}^L(\omega) \le z \le A_{II}(X)_d\sigma_{II}^U(\omega) \end{pmatrix} \tag{18.1}$$

b) Suppose that the partition of F^L, F^U into F_I^L, F_I^U and F_{II}^L, F_{II}^U can be chosen such that $(F_I^L(\omega,X), F_I^U(\omega,X))$, $(F_{II}^L(\omega), F_{II}^U(\omega))$ are stochastically independent. If (17 holds, then

$$\tilde{P}(X,z) := P(F_I^L(\omega,X) \le C_I^{-1}R(\omega) - C_I^{-1}C_{II}z \le F_I^U(\omega,X))$$
$$\times P(F_{II}^L(\omega,X) \le z \le F_{II}^U(\omega,X)).$$

5 FAILURE MODES, LIMIT STATE FUNCTIONS AND UPPER BOUNDS FOR P(X)

According to (8),(10) we have that

$$P(X) = P(\text{There is } F=F(\omega) \text{ such that } CF(\omega)=R(\omega) \text{ and}$$
$$F_j(\omega) - F_j^U(\omega,X) \le 0,\ j=1,\dots,n \tag{19}$$
$$F_j^L(\omega,X) - F_j(\omega) \le 0,\ j=1,\dots,n),$$

where we suppose first that all bounds F_j^L, F_j^U are finite, i.e.

$$-\infty < F_j^L(\omega,X) \le F_j^U(\omega,X) < +\infty \text{ a.s., } 1 \le j \le n, \text{ for all } X \text{ under consideration.} \tag{19.1}$$

Defining

$$t(\omega,F(\omega),X) := \max_{1\le j\le n} \{F_j(\omega)-F_j^U(\omega,X), F_j^L(\omega,X)-F_j(\omega)\}, \tag{19.2}$$

we obtain

$$P(X) = P(\text{There is } F=F(\omega) \text{ such that } CF(\omega)=R(\omega) \text{ and } t(\omega,F(\omega),X)\le 0\}$$
$$= P(\inf\{t(\omega,F(\omega),X) : CF(\omega)=R(\omega)\} \le 0) \tag{20}$$
$$= P(t^*(\omega,X) \le 0),$$

where

$$t^*(\omega,X) := \inf\{t(\omega,F(\omega),X) : CF(\omega)=R(\omega)\} \tag{20.1}$$

is the minimal value of the program

$$\min\ t(\omega,F(\omega),X) \tag{20.2}$$
s.t.
$$CF(\omega) = R(\omega)$$

being equivalent to the **linear program**

minimize t (21)

s.t.

$$F_j - F_j^U(\omega,X) - t \le 0, \; j=1,\dots,n \tag{21.1}$$

$$F_j^L(\omega,X) - F_j - t \le 0, \; j=1,\dots,n \tag{21.2}$$

$$CF(\omega) = R(\omega) \tag{21.3}$$

with the variables $F_1, F_2, \dots, F_n, t$.

Because of condition (19.1), for each (ω,X) we get

$$t(\omega,F(\omega),X) \ge \max_{1\le j\le n} \frac{1}{2}(F_j^L(\omega,X) - F_j^U(\omega,X)) > -\infty \tag{21.4}$$

for arbitrary $F(\omega)$; hence, the objective function of the linear program (21) is bounded from below for all (ω,X). Since the LP (21) always has a feasible solution, for each (ω,X) an optimal solution $\begin{pmatrix} F^* \\ t^* \end{pmatrix}$ of (21) is guaranteed, and we have that

$$t^* = t(\omega,F^*(\omega),X) = t^*(\omega,X). \tag{22}$$

Consequently, by means of duality theory the optimal value $t^*(\omega,X)$ of the equivalent programs (20.2) and (21) can be represented also by the optimal value of the dual program of (21) given by

$$\max R(\omega)'u - F^U(\omega,X)'\tilde{u}^+ + F^L(\omega,X)'\tilde{u}^- \tag{23}$$

s.t.

$$C'u - \tilde{u}^+ + \tilde{v}^- = 0 \tag{23.1}$$

$$1'\tilde{u}^+ + 1'\tilde{u}^- = 1 \tag{23.2}$$

$$\tilde{u}^+ \ge 0, \; \tilde{v}^- \ge 0, \tag{23.3}$$

where $u \in \mathbb{R}^n$ is not restricted. Thus, $t^*(\omega,X)$ reads

$$t^*(\omega,X) = \max\{ \begin{pmatrix} R(\omega) \\ -F^U(\omega,X) \\ F^L(\omega,X) \end{pmatrix}' \delta : \begin{pmatrix} u \\ \tilde{u}^+ \\ \tilde{v}^- \end{pmatrix} =: \delta \in \Delta_o \}, \tag{24}$$

where Δ_o denotes the convex polyhedron in $\mathbb{R}^{m+2n}$ represented by the constraints (23.1)-(23.3) of the LP (23). Taking any subset $\Delta_1 \subset \Delta_o$ of Δ_o, and defining then $t_1^*(\omega,X)$ by

$$t_1^*(\omega,X) := \sup\{ \begin{pmatrix} R(\omega) \\ -F^U(\omega,X) \\ F^L(\omega,X) \end{pmatrix}' \delta : \; \delta \in \Delta_1 \}, \tag{25}$$

we get

$$t^*(\omega,X) \ge t_1^*(\omega,X), \tag{25.1}$$

which yields for P(X) the following upper bound:

$$P(X) = P(t^*(\omega,X) \le 0) \le P(t_1^*(\omega,X) \le 0). \tag{25.2}$$

Moreover, if

$$\delta^{(\ell)} = \begin{pmatrix} u^{(\ell)} \\ \tilde{u}^{+(\ell)} \\ \tilde{u}^{-(\ell)} \end{pmatrix}, \quad \ell=1,\dots,\ell_o,$$

denote the extreme points of the convex polyhedron Z_o, then

$$t^*(\omega,X) = \max_{1\le\ell\le\ell_o} R(\omega)'u^{(\ell)} - F^U(\omega,X)'\tilde{u}^{+(\ell)} + F^L(\omega,X)'\tilde{u}^{-(\ell)}, \tag{26}$$

which shows that $t^*(.,X)$ is measurable. Hence, $S(F^L(.),F^U(.)) = \{\omega\in\Omega: t^*(\omega,X)\le 0\}$ is measurable, cf. (10),(20), and we get

$$P(X) = P(R(\omega)'u^{(\ell)} - F^U(\omega,X)'\tilde{u}^{+(\ell)} + F^L(\omega,X)'\tilde{u}^{-(\ell)} \le 0,\ 1\le\ell\le\ell_o). \tag{26.1}$$

According to (6),(7), the survival, failure, resp., of the underlying structure can be described by the inequality

$$t^*(\omega,X) \le 0,\ t^*(\omega,X) > 0, \text{ respectively.} \tag{27}$$

Thus, the structure fails if

$$R(\omega)'u^{(\ell)} + F^U(\omega,X)'\tilde{u}^{+(\ell)} - F^L(\omega,X)'\tilde{u}^{-(\ell)} > 0 \text{ for at least one } 1\le\ell\le\ell_o; \tag{27.1}$$

obviously, (27.1) represents the different **failure modes** of the structure.

Having a certain number $\ell_1\le\ell_o$ of basic solutions $\delta^{(\ell_\tau)}$, $\tau=1,\dots,\ell_1$, of the LP (23), and defining

$$\tilde{t}_1^*(\omega,X) := \max_{1\le\tau\le\ell_1} R(\omega)'u^{(\ell_\tau)} - F^U(\omega,X)'\tilde{u}^{+(\ell_\tau)} + F^L(\omega,X)'\tilde{u}^{-(\ell_\tau)}, \tag{28}$$

corresponding to (25.1), here we get

$$t^*(\omega,X) \ge \tilde{t}_1^*(\omega,X) \tag{28.1}$$

and therefore

$$P(X) = P(t^*(\omega,X) \le 0) \le P(\tilde{t}_1^*(\omega,X) \le 0). \tag{28.2}$$

6 THE PROBABILITY OF FAILURE p_f

According to (8),(10),(20) and (27), for the probability of failure $p_f:=1-p_s=1-P(X)$ we obtain

$$\begin{aligned} p_f &= P(t^*(\omega,X) > 0) \\ &= P(R(\omega)'u - F^U(\omega,X)'\tilde{u}^+ + F^L(\omega,X)'\tilde{u}^- > 0 \text{ for at least one } \begin{pmatrix} w \\ \tilde{u}^+ \\ \tilde{u}^- \end{pmatrix} \in \Delta_o) \\ &= P(R(\omega)'u^{(\ell)} - F^U(\omega,X)'\tilde{u}^{+(\ell)} + F^L(\omega,X)'\tilde{u}^{-(\ell)} > 0 \text{ for at least one } 1\le\ell\le\ell_o) \\ &= P(\bigcup_{\ell=1}^{\ell_o} \mathbf{F}_\ell(X)), \end{aligned} \tag{29}$$

where $\mathbf{F}_\ell(X)$ denotes the ℓ-th failure domain

$$\mathbf{F}_\ell(X) := \{\omega\in\Omega: R(\omega)'u^{(\ell)} - F^U(\omega,X)'\tilde{u}^{+(\ell)} + F^L(\omega,X)'\tilde{u}^{-(\ell)} > 0\} \tag{29.1}$$
$$= \{\omega\in\Omega: M_\ell(\omega,X) < 0\}$$

with the corresponding **limit state function**

$$M_\ell(\omega,X) := F^U(\omega,X)'\tilde{u}^{+(\ell)} - F^L(\omega,X)'\tilde{u}^{-(\ell)} - R(\omega)'u^{(\ell)}, \tag{29.2}$$

$\ell=1,\dots,\ell_o$; especially, for trusses, cf. (2.2), (18.1), we find

$$M_\ell(\omega,X) := \sigma^U(\omega)'A(X)_d\tilde{u}^{+(\ell)} - \sigma^L(\omega)'A(X)_d\tilde{u}^{-(\ell)} - R(\omega)'u^{(\ell)}. \tag{29.3}$$

Using known inequalities for probabilities, for p_f we find the bounds

$$\max_{1\le\ell\le\ell_o} p_{f,\ell} \le p_f \le \sum_{\ell=1}^{\ell_o} p_{f,\ell}, \tag{30}$$

where $p_{f,\ell}$ is given by

$$p_{f,\ell} := P(\mathbf{F}_\ell(X)) = P(M_\ell(\omega,X) < 0)$$
$$= P(F^U(\omega,X)'\tilde{u}^{+(\ell)} - F^L(\omega,X)'\tilde{u}^{-(\ell)} < R(\omega)'u^{(\ell)} \tag{30.1}$$
$$= 1 - P(R(\omega)'u^{(\ell)} \le F^U(\omega,X)'\tilde{u}^{+(\ell)} - F^L(\omega,X)'\tilde{u}^{-(\ell)}),$$

and sharper bounds can be obtained by using more genral Bonferroni-inequalities for probabilities.

7 REPRESENTATION OF p_s BY USING CONES

According to (9) we have that

$$p_s = P(R(\omega) \in C[F^L(\omega),F^U(\omega)]),$$

where $[F^L,F^U]$ is given by (8.2). Representing therefore the vector F of internal member forces/bending-moments by

$$F = F^L+\Delta F^L = F^U-\Delta F^U$$

with n-vectors $\Delta F^U,\Delta F^L\ge 0$, the condition $R \in C[F^L,F^U]$ can be represented by

$$\begin{aligned} R - CF^U &= - C\Delta F^U \\ -(F^U-F^L) &= - \Delta F^U - \Delta F^L \\ \Delta F^U\ge 0,\ &\Delta F^L\ge 0. \end{aligned} \tag{31}$$

Thus, we consider the cone Y_o $\mathbb{R}^{m+n}$ defined by

$$Y_o = \{\begin{pmatrix} C & 0 \\ I & I \end{pmatrix}\begin{pmatrix} \Delta F^U \\ \Delta F^L \end{pmatrix}: \Delta F^U\ge 0,\ \Delta F^L\ge 0\} \tag{32}$$
$$= \left\{\sum_{k=1}^{2n} \alpha_k\, y_k: \alpha_k\ge 0,\ k=1,\dots,2n\right\},$$

where the cone-generators $y^{(k)}$, $k=1,\ldots,2n$, are given by

$$y^{(k)} := \begin{pmatrix} c_k \\ e_k \end{pmatrix}, \ 1 \le k \le n, \ y^{(k)} := \begin{pmatrix} 0 \\ e_k \end{pmatrix}, \ n < k \le 2n, \tag{32.1}$$

and c_k, e_k denotes the k-th column of C, of the nxn identity matrix I, respectively. Having Y_o, the set $S(F^L(.),F^U(.))$ defined in (8.1) can be described by

$$S(F^L(.),F^U(.)) = \{\omega \in \Omega: \begin{pmatrix} R(\omega)-CF^U(\omega,X) \\ -F^U(\omega,X)+F^L(\omega,X) \end{pmatrix} \in (-1)Y_o\}, \tag{33}$$

which shows again the measurability of $S(F^L(.),F^U(.))$. Moreover, the probability function $P=P(\lambda,X)$ representing (10)-(10.3) can be given by

$$P(\lambda,X) = P(\begin{pmatrix} CF^U(\omega,X) - \lambda R_o(\omega) \\ F^U(\omega,X) - F^L(\omega,X) \end{pmatrix} \in Y_o). \tag{33.1}$$

According to the representation (32) of Y_o, there are a finite number of boundary hyperplanes in $\mathfrak{R}^{m+n}$, represented by vectors $\eta^{(\ell)} = \begin{pmatrix} w^{(\ell)} \\ v^{(\ell)} \end{pmatrix}$, $\ell=1,\ldots,\ell'_o$, such that

$$Y_o = \{y = \binom{w}{v} \in \mathfrak{R}^{m+n}: y'\eta^{(\ell)} = w'w^{(\ell)} + v'v^{(\ell)} \ge 0, \ 1 \le \ell \le \ell'_o\}. \tag{34}$$

Hence, because of (33) and (34), the survival of the structure can be represented also by the inequalities

$$(R(\omega)-CF^U(\omega,X))'w^{(\ell)} + (-F^U(\omega,X) + F^L(\omega,X)'v^{(\ell)} \le 0, \ 1 \le \ell \le \ell'_o,$$

which yields

$$(R(\omega)'w^{(\ell)} - F^U(\omega,X)'(C'w^{(\ell)}+v^{(\ell)}) + F^L(\omega,X)'v^{(\ell)} \le 0, \ 1 \le \ell \le \ell'_o, \tag{35}$$

and therefore

$$p_s = P(R(\omega)'w^{(\ell)} - F^U(\omega,X)'(C'w^{(\ell)})+v^{(\ell)}+F^L(\omega,X)'v^{(\ell)} \le 0, \ 1 \le \ell \le \ell'_o). \tag{35.1}$$

Obviously, the conditions for structural safety given by (27) and (35) coincide.

For arbitrary subset $Y_o^{(\ell)}$, $\ell=1,2$, such that

$$Y_o^{(1)} \subset Y_o \subset Y_o^{(2)}, \tag{36}$$

we get

$$p_s^{(1)} \le p_s \le p_s^{(2)}, \tag{36.1}$$

where the bounds $p_s^{(\ell)}$, $\ell=1,2$, are defined by

$$p_s^{(\ell)} := P\left(\begin{pmatrix} CF^U(\omega,X) - R(\omega) \\ F^U(\omega,X) - F^L(\omega,X) \end{pmatrix} \in Y_o^{(\ell)}\right), \ \ell=1,2. \tag{36.2}$$

7.1 Construction of approximating cones

Approximations $Y_o^{(\ell)}$, $\ell=1,2$, of Y_o can be obtained by spherical or ellipsoidal approximation of the convex polyhedron generated by $y^{(k)}$, $k=1,\dots,2n$.

8 COMPUTATION OF SENSITIVITIES OF PROBABILITIES OF SURVIVAL AND PROBABILITIES OF FAILURE

Sensitivities of p_s,p_f with respect to the load factor λ and the vector X of design variables X_k, $k=1,2,\dots,r$, can be obtained now by applying the differerentiation formulas given in [6],[7],[8],[9] to the probability functions (10),(10.2)-(10.5),(16.1),(18.1),(26.1),(33.1).

REFERENCES

[1] Arnbjerg-Nielsen, Torben: Rigid-ideal plastic model as a reliability analysis tool for ductile structures. Ph.D. Dissertation, Technical University of Denmark, Lyngby 1991

[2] Augusti, G. et al.: Probabilistic Methods in Structural Engineering. Chapman and Hall, London-New York 1984

[3] Haftka, R.T. et al.: Elements of Structural Optimization. Kluwer, Dordrecht-Boston-London 1990

[4] Hodge, P.G.: Plastic Analysis of Structures. Mc Graw-Hill, New York 1959

[5] Kirsch, U.: Structural Optimization. Springer-Verlag, Berlin-Heidelberg-New York 1993

[6] Marti, K.: Differentiation of Probability Functions: The Transformation Method. To appear in "Computers and Mathematics with Applications", 1995

[7] Marti, K.: Differentiation Formulas for Probability Functions: The Transformation Method. To appear in "Mathematical Programming, Series B", 1995

[8] Marti, K.: Differentiation of Probability Functions Arising in Structural Reliability. In: R. Rackwitz (ed.): Reliability and Optimization of Structural Systems. Chapman and Hall, London 1995

[9] Marti, K.: Computation of Probability Functions and its Derivatives by means of Orthogonal Functions Series Expansions. In: K. Marti, P. Kall (eds.): Stoch. Programming: numerical techniques and engineering applications. LNEMS Vol. 423, Springer-Verlag, Berlin-Heidelberg-New York 1995

[10] Nafday, A.M., Corotis, R.B.: Failure Mode Enumeration for System Reliability Assessment by Optimization Algorithms. In: P. Thoft-Christensen (ed.): Reliability and Optimization of Structural Systems. Springer-Verlag, Berlin-Heidelberg-New York 1987

PART TWO

Contributed Papers

Automatic Control

8

Robust stabilization of nonlinear systems by optimal controllers

J. Kabziński
Institute of Automatic Control, Technical University
PL 90-924 Łódź, Stefanowskiego 18/22

Abstract

In this note we investigate robustness of optimally controlled systems in presence of nonlinear state-depending perturbations effecting the input additively together with nonlinear change in the control gain. We derive two types of robustness conditions for optimal controller. The first one can be verified *a priori* and the second - if the optimal control problem is solved. In the next section we propose robust controller design technique without solving optimal control problem. Next we study special systems and give numerical examples.

Keywords

Nonlinear control, robust control, optimal control

1 INTRODUCTION

It is well recognized by control scientists that robustness is a basic desirable feature in the design of control systems. It is also known that optimally controlled systems possesses this property.

Let us consider a system described by differential equation:

$$\dot{x}(t)=f(x(t),u(t)),\quad f(0,0)=0,\quad x(t)\in R^{n},\quad u(t)\in R^{m}. \tag{1}$$

The control $u(\cdot)$ is chosen to minimize the performance index

$$J=\int_{0}^{\infty}(g(x(t))+h(u(t)))\,dt,\qquad g(0)=0,\quad g(x)>0\ \text{ for } x\neq 0,\quad h(0)=0,\quad h(u)>0\ \text{ for } u\neq 0. \tag{2}$$

We will assume that the data of the above problem are sufficiently smooth to guarantee existence of optimal control in the feedback form $k: R^n\to R^m$, $u(t)=k(x(t))$ and that the optimal cost-to-go value function $V(\cdot)$ is continuously differentiable and satisfies Hamilton-Jacobi-Bellman equation:

$$\forall\, x\in R^{n},\ u\in R^{m}\qquad V_x^T f(x,u)+g(x)+h(u)\geq V_x^T f(x,k(x))+g(x)+h(k(x))=0. \tag{3}$$

Under this optimal control law the origin is globally asymptotically stable equilibrium point of the closed loop system (if we assume $g(x)\geq 0$ instead of $g(x)>0$ for $x\neq 0$ local observability condition is necessary to derive asymptotical stability (Glad, 1984)).

The optimality of the control implies also that the closed loop system is robust in presence of perturbations in the feedback loop (Tsitsiklis, Athans 1984, Glad 1984, Geromel, Yamakami 1985). Under some additional constraints the closed loop system has a guaranteed infinite gain margin and 50 % gain reduction margin and 60 degree phase margin in each feedback channel (Tsitsiklis, Athans 1984).

In this note we investigate robustness of optimally controlled systems in presence of nonlinear state-depending perturbations effecting the input additively together with nonlinear change in the control gain. we derive two types of robustness conditions for optimal controller. The first one can be verified *a priori* and the second - if the optimal control problem is solved. In the next section we develop ideas from Kabziński (1991) to propose robust controller design technique without solving optimal control problem. Next we study special systems and give numerical examples.

2 ROBUSTNESS OF OPTIMAL CONTROLLER

Let us consider a nominal system being a special form of (1):

$$\dot{x}(t) = A(x(t)) + B(x(t))u(t), \qquad A(0) = 0 \tag{4}$$

and a special form of a performance index:

$$J = \int_0^\infty \left(g_0(x(t)) + n(x(t)) + \tfrac{1}{2}u(t)^T R u(t)\right) dt, \qquad g_0(0) = 0, \quad g_0(x) > 0 \;\text{ for } x \neq 0, \quad n(0) = 0 \tag{5}$$

where matrix R is positive definite. The optimal feedback control law is given by:

$$k(x) = -R^{-1}B^T(x)V_x(x). \tag{6}$$

We investigate robustness of the optimal controller subject to non-linear, state dependant perturbation effecting the input channel. The perturbed system is described by:

$$\dot{x}(t) = A(x(t)) + B(x(t))[d(x(t)) + p(u(t))], \tag{7}$$

where $d(\cdot)$ and $p(\cdot)$ are memoryless nonlinearities satisfying $d(0)=0$, $p(0)=0$ and such that equation (6) has a unique solution define for all positive t. For $d(\cdot)\equiv 0$ and $p(u)=u$ we get the nominal system.

Theorem 1.

If there exist positive number b such that for every u and x:

$$2p(u)^T R u - (1+b)u^T R u \geq 0, \tag{8}$$

$$2bn(x) - d(x)^T R d(x) \geq 0, \tag{9}$$

then the optimal control law for problem (4, 5) is robust stabilizing control for system (7).

Proof:

We try cost-to-go value function for problem (4,5) as Lyapunov function candidate for (7). It follows from (6) and (3) that the system derivative is given by:

$$\dot{V}(x) = V_x(x)^T[A(x)+B(x)d(x)+B(x)p(k(x))] = V_x(x)^T[A(x)+B(x)k(x)] + V_x(x)^T B(x)[p(k(x)) - k(x)] +$$
$$+ V_x(x)^T B(x)d(x) = - g_0(x) - n(x) - \frac{1}{2} k(x)^T Rk(x) - p(k(x))^T Rk(x) + k(x)^T Rk(x) - k(x)^T Rd(x) = \quad (10)$$
$$= - g_0(x) - n(x) + \frac{1}{2} k(x)^T Rk(x) - k(x)^T Rd(x) - p(k(x))^T Rk(x).$$

After adding and subtracting $bk(x)^T Rk(x) + \frac{1}{b} d(x)^T Rd(x)$ we are able to rewrite (10) in the following form:

$$\dot{V}(x) = \frac{1}{2}\{-[2k(x)^T Rp(k(x)) - (1+b)k(x)^T Rk(x)] - \frac{1}{b}[d(x)+bk(x)]^T R[d(x)+bk(x)] +$$
$$- \frac{1}{b}[2bn(x) - d(x)^T Rd(x)]\} - g_0(x), \quad (11)$$

so under (8, 9) $\dot{V}(\cdot)$ is negative definite. □

Theorem 1 formulates conditions that do not depend on the optimal control problem solution, allowing then *a priori* analysis of the closed loop system stability. The derivation of the robust controller requires however the solution of the nonlinear optimal control problem (NOCP).

If $d(x) \equiv 0$, then assumption (8) is satisfied if

$$2p(u)^T Ru - u^T Ru \geq 0, \quad (12)$$

so (12) generalizes a familiar 50% gain reduction robustness condition.

If $p(u)=u$ then condition (8) is satisfied with $b=1$ and (9) is reduced to:

$$2n(x) - d(x)^T Rd(x) \geq 0, \quad (13)$$

what is familiar from Geromel and Yamakami (1985). If $n(x)$ is chosen to be a quadratic form $n(x) = \frac{1}{2}x^T Qx$ we can replace (9) by a more restrictive condition:

$$\frac{\| d(x) \|^2}{\| x \|^2} \leq \frac{b\,\lambda_{min}(Q)}{\lambda_{max}(R)}, \quad (14)$$

where λ_{min} and λ_{max} are minimal and maximal eigenvalue.

Function $n(x)$ is designed to compensate nonlinear state dependent disturbance $d(x)$, while function $g_0(x)$ influences the system convergence rate under the worst possible disturbance, what is illustrated by the inequality $\dot{V}(x) \leq - g_0(x)$. If $V(x) = x^T Hx$, $g_0(x) \geq x^T Gx$, then for any solution of (7) under control (6) $\| x(t) \|^2 \leq \frac{\lambda_{max}(H)}{\lambda_{min}(H)} \| x(0) \|^2 \exp\left(- \frac{\lambda_{min}(G)}{\lambda_{max}(H)} t \right)$.

Theorem 2.

Assume that $k(x)$ is the optimal control law for problem (4, 5). Assume that condition (8) is satisfied and that in some neighborhood Θ of $x=0$, for $x \neq 0$:

$$2k(x)^T R d(x) + b k(x)^T R k(x) + 2n(x) > 0, \tag{15}$$

then $x=0$ is the asymptotically stable equilibrium point of system (7). The stability domain is given by $\Omega = \{x: V(x) \le inf_{x \in \partial\Theta} V(x)\}$.

Proof:

It follows easily from (11) that (15) is sufficient for $\dot{V}(x) \le -g_0(x)$. in Θ.□

Conditions (8, 15) are less conservative than (8, 9), but to verify (15) it is necessary to solve NOCP and obtain $k(x)$.

3 ROBUST CONTROLLER DESIGN WITHOUT SOLVING OPTIMAL CONTROL PROBLEM

The derivation of the robust controller proposed in section 2 requires the solution of the nonlinear optimal control problem (NOCP). Solving a NOCP is always a difficult task, although there are several methods, which can be helpful in solving (4, 5). Let us investigate if we are able to utilize theorem 1 without solving a NOCP.

Theorem 3.

Assume that condition (8) is satisfied and that there exist positive definite function $V(\cdot)$ and positive numbers a, b, such that in a neighborhood Θ of $x=0$, for $x \neq 0$:

$$[-V_x(x)^T A(x) + \tfrac{1}{2} V_x(x)^T B(x) R^{-1} B(x)^T V_x(x)]\, 2b(1-a) - d(x)^T R d(x) > 0, \tag{16}$$

or:

$$a[-V_x(x)^T A(x) + \tfrac{1}{2} V_x(x)^T B(x) R^{-1} B(x)^T V_x(x)] > 0, \tag{17}$$

$$-V_x(x)^T B(x) d(x) - (1-a) V_x(x)^T A(x) + \tfrac{1}{2}(1-a+b) V_x(x)^T B(x) R^{-1} B(x)^T V_x(x) > 0, \tag{18}$$

then under the control law: $k(x) = -R^{-1} B^T(x) V_x(x)$ $x=0$ is the asymptotically stable equilibrium point of system (7). The stability domain is given by $\Omega = \{x: V(x) \le inf_{x \in \partial\Theta} V(x)\}$.

Proof:

Indeed the control law $k(x)$ is optimal control for system (4) with the performance index (5) where:

$$g(x) = g_0(x) + n(x) = -V_x(x)^T A(x) + \tfrac{1}{2} V_x(x)^T B(x) R^{-1} B(x)^T V_x(x), \quad n(x) = (1-a) g(x), \quad g_0(x) = a g(x). \tag{19}$$

It follows from (16) that condition (9) is satisfied and the thesis follows like in the proof of theorem 1. Condition (18) is equivalent to inequality (15), condition (17) ensures that $g_0(x) > 0$ and the proof proceeds like in theorem 2.□

The design technique introduced by theorem 3 consists in selecting function $V(\cdot)$. This function is connected with the open loop system and it parameterizes the optimal control and the performance index.

The function $V(x)$ satisfying (16) is easy to find if the system without control $\dot{x}(t) = A(x(t))$ is stable. Then there exists Lyapunov function $V(\cdot)$, such that $-V_x(x)^T A(x) > 0$. Parameterization of $V(\cdot)$ allows to satisfy (16) for a given class of disturbances $d(x)$.

In many important cases including bilinear systems $A(x) = Ax$. If matrix A is stable then there exist positive definite matrices P and Q satisfying Lyapunov equation:

$$PA + A^T P = -Q \tag{20}$$

and Lyapunov function is $V(x) = \frac{1}{2} x^T P x$.. Condition (16) takes the form:

$$[x^T Q x + V_x(x)^T B(x) R^{-1} B(x)^T V_x(x)]\, b(1-a) - d(x)^T R d(x) > 0, \tag{21}$$

If matrix A is unstable Q is not positive definite and inequality (21) can be satisfied only locally.

4 EXAMPLES

4.1 Robustness of optimal regulator

Let us consider a scalar NOCP defined by the nominal system and the performance index:

$$\dot{x} = -1.4x - 0.5x^2 + (0.5 - 2.5x)u, \qquad x(0) < 0, \tag{22}$$

$$J = \int_0^\infty (1.2x^2 + \tfrac{1}{2} u^2)\, dt. \tag{23}$$

As it is a scalar problem we are able to solve HJB equation and to find optimal control:

$$k(x) = \frac{(1.4 + 0.5x^2) - \sqrt{(x^2(25.25x^2 - 8.6x + 2.96))}}{(1 - 5x)}. \tag{24}$$

If we take $p(u)=u$ and $g_0(x)=\varepsilon x^2$ we obtain for $\varepsilon \to 0$ $n(x)<1.2x^2$. Sets defined by condition (9) and (15) are plotted in fig. 1, together with disturbances $d(x)=-0.5x^2$, $d(x)=1.5x$, $d(x)=-1.5x$. State histories corresponding to these disturbances, under optimal control $k(x)$, starting from $x=-3.05$ are plotted in fig. 2.

4.2 Robust controller synthesis for nonlinear system

Let us consider nominal system described by equation (4), where:

$$x = \begin{bmatrix} x_1 \\ x_2 \end{bmatrix}, \quad A(x) = 10 \begin{bmatrix} -x_1^3 - x_2 \\ x_1 - x_2^5 \end{bmatrix}, \quad B(x) = 10 \begin{bmatrix} 2x_1 x_2 \\ x_2 + 1 \end{bmatrix}. \tag{25}$$

We look for Lyapunov function for the system $\dot{x} = A(x)$. We can use Krasovski method and choose

$$V(x_1, x_2) = \tfrac{1}{2}(x_1^3 + x_2)^2 + \tfrac{1}{2}(x_1 - x_2^5)^2, \qquad V_x(x) = \begin{bmatrix} 3x_1^2(x_1^3 + x_2)^2 + x_1 - x_2^5 \\ x_1^3 + x_2 - 5x_2^4(x_1 - x_2^5) \end{bmatrix}. \tag{26}$$

The optimal stabilizing feedback is given by:

$$k(x_1,x_2) = -\frac{10}{R}[2x_1x_2 \quad x_2+1]\begin{bmatrix} 3x_1^2(x_1^3+x_2)^2+x_1-x_2^5 \\ x_1^3+x_2-5x_2^4(x_1-x_2^5) \end{bmatrix}. \tag{27}$$

If we choose $V(x_1,x_2)=\frac{1}{2}(x_1^2+x_2^2)$ we obtain respectively:

$$V_x(x)=\begin{bmatrix} x_1 \\ x_2 \end{bmatrix}, \quad k(x_1,x_2) = -\frac{10}{R}[2x_1x_2 \quad x_2+1]\begin{bmatrix} x_1 \\ x_2 \end{bmatrix}. \tag{28}$$

As the input is scalar, condition (16) is reduced to:

$$|d(x)| < \left\{\frac{2b(1-a)}{R}\left[-V_x(x)^T A(x) + \frac{1}{2}V_x(x)^T B(x) R^{-1} B(x)^T V_x(x)\right]\right\}^{½}. \tag{29}$$

The region defined by this condition is plotted for $R=20$, $b=1$, $a=0$ in fig. 3 and 4 for two choices of $V(x)$. The corresponding regions defined by (18) are shown in fig. 5 and 6.

4.3 Global robust stabilizer for bilinear system

Let us consider bilinear system:

$$\dot{x} = Ax + B(x)u = Ax + (B_0 + B_1(x))u, \quad A=\begin{bmatrix} 0 & 1 \\ -5 & 2 \end{bmatrix}, \quad B_0=\begin{bmatrix} 1 \\ 0 \end{bmatrix}, \quad B_1=\begin{bmatrix} 2x_2 \\ x_1 \end{bmatrix}. \tag{30}$$

Linear stabilizing feedback was proposed for this system by Derese and Noldus (1980), but under this control only local stability is possible. Let us consider disturbances d(x) satisfying:

$$d(x)^2 \le x^T Q_1 x, \quad Q_1 = (1-a)Q = 0.7\begin{bmatrix} 1 & 0 \\ 0 & 10 \end{bmatrix}. \tag{31}$$

Then matrix P is calculated from (20) and the stabilizing control is given by $k(x) = -R^{-1}B(x)^T Px$. Condition (21) is (conservatively) satisfied with R=1, b=1, for $d(x)$ satisfying (31). The system was tested for $d(x)=\sqrt{0.7}\,x_1$, $d(x)=\sqrt{7}\,x_2$, $d(x)=0$. State trajectories are plotted in fig 7. All starting points are chosen outside stability regions obtained by Derese and Noldus (1980).

5 CONCLUSIONS

We have investigated robustness of optimally controlled nonlinear systems in presence of nonlinear state-depending perturbations effecting the input additively together with nonlinear

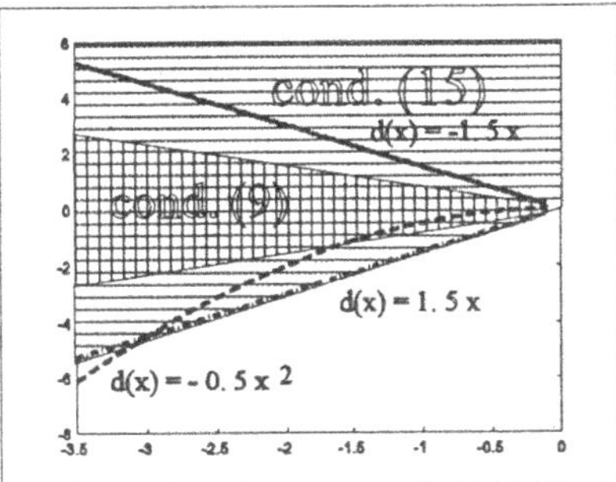

Figure 1 Disturbances and robust stability regions for example 4.1.

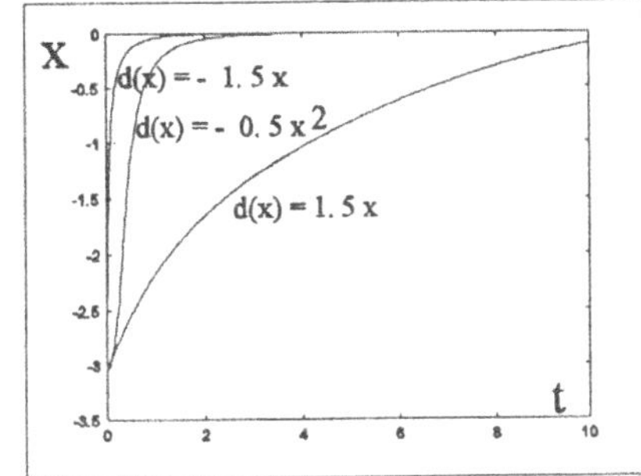

Figure 2 State histories for different d(x) (example 4.1).

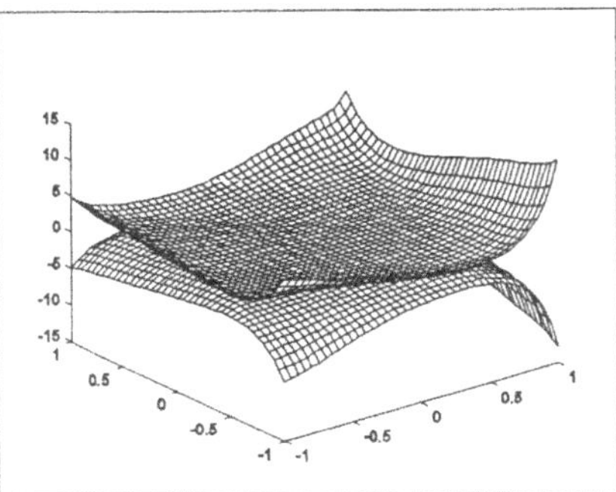

Figure 3 *A priori* disturbance bounds (*V(x)* from Krasovski method, example 4.2).

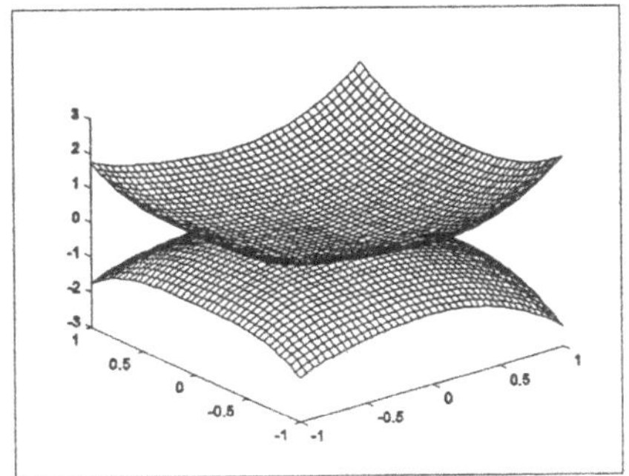

Figure 4 *A priori* disturbance bounds (*V(x)* - quadratic form, example 4.2).

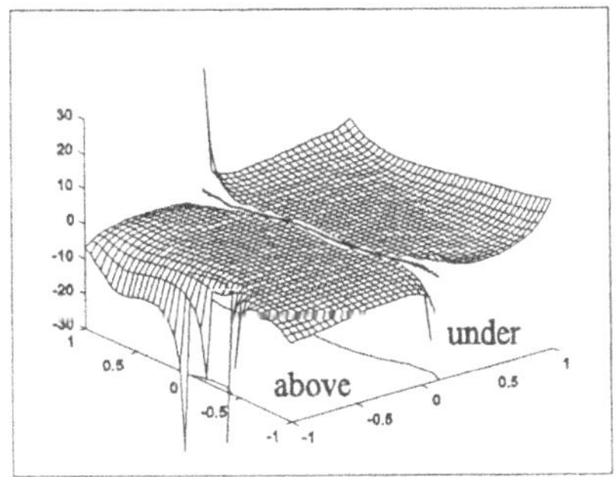

Figure 5 *A posteriori* robustness regions (*V(x)* from Krasovski method, example 4.2).

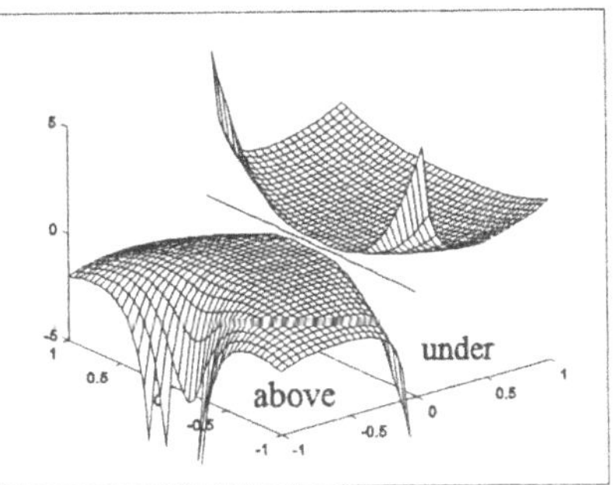

Figure 6 *A posteriori* robustness regions (*V(x)* -quadratic form, example 4.2).

change in the control gain. We derived two types of robustness conditions for optimal controller. The first one can be verified *a priori* and the second - if the optimal control problem is solved. The first condition is far more conservative as it was shown in example 4.1. The designer selects the performance index to obtain satisfactory robustness bonds for disturbances, good system behavior and bounded control signals. Then nonlinear optimal control problem is solved. If it is too difficult to solve NOCP, we can proceed as in section 3. The designer is to select positive definite function V(x) associated with the system without control and matrix R penalizing controls in the performance index. The stabilizing control is then found without solving optimal control problem. This method is effective for design of robust controllers for nonlinear systems stable without control. Example 4.2 demonstrates how robustness regions are effected by design parameters and example 4.3 shows how to obtain the controller for given *a priori* disturbance bounds.

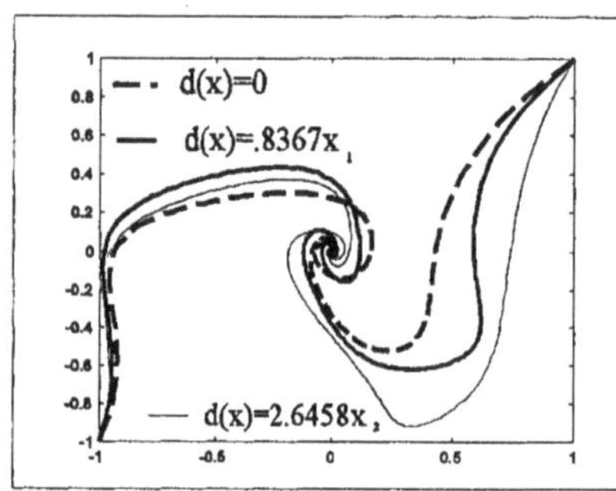

Figure 7 State trajectories under stabilizing control for several disturbances (example 4.3).

6 REFERENCES

Derese L. and Noldus I., (1980) Design of linear feedback laws for bilinear systems, *Int. Jour. Control*, **vol. 31**, 219-237

Geromel, J.C. and Yamakami A.. (1985) On the robustness of nonlinear regulators and its application to nonlinear systems stabilization, *IEEE Trans. Automatic Control*, **vol. AC-30**, 1251-1254.

Glad S.T. (1984) On the gain margin of nonlinear and optimal regulators, *IEEE Trans. Automatic Control*, **vol. AC-29**, 615-620.

Kabziński J. (1991) Optimal control for stabilization of nonlinear systems, in *System Modelling and Optimization*, Kall p. ed., Lecture Notes in Control and Inf. Sc., **vol 180**, 520-529.

Tsitsiklis J.N. and Athans M. (1984) Guaranteed robustness properties of multivariable nonlinear stochastic optimal regulators, *IEEE Trans. Automatic Control*, **vol. AC-29**, 690-696.

9

Weighted H^2 approximation of transfer functions

Juliette Leblond and Martine Olivi
INRIA
BP 93, 06902 Sophia–Antipolis Cedex, France, e-mail :
leblond,olivi@sophia.inria.fr

Abstract

The aim of this work is to generalize to the case of weighted L^2 spaces some results about L^2 approximation by analytic and rational functions which are useful to perform the identification of unknown transfer functions of stable (linear causal time–invariant) systems from incomplete frequency data.

Keywords

Hardy spaces, weighted rational approximation, frequency domain identification.

1 INTRODUCTION

Let $L^2(\mathbb{T})$ be the *real* Hilbert space of square–summable functions on the unit circle $\mathbb{T}$ satisfying the conjugate–symmetry property $f(e^{-i\theta}) = \overline{f(e^{i\theta})}$, or equivalently whose Fourier coefficients are real. The space $L^2(\mathbb{T})$ is endowed with its classical inner product $< \, , \, >$ and associated norm $\| \; \|$. The Hardy space H^2 of the unit disk $\mathbb{D}$ is the closed subspace of $L^2(\mathbb{T})$ which contains functions whose Fourier coefficients of negative index are zero. These functions admit an analytic extension in $\mathbb{D}$ and H^2 is isometric to the space of analytic functions in $\mathbb{D}$ which satisfy the growth condition:

$$\sup_{r<1} \frac{1}{2\pi} \int_{-\pi}^{\pi} |f(r\, e^{i\theta})|^2 d\theta < \infty \,, \tag{1}$$

which provides a norm on this space (see Garnett (1981), II.3). Furthermore, $\bar{H}_0^2$ denotes the orthogonal complement of H^2 in $L^2(\mathbb{T})$. It consists in $L^2(\mathbb{T})$ functions which possesses Fourier coefficients of non–negative index equal to zero or, equivalently, in functions analytic outside the closed unit disk and vanishing at infinity satisfying the growth condition (1) for $r > 1$. The two spaces H^2 and $\bar{H}_0^2$ are isometrical under the map:

$$\begin{aligned} L^2(\mathbb{T}) &\rightarrow L^2(\mathbb{T}) \\ f(z) &\mapsto \frac{f(1/z)}{z} = \check{f}(z) \,. \end{aligned} \tag{2}$$

Functions belonging to $\bar{H}_0^2$ are transfer functions of linear strictly causal stationary discrete time systems of finite variance for a white noise input or, equivalently, which possess an $l^1(\mathbb{N})$–input / $l^2(\mathbb{N})$–output stability property. For an unknown $\bar{H}_0^2$ transfer function, assume that we are only given some of its (possibly noisy) pointwise values at some frequencies belonging to a symmetric subset K of $\mathbb{T}$ (which corresponds to the bandwidth of the associated system). Such measurements may be obtained using harmonic identification procedures. In order to identify the unknown system, we want to find a rational $\bar{H}_0^2$ function of bounded Mac–Millan degree accounting well enough for these data, in a sense to be made precise below.

Let μ be some positive finite measure on the unit circle $\mathbb{T}$, absolutely continuous with respect to the Lebesgue measure λ (for which we write $d\lambda(\theta) = d\theta$) and satisfying $d\mu(-\theta) = -d\mu(\theta)$. In this work, we want to identify the unknown transfer function by solving approximation problems with respect to this measure μ. This will take place in the *real* Hilbert space $L^2(\mu)$ of conjugate–symmetric functions that are square–summable on $\mathbb{T}$ with respect to the measure μ endowed with the inner product defined by:

$$< f , g >_\mu = \frac{1}{2\pi} \int_{-\pi}^{\pi} f(e^{i\theta})\, g(e^{-i\theta})\, d\mu(\theta) = \frac{1}{2i\pi} \int_{\mathbb{T}} f(z)\, \check{g}(z)\, d\mu(z) \,,$$

and with the associated norm $\| \ \|_\mu$. This will be done for every measure μ such that $L^2(\mathbb{T}) = L^2(\mu)$, space equality which we assume to hold during the sequel of this introduction and for which we will provide a necessary and sufficient condition in section 2. This condition is indeed required for problems (P_1) and (P_2) below to make sense.

In the classical stochastic framework, this type of approximation problems come up when minimizing the variance of the output error between the searched model and the "true system", the quantity $d\mu/d\lambda$ being the spectral density of the noisy input (when this input is a white noise, then $\mu = \lambda$). Such a weighting μ may also be present in the criterion when one pursues an identification procedure with the purpose of designing a controller, in which case μ represents the control performance criterion. Moreover, these weighted L^2 criteria may arise in control problems, when computing stable optimal controllers under some parametrization, for example. Such a μ may be simply used to weight some frequencies more than the others and to represent the confidence one has in the available measurements for either identification, filtering or control issues.

For any symmetric subset Γ of $\mathbb{T}$, $L^2(\Gamma)$ stands for the *real* Lebesgue space of square summable conjugate–symmetric functions on Γ. If χ_Γ denotes the characteristic function of Γ, $\|f\|_{\Gamma,\mu}$ is defined to be equal to $\|\chi_\Gamma f\|_\mu$; it induces a norm on the space of square summable functions on Γ with respect to μ.

A preliminary step is to get an interpolant $\phi \in L^2(K)$ for the given experimental data and a function $\kappa \in L^2(J)$, $J = \mathbb{T} \setminus K$, which reflects the behaviour of the system to be identified outside the bandwidth (if nothing is known, one can take $\kappa = 0$). Now, the identification procedure we are concerned with splits in two approximation issues. The first one furnishes an $\bar{H}_0^2$ approximant to the prescribed data and can be formulated as a bounded (dual) extremal problem:

(P_1) *Given $\phi \in L^2(K)$, $\kappa \in L^2(J)$, and $M > 0$, find a function $f_0 \in \bar{H}_0^2$ which minimizes $\|\phi - f\|_{K,\mu}$ among the functions $f \in \bar{H}_0^2$ which satisfy the constraint $\|\kappa - f\|_{J,\mu} \leq M$.*

The second issue is the following rational approximation problem which may be associated to a model reduction step, in the overall identification procedure.

(P_2) *Given $f \in \bar{H}_0^2$ and an integer $n > 0$, find a rational function $r \in \bar{H}_0^2$ of Mac–Millan degree less than n which minimizes $\|f - p/q\|_\mu$, where p/q ranges over the rational functions in $\bar{H}_0^2$ of Mac–Millan degree less than n.*

Problems (P_1) and (P_2) are stated here in the Hardy space $\bar{H}_0^2$ where they may be associated to identification issues for discrete time systems. However, analogous identification questions for continuous time systems, which naturally take place in Hardy spaces of the right half–plane, can also be formulated this way up to a Möbius transformation.

We give now the main statements of our results, the proofs of which are detailed in Leblond and Olivi (1995). We first state a necessary and sufficient condition for $L^2(\mathbb{T}) = L^2(\mu)$ to hold in section 2. We then solve the analytic approximation problem (P_1) and provide an explicit characterization of its solution in section 3. Finally, section 4 is devoted to the rational approximation problem (P_2) for the solution of which we lay the foundation stone of an algorithm.

2 WEIGHTED H^2 SPACES

Let $H^2(\mu)$ and $\bar{H}_0^2(\mu)$ be the weighted *real* Hardy spaces respectively defined to be the $L^2(\mu)$ closure of polynomials and the $L^2(\mu)$ closure of $\{1/z^k,\ k > 1\}$. Let $L^\infty(\mathbb{T})$ be the *real* Banach space of essentially bounded conjugate–symmetric functions. The Hardy space H^∞ is defined to be the $L^\infty(\mathbb{T})$ closure of polynomials; it verifies $H^\infty = H^2 \cap L^\infty(\mathbb{T})$ (Garnett (1981), II.4). We then have:

Theorem 1 *Let μ be a finite positive measure absolutely continuous with respect to the Lebesgue measure and such that $d\mu(-\theta) = -d\mu(\theta)$. Then, $L^2(\mathbb{T}) = L^2(\mu)$ if and only if*

$$d\mu = |\nu|^2 \, d\lambda \tag{3}$$

for a function ν which belongs to H^∞ and is invertible in H^∞. In this case, we also have:

$$H^2 = H^2(\mu) \quad \text{and} \quad \bar{H}_0^2 = H_0^2(\mu) .$$

The proof of theorem 1 relies in particular on Szegö's and Beurling's theorems, see e.g. Garnett (1981), Hoffman (1988).

We assume in the sequel that μ verifies (3). Consequently, the above approximation problems (P_1) and (P_2) deal with and lead to $\bar{H}_0^2$ transfer functions and then respects the usual stability property of the systems; this is required for identification purposes though not for mathematical reasons. Moreover, for any symmetric subset Γ of $\mathbb{T}$, the space of square–summable functions on Γ with respect to μ (that is of finite $\| \|_{\Gamma,\mu}$ norm) coincides with $L^2(\Gamma)$.

For sake of simplicity but without loss of generality, we suppose from now on that the measure μ is such that:

$$\frac{1}{2\pi}\int_{-\pi}^{\pi} d\mu(\theta) = 1 \,,$$

or equivalently that $\|\nu\| = 1$.

3 WEIGHTED H^2 APPROXIMATION

In this section, we get a solution to problem (P_1) by solving the following analogous problem (P_1') in H^2. Indeed, thanks to the isometry (2), problems (P_1) and (P_1') are equivalent.

(P_1') *Given $\varphi \in L^2(K)$, $h \in L^2(J)$, and $M > 0$, find a function $g_0 \in H^2$ which minimizes $\|\varphi - g\|_{K,\mu}$ among the functions $g \in H^2$ which satisfy the constraint $\|h-g\|_{J,\mu} \leq M$.*

In the unweighted case where $\mu = \lambda$, problem (P_1') has been solved in Alpay et al. (1993) and, in a somewhat dual form, in Krein and Nudel'man (1975). Moreover, it has been approached in the general H^p setting, $1 \leq p \leq \infty$, in Baratchart et al. (1994).

Assume further that K is a subset of $\mathbb{T}$ such that both K and its complementary J are of positive Lebesgue measure. Theorem 1 allows us to deduce the following solution to problem (P_1') from results obtained in the unweighted case (see Alpay et al. (1993)). Let

$$C_M^h(\mu) = \{g_{|K} \,,\; g \in H^2 \text{ s.t. } \|h - g\|_{J,\mu} \leq M\} \,.$$

Theorem 2 *If μ is given by* (3) *for some $\nu \in H^\infty$, invertible in H^∞, then, there exists a unique solution $g_0 \in H^2$ to problem (P_1'). Moreover, if $\varphi \notin C_M^h(\mu)$, $\|h - g_0\|_{J,\mu} = M$.*

Note that when $M \to \infty$, problem (P_1') becomes ill–posed. This follows from the fact that $H^2_{|K}$, the space of traces on K of H^2 functions, is dense in $L^2(K)$. Moreover, let (g_n) be a sequence of H^2 such that $\|\varphi - g_n\|_{K,\mu}$ tends to 0; then, if φ is not the trace on K of an H^2 function, then $\lim_{n\to\infty} \|g_n\|_{J,\mu} = \infty$.

Now, the solution g_0 of problem (P_1') can be explicitly characterized. Let $T \,:\, H^2 \to H^2$ be the Toeplitz operator with symbol χ_J:

$$T(g) = P_{H^2}(\chi_J\, g) \,,\; \forall g \in H^2 \,,$$

where P_{H^2} denotes the orthogonal projection from $L^2(\mathbb{T})$ onto H^2. The operator T is bounded, self-adjoint, positive, with norm 1 and spectrum equal to $[0,1]$. We have that:

$$g_0 = \nu^{-1}\,(1 + l\,T)^{-1}\,P_{H^2}\,(\nu(\chi_K\,\varphi + (l+1)\,\chi_J\,h))\ , \tag{4}$$

for some $l \in (-1, +\infty)$ such that $\|g_0 - h\|_{J,\mu} = M$.

This is obtained from the analogous expression of the solution to problem (P_1') for $\mu = \lambda$ which can be recovered from (4) by setting $\nu = 1$ (Alpay et al. (1993), Baratchart and Leblond (1993)). Hence, for given functions φ, h and a measure μ, the solution (4) of problem (P_1') is equal to the product of ν^{-1} by the solution of the analogous unweighted problem for the functions $\varphi\,\nu$ and $h\,\nu$.

It can also be shown that M is decreasing with respect to l so that g_0 can be numerically computed using a dichotomy procedure. Furthermore, as $l \to -1$, then the error $\|\varphi - g_0\|_{K,\mu}$ goes to zero while $M \to \infty$ if $\varphi \notin H^2_{|K}$. However, if $\varphi \in H^2_{|K}$, then it follows from a result of Patil (1972) that, as $l \to -1$, $g_0 \to \varphi$ in H^2.

Finally, thanks to (2), this provides us with the solution to problem (P_1). Indeed, if we take $\varphi = \check{\phi}$, $h = \check{\kappa}$, and then build the associated solution g_0 to (P_1'), then $f_0 = \check{g}_0$ is the solution to (P_1) we are looking for.

4 WEIGHTED RATIONAL APPROXIMATION

About the rational approximation problem (P_2), as in the case of the Lebesgue measure, a result of normality has been proved: unless f is rational of degree less than n, a rational local best approximant to f at order n has degree exactly n.

We assume that f is not rational of degree less than n. In this case, (P_2) becomes:

(P_2) *given $f \in \bar{H}^2_0$ and an integer $n > 0$, find a rational function $r \in \bar{H}^2_0$ of Mac-Millan degree n which minimizes $\|f - p/q\|_\mu$, where p/q ranges over the rational functions in $\bar{H}^2_0$ of Mac-Millan degree n.*

Recall that a rational function p/q lies in $\bar{H}^2_0$ if and only if $d^\circ p < d^\circ q$ and q is a Schur polynomial (i.e. has all its roots in the unit disk). Moreover, we shall assume that q is monic. The first step is to eliminate the numerator p. Now, the set of approximants has a manifold structure and by differentiating the above criterion with respect to the coefficients of p, we get that every critical point (p,q) satisfies:

$$< f - \frac{p}{q}, \frac{z^i}{q} >_\mu = 0\ ,\ i = 0, \cdots, n-1\,.$$

Hence, if V_q denotes the n-dimensional linear subspace of $\bar{H}^2_0$ generated by $\{z^i/q\}$, $i = 0, \cdots, n-1$, the rational approximant p/q is the orthogonal projection of f onto V_q with respect to μ. In this way, p becomes a function of q denoted by $L_\mu(q,f)$ or simply by $L_\mu(q)$ when the dependence on f is clear from the context.

Let now $\{\Phi_j\}$, $j \geq 0$, be the system of orthonormal polynomials on $\mathbb{T}$ for the measure $d\mu/|q|^2$ (see Szegö (1939), XI). By choosing $\{\Phi_j/q\}$, $j = 0, \cdots, n-1$, as a basis of V_q, we get that:

$$L_\mu(q) = \sum_{j=0}^{n-1} < f, \frac{\Phi_j}{q} >_\mu \Phi_j \,. \tag{5}$$

Define the reciprocal polynomial $\tilde{P}$ of a polynomial P of degree k by

$$\tilde{P}(z) = z^k P(1/z) \,.$$

Although $L_\mu(q)$ is a polynomial of degree possibly less than $n-1$, its reciprocal polynomial will still be considered to be:

$$\widetilde{L_\mu(q)}(z) = z^{n-1} L_\mu(q)(1/z) \,.$$

Let g_μ be the orthogonal projection on H^2 of $g\,|\nu|^2$ which can also be expressed as:

$$g_\mu(z) = \frac{1}{2i\pi} \int_{\mathbb{T}} g(\xi) \, \frac{d\mu(\xi)}{\xi - z} \,, \tag{6}$$

Using the Christoffel–Darboux formula (Szegö (1939), XI):

$$\sum_{j=0}^{n-1} \Phi_j(\xi) \, \Phi_j(z) = \frac{\widetilde{\Phi_n}(\xi) \, \widetilde{\Phi_n}(z) - \Phi_n(\xi) \, \Phi_n(z)}{1 - \xi \, z} \,,$$

together with the fact that $\Phi_n = q$, we deduce from (7) the following result.

Proposition 1 *The reciprocal polynomial $\widetilde{L_\mu(q)}$ is given by $\widetilde{L_\mu(q)} = \tilde{q} \, g_\mu - q \, v_\mu(q)$, for some function $v_\mu(q)$ which belongs to H^2. Equivalently, $\widetilde{L_\mu(q)}$ is the remainder of the division in H^2 of $\tilde{q} \, g_\mu$ by q.*

An immediate consequence of proposition 1 is that:

$$L_\mu(q, f) = L_\lambda(q, f_\mu) \,, \tag{7}$$

where $f_\mu = \check{g}_\mu$ and $L_\lambda(q, f_\mu)$ is the numerator associated with q when solving (P_2) for the Lebesgue measure λ and the function f_μ (see Baratchart et al. (1992)).

We are thus led to minimize the function $\Psi_\mu(., f)$ defined on the set Δ_n of monic polynomials q of degree n whose roots belong to $\mathbb{D}$ by:

$$\begin{array}{rccl} \Psi_\mu(., f): & \Delta_n & \longrightarrow & \mathbb{R} \\ & q & \longmapsto & \|f - \dfrac{L_\mu(q)}{q}\|_\mu^2 \,. \end{array} \tag{8}$$

It follows from (7) that

$$\Psi_\mu(q, f) = \Psi_\lambda(q, f_\mu) + \|f\|_\mu^2 - \|f_\mu\|^2 \,, \tag{9}$$

where $\Psi_\lambda(q, f_\mu)$ is the criterion to be minimized in the unweighted case for the function f_μ. Hence, minimizing the weighted criterion for some function f is just the same than minimizing the usual L^2 one for the associated function f_μ. We can thus apply differential tools as in the case of the Lebesgue measure λ for which problem (P_2) has been solved in Baratchart et al. (1991) and (1992).
The procedure goes as follows. From (8), the function $\Psi_\mu(., f)$ is smooth so that its local minima can be found by a gradient algorithm. Moreover, if g and ν are analytic in a disk $D_r = \{ z, |z| < r \}$ for some $r > 1$, then $\Psi_\mu(., f)$ extends to a neighbourhood of Δ_n and possesses recursive properties: when we meet the boundary of Δ_n we are led to solve a problem of lower order and the solution of a problem of order $k < n$ provides a boundary initial point to search a minimum at order $k + 1$. Thus, this procedure continues through different orders until we find a local minimum at order n.

Theoretical questions could also be considered, such as the consistency problem: if f is already rational of degree n, is it the *single* critical point of the problem? Once again, the answer does not come straightfully as in the unweighted case and depends on the measure μ. It would be interesting to link our further results with the ones obtained in a stochastic framework, see Ljung (1987).

To conclude, let us mention that we already got good numerical results in the unweighted case for the identification at order 8 of hyperfrequencies filters from experimental data provided by the french CNES.
As a consequence of the results of sections 3 and 4, the algorithms which solve the weighted approximation problems (P_1) and (P_2) are the same than in the unweighted case, up to changes of variables that link the function f to be approximated to the weight μ. These latter are to be implemented.

REFERENCES

Alpay, D., Baratchart, L. and Leblond, J. (1993) Some extremal problems linked with identification from partial frequency data. In J.L. Lions, R.F. Curtain, A. Bensoussan, editor, *10th conference on analysis and optimization of systems, Sophia-Antipolis 1992, L.N.C.I.S.*, Springer-Verlag, **185**, 563–573.

Baratchart, L., Cardelli, M. and Olivi, M. (1991) Identification and rational L^2 approximation : a gradient algorithm. *Automatica*, **27(2)**:413–418.

Baratchart, L. and Leblond, J. (1993) Characterization of solutions to a class of bounded extremal problems in L^2. Unpublished.

Baratchart, L., Leblond, J. and Partington, J.R. (1994) Hardy approximation to L^p functions on subsets of the circle. INRIA research report **2377**.

Baratchart, L., Olivi, M. and Wielonsky, F. (1992) On a rational approximation problem in the real hardy space H_2. *Theoretical Computer Science*, **94**:175–197.

Cardelli, M. and Saff, E.B. (1992) An algorithm for a certain type of rational approximation in H_2^-. Unpublished.

Garnett, J.B. (1981) *Bounded analytic functions.* Academic Press.

Hoffman, K. (1988) *Banach spaces of analytic functions.* Dover.

Krein, M.G. and Nudel'man, P.Y. (1975) Approximation of $L^2(\omega_1, \omega_2)$ functions by minimum-energy transfer functions of linear systems. *Problemy Peredachi Informatsii*, **11(2)**:37–60. English translation.
Leblond, J. and Olivi, M. (1995) Hardy approximation in weighted L^2 spaces of an arc. In preparation.
Ljung, L. (1987) *System identification : Theory for the user.* Prentice-Hall.
Patil, D.J. (1972) Representation of H^p functions. *Bull. A.M.S.*, **78(4)**.
Szegö, G. (1939) Orthogonal polynomials. *Col. Pub. A.M.S*, **23**.

On design of H_∞ optimal controls for uncertain nonlinear systems

Songlin Tong
Department of Adaptive Systems, Institute of Information Theory and Automation, Academy of Sciences of the Czech Republic
P.O.Box 18, 182 08 Prague, Czech Republic. e-mail: tong@utia.cas.cz

Zhongjun Zhang, Zhengzhi Han
Department of Automation, Shanghai Jiao Tong University
Shanghai, 200030, P.R.China.

Abstract

In this paper, H_∞ optimal control problems for some uncertain nonlinear systems are discussed. By the applications of LSDA-approach and game theory, this kind of optimal problems is evoluted to some convex programmings. Therefore global optimization algorithms for the problems can be obtained.

Keywords

Disturbance attenuation problem (LSDA), game theory, uncertain nonlinear system, robust controller.

1 INTRODUCTION

The true nature of H_∞ optimal control theory is to design a controller for the considered system. This controller should stabilize the closed loop system and meanwhile optimize the H_∞ norm of the gain from disturbance acting the system to a given output (which is often called a penalty) (see Francis, 1987). For details, consider a system:

$$\dot{x} = f(x,u,w), \tag{1}$$

$$z = h(x,u,w), \qquad t \geq t_0, \tag{2}$$

where x, u, w is state vector, control input and disturbance, respectively, $w \in L_2[t_0,\infty)$, f and h are continuous functions with $f(0) = 0, h(0) = 0$, z is the penalty output.

For this system, if there exists a feedback controller, say $u = u^*$, such that,

(1) the closed loop system

$$\dot{x} = f(x, u^*, 0), \qquad t \geq t_0 \tag{3}$$

is asymptotically stable at the equilibrium point $x_0 = 0$;
(2) whenever $w \neq 0$, L_2-norm of the gain from w to z is not bigger than a given positive constant γ , i.e.,

$$\int_{t_0}^{T} \|z\|^2 dt \leq \gamma^2 \int_{t_0}^{T} \|w\|^2 dt, \qquad T \geq t_0. \tag{4}$$

We will call controller $u = u^*$ a solution of the locally stable disturbance attenuation problem with an attenuation parameter (γ-LSDA, or LSDA- problem, for short) for the system Σ.

It is clear that the H_∞ optimal control theory is just to optimize all the number with which the γ-LSDA problems are solvable. So, both the LSDA problem and the optimization of all attenuation parameters are of importance in the theory of H_∞ optimal control.

In recent years, much attention has been paid to γ-LSDA problems (see, *e.g.* , Hill, 1992; Isidori, 1992; Tong, 1994; Van der Schaft, 1992). For linear system, the existence of a state feedback solution to the LSDA problem is equivalent to the solvability of a matrix Riccatti equation (Scherer, 1989). The linearization approach is still the first step in the study of LSDA problems for nonlinear systems (Van der Schaft, 1991). To deal with affine nonlinear systems, differential game theory has been used by several authors, specially, in a recent paper (Tong, 1994), the authors have investigated this problem directly for some much generalized nonlinear systems by means of convex game theory. Some interesting results have been reached.

On the other hand, LSDA problems will become very difficult to deal with when there exist some uncertainties (such as structured uncertainties) involving in the considered systems. Generally speaking, there is no feasible schedule of design to such problems, especially as concerns the global optimization of attenuation parameters, even for uncertain linear systems. Fortunately, papers (Geromel, 1992) and (Peres, 1993) offered a kind of design methods to the H_∞ optimal problems for a class of uncertain linear systems. They provided an algorithm which ensured a global optimization.

In this paper, we deal with both the LSDA problems and the attenuation parameters optimizations for the nonlinear systems with a convex structured uncertainty. We use the results obtained by (Tong, 1994) to solve the LSDA problem for each nominal nonlinear system. Then by some interesting operations, we transfer the optimization of the attenuation parameters for the whole class of the uncertain nonlinear systems into a mathematical programming on some convex sets. Therefore, it is much easy to provide global optimization algorithms for the problems. By the way, such optimizations also have strong robustness. Certainly, when the systems degenerate to linear cases, our results coincident comparatively with that of papers (Peres, 1993 and Geromel, 1992) and moreover, it can be found that, in this case, the reliability of such inference provided here has been strengthened.

The paper is organized in the following manner. In section 2 the features of LSDA problem and uncertain nonlinear system are recalled at first. We also make some conventions in the section. The formulation of the problem we discuss and the main results are presented in section 3. We just give the outlines of the proofs for the results here. In the last section, section 4, we use the obtained results to discuss uncertain linear systems, this can be considered as a comparison of our newly obtained conceptions with the known ones on the optimization problems.

2 FORMULATION OF PROBLEMS

We will explain some necessary conceptions at the beginning of the section. Then describe the considered problems in detail.

2.1 LSDA Problem

Consider nonlinear system

$$\dot{x} = f(x,u,w) \quad (5)$$
$$z = h(x,u,w) \qquad t \geq t_0 \quad (6)$$

where $x \in R^n$ is the state vector, u, w denote the control input and the disturbance input. We suppose that u and w take values in convex bounded closed sets U and W respectively, also, $w(t) \in L_2[t_0, \infty)$. Furthermore, f and h are continuous functions with $f(0) = 0, h(0) = 0$. These assumptions, will, if not said otherwise, be preserved for all the systems considered.

The output z is always called cost or penalty output for the reason of its practical explanation of a certain kind of cost the system pays because of the occurrence of disturbance w.

LSDA problem has been defined in last section ((1)-(4)). It is clear that the objective of the H_∞ control theory is to search a solution for the LSDA problem which with the minimum attenuation parameter. Therefore, the understanding and investigation on LSDA problem is of much importance and value.

2.2 Description of uncertain nonlinear systems

Consider the following group of nonlinear systems Σ:

$$\dot{x} = f(x) + g(x)u + q(x)w \quad (7)$$
$$z = (h(x), u)^\tau \qquad t \geq t_0 \quad (8)$$

where the function vector $(f, q)^\tau$ belongs to the set D:

$$D = \{\textstyle\sum_{i=1}^N \xi_i (f_i, q_i)^\tau \mid \xi_i \geq 0, \sum_{i=1}^N \xi_i = 1\},$$

and the symbol τ means transportion of matrix or vector.

Suppose that function vectors (f_i, q_i) form the following systems Σ_i:

$$\dot{x} = f_i(x) + g(x)u + q_i(x)w \quad (9)$$
$$z = (h(x), u)^\tau \qquad t \geq t_0 \quad (10)$$

we will call systems $\Sigma_i, i = 1, \cdots, N$, the nominal systems of uncertain system Σ. This means that the uncertainty of Σ is formed by a convex combination of Σ_i. Clearly Σ is just a common determined nonlinear system when $N = 1$.

The fixed matrix function $g(x)$ implies that we hope that we can control the system in a certain fixed way.

For simplicity, we make the following conventions:

Given $(f, q)^\tau$, the corresponding form of Σ is denoted as F, and we write $F \in \Sigma$. So, for nominal system Σ_i, we say $F_i \in \Sigma$.

3 MAIN RESULTS

3.1 Robust Controllers and Attenuation Parameters

1. The robust solutions of LSDA problems for Σ

For $F \in \Sigma$, let $\mathcal{S}_F$ represents the set

$$\mathcal{S}_F = \{\mathcal{T}\colon R^n \to U \mid \text{such that closed- loop system } \dot{x} = f(x) + g(x)\mathcal{T}(x) \text{ is asymptotically stable at } 0\}.$$

Define a functional $h_F\colon \mathcal{S}_F \to R^n$ as

$$h_F(\mathcal{T}) = \sup_{w \neq 0} \frac{\|z\|}{\|w\|}, \mathcal{T} \in \mathcal{S}_F,$$

where $\|.\|$ is L_2-norm. Therefore, the functional $h_F(\mathcal{T})$ is actually the value of the L_2-gain of closed-loop system from w to z.

Another set is the epigragh of h_F:

$$\text{epi}h_F = \{(\mathcal{T}, \gamma) \mid h_F(\mathcal{T}) \leq \gamma, \gamma > 0, \mathcal{T} \in \mathcal{S}_F\}.$$

So, if $(\mathcal{T}, \gamma) \in \text{epi}h_F$, the feedback $u = \mathcal{T}(x)$ can stabilize the system F when $w = 0$, and otherwise

$$\|z\| \leq \gamma\|w\|$$

for any feasible disturbance w. This means that $u = \mathcal{T}(x)$ is a solution to the γ-LSDA problem of F.

Define set:

$$\mathcal{P}_1 = \{\gamma > 0 \mid \exists \mathcal{T}\colon (\mathcal{T}, \gamma) \in \text{epi}h_F, \forall F \in \Sigma\}.$$

If $\mathcal{P}_1$ is not empty, then, for $\gamma \in \mathcal{P}_1$, there exists a $\mathcal{T}$ such that $u = \mathcal{T}(x)$ is a solution to LSDA problem for every system $F \in \Sigma$. This means we get a robust LSDA controller for the whole group of systems Σ.

2. H_∞ optimization problem for the uncertain systems

We remember that H_∞ control problem is principally to optimize attenuation parameters with which the LSDA problems are solvable.

Now we discuss the optimization problem for Σ:

$$(P_1) \qquad \min\{\gamma > 0 \mid \gamma \in \mathcal{P}_1\}.$$

It is worthy to make the following remark:

Remark 1. If there exists a solution, say γ_0 , to problem (P_1), then for the case of $N = 1$, there must exist a feedback controller which reaches the objective of the H_∞ optimal control for system F_1. When $N > 1$, according to the meaning of γ_0, we can find a operator T_ε, for every $\varepsilon > 0$, which solves the common $(\gamma_0 + \varepsilon)$-LSDA problems of system group Σ.

It should be pointed out, meanwhile, that problem (P_1) is usually difficult to deal with, especially there is no feasible method to get a global solution for the problem.

3.2 LSDA-approach Based Method to the Optimization

It is helpful to recall some results about the LSDA problem before we start our inference. In papers (Hill 1992, Van der Schaft 1991 and Veillette 1989), one can find more detailed descriptions of LSDA-problems.

For system $F \in \Sigma$, its Hamilton functional is defined as:

$$H_F(x,p,u,w)(\mu) = \\ p^\tau[f(x) + g(x)u + q(x)w] + z^\tau z - \mu^{-1}w^\tau w,$$

where $x, p \in R^n, u \in U, w \in W, \mu > 0$.

Based on the inference of Tong(1994), $\forall x, p \in R^n$, $H_F(x,p,u,w)$ must has a saddle point respecting to every pair of (u, w), moreover, the saddle point can be expressed as $u^* = -\frac{1}{2}g(x)^\tau p$, and $w^* = \frac{\mu}{2}q(x)^\tau p$. Therefore,

$$H_F^*(x,p)(\mu) = H_F(x,p,u^*,w^*)(\mu) \\ = p^\tau f(x) + \tfrac{1}{4}[p^\tau(\mu q(x)q(x)^\tau - g(x)g(x)^\tau)p] + h(x)^\tau h(x).$$

Theorem 1. *For a system $F \in \Sigma$, if there exists a differential positive functional V such that*

$$H_F^*(x, V_x)(\mu) \le 0 \tag{11}$$

for every $x \in R^n$ where V_x stands for $\frac{dV(x)}{dx}$.

Then, feedback controller $u = u^*(x, V_x) = -\frac{1}{2}g(x)^\tau V_x$ is a solution to the LSDA problem of system F, the attenuation parameter of which is $\gamma = \sqrt{\mu^{-1}}$.

Let us make following assumption:

Assumption (A) All the LSDA-problems will be considered for such parameters: A=$\{\gamma \mid \gamma = \sqrt{\mu^{-1}}, \mu q(x)q(x)^\tau - g(x)g(x)^\tau \ge 0, \forall x\}$

Define set $\mathcal{E}_F$ as

$\mathcal{E}_F = \{(V,\mu) \mid \mu > 0,\ V$ satisfying theorem 1 $\}$.

Theorem 2. *For any $F \in \Sigma$, the following statements are valid*
(i). $\mathcal{E}_F$ is convex;
(ii).If $(V, \mu) \in \mathcal{E}_F$, then $u = u^*(x, V_x)$ is a solution to the LSDA problem of system F with an attenuation parameter $\gamma = \sqrt{\mu^{-1}}$.

This conclusion can be verified from the definition of the set $\mathcal{E}_F$ easily.

It is a natural ideal to find a way of describing the H_∞ optimal problems of systems Σ by the use of nominal systems F_i. To do so, the following set is important.

$$\mathcal{E} = \cap_{i=1}^N \mathcal{E}_{F_i} = \{(V, \mu) \mid (V, \mu) \in \mathcal{E}_{F_i}, i = 1, \cdots, N\}.$$

Surely, $\mathcal{E}$ is a convex set.

Theorem 3.

$$\{(u^*(x, V_x), \sqrt{\mu^{-1}}) \mid (V, \mu) \in \mathcal{E}\} \subseteq \cap_{F \in \Sigma} \mathrm{epi} h_F.$$

Proof. for r arbitrary $F = \sum_{i=1}^N \xi_i F_i \in \Sigma$, when $(V, \mu) \in \mathcal{E}$, we have, for $x \in R^n$,

$$H^*_{F_i}(x, V_x)(\mu) \leq 0$$

because $(V, \mu) \in \mathcal{E}_{F_i}$. So,

$$\begin{aligned} H^*_F(x, V_x)(\mu) &= V_x^\tau f(x) + \frac{1}{4}[V_x^\tau(\mu q(x) q(x)^\tau - g(x) g(x)^\tau) V_x] h(x)^\tau h(x) \\ &= V_x^\tau \sum_{i=1}^N \xi_i f_i(x) + \frac{1}{4}[V_x^\tau(\mu(\sum_{i=1}^N \xi_i q_i(x))(\sum_{i=1}^N \xi_i q_i(x))^\tau - g(x) g(x)^\tau) V_x] \\ &\quad + h(x)^\tau h(x). \end{aligned}$$

The above formula is convex with respect to q, so that

$H^*_F(x, V_x)(\mu) \leq \sum_{i=1}^N \xi_i H^*_{F_i}(x, V_x)(\mu) \leq 0,$
So, $(u^*(x, V_x), \sqrt{\mu^{-1}}) \in \mathrm{epi} h_F$ is valid.

Remark. Theorem 3 implies that the set $\mathcal{P}_1$ in nonempty when $\mathcal{E}$ in nonempty.

Let $\mathcal{P}_2$ stands for the set

$$\mathcal{P}_2 = \{\mu > 0 \mid \exists V : (V, \mu) \in \mathcal{E}\}.$$

Theorem 4. *Set $\mathcal{P}_2$ is a convex set and*

$$\{\sqrt{\mu^{-1}} \mid \mu \in \mathcal{P}_2\} \subseteq \mathcal{P}_1.$$

This statement is right clearly.

The convexity of the set $\mathcal{P}_2$ is very important for our inference. Theorem 4 implies that we have constructed a convex subset for set $\mathcal{P}_1$. This makes it possible to discuss optimal problem (P_1) by means of convex programming.

3.3 Global Optimization Algorithms

Consider a convex programming problem:

$$(P_2) \qquad \max\{\mu > 0 \mid \mu \in \mathcal{P}_2\}.$$

Definition. If μ^* is a solution of problem (P_2), then we call number $\gamma^* = \frac{1}{\sqrt{\mu^*}}$ a quasi-optimal attenuation coefficient of systems Σ.

Remark 1.. Programming (P_2) is principally a convex optimal problem, so, generally speaking, it can be easily solved. Unfortunately, in general case, $\mathcal{P}_2$ is just a subset of $\mathcal{P}_1$, hence (P_2) and (P_1) are not equivalent. This is why we present the name of quasi-optimal coefficient here. However, it is worthy to point out that quasi-optimization is rightly the optimization when Σ degenerates to linear systems (See the next section).

The problem (P_2) is a simple convex linear programming. For such a problem, there are many well-developed algorithms, for instance, one can find some detailed discussion in book (Rockafellar,1970). For this reason and for simplicity, we don't list any of the algorithms here.

4 AN EXAMPLE: LINEAR CASE

Consider a family of linear systems Σ_l

$$\dot{x} = Ax + B_1 w + B_2 u \qquad (12)$$
$$z = Cx + Du \qquad t \geq t_0, \qquad (13)$$

where the matrix pair of (A, B_1) belongs to set

$$\{\textstyle\sum_{i=1}^N \xi_i(A_i, B_{1i}) \mid \xi_i \geq 0, \sum_{i=1}^N \xi_i = 1\},$$

and the systems related to (A_i, B_{1i}) are denoted by $F_i \in \Sigma_l$.

If we require the controller for linear system having the linear form of feedback of state, say, $u = -Kx$, then with the using of the same arguments as before we get the following descriptions.

Lemma. *For any system $F \in \Sigma_l$, and $\gamma > 0$, the following statements are equivalent:*
(1) there exists matrix K such that $u = -Kx$ is the solution of γ-LSDA problem of system F;
(2) there exists a positive symmetric matrix $\mathcal{W}$ such that

$$H_F^*(x, \mathcal{W})(\mu) \leq 0, \quad x \in R^n$$

where $H_F^*(x, \mathcal{W}) = H_F^*(x, V_x)$, $V = \frac{1}{2}x^\tau \mathcal{W} x$, and $\mu = \frac{1}{\gamma^2}$.

This conclusion was presented in paper (Scherer, 1989) where it was proved for all linear systems. Now we can get a interpretation of theorem 4 to the linear case:

Theorem 5. *The following statements are valid:*
(1) $\mathcal{P}_2$ is a convex set;
(2) the set $\mathcal{P}_2$ is isomorphic to the set $\mathcal{P}_1$, i.e.,

$\{\frac{1}{\sqrt{\mu}} \mid \mu \in \mathcal{P}_2\} = \mathcal{P}_1$.

It is clear that the implication of theorem 5 can be stated as following.

Theorem 6. *For the uncertain linear systems Σ_l, its H_∞ optimization problems is equivalent to a convex linear problem (P_2).*

Remark 2. This conclusion means that, on the set A, optimization problem (P_2) is equivalent to convex programming (P_1). Infact, whenever $A \cap P_1$ is nonempty, assumption (A) becomes trivial. To the other cases, the relative problems have been investigated in my Ph.D thesis.

REFERENCES

Francis. B.A. (1987) *A Course in H_∞-control Theory.* Springer-verlag Berlin, Heidelberg.
Geromel, J.C., Peres, P.L.D. and Souza, S.R. (1992) *System and Control Letters*, **19**, 23–7.
Hill, D.J. (1992) *Proc. of 31st CDC*, Tuscon, Az., 3259–64.
Isidori, A. et al. (1992) *IEEE Trans. Auto. Contr.*, **37**, 1283–93.
Peres, P.L.D. et al (1993) *System and Control Letters*, **20**, 413–8.
Rockafellar, R.T. (1970) *Convex Analysis.* Princeton Uni. Press, Princeton, NJ.
Scherer, C. (1989) *System and Control Letters*, **12**, 383–91.
Tong, S. et al. (1994) *Proc. of 33rd CDC*, Lake Buena Vista, FL. 3727–8.
Van der Schaft, A.J. (1991) *System and Control Letters*, **16**, 1–8.
Van der Schaft, A.J. (1992) *IEEE Trans. Auto. Contr.*, **37**, 770–84.
Veillette, R.J. et al (1989) *System and Control Letters*, **13**, 193-204.

Biomedical Systems

11

Constrained optimization algorithms and automatic differentiation for parameter estimation with application to granulocytic models

Bernd Tibken, Eberhard P. Hofer
University of Ulm
Department of Measurement, Control and Microtechnology,
University of Ulm, D-89069 Ulm, Germany.
Tel: +49 731 502 6334. Fax: +49 731 502 6301.
e-mail: bernd.tibken@e-technik.uni-ulm.de

Abstract

In this paper a dynamic model of granulocytopoiesis is presented which shall be used to estimate cell numbers after bone marrow transfusion. The effects of bone marrow transfusion are interpreted as a change of the initial conditions of the model. The unknown initial conditions of the model are computed on the basis of real patient data with the help of a suitably constructed optimization problem. This optimization problem is solved numerically using an SQP method to compute a stationary point. The step size is controlled via the Armijo rule and the necessary derivatives of the objective function are computed with the automatic differentiation technique.

Keywords

Biomathematical modeling, parameter estimation, real patient data, simulation, optimization, armijo rule, automatic differentiation, system theory

1 INTRODUCTION

During recent years biomathematical models to describe granulocytopoiesis have been published (Fliedner and Steinbach, 1987, Hofer, Tibken and Fliedner, 1991, Fliedner, Steinbach and Szepesi, 1988). Although the medical research has almost revealed the structure of granulocytopiesis, the application of these models in the medical treatment of individual patients has been investigated only in recent years (Hofer, Fan and Tibken, 1991, Hofer and Tibken, 1994, Hofer, 1995). In this paper, the model originally proposed by Fliedner and Steinbach is used to quantify the individual perturbation to the hemopoietic system due to bone marrow transfusion. This quantification is possible via the estimation of cell numbers for the model from granulocyte measurements from the blood. Based on real patient data the initial numbers of surviving and transplanted cells is calculated, respectively.

2 MODEL OF GRANULOCYTOPOIESIS

The model of granulocytopoiesis used consists of 26 differential equations which describe the time evolution of the system. It describes 6 cell and 2 hormone compartments. At the origin of the production of granulocytes a pool S of stem cells is assumed followed by the progenitor pool CBM, the precursor pool P, the maturation pool M, the reserve pool R, and the functional pool F in the blood. The differential equations describing the model are given by

$$\begin{aligned}
\frac{dS}{dt} &= \lambda_S(1-2\rho)S\,, \\
\frac{dCBM_1}{dt} &= 2(1-\rho)\lambda_S S + (\alpha_C - \lambda_C)CBM_1\,, \\
\frac{dCBM_i}{dt} &= \lambda_C CBM_{i-1} + (\alpha_C - \lambda_C)CBM_i\,, \\
& \quad i = 2,3,\ldots,10\,, \\
\frac{dP_1}{dt} &= \lambda_C CBM_{10} + (\alpha_P - \lambda_P)P_1\,, \\
\frac{dP_i}{dt} &= \lambda_P P_{i-1} + (\alpha_P - \lambda_P)P_i\,, \\
& \quad i = 2,3,\ldots,10\,, \\
\frac{dM}{dt} &= \lambda_P P_{10} - \lambda_M M\,, \qquad (1) \\
\frac{dR}{dt} &= \lambda_M M - \lambda_R R\,, \\
\frac{dF}{dt} &= \lambda_R R - \lambda_F F\,, \\
\frac{dReg.I}{dt} &= \varphi_1 - \lambda_{Reg.I} Reg.I\,, \\
\frac{dReg.II}{dt} &= \varphi_2 - \lambda_{Reg.II} Reg.II\,.
\end{aligned}$$

The pools CBM and P are divided up into 10 subcompartments in order to describe the cell transit time variability. The transit rate λ_S of the stem cells into division is regulated by the content of all cell compartments in the bone marrow. The fraction ρ of cells returns into the stem cell pool after division while the other cells divide further and mature to granulocytes.

The parameters in the differential equations are choosen to reflect the medical knowledge about the cell transit times and cell amplification due to division.

3 ESTIMATION OF CELL NUMBERS

The mathematical model (1) of granulocytopoiesis consists of a system of nonlinear differential equations given in the previous section. The available real patient data after bone marrow transfusion are measurements of the granulocyte concentration in the blood. Thus, only $F(t)$ for non equidistant time instances is available as measurement. Based on this measurements the initial values for the content of the cell compartments are estimated. For the initial values of general systems of nonlinear differential equations with measurements at non equidistant times no general, systematic and effective method is available. A suitable way to circumvent this problem is to use numerical optimization software to minimize a problem specific performance index. The only measurable datum is the concentration of F cells in the blood. Thus it is reasonable to perform the estimation using the performance index

$$J = \sum_{i=1}^{N} (F_i - F(t_i))^2$$

where t_i, F_i, $i = 1, \ldots, N$ are the measurement times and the numbers of granulocytes in the blood, respectively.

The function J is to be minimized with respect to the initial values of the system (1). These initial values are parametrized as follows

$$\begin{aligned} S(0) &= \theta_1 S^* \\ CBM_i(0) &= \theta_2 CBM_i^* \ , \quad i = 1, \ldots, 10 \\ P_i(0) &= \theta_3 P_i^* \ , \quad i = 1, \ldots, 10 \\ M(0) &= \theta_4 M^* \\ R(0) &= \theta_5 R^* \\ F(0) &= \theta_6 F^* \end{aligned} \tag{2}$$

where S^*, CBM_1^*, ..., F^* are the numbers of cells in the respective compartments in the healthy steady state.

Thus the minimization problem

$$J_{min} = \min_{\theta \geq 0} J(\theta)$$

has to be solved in order to estimate the initial cell numbers of a patient based on the observed data. This optimization problem is a constrained optimization with the lower bound 0 on all variables. The numerical methods which have been used and are implemented in MATLAB to solve this problem will be given in the next section.

4 OPTIMIZATION METHOD

The optimization problem described in the last section has been solved using a special SQP method. The basic idea is to solve a sequence of easier problems whose solution converges to the solution of the original problem. We start with an initial estimate $\theta^{(0)}$ for the minimum and will construct a convergent sequence $\theta^{(k)}$ which satisfies $J(\theta^{(k)}) \geq J(\theta^{(k+1)})$ and $\lim_{k\to\infty} J(\theta^{(k)}) = J_{min}$. We compute this sequence with the help of the auxiliary problem

$$\min_{\theta\geq 0}(J(\theta^{(k)}) + (\frac{\partial J}{\partial \theta}(\theta^{(k)}))^T(\theta - \theta^{(k)}) + \frac{1}{2}(\theta - \theta^{(k)})^T(\frac{\partial^2 J}{\partial \theta^2}(\theta^{(k)}))(\theta - \theta^{(k)}))$$

which is constructed from the original problem by a Taylor series expansion of the objective function J at the actual estimate $\theta^{(k)}$ up to the quadratic terms. If the hessian $\frac{\partial^2 J}{\partial \theta^2}(\theta^{(k)})$ is positive definit this auxiliary problem will be solved and the solution is denoted by $\tilde{\theta}$. The next iterate $\theta^{(k+1)}$ is then sought between $\theta^{(k)}$ and $\tilde{\theta}$ on the line segment connecting these two points in the parameter space if the direction from $\theta^{(k)}$ to $\tilde{\theta}$ is a descent direction for J. If either the hessian is not positive definit or the direction is no descent direction the next iterate is sought in the direction of the negative gradient of J at $\theta^{(k)}$ where the search has to be restricted to the components for which the corresponding components of $\theta^{(k)}$ are not on the boundary of the feasible set, *e.g.* , for the components which are strictly positive. The search is carried out using the armijo rule. This results in a convergent sequence which converges (Kosmol, 1989) to a zero of the gradient of J. Thus, if the initial estimate is good enough the convergence to the minimum is guaranteed. The optimization method is implemented using MATLAB and the auxiliary quadratic optimization problem is solved using the corresponding routine in the MATLAB Optimization Toolbox.

The computation of gradient and hessian is carried out using the technique of automatic differentiation. This technique has been implemented in MATLAB using m-files and the necessary computations are done after a call to the corresponding m-files. A special Runge-Kutta method has been implemented which uses automatic differentiation for the calculation of the derivatives of the numerical solution to (1) with respect to the parameters $\theta_1, \ldots \theta_6$.

5 CONCLUSION AND OUTLOOK

In this paper a special optimization method has been presented which has been implemented using MATLAB. It is based on a quadratic approximation of the function to be minimized and takes simple bounds on the variables into account in each step. Thus, infeasible points are never generated during the optimization. This is important because in the application in mind the variables are cell numbers which have to be positive in order to have an biological interpretation. The convergence near the optimum is quadratic and first results look very promising.

The method has been applied to the estimation of cell numbers after severe irradiation and bone marrow transplantation after leukemia treatment.

Future work will concentrate on the application of Quasi Newton Methods which need only gradient information and no hessian. Thus, eventually the necessary computer time to perform an estimation can be reduced drastically.

REFERENCES

Hofer, E.P., Fan, Y. and Tibken, B. (1991) Extraction of Rules for Model Based Estimation of Granulocytopoiesis, in *The 5th German-Japanese Seminar on Nonlinear Problems in Dynamical Systems - Theory and Applications -*. (ed. M. Frik), Duisburg.

Hofer, E.P., Tibken, B. and Fliedner, T.M. (1991) Modern control theory as a tool to describe the biomathematical model of granulocytopoiesis, in *Analyse Dynamischer Systeme in Medizin, Biologie und Ökologie.* (eds. D.P.F. Möller, O. Richter), Informatik Fachberichte 275, Springer, Berlin.

Hofer, E.P. and Tibken, B. (1994) A Clinical Decision Support System for the Treatment of Irradiated Persons Based on a Biomathematical Model of Granulocytopoiesis, in *AUTOMATIC CONTROL 12th Triennial World Congress of IFAC* (eds. G.C. Goodwin, R.J. Evans), Pergamon, 3.

Hofer, E.P. (1995) On the Way to a Decision Support Tool for Treatment of Irradiated Persons Based on Hemopoietic Models. *Journal of SICE*, **34**, 410–7.

Fliedner, T.M., Steinbach, K.-H. and Szepesi, T. (1988) Hematological indicators in the determination of clinical management strategies in radiation accidents, in *International Conference on Biological Effects of Large Dose Ionizing and Non-Ionizing Radiation*, Hangzhou, China.

Fliedner, T.M. and Steinbach, K.-H. (1987) Simulationsmodelle von Perturbationen des granulozytären Zellerneuerungssystems, in *Modelle der Pathologischen Physiologie.* (eds. W. Doerr, H. Schipperges), Springer, Berlin.

Kosmol, P. (1989) *Methoden zur numerischen Behandlung nichtlinearer Gleichungen und Optimierungsaufgaben.* Teubner, Stuttgart.

Expert system for diagnosis of womens' menstrual cycle using natural family planning method

Andrzej Urbaniak
Institute of Computing Science, Poznań University of Technology
ul. Piotrowo 3A, 60-965 POZNAŃ, Poland
Phone: +48 61 790790; fax: +48 61 771525;
E-mail: urbaniak@pozn1v.put.edu.pl

Abstract

In the contribution the algorithm and user-friendly software for fertility diagnosis, are considered. A part of research concerning the algorithms for multi-check methods was presented by Urbaniak (1994a). In this paper will be present the algorithms for expert system supporting the women fertility diagnosis and evaluation of pregnancy achievement.

Keywords

Biological system, expert system, Natural Family Planning (NFP), medical diagnosis

1 INTRODUCTION

People have been interested more and more in ecological aspects of life. This is connected to the application of any natural methods in human life, as well in normal as in sick conditions. These tendencies cause the great interest in natural methods determining fertility and infertility days in the woman menstrual cycles. Methods of Natural Family Planning (NFP) are interesting because of their high effectiveness, comparable to hormone contraception methods.

Some may complain of two inconveniences of the NFP methods. These are the following:

- necessity of daily observations and documentation's of fertility symptoms,
- necessity of well education to determine fertility and infertility days.

---

* Presented on 17th IFIP Conference on **System Modelling and Optimization**, July 10-14, Prague

In many scientific centres research for easier NFP methods application is provided. There are two main approaches. The first one, it is construction of special devices to measure hormone changes during the woman cycle. This research has been developed, for example, by Thornton et al.(1990), Brown et al. (1984, 1992). The enzyme immunoassays for urinary oestrogen detection was elaborated and effectively used to determine fertility days. The method was worked out as a simply procedure which can be performed at home.
Second approach is based on the microprocessor technic. BIOSELF system is first, fully acceptable device. It is an electronic thermometer with the microcomputer (Flynn &Brooks, 1990). This system was programmed according to Temperature-only Method and had been tested in many countries.

Our research it is the second approach which uses standard PC computers with Expert System (ES) to create fertility awareness. The advantage of PC systems application is: possibility of determining fertile days without education about the rules of NFP methods. Of course, the most important is constant observation of woman body symptoms. It is part of the activity in which woman can not be replaced. The rest-part, with rules of NFP methods, takes over the PC software. The software prepared as a user-friendly program is the crucial point of the system.

2 BASES FOR NFP METHODS

Natural Family Planning methods are based on three major facts.

1. There is only one ovulation in the cycle. There is possibility for the second ovum to grow up, but only during the next 24 hours to the first ovulation. Ogino and Knaus have described that ovulation in the menstrual cycle occurs about 14±2 days before the next menstruation.
2. The ovum life time is no longer than 24 hours. The possibility for the conception is usually shorter.
3. The sperm life in the women's vagina is maximum 5 to 6 days. This time depends on the special kind of cervix mucus which is necessary to keep the sperm alive 5-6 days. General conclusions is that conception may occur in the time around ovulation and no longer than during 6 to 7 days in the cycle. During other days of the cycle conception is not possible.

Women menstrual cycle is controlled by the complex interaction of hormones (the chemical messengers of the body) and reproductive organs. This causes that each month a grown up ovum is released from the ovary.

Thus the women menstrual cycle consists of the following phases.

Phase 1 it is time from first day of menstruation to the moment when follicles start to grow up in the ovaries. As length of this part of the cycle varies from cycle to cycle of the individual women, as well as from women to woman, it is sometimes difficult to of make accurate assessments of the beginning of fertility. For this reason it is called as relatively *infertile phase.*

Phase 2 starts from the time when follicles begin to grow up to first signs of any mucus or wetness to 48 hours after to few days after ovulation (time calculated by temperature and mucus observations). (It together takes 24 hours for the lifespan of the ovum, 5-6 days for the livespan of the sperm, because it is impossible to be precise about the actual time of ovulation.) This is *fertile phase.*

Phase 3 is counted from a few days after ovulation (calculated by temperature and mucus observations) until the last day before the next menstruation which is the end of that cycle. This is called *absolutely infertile phase.*
From Ogino and Knaus discovery a lot of NFP methods have been worked out.. Now a lot of these methods have been documented by the wide experimental research provided by world famous scientific centres.

3 MODERN NFP METHODS

Calendar (or Calculation) Method was the earliest method of NFP. This method is based on the fact, that ovulation occurs 14+2 days before the next menstruation. Taking into account the "worst case" (the length of the shortest cycle from the last 12 cycles) the fertility days are calculated according to the simple rule. In the thirties this method was very the important for many couples. In those years women had more regular cycles than now, because they did not work professionally, and ecological conditions were better. Thus then the effectiveness of the Calendar Method was high too. Presently, it is impossible to based on the regularity of women's cycle. Many factors of present life conditions cause irregularity of women's cycles. In consequence, the determination of the fertility or infertility days in women cycles can be possible only when using the every day observations of major fertility signs.
There are the following:

1. basal body temperature (BBT),
2. quantity and quality of the cervix mucus,
3. state of the cervix.

The modern methods of fertility awareness apply one or more of mentioned above signs. Daily observation of these signs gives with the high reliability, possibility to determining fertile and infertile days.

3.1. Temperature-only Method

This method is based on the fact, that woman's basal body temperature(BBT) rises suddenly after the ovulation. The minimal shift of the temperature has to be 0.2 Celsius degrees, and the fertile phase ends when the basal body temperature elevates at least 0.2 ^{o}C for three consecutive days, without any disturbances. The Temperature-only Method allows to determine only post-ovulatory infertile phase. BBT to be used in this method, and in (any other methods with BBT, has to be taken for five or eight minutes at the same waking time each morning. Rules for the Temperature-only Method were presented by Marshal. A lot of researchers underline the high effectiveness of the Temperature-only Method (Donnay, 1991).

3.2. Billings Ovulation Method

The cervical mucus method of NFP relies on the woman self observation of presence and changes cervical mucus throughout the menstrual cycle, and these changes are used to determine fertile and infertile phases. This method is also known as the Billings Ovulation Method or Ovulation Method Billings (Brown et al., 1984, 1987). Two important factors are needed: sensation and appearance of the mucus. The important point is that the cervix produces different types of mucus - fertile-type mucus and infertile-type mucus which appear at different parts of the cycle. Together with ovum growing up in the ovary, the cervical mucus becomes more clear, stretchy, wet and slippy. It depends on rising level of oestrogen. After ovulation the mucus becomes thicker, sticky or disappear - dry days.

3.3. Double-check Method (Sympto-Thermal Method)

The sympto-thermal double-check method is based on the observation of three major signs of the menstrual cycle. These are the following:
1. basal body temperature (BBT);
2. quantity and quality of the mucus;
3. state of the cervix

All changes of these signs are caused by the changes of hormones levels in the blood. From the technical point of view it would be very interesting to measure precisely these hormones levels. But the three signs, mentioned above, are able to measure these parameters indirectly.

Historically, the Sympto-Thermal Method is based on the earlier worked out methods. In general, all of these methods of NFP can be divided into two groups. First one - it is group of the one-symptom methods based on: either the BBT, either the mucus, or the cervix. Second group there is group of the methods which apply two and more fertility signs to crosscheck beginning and end of fertile phase. This group has been intensively developed, especially in the last two decades. The Sympto-Thermal Methods (Double-Check Method) have one of the highest effectiveness rate. Formula devised by Raymond Pearl in the 1930's is standard procedure expressing the effectiveness of any NFP Methods.

The particular rules of the Sympto-Thermal Method depend on the following factors (Flynn&Brooks, 1990):
1. status of women life (ordinary cycle, post-partum cycle, premenopause cycle, and post-pill cycle);
2. length of the history of observations,
3. sort and number of signs observed.

These different rules must be implemented in the engine module of the Expert System. The women needs to declare properly observed mucus sign concerning its **sensation (feeling)** and **appearance.** In the similar way the women needs to check and describe proper features of the cervix (**position, opening, firmness**). Taking into account every day observation of the above two signs and adding the BBT results we obtain sufficient data to estimate fertility stage on every day of women menstrual cycle.

From the algorithm point of view, there is very important to define the **temperature shift**. Temperature shift is connected to the ovulation of the ovulation and is described as a higher BBT level sustaining for at least three consecutive days without any disturbances. . Additionally, to recognize temperature shift, the **cover line** needs to be drawn over the low temperature phase (Flynn&Brooks, 1990).

For the mucus and cervix observation, the **peak day** is defined as a last day of the fertile-type mucus or maximum cervix features. All rules of sympto-thermal method are given by Flynn & Brooks, (1990).

4 CHARACTERISTICS OF AN EXPERT SYSTEM

We must note that systematic and responsible observations are very important factors for adaptation of the NFP methods. This remark especially important when collection of data and chart interpretation is realized automatically using a PC system.
The computer system can fulfil the following expectations concerning with the two main inconveniences of NFP methods:

- necessity of documentation of daily observations,
- good knowledge about the rules of fertility and infertility to determine.

These above mentioned inconveniences can be realized using the well designed DSS or ES with clear and friendly graphical interface (Urbaniak, 1994c). The user ought to only introduce to the system the observed signs of fertility and other functions are realized by algorithms. All data are introduced by the way of answers for the questions which are highlighted on the screen.
There are the following advantages in the well-designed Expert System: the increase availability, the cost reducing, the danger reducing, the permanence, the multiple expertise, the reliability increasing, the explanation, the fast response, the steady, unemotional, and complete response at all times, the intelligent tutor, the intelligent database.

The worked out first implementations of the algorithms concern the basic situations in the woman live only. The special cases demands the different rules to introduce into the algorithms. A study on this problem are begun now (Burdziak, 1994).

5 USER INTERFACES

4.1. BILMETH - Billings Ovulation Method

The worked out friendly user's interface allows to apply the system without special experience in the computer science (Szczepaniak, 1994). There are following rules of the options (Fig.1):

1. Choice of the accessible option is realized by pressing the "hot" key, identified by the highlighting letter or by mouse. This change is presented graphically by simulation of the pressed key.
2. The program has help function. This function is connected with given operation which is under realization.
3. There is unification way of data introducing. All data must be confirmed by *Enter*

Options

Menu USER. This option enables to create a new user *(NEW)* or register of an existing user *(PRESENT)*. The each individual is identified by forename, name and password.
Menu NEW. This instruction is used to create new chart of the cycle observation. The important data about each cycle is at beginning. Additionally we can shortly describe the special feature of the cycle (optionally).
Menu PRESENT. This instruction is prepared to introduce of data the present cycle. The user has to choose proper symptom from given set the and confirm it.
Menu HIST is used to analyze previous cycles (history of the cycles).
Menu dRUK enables to print out the cycle chart.

Menu HELP and INFO there is context help to apply system and gives information about authors and institution.

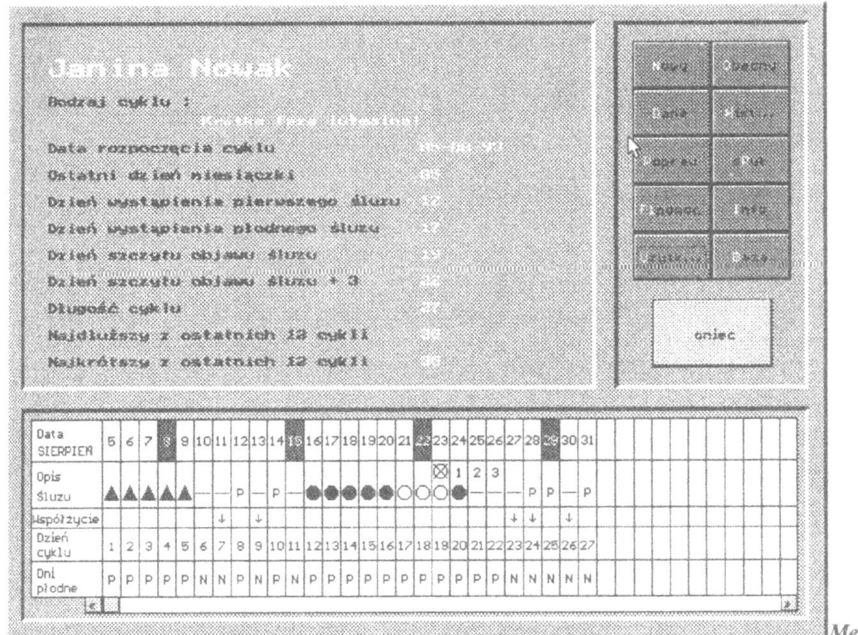

Fig. 1 An example of dialogue window for Billings Ovulation Method

4.2. STBB - Double-check Method

The elaborated software is friendly to the user who knows only basic rules of the keyboard and how to use mouse (Białobłocki, Urbaniak, 1994, Urbaniak, 1994b).

During working phase, the screen is divided into three parts (Fig. 2):

- menu line (first screen line);
- status line (last screen line);
- work part (between menu and status lines).

Menstrual Cycle Chart Creation. Spreading of the CHART menu allows to introduce new chart user or to fulfil existing chart. The user can choose of any chart cycle identified by date of its preparation.

Mark Data On the Chart. All observations are marked in days only. It is impossible to edit a day which edition was finished, nor any of the future days. Edition of each day can be finished or closed. The difference between closed and finished concern the situation when all symptoms are observed already or only one of them. After when the day edition is finishing no changes are possible. In situation when some symptoms were not observed and marked we resign edit it. In the window with the mucus sensation is presented.

Chart Printing. An option allows to print out any chart of menstrual cycle.

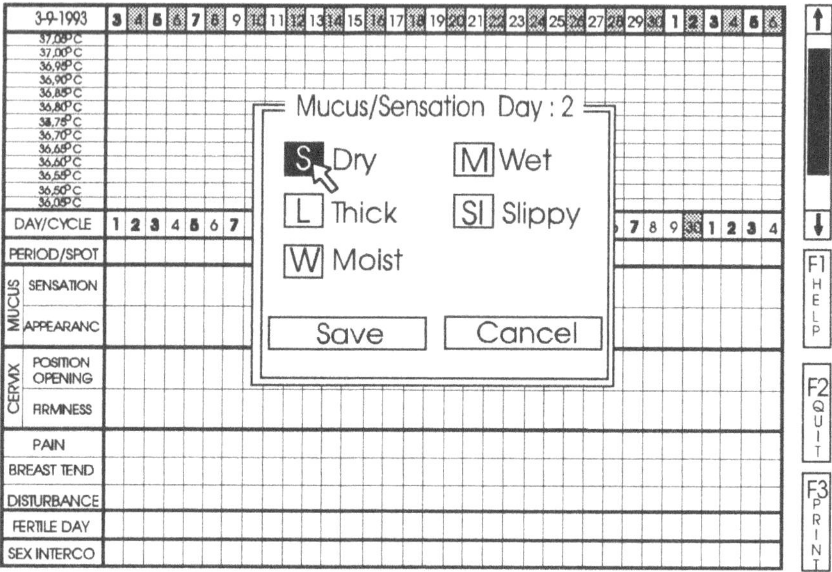

Fig. 2. An example of the dialogue window

5 FUTURE RESEARCH

In many scientific centres research on NFP methods is performed and new methods and modern modification of an existing ones are published. The intensive research expansion in the computer science and especially in the field of expert systems and artificial intelligent allows to apply these results to software systems. The software presented above is based on the scientific background of NFP and now, the software is tested by NFP teachers.
Our next research plans concerns following directions.

First one concern supplementation of the software with rules of the special situations. And after this we will have the basic software product, but only for main situation of the woman life. In the next step we plan to extend the software to the special situations: i.e. after birth, premenopause and post hormonal contraception. These special situations demand the modification of the rules and algorithm to determine fertility and infertility days.

Second direction concerns creation of the software for computer oriented teaching. Presently we have first version of the software which actually has been tested for different NFP methods.

NFP effectiveness study this is third direction. All prepared software generate formatted files with necessary for effectiveness calculations data, according to European NFP Study (Urbaniak, Zmyślony, 1994).

The important results of the research were achieved thanks to good collaboration between computer science engineers and NFP teachers, and feedback under the form of well organized testing procedure.

REFERENCES

Białobłocki, T., Urbaniak, A. (1994) Fertility analysis in the woman cycle using symptothermal method. *Biulletin of Polish National Society of Natural Family Planning Teachers*, **3**, 2–4. (in polish)

Brown, J.B., Blackwell, L.F., Billings, J.J. et al. (1984) Determination of ovarian hormone levels in urine for identifying the fertile and infertile phases of the cycle. *Family Health Internat. Expert Meeting*, North Carolina.

Brown, J.B., Blackwell, L.F., Billings, J.J. et al. (1987) Natural family planning. *American Journal of Obstetrics Gynecology*, **157**, 1082–9.

Brown, J.B., Blackwell, L.F., Holmes, J., Smyth, K. (1992) New assays fir identifying the fertile period. *International Journal of Gynecology & Obstetrics*, **1**, (1).

Burdziak, A. (1994) *NFP - expert system.* Master Degree, Institute of Comp. Science, Poznań University.

Donnay, F. (1991) Maitrese de la fecondite. *L'enfant en milien tropical*, Centre Internationale de l'Enfance, no. 193-194, Paris.

Flynn, A., Brooks, M. (1990) *A manual of Natural Family Planning.* Unwin Hyman Limited, London.

Szczepaniak, P. (1994) *DSS for ovulation method.* Master degree, Institute of Comp. Science.

Thornton, S.J., Pepperell, R.J. Brown, J.B. (1990) Home monitoring of gonadotropin ovulation induction using the Ovarian Monitor. *Fertility and Sterility*, **56**, (6), 1076–82.

Urbaniak, A. (1994a) Computer and Natural Family Planning. *Biulletin of Polish National Society of Natural Family Planning Teachers*, **2**, 6–7. (in polish)

Urbaniak, A. (1994b) Fertility diagnosis and simulation based on the multi-check method of natural family planning, in *Proc. CISS – First Joint Conf. of International Simulation Societas* (eds. J. Halin, W. Karplas), Zurich.

Urbaniak, A. (1994c) Main features of the expert systems for determining the fertile and infertile days in the women menstrual cycle. Plenary lecture on *VI Congress IFFLP/FIDAF*, Lublin, Poland.

Urbaniak, A., Zmyślony, R. (1994) An effectiveness research of NFP methods using NPR system, in *VI Congress IFFLP/FIDAF Lublin* (ed. R. Sikorski), Poland.

13

Metabolic flux determination by stationary 13-C tracer experiments: Analysis of sensitivity, identifiability and redundancy

Wolfgang Wiechert
Institut für Biotechnologie, Forschungszentrum Jülich
52425 Jülich, Germany. e-mail: w.wiechert@kfa-juelich.de

Abstract

Stationary ^{13}C tracer experiments supply a large amount of information related to metabolic fluxes in microorganisms. Unknown intracellular fluxes can be determined from some directly measured metabolic fluxes and the fractional labelling of intracellular carbon atom pools. To this end the algebraic flux and carbon balance equations have to be solved by parameter-fitting. The statistical quality of the results is judged by local (i.e. linearized) sensitivity analysis. On the other hand some generally applicable computer algebraic algorithms are given for global redundancy and identifiability analysis. They enable the dimension of the flux determination problem to be significantly reduced. As an application example some global results for the anaplerotic reaction section from the central metabolism are derived.

Keywords

Metabolic fluxes, isotope labelling, sensitivity, identifiability, redundancy, computer algebra

1 INTRODUCTION

The determination of intracellular metabolic fluxes in microorganisms is of great interest for biotechnology. A complete quantitation of all reaction rates in the central metabolic pathways allows the effect of genetic manipulations on the yield of certain desired metabolic products to be characterized *in vivo*. Moreover such a method will supply valuable information for a systematic approach to increase product formation by microorganisms (Bailey, 1991).

A recently developed method of determining intracellular fluxes is based on ^{13}C NMR labelling data in combination with direct measurements of the fluxes between the cell interior and the surrounding medium (Marx *et al.*, 1995). The organism must therefore be in a metabolic steady state which can be maintained inside a bioreactor. In this situation a tracer substrate which is labelled at a certain carbon atom by ^{13}C is fed into the system. After an isotopically stationary steady state has been reached for all metabolic intermediates, the fractional amount of labelled carbon atoms within intracellular metabolites can be measured (Figure 1). For this purpose the

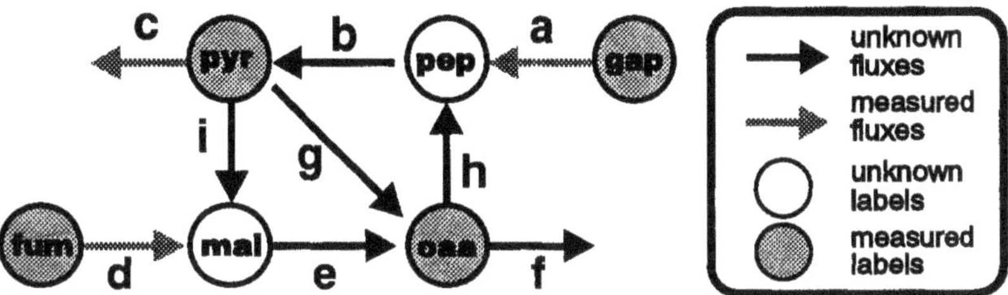

Figure 1 The anaplerotic reaction section in the central metabolism of *C. glutamicum* with the measurement data that can be supplied by a ^{13}C NMR tracer experiment. Upper row: glycolytic metabolites. Lower row: metabolites in the citric acid cycle.

labelled amino acids are extracted from the cell protein, which is the distinguishing feature of the new method.

This new technique supplies a previously unattainable amount of (stationary) measurement data from which the intracellular metabolite fluxes have to be determined. 15 fluxes and 23 fractional labels within *C. glutamicum* could be directly measured by Marx *et al.* (1995)! From this data set nearly all fluxes in cellular metabolism could be determined and additionally several reverse fluxes (Wiechert *et al.*, 1995).

However, since the corresponding model equations are nonlinear and the metabolic networks display a high degree of coupling, it cannot easily be judged whether the measurements contain sufficient information about the unknown fluxes (sensitivity and identifiability problem). On the other hand several time-consuming measurement procedures can be omitted if it turns out that certain data contain no additional information (redundancy problem). This contribution presents some generally applicable algorithms and an application for the treatment of identifiability and redundancy problems corresponding to metabolic carbon isotope labelling systems.

2 EXAMPLE: ANAPLEROTIC REACTIONS

The model equations will be introduced using a very simple example from the central metabolism of *C. glutamicum*. The anaplerotic section of the metabolism (as shown in Figure 1) was chosen because it serves to demonstrate a certain nonidentifiability problem that will be explained later on. More details concerning the formulation of model equations can be taken from (Wiechert *et al.*, 1995).

Firstly, the state vectors are constructed as follows. All metabolic fluxes in the system are combined into the flux vector

$$\mathbf{v} = (a, b, c, d, e, f, g, h, i)^T \ .$$

In our case all participating metabolites have 3 carbon atoms which do not change places in any of the reaction steps. For this reason we can restrict our study to the first carbon atom of each metabolite. The ^{13}C labelling fractions in each carbon atom pool are then given by the vectors

$$\mathbf{x}^{\text{inp}} = (gap, fum)^T \quad \text{and} \quad \mathbf{x} = (pep, pyr, oaa, mal)^T \ .$$

Here $\mathbf{x}^{\text{inp}}$ corresponds to input metabolites with known labelling state while $\mathbf{x}$ corresponds to metabolic intermediates.

Secondly, some material balance equations hold because the system is in a stationary state. The metabolic fluxes connected with an intermediate pool add up to zero (linear stoichiometric equations), which can be used for term elimination:

$$\begin{array}{lrcl} pep: & a+h & = & b \\ pyr: & b & = & c+g+i \\ oaa: & e+g & = & f+h \\ mal: & d+i & = & e \end{array} \quad \Longrightarrow \quad \begin{array}{rcl} b & = & a \qquad\qquad +h \\ e & = & a-c+d-g+h \\ f & = & a-c+d \\ i & = & a-c \qquad -g+h \end{array} \tag{1}$$

Similarly, the isotope labelling balance equations reflect the fluxes of labelled carbon atoms. A metabolic flux $\mathbf{v}_i : \mathbf{x}_j \rightarrow \mathbf{x}_k$ carries the absolute amount of $\mathbf{v}_i \cdot \mathbf{x}_j$ of labelled material per time unit. This gives rise to a system of equations which is bilinear with respect to fluxes and labels:

$$\begin{pmatrix} 0 \\ 0 \\ 0 \\ 0 \end{pmatrix} = \begin{pmatrix} -b & . & h & . \\ b & -c-g-i & . & . \\ . & g & -f-h & e \\ . & i & . & -e \end{pmatrix} \begin{pmatrix} pep \\ pyr \\ oaa \\ mal \end{pmatrix} + \begin{pmatrix} a & . \\ . & . \\ . & . \\ . & d \end{pmatrix} \begin{pmatrix} gap \\ fum \end{pmatrix} . \tag{2}$$

Using (1) the 4×4 system matrix in (2) further reduces to:

$$\begin{pmatrix} -a-h & . & h & . \\ a+h & -a-h & . & . \\ . & g & -a+c-d-h & a-c+d-g+h \\ . & a-c-g+h & . & -a+c-d+g-h \end{pmatrix}$$

$$= a \cdot \begin{pmatrix} -1 & . & . & . \\ 1 & -1 & . & . \\ . & . & -1 & 1 \\ . & 1 & . & -1 \end{pmatrix} + c \cdot \begin{pmatrix} . & . & . & . \\ . & . & . & . \\ . & . & 1 & -1 \\ . & -1 & . & 1 \end{pmatrix} + d \cdot \begin{pmatrix} . & . & . & . \\ . & . & . & . \\ . & . & -1 & 1 \\ . & . & . & -1 \end{pmatrix} + \ldots \tag{3}$$

From now on $\mathbf{v}$ is replaced by the reduced flux vector $\mathbf{v} = (a, c, d, h, g)^T$.

3 GENERAL STRUCTURE OF MODEL EQUATIONS

Collecting all integer coefficients associated with a carbon flux $\mathbf{v}_i$ into the atom transition matrices $\mathbf{P}_i, \mathbf{P}_i^{\text{inp}}$ as has been done in (3) the general algebraic structure of the metabolic carbon isotope labelling system (2) is comprehensively given by:

$$\mathbf{0} = \left(\sum_{i=1}^{\dim \mathbf{v}} \mathbf{v}_i \cdot \mathbf{P}_i\right) \cdot \mathbf{x} + \left(\sum_{i=1}^{\dim \mathbf{v}} \mathbf{v}_i \cdot \mathbf{P}_i^{\text{inp}}\right) \cdot \mathbf{x}^{\text{inp}} \stackrel{def}{=} \mathbf{P}(\mathbf{v}) \cdot \mathbf{x} + \mathbf{P}^{\text{inp}}(\mathbf{v}) \cdot \mathbf{x}^{\text{inp}} \tag{4}$$

This is sufficient to understand the mathematical problems under consideration. For a reasonably complex metabolic network $\dim \mathbf{v} \approx 30$ and $\dim \mathbf{x} \approx 100$ can be reached, which makes clear that identifiability and redundancy are indeed nontrivial problems. For 'nonpathological' systems the matrix $\mathbf{P}(\mathbf{v})$ can be proven to be positively definite (Anderson, 1983) so that we obtain a well defined solution

$$\mathbf{x} = \mathbf{x}(\mathbf{v}) = -\mathbf{P}(\mathbf{v})^{-1} \cdot \mathbf{P}^{\text{inp}}(\mathbf{v}) \cdot \mathbf{x}^{\text{inp}}. \tag{5}$$

Several of the fluxes $\mathbf{v}_i$ as well as the carbon atom labels $\mathbf{x}_j$ can be directly measured (see Figure 1). This is formally described by coordinate projection matrices $\mathbf{M}$ (i.e. matrices composed of unit vectors). The measurement noise ε is assumed to be normally distributed with expectation $\mathbf{0}$ and known covariance matrix $\mathbf{\Sigma}$ giving rise to the measurement equations:

$$\begin{array}{lcl} \mathbf{w} &=& M_{\mathbf{w}} \cdot \mathbf{v} + \varepsilon_{\mathbf{w}} \\ \mathbf{y} &=& M_{\mathbf{y}} \cdot \mathbf{x} + \varepsilon_{\mathbf{y}} \end{array}, \quad \mathbf{\Sigma} = \text{Cov}(\varepsilon) = \text{Cov}\begin{pmatrix} \varepsilon_{\mathbf{w}} \\ \varepsilon_{\mathbf{y}} \end{pmatrix} = \begin{pmatrix} \mathbf{\Sigma}_{\mathbf{w}} & 0 \\ 0 & \mathbf{\Sigma}_{\mathbf{y}} \end{pmatrix} \tag{6}$$

4 FLUX ESTIMATION AND SENSITIVITY ANALYSIS

Usually, there are more measured quantities than unknown fluxes. The problem of flux determination thus is an inverse problem associated with equation (5) and (6). It can be treated numerically by computing a least squares estimate with respect to the appropriate weighted vector norms $||\mathbf{x}||_{\mathbf{\Sigma}} = \mathbf{x}^T \cdot \mathbf{\Sigma}^{-1} \cdot \mathbf{x}$:

$$\hat{\mathbf{v}} = \arg\min_{\mathbf{v}} ||\mathbf{w} - M_{\mathbf{w}} \cdot \mathbf{v}||^2_{\mathbf{\Sigma}_{\mathbf{w}}} + ||\mathbf{y} - M_{\mathbf{y}} \cdot \mathbf{x}(\mathbf{v})||^2_{\mathbf{\Sigma}_{\mathbf{y}}} \tag{7}$$

The statistical quality of the obtained flux estimate $\hat{\mathbf{v}}$ can be judged by computing sensitivity measures and confidence bounds. The key to the solution of these problems is obtained by implicit differentiation of equation (4):

$$\frac{\partial \mathbf{x}}{\partial \mathbf{v}_j}(\mathbf{v}) = -\mathbf{P}(\mathbf{v})^{-1} \cdot (\mathbf{P}_j \cdot \mathbf{x} + \mathbf{P}_j^{\text{inp}} \cdot \mathbf{x}^{\text{inp}}) \tag{8}$$

This shows that the same matrix inversion serves for both the computation of $\mathbf{x}$ and its derivatives with respect to v, which is also of great use for an efficient iterative solution of the least squares problem (7). Using (8) and a standard Gauß-Newton linearization argument (Seber, Wild, 1989), it is merely a technical problem to compute the combined partial derivative $\partial(\mathbf{w}, \mathbf{y})/\partial \mathbf{v}$ (on the assumption that no noise is present, i.e. $\varepsilon = \mathbf{0}$) and from this:

	the weighted output sensitivity	$\text{Sens}_{\mathbf{v}}^{\mathbf{w},\mathbf{y}}(\mathbf{v})$	$=$	$\sqrt{\mathbf{\Sigma}}^{-1} \cdot \frac{\partial(\mathbf{w},\mathbf{y})}{\partial \mathbf{v}}(\mathbf{v})$
	the estimator's covariance matrix	$\text{Cov}(\hat{\mathbf{v}})$	$\approx$	$\left[\text{Sens}_{\mathbf{v}}^{\mathbf{w},\mathbf{y}}(\hat{\mathbf{v}})^T \cdot \text{Sens}_{\mathbf{v}}^{\mathbf{w},\mathbf{y}}(\hat{\mathbf{v}})\right]^{-1}$
and	the weighted estimator's sensitivity	$\text{Sens}_{\mathbf{w},\mathbf{y}}^{\hat{\mathbf{v}}}$	$\approx$	$\text{Cov}(\hat{\mathbf{v}}) \cdot \text{Sens}_{\mathbf{v}}^{\mathbf{w},\mathbf{y}}(\hat{\mathbf{v}})^T$

The convergence of the parameter-fitting algorithm that has been applied to the data set in (Marx *et al.*, 1995) can be taken as an indicator of flux identifiability from the measured data. Sensitivity analysis then helped to detect redundant data sources and identifiability problems (these results will be presented elsewhere). However, they rely on a linearization argument and thus are purely local in nature. Moreover they represent *a posteriori* results in contrast to *a priori* results that make no use of concrete measurement data. To this end some generally applicable computer algebraic methods for global system analysis will now be presented. Finally, they will be applied to solve the already mentioned nonidentifiability problem that occured in the anaplerotic section.

5 IDENTIFIABILITY AND REDUNDANCY ANALYSIS

Firstly, a precise definition of the terms 'identifiability' and 'redundancy' has to be given. Aiming at the application of computer algebraic methods we will assume from now on that $\mathbf{w}, \mathbf{y}$ can be measured without any error (i.e. $\varepsilon_{\mathbf{w}} = \varepsilon_{\mathbf{y}} = \mathbf{0}$). Furthermore it will be assumed for simplicity that the entries of $\mathbf{v}, \mathbf{x}$ have been ordered such that the last components are the measured ones. Introducing the vectors $\mathbf{u}, \mathbf{z}$ of unknown fluxes or labels we get:

$$\mathbf{v} = \begin{pmatrix} \mathbf{u} \\ \mathbf{w} \end{pmatrix} = \begin{pmatrix} \mathbf{u} \\ \mathbf{M_w} \cdot \mathbf{v} \end{pmatrix} \quad \text{and} \quad \mathbf{x} = \begin{pmatrix} \mathbf{z} \\ \mathbf{y} \end{pmatrix} = \begin{pmatrix} \mathbf{z} \\ \mathbf{M_y} \cdot \mathbf{x} \end{pmatrix}.$$

The set of all flux and label vectors producing the observed output $(\mathbf{w}, \mathbf{y})$ then is

$$\mathbf{\Omega}(\mathbf{w}, \mathbf{y}) \stackrel{def}{=} \{(\mathbf{v}, \mathbf{x}) | \, \mathbf{P}(\mathbf{v}) \cdot \mathbf{x} + \mathbf{P}^{\mathrm{inp}}(\mathbf{v}) \cdot \mathbf{x}^{\mathrm{inp}} = \mathbf{0} \ \wedge \ \mathbf{M_w} \mathbf{v} = \mathbf{w} \ \wedge \ \mathbf{M_y} \mathbf{x} = \mathbf{y}\}.$$

Identifiability of $\mathbf{v}$ now precisely means that $\mathbf{\Omega}(\mathbf{w}, \mathbf{y})$ contains exactly one element.

Clearly, if the unknown fluxes $\mathbf{u}$ can be uniquely determined from $(\mathbf{w}, \mathbf{y})$, any further measurement data are redundant with $(\mathbf{w}, \mathbf{y})$ by (5). However, it is desirable to decide redundancy independently from the identifiability problem. By a *redundancy relation* we denote some parametrized equation of type $f_{(\mathbf{w},\mathbf{y})}(\mathbf{v}, \mathbf{x}) = 0$ that holds for all $(\mathbf{v}, \mathbf{x}) \in \mathbf{\Omega}(\mathbf{w}, \mathbf{y})$.

As proven by Wiechert and Schwingenheuer (1995) the identifiability and redundancy problems can be almost completely solved using the Gröbner base algorithm from polynomial ideal theory. However, the computational complexity of this algorithm is much too great to be applied to complex metabolic networks. For this reason, efficient methods are required for further complexity reduction. To this end we will restrict ourselves to linear algebraic methods that enable *linear* redundancy relations of type

$$f_{(\mathbf{w},\mathbf{y})}(\mathbf{v}, \mathbf{x}) = \eta(\mathbf{w}, \mathbf{y}) + \sum_{i=1}^{\dim \mathbf{v}} \eta_i^{\mathbf{v}}(\mathbf{w}, \mathbf{y}) \cdot \mathbf{v}_i + \sum_{i=1}^{\dim \mathbf{x}} \eta_i^{\mathbf{x}}(\mathbf{w}, \mathbf{y}) \cdot \mathbf{x}_i = 0 \quad \forall_{(\mathbf{v},\mathbf{x}) \in \mathbf{\Omega}(\mathbf{w},\mathbf{y})} \tag{9}$$

to be constructed. Clearly, each relation of this type can be considered for dimension reduction of $\mathbf{u}$ or $\mathbf{z}$, which simultaneously helps to treat the identifiability problem. Consequently, a combination of dimension reduction steps with the Gröbner base algorithm has the potential to treat even complex networks.

6 COMPUTER ALGEBRAIC ALGORITHMS

As a preliminary equation (5) can be reformulated using Cramer's rule:

$$\mathbf{x}_i \cdot \det \mathbf{P}(\mathbf{v}) = -\det\left(\mathbf{P}(\mathbf{v}) \boxed{i} \, \mathbf{P}^{\mathrm{inp}}(\mathbf{v}) \cdot \mathbf{x}^{\mathrm{inp}}\right), \quad i = 1, \ldots, \dim \mathbf{x} \tag{10}$$

where $\boxed{i}$ denotes replacement of the ith column by the vector on the right side. All entries of $\mathbf{P}(\mathbf{v})$ and $\mathbf{P}^{\text{inp}}(\mathbf{v}) \cdot \mathbf{x}^{\text{inp}}$ are linear combinations of fluxes $\mathbf{v}_i$. Consequently, both sides of (10) are polynomials of at most degree $\dim \mathbf{x}$ with respect to the $\mathbf{v}_i$. Now separating the unknown part $\mathbf{u}$ from the known part $\mathbf{w}$ of $\mathbf{v}$ and using multi-index notation (i.e. $\mathbf{u}^\alpha = \mathbf{u}_1^{i_1} \cdot \ldots \cdot \mathbf{u}_m^{i_m}$, $\alpha = (i_1, \ldots, i_m)$, $\#\alpha = i_1 + \ldots + i_m$) the left and right sides in equation (10) can be rewritten in the form

$$\mathbf{x}_i \cdot \sum_{\#\alpha \leq \dim \mathbf{x}} c^\alpha(\mathbf{w}) \cdot \mathbf{u}^\alpha = - \sum_{\#\alpha \leq \dim \mathbf{x}} c_i^\alpha(\mathbf{w}, \mathbf{x}^{\text{inp}}) \cdot \mathbf{u}^\alpha, \quad i = 1, \ldots, \dim \mathbf{x} \tag{11}$$

with certain polynomials c^α, c_i^α that do not depend on $\mathbf{u}$. Assume now that a linear relation of type

$$\eta = \sum_{i=1}^{\dim \mathbf{x}} \eta_i^{\mathbf{x}} \cdot \mathbf{x}_i \iff \eta \cdot \det \mathbf{P}(\mathbf{v}) = \sum_{i=1}^{\dim \mathbf{x}} \eta_i^{\mathbf{x}} \cdot \mathbf{x}_i \cdot \det \mathbf{P}(\mathbf{v}) \tag{12}$$

holds for all $\mathbf{u}$. Then we have from (10) and (11):

$$\eta \cdot \sum_{\#\alpha \leq \dim \mathbf{x}} c^\alpha(\mathbf{w}) \cdot \mathbf{u}^\alpha = - \sum_{i=1}^{\dim \mathbf{x}} \eta_i^{\mathbf{x}} \cdot \sum_{\#\alpha \leq \dim \mathbf{x}} c_i^\alpha(\mathbf{w}, \mathbf{x}^{\text{inp}}) \cdot \mathbf{u}^\alpha$$

This is a polynomial equation with respect to $\mathbf{u}$ and henceforth all coefficients corresponding to the same multi-index α must be equal. This leads to

Algorithm 1 *Compute a priori linear redundancy relations with respect to* $\mathbf{x}$.

Step 1: Compute all polynomials $c^\alpha(\mathbf{w}), c_i^\alpha(\mathbf{w}, \mathbf{x}^{\text{inp}})$ *for* $\#\alpha \leq \dim \mathbf{u}$.
Step 2: For each α *a linear equation* $\eta \cdot c^\alpha(\mathbf{w}) = -\sum_{i=1}^{\dim \mathbf{x}} \eta_i^{\mathbf{x}} \cdot c_i^\alpha(\mathbf{w}, \mathbf{x}^{\text{inp}})$ *holds. Use this equation set to determine a basis for the space of all equations of type (12).*

Clearly, steps 1 and 2 should be performed in succession for each new α. Because there are usually many more equations than variables $\eta, \eta_i^{\mathbf{x}}$, most generated equations will be trivial (i.e. $c^\alpha = 0 \wedge \forall_i c_i^\alpha = 0$) or redundant with the already computed solutions. A simple stopping criterion for this process (that cannot be given here for shortness) can be computed numerically using a Monte Carlo method. Moreover, efficient algorithms for symbolic determinant evaluation are discussed by Gielen and Sansen (1991).

The next algorithm determines mixed linear relations for $\mathbf{u}, \mathbf{z}$. To this end we split the atom transition matrices $\mathbf{P}_i$ from equation (4) into two parts $\mathbf{Q}_i, \mathbf{R}_i$ such that

$$\mathbf{P}_i \cdot \mathbf{x} = (\mathbf{Q}_i\ \mathbf{R}_i) \cdot \begin{pmatrix} \mathbf{z} \\ \mathbf{y} \end{pmatrix} = \mathbf{Q}_i \cdot \mathbf{z} + \mathbf{R}_i \cdot \mathbf{y}, \quad i = 1, \ldots, \dim \mathbf{v}.$$

Rearranging all completely unobservable bilinear terms $\mathbf{u}_i \cdot \mathbf{z}_j$ in (4) to the left side we get

$$- \sum_{i=1}^{\dim \mathbf{u}} \mathbf{u}_i \cdot \mathbf{Q}_i \cdot \mathbf{z} = \sum_{i=1}^{\dim \mathbf{w}} \mathbf{w}_i \cdot \mathbf{Q}_{i+\dim \mathbf{u}} \cdot \mathbf{z} + \sum_{i=1}^{\dim \mathbf{v}} \mathbf{v}_i \cdot \mathbf{R}_i \cdot \mathbf{y} + \sum_{i=1}^{\dim \mathbf{u}} \mathbf{v}_i \cdot \mathbf{P}_i^{\text{inp}} \cdot \mathbf{x}^{\text{inp}}. \tag{13}$$

This leads us to

Algorithm 2 *Compute a priori linear redundancy relations with respect to* $\mathbf{u}, \mathbf{z}$.

Step 1: Determine all matrices $\mathbf{Q}_i$, $i = 1, ..., \dim \mathbf{u}$.
Step 2: Determine an annihilation matrix $\mathbf{A}$ *fulfilling* $\mathbf{A}^T \cdot \mathbf{Q}_i = \mathbf{0}$ *for* $i = 1, ..., \dim \mathbf{u}$ *by computing a vector space complement of all column vectors of* $\mathbf{Q}_1, ..., \mathbf{Q}_{\dim \mathbf{u}}$.
Step 3: Multiply equation (4) with $\mathbf{A}$ *from the left thus producing a set of linear equations with respect to* $\mathbf{u}, \mathbf{z}$.

Since the matrices $\mathbf{P}_i, \mathbf{P}_i^{\text{inp}}$ are usually sparsely populated there is a good chance that this equation set is nontrivial. Clearly, this chance increases with the number of measured fluxes and labels.

7 APPLICATION EXAMPLE

Both algorithms will now be applied to the anaplerotic example. In (Marx *et al.*, 1995) the peripheral fluxes a, c, d could be indirectly estimated from the overall data set. Additionally, the fractional intermediary labels *pyr*, *oaa* and the input labels *gap*, *fum* could be measured (see Figure 1). By equation (1) this leaves only 2 degrees of freedom (represented by g, h) in the flux system.

Algorithm 1 is now used to compute all linear redundancy equations of type (12), i.e. $\eta = \eta_{pep} \cdot pep + \eta_{pyr} \cdot pyr + \eta_{oaa} \cdot oaa + \eta_{mal} \cdot mal$ with coefficients depending only on measured parameters. This vector space turns out to be 2-dimensional as determined by the equations

$$0 = \eta_{mal}, \quad \eta = -\frac{a \cdot gap + d \cdot fum}{a - c + d} \cdot \eta_{oaa}, \quad 0 = \eta_{pep} + \eta_{pyr} + \frac{c}{a - c + d} \cdot \eta_{oaa} \,.$$

As can easily be verified these conditions are fulfilled by the linearly independent equations

$$\begin{array}{rcl} 0 & = & pep \quad -pyr \\ a \cdot gap + d \cdot fum & = & \quad c \cdot pyr \quad +(a - c + d) \cdot oaa \end{array} \tag{14}$$

The first equation holds independently of the parameters a, c, d. This shows that it is never useful to measure both *pyr* and *pep*. The second equation turns out to be the systems overall label balance (remember that $a - c + d = f$ by (1)). As a result it is superfluous to measure *oaa* too. For dimensional reasons it follows that f, g cannot be simultaneously determined without further measurements.

Applying Algorithm 2 requires to compute the left side matrix of equation (13)), which is

$$-\sum_{i=1}^{\dim \mathbf{u}} \mathbf{u}_i \cdot \mathbf{Q}_i = h \cdot \begin{pmatrix} -1 & . \\ 1 & . \\ . & 1 \\ . & -1 \end{pmatrix} + g \cdot \begin{pmatrix} . & . \\ . & . \\ . & -1 \\ . & 1 \end{pmatrix} \implies \mathbf{A}^T = \begin{pmatrix} 1 & 1 & 0 & 0 \\ 0 & 0 & 1 & 1 \end{pmatrix}.$$

Multiplication of (2) with $\mathbf{A}^T$ yields (use (3)):

$$0 = \begin{pmatrix} \cdot & -a-h & h & \cdot \\ \cdot & a-c+h & -a+c-d-h & \cdot \end{pmatrix} \begin{pmatrix} pep \\ pyr \\ oaa \\ mal \end{pmatrix} + \begin{pmatrix} a & \cdot \\ \cdot & d \end{pmatrix} \begin{pmatrix} gap \\ fum \end{pmatrix}$$

From this two alternative solutions for h are obtained (which of course must be equal):

$$h_1 = a \cdot \frac{pyr - gap}{oaa - pyr}, \quad h_2 = (c - a) - d \cdot \frac{oaa - fum}{oaa - pyr}$$

Consequently, g is not identifiable from the measured data, which explains the identifiability problem that occurred with the data in (Marx *et al.*, 1995). It should be pointed out that the equality $h_1 = h_2$ produces exactly the already mentioned overall label balance in (14). On the other hand $pep = pyr$ is not reproduced with Algorithm 2. This shows that both algorithms have a different scope.

Although *mal* is currently not measurable it should be mentioned that g is identifiable in this situation. Algorithm 2 then produces: $mal = (h + a - c + d) \cdot (oaa - mal)/(pyr - mal)$

REFERENCES

Anderson, D.H. (1983) *Compartmental Modelling and Tracer Kinetics*, volume 50 of *Lecture notes in Biomathematics*. Springer.

Bailey, J.E. (1991) Towards a science of metabolic engineering. *Science*, **252**, 1668–75.

Gielen, G. and Sansen, W. (1991) *Symbolic Analysis for Automated Design of Analog Integrated Circuits*. Kluwer Academic Publishers.

Marx A., de Graaf A.A., Wiechert W., Eggeling L. and H. Sahm (1995). Determination of the Fluxes in Central Metabolism by ^{13}C NMR Combined with Metabolite Balancing. Submitted.

Seber, G.A.F. and Wild, C.J. (1989) *Nonlinear Regression*. Wiley.

Wiechert W., de Graaf A.A. and Marx A. (1995) In Vivo Stationary Flux Determination Using ^{13}C NMR Isotope Labelling Experiments, in *CAB 6, Computer Applications in Biotechnology* (ed. K. Schügerl and A. Munack), Pergamon. In Press.

W. Wiechert and V. Schwingenheuer. Algebraic methods for the analysis of redundancy and identifiability in metabolic ^{13}C labelling systems. 1995. in *Bioinformatik 1995* (ed. D. Schomburg). In Press.

Discrete Event Systems

14

Binding-time analysis applied to mathematical algorithms

Robert Glück, Ryo Nakashige*, Robert Zöchling°*
**DIKU, Dept. of Computer Science, University of Copenhagen*
Universitetsparken 1, DK-2100 Copenhagen Ø, Denmark.
°Inst. für Computersprachen, Vienna University of Technology,
Argentinierstr. 8, A-1040 Vienna, Austria.
e-mail: {glueck,ryon}@diku.dk, e1802gab@vm.univie.ac.at

Abstract

Our goal is to incorporate state-of-the-art partial evaluation in a library of general-purpose algorithms - in particular, mathematical algorithms - in order to allow the automatic creation of efficient, special-purpose programs. The main goal is efficiency: a specialized program often runs significantly faster than its generic version.

This paper shows how a binding-time analysis can be used to identify potential sources for specialization in mathematical algorithms. The method is surprisingly simple and effective. To demonstrate the effectiveness of this approach we used an automatic partial evaluator for Fortran that we developed. Results for five well-known algorithms show that some remarkable speedup factors can be obtained on a uniprocessor architecture.

Keywords

Scientific computing, numerical algorithms, partial evaluation, binding-time analysis.

1 INTRODUCTION

The application of partial evaluation to mathematical algorithms seems especially promising for several reasons. A large body of general-purpose algorithms is available (*e.g.* the NAG library contains more than 1000 mathematical algorithms). Their motivation clearly comes from the practical world of scientific computing, which has been dictating their development over a long time. High performance of mathematical algorithms is a key issue in most scientific and engineering applications.

Our goal is to incorporate state-of-the-art partial evaluation in a library of general-purpose mathematical algorithms in order to allow the automatic generation of fast, special-purpose programs from generic algorithms. This work is an attempt to capitalize on partial evaluation's ability to identify and extract static computations automatically from mathematical algorithms (Berlin &Weise, 1990; Baier et al., 1994; Andersen, 1995). Early examples of specializing numerical algorithms are provided by (Gustavson et al., 1970) and (Goad, 1982). Interprocedural constant propagation was applied to scientific applications in (Metzger & Stroud, 1993). They do not associate themselves with the partial evaluation paradigm, however.

We demonstrate how a binding-time analysis can be used to identify sources for specialization in general-purpose mathematical algorithms. To demonstrate the effectiveness of this approach we used an automatic partial evaluator for a subset of Fortran 77 which we developed (Kleinrubatscher et al., 1995). Our results show that this approach is strong enough to improve the efficiency of a certain class of mathematical problems.

2 PARTIAL EVALUATION AND BINDING-TIME ANALYSIS

Program Specialization. Assume that P is a general program with two arguments and that its first argument x is known (*static*) while its second argument y is unknown (*dynamic*). A program specializer produces a specialized program P_x that returns the same result when applied to the remaining input y as the original program P when applied to the input x and y, but potentially much faster.

Partial Evaluation is an automatic method for program specialization. In *offline* partial evaluation the transformation process is guided by a *binding-time analysis* performed prior to the specialization phase (Jones et al., 1993). The result of the binding-time analysis is a program in which all expressions are annotated as either static or dynamic. Operations annotated as static are performed at specialization time, while operations annotated as dynamic are delayed until run time (*i.e.* residual code is generated). Partial evaluation differs from ordinary optimizing compilers since it takes the *static input* of programs into account. Optimizing compilers lack binding-time information, thus it is unreasonable to expect a compiler to execute static statements and generate specialized programs.

Binding-TimeAnalysis. The analysis computes a division B of all variables X in a program P given an initial classification of the input variables as either static or dynamic. Variables classified as static depend only on static input variables. Variables classified as dynamic may depend on dynamic input variables.

Algorithm. (Monovariant Binding-Time Analysis) *Call the program variables* $X_1, \ldots, X_N$ *and assume that the input variables are* $X_1, \ldots, X_n$, *where* $1 \leq n \leq N$. *Assume that the binding-times* $\bar{b}_1, \ldots, \bar{b}_n$ *for the input variables are given, where* $\bar{b}_i$ *is either* S *(static) or* D *(dynamic). The task is to compute a congruent division for all program variables:* $B = (b_1, \ldots, b_N)$ *which satisfies* $\bar{b}_i = \mathrm{D} \Rightarrow b_i = \mathrm{D}$ *for the input variables. The analysis is done by the following algorithm:*

1. *Construct the initial division* $\overline{B} = (\bar{b}_1, \ldots, \bar{b}_n, \mathrm{S}, \ldots, \mathrm{S})$ *and set* $B = \overline{B}$.
2. *If the program contains an assignment* $X_k \leftarrow \mathrm{exp}$ *where the variable* X_j *appears in* exp *and* $b_j = \mathrm{D}$ *then set* $b_k = \mathrm{D}$ *in* B.
3. *Repeat step 2 until* B *does not change any longer. Then the algorithm terminates with congruent division* B.

3 PARTIAL EVALUATION OF MATHEMATICAL ALGORITHMS

The algorithms we studied can be classified roughly into one of the following categories (for a given S/D classifications of the input variables). A sequence of operations is *data-independent* if the control flow can be determined at specialization time and does not depend on numeric data. If the control flow is dynamic, then only few computations will be static in a loop since many of the operations will depend on the iteration variable. If the entire control flow and all computations are static, then specialization reduces to ordinary computation.

- Dynamic control flow / dynamic computations; *e.g.* Newton iteration.
- Partially static control flow / dynamic computations; *e.g.* PEQ (Section 4.3).
- Static control flow / dynamic computations; *e.g.* CSI with n static (Section 4.2).
- Static control flow / partially static computations; *e.g.* FFT, CC (Section 4.4, 4.5).

Effects and Limitations. The specialization effects we observed are typically due to *unfolding* (unrolling) of loops and *elimination* of conditionals, *interprocedural constant propagation* (as opposed to intraprocedural), *procedure specialization* (cloning), *precomputation* of *indices* and *coefficients* (*e.g.* involving trigonometric functions). Our experience shows that specialized programs often enable further compiler optimizations (*e.g.* when array indices become known); thus, the choice of the compiler optimization level affects the speedup.

The gain in efficiency has its price (program size) and it is not always desirable to fully unfold loops since the number of iteration may be extremely large; *e.g.* due to the stability condition for solving parabolic equations the number of time steps M must be very large and unfolding the M-bound loop is unacceptable (Section 4.3). Other numerical methods, such as the Romberg integration (Section 4.1) or the Chebyshev approximation (Section 4.5), converge much faster; hence unfolding may be practical.

Partial evaluation is not very effective when computations are extremely data-dependent. For example, in techniques for linear programming the choice of the pivot is not known before run-time, but depending on this choice different computations have to be performed. However, specialization may still remove boundary checks and precompute indices.

4 BINDING-TIME ANALYSIS OF MATHEMATICAL ALGORITHMS

To demonstrate the effectiveness of the approach we chose five well-known algorithms from different subject areas: numerical integration, partial differential equations, function approximation and interpolation.[1] We refer to the literature for a description of the mathematical methods; the algorithms were taken from (Kincaid & Cheney, 1991), (Press et al., 1993). Details about the Fortran partial evaluator can be found in (Kleinrubatscher et al., 1995); a similar system exists for C (Andersen, 1994).

4.1 Romberg integration

The Romberg integration (RI) approximates the integral of a function f in an interval $[a, b]$ using trapezoidal estimates. The input of the RI is the lower and upper limit a,b of the interval, the number of iterations M and values for the function f.

[1]The algorithms were written in Fortran 77. The run times (= user time + system time) are given in seconds using the HP-UX Fortran 77 compiler (optimization option: +02 for CSI and FFT, +01 otherwise) and a HP workstation (Apollo 9000, 99 MHz PA-RISC 7100). The size of the programs is lines of 'pretty-printed' text and KBytes of executable code.

Analysis. We consider the S/D classification where M is static. The control-flow is entirely data-independent where all index computations and the weight factor $(4^m - 1)$ are static. The numerical computations are dominated by the computation of the quantities $R(n,0)$ from f. In the remainder of the algorithm additional quantities $R(n,m)$ are to be computed iteratively $(1 \leq n \leq M, 1 \leq m \leq n)$. The interval boundaries a,b are only required for computing the initial quantities $R(n,0)$.

Static	Computations saved	Size estimation
M	test of 3 iterations index computations divisor $(4^m - 1)$	unfolding 3 iterations $O(2^M)$

Results. The results show speedup factors between 1.16 and 1.46 for static M (Table 1), where M ranges from 2 to 10 (run-times for 10000 repetitions). The code size grows exponentially $O(2^M)$, but the Romberg integration converges fast to the integral of f; hence only a moderate value of M isneeded.

4.2 Cubic splines interpolation

The cubic splines interpolation (CSI) approximates a function using a cubic polynomial. The input of the CSI is the number of points n, their x-coordinates $x[\,]$ and their y-coordinates $y[\,]$. The output is the second derivative for each point (the computation of the y-coordinate for an arbitrary point is then straightforward). The algorithm uses a natural cubic spline.

Analysis. We consider two S/D classifications of the input variables. Motivation: the number of points is often known beforehand and the x-coordinates may be fixed when regular measurements are taken. The control-flow is entirely data-independent and all iterations can be unfolded. If the x-coordinates $x[\,]$ are static, a series of coefficients can be precomputed.

Static	Computations saved	Size estimation
n	tests of 3 iterations index computations	unfolding 3 iterations $O(n) \approx n(\lvert\mathrm{for}_{i_1}\rvert + \lvert\mathrm{for}_{i_2}\rvert + \lvert\mathrm{for}_{i_3}\rvert)$
$n, x[\,]$	tests of 3 iteration index computations coefficient computations	unfolding 3 iterations $O(n) \approx n(\lvert\mathrm{for}_{i_1}\rvert + \lvert\mathrm{for}_{i_2}\rvert + \lvert\mathrm{for}_{i_3}\rvert)$

Results. The results for the CSI show speedup factor between 1.16 and 1.63 for n static (Table 3), and 1.63 and 2.19 for n and $x[\,]$ static (Table 4), where n ranges from 10 to 1000 (run-times for 100000 repetitions). Note how an additional static $x[\,]$ reduces the size of the residual programs while increasing the speedup. The code size grows linearly $O(n)$.

4.3 Partial differential equations of parabolic type

Partial differential equations are usually solved using the finite difference method. The approximative values of the solution function are computed at so-called *mesh points* (x_i, t_j) and the numerical solution is advanced step by step in the time-direction. The input for solving parabolic equations (PEQ) with the explicit method is the number of mesh points n (x-direction), the step size k and the number of steps M (t-direction), as well as the values $v_{i,0}$ of the initial profile and the boundary values $v_{0,j}$, $v_{n+1,j}$ $(1 \leq i \leq n, 0 \leq j \leq M)$. The output are the values $v_{i,M}$ of the last iteration.

For the algorithm to be stable, it is necessary to assume $k \leq \frac{1}{2}(\frac{1}{n+1})^2$. For example, if $n = 99$ then the largest permissible value for step size $k = \frac{1}{2}10^{-4}$. A solution for $0 \leq t \leq 10$ then requires $M \geq \frac{1}{5}10^6$. Unfolding the M-bound iteration is therefore unacceptable.

Analysis. We consider the S/D classification where the number n is static. The control-flow is entirely data-independent, all numerical operations are dynamic. Making more parameters static will not contribute to the speedup: the initial profile is used only in the first iteration and the boundary values only for computing two new values in each iteration (moreover, we do not want to unfold the M-bound iteration).

Static	Computations saved	Size estimation
n	tests of 3 iterations index computations	unfolding 3 iterations $O(n) \approx n(\lvert\text{for}_\text{l}\rvert + \lvert\text{for}_{\text{i}_1}\rvert + \lvert\text{for}_{\text{i}_2}\rvert)$

Results. The results show speedup factors between 1.63 and 1.85 for static n (Table 2), where n ranges from 10 to 1000 (run-times for 10000 repetitions). The computation of new mesh points is straightforward involving only arithmetic operations; this is reflected in the speedup. The code size grows linearly $O(n)$.

4.4 Fast Fourier transformation

The fast Fourier transformation (FFT) approximates the Fourier transformation using N points. The FFT is the fastest known algorithm for calculating a discrete Fourier Transformation. The input is a sequence of points (given as a function f) and a variable N indicating the number of points. The output is the Fourier coefficient at each point.

Analysis. We consider the S/D classification where $N(= 2^m)$ is static. The FFT is entirely data-independent with a significant number of numerical operations depending only on the number of points.

Static	Computations saved	Size estimation
$N(= 2^m)$	tests of 5 iterations computations of w, Z index computations using j, k, n index computations for $C[\,]$, $D[\,]$	unfolding 5 iterations $O(N \log_2 N) \ll O(N^2)$

Results. The results for the FFT show speedup factors between 1.83 and 5.05 for static N (Table 5), where N ranges from 16 to 512 (run-times for 10000 repetitions). The computational costs of the FFT are mirrored in the growth of code size $O(N \log_2 N)$. Note the good speedup for larger N despite the growth in code size.

4.5 Chebyshev approximation

Chebyshev polynomials approximate continuous functions in a given interval. Once the Chebyshev coefficients (CC) are determined the approximation of $f(x)$ for arbitrary x is straightforward. They yield the 'most accurate' approximation of degree n. The input of the algorithm computing the CC is the lower and upper limit a,b of the interval, the maximum degree n of the Chebyshev polynomials, and f. The output are the Chebyshev coefficients.

Analysis. We consider the S/D classification where n is static. The CC is entirely data-independent with a significant number of numerical computations depending only on the degree n (*cf.* Section 4.4).

Static	Computations saved	Size estimation
n	tests of 3 iteration index computations coefficients (cos)	unfolding 3 iterations $O(n^2) \approx n^2(\lvert \mathrm{for_j} \rvert \lvert \mathrm{for_k} \rvert)$

Results. The results show remarkable speedup factors between 8.50 and 14.8 for static n (Table 6), where n ranges from 10 to 100 (run-time for 10000 repetitions). The speedup is mostly due to the precomputation of the trigonometric coefficients. To determine their influence, we removed their computation: the speedup was still 4.2. This shows that the specialization effect depends strongly on the efficiency of the standard functions in the mathematical library, but also gives a surprisingly good lower bound for the speedup. The code size grows $O(n^2)$, but the approximation error decreases rapidly; hence only a moderate degree n of the polynomial is needed (*e.g.* 30 or 50).

5 FURTHER WORK

Further work is desirable in several directions. Partial evaluators for numerical programs could take advantage of additional knowledge about mathematical and scientific functions, and exploit algebraic simplifications. Additional precision could be gained by online partial evaluation techniques and exploiting information about partially static data structures (*e.g.* sparse matrices). A combination of partial evaluation and traditional compiler optimizations seems promising; more should be known about their interaction. We expect that these techniques will improve the results presented in this paper. Another interesting direction is to exploit the parallelism exposed by partial evaluation on parallel computers.

REFERENCES

Andersen, L.O. (1994) *Program Analysis and Specialization for the C Programming Language.* DIKU, Department of Computer Science, University of Copenhagen. DIKU Report 94/19.

Andersen, P.H. (1995) *Partial Evaluation Applied to Ray Tracing.* DIKU, Department of Computer Science, University of Copenhagen. DIKU Report 95/2.

Baier, R., Glück, R., Zöchling, R. (1994) Partial evaluation of numerical programs in Fortran, in *ACM SIGPLAN Workshop on Partial Evaluation and Semantics-Based Program Manipulation.* University of Melbourne, Technical Report 94/9.

Berlin, A., Weise, D. (1990) Compiling scientific code using partial evaluation. *IEEE Computer,* **23**(12), 25–37.

Goad, C. (1982) Automatic construction of special purpose programs, in *6th Conference on Automated Deduction* (ed. D.W. Loveland), LNCS, vol. 138, Springer-Verlag.

Gustavson, F.G., Linige,r W., Willoughby, A.R. (1970) Symbolic generation of an optimal Crout algorithm for sparse systems of linear equations. *J. of the ACM,* **17**(1), 87–109.

Jones, N.D., Gomard, C.K., Sestoft, P. (1993) *Partial Evaluation and Automatic Program Generation.* International Series in Computer Science. Prentice Hall.

Kincaid, D., Cheney, W. (1991) *Numerical Analysis: Mathematics of Scientific Computing.* Brooks/Cole.

Kleinrubatscher, P., Kriegshaber, A., Zöchling, R., Glück, R. (1995) Fortran program specialization. *SIGPLAN Notices,* **30**(4), 61–70.

Metzger, R., Stroud, S. (1993) Interprocedural constant propagation: an empirical study. *ACM Letters on Programming Languages and Systems,* **2**(1–4), 213–32.

Press, W.H., Teukolsky, S.A., Vetterling, W.T., Flannery, B.P. (1993) *Numerical Recipes in Fortran: the Art of Scientific Computing.* 2nd ed., Cambridge University Press.

APPENDIX 1 RESULTS

Table 1 Romberg (M static)

RI	Time	Ratio	Lines	KB
Src	0.16		99	20.5
Res, M=2	0.12	**1.33**	52	20.5
Src	0.44		99	20.5
Res, M=4	0.33	**1.33**	112	20.5
Src	1.24		99	20.5
Res, M=6	0.85	**1.46**	252	24.6
Src	3.98		99	20.5
Res, M=8	3.09	**1.29**	688	36.9
Src	14.37		99	20.5
Res, M=10	12.24	**1.17**	2284	86.0
Src	54.79		99	20.5
Res, M=12	47.42	**1.16**	8496	282.6

Table 2 Parabolic equation (n static)

PEQ	Time	Ratio	Lines	KB
Src	6.69		86	20.5
Res, n=10	4.11	**1.63**	115	20.5
Src	29.14		86	20.5
Res, n=50	16.76	**1.74**	315	28.7
Src	57.45		86	20.5
Res, n=100	32.44	**1.77**	565	36.9
Src	113.26		86	20.5
Res, n=200	63.98	**1.77**	1065	57.3
Src	169.98		86	20.5
Res, n=300	96.70	**1.76**	1565	77.8
Src	298.62		86	20.5
Res, n=500	161.82	**1.85**	2565	114.7
Src	587.82		86	20.5
Res, n=1000	319.02	**1.84**	5065	217.1

Table 3 Cubic splines (n static)

CSI	Time	Ratio	Lines	KB
Src	0.57		74	20.5
Res, $n = 10$	0.35	**1.63**	94	20.5
Src	3.16		74	20.5
Res, $n = 50$	2.15	**1.47**	457	28.7
Src	5.12		74	20.5
Res, $n = 100$	4.35	**1.18**	907	36.9
Src	10.25		74	20.5
Res, $n = 200$	8.66	**1.18**	1807	57.3
Src	15.46		74	20.5
Res, $n = 300$	13.28	**1.16**	2707	77.8
Src	25.60		74	20.5
Res, $n = 500$	22.03	**1.16**	4507	118.8
Src	52.28		74	20.5
Res, $n = 1000$	45.02	**1.16**	9007	262.1

Table 4 Cubic splines ($n, x[]$ static)

CSI	Time	Ratio	Lines	KB
Src	0.57		74	20.5
Res, $n = 10, x$	0.30	**1.90**	74	20.5
Src	3.16		74	20.5
Res, $n = 50, x$	1.44	**2.19**	354	24.6
Src	5.12		74	20.5
Res, $n = 100, x$	3.02	**1.70**	704	28.7
Src	10.25		74	20.5
Res, $n = 200, x$	6.28	**1.63**	1404	41.0
Src	15.46		74	20.5
Res, $n = 300, x$	9.29	**1.66**	2104	49.2
Src	25.60		74	20.5
Res, $n = 500, x$	15.73	**1.63**	3504	73.7
Src	52.28		74	20.5
Res, $n = 1000, x$	30.59	**1.71**	7004	122.9

Table 5 FFT (N static)

FFT	Time	Ratio	Lines	KB
Src	2.17		151	20.5
Res, N=16	0.43	**5.05**	701	24.6
Src	5.71		151	20.5
Res, N=32	1.37	**4.17**	1469	36.9
Src	15.27		151	20.5
Res, N=64	4.29	**3.56**	3069	57.3
Src	40.68		151	20.5
Res, N=128	14.22	**2.86**	6397	102.4
Src	115.45		151	20.5
Res, N=256	50.31	**2.29**	13309	192.5
Src	1183.78		151	20.5
Res, N=512	647.72	**1.83**	27645	720.9

Table 6 Chebyshev (n static)

CC	Time	Ratio	Lines	KB
Src	1.70		72	20.5
Res, n=10	0.20	**8.50**	170	20.5
Src	6.21		72	20.5
Res, n=20	0.55	**11.3**	520	28.7
Src	13.70		72	20.5
Res, n=30	1.22	**11.2**	1070	41.0
Src	23.96		72	20.5
Res, n=40	2.38	**10.1**	1820	57.3
Src	38.75		72	20.5
Res, n=50	2.62	**14.8**	2770	81.9
Src	145.14		72	20.5
Res, n=100	14.74	**9.85**	10520	335.8

APPENDIX 2 ANNOTATED ALGORITHMS

The algorithms are presented in a pseudo code containing additional details beyond the pure mathematical formula; see Kincaid & Cheney (1991), Press et al. (1993). The program annotations were determined automatically by the monovariant binding-time analysis (Section 2).

Notation. The for $i = e_1, e_2, e_3$ do ... end denotes an iteration from e_1 to e_2 with a step of e_3 (if e_3 is omitted then 1 is assumed). Dynamic variables and operations are annotated as ***op***; all other operations *op* are static (*i.e.* they can be precomputed at specialization time).

Romberg integral (M static)

```
input a, b, M, f
integer M, n, m, i
real h, s
real function f
real array r[ , ]
h     ← b−a
r[0,0] ← (f(a)+f(b))*h/2
for n = 1, M do
   h    ← h/2
   s    ← 0
   for i = 1, 2**(n − 1) do
      s    ← s+f(a+(2 * i − 1)*h)
   end
   r[n, 0] ← r[n − 1, 0]/2+h*s
   for m = 1, n do
      r[n, m] ← r[n, m − 1]+ 1/(4**m−1) *
                  (r[n, m − 1]−r[n − 1, m − 1])
   end
end
output r[M, M]
```

Cubic splines interpolation ($n, x[\,]$ static)

```
input n, x[ ], y[ ]
integer i, n
real array b, h, x, u, v, y, z
for i = 0, n − 1 do
   h[i] ← x[i + 1] − x[i]
   b[i] ← (6/h[i])*(y[i + 1]−y[i])
end
u[1]← 2 * (h[0] + h[1])
v[1]← b[1]−b[0]
for i = 2, n − 1 do
   u[i]← 2 * (h[i] + h[i − 1])−h[i − 1]**2/u[i − 1]
   v[i] ← b[i]−b[i − 1]−h[i − 1]*v[i − 1]/u[i − 1]
end
z[n]← 0
for i = n − 1, 1, −1 do
   z[i] ← (v[i]−h[i]*z[i + 1])/u[i]
end
z[0]← 0
output z[ ]
```

Parabolic equation (n static)

```
input n, k, M
integer n, M, l, j
real k, h, s, ss, t
real function g, a, b
real array v, w
h     ← 1/(n + 1)
s     ← k/(h**2)
ss    ← 1−2*s
for l = 0, n + 1 do
   w[l] ← g(l)
end
t   ← 0
for j=1, M do
   t     ← t+k
   v[0]← a(t)
   v[n + 1] ← b(t)
   for i = 1, n do
      v[i] ← s*w[i − 1]+ss*w[i]+s*w[i + 1]
   end
   for i = 0, n + 1 do
      w[i] ← v[i]
   end
end
output v[ ]
```

FFT ($N = 2^m$ static)

```
input m, f
integer j, k, m, n, N
real const π = 3.14...
complex u, v, w
complex array C, D, Z
complex function f
N  ← 2**m
w  ← exp(−2 * π * √−1/N)
for k = 0, N − 1 do
   Z[k] ← w**k
   C[k] ← f(2 * π * k/N)
end
for n = 0, m − 1 do
   for k = 0, 2**(m − n − 1) − 1 do
      for j = 0, 2**n − 1 do
         u  ← C[2**n * k + j]
         v  ← Z[j * 2**(m − n − 1)]*
                C[2**n * k + 2**(m − 1) + j]
         D[2**(n + 1) * k + j] ← (u + v)/2
         D[2**(n + 1) * k + j + 2**n]
                              ← (u − v)/2
      end
   end
   for j = 0, N − 1 do
      C[j] ← D[j]
   end
end
output C[]
```

Chebyshev approximation (n static)

```
input n, xa, xb, func
integer n, k, j
real xa, xb, xm, xp, sm
real function func
real array c, f
real const π = 3.14...
xp ← (xb+xa)/2
xm← (xb−xa)/2
for k = 1, n do
   f[k] ← func(xp+xm* cos(π * (k − 0.5)/n))
end
for j = 0, n − 1 do
   sm ← 0
   for k = 1, n do
      sm ← sm+f[k]* cos(π * j * (k − 0.5)/n)
   end
   c[j] ← (2/n)*sm
end
output c[]
```

Invariant state progress and relation modelling of DEDS

Gabriel Juhás and Miroslav Kocian
Institute of Control Theory and Robotics, Slovak Academy of Sciences, Dúbravská cesta 9, 842 37 Bratislava, Slovak Republic.
Tel: 42-7-378 3216. Fax: 42-7-376045.
e-mail: `utrrjuhi@savba.sk`

Abstract

The states are used as the fundamental concept (instead of events) and the discrete state change systems (DSCSs) are defined. The possibility of DSCSs modelling, control and description of behaviour specifications using the presented state relation approach is investigated. The generalized T-invariants of Petri nets are defined as a special case of the presented approach.

Keywords

Discrete event system, relation, Petri nets, T-invariants

1 INTRODUCTION

From a historical general system theory point of view, the DEDSs follow the state/transition systems formally defined by a state set and binary relation on this set (Klir, 1972). In the paper it is tried to use the notion of (binary) relation (between states) in the DEDS modelling and control in similar way like predicate calculus is used (Kumar et al., 1993). The simple set structure given in (Klir, 1972) is used and related calculus is developed. The state space and state changes (instead of events) are used as fundamental concepts, consequently a discrete state change system (DSCS) is defined instead of DEDS. An example of the approach is shown on the cat and mouse maze problem (Ramadge and Wonham, 1989; Capkovic, 1993).

Generalized T-invariants of Petri nets can be considered as the special case of the presented theory, if the system is modeled by a Petri net. By means of generalized T-invariants it is possible to describe the relevant kind of discrete processes, where the cycle is understood to repeatedly reach states that are in the given invariance relation. The behaviour according to this relation determines the state progress. The presented definition of T- invariants is more global than the standard definition that can be included in it as the special case.

2 RELATION MODELLING OF DEDS

2.1 Discrete state change systems

Let D is the state set of a given system *D*. The notion direct discrete state change (or shortly direct state change) of a state $d \in D$ to a state $d' \in D$ expresses the fact, that during this state change the system is not found in any state $d'' \in D$.

A discrete state change system (DSCS) is a system, where each state change of the system is arisen from a sequence of direct state changes. If the time is ignored and only the order of states is considered, then each possible state trajectory of DSCS can be reduced to a sequence of states that for each pair of neighbours in this sequence holds that direct state change from the first element of this pair to the second element is possible. As follows, these sequences are called the logical state trajectories of the DSCS. A (logical) state trajectory that there is considered to run the time to infinity, is called total (logical) state trajectory. The notion (logical) behaviour means the set of all possible (logical) state trajectories of the given DSCS.

Similarly to (Ramadge and Wonham, 1989), in control of DSCS it is premised that certain direct discrete state changes of the system can be disabled when it is desired.

2.2 Relation state change generator

Let S is an arbitrary set and L is an arbitrary binary relation on S x S.

Let $L(s,.)$ is a relation given so that $L(s,.) = L \cap (\{s\} x\ S)$ and let $L(.,s) = L \cap (S\ x\ \{s\})$.

The support of a binary relation L is given by the union of the first and the second domain of it and denoted by L_{supp} ($L_{supp} = \{s;\ \exists\ s': [s, s'] \in L \vee [s',s] \in L\}$).

The image of an element s in a relation L is denoted by $L(s)$ ($L(s) = \{s';\ [s, s'] \in L\}$).

Binary relation $L^k = \{\ [s_1, s_{k+1}]\ ;\ \exists \{s_i\}_{i=1}^{k+1}: [s_i, s_{i+1}] \in L$ for $\forall\ i \in \{1, ..., k\}\}$ (k is a positive integer), is called k-power of L. The symbol $L^{\otimes}$ denotes the binary relation that is the union of all powers of the relation L ($L^{\otimes} = \bigcup_{k \in Z^+} L^k$, where Z^+ is denoting the set of positive integers).

The notion L-sequence denotes a sequence of elements from S $\{s_i\}_{i=1}^{N}$, where $N \in Z^+ \cup \{\infty\}$ that next two conditions are satisfied:

1. If $N > 1$ then $\forall\ i \in \{2, .. N\}$: $[s_{i-1}, s_i] \in L$. 2. If $N \neq \infty$ then $L(s_N,.)$ is the empty set.

The sequence for which holds only the first condition is called part of L-sequence.

Let $s = \{s_i\}_{i=1}^{N}$ is a sequence of elements from S so that there exists some L-sequence $q = \{s_{k_j}\}_{j=1}^{M}$, $M \in Z^+ \cup \{\infty\}$, q is selected from s and next three conditions hold:

1. $k_1 = 1$. 2. If $N = \infty$ then $M = \infty$. 3. If $N \neq \infty$ then $k_M = N$.

Then s is denoted by symbol (L)-sequence. It is said that s is containing q and on the contrary q is contained in s. If q is a part of L-sequence, then s is called part of (L)-sequence.

A binary relation ${}^{\{s}L = L(s,.) \cup \{[s',s'']; [s',s''] \in L \wedge [s,s'] \in L^{\otimes}\}$ is represented in the graph of the binary relation L by the tree with the root labelled by an element $s \in S$.

Definition: Relation state change generator is a two tuple $S = (S, F)$ where S is a set of states and F is a direct discrete state change relation on S x S. Elements of the relation F are called possible direct discrete state changes of S. F-sequences are called total state trajectories of the generator S and parts of F-sequences are called state trajectories the generator S. By the notion the behaviour of generator S is understood the set of its all state trajectories.

A state s' is reachable from a state s in S if and only if $[s, s'] \in F^{\otimes}$ (i.e. iff $\exists\, k \in Z^+$: $[s, s'] \in F^k$). The binary relation $F^{\otimes}$ is called the reachability relation.

Definition: Let $S = (S, F)$ is a generator. A state $s' \in S$ is live according to a state $s \in S$ in generator S iff for each F-sequence $\{s_i\}_{i=1}^{N}$, where $N \in Z^+ \cup \{\infty\}$, holds the statement: If $\exists$ k $\leq N$: $s_k = s \Rightarrow \exists\, l, k < l \leq N$: $s_l = s'$. If s' is live according to each state from S, then s' is live in S and if each state of S is live in S, then it is said that the generator S is live.

Definition: Let $P = E \times E^*$, where E is an arbitrary set and E* is the set of all sequences of elements from E. Let G is a relation on P x P so, that for each $[[e, q], [e', q']] \in G$ holds that if $q = [\{e_i\}_{i=1}^{N}]$ and $N \in Z^+$, then $q' = \{e_i\}_{i=1}^{N+1}$, where $e_{N+1} = e'$. Then the relation state change generator $P = (P, G)$ is called relation state change generator with memory, the set E is called pure state set of P and the set E* is called memory state set of P. A state trajectory of the generator P with first element $[e,e] \in ExE$ is called the state trajectory starting from the state e with the reseted memory.

2.3 Modelling and control of DSCS via relation generators

To model the behaviour of a DSCS it is used a generator, where states of the given system are represented by states of the generator, and the possible direct changes of the modelled system are represented by elements of direct discrete state change relation of the generator (direct modelling). If the possibility of direct changes of an actual state depends on a state history of the system, then it is used as the model a generator with memory, where the state history of the modelled system is represented by a state of memory of the generator (memorial modelling). The memorial modelling can be used to model logical behaviour and the control of each DEDS that can be modelled by a formal language.

The relation between a generator (S, F) modelling some system D and a generator (S, C) modelling this system D controlled by some supervisor is briefly formulated as follows: for each $s \in C_{supp}$ holds that $F_u(s) \subseteq C(s) \subseteq F(s)$, where elements of F_u represent uncontrollable direct state changes of the system D.

Definition: Let $S = (S, F)$ is a generator. Let relation $F_u \subseteq F$. Let L, $L \subseteq F$ is a binary relation on S x S and let $s \in L_{supp}$. It is said that L is controllable in the point s in regard to F_u if and only if $F_u(s,.) \subseteq L(s,.)$. If L is controllable in each point of its support in regard to F_u, then it is called controllable in regard to F_u.

Definition: Let $S = (S, F)$ is a generator. Let relation $F_u \subseteq F$. Let s, $s' \in S$. Iff exists a controllable relation C in regard to F_u that state $s' \in S$ is live according to the state s in generator (S, C), then s' is controllably reachable from s in regard to F_u in the generator S = (S, F). If, in addition, holds that s' is the end point of C_{supp} (C(s') is the empty set) then s' is controllably stabilizable from the state s in regard to F_u in the generator S.

2.4 Behavioural specifications and relation calculus for DSCS

The main elements of possible problem settings are properties of the system model, relations and their properties, relationships between the system model and relations and mutual relationships between relations.

Some relationships between relations can be formulated as follows. Let L and K are binary relations on some cartesian product S x S of some set S.

1. $L \subseteq K^{\otimes}$ (informally it means that K can go through L)
1.1. $L \subseteq K$ (K can detailly go through K)
1.1.1. $L = K$ (K have to go detailly through L)
1.2. $L \subseteq K^{\otimes} \wedge \forall$ K-sequence $\{s_i\}_{i=1}^{N}$ so that $s_1 \in L_{supp}$ is (L)-sequence (K starting in the support of L have to go through L)
1.2.1. $L = K^{\otimes}$ (K have to go through L)

The other relevant problems consist in an investigation of relationships between some relations L, H that both are subsets of any relation $K^{\otimes}$.

If above mentioned relation K is the relation of direct state changes of some system modelling generator, then the relationships between L and K represent relationships between relation L and the system model. The considered system can or have to go (detailly) through the relation L. The relation L with some mentioned relationship to the system can characterise a possible and/or required behaviour. From this point of view the behaviour of the system given by relation L fulfilling the criteria 1.2. can be understood as partial observable.

As follows, let L, K are arbitrary relations on S x S. If for each L-sequence q exists K-sequence s that s is containing q, then $L \subseteq K^{\otimes}$ (K can go through L). If, in addition, K is transitive, then $L \subseteq K$. Let $\mathcal{S}$ = (S, F) is a generator. If for each $s \in F_{supp}$ holds that the cardinal number of F(s) is equal to one, F is coherent and $F^{\otimes}$ is reflexive, then the generator is live.

Logical expressions about properties of a relation K can be divided into two classes. Let V(K) denotes a logical expression about a relation K.

1. If V(K) = 1 (true) and for each $L \subseteq K$ holds V(L) = 1 then it is said that V(K) is an expression of the class one.
2. If V(K) = 1 and there exists $L \subseteq K$ that V(L) = 0 (false) then V(K) is an expression of the class two.

The following assertion can be use in the modular problem solving.

Let $H_i = \{1, .., n\}$, $n \in Z^+$ are binary relations on S x S. Let $V_i(H_i)$ are expressions of the class one for all $i \in \{1,..,n\}$, then $V(\bigcap_{i=\{1,..,n\}} H_i) = V_1(\bigcap_{i=\{1,..,n\}} H_i) \wedge .. \wedge V_n(\bigcap_{i=\{1,..,n\}} H_i) = 1$.

The notion introduced in the paper can be used to derive a proof methodology to examine properties of relations if the other properties (of these relations and/or other relations) are given. The proof methodology is based on the definitions of basic proofs rules so that by the combinations of them more complex proofs can be constructed. Assumptions of proof rules can be given and/or derived by previous applications of proof rules and/or derived from the detail modelling level. The nature of basic proof rules can be shown as follows. Let H, K, L are some relations on S x S, so that $K \subseteq H^{\otimes}$. Let holds: if $K \subseteq H^{\otimes}$, then $L \subseteq H^{\otimes}$. Then $L \subseteq H^{\otimes}$.

Let L is an arbitrary relation on S x S. Let $s_1, s_2, s_3 \in L_{supp}$ and $(s_1, s_2) \in L^{\otimes}$. Let for each $s \in L_{supp}$ holds: if $(s, s_2) \in L^{\otimes}$ then $(s, s_3) \in L^{\otimes}$. Then $(s_1, s_3) \in L^{\otimes}$.

Let H, K, L are some relations on S x S, so that $K \subseteq H^{\otimes}$ (H can go through K) and $L \subseteq K^{\otimes}$ (K can go through L). Then $L \subseteq H^{\otimes}$. This assertion can be used to introduce some type of hierarchy of relations on S x S and hierarchy of the behaviour description.

2.5 The cat and mouse maze example

A cat and mouse are placed in the maze on the Figure 1. Let γ is a function, that maps rooms of the maze onto a set of numbers $A = \{0,1,2,3,6\}$ in accordance with Figure 1.

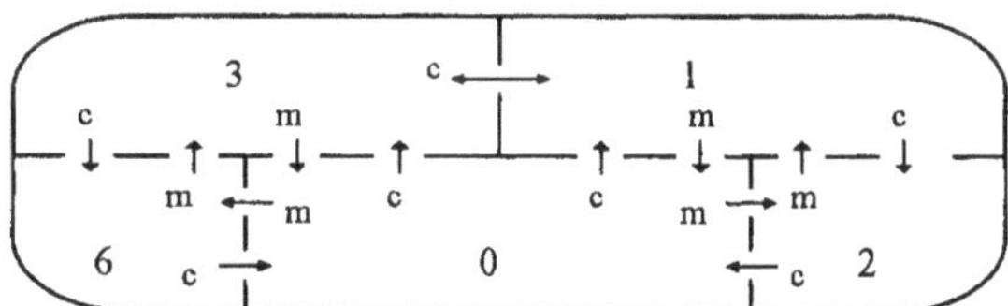

Figure 1 The cat and mouse maze.

Let set $B = A \cup \{7\}$. Let set $S = A \times B$ so, that each pair $[a,b] \in S$ represents the state, in which the cat is in the room mapped by γ onto a, and the free mouse is in the room mapped by γ onto b (if $b \in A$) or the cat has caught the mouse (if $b = 7$).

Let F_u, F_c and F are binary relations on S x S given as follows.

$F_u = \{[[a,b], [a',b']];\ (((a = 1 \wedge a' = 3) \vee (a = 3 \wedge a' = 1)) \wedge b' = b \wedge b \in B) \vee (b = a \wedge a' = a \wedge b' = 7 \wedge a \in A)\}$

$$\begin{aligned} F_c = \{[[a,b], [a',b']];\ & ([a',b'] = [(a+1)_{mod\,3}, b] \wedge a \in \{0,1,2\} \wedge b \in B) \vee \\ & ([a',b'] = [a, (b-1)_{mod\,3}] \wedge a \in A \wedge b \in \{0,1,2\}) \vee \\ & ([a',b'] = [(a+3)_{mod\,9}, b] \wedge a \in \{0,3,6\} \wedge b \in B) \vee \\ & ([a',b'] = [a, (b-3)_{mod\,9},] \wedge a \in A \wedge b \in \{0,3,6\})\} \end{aligned}$$

and $F = F_u \cup F_c$.

Then the possible movement of the animals can be directly modelled by the behaviour of the generator $\mathcal{S} = (S, F)$ starting in the initial state [2,6], where elements of F_u represent direct state changes, that cannot be disabled (the door between rooms 1 and 3 cannot be closed).

The required behaviour specification is expressed as follows.

1. The animals initially placed in these rooms (represented by pair of number [2,6]) will never occupy the same room simultaneously.
2. It is always possible for the cat and the mouse to return to the initial state.
3. The greatest possible freedom of animal movement is demanded.

The task is to determine, whether the behaviour satisfying these requirements can be reached by a supervision, and if it is possible, then to determine the control properties, that must be fulfilled by the supervisor. To translate this task to modelling formalism notions the following task is obtained. The relation C on S x S satisfying under mentioned requirements is sought.

1. ${}^{\langle[2,6]}C = C \subseteq F$
2. L is controllable

3. $\forall$ [a,a] $\in$ S: [[2,6], [a,a]] $\notin C^{\otimes}$
4. $\forall$ [a,b] $\in C^{\otimes}([2,6])$: [[a,b],[2,6]] $\in C^{\otimes}$
5. There exists no relation K on S x S for which all above mentioned requirements are fulfilled and holds: ${}^{\langle[2,6]}C \subset {}^{\langle[2,6]}K$.

The relation C satisfying these all 5 requirements is formally given as follows.

$$C = \{[[a,b], [a',b']];\ ([a',b'] = [(a+1)_{mod\,3}, b] \wedge a \in \{0,1,2\} \wedge b = 6) \vee$$
$$([a',b'] = [a, (b-3)_{mod\,9},] \wedge a = 2 \wedge b \in \{0,3,6\}) \vee$$
$$(a = 0 \wedge a' = 3 \wedge b' = b = 6]) \vee$$
$$(((a = 1 \wedge a' = 3) \vee (a = 3 \wedge a' = 1)) \wedge b' = b = 6)\}$$

Then, by comparison of the relation F and the relation C in each point of s $\in C_{supp}$ it is obtained the information about direct state changes from the state represented by the s, which must be enabled by a supervisor (direct state changes represented by elements of C(s,.)) and disabled (prohibited) by a supervisor (direct state changes represented by elements of (F(s,.) - C(s,.))). Then the behaviour of this supervised system starting in the initial state can be modelled by the behaviour of the generator *H* = (S, C) starting in the initial state [2,6].

3 INVARIANT STATE PROGRESS IN PETRI NETS

The state relation is defined by the state difference function Δ, so the states are in this state relation, if the difference between them is equal to given Δ. The behaviour according to this relation is examined. Such behaviour (state progress) is understood as to reach the states that are in this relation. There are many issues as the existence of infinite behaviour according to this relation, determination of event sequences assuring this behaviour (state progress).

The problem is investigated by means of Petri nets theory (Reisig, 1985) that enable to use analytic algebraic methods. As follows, m x n matrix is denoted by [row 1; ..; row m].

Definition: A Petri net is a 4-tuple PN = (P, T, I, O) such that : P = $\{p_1, .., p_m\}$ is a finite and nonempty set of places, T = $\{t_1, .., t_n\}$ is a finite and nonempty set of transitions P $\cap$ T is the empty set, I(p,t): P x T $\rightarrow$ N is an input function, N = $Z^+ \cup \{0\}$), O(t,p): T x P $\rightarrow$ N is an output function (**I** and **O** can be understood as an integer m x n and n x m matrices).

In a Petri net the system states are characterised by markings $\boldsymbol{m}$: P $\rightarrow$ N, $\boldsymbol{m} = (m(p_1); .. ; m(p_m))$. The set of all markings is denoted by M. In the equation $\boldsymbol{m}' = \boldsymbol{m} + \mathbf{C} \cdot \boldsymbol{y}$, where **C** is the structure (incidence) matrix of the given Petri net ($\mathbf{C} = \mathbf{O}^t - \mathbf{I}$), vectors $\boldsymbol{y}$ determine the firing numbers of each transition that are necessary to reach the state $\boldsymbol{m}'$ from the state $\boldsymbol{m}$. If the difference Δ is given, then vectors $\boldsymbol{y}$, that are integer solutions of the equation $\mathbf{C} \cdot \boldsymbol{y} = \Delta$, determine the firing numbers of each transition in sequences of events, that assure the state progress (to reach state $\boldsymbol{m}'$ from state $\boldsymbol{m}$ that $(\boldsymbol{m},\boldsymbol{m}')$ belongs to the relation given by the difference Δ). Classic T - invariants are solutions of equation $\mathbf{C} \cdot \boldsymbol{y} = \mathbf{0}$ (Lautenbach, 1986). The solutions of the equation $\mathbf{C} \cdot \boldsymbol{y} = \Delta$ (when Δ is understood as linear function) have the similar geometric interpretation as the solution of equation $\mathbf{C} \cdot \boldsymbol{y} = \mathbf{0}$. That is the reason, that the algorithms for determination of T - invariants can be simply modified to solve the equation $\mathbf{C} \cdot \boldsymbol{y} = \Delta$. The properties of solutions of both equations ($\mathbf{C} \cdot \boldsymbol{y} = \mathbf{0}$, $\mathbf{C} \cdot \boldsymbol{y} = \Delta$) are the same in special cases, where difference Δ is a linear function without absolute coeficients. There is the possibility to define generalized T - invariants as solutions of the equation $\mathbf{C} \cdot y = \Delta$, where Δ is understood as invariant function and the states in the relations given by Δ are invariant states. The behaviour according to the relation can be formally described by the relation reachability graph, where the neighbouring nodes belong to this relation (given by difference Δ) and the arcs represent realizable solutions of the above mentioned equation ($\mathbf{C} \cdot y = \Delta$). The generalized T-invariants can be understood as a special case of the state relation. By means of generalized T-invariants it is possible to describe the relevant and interesting kind of discrete processes i.e. real cyclic processes, where the cycle is understood to repeatedly reach some similar (not exactly the same) state to the initial state (e.g. systems with counters).

Definition (Generalized T-invariants): Let PN = (P, T, I, O) is a Petri net, let M is the set of all functions (markings) $\boldsymbol{m}$: P $\rightarrow$ N, let C = O^t - I, $\boldsymbol{m}$, $\boldsymbol{m}' \in M$ and $\boldsymbol{y}$ is a firing vector. Let D = $\{[\boldsymbol{m},\boldsymbol{m}'], [\boldsymbol{m},\boldsymbol{m}'] \in M \times M\}$ is the invariance relation. Then the integer solutions of the system $\mathbf{C} \cdot \boldsymbol{y} = \boldsymbol{m}' - \boldsymbol{m}$ are called T-invariants in $\boldsymbol{m}$ in regard to the relation D.

Let $\Delta(M,\mathrm{H})$, where H is an arbitrary set, is a function Δ: $M \times \mathrm{H} \rightarrow Z^m$ (m is the number of places in the Petri net) such that for each h $\in$ H $\exists$ $[\boldsymbol{m},\boldsymbol{m}'] \in$ D that $\boldsymbol{m}' = \boldsymbol{m} + \Delta(\boldsymbol{m},\mathrm{h})$. The function Δ is called the invariant function and the equation $\mathbf{C} \cdot \boldsymbol{y} = \Delta(M,\mathrm{H})$ is called the invariance relation equation. Evidently, integer solutions of the invariance relation equation in $\boldsymbol{m}$ are T-invariants in $\boldsymbol{m}$ in regard to the relation D. Classic T-invariants represent the special case with the difference function $\Delta(M,\mathrm{H}) = \Delta(\mathrm{H}) = \mathbf{0}$.

The further formal definitions of above mentioned notions, solutions of the equations and the investigation of relation reachability graph properties, are presented in the (Juhás and Kocian, 1994, 1995). A motivation of generalized T-invariant definition is shown on an example of the simple production line.

Example: The line (modelled by Petri net in Figure 2) is formed by the input buffer, the numeric control (NC) machine, the output buffer and the counters. In the system there is running repeatedly the cyclic process, where the product is worked up by the NC machine, the quality test is performed and according to the result the related product counter is incremented.

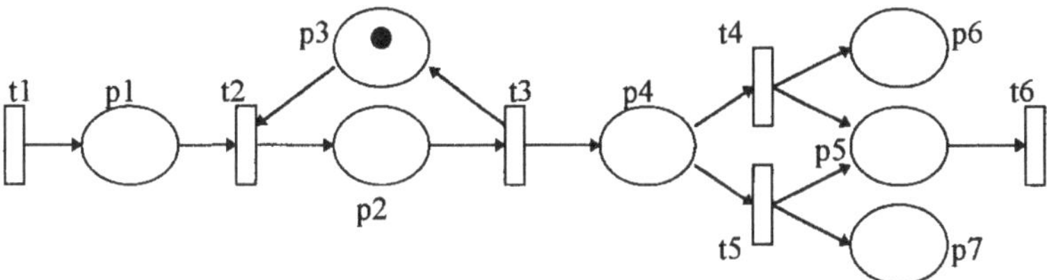

$$\mathbf{C} = [\ 1,-1,0,0,0,0;\ \ 0,1,-1,0,0,0;\ \ 0,-1,1,0,0,0;\ \ 0,0,1,-1,-1,0;\ \ 0,0,0,1,1,-1;\ \ 0,0,0,1,0,0;\ \ 0,0,0,0,1,0]$$

Figure 2 The Petri net model of a simple product line

Transition t1 - the arrival of the product, place p1 - waiting in the input buffer, t2 - the start of the working up, p2 - the state of working up, t3 - the finish of the working up and the quality test start, p3 - the state of the NC machine (ready or not), p4 - the quality test, t4 and t5 - the quality test finish and increment of the related counter, p5 - waiting in the output buffer, p6 and p7 - state of the correct and uncorrect product counter, t6 - the departure of the product.

The invariance relation is given as follows: D = {[$\boldsymbol{m},\boldsymbol{m}'$]; $\boldsymbol{m}'(p_i) - \boldsymbol{m}(p_i) = 0$ for $i = 1 .. 5$, $(\boldsymbol{m}'(p_6) - \boldsymbol{m}(p_6)) + (\boldsymbol{m}'(p_7) - \boldsymbol{m}(p_7)) = u \wedge u$ is an integer}. This state progress leads to the solving of the system $\mathbf{A} . \boldsymbol{x} = \mathbf{B} . u$, where $\mathbf{A}$ is 6 x 6 matrix and $A_i = C_i$ for $i = 1 .. 5$, $A_6 = C_6 + C_7$, A_i and C_i are rows of the matrixes $\mathbf{A}$ and $\mathbf{C}$, and $\mathbf{B} = [0;0;0;0;0;1]$ and u is an integer parameter. The T-invariants: $y = k . [1;1;1;1;0;1] + v . [1;1;1;0;1;1]$, k, v are integers. The nonnegative T-invariants indicate numbers of the event executions that are necessary to perform complete pass of the products trough the production line.

4 CONCLUSION

DSCSs are defined by using of state changes instead of events. The relation state change generator is used to model DSCS and related calculus is developed. The presented relation approach can be used to describe the possible behaviour as well as desirable behaviour of the system (control specification). The presented approach is illustrated by an example. The generalized T-invariants of Petri nets are defined as a special case of the presented state relation approach. A building of the relation calculus for DSCSs including algorithms for investigation of behavioural specifications represents the main subject of the further research.

REFERENCES

Capkovic, F (1993) Petri Net-based Approach to the Maze Problem Solving. In Balemi, S., Kozák, P. and Smedinga, R. (Ed.) *Discrete Event Systems: Modelling and Control,. Proc. of a Joint Workshop held in Prague, 1992,* Birkhäuser, 173-179.

Juhás, G. and Kocian, M. (1994) Generalized T-Invariants of Petri Nets and Control of DEDS. In *Preprints of 1st IFAC Workshop on New Trends in Design of Control Systems 1994,* Smolenice, Slovakia, 408-413.

Juhás, G. and Kocian, M. (1995) Invariant Relation Behaviour in Petri Nets. In Snorek, Sujansky, and Verbraeck (Ed.) *Modelling and Simulation 1995, ESM95, Prague, Czech Republic,* A publication of the [SCS], 160-164.

Klir, G. J. (1972) (Ed.)Trends in General System Theory, Wiley - Interscience, New York.

Kumar, R., Garg, V., and Marcus, S.I. (1993) Predicate and Predicate Transformers for Supervisory Control of Discrete Event Dynamic Systems. *IEEE tr. on Automatic Control,* Vol.38, N. 2, 232-247.

Lautenbach, K. (1986) Linear Algebraic Techniques for Place/Transitions Nets. In: *Proceedings of an Advanced Course on Petri Nets,* LNCS 254, 142-167.

Ramadge, P. and Wonham, W.M. (1989). The Control of Discrete Event Dynamic Systems. *Proceedings of the IEEE,* Vol. 77, N. 1, 81-98.

Reisig, W. (1985) *Petri Nets.* Springer-Verlag, Berlin.

Discrete Time Systems

16

Remarks on the observability of nonlinear discrete time systems

Francesca Albertini
Dipartimento di Matematica
Via Belzoni 7, 35131 Padova, Italy. Tel: +39(49)8275966.
e-mail: albertini@pdmat1.math.unipd.it

Domenico D'Alessandro
Universita' di Padova, Dipartimento di Elettronica ed Informatica
Via Gradenigo 6a, 35100 Padova, Italy. Tel: +39(49)8287791.
e-mail: daless@maya.dei.unipd.it

Abstract

The observability of discrete time nonlinear systems is studied. Criteria of observability are given in terms of codistributions. This leads naturally to decompositions similar to the ones known in the continuous time case. Some observability properties of invertible systems are also investigated. In particular, it is shown that, under regularity hypotheses, the weaker notion of forward-backward observability is equivalent to the one of (forward) observability, for these systems.

Keywords

Discrete-time, nonlinear systems, observability.

1 INTRODUCTION

We deal with observability questions for nonlinear discrete-time systems of the form

$$\Sigma \qquad \begin{array}{rcl} x(t+1) & = & f(x(t), u(t)), \qquad t = 0, 1, 2, \ldots \\ y(t) & = & h(x(t)). \end{array}$$

We consider single input single output systems, since the general case involves only notational changes. In Σ we assume that $x(t) \in M, y(t) \in Y$ and $u(t) \in U$, with M and Y connected, second countable, Hausdorff, differentiable manifolds, of dimensions n and 1, respectively. We also assume that the control space U is an open interval of $\mathbb{R}$, such that $0 \in U$. Such a system is said to be of class C^k, if the manifolds M and Y are of class C^k, and the functions $f: M \times U \to M$ and $h: M \to Y$, are of class C^k. We shall often use the abbreviated notation $f_u(x) := f(x, u)$.

Definition 1 Two states $x_1, x_2 \in M$ are said to be indistinguishable, and we write $x_1 I x_2$, if for each sequence of controls, $u_1, ..., u_j$, with $j \geq 0$, we have

$$h(f_{u_j} \circ \cdots \circ f_{u_1}(x_1)) = h(f_{u_j} \circ \cdots \circ f_{u_1}(x_2)).$$

Analogously x_1, x_2 are said to be k–indistinguishable, and we write $x_1 I^k x_2$, if the previous condition holds for each $0 \leq j \leq k$.

Definition 2 One state x_0 is said to be observable (k-observable), if, for each $x_1 \in M$, $x_0 I x_1$ ($x_0 I^k x_1$) implies $x_0 = x_1$.

Definition 3 One state x_0 is said to be locally observable (k-observable), if there exists a neighborhood W_{x_0} of x_0, such that, for each $x_1 \in W_{x_0}$, $x_0 I x_1$ ($x_0 I^k x_1$) implies $x_0 = x_1$.

Definition 4 A system Σ is (locally) (k-) observable, if each state $x \in M$ enjoys this property.

In the following, if we say that x_e is an equilibrium point, we always mean that $f(x_e, 0) = x_e$. We say that a subset of M is *generic* if its complement is contained in a proper analytic subset of M. Given a set L of C^∞ functions, defined on M, we shall denote by dL the codistribution spanned by all the differentials of these functions. By definition, these are exact differentials.

A previous study on the observability of discrete time nonlinear systems can be found in Nijmeijer (1982), where the case of systems without controls is considered. The use of the differential geometric concepts of invariant distributions and codistributions (see below), in the discrete time setting, is introduced in Monaco and Normand-Cyrot (1986). The classical paper Hermann and Krener (1977) deals with questions of nonlinear observability in the continuous time context. A complete treatment of this case is given in the books Isidori (1989) and Nijmeijer and Van der Schaft (1990).

2 OBSERVABILITY CRITERIA

From now on, assume that a C^∞ system Σ is given. Define

$$\Theta = \{h(f_{u_j} \circ \cdots \circ f_{u_1}(\cdot)) \mid u_1, ..., u_j \in U^j \, j \geq 0\}. \tag{1}$$

The previous set of functions will become the main object of our study. The following result holds:

Theorem 5 *If dim $d\Theta(x_0) = n$, then x_0 is a locally observable state for Σ. Conversely if Σ is locally observable, then there exists an open subset of M, where this condition is verified. If in addition the system is analytic and locally observable, then this condition is verified in a generic subset of M.*

Proof. Assume *dim* $d\Theta(x_0) = n$. Then, there exist n functions in Θ, $H_i(\cdot) := h(f_{u^i_{j_i}} \circ \cdots \circ f_{u^i_1}(\cdot))$, $i = 1, ..., n$, whose differentials are linearly independent at x_0. By continuity, they also are independent in a neighborhood W_{x_0} of x_0. Therefore, $H_i(\cdot)$, $i = 1, ..., n$, define a smooth mapping from M to Y^n, which, restricted to W_{x_0}, is injective. If, for $x_1 \in W_{x_0}$, it is $x_1 I x_0$, in particular, for all $i = 1, ..., n$, it must hold $H_i(x_0) = H_i(x_1)$. By the injectivity of $H_i(\cdot)$, $i = 1, ..., n$, it follows that $x_0 = x_1$.

For the converse implication, assume that Σ is locally observable and it does not exist an open subset of M where $dim\, d\Theta(x) = n$, which is equivalent to say that $dim\, d\Theta(x) < n$ for all $x \in M$. Let $r = \max_{x \in M} dim\, d\Theta(x) (< n)$, and choose $x_0 \in M$, such that $dim\, d\Theta(x_0) = r$. By continuity, there exists an open neighborhood W_{x_0} of x_0, such that $dim\, d\Theta(x) = r$, for all $x \in W_{x_0}$. Therefore there exist $H_1(\cdot), ..., H_r(\cdot)$, in Θ, whose differentials in W_{x_0} are linearly independent. We may take these functions, $H_1(\cdot), ..., H_r(\cdot)$, along with a set of complementary independent functions, as partial coordinates in W_{x_0}. Since every function in Θ only depends on the first $r < n$ coordinates, points in W_{x_0}, with the last $n - r$ coordinates equal, cannot be distinguished. This contradicts the hypothesis of the local observability of the system, and shows that there exists an open subset in M, such that $dimd\Theta(x) = n$ is true. The last sentence of the theorem follows easily from the fact that, if the above is true in an open subset of M, and the system is analytic, then, it is true everywhere except for the set of zeros of an analytic function, namely an analytic set. It is therefore true in what we have called a generic subset of M. □

If $d\Theta$ is constant dimensional in a neighborhood of x_0, then a stronger result about local observability can be derived (the result is easily proved by specializing to x_0 the proof of Theorem 5).

Corollary 6 Assume $d\Theta$ is constant dimensional in a neighborhood W_{x_0} of x_0. Then, x_0 is locally observable if and only if $dimd\Theta(x_0) = n$.

Remark 7 Analogous criteria can be given for local k-observability, by considering, in the previous statements, the following set of functions

$$\Theta_k = \{h(f_{u_j} \circ \cdots \circ f_{u_1}(\cdot)) \mid u_1, ..., u_j \in U^j\, k \geq j \geq 0\}.$$

The proofs follow the same lines as above.

Remark 8 In view of Corollary 6, it is of interest to give criteria for the constant dimensionality of the codistribution $d\Theta$, in a neighborhood of a given point. This is also important because, as we will see in the sequel, it is possible, in this situation, to obtain a local state space decomposition (see Section 3 below). Moreover, under this assumption, it is also possible to prove the equivalence between *forward* and *forward-backward* observability for invertible systems (see Section 4 below). In the continuous-time setting, the class of functions, used to characterize observability, gives raise to a constant dimensional codistribution if analyticity and accessibility are verified (see Nijmeijer and Van der Schaft (1990) Proposition 3.38). This is not true in general for $d\Theta$ or $d\Theta_k$, in the discrete-time case. The following example illustrates this issue and clarifies what we mean for accessibility in our context. Later, we shall give a sufficient condition for local constant dimensionality of $d\Theta$ (see Proposition 13).

Example 9 We recall (Jakubczyk and Sontag (1990) and Albertini and Sontag (1993)) that a point $x_0 \in M$ is said to be *forward accessible*, if it is possible to reach from it an open subset of M. For the following system, 0 is a forward accessible equilibrium point

$$\begin{aligned} x_1(t+1) &= x_1(t) + x_2^2(t) + u^2(t) \\ x_2(t+1) &= x_1(t) \\ y(t) &= x_1(t) \end{aligned}$$

In fact, if $x_1(0) = 0$, and $x_2(0) = 0$, it is $x_1(2) = u^2(0) + u^2(1)$ and $x_2(2) = u^2(0)$. Therefore, it is possible to reach, in two steps, any point such that $x_1 \geq 0, x_2 \geq 0$. The codistribution $d\Theta$ has not constant dimension in a neighborhood of 0. In fact, it is $dh(x) = [1,0]$, $dh(f_{u_1}(x)) = [1, 2x_2]$, and for $j \geq 2$, defined $\bar{H}(\cdot) := h(f_{u_j} \circ \cdots \circ f_{u_2}(\cdot))$, it is

$$\frac{\partial h(f_{u_j} \circ \cdots \circ f_{u_1}(x))}{\partial x_2} = \frac{\partial \bar{H}(f_{u_1}(x))}{\partial x} \frac{\partial f_{u_1}(x)}{\partial x_2}$$

$$= [\frac{\partial \bar{H}(f_{u_1}(x))}{\partial x_1}, \frac{\partial \bar{H}(f_{u_1}(x))}{\partial x_2}][2x_2, 0]^T = \frac{\partial \bar{H}(f_{u_1}(x))}{\partial x_1} 2x_2.$$

This shows that $d\Theta$ has dimension 1 for any point such that $x_2 = 0$, and 2 elsewhere.

3 STATE SPACE DECOMPOSITION

The study of the observability for nonlinear discrete time systems, using standard differential geometric tools, leads naturally to consider the notion of invariant distribution, as a natural generalization of the concept of invariant subspace, used in the study of linear systems. This concept has been widely used in characterizing structural properties of nonlinear continuous time systems and its importance, in the study of the discrete time case, seems to have been first pointed out in Monaco and Normand-Cyrot (1984). We will use the dual concept of invariant codistribution to perform a state space decomposition, in a neighborhood of an equilibrium point. The theory very much resembles the continuous time one presented in Isidori (1989).

Recall that, given a covector field ω and a mapping $f(\cdot)$, on a manifold M, $f^*\omega := \omega(f(\cdot))\frac{\partial f}{\partial x}(\cdot)$. Notice that, in particular, if ω is an exact differential, i.e. $\omega = d\lambda$, for a function $\lambda(\cdot)$, it is $f^*\omega = d(\lambda \circ f(\cdot))$. We have the following definition:

Definition 10 A codistribution Ω is said to be invariant under f_u, if

$$f_u^*\Omega \subseteq \Omega, \quad \text{for each} \quad u \in U.$$

Theorem 11 *Suppose that there exists a constant dimensional f_u- invariant codistribution of dimension r, containing dh, and spanned by exact differentials, in a neighborhood* *of an equilibrium point x_e. Then there exists a coordinate change, such that for (x,u) in a suitable neighborhood $\tilde{W}_{x_e} \times U_0$ of $(x_e, 0)$, the system Σ reads as*

$$\Sigma' \quad \begin{array}{rcl} z_1(t+1) &=& \hat{f}(z_1(t), z_2(t), u(t)), \qquad t = 0,1,2,\ldots, \\ z_2(t+1) &=& \hat{f}(z_2(t), u(t)) \\ y(t) &=& \hat{h}(z_2(t)), \end{array}$$

where z_1 and z_2 have dimensions $n-r$ and r respectively.

Proof. Under the stated hypotheses, there exist r functions $\lambda_1(\cdot), \ldots, \lambda_r(\cdot)$ on M, such that $d\lambda_1(x_e), \ldots, d\lambda_r(x_e)$ form a basis of $\Omega(x_e)$. These functions, along with a complementary set of $n-r$ linearly independent functions, give a coordinate change, $T(x)$, in a neighborhood W_{x_e} of x_e. We write $\hat{h}(\cdot), \hat{f}(\cdot), \hat{\lambda}_i(\cdot)$, for $h(\cdot), f(\cdot)$ and $\lambda_i(\cdot)$, $i = 1, \ldots, r$, respectively, in these coordinates. In particular, if $z = T(x)$ denotes the new coordinates, we can as well assume that $\hat{\lambda}_i = z_{n-r+i}$, $i = 1, \ldots, r$. Notice that, since dh is in Ω, $\hat{h}(\cdot)$ only depends on the last r coordinates. By continuity, we can choose a neighborhood $\tilde{W}_{x_e} \times U_0$ of $(x_e, 0)$, such that $f_u(x) \in W_{x_e}$, for all pairs $(x,u) \in \tilde{W}_{x_e} \times U_0$. Choosing x and u in this way, and remembering that, by the invariance property, $d\lambda_i(f_u)$, for $i = 1, \ldots, r$, is still in Ω, we also have

$$0 = \frac{\partial}{\partial z_j} \hat{\lambda}_i(\hat{f}_u(\cdot)) = \frac{\partial \hat{f}_u^{n-r+i}}{\partial z_j}(\cdot),$$

$i = 1, \ldots, r$, $j = 1, \ldots, n-r$. This shows that the last r components of $\hat{f}(\cdot)$ are independent of the first $n-r$ components of z. Therefore, locally the system can be written in the form Σ'. □

If $d\Theta$ is constant dimensional in a neighborhood of x_e, it is a good candidate to be used to perform the change of coordinates described above. $d\Theta$ is, in fact, f_u-invariant, it is spanned by exact differentials, and it contains dh. Moreover $d\Theta$ is the smallest codistribution which enjoys these properties. The proof of this fact follows the same lines of the one given for the observability codistribution in the continuous time case Isidori (1989). The only modification consists in replacing the definition of invariance, given in the continuous time context, with the one given here. Notice that the only if part of Corollary 6 also holds if we consider an arbitrary constant dimensional codistribution which properly contains $d\Theta$. In view of this fact, in the hypotheses of Theorem 11, since Ω must contain $d\Theta$, if $r < n$, x_e is not locally observable.

4 PROPERTIES OF INVERTIBLE SYSTEMS

In this section, we will deal with a particular class of discrete time nonlinear systems.

Definition 12 A system Σ is said to be *invertible*, if for all $u \in U$, the function $f_u : M \to M$, is a diffeomorphism (we denote by f_u^{-1} the inverse function of f_u).

Invertible systems arise, for example, when a continuous-time model is controlled under digital control, via *sampling*. Further motivations for the study of this class of systems are given in Jakubczyk and Sontag (1990). For invertible systems, it is possible to define an *inverse* system by

$$\Sigma^- \qquad \begin{array}{lcl} x(t+1) & = & f^{-1}(x(t), u(t)), \quad t = 0, 1, 2, \dots \\ y(t) & = & h(x(t)). \end{array}$$

Using this system one can define *backward* indistinguishability and observability, following the same lines of Definitions 1-4. It is also possible to define (see example 9 and Jakubczyk and Sontag (1990), Albertini and Sontag (1993)) *backward* accessibility. These definitions extend to *forward-backward* indistinguishability, observability and accessibility in an obvious manner. We state now a sufficient condition for the constant dimensionality of the codistribution $d\Theta$ defined in (1)

Proposition 13 Consider an invertible system Σ and an equilibrium point x_e. Assume that the following rank condition is verified

$$\sup_{j \geq n} \left\{ rank \left. \frac{\partial}{\partial u} \right|_{u_1 = \dots = u_j = 0} f_{u_j} \circ \cdots \circ f_{u_1}(x_e) \right\} = n. \tag{2}$$

Then, the codistribution $d\Theta$ is constant dimensional, in a neighborhood of x_e.

Proof. Using the rank condition (2), it is easy to show that there exist an open subset of M, $F(x_e)$, which contains x_e, and such that each point of $F(x_e)$ can be reached by x_e. Also it can be shown that there exists an open subset $B(x_e)$, which contains x_e, and such that from any point in it, it is possible to reach x_e. (The proof of these facts follows from the one of the accessibility criterion given in Proposition 2.3 of Jakubczyk and Sontag (1990).) We consider the set $L(x_e) = B(x_e) \cap F(x_e)$. $L(x_e)$ is an open set and contains x_e. Moreover, for a point x_1 in $L(x_e)$, there exist two sequences of controls such that

$$x_1 = f_{u^F_{j_F}} \circ \cdots \circ f_{u^F_1}(x_e), \qquad \text{and} \qquad x_e = f_{u^B_{j_B}} \circ \cdots \circ f_{u^B_1}(x_1). \tag{3}$$

Consider now k_F functions, $h^i(\cdot)$, $i = 1, \dots, k_F$, such that $dh^i(x_1)$, $i = 1, \dots, k_F$, form a basis for $d\Theta(x_1)$. The functions $h^i(f_{u^F_{j_F}} \circ \cdots \circ f_{u^F_1}(\cdot))$, $i = 1, \dots, k_F$ are in Θ; moreover, using (3) (left hand side),

$$dh^i(f_{u^F_{j_F}} \circ \cdots \circ f_{u^F_1}(x_e)) = dh^i(x_1) \frac{\partial f_{u^F_{j_F}} \circ \cdots \circ f_{u^F_1}(x_e)}{\partial x}.$$

By the invertibility of $f_{u^F_{j_F}} \circ \cdots \circ f_{u^F_1}(\cdot)$, $\frac{\partial f_{u^F_{j_F}} \circ \cdots \circ f_{u^F_1}(x_e)}{\partial x}$ is nonsingular, so, by the linear independence of $dh^i(x_1)$, we have that $dh^i(f_{u^F_{j_F}} \circ \cdots \circ f_{u^F_1})(x_e)$ also are linearly independent. Therefore they can be included in a basis of $d\Theta(x_e)$, and this shows that

$$dim d\Theta(x_1) \leq dim d\Theta(x_e). \tag{4}$$

Analogously, if we choose a set of functions $h^i(\cdot)$, $i = 1, ..., k_B$, such that $dh^i(x_e)$ form a basis for $d\Theta(x_e)$, we have that $h^i(f_{u^B_{jB}} \circ \cdots \circ f_{u^B_1}(\cdot))$, $i = 1, ..., k_B$, is still in Θ. By the invertibility of $f_{u^B_{jB}} \circ \cdots \circ f_{u^B_1}(\cdot)$ and using (3) (right hand side), we have, as above,

$$dim d\Theta(x_e) \leq dim d\Theta(x_1). \tag{5}$$

Combining (4) and (5), we have that Θ is constant dimensional in $L(x_e)$. □

We conclude showing that, under regularity hypotheses, locally forward observability (l.f.o) and locally forward-backward observability (l.f.b.o) are equivalent for equilibrium points of invertible systems.

Theorem 14 *Let Σ be an analytic invertible system, and $x_e \in M$ be an equilibrium point. Assume that $d\Theta$ has constant dimension in a neighborhood W_{x_e} of x_e. Then, x_e is l.f.o. if and only if it is l.f.b.o.*

Proof. It is obvious that if x_e is l.f.o. it is also l.f.b.o. Conversely, assume that x_e is not l.f.o. Under the stated hypotheses and using Corollary 6, we know that $dim d\Theta(x) = r < n$, in a neighborhood W_{x_e} of x_e. We can therefore perform a change of coordinates as in Theorem 11. System Σ reads as Σ', as long as, x is in a suitable neighborhood $\tilde{W}_{x_e} \subseteq W_{x_e}$ and u is in a suitable neighborhood U_0 of 0. It is straightforward to verify that also the inverse system Σ^- can be put in a triangular form, again for $x \in \tilde{W}_{x_e}$, and $u \in U_0$.

Assume, by contradiction, that x_e is l.f.b.o, and let $V_{x_e} \subseteq \tilde{W}_{x_e}$, be a neighborhood of x_e, where the l.f.b.o. holds. Then choose any $\bar{x} \in V_{x_e}$ such that, in the z-coordinate, the last r components of x_e and of $\bar{x}$ are equal. Since $\bar{x} \in V_{x_e}$, there exists k, $\tilde{u}_1, \ldots, \tilde{u}_k \in U$, $\epsilon_1, \ldots, \epsilon_k$, with $\epsilon_i = \pm 1$, $i = 1, ..., k$, such that:

$$h(f^{\epsilon_k}_{\tilde{u}_k} \circ \cdots \circ f^{\epsilon_1}_{\tilde{u}_1}(x_e)) \neq h(f^{\epsilon_k}_{\tilde{u}_k} \circ \cdots \circ f^{\epsilon_1}_{\tilde{u}_1}(\bar{x})). \tag{6}$$

By continuity, there exists $\tilde{U}_0 \subseteq U_0$ neighborhood of 0 such that, for $i = 1, \ldots, k$:

$$u_1, \ldots, u_i \in \tilde{U}_0 \Rightarrow f^{\epsilon_i}_{u_i} \circ \cdots \circ f^{\epsilon_1}_{u_1}(x_e) \in \tilde{W}_{x_e}, \quad \text{and} \quad f^{\epsilon_i}_{u_i} \circ \cdots \circ f^{\epsilon_1}_{u_1}(\bar{x}) \in \tilde{W}_{x_e}. \tag{7}$$

From (6), and by analyticity, there exists $\bar{u}_1, \ldots, \bar{u}_k$, with $\bar{u}_i \in \tilde{U}_0$, such that

$$h(f^{\epsilon_k}_{\bar{u}_k} \circ \cdots \circ f^{\epsilon_1}_{\bar{u}_1}(x_e)) \neq h(f^{\epsilon_k}_{\bar{u}_k} \circ \cdots \circ f^{\epsilon_1}_{\bar{u}_1}(\bar{x})). \tag{8}$$

However, since $\bar{u}_i \in \tilde{U}_0$, equation (7) implies that, $\forall i = 1, ..., k$, $f^{\epsilon_i}_{\bar{u}_i} \circ \cdots \circ f^{\epsilon_1}_{\bar{u}_1}(x_e) = x^i_e \in \tilde{W}_{x_e}$ and also $f^{\epsilon_i}_{\bar{u}_i} \circ \cdots \circ f^{\epsilon_1}_{\bar{u}_1}(\bar{x}) = \bar{x}^i \in \tilde{W}_{x_e}$. Thus, by the triangular form, in the z-coordinate, the last r components of x^i_e and of $\bar{x}^i$ are equal. So, also $h(x^k_e) = h(\bar{x}^k)$, which contradicts equation (8). □

REFERENCES

Albertini, F. and Sontag, E.D. (1994) Discrete-time transitivity and accessibility: analytic systems. *SIAM J. Control and Opt.*, **31**, 1599–622.

Hermann, R. and Krener, A.J. (1977) Nonlinear controllability and observability. IEEE *Trans. Automat. Control.*, **AC-22**, 728–40.

Isidori, A. (1989) *Nonlinear Control Systems.* 2nd edition, Springer Verlag, Berlin.

Jakubczyk, B. and Sontag, E.D. (1990) Controllability of nonlinear discrete-time systems: A Lie-algebraic approach, *SIAM J. Control and Opt.*, **28**, 1–33.

Monaco, S. and Normand-Cyrot, D. (1984) Invariant distributions for discrete time nonlinear systems. *Systems and Control Letters*, **5**, 191–6.

Nijmeijer, H. (1982) Observability of autonomous discrete time non-linear systems: a geometric approach. *Int.J.Control*, **36**, 867–74.

Nijmeijer, H. and Van der Schaft, A. J. (1990) *Nonlinear Dynamical Control Systems.* Springer-Verlag, New York.

17

Risk-sensitive control and dynamic games: The discrete-time case

Paolo Dai Pra, Cristina Rudari
Universita' di Padova
Dipartimento di Matematica, Via Belzoni 7, 35131 Padova, Italy.
Tel: +39(49)8275966. e-mail: daipra@pdmat1.math.unipd.it

Abstract
We show that a suitable small parameter limit of a family of (possibly partially observed) risk-sensitive stochastic control problems is a deterministic dynamic game. We identify the corresponding cost functional through large deviations techniques.

Keywords
Risk-sensitive control, dynamic games, large deviations.

1 INTRODUCTION

The aim of this paper is to investigate connections between risk-sensitive stochastic optimal control and dynamic games. The existence of a link between these two families of optimization problems was first noticed by Jacobson (1973, 1977) in the case of linear systems with quadratic cost, where the explicit solutions are available. In the nonlinear setting the relation is more subtle: a deterministic game is recovered from a risk-sensitive stochastic control problem through a limiting procedure, consisting in letting the noise variance and the risk parameter going to zero and infinity respectively, at the same rate. As pointed out by Whittle (1990, 1991), large deviation theory is a natural probabilistic tool to study small-noise high-risk limit of stochastic control problems, providing rather convincing heuristic arguments but, so far, no rigorous proof.

In the case of continuous time diffusion processes, a different approach to the limit of risk-sensitive control problems was proposed by Fleming and McEneaney (1992, 1995). Their method relies on the theory of viscosity solutions for Hamilton-Jacobi (Dynamic Programming) equations, and has been applied to both finite time horizon and infinite time horizon problems.

When mimicking Fleming's method for discrete time models, one has to deal with a Dynamic Programming equation, for the risk-sensitive control problem, having the form of an integral difference equation, in which the noise distribution enters explicitely. The structure of this difference equation naturally suggests, in the analysis of its small-noise high-risk limit, the use of large deviation techniques, exploiting thus some of Whittle's ideas. Such large deviation analysis of the Dynamic Programming equation has been completed by James, Baras and Elliott (1994) for nonlinear discrete time systems driven by a Gaussian additive noise.

The aim of this paper is to extend the results of James, Baras and Elliott by making a systematic use of large deviation methods. We give here a rather wide class of systems whose small noise-high risk limit can be identified as a dynamic game. The models in James, Baras and Elliott (1994) are covered by our approach, that in turn does not rely on any specific assumption on the distribution of the noise. In particular, we can treat nontrivial models with non-Gaussian and/or non-additive noise.

We finally mention that different, but close in spirit, limit theorems for risk-sensitive control problems have been obtained by Barron and Jensen (1989), and Runolfsson (1994).

The paper is organized as follows. In Section 2 we introduce the (new) concept of Uniform Large Deviation Family, that is used in Section 3 to analyze the limit of totally observed risk-sensitive control problems. Section 4 is devoted to the partially observed case, and to the discussion of a relevant example.

2 UNIFORM LARGE DEVIATION FAMILIES

We introduce here the main technical tool of this paper. In what follows we let X be a Polish (metric, separable and complete) space, and Θ be a set.

Definition 1 We say that a family $\{P^\epsilon(dx;\theta) : \epsilon > 0, \theta \in \Theta\}$ of probability measures on X is a Uniform Large Deviation Family (ULDF) if there exists a function $H : X \times \Theta \longrightarrow [0,+\infty]$ such that
i) $x \longrightarrow H(x;\theta)$ is lower semicontinuous (l.s.c.), for every fixed $\theta \in \Theta$;
ii) for $C \subset X$ closed

$$\limsup_{\epsilon\to 0} \sup_{\theta\in\Theta} \left[\epsilon \log P^\epsilon(C;\theta) + \inf_{x\in C} H(x;\theta)\right] \leq 0; \tag{1}$$

iii) for $A \in X$ open

$$\liminf_{\epsilon\to 0} \inf_{\theta\in\Theta} \left[\epsilon \log P^\epsilon(A;\theta) + \inf_{x\in A} H(x;\theta)\right] \geq 0. \tag{2}$$

The function H is called the *rate function*. The following result is a version of Varadhan's Lemma (Varadhan (1984)); its proof requires a modification of the classical argument in Varadhan (1984), and can be found in Dai Pra and Rudari (1995a).

Lemma 2 Let $F_\epsilon, F : X \longrightarrow \mathbb{R}$ be bounded continuous functions such that $F^\epsilon \longrightarrow F$ uniformly as $\epsilon \longrightarrow 0$. Then

$$\lim_{\epsilon\to 0} \epsilon \log \int e^{\epsilon^{-1}F_\epsilon(x)} P^\epsilon(dx;\theta) = \sup_{x\in X}[F(x) - H(x;\theta)] \tag{3}$$

uniformly in $\theta \in \Theta$.

3 SMALL-NOISE HIGH-RISK LIMIT: THE CASE OF COMPLETE OBSERVATION

Let $x^{\epsilon,\mathbf{u}} = (x_n^{\epsilon,\mathbf{u}})_{n=0}^N$ be a stochastic process, defined on a probability space $(\Omega, \mathcal{F}, P)$, taking values in a Polish space X and adapted to a filtration $(\mathcal{F}_n)_{n=0}^N$. The process is indexed by a parameter $\epsilon > 0$ and by a $(\mathcal{F}_n)$-adapted process $\mathbf{u} = (u_0, \ldots, u_{N-1})$, taking values in a metric space U. The law of the process $x^{\epsilon,\mathbf{u}}$ is determined by a $\mathcal{F}_0$-measurable initial condition x_0, and by the following transition probabilities:

$$P(x_{n+1}^{\epsilon,\mathbf{u}} \in \cdot|\mathcal{F}_n) = P_n^\epsilon(\cdot; x_n^{\epsilon,\mathbf{u}}, u_n); \quad n = 0, \ldots, N-1 \tag{4}$$

where the probability kernels P_n^ϵ satisfy the following

Assumption A.

1. For fixed $\epsilon > 0$, $n = 0, \ldots, N-1$, the map $(x, u) \to P_n^\epsilon(\cdot; x, u)$, from $X \times U$ to $\mathcal{M}_1(X)$ (space of probability measures on X) is continuous in x, uniformly for $u \in U$, with respect to the weak topology in $\mathcal{M}_1(X)$;
2. For n fixed, $\{P_n^\epsilon(\cdot; x, u) : \epsilon > 0, (x, u) \in X \times U\}$ is a ULDF with rate function $H_n(z; x, u)$.

We denote by $\mathcal{U}$ the class of *admissible controls*, i.e. the $(\mathcal{F}_n)$-adapted, U-valued stochastic processes. The aim of the controller is to minimize over $\mathcal{U}$ the cost functional

$$J^\epsilon(\mathbf{u}) = E\Big\{\exp\Big[\epsilon^{-1}\Big(\sum_{n=0}^{N-1} g_n(x_n^{\epsilon,\mathbf{u}}, u_n) + g_N(x_N^{\epsilon,\mathbf{u}})\Big)\Big]\Big\} \tag{5}$$

where the g_n's are bounded functions, continuous in x, uniformly for $u \in U$.

The *value* $J_*^\epsilon = \inf_{\mathbf{u}} J^\epsilon(\mathbf{u})$ can be obtained through the Dynamic Programming equation (Bertzekas (1976))

$$\left\{\begin{array}{lcl} V_n^\epsilon(x) & = & \inf_{u\in U}\left[e^{\epsilon^{-1}g_n(x,u)}\int V_{n+1}^\epsilon(x')P_n^\epsilon(dx'; x, u)\right] \\ V_N^\epsilon(x) & = & e^{\epsilon^{-1}g_N(x)} \end{array}\right. \tag{6}$$

which yields $J_*^\epsilon = E(V_0^\epsilon(x_0))$. The function $(n, x) \to V_n^\epsilon(x)$ is called the *value function* for the risk-sensitive stochastic control problem (4)(5), where ϵ^{-1} plays the role of a risk parameter. Under Assumption A 1., it is easily proved that V_n^ϵ is a bounded continuous function of x.

To the above risk-sensitive stochastic control problem we will associate a deterministic dynamic game of the following type. The dynamics are given by the difference equations

$$\begin{array}{ll} x_{n+1} = f_n(x_x, u_n, w_n) & n = 0, \ldots, N-1 \\ x_0 \ \text{given} & \end{array} \tag{7}$$

with $x_n \in X, u_n \in U, w_n \in W$, W being an arbitrary set. The controls $\mathbf{u} = (u_0, \ldots, u_{N-1})$ are such that u_n is a function of $(x_0, \ldots, x_n)$. Letting $\hat{\mathcal{U}}$ denote this class of controls, and $\mathbf{w} = (w_0, \ldots, w_{N-1})$, the aim of the controller is to minimize over $\hat{\mathcal{U}}$ the cost functional

$$K(\mathbf{u}) = \sup_{\mathbf{w}}\Big\{\sum_{n=0}^{N-1} G_n(x_n, u_n, w_n) + G_N(x_N)\Big\} \tag{8}$$

where the G_n's are given functions. The quantity $K_* = \inf_{\mathbf{u}} K(\mathbf{u})$ is the (upper) value for the game. The dynamic programming equation for this game is

$$\begin{aligned} S_n(x) &= \inf_{u\in U}\sup_{w\in W}\left\{G_n(x,u,w)+S_{n+1}(f_n(x,u,w))\right\} \\ S_N(x) &= G_N(x). \end{aligned} \tag{9}$$

Also in this context the map $(n,x) \to S_n(x)$ is called the (upper) *value function* for the game, and $K_* = S_0(x_0)$. The main result of this Section is the following Theorem, which identifies the $\epsilon \to 0$ limit of the stochastic control problem (4)(5).

Theorem 3 *Suppose Assumption A is satisfied and, for $n = 0,\ldots,N-1$, let $f_n : X \times U \times W \to X$ be functions such that for each $(x,u) \in X \times U$ the map $w \to f_n(x,u,w)$ is onto. Then the value function $V_n^\epsilon(x)$ defined in (6) is such that*

$$\lim_{\epsilon\to 0}\epsilon\log V_n^\epsilon = S_n \tag{10}$$

uniformly on X, where S_n is the value function of the dynamic game (7)(8) with

$$\begin{aligned} G_N(x) &= g_N(x) \\ G_n(x,u,w) &= g_n(x,u) - H_n(f_n(x,u,w);x,u) \quad \text{for } n=0,\ldots,N-1. \end{aligned} \tag{11}$$

Proof. The proof consists in a simple use of Lemma 2. Indeed, we define $S_n^\epsilon = \epsilon\log V_n^\epsilon$, and write (6) in the equivalent form

$$S_n^\epsilon(x) = \inf_{u\in U}\left\{g_n(x,u)+\epsilon\log\int e^{\epsilon^{-1}S_{n+1}^\epsilon(x')}P_n^\epsilon(dx';x,u)\right\}. \tag{12}$$

Since $S_N^\epsilon \equiv g_N = S_N$, we can proceed by induction, and assume $S_{n+1}^\epsilon \to S_{n+1}$ uniformly, as $\epsilon \to 0$. Then, by Lemma 2,

$$\epsilon\log\int e^{\epsilon^{-1}S_{n+1}^\epsilon(x')}P_n^\epsilon(dx';x,u) \to \sup_{x'}[S_{n+1}(x')-H_n(x';x,u)] \tag{13}$$

uniformly. Thus

$$\begin{aligned} \lim_{\epsilon\to 0} S_n^\epsilon(x) &= \inf_{u\in U}\sup_{x'\in X}[g_n(x,u)-H_n(x';x,u)+S_{n+1}(x')] \\ &= \inf_{u\in U}\sup_{w\in W}[g_n(x,u)-H_n(f(x,u,w);x,u)+S_{n+1}(x')] \\ &= \inf_{u\in U}\sup_{w\in W}[G_n(x,u,w)+S_{n+1}(x')] \\ &= S_n(x) \end{aligned}$$

where the limit is uniform. □

4 THE CASE OF PARTIAL OBSERVATION

Consider a risk-sensitive stochastic control problem of type (4)(5). When the state process (4) is only *partially observed* one changes the class of admissible controls to the set of those control processes $\mathbf{u}$ that are adapted to the filtration $(\mathcal{F}_n^y)_{n=0}^N$ generated by an *observation process* $y^{\epsilon,\mathbf{u}}$. This observation process, that takes value in a Polish space Y, is assumed to obey to the following requirements:

$$P\{y_n^{\epsilon,\mathbf{u}} \in \cdot | \mathcal{F}_n \vee \mathcal{F}_{n-1}^y\} = Q_n^\epsilon(\cdot; x_n^{\epsilon,\mathbf{u}}) \quad n = 0, \ldots, N-1, \tag{14}$$

$$P\{x_{n+1}^{\epsilon,\mathbf{u}} \in \cdot | \mathcal{F}_n \vee \mathcal{F}_n^y\} = P_n^\epsilon(\cdot; x_n^{\epsilon,\mathbf{u}}, u_n) \tag{15}$$

where the Q_n^ϵ are given probability kernels. Clearly, (4),(14) and (15) together determine the joint law of the process $(x^{\epsilon,\mathbf{u}}, y^{\epsilon,\mathbf{u}})$. Denoting by $\mathcal{U}^y$ the class of control processes that are adapted to the filtration $(\mathcal{F}_n^y)$, the aim of the controller is to minimize over $\mathcal{U}^y$ the cost functional $J^\epsilon(u)$ given by (5).

The following (rather long!) list of requirements make the problem tractable with our method.

Assumption B.

1. For fixed $\epsilon > 0$, $n = 0, \ldots, N-1$ the map $(x, u) \to P_n^\epsilon(dx'; x, u)$ is continuous in x uniformly for $u \in U$, and the map $x \to Q_n^\epsilon(dy; x)$ is continuous.
2. For any n, $\{P_n^\epsilon(dx'; x, u) : \epsilon > 0, (x, u) \in X \times U\}$ and $\{Q_n^\epsilon(dy; x); \epsilon > 0, x \in X\}$ are ULDF's with rate functions $H_n^P(x'; x, u)$, $H_n^Q(y; x)$ respectively.
3. The spaces X, Y are locally compact Polish spaces, and U is a compact metric space.
4. The functions g_n, $n = 0, \ldots, N$, in the cost function (5) are uniformly continuous.
5. There are sequences $(C_m)_{m \geq 0}$, $(K_m)_{m \geq 0}$ of compact sets in X and Y respectively, such that, for $n = 0, \ldots, N-1$

$$\limsup_{m \to \infty} \limsup_{\epsilon \to 0} \sup_{(x,u) \in X \times U} \epsilon \log P_n^\epsilon(C_m^c; x, u) = -\infty$$

$$\limsup_{m \to \infty} \limsup_{\epsilon \to 0} \sup_{x \in X} \epsilon \log Q_n^\epsilon(K_m^c; x) = -\infty.$$

6. Let C_1, C_2 be arbitrary compact subsets of X, and $A_n(C_1, C_2) = \{(x', x, u) : H_n^P(x'; x, u) < \infty\}$. Then H_n^P is uniformly continuous in $A_n(C_1, C_2)$, for any choice of C_1, C_2.
7. The rate function $H_n^Q(y; x)$ is uniformly continuous in $K \times X$, for any $K \subset Y$ compact.

For simplicity, we also assume from now on that the initial condition x_0 for the state dynamics is deterministic; this requirement can be considerably weakened.

The $\epsilon \to 0$ limit of the partially observed stochastic control problems defined above will be described as a dynamic game of the following type. The state-observation process is determined by the difference equations

$$\begin{cases} x_{n+1} &= f_n(x_n, u_n, w_n) \\ x_0 & \text{given} \\ y_n &= h_n(x_n, v_n) \end{cases} \tag{16}$$

with $x_n \in X, y_n \in Y, w_n \in W, v_n \in V$, W and V being arbitrary sets. The class $\hat{\mathcal{U}}^y$ of admissible controls is given by those controls $\mathbf{u}$ such that u_n is a function of $(y_0, \ldots, y_n)$. The aim of the controller is to minimize over $\hat{\mathcal{U}}^y$ a cost functional of the form

$$K(\mathbf{u}) = \sup_{\mathbf{w},\mathbf{v}} \Big\{ \sum_{n=0}^{N-1} G_n(x_n, u_n, w_n, v_n) + G_N(x_N) \Big\}. \tag{17}$$

In order to use the machinery introduced in Section 3, we transform the problem (14)(5) into an equivalent (i.e. having the same cost functional) problem with complete observation.

The transformation consists in defining the process $z^{\epsilon,\mathbf{u}} = (z_0^{\epsilon,\mathbf{u}}, \ldots, z_N^{\epsilon,\mathbf{u}})$, with $z_n^{\epsilon,\mathbf{u}} = (y_0^{\epsilon,\mathbf{u}}, \ldots, y_n^{\epsilon,\mathbf{u}}, u_0, \ldots, u_{n-1})$. The process $z^{\epsilon,\mathbf{u}}$ is adapted to the filtration $(\mathcal{F}_n^y)$, and it can be shown that there are probability kernels $P_n^{z,\epsilon}(dz_{n+1}; z_n, u_n)$ such that

$$P\{z_{n+1}^{\epsilon,\mathbf{u}} \in \cdot | \mathcal{F}_n^y\} = P_n^{z,\epsilon}(\cdot; z_n^{\epsilon,\mathbf{u}}, u_n) \tag{18}$$

for any admissible $\mathbf{u} \in \hat{\mathcal{U}}^y$. Moreover, the cost functional $J^\epsilon(\mathbf{u})$ can be rewritten in terms of the process $z^{\epsilon,\mathbf{u}}$ as

$$J^\epsilon(\mathbf{u}) = E\{e^{\epsilon^{-1}\gamma^\epsilon(z_N^{\epsilon,\mathbf{u}})}\} \tag{19}$$

with

$$\gamma^\epsilon(z_N^{\epsilon,\mathbf{u}}) = \epsilon \log E\Big\{ \exp\Big[\epsilon^{-1}\Big(\sum_{n=0}^{N-1} g_n(x_n^{\epsilon,\mathbf{u}}, u_n) + g_N(x_N)\Big)\Big] | z_N^{\epsilon,\mathbf{u}} \Big\}.$$

We have therefore transformed the partially observed problem (14)(5) into the equivalent totally observed (18)(19). The value function $V_n^\epsilon(z_n)$ for this last problem will be referred to as the value function for (14)(5).

It turns out that Theorem 3 cannot immediately applied to this transformed problem, since the probability kernels $P_n^{z,\epsilon}$ do not necessarily satisfy Assumption A. However, using Assumption B and a refinement of Lemma 2, the proof can be repeated in its essential lines, yielding the Theorem that follows. The details are, however, rather long and technical, and are omitted. The complete analysis of the partially observed case is the subject of a forthcoming paper (Dai Pra and Rudari (1995b)).

Theorem 4 *Suppose Assumption B is satisfied and, for $n = 0, \ldots, N-1$, let $f_n : X \times U \times W \to X$, $h_n : X \times V \to Y$ be functions such that for each $(x, u) \in X \times U$ the maps $w \to f_n(x, u, w)$ and $v \to h_n(x, v)$ are is onto. Then the value function $V_n^\epsilon(z_n)$ for the problem (14)(5) is such that*

$$\lim_{\epsilon \to 0} \epsilon \log V_n^\epsilon = S_n \tag{20}$$

exists, uniformly on $Y^n \times U^{n-1}$. Moreover, $\sup_{z_0} S_0(z_0)$ is the value function of the dynamic game (16)(17) with

$$\begin{aligned}
G_N(x) &= g_N(x) \\
G_n(x,u,w,v) &= g_n(x,u) - H_n^P(f_n(x,u,w);x,u) - H_n^Q(h_n(x,v);x) \\
&\quad \text{for } n = 1,\dots,N-1 \\
G_0(u,w) &= g_0(x_0,u) - H_0^P(f_0(x_0,u,w);x_0,u).
\end{aligned} \tag{21}$$

Example 5 Consider the risk-sensitive stochastic control problems with cost functional given by (6), and dynamical equations

$$\begin{aligned}
x_{n+1}^{\epsilon,\mathbf{u}} &= F\Big(a_n(x_n^{\epsilon,\mathbf{u}},u_n) + c_n(x_n^{\epsilon,\mathbf{u}},u_n)w_n^\epsilon\Big) \\
x_0^{\epsilon,\mathbf{u}} &= x_0 \\
y_n^{\epsilon,\mathbf{u}} &= G\Big(b_n(x_n^{\epsilon,\mathbf{u}}) + d_n(x_n^{\epsilon,\mathbf{u}})v_n^\epsilon\Big)
\end{aligned} \tag{22}$$

where $x_n^{\epsilon,\mathbf{u}}, w_n^\epsilon \in \mathbb{R}^d$, $y_n^{\epsilon,\mathbf{u}}, v_n^\epsilon \in \mathbb{R}^k$, $u_n \in U \subset \mathbb{R}^q$ compact, a_n, b_n are bounded uniformly continuous functions, c_n, d_n are matrix valued functions, uniformly continuous, bounded, and such that $c_n c_n^T, d_n d_n^T$ have a spectrum which is uniformly bounded away from zero. Moreover F, G are injective, uniformly continuous functions, and $w_0^\epsilon, \dots, w_{N-1}^\epsilon, v_0^\epsilon, \dots, v_N^\epsilon$ are two sequences of i.i.d. random variables with the following distribution:

$$w_n^\epsilon \sim C_1 \sum_{m \in M} \alpha_m \exp[\frac{1}{2\epsilon}\|w - m\|^2]$$

$$v_n^\epsilon \sim C_2 \sum_{m \in N} \beta_m \exp[\frac{1}{2\epsilon}\|v - m\|^2]$$

where $\alpha_m, \beta_m > 0$, M, N are finite subsets of $\mathbb{R}^d$ and $\mathbb{R}^k$ respectively, and C_1, C_2 are normalization factors. The model studied in James, Baras and Elliott (1994) corresponds to $F = id$, $G = id$, c_n, d_n scalar matrices, M, N consisting of a single element. By using the results in Dai Pra and Rudari (1995a), we have that the transition probabilities corresponding to (22) form ULDF's with rate functions

$$H_n^P(x';x,u) = \frac{1}{2}\min_{m \in M} \|F_n^{-1}(x') - m - a_n(x,u)\|^2$$

$$H_n^Q(y;x) = \frac{1}{2}\min_{m \in N} \|G_n^{-1}(x) - m - b_n(x)\|^2.$$

If we assume the functions g_n in (6) to be uniformly continuous and bounded, then it can be shown that Assumption B is satisfied (see Dai Pra and Rudari (1995b)). By using Theorem 4 one obtain, among all possible choices, the following dynamic game:

$$\begin{aligned}
x_{n+1} &= F\Big(a_n(x_n,u_n) + c_n(x_n,u_n)w_n\Big) \\
y_n &= G\Big(b_n(x_n) + d_n(x_n)v_n\Big)
\end{aligned}$$

$$K(\mathbf{u}) = \sup_{\mathbf{w},\mathbf{v}} \Big[g_0(x_0,u_0) - \tfrac{1}{2}\,\text{dist}^{\,2}(w_0,M) + \sum_{n=1}^{N-1} \Big(g_n(x_n,u_n) - \tfrac{1}{2}\,\text{dist}^{\,2}(w_n,M) - \tfrac{1}{2}\,\text{dist}^{\,2}(v_n,N) \Big) + g_N(x_N) \Big].$$

REFERENCES

Barron, E.N. and Jensen, R. (1989) Total risk aversion, stochastic optimal control and differential games. *Appl.Math.Optimiz.* **19**, 313–27.

Bertzekas, D.P. (1976) *Dynamic Programming and Stochastic Control.* Academic Press, London.

Dai Pra, P. and Rudari, C. (1995a) Large deviation limit for discrete-time totally observed stochastic control problems with multiplicative cost. To appear in *Appl.Math.Optimiz.*.

Dai Pra, P. and Rudari, C. (1995b) Large deviation limit for discrete-time partially observed stochastic control problems with multiplicative cost. In preparation.

Fleming, W.H. and McEneaney, W.M. (1992) *Risk Sensitive Control and Differential Games.* Springer LN in Control and Inf. Sci., No.184, 185–97.

Fleming, W.H. and McEneaney, W.M. (1995) Risk sensitive control on an infinite time horizon, to appear on *SIAM J. on Control and Optimiz.*

Jacobson, D.H. (1973) Optimal Stochastic Linear Systems with Exponential Performance Criteria and their relation to Deterministic Differential Games. *IEEE Trans.Autom. Control*, AC-18 (2), 124–31.

Jacobson, D.H. (1977) *Extensions of Linear-Quadratic Control, Optimization and Matrix Theory.* Academic Press, London.

James, M.R., J.S.Baras, J.S. and Elliott, R.J. (1994) Risk-Sensitive Control and Dynamic Games for Partially Observed Discrete-Time Nonlinear Systems. *IEEE Trans.Aut.Control.* **39**, 780–92.

Runolfsson, T. (1994) The equivalence between infinite-horizon optimal control of stochastic systems with exponential-of-integral performance index and stochastic differential games, *IEEE Trans.Aut. Control* **39**, 1551–63.

Varadhan, S.R.S. (1984) *Large Deviations and Applications.* Society for Industrial and Applied Mathematics, Philadelphia.

Whittle, P. (1990) A risk-sensitive maximum principle, *System and Control Letters*, **15**, 183–92.

Whittle, P. (1991) A risk-sensitive maximum principle: the case of imperfect state observation. *IEEE Trans. Autom. Control*, AC-36 (7), 793–801.

18

Dynamic portfolio optimization based on reference trajectories

Andrzej M.J. Skulimowski
Institute of Automatic Control, University of Mining & Metallurgy
Cracow, Poland

Abstract

We propose a multicriteria model for the dynamic multi-stage portfolio optimization. The information available at each decision step is applied to attain simultaneously the maximal rate of return, minimal risk, and the maximal investment flexibility by an appropriate selection of an optimal compromise decision. The input data concerning the portfolio structure and the state of economical environment at each moment of time is provided in form of the data matrices, transformed then to a state vector, only a part of it being known to the decision-maker. The evolution of the portfolio structure is described as a discrete-time control system, with the sell/buy transactions as controllable events. Missing coefficients are estimated using the Kalman filtering techniques. The investor's goals are modelled as reference points in the space of the criteria values for different planning horizons. To find the best-compromise allocation strategy we propose a decision-making method applying multiple reference points in dynamic multicriteria problems. This method can be implemented as an on-line interactive decision support system.

Keywords

Multicriteria optimization, portfolio theory, dynamic programming, discrete-time control systems.

1 INTRODUCTION

This paper addresses the problem of dynamic selection of the portfolio structure so that the criteria of maximal rate of return, minimal risk, maximal investment flexibility, and the demanded consumption curve were simultaneously taken into account while looking for an investment strategy. We apply consequently the multicriteria approach, so that the method proposed should be flexible enough to model the preferences of a possibly large class of investors, being thus a universal tool for a portfolio manager dealing with investors characterized by diversified risk attitudes. The desired consumption distribution over time serves at the same time as a trajectory objective and a reference trajectory in the multicriteria decision-making problem. As a special feature of the decision support model here presented, we allow to calculate the current value of portfolio and those of returns in several currencies simultaneously. Further, the user of the system is allowed to create his/her own artificial currencies fitting best his/her particular utilities.

Research supported in part by Swiss National Fund for Scientific Research, Grants No.12-30240.90 and 21-0377738.93/1.

In the sequel we will admit no special limitation regarding the portfolio structure under consideration, which may contain stocks, corporate bonds, government bonds, treasury bills, precious metals, real estate, and liquidity, as well as options and futures on stocks, currencies, composite indices, bonds, and metals. Nevertheless, we will concentrate our attention on pure security portfolios, especially stocks and options on stocks and currencies. Each of the above will be referred to as a portfolio category, which in the sequel may be subdivided into the country of issue, place of trade, the currency of trade, and the physical place of storing or the bank account where it is deposited.

The portfolio structure for each moment from the time interval considered is described in form of the state vector in a discrete-time control system. The data concerning the economical environment used in the optimization process are stored in the data matrices, being known only partly to the decision-maker. Missing coefficients are estimated using the Kalman filtering techniques. The investor's goals are modelled as reference points in the space of the criteria values for different planning horizons. To find the best compromise allocation strategy we propose a decision-making method applying multiple reference points in dynamic multicriteria problems. The method can be implemented as an on-line interactive decision support system.

Summarizing, the main task of the decision support system proposed is to provide the decision-maker with a possibly best compromise investment strategy, taking into account its investment goals, risk aversion (or attitude), an attitude to make flexible investment decisions, and the possibility of adaptively modifying the selected investment strategy in a response to the changes of the environment. Such strategy will be derived based on economic data such as past and current prices, price expectations, turnover *etc.*, provided by the decision-maker from external knowledge bases and evaluated within the system.

2 PRINCIPLES OF PORTFOLIO MODELLING

Following the above remarks, the physical structure of portfolio at a moment t will be described by the vector $x(t) \in \mathbb{R}^n$. The unit prices for each portfolio category are expressed in m different currency units and are contained in the real matrix $P(t) \in M_{n(t),m(t)}$, i-th row of $P(t)$ containing the prices in the i-th currency for $i = 1, ..m(t)$.

By the portfolio transformations we will mean the sell/buy transactions, the generation of options, the expiration and exercising of options and futures, the conversion of convertible bonds and warrants, and some other processes. The portfolio transformations are described by the equation

$$x(t+1) = A(t)x(t) + B(t)u(t) + \xi(t) \quad (1)$$

with the matrix $A(t)$ determined by the transformations which cannot be influenced by the decision-maker, while the control variables $u(t)$ can be fully determined, perhaps subject to a set of constraints. The matrix $B(t)$ models the control laws, and $\xi(t)$ is the stochastic disturbance which can be interpreted as unexpected gains or losses. The evolution of prices may be described by the following rule

$$P(t+1) = \psi(P(t), \varepsilon(t), \zeta(t)) \quad (2)$$

where ψ is the price formation mechanism (usually known only partly), $\varepsilon(t)$ is the state vector of economic environment, which may also be at least in part unknown, and $\zeta(t)$ is the random disturbance factor. Thus the increase $\Delta V(t,t+1)$ of the portfolio value $V(t)$ on the time interval $[t,t+1)$, is given by the formula

$$\Delta V(t,t+1) \; := \; V(t+1) - V(t) \; = \tag{3}$$
$$= \; [P^T(t+1)A(t) - P^T(t)]x(t) + R(t,t+1) + P^T(t)b(t) - C(t-1,t) - P^T(t)s(t),$$

where $b(t)$ is the vector of buys, $s(t)$ - the vector of sales, and $R(t-1,t)$ and $C(t-1,t)$ are the direct net revenue (such as dividends or interests), and consumption vectors for the period $(t-1,t]$, respectively. Throughout the paper the symbol A^T will denote the transposition of the matrix A.

Each of m rows of $P(t), p_j(t)$, as well as each coordinate $r_j(t)$ of $R(t)$ corresponds to a monetary unit the current value may be expressed in. As the monetary units one can use real currencies, each one of portfolio categories, artificial currencies such as ECU, *etc.* In the most common case of one privileged monetary unit, $P(t)$ consists of a single price vector $p(t)$. The coordinates of the vector $x(t)$ may be interpreted either as physical quantities of each portfolio category, *e.g.* number of stocks of specific companies, weight of gold, amount of cash *etc.*, in this paper called items, or as estimated values of real estate.

For the sake of simplicity, in the sequel we will assume that the decision-maker is able to choose one of the currencies, or define own value function, so that one can employ the scalar measure v for the portfolio value. Hence (3) can be rewritten as

$$\Delta v(t,t+1) := v(t+1) - v(t) =$$
$$= [p^T(t+1)A(t) - p^T(t)]x(t) + R(t,t+1) + p^T(t)b(t) - C(t-1,t) - p^T(t)s(t),$$

where $v(t) := \phi(V(t)),\;\; r(t-1,t) := \phi(R(t-1,t)),\;\; c(t-1,t) := \phi(C(t-1,t))$ for certain aggregating currency function $\phi : I\!R^n \to I\!R$. As an example of ϕ may serve the projection on one of the coordinates corresponding to the selection of a single currency, or a linear combination of the coordinates which imposes the book-keeping in an artificial currency such as ECU.

2.1 Modelling the portfolio transformations

The revenue vector R may be represented as the sum

$$R(t-1,t) = \sum_{i=1}^{n} r_k(t-1,t), \tag{4}$$

where $r_k,\;\; k = 1,..n$, is the revenue vector coming from the k-th portfolio category. On the other hand, if the portfolio structure changes according to

$$x(t) = A(t-1,t)x(t-1) + \sum_{k=1}^{n}[x_{kj+}(t-1,t) - x_{kj-}(t-1,t)],$$

where $A(t-1,t)$ is the *growth rate matrix*, then the total revenue on the interval $[t-1,t)$ equals to

$$R(t-1,t) = P^T(t)x(t) - P^T(t-1)A(t-1,t)x(t-1).$$

The values of

$$s_{kj}(t-1,t) := r_{kj}(t-1,t)/(M_{kj}(t-1)x_k(t-1)) \tag{5}$$

express the net rate of return of k-th portfolio category expressed in the j-th monetary unit, for $k = 1,..n,\;\; j = 1,..m$. At each moment of time t the structure of the portfolio vector $x(t)$ may be transformed according to the constraints

$$\sum_{k=1}^{n} x_{kj+}(t-1,t)p_{kj+}(t-1,t) \; \leq \tag{6}$$

$$\leq \sum_{k=1}^{n} x_{kj-}(t-1,t)p_{kj-}(t-1,t) + r_j(t-1,t) + d_j(t-1,t), \qquad \text{for } j = 1,..m,$$

where $P_+(t-1,t)$ and $P_-(t-1,t)$ are the matrices of minimal buy, and maximal sale prices for each portfolio category over the period $[t-1,t]$, respectively, $d_j(t-1,t)$ is the debt in the j-th monetary unit and $x_{kj+}(t-1,t)$ and $x_{kj-}(t-1,t)$ are the amounts of each items bought, or resp. sold over the period $[t-1,t]$ and fulfilling the balance equations

$$x_k(t) = x_k(t-1) + \sum_{j=1}^{m}(x_{jk+}(t-1,t) - x_{jk-}(t-1,t)), \quad \text{for } k = 1,..n. \tag{7}$$

A simple exchange transaction consisting in selling the quantity $x_{kj-}(t-1,t)$ from the k-th portfolio category at price $p_{kj-}(t-1,t)$ and buying an equivalent amount of the i-th item, $x_{ij+}(t-1,t)$, at the price $p_{ij+}(t-1,t)$ expressed in the same j-th monetary unit may be written as

$$x_{ij+}(t-1,t) = p_{kj-}(t-1,t)/p_{kj+}(t-1,t)x_{kj-}(t-1,t).$$

The mixed exchange transaction involving the payment in various monetary units and investing a part of the revenue may be expressed as

$$x_{k+}(t-1,t) = \sum_{j=1}^{n}[\sum_{i=1}^{m} p_{ij-}(t-1,t)x_{ikj-}(t-1,t)/p_{kj+}(t-1,t) + r_j(t-1,t)/p_{kj+}(t-1,t)], \tag{8}$$

where $x_{ikj+}(t-1,t)$ denotes the quantity of i-th item sold for the j-th monetary unit and used for buying k-th item.

Arbitrary change of the structure vector between the moments $t-1$ and t according to the constraints (8) will be called *investment*. It is easy to see that each investment can be represented as the superposi- tion of mixed exchange transactions.

In the sequel we will assume that the structure of the portfolio vector at moment t may be influenced by the values of $x(t-1)$, which may be expressed *e.g.* as the *growth rate constraints*

$$x_i(t) - x_i(t-1) \leq \sum_{j \in J} \lambda_j(t)x_j(t-1) \quad \text{for certain} \quad J \subset \{1,...n\}. \tag{9}$$

The coefficients $\lambda_j(t)$ represent the relative amounts of j-th items in the portfolio which can be used for buying the i-th items at the moment t.

2.2 Description of portfolio evolution as a discrete-time control system

An r-th complex transaction protocol may be available as the *transition form* $\Gamma_r(t,t+1)$, i.e the matrix

$$[q^r_{ij}]_{1 \leq i \leq N(t), 1 \leq j \leq N(t+1)}$$

The value of q^r_{ij} denotes the (physical) amount of the i-th portfolio category changed for the j-th category on the interval $[t,t+1)$ within the r-th transaction, and $N(t)$ denotes the total amount of portfolio classes considered. Usually, by selling the various portfolio catego- ries an amount $q_{\beta\nu}(t,t+1)$ of the β-th category will be exchanged for liquidity $b_\nu(t+\tau)$, and an amount of $p_{\beta\nu}(t)q_{\beta\nu}(t,t+1)$ will be located on the ν-th account after the time lapse τ. Analogously, a buy consists in replacing the liquidity $b_\nu(t+\tau)$ by a specified amount of the μ-th portfolio category. For the sake of simplicity we will consider immediate buys only.

The *aggregated transaction matrix* is the matrix

$$\Theta(t,t+1) = [q_{ij}(t,t+1)]_{1\leq i\leq N(t), 1\leq j\leq N(t+1)} \quad (10)$$

derived from the matrices $\Gamma_r(t,t+1)$, $r = 1,...p$. It contains the absolute values of changes of each portfolio category on [t,t+1).

The relative values $r_{ij}(t,t+1) := q_{ij}(t,t+1)/x_i(t)$ will be stored in the matrix $A_2(t)$ of the same dimensions that Θ. It is easy to see that then we get the equation

$$x(t+1) = A_1(t)x(t) + A_2(t)\Theta(t,t+1)e + \xi(t) \quad (11)$$

where $e = (1,..1) \in I\!R^{N(t)}$ and $\xi(t)$ is the random perturbance representing the unexpected losses or gains, which constitutes the simplest form of the state-space portfolio model. Observe that the matrix A_1 need not be quadratic as the dimensions $N(t+1)$ and $N(t)$ of $x(t+1)$ and $x(t)$, respectively, may be different in case where new portfolio category is appended to the portfolio vector between t and $t+1$, or if another one is not to be considered any more. The next step consists in distinguishing a subset of controlled actions and transforming the system to the form

$$x(t+1) = A(t)x(t) + B(t)u(t) + w(t) + \xi(t), \quad (12)$$

where $u(t)$ are controls and $B(t)$ is the control matrix describing the control laws. As non-controlled events $w(t)$ should be regarded those transactions which are imposed to the decision-maker. as *e.g.* the obligatory future realization after assignment, while all those being freely chosen are controls.

3 FORMULATION OF THE MULTICRITERIA OPTIMIZATION PROBLEM

Below we will show an application of reference points in the criteria space to select a satisfactory solution to the portfolio management problem at each step of the optimization process.

The problem of choice of an optimal portfolio structure at a moment t will be modelled as the multicriteria optimization problem

$$(F_t : U \rightarrow I\!R^N) \rightarrow \min, \quad (13)$$

where U is the decision space, $I\!R^N$ the criteria space, and $F_t(u) = (F_1(t,u), F_2(t,u), ...,$ $F_N(t,u))$ is a vector objective function for the decision step [t,t+1). Let us recall that a solution u_0 to the multicriteria optimization problem (13) is called *Pareto-optimal*, or nondominated, iff the admission of no other control results in the values of $F_1,...F_N$ being all better than those achieved for u_0 , i.e. for the simultaneous minimization of F_t for fixed moment of time t :

$$\forall u \in U \ [\forall 1 \leq i \leq N \ \ F_i(t,u) \leq F_i(t,u_0)] \ \Rightarrow \ u = u_0.$$

Here, we will consider the following three most fundamental criteria for each single decision step :

- the total yields on the interval $[t-1,t)$, $F_1(t,u)$,
- the rate of risk, $(F_2(t,u)$,

and

- the amount of liquidity available immediately at time t, $F_3(t,u)$.

To find an optimal investment strategy for the period of time [0,T] we will request the decision-maker to formulate the additional information about his preferences in form of *reference trajectories*. Then we will interactively generate the nondominated solutions to (13) which fulfill the decision-maker's demands. The compromise strategy thus being found may be regarded as a result of the maximization of an a priori unknown utility function over the set of nondominated points.

A reference point is an element of the criteria space which represents the values of criteria being of a special importance to the decision-maker. A *reference trajectory* consists of reference points defined over certain planning period. In the above decision model the reference points may be results of investor's presumptions regarding the investment process, *e.g.* one can assume that 8% rate of return at a 3% risk rate with the confidence level 90% is a satisfactory solution, while the parameters (12%, 1%, 99%) would be a desired solution. The methods of simultaneously applying multiple reference points to select a compromise decision in a multicriteria decision making problem has been studied in the recent years by the author Górecki, Skulimowski (1986), Górecki, Skulimowski (1988), Skulimowski, Schmid (1992).

Besides of the well-known desired values of criteria in the multiple reference points method one can consider also anti-ideal reference points, which achievement may be regarded as the failure, the solutions available at the pre-decision stage (or status-quo reference points), and the lower limits of optimality which determine the limits of the optimal values (if any). The reference points come as results of experts judgments independently from the statement of the multicriteria problem (13). The resulting model of decision-maker preferences (see Górecki, Skulimowski (1986) for more details) consists of two subsets of the criteria space R_{-1} and R_1 , aggregated as convex hulls of the neighboring classes of reference points, and two functions g_1 and g_{-1} modelling the decision-maker's wish to reach the set R_1 and to avoid R_{-1}.

To select a solution, taking into account all above preference structure, we find the set D which contains all solutions nondominated in the sense of the maximization of g_{-1} and minimization of g_1 . Applying this method we reduce the decision problem to the bicriteria trade-off between the measures of proximity to the sets R_1 and R_{-1} . Thus we confine our interactive search for a compromise solution to the intersection of $D$and of the set of nondominated points to the multicriteria optimization problem (13). This procedure will be repeated on-line for partial optimization problems ending at each moment $t, t \in [t_1, T], t_1 > 0$.

3.1 The trajectory optimization problem

A general formulation of the multicriteria optimal control problem for the dynamical system described by difference equations may be presented as follows:

$$x(t+1) = f(x(t), u(t)), \quad x_0 \in X_0, \quad u(t) \in U(t) \quad \text{for } t_0 \leq t \leq T, \tag{14}$$

where $x(t) \in I\!R^n, u(t) \in I\!R^k$, and the constraint set $U(t)$ is a closed and convex subset of $I\!R^k$. The classical multicriteria optimization task for the above system consists in simultaneous minimization of the functions $F(u) := (F_1(u), ..., F_N(u))$ for the terminal time T. This formulation leads usually to finding a Pareto-optimal control u_{opt} satisfying some additional conditions resulting from the consideration of the decision-maker's preferences which are not included in the preliminary problem formulation.

In contradistinction to (13)-(14), a *trajectory optimization problem* (see Wierzbicki (1988)) involves the consideration of the values of $F(x(t), u(t))$ for the intermediary values of t, in the sense that they should be optimal for each moment of time. Thus this problem is transformed in fact into the multicriteria problem with an number of objectives equal to the number of time moments considered. The additional preference information may also be given in form of conditions concerning a subinterval of $[t_0, T]$, as *e.g.*

$$F(u(t)) \in Q(t) \quad \text{for} \quad t \in [t_1, T] \tag{15}$$

where Q is so-called reference multifunction for (13) - (15) and $t_1 \in [t_0, T]$. An approach to solve a trajectory optimization problem, which has been admitted for portfolio optimization, consists in finding a finite sequence of time instances $t_i \in [0, T]$, $i = 1, ...k$, such that the original problem is equivalent to the multicriteria optimization problem with the objectives being the momentary values of $F(x(t), u(t))$ for $t := t_i, i = 1, ..k$. Further, in the reference output trajectory problem considered here to model the consumption demands, we have explored the dependence of the values of the remaining criteria on the consumption curve to reduce the dimensionality of the problem.

3.2 Multiple reference point and reference trajectory method applied for dynamic portfolio optimization problems

The solution of a portfolio optimization problem based on the above presented method can be undertaken in the following steps.

I. The modelling stage.

M1. The portfolio structure must be described symbolically using the prescribed format (cf. Skulimowski, Schmid (1992)). The system will automatically derive the state vector. Each state variable will have a fixed meaning available to the user as the interpretation vector.

M2. The transaction tables have to entered into the system. The transaction matrix will be created automatically.

M3. The price information must be supplied to the system.

M4. The planning period, the past data, and transaction constraints are to be specified by the user.

II. The goal setting and interactive solution procedure.

G1. The level of risk it a, desired revenues, and the amount of liquidity are to be predefined for the next series of interactive steps.

G2. The user defines its desired consumption curve as a reference trajectory over the planning period

G3. Step G2 is repeated for different consumption curves so as the revenues desired in step G1 were achieved at a risk level it a. If these goals cannot be achieved, the system automatically points out the time moments to update the consumption in the next interactive step.

G4. If no compromise solution can be found : step G1 is repeated for interactively modified values of risk, revenues, and liquidity levels.

III. The implementation of the decision suport system.

S1. Once a compromise decision is made, a corresponding sequence of optimal transactions is calculated by the system and communicated to the decision-maker. Missing states of environment are estimated using the generalized Kalman filtering techniques.

S2. The optimal values of all performance criteria are calculated and the results visualized on the screen.

4 DISCUSSION

The portfolio modeling and optimization methodology here proposed may be regarded as a practical supplement of theoretical results on , compare *e.g.* the results presented in Östermark (1991), Perold, Sharpe (1988) and in the references cited therein. The presented portfolio optimization model would be especially well suited for short- and middle-range optimal strategy planning. It is expected that the best results can be achieved for calculating single-session strategies in case of securities trading. An implementation of long-term economical forecasts into the model, as *e.g.* the estimations of long-term price trends would provide a base for strategic decision support but would far exceed the framework of this study. The underlying idea while deriving this model has been a possibility of implementing it as a integrated interactive portfolio management and optimization package, in several versions fitting the needs of small through large investors. Some of the main elements of such package have been already implemented in the prototype versions:

- the description and coding of the portfolio structure in the state vector can be done with the tools provided by the system for the management and evaluation of quantitative information developed at the Institute of Computer Science, University of St. Gallen;
- the prototype of the decision support system based on multiple reference points, making also possible an application of reference trajectories and trajectory objectives, has been implemented including additional features especially suited for the portfolio management purposes;
- the databases, user interfaces, and the interface of the multicriteria decision support system to the implementation of the discrete-time stochastic control system has been established using standard software tools.

The database storing the sell-buy transactions, and determining the transaction plausibility by finding the demanded value of the margin requirement, which is especially important for options trading, has been coded as a part of a package developed by a commercial software company. The software package here presented can also be regarded as a supplement to other financial analysis products offered on the market.

REFERENCES

Górecki, H., Skulimowski, A.M.J. (1986) A joint consideration of multiple reference points in multicriteria decision-making. *Found. Contr. Engrg.*, **11**, No.2.

Górecki, H., Skulimowski, A.M.J. (1988) Safety Principle in Multiobjective Decision Support in the Decision Space Defined by Availability of Resources. *Arch. Automatyki i Telemech.*, **17**, 68–87.

Östermark, R. (1991) Vector forecasting and dynamic portfolio selection : Empirical efficiency of recursive multiperiod strategies. it EJOR, bf 55, 46–56.

Perold, A.F., Sharpe, W.F. (1988) Dynamic Strategies for Asset Allocation. it Financial Analysts Journal, 16–27.

Skulimowski, A.M.J., Schmid, B.F. (1992) Redundance-free Description of Partitioned Complex Systems. *Mathl Comput. Modelling*, **16**, 71–92.

Wierzbicki, A.P. (1988) Dynamic Aspects of Multi-Objective Optimization, in *Multiobjective Problems of Mathematical Programming*, held in Yalta, Oct. 26 - Nov. 2, 1988, *Lecture Notes in Economics and Math. Systems*, **351**, Springer-Verlag.

19

Stability analysis of time–varying discrete interval systems

Karel Sladký
Academy of Sciences of the Czech Republic,
Institute of Information Theory and Automation
P. O. Box 18, CZ–182 08 Prague 8, Czech Republic. Tel: 42-2-66414906.
Fax: 42-2-66414903. e-mail: sladky@utia.cas.cz

Abstract
In this article necessary and sufficient conditions for the stability of time-varying discrete interval systems are studied and the corresponding stability margins are obtained. Our analysis heavily employs the Perron–Frobenius theorem for nonnegative matrices and its extensions to a class of interval matrices. We show that by combining the approaches suggested in Bauer *et al.* (1993) and Han and Lee (1994) the existing tests on stability and stability margins both of time-varying and time-invariant discrete interval systems can be considerably improved. The obtained results are tested on several numerical examples.

Keywords
stability of linear systems, interval matrices, nonhomogeneous matrix products

1 INTRODUCTION

An interval matrix is a real matrix in which all the elements are known only to the extent that each belongs to specified closed interval. In particular, an $r \times r$ interval matrix A_I is in fact a set of real matrices

$$A_I = \{A = [a_{ij}] : a_{ij} \in [b_{ij}, c_{ij}], \quad i,j = 1,\ldots,r\}$$

where $b_{ij} \le c_{ij}$ are given real numbers. Let $B = [b_{ij}]$, $C = [c_{ij}]$, and hence $A_I = [B \quad C]$.

The time-varying discrete system is given for $k = 0, 1, \ldots,$ by

$$x(k+1) = A(k)\,x(k), \quad x(0) = x_0 \tag{1}$$

where $x(k) = [x_i(k)]$ is an r-dimensional state vector and $A(k) \in A_I$.

The system (1) is (globally) asymptotically stable if for every $x(0) = x_0$ and every matrix sequence $\{A(k), k = 0, 1, \ldots : A(k) \in A_I\}$ $\lim_{k\to\infty} x(k) = 0$. The time-varying system (1) is said to be stable with stability margin h, where $0 \le h < 1$ (or to have the degree of stability h), if there exists a number $c \geq 0$ (depending on x_0) such that for an arbitrary matrix sequence $A(k) \in A_I$ it holds that $|x_i(k)| \leq c(1-h)^k$ for $i = 1, \ldots, r$. Observe that from our definition of the stability margin we immediately conclude that the time-invariant discrete system (1) (i.e. if in (1) $A(k) \equiv A \in A_I$, $\forall k = 0, 1, \ldots$) is stable with stability margin h if and only if the modulus of every eigenvalue of any $A \in A_I$ is less than $1-h$. A matrix $A \in A_I$ is called extreme (vertex) if for every $i, j = 1, 2, \ldots, r$ either $a_{ij} = b_{ij}$ or $a_{ij} = c_{ij}$, and is (asymptotically) stable iff the system $x(k+1) = A\,x(k)$, $\forall k$ is (asymptotically) stable. Considering some $A \in A_I$, the symbol $|A|$ is reserved for the corresponding modulus matrix, i.e. $|A| = [|a_{ij}|]$, where $|a_{ij}|$ is the absolute value of a_{ij}. In the sequel, we denote by $\bar{A}_I$ the (finite) set of all extreme matrices. Obviously, $\bar{A}_I \subset A_I$ and $\bar{A}_I$ possesses not more than 2^{r^2} different matrices. An extreme matrix $\hat{A} \in \bar{A}_I$ is called *dominating* if $|\hat{A}| \geq |A|$ for any $A \in \bar{A}_I$ (obviously then also $|\hat{A}| \geq |A|$ for any $A \in A_I$).

The stability analysis of dynamic interval systems is very important in the robust controller design. In recent years, stability of dynamic interval systems has been studied by many authors and some sufficients conditions for the stability have been obtained mostly for time-invariant interval systems. In particular, Jiang (1988) has attempted to show that the time-invariant discrete system (1) is asymptotically stable, if and only if all extreme matrices are asymptotically stable. This result was shown to be incorrect (cf. *e.g.* Kolla and Farison (1988) and the Correspondence in *International Journal of Control* (1989)). However, it was shown in Mori and Kokame (1987) and in Juang *et al.* (1989a) that the time-invariant system (1) is asymptotically stable if the norm of the extreme matrices is restricted to be less than unity.

In the present contribution we focus our attention on the stability of time-varying interval systems. This topic was recently studied in Bauer *et al.* (1993) and in Han and Lee (1994). In Bauer *et al.* (1993) a necessary and sufficient condition for the stability of time-varying discrete interval matrices was presented. The condition is based on checking and evaluating the products of extreme matrices using standard matrix norms. Even the authors presented an algorithmic procedure for generating and checking these products, there are severe computational limits of an actual implementation of the algorithm since the computational complexity increases exponentially with the size of the considered interval matrix (recall that the number of extreme matrices from $\bar{A}_I$ equals 2^{r^2}). On the other hand in Han and Lee (1994) a simple sufficient condition for the stability of time-varying discrete interval systems was presented. We show that the sufficient condition on stability of time-varying discrete interval systems presented in Han and Lee (1994) can be easily obtained from the Perron-Frobenius theorem on maximal eigenvalue of a nonnegative matrix. The simple test on stability suggested in Han and Lee (1994) can be further amplified by combining this test with the approach presented in Bauer *et al.* (1993) and by evaluating the resulting matrices by weighted matrix norms. The obtained results are also useful for improving existing stability results on time-invariant systems.

2 PRELIMINARIES

Recall that, according to the well-known Perron–Frobenius theorem (cf. *e.g.* Gantmacher (1966), Berman and Plemmons (1979)), the spectral radius, denoted $\rho(\cdot)$, of a nonnegative matrix is equal to its largest positive eigenvalue (called the Perron eigenvalue) and the corresponding left, right eigenvector (denoted $v(\cdot)$, $u(\cdot)$ and called the left, right Perron eigenvector respectively) can be selected nonnegative. In case that the matrix is irreducible, the Perron eigenvalue is simple, the Perron eigenvector is unique up to a multiplicative constant and can be selected positive.

In particular, for the dominating matrix $\hat{A}$ it holds (remember that symbols $\geq$, $>$ in a matrix relation are considered componentwise)

$$\rho(|\hat{A}|)\,u(|\hat{A}|) = |\hat{A}|\,u(|\hat{A}|) \geq |A|\,u(|\hat{A}|), \quad \rho(|\hat{A}|)\,v(|\hat{A}|) = v(|\hat{A}|)\,|\hat{A}| \geq v(|\hat{A}|)\,|A| \tag{2}$$

where the left, resp. right, eigenvector $u(|\hat{A}|) \geq 0$, resp. $v(|\hat{A}|) \geq 0$. Observe that on premultiplying the first relation of (2) by $v(|A|)$, we immediately conclude that $\rho(|\hat{A}|) \geq \rho(|A|)$ for every $|\hat{A}| \neq |A|$.

Recall that the one norm (denoted $\|\cdot\|_1$) and the maximum (or infinity) norm (denoted $\|\cdot\|_\infty$) of the matrix $A = [a_{ij}]$ is given by $\max_j \sum_{i=1}^{r} |a_{ij}|$ and by $\max_i \sum_{j=1}^{r} |a_{ij}|$ respectively. Similarly, for every $u \in \mathbb{R}_+^r$ (i.e. $u > 0$, where $u = [u_1, \ldots, u_r]$) the u-weighted maximum norm of the matrix A is given by $\|A\|_u = \max_{i=1,\ldots,r} \sum_{j=1}^{r} |a_{ij}|\frac{u_j}{u_i}$, and hence the u-weighted maximum norm of a (column) vector $d \in \mathbb{R}^r$, $\|d\|_u = \max_{i=1,\ldots,r} |d_i|$. It can be easily shown that $\|\cdot\|_u$ is indeed a norm, in particular, $\|A\|_u \geq 0$ with $\|A\|_u = 0$ iff $A = 0$; $\|\lambda A\|_u = |\lambda|\|A\|_u$ for every $\lambda \in \mathbb{R}$; and if $A, B \in \mathbb{R}^{r \times r}$, then $\|A+B\|_u \leq \|A\|_u + \|B\|_u$. Moreover, supposing that $A = [a_{ij}] \in A_I$ is irreducible, let $u > 0$ be selected such that $\rho(|A|)\,u = |A|\,u$. Then $\|A\|_u = \rho(|A|)$ (observe that $\sum_{j=1}^{r} |a_{ij}|\frac{u_j}{u_i} = \rho(|A|) \quad \forall i = 1, \ldots, r$). In case of reducible A, for an arbitrary small $\varepsilon > 0$ we can find $u^\varepsilon > 0$ such that $\rho(|A|) + \varepsilon \geq \|A\|_{u^\varepsilon} \geq \rho(|A|)$.

Observe that if $A \in A_I$ is stable, it may happen that $\|A\| > 1$ both for the one and the maximum norm. Consider *e.g.* $A_I = \begin{bmatrix} 0.8 & 9.2 \\ 0.001 & 0.8 \end{bmatrix}$, then $\|A\|_1 = \|A\|_\infty = 10$; however, calculating the eigenvalues we can conclude that the Perron eigenvalue is equal to $\rho(A) = 0.8959$ and the corresponding right Perron eigenvector can be set to $u = [85.92 \;\; 1]$. Hence the weighted u-maximum norm of the matrix $\|A\|_u = 0.8959$.

Stability criteria for time-invariant discrete systems based on examination of extreme matrices are summarized in the following proposition (unless otherwise stated, $\|\cdot\|$ denotes an arbitrary matrix norm):

Proposition 1 (cf. Mori and Kokame (1987), Juang *et al.* (1989a)).
The time-invariant discrete system (1) is globally asymptotically stable if there exists a matrix norm such that the norm of every extreme matrix in A_I is less than one.

Now we present a stability criterion similar to that in Proposition 1 that is based on the dominating matrix $\hat{A}$ instead of matrix norms of extreme matrices. Recalling that $|\hat{A}| \geq |A|$ for every $A \in A_I$ the following Proposition 2 can be easily verified by iterating (2).

Proposition 2 *The time-invariant discrete system (1) is globally asymptotically stable if* *the modulus matrix of the dominating matrix $\hat{A}$ is asymptotically stable, i.e. if $\rho(|\hat{A}|) < 1$. Then $h = 1 - \rho(|\hat{A}|)$ is the stability margin of (1) and there exists a matrix norm such that the norm of every matrix $A \in A_I$ (and hence also every extreme matrix in A_I) is less than unity, i.e. the stability conditions due to Proposition 1 are fulfilled.*

The following proposition slightly extends some of the results reported in Han and Lee (1994), in particular, the part concerning the stability margins is believed to be new. On the contrary to the original proof, observe that Proposition 3 also immediately follows by iterating (2) where time-varying matrix $A(k)$ replaces the matrix A, Corollary 1 and Corollary 2 follow then trivially.

Proposition 3 (cf. Han and Lee (1994) Theorem 1 and Lemma 1).
The time-varying discrete system (1) is globally asymptotically stable if $\rho(|\hat{A}|) < 1$. Then $h = 1 - \rho(|\hat{A}|)$ is the stability margin of (1).

Corollary 1 (cf. Han and Lee (1994) Corollary 1 and Corollary 2).
(i) Let $C \geq |B|$. Then (1) is asymptotically stable if and only if $\rho(C) < 1$.
(ii) Let $|B| \geq |C|$. Then (1) is asymptotically stable if $\rho(|B|) < 1$.

From Corollary 1 we immediately obtain necessary and sufficient conditions for stability of nonnegative time-invariant discrete interval systems studied in Shafai *et al.* (1991) and in Chen (1993).

Corollary 2 (cf. Shafai *et al.* (1991) and Chen (1993) Theorem 1).
Let $B \geq 0$. Then (1) is globally asymptotically stable if and only if $\rho(C) < 1$. In particular, also the time-invariant system given by (1) is globally asymptotically stable.

Of course, the obtained results are also closely connected with the necessary and sufficient conditions for the stability of time-varying interval matrices reported in Bauer *et al.* (1993). These results can be summarized as follows.

Proposition 4 (cf. Bauer *et al.* (1993) Theorem and Lemma).
The time-varying discrete system (1) is globally asymptotically stable if and only if there exists a finite n_0 such that $\| \prod_{k=0}^{n_0} A(k) \| < 1 \quad \forall A(k) \in A_I,\ k = 0, 1, \ldots, n_0$, where $\| \cdot \|$ is reserved for one norm or for maximum norm. The above condition is fulfilled if and only if $\| \prod_{k=0}^{n_0} A(k) \| < 1 \quad \forall A(k) \in \bar{A}_I,\ k = 0, 1, \ldots, n_0$.

In what follows, we extend the results of Propositions 1, 2, 3, 4 and present some useful connections between similarity transformations, matrix norms and spectral radii. The obtained results will enable to test stability using spectral radii of the modulus matrices arising by multiplying extreme matrices of the considered interval system. In particular, we suggest an algorithmic procedure for evaluating products of extreme matrices to obtain sufficient stability conditions for the time-varying discrete interval system (1) along with the corresponding stability margins. The obtained results and the algorithmic procedure are tested on five numerical examples taken from the literature.

3 MAIN RESULTS

For the sake of brevity sometimes we shall set $P(n) \equiv \prod_{k=0}^{n-1} A(k)$ where $A(k) \in A_I$. Observe that $P(1) \in A_I$, denote by $\hat{P}(1) \equiv \hat{A}$ the dominating matrix and recall that $\rho(|P(\cdot)|)$, resp. $u(|P(\cdot)|)$, is reserved for the spectral radius of $|P(\cdot)|$, resp. its right Perron eigenvector.

Lemma 1 *Let $n_0 = 1, 2, \ldots$ and $u \in \mathbb{R}_+^r$ be fixed. Then there exists a matrix sequence $\{\hat{A}(k) : \hat{A}(k) \in \bar{A}_I,\ k = 0, 1, \ldots, n_0 - 1\}$ (called* (n_0, u)-dominating matrix sequence*) such that $\|\hat{P}(n_0, u)\|_u \equiv \|\prod_{k=0}^{n_0-1} \hat{A}(k)\|_u \geq \|P(n_0)\|_u \equiv \|\prod_{k=0}^{n_0-1} A(k)\|_u$ for every matrix sequence $\{A(k) : A(k) \in A_I, k = 0, \ldots, n_0 - 1\}$.*

Lemma 2 *Let for some $n_0 = 1, 2, \ldots$ there exists a matrix sequence $\{\hat{A}(k) : \hat{A}(k) \in \bar{A}_I,\ k = 0, 1, \ldots, n_0 - 1\}$ (called* n_0-dominating matrix sequence*) such that $u(|\hat{P}(n_0)|) > 0$ and $\rho(|\hat{P}(n_0)|)\, u(|\hat{P}(n_0)|) \geq |P(n_0)|\, u(|\hat{P}(n_0)|)$ for any $P(n_0) \equiv \prod_{k=0}^{n_0-1} A(k)$, $A(k) \in \bar{A}_I$. Then $\rho(|\hat{P}(n_0)|) \geq \rho(|P(n_0)|)$ for an arbitrary matrix sequence $\{A(k) : A(k) \in A_I, k = 0, \ldots, n_0 - 1\}$. Moreover, $\rho(|\hat{P}(n_0)|) = \|\hat{P}(n_0)\|_{\hat{u}}$ for $\hat{u} = u(|\hat{P}(n_0)|)$.*

Theorem 1 *The time-varying discrete system (1) is asymptotically stable if there exists a finite $n_0 = 1, 2, \ldots$ such that $\rho(|\hat{P}(n_0)|) < 1$ where $\hat{P}(n_0) \equiv \prod_{k=0}^{n_0-1} \hat{A}(k)$, $\hat{A}(k) \in \bar{A}_I$ is an n_0-dominating matrix sequence. Then $h = 1 - (\rho(|\hat{P}(n_0)|))^{\frac{1}{n_0}}$ is the stability margin of (1).*

Remark 1 Observe that the "if" part of Theorem 1 for $n_0 = 1$ is identical with Proposition 3 and if in Lemma 1 the u-weighted maximum norm is replaced by one or maximum norm, Lemma 1 is identical with the Lemma in Bauer *et al.* (1993).

As a simple example shows, it need not hold that $(\rho(|\hat{P}(n+1)|))^{\frac{1}{n+1}} \leq (\rho(|\hat{P}(n)|))^{\frac{1}{n}}$, however, some monotonicity properties of suitable subsequences of $\{(\rho(|\hat{P}(n)|))^{\frac{1}{n}}\}$ are presented in Proposition 5.

Example. Let $A_I = A = \begin{bmatrix} -0.5 & 0.5 \\ 0.5 & 0.5 \end{bmatrix}$. Obviously $\rho(A) = \sqrt{0.5} = 0.7071$.

Then $|\hat{A}| = \begin{bmatrix} 0.5 & 0.5 \\ 0.5 & 0.5 \end{bmatrix}$ with $\rho(|\hat{A}|) = 1$, $\hat{P}(2) = |\hat{P}(2)| = \begin{bmatrix} 0.5 & 0 \\ 0 & 0.5 \end{bmatrix}$ where $\rho(|\hat{P}(2)|) = 0.5$, and $(\rho(|\hat{P}(2)|))^{\frac{1}{2}} = \sqrt{0.5} = 0.7071$, however,

$$\hat{P}(3) = \begin{bmatrix} -0.25 & 0.25 \\ 0.25 & 0.25 \end{bmatrix} \implies |\hat{P}(3)| = \begin{bmatrix} 0.25 & 0.25 \\ 0.25 & 0.25 \end{bmatrix}$$

and hence $\rho(|\hat{P}(3)|) = 0.5 \implies (\rho(|\hat{P}(3)|))^{\frac{1}{3}} = 0.79.$

Proposition 5 *For every integer k, $n > 1$ $\rho(|\hat{P}(1)|) \geq (\rho(|\hat{P}(n)|))^{\frac{1}{n}}$, and $\rho(|\hat{P}(n)|) \geq (\rho(|\hat{P}(n \cdot k)|))^{\frac{1}{k}}$.*

Now we present useful connections between matrix norms and spectral radii of matrix sequences. The symbol $\|\cdot\|$ is reserved for one norm or for maximum norm.

Theorem 2 *For every $\varepsilon > 0$, $n = 1, 2, \ldots$ and $P(n_0) \equiv \prod_{k=0}^{n_0-1} A(k)$, where $A(k) \in A_I$, there exists a nonsingular diagonal matrix $T^{\varepsilon}(n)$ such that $\|(T^{\varepsilon}(n))^{-1}P(n)T^{\varepsilon}(n)\| - \varepsilon \leq \rho(|P(n)|) \leq \|(T^{\varepsilon}(n))^{-1}P(n)T^{\varepsilon}(n)\|$. Furthermore, if there exists a strictly positive Perron eigenvector of $|P(n)|$, say $u = [u_i]$ then $\|(T(n))^{-1}P(n)T(n)\| = \rho(|P(n)|)$ for $T(n) = \operatorname{diag}\{u_i\}$. In particular, $\rho(|P(n)|) < 1$ if and only if there exists a nonsingular, diagonal matrix $T(n)$ such that $\|(T(n))^{-1}P(n)T(n)\| < 1$.*

Now we are in a position to present an algorithmical procedure for checking stability of time-varying interval systems.

Algorithm 1

Step 0. Set $n = 1$ and generate the set $\mathcal{P}(1) \equiv \bar{A}_I$ of all extreme matrices.

Step 1. Find in the set $\mathcal{P}(1) \equiv \bar{A}_I$ the dominating matrix $\hat{P}(1) \equiv \hat{A}$ and calculate the spectral radius $\rho(|\hat{P}(1)|)$ of the modulus of $\hat{P}(1)$. If $\rho(|\hat{P}(1)|) < 1$ terminate with conclusion of stability and calculate the stability margin $h(1) = 1 - \rho(|\hat{P}(1)|)$ (if $\hat{P}(1) = |\hat{P}(1)|$ this stability margin is the tightest possible one). In case that $\rho(|\hat{P}(1)|) \geq 1$ then check if also $|\hat{P}(1)| \in \bar{A}_I$ (i.e. $|\hat{P}(1)| = \hat{P}(1)$). If it is the case then terminate with conclusion of instability, else proceed to Step 2.

Step 2. Set $n = n + 1$ and generate the set $\mathcal{P}(n)$ of all products of n extreme matrices. Proceed to Step 3.

Step 3. Calculate the spectral radius $\rho(|P(n)|)$ of every matrix $|P(n)|$ where $P(n) \in \mathcal{P}(n)$. Let $\hat{P}(n) = \prod_{k=0}^{n-1} \hat{A}(k) \in \mathcal{P}(n)$ be such that $u(|\hat{P}(n)|) > 0$ and $\rho(|\hat{P}(n)|)\,u(|\hat{P}(n)|) \geq |P(n)|\,u(|\hat{P}(n)|)$ for every $P(n) \in \mathcal{P}(n)$. If $\rho(|\hat{P}(n)|) < 1$ then terminate with conclusion of stability and calculate the stability margin $h(n) = 1 - (\rho(|\hat{P}(n)|)^{\frac{1}{n}}$ (if $\hat{P}(n) = |\hat{P}(n)|$ this stability margin is the tightest possible one), else proceed to Step 2.

4 ILLUSTRATIVE EXAMPLES

Finally, we shall test Algorithm 1 on five numerical examples of linear interval systems borrowed from Bauer *et al.* (1993), Han and Lee (1994), Kolla *et al.* (1989) and of Juang and Shao (1989).

Example 1 (see Example 1 in Bauer *et al.* (1993)).
Consider a discrete-time dynamic system with the following interval matrix

$$A_I = \begin{bmatrix} [-0.8 \quad 0.0] & [0.05 \quad 0.35] \\ [0.05 \quad 0.35] & [0.0 \quad 0.8] \end{bmatrix}. \qquad \text{Since} \qquad |\hat{A}| \equiv |\hat{P}(1)| = \begin{bmatrix} 0.8 & 0.35 \\ 0.35 & 0.8 \end{bmatrix}$$

is not stable and $|\hat{A}| \notin \bar{A}_I$, we calculate $P(2)$'s (products of two extreme matrices) and find that $|\hat{P}(2)| = \begin{bmatrix} 0.0 & 0.35 \\ 0.35 & 0.8 \end{bmatrix}^2 = \begin{bmatrix} 0.1225 & 0.28 \\ 0.28 & 0.7625 \end{bmatrix}$, $u(|\hat{P}(2)|) = \begin{bmatrix} 0.375 \\ 1.000 \end{bmatrix}$, where $\hat{P}(2)$ is a dominating matrix sequence. Even if $\|\hat{P}(2)\|_1 = \|\hat{P}(2)\|_\infty = 1.043$, for the spectral radius of $|\hat{P}(2)|$ we get $\rho(|\hat{P}(2)|) = 0.8676$. Hence the system is stable and the resulting stability margin $h = 1 - (0.8676)^{\frac{1}{2}} = 0.0685$.

Example 2 (see Example 2 in Bauer *et al.* (1993)).
Consider a discrete-time dynamic system with the following interval matrix

$$A_I = \begin{bmatrix} 0.7 & [-0.2 \quad 0.45] \\ 0.45 & [-0.1 \quad 0.2] \end{bmatrix} \qquad \text{Obviously,} \qquad \hat{A} \equiv \hat{P}(1) = \begin{bmatrix} 0.7 & 0.45 \\ 0.45 & 0.2 \end{bmatrix},$$

with $\|\hat{P}(1)\|_\infty = \|\hat{P}(1)\|_1 = 1.15$ and moreover $|\hat{P}(1)| = \hat{P}(1) \in \bar{A}_I$ with $\rho(|\hat{P}(1)|) = 0.967$. Hence the considered system is stable with stability margin 0.032 that is the tightest possible one.

Example 3 (see Example 3 in Juang and Shao (1989)).
Consider a discrete-time dynamic system with the following interval matrix

$$A_I = \begin{bmatrix} [-0.20 \quad 0.16] & [-0.34 \quad 0.02] \\ [-0.24 \quad 0.12] & [-0.16 \quad 0.20] \end{bmatrix}. \qquad \text{Then} \qquad |\hat{A}| \equiv |\hat{P}(1)| = \begin{bmatrix} 0.2 & 0.34 \\ 0.24 & 0.2 \end{bmatrix},$$

$\|\hat{P}(1)\|_1 = \|\hat{P}(1)\|_\infty = 0.54$ and $\rho(|\hat{P}(1)|) = 0.485$. Hence the considered system is stable with stability margin at least 0.515 (since $|\hat{P}(1)| \notin \bar{A}_I$ this stability need not be the tightest possible one).

Example 4 (see Example 1 in Han and Lee (1994)).
Consider a discrete-time dynamic system with the following interval matrix

$$A_I = \begin{bmatrix} [-0.7 \quad 0.6] & 0.6 \\ 0.1 & 0.5 \end{bmatrix}. \qquad \text{Then} \qquad |\hat{A}| \equiv |\hat{P}(1)| = \begin{bmatrix} 0.7 & 0.6 \\ 0.1 & 0.5 \end{bmatrix},$$

$\|\hat{P}(1)\|_1 = 1.1$, $\|\hat{P}(1)\|_\infty = 1.3$ however $\rho(|\hat{P}(1)|) = 0.8646$, $|\hat{P}(1)| \notin \bar{A}_I$. Hence the considered system is stable with stability margin at least 0.1354.

Example 5 (see Example 3 in Kolla *et al.* (1989)).
Consider a discrete-time dynamic system with the following interval matrix

$$A_I = \begin{bmatrix} [-0.5 \quad 0.5] & [0 \quad 0.6] \\ [-0.25 \quad 0.75] & 0 \end{bmatrix}. \qquad \text{Then} \qquad \hat{A} \equiv |\hat{P}(1)| = \hat{P}(1) = \begin{bmatrix} 0.5 & 0.6 \\ 0.75 & 0 \end{bmatrix},$$

$\|\hat{P}(1)\|_1 = 1.25$, $\|\hat{P}(1)\|_\infty = 1.1$, but $\rho(|\hat{P}(1)|) = 0.966$. Hence the considered system is stable with stability margin 0.034 that is the tightest possible one.

REFERENCES

Bauer, P.H., Premaratne, K., and Durán, J. (1993) A necessary and sufficient condition for robust asymptotic stability of time-variant discrete systems. *IEEE Trans. Automat. Control*, **AC-38**, 1427–30.

Berman, A. and Plemmons, R.J. (1979) *Nonnegative Matrices in the Mathematical Sciences*. Academic Press, New York.

Chen, J. (1993) Comments on 'A necessary and sufficient conditions for the stability of nonnegative interval discrete systems.' *IEEE Trans. Automat. Control*, **AC-38**, 189.

Gantmakher, F.R. (1966) *Teoriya matric*. Second edition. Nauka, Moscow. English translation: *The Theory of Matrices, Volume I, II*. Chelsea, New York 1959.

Han, H.S. and Lee, J.G. (1994) Necessary and sufficient conditions for stability of time-varying discrete interval matrices. *Internat. J. Control*, **59**, 1021–9.

Internat. J. Control (1989) **49**, 1095–1108, Correspondence.

Jiang, C.L. (1988) Sufficient and necessary condition for the asymptotic stability of discrete linear interval systems. *Internat. J. Control*, **47**, 1563–5.

Juang, Y.-T. and Shao, C.-S. (1989) Stability analysis of dynamic interval systems. *Internat. J. Control*, **49**, 1401–8.

Juang, Y.-T., Tung, S.-H., and Ho, T.-C. (1989a) Sufficient condition for asymptotic stability of discrete interval systems. *Internat. J. Control*, **49**, 1799–803.

Kolla, S.R. and Farison, J.B. (1988) Counterexamples to 'Sufficient and necessary condition for the asymptotic stability of discrete linear interval systems.' *Internat. J. Control*, **48**, 1751–2.

Kolla, S.,R., Yedavalli, R.K., and Farison, J.B. (1989) Robust stability bounds on time-varying perturbations for state-space models of linear discrete-time systems. *Internat. J. Control*, **50**, 151–9.

Mori, T. and Kokame, H. (1987) Convergence properties of interval matrices and interval polynomials. *Internat. J. Control*, **45**, 243–8.

Shafai, B., Perev, K., Cowley, J., and Chehab, Y. (1991) A necessary and sufficient conditions for the stability of nonnegative interval discrete systems. *IEEE Trans. Automat. Control*, **AC-36**, 742–6.

Distributed Parameter Systems

20

The relaxation theory applied to optimal control problems of semilinear elliptic equations

Eduardo Casas
Dpto. de Matemática Aplicada y Ciencias de la Computación, E.T.S.I. Industriales y de Telecomunicación, Universidad de Cantabria
Av. Los Castros s/n, 39071 Santander, Spain. Tel: 34-42-201427.
Fax: 34-42-201829. e-mail: casas@macc.unican.es

Abstract
The relaxation methods, such as described by Warga (1972), are applied to the study of state-constrained optimal control problems governed by semilinear elliptic equations. The main issue is to prove the convergence of the solutions of the discretized control problems to optimal controls of the relaxed continuous problem. In order to obtain this result we make a stability assumption of the optimal cost functional with respect to small perturbation of the feasible state set.

Keywords
Relaxed controls, state constraints, elliptic equations, finite element method

1 INTRODUCTION

In this paper, we are concerned with an optimal control problem of a semilinear elliptic equation, with control and state constraints. No convexity assumption is made on the feasible control set or the cost functional, therefore the existence of an optimal control can not be proved. However, after the discretization of the continuous problem (by using finite elements), we usually obtain a finite dimensional optimization problem having a solution. Then some questions arise: can we state any relation between these solutions and the continuous control problem?, do they converge to something in some topology when the dimension of the discretized problem is increased?

The reader is referred to Chryssoverghi (1985), (1986) and Chryssoverghi and Kokkinis (1994) for some related papers. We try to improve the existing theory by assuming less restrictive conditions on the problem (mainly on the state equation) and by including pointwise constraints in the control problem. On the other hand, in order to prove the convergence of the discretization of a state-constrained optimal control problem, it is usually required to enlarge the set of feasible states. Hereafter we will give a stability condition for the control problem in such a way that for the stable problems is not necessary to enlarge the set of feasible states to achieve the convergence of the discretizations. Moreover, we prove that almost all problems are stable; see Casas (1992).

2 THE OPTIMAL CONTROL PROBLEM

Let Ω be an open bounded subset of $\mathbb{R}^n$, $n = 2$ or 3, with a Lipschitz boundary. To simplify the presentation we will also assume that Ω is convex. However this assumption can be removed if we assume $\Gamma = \partial\Omega$ to be of class $C^{1,1}$, we will come back to this issue below. In Ω we consider the operator

$$Ay = -\sum_{i,j=1}^{n} \partial_{x_j}[a_{ij}(x)\partial_{x_i}y],$$

where $a_{ij} \in C^{0,1}(\bar{\Omega})$, $1 \leq i, j \leq n$ and satisfy the usual ellipticity assumption

$$\exists\lambda > 0 \text{ such that } \sum_{i,j=1}^{n} a_{ij}(x)\xi_i\xi_j \geq \lambda\|\xi\|^2 \quad \forall x \in \Omega \text{ and } \forall\xi \in \mathbb{R}^n.$$

Let $\mathcal{K}$ be a compact subset of $\mathbb{R}^m$, $m \geq 1$, and let $f : \Omega \times (\mathbb{R} \times \mathcal{K}) \longrightarrow \mathbb{R}$ be a Carathèdory function, monotone non increasing with respect to the second variable and satisfying

$$\begin{cases} \exists\psi_0 \in L^2(\Omega) \text{ such that } |f(x,0,u)| \leq \psi_0(x) \quad \forall u \in \mathcal{K} \text{ and a.e. } x \in \Omega, \\ \forall M > 0\ \exists\psi_M \in L^2(\Omega) \text{ such that} \\ |f(x,y_2,u) - f(x,y_1,u)| \leq \psi_M(x)|y_2 - y_1| \text{ a.e. } x \in \Omega,\ \forall|y_1|,|y_2| \leq M,\ \forall u \in \mathcal{K}. \end{cases} \tag{1}$$

Let us denote by $\mathcal{U}$ the set of measurable functions $u : \Omega \longrightarrow \mathcal{K}$. For each $u \in \mathcal{U}$ we consider the state equation

$$\begin{cases} Ay = f(x,y(x),u(x)) & \text{in } \Omega, \\ y = 0 & \text{on } \Gamma. \end{cases} \tag{2}$$

The next theorem claims the well posedness of the state equation.

Theorem 1 *Under the previous assumptions* (2) *has a unique solution* y_u *in* $H_0^1(\Omega) \cap H^2(\Omega)$. *Moreover there exists a constant* $C_{\mathcal{K}} > 0$ *such that*

$$\|y_u\|_{H^2(\Omega)} \leq C_{\mathcal{K}} \quad \forall u \in \mathcal{U}. \tag{3}$$

This theorem can be proved by classical arguments. First we can consider f bounded and use Schauder's fixed point theorem to deduce the existence of a solution in $H^1(\Omega)$, the uniqueness being a consequence of the monotonicity of f w.r.t. the second variable. If f is not bounded, we can define

$$f_M(x,y,u) = \begin{cases} +M & \text{if } y > +M, \\ -M, & \text{if } y < -M \\ f(x,y,u) & \text{otherwise.} \end{cases}$$

Thus we have a unique solution y_M for every M. Then using again the monotonicity of f and (1), we can obtain the boundedness y_M in Ω independently of M by Stampachia's method (1965). Thus $f_M(x, y_M(x), u(x)) = f(x, y_M(x), u(x))$ for M large enough, which implies that y_M is a solution of (2) in $H^1(\Omega) \cap L^\infty(\Omega)$, the uniqueness in this space is once more a consequence of the monotonicity of f w.r.t. y. Finally, the $H^2(\Omega)$-regularity follows from the convexity of Ω and the Lipschitz regularity of the coefficients a_{ij}; see Grisvard (1985).

Remark 1 *In the case of a nonconvex set Ω, the previous theorem remains to be true under the $C^{1,1}$-regularity of Γ. Indeed the unique modification in the previous argumentation arises in the proof of the $H^2(\Omega)$-regularity of the solution. This regularity is still true when the convexity assumption of Ω is replaced by the $C^{1,1}$-regularity of Γ; see Grisvard (1985). It is known that the C^1-regularity of Γ is not enough to assure the $H^2(\Omega)$-regularity of y; see Jerison and Kenig (1994).*

Now we consider a Carathéodry function $L : \Omega \times (\mathbb{R} \times \mathcal{K}) \longrightarrow \mathbb{R}$ satisfying

$$\forall M > 0 \;\; \exists \phi_M \in L^1(\Omega) \quad \text{such that } |L(x,y,u)| \le \phi_M(x) \;\text{ a.e. } x \in \Omega, \forall |y| \le M, \; \forall u \in \mathcal{K}. \tag{4}$$

Now we can state the optimal control problem as follows

$$(\mathrm{P}_\delta) \begin{cases} \text{Minimize } J(u) = \displaystyle\int_\Omega L(x, y_u(x), u(x))dx, \\ u \in \mathcal{U}, \; g(x, y_u(x)) \le \delta \; \forall x \in \bar{\Omega}, \end{cases}$$

where $\delta \in \mathbb{R}$ and $g : \bar{\Omega} \times \mathbb{R} \longrightarrow \mathbb{R}$ is a continuous function.

We finish this section introducing the stability concept of (P_δ) w.r.t. perturbations of the set of feasible states.

Definition 1 *We will say that (P_δ) is stable to the right if*

$$\lim_{\delta' \searrow \delta} \inf (\mathrm{P}_{\delta'}) = \inf (\mathrm{P}_\delta). \tag{5}$$

Analogously, (P_δ) is stable to the left if

$$\lim_{\delta' \nearrow \delta} \inf (\mathrm{P}_{\delta'}) = \inf (\mathrm{P}_\delta). \tag{6}$$

(P_δ) is said stable if it is stable to the left and to the right simultaneously.

The next theorem establishes how often this stability condition is satisfied.

Theorem 2 *There exists $\delta_0 \in \mathbb{R}$ such that (P_δ) has no feasible control for $\delta < \delta_0$. For every $\delta > \delta_0$, except at most a countable number of them, problem (P_δ) is stable.*

Proof. From Theorem 1 we know the existence of a constant $M > 0$ such that $|y_u(x)| \leq M$ for all $x \in \Omega$ and $u \in \mathcal{U}$. Let us set λ_M and Λ_M the minimum and maximum of g over $\bar{\Omega} \times [-M, +M]$. Then it is obvious that (P_δ) has no feasible control for $\delta < \lambda_M$, while every element of $\mathcal{U}$ is a feasible control for every $\delta \geq \Lambda_M$. Let us set $\delta_0 = \inf\{\delta : (P_\delta)$ has at least one feasible control$\}$. Then $\lambda_M \leq \delta_0 \leq \Lambda_M$.

Finally we prove that (P_δ) is stable for almost all $\delta > \delta_0$. Let us consider the function $h : (\delta_0, +\infty) \longrightarrow \mathbb{R}$ defined by $h(\delta) = \inf(P_\delta)$. Then h is a monotone non increasing function and then continuous at every point δ, except at most a countable number of them. The theorem follows from the fact that the continuity of h in δ is equivalent to the stability of (P_δ). □

3 THE RELAXED CONTROL PROBLEM

In this section we apply the relaxation theory to immerse the control set $\mathcal{U}$ in a bigger class of controls such that the new control problem has at least one solution. To do this we follow the approach described by Warga (1972). As usual $C(\mathcal{K})$ denotes the space of continuous functions endowed with the maximum norm and $M(\mathcal{K}) = C(\mathcal{K})^*$ is the space of real regular Borel measures in $\mathcal{K}$. We also set $\mathcal{P}(\mathcal{K})$ equal to the subset of $M(\mathcal{K})$ formed by the probability measures in $\mathcal{K}$. With $\mathcal{R}$ we denote the subset of the Banach space $L^\infty(\Omega, M(\mathcal{K})) = L^1(\Omega, C(\mathcal{K}))^*$ formed by the functions r such that $r(x) \in \mathcal{P}(\mathcal{K})$ for almost all $x \in \Omega$. Given a Carathéodory function $\phi : \Omega \times \mathcal{K} \longrightarrow \mathbb{R}$, we denote by $\phi_R : \Omega \times M(\mathcal{K}) \longrightarrow \mathbb{R}$ the function defined by

$$\phi_R(x, \mu) = \int_{\mathcal{K}} \phi(x, k)\, d\mu(k).$$

In some sense, ϕ_R can be considered as an extension of ϕ, simply by taken $\phi(x, k) = \phi_R(x, \delta_{[k]})$, where $\delta_{[k]}$ is the Dirac measure centered at the point k of $\mathcal{K}$. Consequently, we can define the functions $f_R, L_R : \Omega \times \mathbb{R} \times M(\mathcal{K}) \longrightarrow \mathbb{R}$ by

$$f_R(x, y, \mu) = \int_{\mathcal{K}} f(x, y, k)\, d\mu(k) \quad \text{and} \quad L_R(x, y, \mu) = \int_{\mathcal{K}} L(x, y, k)\, d\mu(k).$$

Now we define the relaxed control problem in the following way

$$(\mathrm{RP}_\delta) \begin{cases} \text{Minimize } J_R(r) = \displaystyle\int_\Omega L_R(x, y_r(x), r(x))\, dx \\ r \in \mathcal{R},\ g(x, y_r(x)) \leq \delta\ \forall x \in \bar{\Omega}, \end{cases}$$

with $y_r \in H_0^1(\Omega) \cap H^2(\Omega)$ being the solution of the problem

$$\begin{cases} Ay = f_R(x, y(x), r(x)) & \text{in } \Omega, \\ y = 0 & \text{on } \Gamma. \end{cases} \tag{7}$$

Let us remark that $\mathcal{U}$ can be considered as a subset of $\mathcal{R}$ by identifying $u(x)$ and $r(x) = \delta_{[u(x)]}$. Moreover, with this identification we have $J_R(r) = J(u)$, $y_r = y_u$ and $f_R(x, y_r(x), r(x)) = f(x, y_u(x), u(x))$. On the other hand $\mathcal{U}$ is dense in $\mathcal{R}$; see Warga (1972). Therefore (RP_δ) can be considered as an extension of (P_δ). Furthermore (RP_δ) has at least one solution as we will prove below. Then a natural question is whether $\inf(\mathrm{RP}_\delta) = \inf(\mathrm{P}_\delta)$. Here we have the answer

Theorem 3 *Let δ_0 be as in Theorem 2.* (RP_δ) *has at least one solution for every $\delta > \delta_0$. Moreover* $\inf(\mathrm{RP}_\delta) = \inf(\mathrm{P}_\delta)$ *if and only if* (P_δ) *is stable to the right.*

Proof. It is well known that $\mathcal{R}$ is convex, metrizable (with coinciding metric and weak* topologies) and compact. Then the set of feasible relaxed controls is compact too. Moreover this set is not empty due to the fact that $\delta > \delta_0$. Finally, the continuity of J_R on $\mathcal{R}$ allows to conclude the existence of an optimal relaxed control.

To prove the second part of the theorem we first state the following inequalities

$$\inf(\mathrm{RP}_{\delta'}) \le \inf(\mathrm{P}_{\delta'}) \le \inf(\mathrm{RP}_\delta) \le \inf(\mathrm{P}_\delta) \quad \text{for every } \delta' > \delta.$$

The first and the last inequalities are a consequence of the identification of every feasible control for (P_δ) (resp. (P'_δ)) with a feasible control for (RP_δ) (resp. (RP'_δ)). Let us prove the second inequality. If r is a feasible control for (RP_δ), then we can take a sequence of controls $\{u_k\}_{k=1}^\infty$ such that $r_k(x) = \delta_{[u_k(x)]} \to r(x)$ weakly* in $L^\infty(\Omega, M(\mathcal{K}))$. Since $y_{u_k} \to y_r$ uniformly in $\bar{\Omega}$, then $\delta \ge g(x, y_r(x)) = \lim_{k\to\infty} g(x, y_{u_k}(x))$, therefore $g(x, y_{u_k}(x)) \le \delta'$ for every $x \in \bar{\Omega}$ and k bigger than a certain k_0, only depending on δ'. Thus the controls $\{u_k\}_{k\ge k_0}$ are feasible for problem $(\mathrm{P}_{\delta'})$ and

$$J_R(r) = \lim_{k\to\infty} J(u_k) \ge \inf(\mathrm{P}_{\delta'}),$$

which leads to the desired inequality.

Finally, the proof will be concluded by proving that $\lim_{\delta'\searrow\delta} \inf(\mathrm{RP}_{\delta'}) = \inf(\mathrm{RP}_\delta)$. Let $r_{\delta'}$ be a solution of $(\mathrm{RP}_{\delta'})$ for every $\delta' > \delta$. Using the compactness of $\mathcal{R}$, we can take a sequence $\{r_{\delta_j}\}_{j=1}^\infty$, with $\delta_j \searrow \delta$, such that $r_{\delta_j} \to r$ weakly* for some $r \in \mathcal{R}$. Using the uniform convergence $y_{r_{\delta_j}} \to y_r$ in $\bar{\Omega}$, we obtain

$$g(x, y_r(x)) = \lim_{j\to\infty} g(x, y_{r_{\delta_j}}(x)) \le \lim_{j\to\infty} \delta_j = \delta,$$

for every $x \in \bar{\Omega}$, Therefore r is a feasible control for (RP_δ). Thus we have

$$\inf(\mathrm{RP}_\delta) \le J_R(r) = \lim_{j\to\infty} J_R(r_{\delta_j}) = \lim_{\delta'\searrow\delta} \inf(\mathrm{RP}_{\delta'}) \le \inf(\mathrm{RP}_\delta).$$

□

Corollary 1 *If problem* (P_δ) *is stable to the right and it has a solution* $\bar{u}$, *then* $\bar{r}(x) = \delta_{[\bar{u}(x)]}$ *is also a solution of* (RP_δ).

4 NUMERICAL APPROXIMATION OF THE CONTROL PROBLEM

In this section we consider the numerical discretization of problem (P_δ). As we will see below, the discrete problems have at least one solution, the natural question is to relate these optimal discrete controls with the continuous problem. Since the existence of a solution of (P_δ) can not be stated under our assumptions, we look at the relaxed control problem, which has a solution. We will prove that the optimal discrete controls converge to optimal relaxed controls in some topology.

Let $\{\mathcal{T}_h\}_{h>0}$ be a regular family of triangulations in $\bar{\Omega}$ satisfying the inverse assumption; see Ciarlet (1978). Let us take $\bar{\Omega}_h = \cup_{T\in\mathcal{T}_h} T$, Ω_h its interior and Γ_h its boundary. Then we assume that $\bar{\Omega}_h$ is convex and the vertices of $\mathcal{T}_h$ placed on the boundary Γ_h are points of Γ. To every boundary triangle T of $\mathcal{T}_h$ we associate another triangle $\tilde{T} \subset \bar{\Omega}$ with two interior sides to Ω coincident with two sides of T and the third side is the curvilinear arc of Γ limited by the other two sides. We denote by $\tilde{\mathcal{T}}_h$ the family formed by these boundary triangles with a curvilinear side and the interior triangles to Ω of $\mathcal{T}_h$, so $\bar{\Omega} = \cup_{T\in\tilde{\mathcal{T}}_h} T$. Now let us consider the spaces

$$\mathcal{U}_h = \{u_h \in \mathcal{U} : u_h|_T \text{ is constant } \forall T \in \tilde{\mathcal{T}}_h\}$$

$$V_h = \{y_h \in C(\bar{\Omega}) : y_h|_T \in \mathcal{P}_1\ \forall T \in \mathcal{T}_h \text{ and } y_h(x) = 0\ \ \forall x \in \bar{\Omega}\backslash\Omega_h\},$$

where $\mathcal{P}_1$ is the space of the polynomials of degree less than or equal to 1. It is obvious that $V_h \subset H_0^1(\Omega)$. For each $u_h \in \mathcal{U}_h$ we denote by $y_h(u_h)$ the unique element of V_h that satisfies:

$$\int_{\Omega_h} \sum_{i,j=1}^{n} a_{ij}(x)\partial_{x_i} y_h(u_h)(x)\partial_{x_j} z_h(x)dx = \int_{\Omega_h} f(x, y_h(u_h)(x), u_h(x))z_h(x)dx \quad \forall z_h \in V_h.$$

Now we formulate the finite dimensional optimal control problem:

$$(\mathrm{P}_{\delta h}) \begin{cases} \text{minimize } J_h(u_h) = \displaystyle\int_{\Omega_h} L(x, y_h(u_h)(x), u_h(x))dx \\ \\ \text{subject to } u_h \in \mathcal{U}_h \text{ and } g(x_j, y_h(u_h)(x_j)) \leq \delta \ \ 1 \leq j \leq n(h), \end{cases}$$

where $\{x_j\}_{j=1}^{n(h)}$ is the set of vertices of $\mathcal{T}_h$. The following theorem state the existence of a solution for $(\mathrm{P}_{\delta h})$.

Theorem 4 *For every* $\delta > \delta_0$ *there exists* $h_\delta > 0$ *such that* $(\mathrm{P}_{\delta h})$ *has at least one solution* $\bar{u}_h$ *for all* $h \leq h_\delta$.

Proof. Since $\mathcal{U}_h$ is a compact set and J_h is continuous, the existence of a solution of $(\mathrm{P}_{\delta h})$ will be assured if we prove that the set of feasible controls is nonempty. Let $u_0 \in \mathcal{U}$ be a feasible control for (P_{δ_0}) and let us take $u_{0h} \in \mathcal{U}_h$ such that $u_{0h}(x) \to u_0(x)$ for almost every point $x \in \Omega$ as $h \to 0$. Then $y_h(u_{0h}) \to y_{u_0}$ uniformly in $\bar{\Omega}$. Since $g(x, y_{u_0}(x)) \leq \delta_0$ for every $x \in \bar{\Omega}$, from the uniform convergence and the inequality $\delta > \delta_0$ we deduce the existence of $h_\delta > 0$ such that the inequality $g(x, y_h(u_{0h})(x)) \leq \delta$ holds for all $x \in \bar{\Omega}$ and each $h \leq h_\delta$. Therefore u_{0h} is a feasible control for $(\mathrm{P}_{\delta h})$, which completes the proof. □

Finally we prove the convergence result of the numerical approximation.

Theorem 5 *Let us assume that* (P_δ) *is stable and let* $h_\delta > 0$ *be as in Theorem 4. Given a family of controls* $\{\bar{u}_h\}_{h<h_\delta}$, $\bar{u}_h$ *being a solution of* $(\mathrm{P}_{\delta h})$, *there exist subsequences* $\{\bar{u}_{h_k}\}_{k\in N}$, *with* $h_k \to 0$ *as* $k \to \infty$, *and elements* $\bar{r} \in \mathcal{R}$ *such that* $\bar{r}_{h_k}(x) = \delta_{[\bar{u}_{h_k}(x)]} \to \bar{r}$ *in the weak* topology of* $L^\infty(\Omega, M(\mathcal{K}))$. *Each one of these limit points is a solution of* (RP_δ). *Moreover we have*

$$\lim_{h\to 0} J_h(\bar{u}_h) = \inf(\mathrm{RP}_\delta) = \inf(\mathrm{P}_\delta).$$

Proof. Let $\bar{y}_h$ be the state associated to $\bar{u}_h$ and let us write $\bar{r}_h(x) = \delta_{[\bar{u}_{h_k}(x)]}$. Since $\{\bar{r}_h\}_{h\leq h_\delta} \subset \mathcal{R}$ and $\mathcal{R}$ is a weak* compact metrizable subset of the space $L^\infty(\Omega, M(\mathcal{K}))$, we can extract a subsequence $\{\bar{r}_{h_k}\}$ such that $h_k \to 0$ and $\bar{r}_{h_k}(x) = \delta_{[\bar{u}_{h_k}(x)]} \to \bar{r}$ weakly* in $L^\infty(\Omega, M(\mathcal{K}))$ for some element $\bar{r} \in \mathcal{R}$. Now we prove that $\bar{r}$ is a solution of (RP_δ). Let $\bar{y}$ be the state associated to $\bar{r}$. Since $\bar{y}_{h_k} \to \bar{y}$ uniformly in $\bar{\Omega}$ and $g(x_j, \bar{y}_{h_k}(x_j)) \leq \delta$ for every $1 \leq j \leq n(h)$, it follows that $g(x, \bar{y}(x)) \leq \delta$ and therefore $\bar{r}$ is a feasible control for the problem (RP_δ).

Let $\delta' \in [\delta_0 + \epsilon, \delta)$, with $0 < \epsilon < \delta - \delta_0$ fixed, and let $r_{\delta'}$ be a solution of $(\mathrm{RP}_{\delta'})$. Since $\mathcal{U}$ is dense in $\mathcal{R}$, we can take sequence $\{u^j\}_{j=1}^\infty \subset \mathcal{U}$ such that $u^j \to r_{\delta'}$ weakly* in $L^\infty(\Omega, M(\mathcal{K}))$. Taking into account the uniform convergence $y_{u^j} \to y_{r_{\delta'}}$, we deduce the existence of $j_{\delta'} \in N$ such that $g(x, y_{u^j}(x)) \leq \delta' + \epsilon/2$ for every $x \in \bar{\Omega}$ and $j \geq j_{\delta'}$. For each j fixed we take a sequence $\{u_h\}_{h>0}$, with $u_h \in \mathcal{U}_h$ and such that $u_h(x) \to u^j(x)$ for almost all point $x \in \Omega$. From the uniform convergence $y_h(u_h) \to y_{u^j}$ and the inequality $g(x, y_{u^j}(x)) \leq \delta' + \epsilon/2 < \delta$ for all $x \in \bar{\Omega}$ we deduce the existence of h^j such that $g(x, y_h(u_h)(x)) \leq \delta\ \forall x \in \bar{\Omega}$ and $\forall h \leq h^j$. Hence u_h is a feasible control for $(\mathrm{P}_{\delta h})$ always that $h \leq h^j$. Then, we get $J_{h_k R}(\bar{r}_{h_k}) = J_{h_k}(\bar{u}_{h_k}) \leq J_{h_k}(u_{h_k})$ whenever $h_k \leq h^j$. Thus we have

$$J_R(\bar{r}) = \lim_{k\to\infty} J_{h_k R}(\bar{r}_{h_k}) \leq \lim_{k\to\infty} J_{h_k}(u_{h_k}) = J(u^j).$$

Now taking the limit when $j \to \infty$, we obtain that $J_R(\bar{r}) \leq J_R(r_{\delta'})$. Finally, the feasibility of $\bar{r}$ for (RP_δ) and the stability condition (Definition 1) enables us to conclude that

$$\inf(\mathrm{RP}_\delta) \leq J_R(\bar{r}) \leq \lim_{\delta'\nearrow\delta} J_R(r_{\delta'}) = \lim_{\delta'\nearrow\delta} \inf(\mathrm{RP}_{\delta'}) \leq \lim_{\delta'\nearrow\delta} \inf(\mathrm{P}_{\delta'}) = \inf(\mathrm{P}_\delta) = \inf(\mathrm{RP}_\delta),$$

which proves that $\bar{r}$ is a solution of (RP_δ). The rest of theorem is immediate. □

ACKNOWLEDGEMENT

The author would like to thank Dirección General de Investigación Científica y Técnica (Spain) for its support to this research.

REFERENCES

Casas, E. (1992) Finite element approximations for some state-constrained optimal control problems, in *Mathematics of the Analysis and Design in Process Control* (eds. P. Borne, S. Tzafestas, and N. Radhy), Amsterdam, North Holland.

Chryssoverghi, I. (1985) Numerical approximation of nonconvex optimal control problems defined by parabolic equations. *J. Optim. Theory Appl.*, **45**, 73–88.

Chryssoverghi, I. (1986) Nonconvex optimal control of nonlinear monotone parabolic systems. *Systems Control Lett.*, **8**, 55–62.

Chryssoverghi, I. and Kokkinis, B. (1994) Discretization of nonlinear elliptic optimal control problems. *Systems Control Lett.*, **22**, 227–34.

Ciarlet, P.G. (1978) *The Finite Element Method for Elliptic Problems.* North-Holland, Amsterdam.

Grisvard, P. (1985) *Elliptic Problems in Nonsmooth Domains.* Pitman, Boston-London-Melbourne.

Jerison, D. and Kenig, C. (1995) The inhomogeneous Dirichlet problem in Lipschitz domains. To appear.

Stampacchia, G. (1965) Le problème de Dirichlet pour les équations elliptiques du second ordre à coefficients discontinus. *Ann. Inst. Fourier (Grenoble)*, **15**, 198–258.

Warga, J. (1972) *Optimal control of differential and functional equations.* Academic Press, New York.

On the use of space invariant imbedding to solve optimal control problems for second order elliptic equations

Jacques Henry
INRIA, B.P. 105, 78153 Le Chesnay Cedex, France.
e-mail: Jacques.Henry@inria.fr

J. P. Yvon
UTC, Dept. GI, BP 649, 60206 Compiegne Cedex, France.
e-mail: jpyvon@dma.univ-compiegne.fr

Abstract

This paper deals with the application of invariant imbedding to the solution of a control problem of a system governed by a second order elliptic equation. The basic idea is to take advantage of the geometry of the domain (for instance a cylinder or a rectangle) to consider that one space variable plays the role of time for dynamical systems. Then it is possible to decouple the system of optimality in order to get the explicit dependence of the optimal control with respect to the desired state.

Keywords

Optimal control of distributed parameter systems, invariant imbedding.

1 INTRODUCTION

The invariant imbedding technique has been popularized by R. Bellman (Bellman, Dreyfus, 1962) and his coworkers. It allows to decouple systems of coupled first order forward and backward differential equations. It has been widely used to derive the Riccati equation in the context of linear quadratic optimal control problems. It is less known that it can be used to solve second order elliptic boundary value problems with an imbedding variable related to space.

We present here an application of invariant imbedding to solve an optimal control problem governed by an elliptic equation. The method is most easily applied in the case of a cylindrical domain but it can be generalized, for example, to domains derived from cylinders by conformal mapping.

For the sake of simplicity, the method is presented for a model problem of a Laplace equation in a square domain Ω of $\mathbb{R}^2$ for special boundary conditions.

2 SPACE FACTORIZATION OF THE STATE EQUATION

In order to present the method we start with the space decoupling applied to the state equation, then the same method will be applied to an associated control problem. Let us consider the model problem in $\mathbb{R}^2$:

$$\begin{cases} -\Delta y = f & \text{in} \quad \Omega, \\ y = 0 & \text{on} \quad S, \\ y = 0 & \text{on} \quad \Gamma_0, \\ \frac{\partial y}{\partial x_1} = u & \text{on} \quad \Gamma_1, \end{cases} \tag{1}$$

with $\Omega =]0,1[\times]0,a[$, $S = \{\partial\Omega \cap \{x_2 = 0\}\} \cup \{\partial\Omega \cap \{x_2 = a\}\}$, $\Gamma_0 = \partial\Omega \cap \{x_1 = 0\}$ and $\Gamma_1 = \partial\Omega \cap \{x_1 = 1\}$. The function u is given in $H^{-1/2}(\Gamma_1)$ and f is given in $L^2(\Omega)$.

If we pay a special attention to the variable x_1, in a similar way to evolution equations we may consider y as : $y \in L^2(0,1;H_0^1(0,a)) \cap H^1(0,1;L^2(0,a))$. It is solution of the coupled system

$$\begin{cases} \dfrac{\partial y}{\partial x_1} = z, \\ y = 0 \text{ on } \Gamma_0, \\ y = 0 \text{ on } S, \end{cases} \qquad \begin{cases} -\dfrac{\partial z}{\partial x_1} - \dfrac{\partial^2 y}{\partial x_2^2} = f, \\ z = u \text{ on } \Gamma_1. \end{cases} \tag{2}$$

We decouple this system by an invariant imbedding with respect to x_1. Let us consider the family of problems, indexed by s, $0 \leq s \leq 1$, defined on $\Omega_s =]0,s[\times]0,a[$:

$$\begin{cases} -\dfrac{\partial z^s}{\partial x_1} - \dfrac{\partial^2 y^s}{\partial x_2^2} = f, \\ z^s = h \text{ on } \Gamma_s, \end{cases} \tag{3}$$

where $\Gamma_s = \Omega \cap \{x_1 = s\}$ and h is a function in $H^{-1/2}(0,a)$, the left equation of (2) being unchanged but restricted to Ω_s.

One can show Ramos that: $h \mapsto y^s|_{\Gamma_s}$ defines an affine mapping :

$$y^s|_{\Gamma_s} = P(s)h + r(s),$$

with $r(s) \in H^{1/2}(0,a)$, $P(s) \in \mathcal{L}(H^{-1/2}(0,a);H^{1/2}(0,a))$. As a consequence of the previous notations one has $y = y^1$ and if we choose h to be $\dfrac{\partial y^1}{\partial x_1}|_{\Gamma_s}$ and dropping the index s one gets :

$$y(s) = P(s)z(s) + r(s), \ \forall s \in [0,1]. \tag{4}$$

Let D^2 be the operator

$$D^2 : y \in H_0^1(0,a)) \mapsto \frac{\partial^2 y^s}{\partial x_2^2} \in H^{-1}(0,a).$$

By differentiating (4) with respect to $x_1 = s$, one obtains the equations satisfied by P and r, and the result is summarized in :

Result 1 *The solution of the state equation (1) given by the solution of the Riccati equations*

$$\begin{cases} \frac{dP}{dx_1} &= PD^2P + I \quad , \quad P(0) = 0, \\ \frac{dr}{dx_1} &= PD^2r + Pf \quad , \quad r(0) = 0. \end{cases} \tag{5}$$

Then z is solution of

$$-\frac{dz}{dx_1} = D^2Pz + D^2r + f, \ z(1) = u. \tag{6}$$

and y is obtained by (4). □

Remark 21 So the solution of the second order boundary value problem (1) can be obtained by solving a system of uncoupled first order initial value problems : the system (5) is to be solved forwards from 0 to 1, the first equation of (5) being an operator Riccati differential equation. Then equation (6) is to be solved backwards from 1 to 0. □

It can be shown that the operator P defined by (4) is self-adjoint and positive definite.

The same problem can be solved in a symmetric way using invariant imbedding in the domain $\tilde{\Omega}_s :=]s,1[\times]0,a[$. This leads to a "dual" method of decoupling :

$$z(x_1) = Q(x_1)y(x_1) + t(x_1),$$

with

$$\begin{cases} -\frac{dQ}{dx_1} = Q^2 + D^2 \quad , \quad Q(1) = 0, \\ -\frac{dt}{dx_1} = Qt + f \quad , \quad t(1) = 0, \end{cases} \tag{7}$$

the solution y being solution of

$$\frac{dy}{dx_1} = Qy + t, \ y(0) = 0. \tag{8}$$

Remark 22 When considering the finite difference discretization of this problem, the same method applied to the resulting linear system leads to the Gauss block factorization of the block tridiagonal matrix representing the Laplace operator on the rectangle.

3 DECOUPLING OF THE CONTROL PROBLEM (I)

Let us consider the optimal control problem associated to the elliptic equation :

$$\begin{cases} -\Delta y = f & \text{in} \quad \Omega, \\ y = 0 & \text{on} \quad S, \\ y = 0 & \text{on} \quad \Gamma_0, \\ \frac{\partial y}{\partial x_1} = u & \text{on} \quad \Gamma_1, \end{cases} \tag{9}$$

with the criterion to be minimized :

$$J(u) = \frac{1}{2}\int_0^a |y(1,x_2) - y_d(x_2)|^2 \, dx_2 + \frac{\nu}{2}\int_0^a u^2(x_2)\, dx_2 \quad (\nu > 0). \tag{10}$$

If we introduce the adjoint state p which is solution of the adjoint equations

$$\begin{cases} -\Delta p = 0 & \text{in } \Omega, \\ p = 0 & \text{on } S, \\ p = 0 & \text{on } \Gamma_0, \\ \frac{\partial p}{\partial x_1} = y(1,.) - y_d & \text{on } \Gamma_1, \end{cases} \tag{11}$$

the optimality condition takes the form (Lions, 1968):

$$p(1,.) + \nu u = 0 \text{ on } \Gamma_1. \tag{12}$$

The basic idea is to decouple the previous optimality system by an invariant imbedding similar to the one presented in the previous section. The family of optimality systems depends on a parameter s , $0 \leq s \leq 1$ and is defined on $\tilde{\Omega}_s :=]s,1[\times]0,a[$ *. The optimality system takes the form :

$$\begin{cases} -\Delta y^s = f & \text{in } \Omega_s, \\ y^s = 0 & \text{on } S, \\ y^s = h & \text{on } \Gamma_s, \\ \frac{\partial y^s}{\partial x_1} = u & \text{on } \Gamma_1, \end{cases} \qquad \begin{cases} -\Delta p^s = 0 & \text{in } \Omega_s, \\ p^s = 0 & \text{on } S, \\ p^s = g & \text{on } \Gamma_s, \\ \frac{\partial p^s}{\partial x_1} = y^s(1,.) - y_d & \text{on } \Gamma_1, \end{cases} \tag{13}$$

where the function g and h are defined in such a way that y^s and p^s are independent of s. The optimality condition (12) is unchanged. This set of equations is the system of optimality corresponding to the new criterion :

$$J_s(u) = \frac{1}{2}\int_0^a |y^s(1,x_2) - y_d(x_2)|^2\, dx_2 + \frac{\nu}{2}\int_0^a u^2(x_2)\, dx_2 + \int_0^a g(x_2)\frac{\partial y^s}{\partial x_1}(s,x_2)\, dx_2.$$

As previously, considering that the variable x_1 is similar to a "time" variable, equations (13) can be written as first order system with respect to x_1 :

$$\begin{cases} \frac{\partial y^s}{\partial x_1} = z^s & \text{on }]s,1[, \\ y^s(s) = h, \end{cases} \qquad \begin{cases} -\frac{\partial z^s}{\partial x_1} - D^2 y^s = f & \text{on }]s,1[, \\ z^s(1) = u = -\frac{1}{\nu}p^s(1), \end{cases} \tag{14}$$

the adjoint system being :

$$\begin{cases} \frac{\partial p^s}{\partial x_1} = q^s & \text{on }]s,1[, \\ p^s(s) = g, \end{cases} \qquad \begin{cases} -\frac{\partial q^s}{\partial x_1} - D^2 p^s = 0 & \text{on }]s,1[, \\ q^s(1) = y^s(1) - y_d. \end{cases} \tag{15}$$

Then it is possible to express $z^s(s)$ and $q^s(s)$ as an affine function of h and g :

$$\begin{cases} z^s(s) = P_{11}(s)h + P_{12}(s)g + r(s), \\ q^s(s) = P_{21}(s)h + P_{22}(s)g + t(s), \end{cases} \tag{16}$$

*We still denote by S the restriction of S to $[s,1]$

where the functions r and t depend upon f and y_d. One can choose $h = y^0|_{\Gamma_s}$ and $g = p^0|_{\Gamma_s}$ and the relation (16) being valid for any s it is possible replace s by x_1. Dropping the superscript s, the formal calculations go along these lines :

$$\begin{cases} \frac{dz}{dx_1} = -D^2 y - f = \frac{dP_{11}}{dx_1} y + P_{11} z + \frac{dP_{12}}{dx_1} p + P_{12} q + \frac{dr}{dx_1}, \\ \frac{dq}{dx_1} = -D^2 p = \frac{dP_{21}}{dx_1} y + P_{21} z + \frac{dP_{22}}{dx_1} p + P_{22} q + \frac{dt}{dx_1}. \end{cases} \tag{17}$$

The elimination of z between (16) and (17) gives

$$-D^2 y - f = \frac{dP_{11}}{dx_1} y + P_{11}\left(P_{11} y + P_{12} p + r\right) + \frac{dP_{12}}{dx_1} p + P_{12}\left(P_{21} y + P_{22} p + t\right) + \frac{dr}{dx_1}.$$

Formally this allows to write three differential equations verified by P_{11}, P_{12} and r :

$$\begin{cases} \frac{dP_{11}}{dx_1} &= -P_{11}^2 - P_{12} P_{21} - D^2, \\ \frac{dP_{12}}{dx_1} &= -P_{11} P_{12} - P_{12} P_{22}, \\ \frac{dr}{dx_1} &= -P_{11} r - P_{12} t - f. \end{cases} \tag{18}$$

Similarly the elimination of q between (16) and (17) leads to

$$-D^2 p = \frac{dP_{21}}{dx_1} y + P_{21}\left(P_{11} y + P_{12} p + r\right) + \frac{dP_{22}}{dx_1} p + P_{22}\left(P_{21} y + P_{22} p + t\right) + \frac{dt}{dx_1}$$

and the differential equations verified by P_{21}, P_{22} and t are :

$$\begin{cases} \frac{dP_{21}}{dx_1} &= -P_{21} P_{11} - P_{22} P_{21}, \\ \frac{dP_{22}}{dx_1} &= -P_{21} P_{12} - P_{22}^2 - D^2, \\ \frac{dt}{dx_1} &= -P_{21} r - P_{22} t. \end{cases} \tag{19}$$

The initial conditions for these previous equations are consequences of

$$z(1) = -\frac{1}{\nu} p(1),$$

which, taking into account (16), gives

$$P_{11}(1) = 0, \;\; P_{12}(1) = -\frac{1}{\nu} I, \;\; r(1) = 0. \tag{20}$$

And the condition

$$q(1) = y(1) - y_d,$$

implies

$$P_{21}(1) = I, \;\; P_{22}(1) = -\frac{1}{\nu} I, \;\; t(1) = y_d. \tag{21}$$

As a consequence of the form of Riccati equations (18)(19) and initial conditions (20)(21), it is clear that $P_{11} = P_{22}$ and $P_{21} = -\nu P_{12}$, then if we set $P = P_{11}$ and $Q = P_{12}$, we obtain the following :

Result 2 *The solution of the control problem is given by the set of equations :*

$$\left\{ \begin{array}{rclcrcl} \frac{dP}{dx_1} & = & -P^2 + \nu Q^2 - D^2 & , & P(1) & = & 0, \\ \frac{dQ}{dx_1} & = & -PQ - QP & , & Q(1) & = & -\frac{1}{\nu} I, \\ \frac{dr}{dx_1} & = & -Pr - Qt - f & , & r(1) & = & 0, \\ \frac{dt}{dx_1} & = & \nu Qr - Pt & , & t(1) & = & -y_d, \end{array} \right. \tag{22}$$

$$\left\{ \begin{array}{l} z = Py + Qp + r, \\ q = -\nu Qy + Pp + t, \end{array} \right. \tag{23}$$

the optimal solution $\{y, p\}$ being solution of

$$\left\{ \begin{array}{l} \frac{\partial y}{\partial x_1} = Py + Qp + r, \\ y(0) = 0, \end{array} \right. \qquad \left\{ \begin{array}{l} \frac{\partial p}{\partial x_1} = -\nu Qy + Pp + t, \\ p(0) = 0. \end{array} \right. \tag{24}$$

□

4 DECOUPLING OF THE CONTROL PROBLEM (II)

There is an other way to decouple the optimality system (9), (11), (12) which amounts to taking its restriction on $\Omega_s =]0, s[\times]0, a[$. This situation is more complicated than the previous one because the control acts precisely on the *moving boundary*. In particular one cannot take the family of problems derived from (9), (11), (12) by restriction to Ω_s with arbitrary values of z and q on Γ_s, because it does not represent the optimality system of a family of non trivial control problems. So we consider the family of systems † :

$$\left\{ \begin{array}{ll} \frac{\partial y}{\partial x_1} = z & \text{on} \ \ ]0, s[, \\ y(0) = 0, & \end{array} \right. \qquad \left\{ \begin{array}{ll} -\frac{\partial z}{\partial x_1} - D^2 y = f & \text{on} \ \ ]0, s[, \\ z(s) + \frac{1}{\nu} p(s) = \psi(s), & \end{array} \right. \tag{25}$$

with the adjoint system

$$\left\{ \begin{array}{ll} \frac{\partial p}{\partial x_1} = q & \text{on} \ \ ]0, s[, \\ p(0) = 0, & \end{array} \right. \qquad \left\{ \begin{array}{ll} -\frac{\partial q}{\partial x_1} - D^2 p = 0 & \text{on} \ \ ]0, s[, \\ q(s) = y(s) - \varphi(s). & \end{array} \right. \tag{26}$$

One can check that it represents the optimality system for the problem with state equation (9) restricted to Ω_s, boundary condition :

$$\frac{\partial y}{\partial x_1} = u + \psi(s) \quad \text{on} \quad \Gamma_s \tag{27}$$

† In fact y, z, ...should have a superscript s which is omitted for the sake of simplicity.

and with the criterion :

$$J(u) = \frac{1}{2}\int_0^a |y(s,x_2) - \varphi(s,x_2)|^2\,dx_2 + \frac{\nu}{2}\int_0^a u^2(x_2)\,dx_2. \tag{28}$$

In particular one checks that $\varphi(1) = y_d(1)$ and $\psi(1) = 0$. Then it is possible to express $y(s)$ and $p(s)$ as an affine function of φ and ψ :

$$\left\{\begin{array}{lcl} y(s) &=& P_{11}(s)\psi + P_{12}(s)\varphi + r(s), \\ p(s) &=& P_{21}(s)\psi + P_{22}(s)\varphi + t(s), \end{array}\right. \tag{29}$$

where the functions r and t depend upon f and y_d. We will not give all the details of the calculations but let us show the starting point. If one takes the derivative with respect to x_1 of y we have

$$z(x_1) = \frac{dy}{dx_1} = \frac{dP_{11}}{dx_1}\psi + P_{11}\left(\frac{1}{\nu}q - D^2y - f\right) + \frac{dP_{12}}{dx_1}\varphi + P_{12}\left(z + D^2p\right) + \frac{dr}{dx_1}.$$

And, on the other hand,

$$q(x_1) = \frac{dp}{dx_1} = \frac{dP_{21}}{dx_1}\psi + P_{21}\left(\frac{1}{\nu}q - D^2y - f\right) + \frac{dP_{22}}{dx_1}\varphi + P_{22}\left(z + D^2p\right) + \frac{dt}{dx_1}.$$

Then, as we have

$$q(x_1) = y(x_1) - \varphi(x_1) = P_{11}\psi + (P_{12} - I)\varphi + r,$$

it is possible to eliminate q in order to get an equation containing only the functions φ, ψ, r and t. A formal identification gives the following set of equations :

$$\left\{\begin{array}{lcl} \frac{dP_{11}}{dx_1} &=& -\frac{1}{\nu}P_{11}^2 + P_{11}D^2P_{11} - P_{12}D^2P_{21} + P_{12}P_{21} - P_{12} - P_{21} + I, \\ \frac{dP_{12}}{dx_1} &=& -\frac{1}{\nu}P_{11}P_{12} + \frac{1}{\nu}P_{12}P_{22} + P_{11}D^2P_{12} + P_{12}D^2P_{22} - \frac{1}{\nu}P_{11} - \frac{1}{\nu}P_{22}, \\ \frac{dr}{dx_1} &=& -\frac{1}{\nu}P_{11}r + \frac{1}{\nu}P_{12}t + P_{11}D^2r + P_{12}D^2t + P_{11}f. \end{array}\right. \tag{30}$$

It is possible to derive similar equations verified by P_{21} , P_{22} and t. As previously we observe that $P_{11} = -P_{22} = P$ and $P_{21} = \nu P_{12} = Q$. With these notations the final Riccati system of equations is

$$\left\{\begin{array}{lclcl} \frac{dP}{dx_1} &=& -\frac{1}{\nu}P^2 + Q^2 + PD^2P - \nu QD^2Q - 2Q + I &,& P(0) = 0, \\ \frac{dQ}{dx_1} &=& -\frac{1}{\nu}PQ - \frac{1}{\nu}QP + PD^2Q + QD^2P + \frac{2}{\nu}P &,& Q(0) = 0, \\ \frac{dr}{dx_1} &=& PD^2r - QD^2t - \frac{1}{\nu}Pr - \frac{1}{\nu}(I-Q)t + Pf &,& r(0) = 0, \\ \frac{dt}{dx_1} &=& PD^2t + \nu\, QD^2r - (Q-I)r - \frac{1}{\nu}Pt + \nu Qf &,& t(0) = 0. \end{array}\right. \tag{31}$$

Unfortunately the solution of system (31) is not sufficient to determine y and p because these two functions are expressed, by means of (29) in term of the unknown functions φ

and ψ. If we replace φ and ψ by their expressions in the initial conditions of (25), (26) we get the following system of equations :

$$\left\{\begin{array}{lcl} y & = & Pz + \frac{1}{\nu}Pp + Qy - Qq + r, \\ p & = & \nu Qz + Qp - P(y-q) + t. \end{array}\right. \tag{32}$$

It is possible to solve this system and we finally obtain :

$$\left\{\begin{array}{lcl} y & = & Az + Bq + Rr + Tt, \\ p & = & \tilde{A}z + \tilde{B}q + \tilde{R}r + \tilde{T}t. \end{array}\right. \tag{33}$$

where

$$\left\{\begin{array}{lcl} A & = & \left(P^{-1}(I-Q) + \frac{1}{\nu}(I-Q)^{-1}P\right)^{-1} + \left(\frac{1}{\nu}Q^{-1}P + (Q^{-1}-I)P^{-1}(I-Q)\right)^{-1}, \\ B & = & \left(I - Q^{-1} - \frac{1}{\nu}Q^{-1}P(I-Q)^{-1}P\right)^{-1} + \left(I + \nu P^{-1}(I-Q)P^{-1}(I-Q)\right)^{-1}, \\ R & = & \left(I - Q + \frac{1}{\nu}P(I-Q)^{-1}P\right)^{-1}, \\ T & = & \left(P + \nu(I-Q)P^{-1}(I-Q)\right)^{-1}, \\ \tilde{A} & = & -\left(\frac{1}{\nu}I + P^{-1}(I-Q)P^{-1}(I-Q)\right)^{-1} + \nu\left(Q^{-1} - I + \frac{1}{\nu}Q^{-1}P(I-Q)^{-1}P\right)^{-1}, \\ \tilde{B} & = & \left(Q^{-1}P + (Q^{-1}-I)P^{-1}(I-Q)\right)^{-1} + \left(P^{-1}(I-Q) + \frac{1}{\nu}(I-Q)^{-1}\right)^{-1}, \\ \tilde{R} & = & -\left(\frac{1}{\nu}P + (I-Q)P^{-1}(I-Q)\right)^{-1}, \\ \tilde{T} & = & \left(I - Q + \frac{1}{\nu}P(I-Q)^{-1}P\right)^{-1}. \end{array}\right. \tag{34}$$

Finally z and q are solution of

$$\left\{\begin{array}{rcl} -\frac{dz}{dx_1} - D^2(Az + Bq) & = & f + D^2(Rr + Tt), \\ z(1) & = & \frac{1}{\nu}(P(1)y_d - t(1)), \end{array}\right. \tag{35}$$

$$\left\{\begin{array}{rcl} -\frac{dq}{dx_1} - D^2(\tilde{A}z + \tilde{B}q) & = & D^2(\tilde{R}r + \tilde{T}t), \\ q(1) & = & (P(1) - I)y_d + r(1)). \end{array}\right. \tag{36}$$

Result 3 *The optimal control u is given by :*

$$u = -\frac{1}{\nu^2}((\tilde{A}(1)P(1) + \nu\tilde{B}(1)(Q(1)-I))y_d + \nu(\tilde{B}(1) + \tilde{R}(1))r(1) + (\nu\tilde{T}(1) - \tilde{A}(1))t(1)), \tag{37}$$

where P, Q, r and t are solution of (31). *As r and t do not depend on y_d this gives explicitly the linear dependence of u on y_d.*

□

REFERENCES

Bellman, R., Dreyfus S. (1962) *Applied Dynamic Programming.* Princeton University Press.

Lions, J.L., (1968) *Contrôle optimal de systèmes gouvernés par des équations aux dérivées partielles.* Dunod.

Ramos, A.M. to appear.

Semismoothness in parametrized quasi-variational inequalities

Jiří V. Outrata
Institute of Information Theory and Automation
Academy of Sciences of the Czech Republic
Pod vodárenskou věží 4, 182 08 Prague, Czech Republic.
e-mail: outrata@utia.cas.cz

Abstract

The paper deals with stability and sensitivity analysis of a class of parameter-dependent quasi-variational inequalities. By using the stability theory of Robinson for generalized equations, we compute the directional derivative and the generalized Jacobian of the map which assigns to the parameter the (locally unique) solution of the quasi-variational inequality. This enables to prove that this map is in fact semismooth and, consequently, to apply various methods of nonsmooth analysis to the numerical treatment of an interesting class of equilibrium problems.

Keywords

Quasi-variational inequality, directional derivative, generalized Jacobian, semismoothness.

1 INTRODUCTION

Quasi-Variational Inequalities (QVIs), introduced in the seventies in connection with some stochastic impulse control problems, represent an important class of implicitly constrained variational inequalities. They provide a useful tool for modelling of various complicated equilibria (Baiocchi and Capelo, 1984). So far, the basic existence and uniqueness questions have been answered and the mathematicians have turned among others to the stability and sensitivity of their solutions with respect to parameter (Kyparisis and Ip, 1992, Kočvara and Outrata, 1994). In the present paper we use the stability theory of Robinson for generalized equations (GEs) to ensure the lipschitzian behaviour of the (locally unique) solution of our QVI with respect to the parameter and to compute the appropriate directional derivative (Section 2). Then we turn our attention to the generalized Jacobians of this mapping, whereby we use the sensitivity result of Kyparisis (1990). The comparison of directional derivatives and generalized Jacobians enables to prove the main result – the semismoothness of the investigated mapping. This is done in Section 3. The semismoothness property plays a crucial role in a number of important numerical approaches of nonsmooth analysis, in particular in bundle methods of nonsmooth optimization (Schramm and Zowe, 1992) and in a nonsmooth variant of the Newton method

(Qi, 1993). In Section 4 we show, how these methods may be applied to the numerical solution of

- inverse problems generated by parameter-dependent QVIs, and
- optimization problems with QVIs as constraints.

The paper presumes a certain basic knowledge of some concepts of nonsmooth analysis. For reader's convenience we state here at least the definition of a semismooth map.

Definition 1 (Qi and Sun, 1993). An operator $F[\mathbb{R}^n \rightarrow \mathbb{R}^m]$ is said to be *semismooth* at $x_0 \in \mathbb{R}^n$ if it is Lipschitz near x_0 and

$$\lim_{\substack{V \in \partial F(x_0 + th') \\ h' \to h,\, t \downarrow 0}} \{V h\}$$

exists for each $h \in \mathbb{R}^n$.

The following notation is employed: x^i is the ith component of a vector $x \in \mathbb{R}^n$, E is the unit matrix and B is the unit ball. If Ω is a closed convex set, then

$$N_\Omega(x) = \begin{cases} \text{normal cone to } \Omega \text{ at } x, \text{ provided } x \in \Omega \\ \emptyset \text{ otherwise.} \end{cases}$$

2 STABILITY ANALYSIS AND DIRECTIONAL DERIVATIVES

Let $\mathcal{A}$ be an open set in $\mathbb{R}^n$, $F[\mathcal{A} \times \mathbb{R}^n \rightarrow \mathbb{R}^m]$ be a continuously differentiable operator and $g^i[\mathcal{A} \times \mathbb{R}^m \times \mathbb{R}^m \rightarrow \mathbb{R}]$, $i = 1, 2, \ldots, s$, be twice continuously differentiable functions, convex in the third variable. With these data we define the parameter-dependent QVI:

For a given $x \in \mathcal{A}$ find a vector $y \in \Gamma(x, y)$ such that

$$\langle F(x,y), y' - y \rangle \geq 0 \quad \text{for all } y' \in \Gamma(x,y); \tag{1}$$

thereby $\Gamma(x,y) = \{z \in \mathbb{R}^m | g^i(x,y,z) \leq 0,\ i = 1, 2, \ldots s\}$.

Let S be the (set-valued) map assigning to $x \in \mathcal{A}$ the (sets of) solutions of (1). To analyse the local behaviour of this map, we fix a parameter $x_0 \in \mathcal{A}$ and assume that $y_0 \in S(x_0)$. Throughout the first two sections it is supposed that

(LI) the partial gradients $\nabla_3 g^i(x_0, y_0, y_0)$ for $i \in I(x_0, y_0) := \{j \in \{1, 2, \ldots, s\} | g^j(x_0, y_0, y_0) = 0\}$ are linearly independent.

It is well-known that the QVI (1) can be equivalently written as a nonsmooth equation

$$y = \mathrm{Proj}_{\Gamma(x,y)}(y - F(x,y)). \tag{2}$$

Evidently, under a suitable constraint qualification, $z = \mathrm{Proj}_{\Gamma(x_0,y_0)}(y_0 - F(x_0, y_0))$ if and only if there exists a Karush - Kuhn - Tucker (KKT) vector $\lambda_0 \in \mathbb{R}^s_+$ such that

$$0 \in \begin{bmatrix} z - y_0 + F(x_0, y_0) + \sum_{i=1}^s \lambda_0^i \nabla_3 g^i(x_0, y_0, z) \\ -G(x_0, y_0, z) \end{bmatrix} + \begin{bmatrix} 0 \\ N_{\mathbb{R}^s_+}(\lambda_0) \end{bmatrix}; \tag{3}$$

thereby

$$G(x,y,z) = \begin{bmatrix} g^1(x,y,z) \\ g^2(x,y,z) \\ \vdots \\ g^s(x,y,z) \end{bmatrix}.$$

Hence it is clear that under (LI), $y_0 \in S(x_0)$ if and only if y_0 solves together with a $\lambda_0 \in \mathbb{R}^s_+$ the GE

$$0 \in \begin{bmatrix} \mathcal{L}(x_0,y,\lambda) \\ -G(x_0,y,y) \end{bmatrix} + \begin{bmatrix} 0 \\ N_{\mathbb{R}^s_+}(\lambda) \end{bmatrix}, \tag{4}$$

where $\mathcal{L}(x,y,\lambda) = F(x,y) + \sum_{i=1}^{s} \lambda^i \nabla_3 g^i(x,y,y)$ is the *Lagrangian* related to the QVI (1). Due to (LI) to each $y_0 \in S(x_0)$ the KKT vector λ_0 is uniquely determined. Therefore we may introduce another index set

$J(x_0,y_0) := \{i \in I(x_0,y_0) | \lambda_0^i > 0\}.$

The constraints corresponding to $i \in I(x_0,y_0)$, $i \in J(x_0,y_0)$ and $i \in I(x_0,y_0) \setminus J(x_0,y_0)$ are termed *active, strongly active* and *semiactive* at (x_0,y_0), respectively. To shorten the notation, for an index set $K \subset \{1,2,\ldots,s\}$ and a vector $d \in \mathbb{R}^s$, d_K denotes the subvector composed from the components d^i, $i \in K$. Analogously, for a matrix D with s rows, D_K is the submatrix composed from the rows D^i, $i \in K$ and $[D]_n$ is the submatrix with the rows D^i, $i = 1,2,\ldots,n$ $(n \leq s)$. At the index sets $I(x,y)$, $J(x,y)$ the arguments will be dropped whenever it cannot lead to a confusion.

The theory of Robinson (1980) leads to the following important statement, in which Ψ is the map assigning to $x \in \mathcal{A}$ the (sets of) solutions of the GE (4).

Proposition 1. Let

$$Q := \begin{bmatrix} \nabla_2 \mathcal{L}(x_0,y_0,\lambda_0) & (\nabla_3 G(x_0,y_0,y_0))^{\mathrm{T}} \\ -\nabla_2 G(x_0,y_0,y_0) - \nabla_3 G(x_0,y_0,y_0) & 0 \end{bmatrix}$$

and consider the linear GE

$$\xi \in Q \begin{bmatrix} v \\ u \end{bmatrix} + \begin{bmatrix} 0 \\ N_{\mathbb{R}^s_+}(u) \end{bmatrix}. \tag{5}$$

Assume that the GE (4) is strongly stable at (x_0,y_0,λ_0), i. e. the GE (5) possesses a unique solution $(v,u) \in \mathbb{R}^m \times \mathbb{R}^s_+$ for each $\xi \in \mathbb{R}^{m+s}$, cf. Kyparisis (1990). Then there exist neighbourhoods $\mathcal{O}$ of x_0, $\mathcal{M}$ of y_0 and $\mathcal{N}$ of λ_0 such that the map $\widetilde{\Psi}(\cdot) := \Psi(\cdot) \cap (\mathcal{M} \times \mathcal{N})$ is single-valued and Lipschitz on $\mathcal{O}$.

In such a case, evidently, $\widetilde{\Psi}$ splits on $\mathcal{O}$ into two operators $\widetilde{S}$ and $\widetilde{\Lambda}$ which assign to $x \in \mathcal{O}$ the y- and λ-components of the solution $\widetilde{\Psi}(x)$ to the GE (4).

By using results from the theory of complementarity problems, it has been proved in Robinson (1980) that the condition of Prop. 1 is equivalent to the following requirement: The matrix

$$R = \begin{bmatrix} \nabla_2 \mathcal{L}(x_0,y_0,\lambda_0) & (\nabla_3 G_J(x_0,y_0,y_0))^{\mathrm{T}} \\ -\nabla_2 G_J(x_0,y_0,y_0) - \nabla_3 G_J(x_0,y_0,y_0) & 0 \end{bmatrix}$$

is nonsingular and its Schur complement to

$$Q_{\text{red}} = \begin{bmatrix} \nabla_2 \mathcal{L}(x_0, y_0, \lambda_0) & (\nabla_3 G_I(x_0, y_0, y_0))^{\mathrm{T}} \\ -\nabla_2 G_I(x_0, y_0, y_0) - \nabla_3 G_I(x_0, y_0, y_0) & 0 \end{bmatrix}$$

is a P-matrix (has positive principal minors). However, it is not easy to ensure these properties in terms of the original problem data. In fact, we have succeeded to get such a statement only in the case if (1) corresponds to an implicit complementarity problem (ICP), i. e. if $s = m$ and the constraint functions g^i have a special structure

$$g^i(x, y, z) = \varphi^i(x, y) - z^i, \quad i = 1, 2, \ldots, m. \tag{6}$$

Proposition 2. Let $\nabla_2 F(x_0, y_0)$ be positive definite and $\nabla_2 \Phi(x_0, y_0) (\nabla_2 F(x_0, y_0))^{-1}$ be negative semidefinite, whereby

$$\Phi(x, y) = \begin{bmatrix} \varphi^1(x, y) \\ \varphi^2(x, y) \\ \vdots \\ \varphi^m(x, y) \end{bmatrix}.$$

Then the assertion of Prop. 1 holds true.

Proof. The GE (5) attains the form

$$\begin{bmatrix} \xi_1 \\ \xi_2 \end{bmatrix} \in \begin{bmatrix} \nabla_2 F(x_0, y_0) & -E \\ -\nabla_2 \Phi(x_0, y_0) + E & 0 \end{bmatrix} \begin{bmatrix} v \\ u \end{bmatrix} + \begin{bmatrix} 0 \\ N_{\mathbb{R}^m_+}(u) \end{bmatrix}, \tag{7}$$

where $\xi = (\xi_1, \xi_2) \in \mathbb{R}^m \times \mathbb{R}^m$. As $\nabla_2 F(x_0, y_0)$ is positive definite, it is nonsingular and its inverse is also positive definite. Thus, (7) amounts to the equation

$$v = (\nabla_2 F(x_0, y_0))^{-1} (\xi_1 + u)$$

and the linear complementarity problem

$$\xi_2 + (\nabla_2 \Phi(x_0, y_0) - E) (\nabla_2 F(x_0, y_0))^{-1} \xi_1 \in (E - \nabla_2 \Phi(x_0, y_0)) (\nabla_2 F(x_0, y_0))^{-1} u + N_{\mathbb{R}^m_+}(u).$$

Under the imposed assumptions this linear complementarity problem possesses a unique solution u for each ξ_1, ξ_2 and then also the v-component of the solution of (7) is uniquely determined. In this way the suppositions of Prop. 1 have been verified. □

Henceforth it is assumed that the GE (4) corresponding to the original QVI fulfills the assumption of Prop. 1, whereby we are aware that the verification may be quite complicated. Then the results of Robinson (1980, 1991) imply also the directional differentiability of the map $\widetilde{S}$ at x_0. On denoting by $\widetilde{S}'(x_0; h)$ this directional derivative in the direction

$h \in I\!R^n$, $\widetilde{S}'(x_0; h) = v$ satisfies with a vector $u \in I\!R^s$ the system of equations and inequalities

$$\begin{array}{rcl}
\nabla_2\mathcal{L}(x_0,y_0,\lambda_0)v + (\nabla_3 G_I(x_0,y_0,y_0))^{\mathrm{T}}u_I & = & -\nabla_1\mathcal{L}(x_0,y_0,\lambda_0)h \\
\left[\nabla_3 G_J(x_0,y_0,y_0) + \nabla_2 G_J(x_0,y_0,y_0)\right] v & = & -\nabla_1 G_J(x_0,y_0,y_0)h \\
\left[\nabla_3 G_{I\setminus J}(x_0,y_0,y_0) + \nabla_2 G_{I\setminus J}(x_0,y_0,y_0)\right] v & \leq & -\nabla_1 G_{I\setminus J}(x_0,y_0,y_0)h \\
u^i & = & 0 \quad \text{for } i \notin I(x_0,y_0) \\
u_{I\setminus J} & \geq & 0
\end{array} \tag{8}$$
$$(\langle\nabla_1 g^i(x_0,y_0,y_0),h\rangle + \langle\nabla_2 g^i(x_0,y_0,y_0)+\nabla_3 g^i(x_0,y_0,y_0),v\rangle)u^i=0 \text{ for } i\in I(x_0,y_0)\setminus J(x_0,y_0).$$

This system will be used in the next section.

3 GENERALIZED JACOBIANS AND SEMISMOOTHNESS

The local behaviour of the map $\widetilde{S}$ near x_0 may also be characterized by so-called generalized Jacobian, cf. Clarke (1983). Consider the family $\mathcal{P}(I(x_0,y_0)\setminus J(x_0,y_0))$ of all subsets of $I(x_0,y_0)\setminus J(x_0,y_0)$ and denote the single sets from this family by T_i, where i runs through a suitably chosen index set $I\!K(x_0,y_0)$. Denote for $i \in I\!K(x_0,y_0)$

$$D_{J\cup T_i}(x_0,y_0,\lambda_0) = \begin{bmatrix} \nabla_2\mathcal{L}(x_0,y_0,\lambda_0) & \nabla_3 G_{J\cup T_i}(x_0,y_0,y_0)^{\mathrm{T}} \\ -\nabla_2 G_{J\cup T_i}(x_0,y_0,y_0) - \nabla_3 G_{J\cup T_i}(x_0,y_0,y_0) & 0 \end{bmatrix},$$
$$B_{J\cup T_i}(x_0,y_0,\lambda_0) = \begin{bmatrix} -\nabla_1\mathcal{L}(x_0,y_0,\lambda_0) \\ \nabla_1 G_{J\cup T_i}(x_0,y_0,y_0) \end{bmatrix}.$$

From the analysis below Prop. 1 it is clear that each matrix $D_{J\cup T_i}$ is nonsingular (R is nonsingular and each submatrix of a P-matrix is again a P-matrix and hence nonsingular).

Proposition 3. The generalized Jacobian of the map $\widetilde{S}$ at x_0, denoted $\partial\widetilde{S}(x_0)$, admits an upper estimate

$$\partial\widetilde{S}(x_0) \subset \operatorname{conv}\left\{\left[(D_{J\cup T_i}(x_0,y_0,\lambda_0))^{-1}B_{J\cup T_i}(x_0,y_0,\lambda_0)\right]_m \middle| i \in I\!K(x_0,y_0)\right\}. \tag{9}$$

Proof. The proof is essentially given in Outrata (1994). However, as our assumptions are slightly weaker, we repeat briefly its main idea. The strong stability implies due to Kyparisis (1990) the existence of a neighbourhood $\mathcal{U}$ of x_0 such that whenever $x \in \mathcal{U}$ and the maps $\widetilde{S}$, $\widetilde{\Lambda}$ are differentiable at x, then there exist an index $i \in I\!K(x_0,y_0)$ such that

$$D_{J\cup T_i}(x,\,\widetilde{S}(x),\,\widetilde{\Lambda}(x))\begin{bmatrix}\nabla\widetilde{S}(x) \\ \nabla\widetilde{\Lambda}_{J\cup T_i}(x)\end{bmatrix} = B_{J\cup T_i}(x,\,\widetilde{S}(x),\,\widetilde{\Lambda}(x))$$

and the matrix $D_{J\cup T_i}(x,\widetilde{S}(x),\widetilde{\Lambda}(x))$ is nonsingular. By the definition of the generalized Jacobian we just need to construct all possible limits of such derivatives if x tends to x_0. However, all these limits belong to the set $\{(D_{J\cup T_i}(x_0,y_0,\lambda_0))^{-1}B_{J\cup T_i}(x_0,y_0,\lambda_0)|i \in I\!K(x_0,y_0)\}$ and so incl. (9) follows. □

We are now ready to state the main result:

Theorem 4. The map $\widetilde{S}$ is semismooth at x_0.

Proof. By Qi and Sun (1993) it suffices to show that

$$\widetilde{S}'(x_0; h) = \lim_{\substack{V \in \partial\widetilde{S}(x_0+th) \\ t \downarrow 0}} \{V\, h\}$$

and that the convergence is uniform for all $h \in B$. Assume by contradiction that there exists an $\varepsilon > 0$, sequences $\{h_i\} \subset B$, $t_i \downarrow 0$ and matrices $V_i \in \partial\widetilde{S}(x_0 + t_i h_i)$ for which

$$\|V_i\, h_i - \widetilde{S}'(x_0; h_i)\| > \varepsilon \quad \text{for } i = 1, 2, \ldots$$

As the generalized Jacobian mapping is uniformly compact and $\{h_i\} \subset B$, we can select convergent subsequences $h_{i'} \to h$, $V_{i'} \to V$, whereby $h \in B$ and $V \in \partial\widetilde{S}(x_0)$ (due to the closedness of $\partial\widetilde{S}$). The map $\widetilde{S}'(x_0; \cdot)$ is lipschitzian and so

$$\|V\, h - \widetilde{S}'(x_0; h)\| \geq \varepsilon. \tag{10}$$

From Prop. 3 we know that

$$V = \sum_{\ell \in K(x_0, y_0)} \alpha_\ell\, P^\ell(x_0),$$

where $\alpha_\ell \geq 0$, $\sum_{\ell \in K(x_0,y_0)} \alpha_\ell = 1$, and $P^\ell(x_0)$ consists from the first m rows of the matrix $(D_{JUT_\ell}(x_0, y_0, \lambda_0))^{-1}\, B_{JUT_\ell}(x_0, y_0, \lambda_0)$. Ineq. (10) implies the existence of a $j \in I\!K(x_0, y_0)$ such that $\alpha_j > 0$ and $P^j(x_0)\, h \neq \widetilde{S}'(x_0; h)$. However, by (8) this is possible only if

$$\left[\nabla_3\, g^k(x_0, y_0, y_0) + \nabla_2\, g^k(x_0, y_0, y_0)\right] S'(x_0; h) < -\nabla_1\, g^k(x_0, y_0, y_0)\, h \tag{11}$$

for some $k \in T_j$. Let $M(x_0, y_0)$ denote the subset of $I(x_0, y_0) \setminus J(x_0, y_0)$ consisting of those indices k for which ineq. (11) holds. Furthermore, let $\widetilde{I\!K}(x_0, y_0)$ be an index set specifying the subsets of $\mathcal{P}(I(x_0, y_0) \setminus J(x_0, y_0) \setminus M(x_0, y_0))$ in the same way as $I\!K(x_0, y_0)$ specifies the subsets of $\mathcal{P}(I(x_0, y_0) \setminus J(x_0, y_0))$. Clearly, $P^\ell(x_0)\, h = \widetilde{S}'(x_0; h)$ for all $\ell \in \widetilde{I\!K}(x_0, y_0)$.

Due to the continuity of $S'(x_0; \cdot)$, inequalities (11) imply that

$$g^k\left(x_0 + t_{i'} h_{i'}, \widetilde{S}(x_0 + t_{i'} h_{i'}), \widetilde{S}(x_0 + t_{i'} h_{i'})\right) < 0 \quad \text{for } k \in M(x_0, y_0)$$

whenever i' is sufficiently large. Therefore, by Prop. 3, for such i' one has

$$V_{i'} \in \operatorname{conv}\left\{P^\ell(x_0 + t_{i'} h_{i'}) | \ell \in \widetilde{I\!K}(x_0, y_0)\right\}. \tag{12}$$

If we now take also ineq. (10) into account, we observe that there exists a natural number n_0 such that for all $i' > n_0$

$$\|V_{i'} h - \widetilde{S}'(x_0; h)\| > \frac{\varepsilon}{2},$$

where $V_{i'}$ is given by (12). However, this is impossible due to the continuity of the maps $x \mapsto P^\ell(x)$, $\ell \in \widetilde{K}(x_0, y_0)$, in a neighbourhood of x_0. The assertion has been proved. □

Remark. If the functions g^i do not depend on y, the QVI (1) becomes just an ordinary parameter-dependent variational inequality and the above result reduces to (Outrata and Zowe, 1995, Prop. 4).

4 APPLICATIONS

The results of the previous sections may be applied in the numerical solution of different equilibrium problems. Let $m = n$ and consider the inverse problem to the QVI (1):

For a given $y_0 \in \mathbb{R}^m$ find $x \in \mathcal{A}$ such that $y_0 \in S(x)$. Assume that this problem possesses a solution x_0 and that the (LI) constraint qualification as well as the strong stability from Section 2 hold.

Then one could apply the nonsmooth Newton variant of Qi (1993) to the numerical solution of this problem. The Newton iteration attains the form

$$x_{k+1} = x_k - V_k^{-1}(\widetilde{S}(x_k) - y_0), \quad k = 1, 2, \ldots, \tag{13}$$

where $V_k \in \partial_B \widetilde{S}(x_k) := \left\{ \lim \nabla \widetilde{S}(y) | y \to x_k, \ \widetilde{S} \text{ is differentiable at } y \right\}$.

Proposition 5. Assume that all matrices from $\partial_B \widetilde{S}(x_0)$ are nonsingular. Then the method (13) is locally superlinearly convergent.

Proof. As $\widetilde{S}$ is semismooth at x_0, it is an easy consequence of (Qi, 1993, Thm. 3.1).

Of course, under the above mild assumptions we may need a starting point very close to the solution. The reason is that even if the neighbourhood $\mathcal{O}$ (on which $\widetilde{S}$ is defined) is large enough, we actually need the strong stability on the whole $\mathcal{O}$ to dispose with the appropriate estimates of $\partial_B \widetilde{S}(x_k)$. In fact, at nonsmooth points of $\widetilde{S}$ the appropriately modified incl. (9) does not directly lead to the desired matrix V_k, but such points are met extremely rarely during the iteration process. To illustrate this sort of problems, we can take a discretized obstacle problem with a membrane (plate) and a complient obstacle (Kočvara and Outrata, 1994). The task is to compute the load imposed on the membrane (plate) such that a desired deflection is achieved.

Another class of equilibrium problems, where the results of previous sections may well be utilized, are the optimization problems with quasi-variational inequality constraints. Assume that S admits a lipschitzian selection $\widetilde{S}$, defined on a given compact subset of $\mathcal{A}$, denoted U_{ad}. Let $J[\mathcal{A} \times \mathbb{R}^m \to \mathbb{R}]$ be a continuously differentiable objective and consider the nonsmooth optimization problem

$$\begin{aligned} &\text{minimize} && J(x, y) \\ &\text{subject to} && y = \widetilde{S}(x) \\ &&& x \in U_{\text{ad}}. \end{aligned} \tag{14}$$

This problem clearly possesses a solution. It amounts to the minimization of the locally Lipschitz composite function $\Theta(x) := J(x, \widetilde{S}(x))$ over U_{ad} to which a suitable bundle method may be applied. However, one has to ensure that Θ is at least weakly semismooth (Schramm and Zowe, 1992) and be able to compute at each $x \in U_{\text{ad}}$ one arbitrary vector from $\partial\Theta(x)$. As J is continuously differentiable, the semismoothness of $\widetilde{S}$ directly implies the semismoothness of Θ. Moreover, knowing a matrix from $\partial\widetilde{S}(x)$, one can easily compute a desired vector from $\partial\Theta(x)$ by the standard chain rule. Thus, we merely need to require the satisfaction of the (LI) constraint qualification at all pairs $(x, \widetilde{S}(x))$ and the appropriate strong stability assumption for all x from an open set containing U_{ad}. A distinguished example representing this class of problems is the design of robot manipulators, whereby we optimize the stress distribution of each gripper. The constraining QVI results from the reciprocal formulation of the appropriate contact problem with Coulomb friction.

REFERENCES

Baiocchi, C. and Capelo, A. (1984) *Variational and Quasi-Variational Inequalities. Applications to Free Boundary Problems.* Wiley, New York.

Clarke, F. H. (1983) *Optimization and Nonsmooth Analysis.* Wiley, New York.

Kočvara, M. and Outrata J. V. (1994) On optimization of systems governed by implicit complementarity problems. *Numerical Functional Analysis and Optimization*, **15**, 869–87.

Kyparisis, J. (1990) Solution differentiability for variational inequalities. *Mathematical Programming*, **48**, 285–301.

Kyparisis, J. and Ip, Ch. M. (1992) Solution behaviour for parametric implicit complementarity problems. *Mathematical Programming*, **56**, 65–70.

Outrata, J. V. (1994) On optimization problems with variational inequality constraints. *SIAM J. Optimization*, **4**, 340–57.

Outrata, J. V. and Zowe, J. (1995) A Newton method for a class of quasi-variational inequalities. *Computational Optimization and Applications*, **4**, 5–21.

Qi, L. (1993) Convergence analysis of some algorithms for solving nonsmooth equations. *Mathematics of Operations Research*, **18**, 227–44.

Qi, L. and Sun, J. (1993) A nonsmooth version of Newton's method. *Mathematical Programming*, **58**, 353–67.

Robinson, S. M. (1980) Strongly regular generalized equations. *Mathematics of Operations Research*, **5**, 43–62.

Robinson, S. M. (1991) An implicit function theorem for a class of nonsmooth functions. *Mathematics of Operations Research*, **16**, 282–309.

Schramm, H. and Zowe, J. (1992) A version of the bundle idea for minimizing a nonsmooth function: Conceptual idea, convergence analysis, numerical results. *SIAM J. Optimization*, **2**, 121–252.

23

Optimal control problem governed by a semilinear parabolic equation

Jean-Pierre Raymond, Housnaa Zidani
Université Paul Sabatier, UMR CNRS MIP
31062 Toulouse Cedex France. Fax: (33) 61 55 83 85.
e-mail: raymond@cict.fr, zidani@cict.fr

Abstract
This paper deals with optimal control problems governed by semilinear parabolic equations. We obtain necessary optimality conditions in the form of an exact and an approximate Pontryagin's minimum principle for distributed controls and boundary controls.

Keywords
Nonlinear boundary controls, Semilinear parabolic equations, Pontryagin's minimum principle.

1 INTRODUCTION

Let Ω be a smooth bounded open subset in R^N ($N \geq 2$), and Γ its boundary. Let A be a second order symmetric uniformly elliptic operator with regular coefficients ($Ay = -\Sigma_{i,j} D_j(a_{ij}(x)D_i y)$) and let T be a positive number. We consider the following control problem.

Minimize $\qquad (P)$

$$J(y,u,v) = \int_Q F(y(x,t),u(x,t))\,dxdt + \int_\Sigma G(y(s,t),v(s,t))\,dsdt + \int_\Omega \ell(y(x,T))\,dx, \quad (1)$$

where y is the solution of the initial-boundary value problem

$$\begin{aligned} &\frac{\partial y}{\partial t}(x,t) + Ay(x,t) + f(y(x,t),u(x,t)) = 0 && \text{in } Q = \Omega\times]0,T[, \\ &\frac{\partial y}{\partial n_A}(s,t) + g(y(s,t),v(s,t)) = 0 && \text{on } \Sigma = \Gamma\times]0,T[, \quad (2) \\ &y(x,0) = y_0(x) && \text{in } \Omega, \end{aligned}$$

and the control variables u and v satisfy the constraints:

$$u \in U_{ad} = \{u \in L^q(Q) \mid u(x,t) \in K_U \text{ for a.e. } (x,t) \in Q\},$$

$$v \in V_{ad} = \{v \in L^r(\Sigma) \mid v(s,t) \in K_V \text{ for a.e. } (s,t) \in \Sigma\},$$

K_U and K_V are closed subsets of R, $y_0 \in C(\bar{\Omega})$ and

$$q > \frac{N}{2} + 1, \quad r > N + 1.$$

Throughout the paper, the following assumptions are in force.

(A1) - f and F are continuous on R^2, for every $u \in R$, $f(.,u)$ and $F(.,u)$ are of class C^1 on R and we have the following estimates

$$|f(y,u)| + |F(y,u)| + |F'_y(y,u)| \leq M_1(1 + |u|)\eta(|y|),$$

$$C_0 \leq f'_y(y,u) \leq M_1(1 + |u|)\eta(|y|),$$

where η is a nondecreasing function from R^+ into R^+, $M_1 \in R^+$ and $C_0 \in R$. (We have denoted by F'_y and f'_y the partial derivatives of F and f with respect to y, in all the sequel we adopt the same kind of notation for other functions).

(A2) - g and G are continuous on R^2, for every $v \in R$, $g(.,v)$ and $G(.,v)$ are of class C^1 on R and we have the following estimates

$$|g(y,v)| + |G(y,v)| + |G'_y(y,v)| \leq M_1(1 + |v|)\eta(|y|),$$

$$C_0 \leq g'_y(y,v) \leq M_1(1 + |v|)\eta(|y|),$$

where η, M_1 and C_0 are as in (A1).

(A3) - ℓ is of class C^1 on R.

(A4) - The infimum of (P) is finite.

We first prove that the state equation (2) admits a unique weak solution in the space $W(0,T; H^1(\Omega), (H^1(\Omega))') \cap C(\bar{Q})$, this new regularity result (continuity of the state on $\bar{Q}$) is particularly interesting to next consider control problems with pointwise state constraints.

We obtain optimality conditions for (P) in the form of pointwise Pontryagin's principles. For this we follow the method developed in (Bonnans and Casas, 1991) for elliptic problems. This method requires C^0-regularity results for linearized equations. In our knowledge such results are not known in our case (indeed some coefficients of the equation are not bounded and can be negative, see Proposition 1) and we state them in Section 2.

2 EXISTENCE RESULT FOR THE STATE EQUATION

2.1 Linear equation

We first give a new regularity result for some linear equations. This result will be used to prove the existence of a bounded weak solution of (2) and will be the main tool to establish convergence results in the proof of Pontryagin's principles.

Proposition 1.
Let a be in $L^q(Q)$, let b be in $L^r(\Sigma)$ verifying $a(x,t) \geq C_0$ in Q and $b(s,t) \geq C_0$ on Σ. There exists a positive constant $C_1 = C_1(N, q, r, \Omega, T, C_0)$ (independent of a and b), such that for every $\phi \in L^q(Q)$, every $\psi \in L^r(\Sigma)$ and every $y_0 \in C(\bar{\Omega})$, the weak solution $y \in C([0,T]; L^2(\Omega)) \cap L^2(0,T; H^1(\Omega))$ of the equation

$$\begin{aligned} &\frac{\partial y}{\partial t}(x,t) + Ay(x,t) + a(x,t)y(x,t) = \phi(x,t) && \text{in } Q, \\ &\frac{\partial y}{\partial n_A}(s,t) + b(s,t)y(s,t) = \psi(s,t) && \text{on } \Sigma, \\ &y(x,0) = y_0(x) && \text{in } \Omega, \end{aligned} \tag{3}$$

belongs to $C(\bar{Q})$ and satisfies the estimate:

$$\|y\|_{L^2(0,T;H^1(\Omega))} + \|y\|_{L^\infty(Q)} \leq C_1(\|\phi\|_{L^q(Q)} + \|\psi\|_{L^r(\Sigma)} + \|y_0\|_{L^\infty(\Omega)}). \tag{4}$$

Moreover, this solution belongs to $W(0,T; H^1(\Omega), (H^1(\Omega))')$. □

The proof is given in (Raymond and Zidani, 1995) and is based on duality arguments (to estimate y thanks to estimate on the solution of a Cauchy problem), comparison principle and estimates on analytic semigroups.

In the particular case when $a \in C^\infty(\bar{\Omega})$, $b(s,t) = \alpha \geq 0$, $\phi \in L^\infty(Q)$ and $\psi \in L^\infty(\Sigma)$ such a result is given in (Mackenroth, 1982).

2.2 State equation

Theorem 1.
If f and g verify the above assumptions, if $y_0 \in C(\bar{\Omega})$, $u \in L^q(Q)$ and $v \in L^r(\Sigma)$, then the equation (2) admits a unique weak solution y in $W(0,T; H^1(\Omega), (H^1(\Omega))') \cap C(\bar{Q})$, this solution satisfies the estimate

$$\|y\|_{L^\infty(Q)} \leq C_2(\|u\|_{L^q(Q)} + \|v\|_{L^r(\Sigma)} + \|y_0\|_{L^\infty(\Omega)} + 1),$$

where $C_2 = C_2(T, \Omega, N, q, r, C_0)$ is a positive constant. □

Idea on the proof.
Since we cannot directly apply a Faedo Galerkin method to equation (2), we first replace $f(.,u)$ and $g(.,v)$ by truncated functions $f_k(.,u)$ and $g_k(.,v)$ (such that $|f'_{ky}(y,u)|$ and $|g'_{ky}(y,v)|$ are bounded with respect to y).

We next regularize $f_k(y,u(.,.))$ and $g_k(y,v(.,.))$ in order that the derivatives of the regularized functions $(f_k^\varepsilon)'_y(y,u(.,.))$ and $(g_k^\varepsilon)'_y(y,v(.,.))$ be bounded respectively in Q and Σ.

We prove the existence of a bounded weak solution y_k^ε for the equation corresponding to f_k^ε and g_k^ε.

Thanks to Proposition 1, we get estimates on $(y_k^\varepsilon)_\varepsilon$ in $L^2(0,T;H^1(\Omega))\cap C(\bar{Q})$. We pass to the limit when ε tends to zero and we prove that $(y_k^\varepsilon)_\varepsilon$ converges in $L^2(0,T;H^1(\Omega))\cap C(\bar{Q})$ to the solution y_k of the equation corresponding to f_k and g_k. Next we prove that if k is big enough, then y_k is also a solution of (2). The uniqueness is proved thanks to a comparison principle. Moreover, if we denote by y the solution of (2), we can see that y is also a solution of

$$\begin{aligned}
&\frac{\partial z}{\partial t}(x,t) + Az(x,t) + a(x,t)z(x,t) = f(0,u(x,t)) && \text{in } Q,\\
&\frac{\partial z}{\partial n_A}(s,t) + b(s,t)z(s,t) = g(0,v(s,t)) && \text{on } \Sigma, \qquad (5)\\
&z(x,0) = y_0(x) && \text{in } \Omega,
\end{aligned}$$

with $a(x,t) = \int_0^1 f'_y(\theta y(x,t),u(x,t))\,d\theta \geq C_0$ in Q and $b(s,t) = \int_0^1 g'_y(\theta y(s,t),v(s,t))\,d\theta \geq C_0$ on Σ. From (A1), (A2), we deduce that $a \in L^q(Q)$ and $b \in L^r(\Sigma)$. Thanks to Proposition 1, it yields $y \in C(\bar{Q})$ and

$$||y||_{C(\bar{Q})} \leq C_1(||f(0,u)||_{L^q(Q)} + ||g(0,v)||_{L^r(\Sigma)} + ||y_0||_{L^\infty(\Omega)}),$$

$$||y||_{C(\bar{Q})} \leq C_2(||u||_{L^q(Q)} + ||v||_{L^r(\Sigma)} + ||y_0||_{L^\infty(\Omega)} + 1).$$

2.3 Pontryagin principles

We define a distributed Hamiltonian function and a boundary Hamiltonian function by:

$$H_Q(y,u,p) = F(y,u) - pf(y,u)$$

for every $(y,u,p) \in R \times R \times R$,

$$H_\Sigma(y,v,p) = G(y,v) - pg(y,v)$$

for every $(y,v,p) \in R \times R \times R$.

With the regularity results and the estimates stated in Theorem 1, we prove optimality conditions for (P) in the form of two decoupled Pontryagin principles.

Theorem 2.
If $(\bar{y}, \bar{u}, \bar{v})$ is a solution of (P), there then exists $\bar{p} \in W(0,T; H^1(\Omega), (H^1(\Omega))')$ satisfying the equation

$$-\frac{\partial p}{\partial t}(x,t) + Ap(x,t) + f'_y(\bar{y}(x,t), \bar{u}(x,t))p(x,t) = F'_y(\bar{y}(x,t), \bar{u}(x,t)) \quad \text{in } Q,$$

$$\frac{\partial p}{\partial n_A}(s,t) + g'_y(\bar{y}(s,t), \bar{v}(s,t))p(s,t) = G'_y(\bar{y}(s,t), \bar{v}(s,t)) \quad \text{on } \Sigma,$$

$$p(x,T) = \ell'_y(\bar{y}(x,T)) \quad \text{in } \Omega,$$

and such that:

$$H_Q(\bar{y}(x,t), \bar{u}(x,t), \bar{p}(x,t)) = \min_{u \in K_U} H_Q(\bar{y}(x,t), u, \bar{p}(x,t)) \text{ for a.e. } (x,t) \in Q,$$

$$H_\Sigma(\bar{y}(s,t), \bar{v}(s,t), \bar{p}(s,t)) = \min_{v \in K_V} H_\Sigma(\bar{y}(s,t), v, \bar{p}(s,t)) \text{ for a.e. } (s,t) \in \Sigma. \qquad \square$$

This kind of results has been already proved by Wu and Teo (1983) for problems with a linear state equation and a convex cost criterion and by Bonnans and Casas (1991) for problems governed by semilinear elliptic equations with homogeneous Dirichlet boundary conditions.

Very recently, Fattorini and Murphy (1994) have obtained a Pontryagin principle for problems with a boundary control in a Robin boundary condition and a terminal state constraint. Here our assumptions and our proof are completely different.

Sketch of the proof of Theorem 2.
We only give the proof of the boundary Pontryagin's principle, the distributed Pontryagin's principle can be obtained in similar way.

We first write the variation of the cost functional in the form of an Hamiltonian variation. (For this, we introduce an intermediate adjoint state as in (Bonnans and Casas, 1991)). We next use spike perturbations.

i) Thanks to Taylor formula we define $\tilde{G}_y(y_1, y_2, v)$ by:

$$\begin{aligned} G(y_1, v) - G(y_2, v) &= \left(\int_0^1 G'_y(y_2 + \theta(y_1 - y_2), v)\, d\theta\right)(y_1 - y_2) \\ &= \tilde{G}_y(y_1, y_2, v)(y_1 - y_2), \end{aligned}$$

for every $v \in R$, $y_1 \in R$ and $y_2 \in R$. We define $\tilde{F}_y(y_1, y_2, u)$, $\tilde{f}_y(y_1, y_2, u)$, $\tilde{g}_y(y_1, y_2, v)$ and $\tilde{\ell}_y(y_1, y_2)$ in a similar manner.
Let v_1 be in V_{ad}, we denote by y_1 the weak solution of (2) corresponding to $(\bar{u}, v_1)$ and we define the intermediate adjoint state p_1 as the weak solution of the equation:

$$\begin{aligned} &-\frac{\partial p}{\partial t} + Ap + \tilde{f}_y(y_1, \bar{y}, \bar{u})p = \tilde{F}_y(y_1, \bar{y}, \bar{u}) && \text{in } Q, \\ &\frac{\partial p}{\partial n_A} + \tilde{g}_y(y_1, \bar{y}, v_1)p = \tilde{G}_y(y_1, \bar{y}, v_1) && \text{on } \Sigma, \qquad (6) \\ &p(T) = \tilde{\ell}_y(y_1(T), \bar{y}(T)) && \text{in } \Omega. \end{aligned}$$

Let us notice that if $v_1 = \bar{v}$ then $y_1 = \bar{y}$ and the function $p_1 = \bar{p}$ is the so-called adjoint state associated with $(\bar{u}, \bar{v})$.
By a straightforward calculation, we get

$$J(y_1, \bar{u}, v_1) - J(\bar{y}, \bar{u}, \bar{v}) = \int_{\Sigma} \{H_{\Sigma}(\bar{y}, v_1, p_1) - H_{\Sigma}(\bar{y}, \bar{v}, p_1)\}\, ds\, dt.$$

ii) We define

$$S_k(s_0, t_0) = \{(s,t) \in \Sigma \mid |(s_0, t_0) - (s,t)|_{R^{N+1}} \leq 1/k\}.$$

Let v be in K_V, we consider a sequence $(v_k)_k$ of spike perturbations in V_{ad}, given by:

$$v_k(s,t) = \begin{cases} \bar{v}(s,t) & on\ \Sigma \setminus S_k(s_0,t_0), \\ v & on\ S_k(s_0,t_0), \end{cases} \tag{7}$$

and we denote by y_k (respectively p_k) the state associated with $(\bar{u}, v_k)$ (respectively the intermediate adjoint state associated with $(\bar{u}, v_k)$).
We first prove that $(y_k)_k$ converges to $\bar{y}$ in $C(\bar{Q})$. For this, we remark that $y_k - y$ is the weak solution of the equation:

$$\frac{\partial z}{\partial t}(x,t) + Az(x,t) + a_k(x,t)z(x,t) = 0 \qquad \text{in } Q,$$

$$\frac{\partial z}{\partial n_A}(s,t) + b_k(s,t)z(s,t) = g(\bar{y}(s,t), \bar{v}(s,t)) - g(\bar{y}(s,t), v_k(s,t)) \qquad \text{on } \Sigma,$$

$$z(x,0) = 0 \qquad \text{in } \Omega,$$

where

$$a_k(x,t) = \int_0^1 f'_y((\bar{y} + \theta(y_k - \bar{y}))(x,t), \bar{u}(x,t))\, d\theta \geq C_0,$$

$$b_k(s,t) = \int_0^1 g'_y((\bar{y} + \theta(y_k - \bar{y}))(s,t), v_k(s,t))\, d\theta \geq C_0.$$

Thus, from Proposition 1 it yields

$$\|y_k - \bar{y}\|_{L^\infty(Q)} \leq C_1 \|g(\bar{y}, \bar{v}) - g(\bar{y}, v_k)\|_{L^r(\Sigma)}.$$

We get convergence result from assumption (A2). In a similar manner we prove that $(p_k)_k$ converges to $\bar{p}$ in $C(\bar{Q})$.

iii) If (s_0, t_0) is a Lebesgue point for the functions:

$$(s,t) \longmapsto g(\bar{y}(s,t), v),$$

$$(s,t) \longmapsto g(\bar{y}(s,t), \bar{v}(s,t)),$$

$$(s,t) \longmapsto H_{\Sigma}(\bar{y}(s,t), v, \bar{p}(s,t)),$$

$$(s,t) \longmapsto H_{\Sigma}(\bar{y}(s,t), \bar{v}(s,t), \bar{p}(s,t)),$$

it yields :

$$0 \geq \lim_k \frac{1}{meas(S_k(s_0,t_0))}[J(y_k,\bar{u},v_k) - J(\bar{y},\bar{u},\bar{v})] =$$

$$H_\Sigma(\bar{y}(s_0,t_0),v,\bar{p}(s_0,t_0)) - H_\Sigma(\bar{y}(s_0,t_0),\bar{v}(s_0,t_0),\bar{p}(s_0,t_0)).$$

Let us notice that the set of Lebesgue points of the four above functions is of full measure.

iv) Thus we have proved that, for every $v \in K_V$, there exists a set $\Sigma(v,\bar{v})$ of full measure for the N- dimensional Lebesgue measure on Σ, such that

$$H_\Sigma(\bar{y}(s,t),\bar{v}(s,t),\bar{p}(s,t)) \leq H_\Sigma(\bar{y}(s,t),v,\bar{p}(s,t))$$

for all $(s,t) \in \Sigma(v,\bar{v})$. Now we conclude as in (Bonnans and Casas, 1991).

Assumptions (A1)-(A3) are not in general sufficient to prove existence of an optimal control for (P). However, since the infimum of (P) is finite (see $(A4)$), then, for every $\varepsilon > 0$, (P) admit ε^2−solutions that we can characterize when $q = r = \infty$. Moreover this kind of result will be useful to consider problems with state constraints.

Theorem 3.

For every $\epsilon > 0$ and every $\alpha > 0$, there exists an ϵ^2-solution for (P), denoted by $(y_{\epsilon,\alpha}, u_{\epsilon,\alpha}, v_{\epsilon,\alpha})$ such that:

$$H_Q(y_{\epsilon,\alpha}(x,t),u_{\epsilon,\alpha}(x,t),p_{\epsilon,\alpha}(x,t)) \leq H_Q(y_{\epsilon,\alpha}(x,t),u,p_{\epsilon,\alpha}(x,t)) + \epsilon$$

a.e. $(x,t) \in Q$, for every $u \in K_U$ verifying $|u| \leq \alpha$ and

$$H_\Sigma(y_{\epsilon,\alpha}(s,t),v_{\epsilon,\alpha}(s,t),p_{\epsilon,\alpha}(s,t)) \leq H_\Sigma(y_{\epsilon,\alpha}(s,t),v,p_{\epsilon,\alpha}(s,t)) + \epsilon$$

a.e. $(s,t) \in \Sigma$, for every $v \in K_V$ verifying $|v| \leq \alpha$. □

REFERENCES

Bonnans, J.F. and Casas, E. (1991) Un Principe de Pontryagine pour le Contrôle des Systèmes Semilinéaires Elliptiques. *J. Diff. Eq.*, **90**, 288-303.

Fattorini, H.O. and Murphy, T. (1994) Optimal Problems for Nonlinear Parabolic Boundary Control Systems. *SIAM J. Control and Optim.*, **32**, 1577-96.

Mackenroth, U. (1982) Convex Parabolic Boundary Control Problems with Pointwise State Constraints. *J. Math. Anal. Appli.*, **87**, 256-77.

Schmidt, E.J. (1989) Boundary Control for the Heat Equation with Nonlinear Boundary Condition. *J. Diff. Equat.*, **78**, 89-121.

Raymond, J.P. and Zidani, H. (1995) Hamiltonian Pontryagin's Principles for Control Problems Governed by Semilinear Parabolic Equations. *preprint.*

Wu, Z.S. and Teo, K.L. (1983) Optimal Control Problems Involving Second Boundary Value Problems of Parabolic Type. *SIAM J. Control and Optim.*, **21**, 729-57.

24

Shape optimization of hyperelastic rod

B. Rousselet
Universite de Nice
Departement de Mathematiques, Parc Valrose B.P. 71, 06108 Nice Cedex 2, France.
J. Piekarski
Fundamental Technological Research Institute
ul. Swietokrzyska 21, 00-950 Warsaw, Poland.
A. Myslinski
System Research Institute
ul. Newelska 6, 01-447 Warsaw, Poland.

Abstract
The paper is concerned with an optimal design problem for hyperelastic rod. The function describing the position of a point at the line of rod cross-sections centroids in its reference configuration is variable subject to optimization. The necessary optimality condition is formulated. The continuation method with gradient descent method were employed to solve this problem numerically. The numerical results are provided.

Keywords
shape optimization, optimality condition, numerical methods

1 INTRODUCTION

The paper deals with the formulation of a necessary optimality condition for an optimal design problem of a hyperelastic rod as well as with the numerical solution of this problem. The rod is subjected to large displacement involving flexion, shear, torsion and longitudinal extension (see Antman and Kenney (1981)). The equilibrium state of the rod is described by a system of two nonlinear coupled ordinary differential equations of the first order . In this model the function describing the position of the point at the line of centroids of cross-sections of the rod as well as the orthonormal directors are used to characterize the motion and deformation of the rod in an objective and intrinsic framework (see Antman and Kenney (1981)) . Moreover the nonlinear geometry of the rod is taken into account exactly . The existence of solutions to the rod problem was studied by Antman and Kenney (1981), Ciarlet (1986), LeTallec, Mani and Rochinha (1992).

The optimal design problem for hyperelastic rod consists in finding such reference configuration of the rod to minimize its compliance (see Haug, Choi and Komkov(1986),

Rousselet(1992)) measured as a distance between current and reference configurations occupied by the rod loaded by a given body force. The function describing the position of a point at the line of centroids of cross–sections of the rod in its reference configuration is variable subject to optimization. It is assumed that this function is bounded and has bounded derivative.

The aim of this paper is to formulate the shape optimization problem for the highly nonlinear rod model and to derive a necessary optimality condition. We shall show Lipschitz continuity of the solution to the rod model with respect to the optimized variable. The design sensitivity analysis of the solution to the state system is performed and the directional derivative of the cost functional is calculated. Using this derivative a necessary optimality condition is formulated. The calculated derivative is employed as an element of descent direction finding procedure in numerical algorithm for solving this optimization problem. Numerical cxample is provided.

2 KINEMATIC DESCRIPTION

Consider a rod in a reference configuration. This configuration is specified by enough regular mapping $\varphi_0(\xi)$ and an orthonormal pair of vector functions $E_1(\xi)$ and $E_2(\xi)$ of the real variable $\xi \in [0, L]$. The mapping :

$$\varphi_0 : R \ni \xi \to \varphi_0(\xi) \in R^3 \tag{1}$$

defines the line of centroids of cross–sections of a slender three dimensional body called a rod Antman and Kelly (1981). The line of centroids is assumed to be an arbitrary curve. The cross–section $A(\xi)$ of the rod is defined as a plane passing through a point $\varphi_0(\xi) \in R^3$, $\xi \in [0, L]$ and normal to the vector $E_3(\xi) = E_1(\xi) \times E_2(\xi)$. $\varphi_0(\xi)$ is interpreted as the position in a reference configuration of the material point at the centroid line of the section φ (see Antman and Kelly (1981), LeTallec, Mani and Rochinha (1992)). $E_1(\xi)$ and $E_2(\xi)$ characterize the reference configuration of the section ξ. In the reference configuration the rod occupies the domain : $\Omega_0 = \{X \in R^3 \ : \ X = \varphi_0(\xi) + \sum_{i=1}^{2} \tilde{X}_i E_i(\xi),\ 0 \leq \xi \leq L,\ (\tilde{X}_1, \tilde{X}_2) \in A(\xi)\}$ Note that $\{E_i(\xi)\}$, $i = 1, 2, 3$ is a moving orthonormal frame in the reference configuration.

The rod can suffer flexure, torsion, shear and longitudinal extension (see Antman and Kelly (1981), LeTallec, Mani and Rochinha(1992)). After deformation the current configuration of the rod is characterized by a position vector function $\varphi(\xi)$ of the line of centroids and an orthonormal pair of vectors $t_1(\xi)$ and $t_2(\xi)$ defining the cross–section of the deformed rod. The functions φ, t_1, t_2 depending on $\xi \in [0, L]$ have similar interpretation as φ_0, E_1, E_2 (see Antman and Kelly(1981)). The cross–section $A(\xi)$ of the deformed rod is defined as a plane passing through a point $\varphi(\xi) \in R^3$, $\xi \in [0, L]$ and normal to the vector $t_3(\xi) = t_1(\xi) \times t_2(\xi)$. In the current configuration each material point of the deformed rod occupies the position x(X). The set of points x is called the set of admissible configurations and is determined by : $\Omega = \{x \in R^3 \ : \ x = \varphi(\xi) + \sum_{i=1}^{2} X_i t_i(\xi),\ 0 \leq \xi \leq L, (X_1, X_2) \in A(\xi)\}$ where $\varphi : [0, L] \ni \xi \to \varphi(\xi) \in R^3$ defines the position of the line of centroids after deformation and $\{t_i(\xi)\}$, $i = 1, 2, 3$ is an orthonormal moving frame in the current configuration.

We shall assume that the deformed images of the two cross-sections are incapable of touching within the rod (see Ciarlet(1988)) , i.e., $det(t_1(\xi), t_2(\xi), t_3(\xi)) > 0 \ \ \forall \xi \in [0, L]$.

Moreover the deformation $x(X)$ is assumed to satisfy the kinematic boundary conditions imposed to the rod. Assume these conditions are :

$$\varphi(0) = \varphi_0 \quad \varphi(L) = \varphi_0(L) = \varphi_L, \quad t_i(0) = E_i(0) \text{ for } i \text{ given in } \{1,2,3\} \tag{2}$$

where φ_0, φ_L and $E_i(0)$, $i = 1,2,3$ are given vectors. The condition (2) imply the rod is fixed at both ends as well as the rod can rotate around $E_i(0)$.

3 VARIATIONAL FORMULATION

Let $E = R^3$ and $L^p(0,L)$, $1 \le p < \infty$, denotes the space of integrable functions (see Ciarlet(1988)). We denote by $W^{m,p}(0,L)$, $m \ge 1$, $1 \le p < \infty$ the Sobolev space (see Ciarlet(1988)). By K we denote the set of kinematically admissible deformation fields :

$$\begin{aligned} K = \{\{\varphi, t_1, t_2, t_3\} = \{\varphi, t_i\} \in W^{1,p}(0,L;E^4) \ : \varphi(0) = \varphi_0, \varphi(L) = \varphi_L, t_i(0) = E_i(0) \\ \text{for } i \in \{1,2,3\}; \ t_i(\xi)t_j(\xi) = \delta_{ij} \ \forall i,j = 1,2,3, \ \forall \xi \in [0,L]; \\ det(t_1(\xi), t_2(\xi), t_3(\xi)) > 0 \ \forall \xi \in [0,L]\} \end{aligned} \tag{3}$$

By $E(.,.) \ : \ K \to R$ we denote potential energy functional given by Antman and Kelly(1981), Ciarlet(1988), LeTallec, Mani and Rochinha(1992) :

$$E(\varphi, t_i) = \int_0^L w(\xi, \chi^j(t_i), \Gamma^k(\varphi, t_i))d\xi - \int_0^L (\bar{n}\varphi + f t_i)d\xi \tag{4}$$

where $\bar{n} \in L^{p^*}(0,L;E)$ and $f \in L^{p^*}(0,L;E^3)$ are given, $p^* = \frac{p}{p-1}$ and w denotes the stored energy function (see Rousselet, Piekarski, Myslinski(1995)). χ and Γ denote the strain measures (see Antman and Kelly(1981)), LeTallec, Mani and Rochinha(1992)). The equilibrium state of the rod is characterized by :

$$\textit{Find } \{\varphi, t_i\} \in K \textit{ such that : } \quad E(\varphi, t_i) \le E(z, d_i) \ \forall \{z, d_i\} \in K \tag{5}$$

Let us denote by $dK(\varphi, t_i)$ the space of kinematically admissible variations (see LeTallec, Mani and Rochinha(1992)).

$$\begin{aligned} dK(\varphi, t_i) = \{\{\delta\varphi, \delta t_i\} \in W^{1,p}(0,L;E^4) : \delta\varphi(0) = \delta\varphi(L) = 0, \delta t_i(0) = 0 \\ \text{for } i \in \{1,2,3\}; \ \exists \theta \in W^{1,p}(0,L;E), \ \delta t_i = \theta \times t_i \ \forall i = 1,2,3\} \end{aligned} \tag{6}$$

$dK(\varphi, t_i)$ is the space of all tangent vectors to the set K at a point $\{\varphi, t_i\}$. $\delta(.)$ may be interpreted as a differentiation operator on K into a tangent direction to K (see Antman and Kelly(1981), LeTallec, Mani and Rochinha(1992)). Let us introduce the following forms :

$$[W^{1,p}(0,L;E^4)]^2 \ni \tilde{b}(\varphi, t_i, \delta\varphi, \delta t_i) = \int_0^L \{n[(\delta\varphi)' - \theta \times (\varphi)'] + m\theta'\}ds \to R \tag{7}$$

$$W^{1,p}(0,L;E^4) \ni l(\delta\varphi, \delta t_i) = \int_0^L (\bar{n}\delta\varphi + f_i\delta t_i)ds \rightarrow R \tag{8}$$

wher n and m denote resultant contact force and moment respectively. Assuming the solutions to (5) are smooth enough the problem (5) is formally equivalent to the following system (see Ciarlet(1988)) : *Find* $\{\varphi, t_i\} \in K$ *satisfying*

$$\tilde{b}(\varphi, t_i, \delta\varphi, \delta t_i) = l(\delta\varphi, \delta t_i) \ \ \forall\{\delta\varphi, \delta t_i\} \in dK(\varphi, t_i) \tag{9}$$

Problem (9) is formulated in the current configuration. From numerical point of view (see LeTallec, Mani and Rochinha(1992)) it is convenient to consider this problem transformed to the reference configuration. Using the mapping transforming $\{E_i\}$ onto $\{t_i\}$ (see Rousselet, Piekarski and Myslinski(1995)) we can write problem (9) in the reference configuration.

LEMMA **31** *Problem (9) is equivalent to the following variational problem : Find* $\{\varphi, t_i\} \in K$ *satisfying*

$$b(\varphi, t_i, \delta\varphi, \delta t_i) = l(\delta\varphi, \delta t_i) \ \ \forall\{\delta\varphi, \delta t_i\} \in dK(\varphi, t_i) \tag{10}$$

where the form $b(.,.,.,.) : [W^{1,p}(0,L;E^4)]^2 \rightarrow R$ *is given by* $b(\varphi, t_i, \delta\varphi, \delta t_i) = \int_0^L (N\delta\Gamma + M\delta\chi)ds$ *where* N *and* M *denote the contact force and moment in the reference configuration and* $\delta\Gamma$ *and* $\delta\chi$ *denote the increments of strains in the reference configuration.*

Proof. Proof is given in Rousselet, Piekarski and Myslinski(1995). □

The existence of global and local solutions to the state system (10) was studied by Ciarlet(1988).

4 FORMULATION OF THE OPTIMIZATION PROBLEM

Let $\varphi_0 = \varphi_0(\xi)$ defined by (1) be the variable subject to optimization. φ_0 is assumed to satisfy :

$$c_1 \leq \varphi_0(\xi) \leq c_2, \ \left| \frac{d\varphi_0}{d\xi} \right| \leq c_3, \ \left| \frac{d^2\varphi_0}{d\xi^2} \right| \leq c_4, \ \int_{A(\xi)} \varphi_0(\xi)d\xi = c_5 \tag{11}$$

where c_1, c_2, c_3, c_4, c_5 are given positive constants. By U_{ad} we denote the set of admissible designs :

$$U_{ad} = \{\varphi_0 \in C^{2,1}(0,L;E) \ : \ \varphi_0 \text{ satisfies (11) }\} \tag{12}$$

$C^{2,1}$ denotes a space of Lipschitz continuous functions having Lipschitz continuous first and second order derivatives (see Ciarlet(1988)). U_{ad} is assumed to be nonempty. In order to underline the dependence of the solution $\{\varphi, t_i\} \in K \cap W^{2,p}(0,L)$ to the system (10) on $\varphi_0 \in U_{ad}$ we shall write : $b(\varphi_0, \varphi, t_i, \delta\varphi, \delta t_i)$ for $b(\varphi, t_i, \delta\varphi, \delta t_i)$, $l(\varphi_0, \delta\varphi, \delta t_i)$ for $l(\delta\varphi, \delta t_i)$.

From (see Rousselet, Piekarski, and Myslinski(1995)) it follows that for a given point $\varphi_0 \in U_{ad}$, problem (10) has a unique solution $\{\varphi, t_i\} \in K \cap W^{2,p}(0,L)$, $p \geq 2$ in a neighborhood of the linearization point $\{\varphi_0, E_i\} \in K$. We shall consider the following optimization problem : *Find* $\varphi_0 \in U_{ad}$ *minimizing the cost functional :*

$$J(\varphi_0; \varphi, t_i) = \int_0^L \{(\varphi - \varphi_0)^2 + (t_i - E_i)^2\} ds \tag{13}$$

on the set U_{ad}. $\{\varphi, t_i\} \in K \cap W^{2,p}(0,L)$ *is a solution to the system (10).*

The cost functional (13) approximates the compliance of the rod (see Haug, Choi and Komkov(1986)). If $\varphi, t_i, i = 1,2,3$ *are Lipschitz continuous functions with respect to* $\varphi_0 \in U_{ad}$ then from Weierstrass Theorem follows the existence of optimal solution $\hat{\varphi}_0 \in U_{ad}$ to the problem (13).

We shall calculate the directional derivative of the cost functional (13). The directional derivative $dJ(\varphi_0, \delta\varphi_0)$ of the cost functional (13) at a point $\varphi_0 \in U_{ad}$ in the direction $\delta\varphi_0 \in U_{ad}$ is given by :

$$dJ(\varphi_0, \delta\varphi_0) = 2\int_0^L \{(\varphi - \varphi_0)(\frac{\partial \varphi}{\partial \varphi_0} - 1)\delta\varphi_0 + (t_i - E_i)(\frac{\partial t_i}{\partial \varphi_0} - \frac{\partial E_i}{\partial \varphi_0})\delta\varphi_0\} ds + \int_0^L [(\varphi - \varphi_0)^2 + (t_i - E_i)^2] E_3 \frac{d\delta\varphi_0}{ds} ds \tag{14}$$

5 DESIGN SENSITIVITY ANALYSIS OF SOLUTIONS TO THE STATE SYSTEM

Recall from Rousselet, Piekarski and Myslinski(1995) that $\{\varphi, t_i\} \in K \cap W^{2,p}(0,L)$ is a unique local solution to the nonlinear system (10) in a neighborhood of a linearization point $\{\varphi_0, E_i\} \in K$.

Assuming that the solution $\{\varphi, t_i\} \in K \cap W^{2,p}(0,L)$ is sufficiently close to the linearization point $\{\varphi_0, E_i\} \in K$, from Continuity Theorem (see Ciarlet(1988)) it follows that the coercivity condition for the linearized form a of the form b still holds at a point $\{\varphi, t_i\} \in K \cap W^{2,p}(0,L)$, i.e., we have : there exists constant $\alpha > 0$ such that for all $\{\delta\varphi, \delta t_i\} \in dK(\varphi_0, E_i)$ the following condition holds :

$$a(\varphi_0, \varphi, t_i, \delta\varphi, \delta t_i, \delta\varphi, \delta t_i) \geq \alpha(\| \delta\varphi \|^2_{W^{1,2}(0,L)} + \| \delta t_i \|^2_{W^{1,2}(0,L)}) \tag{15}$$

where a denoting the linearized form of the form b is given by Rousselet, Piekarski and Myslinski(1995). Using (15) we are able to prove :

LEMMA **51** *If* $\{\varphi, t_i\} \in K \cap W^{2,p}(0,L)$ *is a unique local solution to the system (10) then there exists the Frechet dertivative of the mapping*

$$U_{ad} \ni \varphi_0 \rightarrow (\varphi, t_i) \in K \tag{16}$$

at a point $\{\varphi, t_i\} \in K \cap W^{2,p}(0,L)$ *in a direction* $\delta\varphi_0 \in U_{ad}$.

Proof. From Rousselet, Piekarski and Myslinski(1995) it follows that a point $\{\varphi, t_i\} \in K \cap W^{2,p}(0,L)$ is a unique solution to the system (10). Let us calculate the derivatives of the form (10) with respect to φ and t_i, respectively, at a point $\{\varphi, t_i\} \in K \cap W^{2,p}(0,L)$ in a direction $\{\tilde{\delta}\varphi, \tilde{\delta}t_i\} \in dK(\varphi, t_i)$. These derivatives are given by :

$$\frac{\partial b}{\partial \varphi}(\varphi_0, \varphi, t_i, \delta\varphi, \delta t_i)\tilde{\delta}\varphi = \int_0^L \{c_i(\frac{d\tilde{\delta}\varphi}{ds}t_i)[t_i(\frac{d\delta\varphi}{ds} - \theta \times \frac{d\varphi}{ds})][c_i(\frac{d\varphi}{ds}t_i) - \bar{c}_i]\frac{d\tilde{\delta}\varphi}{ds}\delta t_i\}ds \quad (17)$$

$$\frac{\partial b}{\partial t_i}(\varphi_0, \varphi, t_i, \delta\varphi, \delta t_i)\tilde{\delta}t_i = \int_0^L \{c_i(\frac{d\varphi}{ds}\tilde{\delta}t_i)[t_i(\frac{d\delta\varphi}{ds} - \theta \times \frac{d\varphi}{ds})] + d_i(\frac{d\tilde{\theta}}{ds}t_i)(\frac{d\theta}{ds}t_i)\} +$$
$$\int_0^L \{[c_i(\frac{d\varphi}{ds}t_i) - \bar{c}_i][\frac{d\delta\varphi}{ds}\tilde{\delta}t_i - (\theta \times \frac{d\varphi}{ds})\tilde{\delta}t_i] + \frac{1}{2}e_{ikl}d_i(\frac{dt_k}{ds}t_l - \frac{dE_k}{ds}E_l)(\frac{d\theta}{ds}\tilde{\delta}t_i)\}ds \quad (18)$$

The derivatives (17) and (18) are linear and bounded. Moreover from (15) it follows positiv definiteness of the derivatives (17) and (18) in the space $W^{1,2}(0,L)$. Hence by implicit function theorem (see Ciarlet(1988)) as well as the regularity of solutions to the system (10) follows the Frechet differentiability of the mapping (16). □

Assume that the curve $\varphi_0 \in U_{ad}$ is perturbed into a curve $(\varphi_0 + \delta\varphi_0) \in U_{ad}$. We derive the design sensitivity of the state system (10). Using Lemma 5.1 and differentiating the system (10) with respect to $\varphi_0 \in U_{ad}$ at a point $\{\varphi, t_i\} \in K$ in a direction $\delta\varphi_0 \in U_{ad}$ we obtain :

$$\frac{\partial b}{\partial \varphi_0}(\varphi_0, \varphi, t_i, \delta\varphi, \delta t_i)\delta\varphi_0 + \frac{\partial b}{\partial \varphi}(\varphi_0, \varphi, t_i, \delta\varphi, \delta t_i)\frac{\partial \varphi}{\partial \varphi_0}\delta\varphi_0 +$$
$$\frac{\partial b}{\partial t_i}(\varphi_0, \varphi, t_i, \delta\varphi, \delta t_i)\frac{\partial t_i}{\partial \varphi_0}\delta\varphi_0 = \frac{\partial l}{\partial \varphi_0}(\varphi_0, \delta\varphi, \delta t_i)\delta\varphi_0 \quad \forall\{\delta\varphi, \delta t_i\} \in dK(\varphi, t_i) \quad (19)$$

The derivatives $\frac{\partial b}{\partial \varphi}(\varphi_0, \varphi, t_i, \delta\varphi, \delta t_i)$ and $\frac{\partial b}{\partial t_i}(\varphi_0, \varphi, t_i, \delta\varphi, \delta t_i)$ are given by (17) and (18) respectively. The derivatives with respect to φ_0 are calculated in Rousselet, Piekarski and Myslinski(1995).

6 NECESSARY OPTIMALITY CONDITION

Using (19) and formulae in Rousselet, Piekarski and Myslinski(1995) we shall calculate the derivative (14) explicitly. Let us introduce an adjoint state $\{\delta\lambda, \delta\varepsilon_i\} \in dK(\varphi, t_i)$ defined as the solution to the following system .

$$\frac{\partial b}{\partial \varphi}(\varphi_0, \varphi, t_i, \delta\lambda, \delta\varepsilon_i)\delta\varphi + \frac{\partial b}{\partial t_i}(\varphi_0, \varphi, t_i, \delta\lambda, \delta\varepsilon_i)\delta t_i = \quad (20)$$
$$-2\int_0^L \{(\varphi - \varphi_0)\delta\varphi + (t_i - E_i)\delta t_i\}ds \quad \forall\{\delta\varphi, \delta t_i\} \in dK(\varphi, t_i)$$

The derivatives $\frac{\partial b}{\partial \varphi}$ and $\frac{\partial b}{\partial t_i}$ in (20) are determined by (17) and (18) respectively. From (15) and Lax–Milgram Lemma (see Ciarlet(1988)) follows the existence of a unique solution $\{\delta\lambda, \delta\varepsilon_i\} \in dK(\varphi, t_i)$ to the adjoint system (20).

Using (14), (17), (18), (20) as well as Rousselet, Piekarski and Myslinski(1995) we can calculate the directional derivative of the cost functional (13) :

$$dJ(\varphi_0, \delta\varphi_0) = \frac{\partial b}{\partial \varphi_0}(\varphi_0, \varphi, t_i, \delta\lambda, \delta\varepsilon_i)\delta\varphi_0 - \frac{\partial l}{\partial \varphi_0}(\varphi_0, \delta\lambda, \delta\varepsilon_i)\delta\varphi_0 + \tag{21}$$
$$2\int_0^L \{(\varphi - \varphi_0)\delta\varphi_0 + (t_i - E_i)\frac{\partial E_i}{\partial \varphi_0}\delta\varphi_0\}ds + \int_0^L [(\varphi - \varphi_0)^2 + (t_i - E_i)^2]E_3\frac{d\delta\varphi_0}{ds}ds$$

where the derivatives with respect to φ_0 are determined in Rousselet , Piekarski and Myslinski(1995).

LEMMA 61 *If $\hat{\varphi}_0 \in U_{ad}$ is an optimal solution to the problem (13), then there exist Lagrange multipliers $\mu_1 \in C^{1,1}(0, L)$, $\mu_2 \in C^{2,1}(0, L)$, $\mu_3 \in [C^{2,1}(0, L)]^*$ such that for all $\delta\varphi_0 \in C^{2,1}(0, L; E)$ satisfying $c_1 \leq \hat{\varphi}_0 + \delta\varphi_0 \leq c_2$ where c_1, c_2 are the constants as in (11), the following condition holds :*

$$\frac{\partial J}{\partial \varphi_0}(\hat{\varphi}_0)\delta\varphi_0 + sgn(\frac{d\hat{\varphi}_0}{d\xi})\int_0^L \mu_1\frac{d\delta\varphi_0}{d\xi}d\xi + sgn(\frac{d^2\hat{\varphi}_0}{d\xi^2})\int_0^L \mu_2\frac{d^2\delta\varphi_0}{d\xi^2}d\xi + \int_{A(\xi)} \mu_3\delta\varphi_0 d\xi \geq 0 \tag{22}$$

where the derivative $\frac{\partial J}{\partial \varphi_0}(\varphi_0)\delta\varphi_0$ is given by (21) and the function sgn(.) is defined by : $sgn(x) = 1$ if $x > 0$, $sgn(x) = 0$ if $x = 0$, $sgn(x) = -1$ if $x < 0$.

Proof. Proof is standard (see Ciarlet (1988)). □

7 NUMERICAL RESULTS

The optimization problem (13) was solved using the descent method (see Haug, Choi and Komkov(1986)). The centroid line φ_0, the design parameter, was interpolated by cubic Hermit spline. The calculated directional derivative (21) was used in the descent algorithm. The state problem (10) is solved using the finite element method. The line of centroids φ as well as directors t_i were discretized using the linear finite elements (see Rousselet, Piekarski and Myslinski(1995), LeTallec, Mani and Rochinha(1992)). The one point Gauss quadrature formula was used.

The continuation method was used to solve the nonlinear system (10) where the predictor colinear with the variable increment in the previous iteration was used and the orthogonal plane method as the correction method was employed.

As an example we considered 2D arch with rectangular, constant cross–section. The centroid line of this arch is assumed to occupy in the initial undeformed configuration the line given by parabola

$$y(\xi) = -(\xi - 0.5)(\xi + 0.5) \quad \text{where } \xi \in [a, b] \tag{23}$$

The numerical data are : $a = -0.5$, $b = 0.5$. The line (23) was interpolated using 5

Table 1 Initial and optimal shapes of arch. *1 - initial non-loaded form, 2 - optimal non-loaded form for* $f_2 = 2.5e + 04$,*3 - optimal non-loaded form for* $f_1 = 10$

ξ	-0.50	-0.40	-0.20	0.00	0.20	0.40	0.50
1	0.00	0.08	0.20	0.25	0.20	0.08	0.00
2	0.00	0.11	0.36	0.45	0.36	0.11	0.11
3	0.00	0.39	0.61	0.75	0.61	0.39	0.11

nodes with coordinates : $x = -0.5, -0.3, 0.0, 0.3, 0.5$. The nodal points of the discretized arch are allowed to move vertically only (see Haug, Choi and Komkov(1986)). To solve the system (10) the segment $[a, b]$ has been divided into 20 elements. The performance of the optimization algorithm is measured by : $w_1 = J(\hat{\varphi}_0)/J(\varphi_0^{ini})$, $w_2 = \| dJ(\hat{\varphi}_0) \| / \| dJ(\varphi_0^{ini}) \|$ where $\hat{\varphi}_0$ and φ_0^{ini} are optimal and initial configuration of the arch centroid line, respectively, J is a cost functional defined by (13) , $dJ(.)$ is determined by (21). In the formula for w_2 the euclidean norm is used.

The obtained result is presented in Table 1. The computations were performed for the following values of the concentrated force f : $f_1(\xi) = 10$ *linear case*, $f_2(\xi) = 2.5 \cdot 10^4$ *nonlinear case*. Table 1 presents the initial, non-loaded form of the arch in the reference configuration (23) as well as the optimal non-loaded forms of the arch with the f_1 and f_2 value of forces, respectively. Note that the maximal height of the optimal shape arch for the force f_2 is lower than the arch height for the load f_1. Moreover the shape near the end points for both loads has been significantly changed in comparison to the initial form. In the linear case : $w_1 = 0.139$, $w_2 = 4.1 \cdot 10^{-2}$. For the nonlinear arch these values are : $w_1 = 0.385$, $w_2 = 2.61 \cdot 10^{-4}$.

REFERENCES

Antman, S.S. and Kenney, C.S. (1981) Large Buckled States of Nonlineary Elastic Rods under Thorsion, Thrust and Gravity. *Archive for Rational Mechanics and Analysis*, **76**, 289 - 338.

Ciarlet, Ph. (1988) *Mathematical Elasticity. Vol 1: Three Dimensional Elasticity.* North–Holland, Amsterdam.

Haug, E.J., Choi, K.K. and Komkov, V. (1986) *Design Sensitivity Analysis of Structural Systems.* Academic Press, New York.

LeTallec, P., Mani, S. and Rochinha, F.A. (1992) Finite Element Computation of Hyperbolic Rods in Large Displacement. *Mathematical Modelling and Numerical Analysis*, **26**, 595 - 625.

Rousselet, B. (1992) A Finite Strain Rod Model and its Design Sensitivity. *Mechanical Structures and Machinery*, **20**, No 4.

Rousselet, B., Piekarski, J. and Myslinski, A. (1995) Design Sensitivity for Hyperelastic Rod in Large Displacements with respect to its Midcurve. *Report.*

Engineering Applications

Dynamic modelling and optimal hierarchical control of a multiple-effect evaporator - superconcentrator plant

P. Gil [1], *H. Duarte-Ramos* [1], *A. Dourado Correia* [2]

[1]) *Grupo de Engenharia Sistémica*
Departamento de Engenharia Electrotécnica, Universidade Nova de Lisboa, Quinta da Torre - 2825 Monte da Caparica, Portugal.
Tel: +351 1 2957787/2954464. Fax: +351 1 2954461/2940965.
e-mail: psg@mail.fct.unl.pt *(P. Gil)*

[2]) *CISUC - Centro de Informática e Sistemas*
Departamento de Engenharia Informática, Universidade de Coimbra, Largo Marquęs de Pombal - 3000 Coimbra, Portugal.
Tel: +351 39 7000000/7000006. Fax: +351 39 701266.
e-mail: adourado@gemini.ci.uc.pt

Abstract

This paper presents new dynamic nonlinear mathematical models for a long tube vertical evaporator and for a plate-type falling film concentrator. Simulation results for a real evaporation system comprising 6 evaporators and one three-section concentrator are presented. In order to prevent from unacceptable deviations in outlet black liquor concentration due to measurable disturbances in weak black liquor, an optimal hierarchical infinite time linear regulator is designed.

Keywords

Evaporator, concentrator, modelling, simulation, hierarchical control.

1 INTRODUCTION

In both economics and environmental issues of wood pulping industry, it is essential that the spent chemicals from the cooking process be recovered for reuse. The very first step in the recovery cycle is to concentrate by an evaporation process the weak black liquor leaving the digesters. For good combustion performance and smooth operation of the boiler, it is quite important that solids content of strong black liquor be permanently high and, simultaneously, steady.

The present paper proposes new mathematical nonlinear models, at unsteady state operation, for a long tube vertical evaporator (LTV) and for a plate-type falling film concentrator, on the basis of fundamentals of heat transfer. In both models, the evaluation of heat transfer rate takes into account the flow regimes outside and inside the evaporator tubes and concentrator plates.

To investigate the feasibility of hierarchical control methodology for regulating black liquor concentration leaving the evaporation plant, it is formulated and implemented an optimal infinite time linear regulator within a two-level structure.

2 MATHEMATICAL MODELLING

2.1 Long Tube Vertical Evaporator

In a LTV evaporator (Fig. 1.a) the liquor enters the main liquor box at the bottom of the system, rises through the tubes and leaves out with higher solids content. In this path, the liquor is heated by the condensing steam located outside the tubes. At the same time, the static head liquor above the tube decreases and thus, at some point along the tube, the liquor starts to boil. When preheater or afterheater sections are incorporated, the liquor is previously heated before entering the respective evaporating section or the main liquor box of next effect.

2.1.1 Modelling of a Single LTV Evaporator

For the purposes of its modelling, the LTV evaporator with afterheater section is broken up into the following three parts: main liquor box, heating body including vapour space and afterheater section. Each one of these subsystems is delimited by a control volume and transient balances are written for total mass, solute and energy, taking into account some simplifying assumptions.

This approach leads to the following system of three nonlinear differential equations (Gil, 1995): black liquor concentration and temperature rates at the tube inlet and black liquor concentration rate at the exit plus one nonlinear algebraic equation describing the temperature gain in the afterheater section, as follows

$$\frac{\mathrm{d}S_{L,it}}{\mathrm{d}t} = \dot{m}_{L,ir} \frac{S_{L,id} - S_{L,it}}{V_r \rho_{L,it}} \tag{1}$$

$$\frac{\mathrm{d}T_{L,it}}{\mathrm{d}t} = \left(\frac{\dot{m}_{L,it}}{V\rho_{L,it}}\right)\left[\left(h_{L,ir} - h_{L,it}\right) - \left(S_{L,ir} - S_{L,it}\right)\frac{\partial h_{L,it}}{\partial t}\right]\left(\frac{\partial h_{L,it}}{\partial T_{L,it}}\right)^{-1} \tag{2}$$

$$\frac{\mathrm{d}S_{L,ot}}{\mathrm{d}t} = \frac{2}{M_{L,t}}\left[\dot{m}_{L,it}S_{L,it} - \xi(\cdot) - \frac{\left(\frac{\partial h_{L,ot}}{\partial S_{L,ot}}\right)\left(\dot{m}_{L,it}S_{L,it} - \xi(\cdot)\right) + \dot{m}_{L,it}\left(h_{vap,ot} - h_{L,it}\right) - \dot{Q} + \gamma(\cdot)}{h_{vap,ot} - h_{L,ot} + S_{L,ot}\frac{\partial h_{L,ot}}{\partial S_{L,ot}}} S_{L,ot}\right] \tag{3}$$

$$T_{L,oAft} = \frac{\dot{Q}_{Aft}}{\dot{m}_{L,iAft}cp_{L,iAft}} + T_{L,iAft} \tag{4}$$

with

$$\xi(\cdot) = \frac{1}{2}M_{L,t}\dot{m}_{L,ir}\frac{S_{L,ir} - S_{L,it}}{\rho_{L,it}V} \qquad \gamma(\cdot) = \frac{M_{L,t}}{2}\left(\frac{\mathrm{d}h_{L,it}}{\mathrm{d}t}\right) \tag{5}$$

As the above equations show, the mathematical model for the LTV evaporator is characterised by a strong interdependence on the descriptive main liquor box variables. In addition, as it was expected, main liquor box dynamics has a relevant influence on the heating section dynamics, without any reciprocity.

2.1.2 Heat Transfer Rate in the Heating Section

In most classical models found in literature, e.g. Niemi and Koistinen (1972), heat transfer rate in the heating section is determined assuming a constant overall heat transfer coefficient or evaluated according some specific empirical expressions. In our approach instead, since the overall heat transfer coefficient is a local parameter, the evaluation of heat transfer rate in heating section is based on a cylindrical discretization of the interface (see Equation 6). The evaluation of local heat transfer coefficient takes into account liquor flow pattern (bubbly, slug, churn, annular or foaming) and condensate flow regime (laminar or turbulent). Moreover, in each section the overall heat transfer coefficient and liquor temperature are supposed uniform and the axial heat flux is assumed negligible, so that

$$\dot{Q} = \pi D\Delta l\sum_{k=1}^{N}\left[\overline{U}_k\left(T_{vap} - T_{L,k}\right)\right] \tag{6}$$

where

$$\overline{U}_k = \left[\frac{r_{t,i}}{r_{I,i}}\frac{1}{\overline{h}_{cond,k}} + \frac{r_{t,i}}{K_I}\ln\frac{r_{I,i}}{r_{I,i}} + \frac{r_{t,i}}{K_t}\ln\frac{r_{t,e}}{r_{t,i}} + \frac{r_{t,i}}{r_{t,e}}\frac{1}{\overline{h}_{conv,k}}\right]^{-1} \tag{7}$$

Relationships for evaluating black liquor heat transfer coefficients can be found, for instance, in Olauson (1981).

2.2 Plate - Type Falling Film Concentrator

Plate-type falling film concentrator (Fig. 1.b) comprises, basically, a liquor distribution system, heating plates, vapour body, entrainment separator system for evaporated vapour and a collecting box. In this system, the heating fluid (steam) flows vertically upward inside the heating elements, while the liquor is distributed from the top over the plates and flows downward to the collecting liquor box.

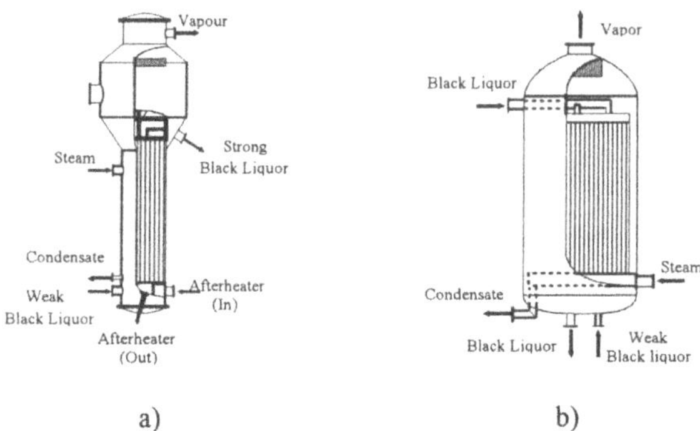

Figure 1 a) Long tube vertical evaporator; b) Plate-type falling film concentrator.

2.2.1 *Modelling of a Single Plate-Type Falling Film Concentrator*

For modelling purposes, the plate-type concentrator comprises the following parts: bank of heating plates, distribution (top) and collecting (bottom) boxes. For each subsystem, energy and mass balance equations are written, assuming some symplifying assumptions. Mathematical expressions describing the concentrator dynamics are given bellow (Gil, 1995).

$$\frac{\mathrm{d}S_{L,op}}{\mathrm{d}t}=\frac{2}{M_{L,p}}\left[\dot{m}_{L,ip}S_{L,ip}-\frac{S_{L,op}\,\dot{m}_{L,ip}}{h_{L,op}-h_{vap,o}-S_{L,op}\dfrac{\partial h_{L,op}}{\partial S_{L,op}}}\left(h_{L,ip}-h_{vap,o}-S_{L,op}\frac{\partial h_{LN,op}}{\partial S_{LN,op}}+\dot{Q}\right)\right] \quad (8)$$

$$\frac{\mathrm{d}S_{L,B}}{\mathrm{d}t}=\frac{\dot{m}_{L,op}\left(S_{L,op}-S_{L,B}\right)}{V_B\,\rho_{L,B}} \quad (9)$$

$$\frac{\mathrm{d}T_{L,B}}{\mathrm{d}t}=\left(\frac{\dot{m}_{L,op}}{V_B\rho_{L,B}}\right)\left[h_{L,op}-h_{L,B}-\frac{\partial h_{L,B}}{\partial S_{L,B}}\left(S_{L,op}-S_{L,B}\right)\right]\left(\frac{\partial h_{L,B}}{\partial T_{L,B}}\right)^{-1} \quad (10)$$

Equation (8) represents black liquor concentration rate at the plates exit, while equations (9, 10) describe, respectively, black liquor concentration and temperature rates within the collecting box.

2.2.2 *Heat Transfer Rate in the Heating Plates*

For evaluating heat transfer rate in the vertical heating plates, by the same reasons mentioned for LTV evaporators, the transfer interface is discretized applying a rectangular mesh. In each elemental section, overall heat transfer coefficient and flow temperature are taken to be constants and heat transfer among adjacent sections is considered negligible. Local overall heat transfer coefficients are evaluated taking into account flow regimes (laminar or turbulent) inside and outside heating elements. The relationship for evaluating heat transfer rate is given bellow.

$$\dot{Q} = N_p b \Delta\ell \sum_{k=1}^{N} \left[\overline{U}_k \left(T_{vap} - T_{L,k} \right) \right] \tag{11}$$

where

$$\overline{U}_k = \left[\frac{1}{\overline{h}_{cond,k}} + \frac{\Delta e_p}{K_p} + \frac{\Delta e_l}{K_l} + \frac{1}{\overline{h}_{conv,k}} \right]^{-1} \tag{12}$$

Relationships for evaluating $\overline{h}_{cond}$, and $\overline{h}_{conv}$ can be found, for instance, in Gil (1995).

3 EVAPORATION PLANT SIMULATION

Black liquor evaporation plant (Fig. 2) consists of six LTV evaporators, incorporating afterheater section, excepting for IIA effect, and one three-section falling film plate-type concentrator. Weak black liquor is fed in parallel to the IV (70%) and V (30%), while fresh steam is fed to the concentrator.

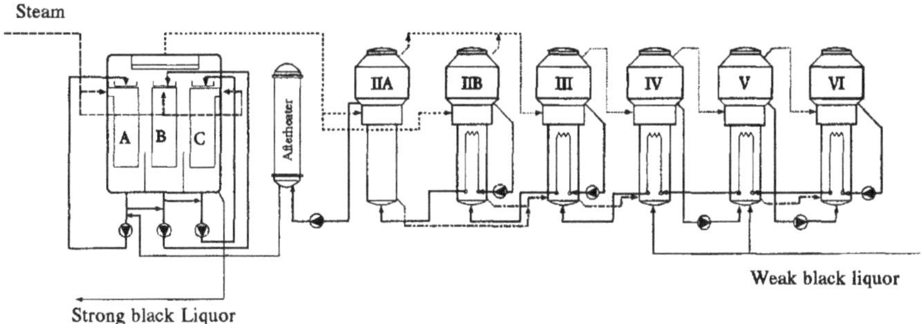

Figure 2 Simplified flow diagram for the evaporation plant.

Based on the proposed nonlinear models, a partially linear model, obtained by Taylor expansion, was derived for simulating the evaporation plant dynamics. In figure 3 it is plotted the transient response to a step change of 10% in weak black liquor concentration.

For that disturbance, the system shows a slight drop in the concentration of black liquor leaving the concentrator. This reduction is mainly explained by the influence of black liquor concentration upon heat transfer rate. The increase in concentration causes a reduction in heat transfer rate which impose a decreasing in concentration gain. Such a drop leads to an augmentation of black liquor mass flux to the forward stages, thus contributing for a gradual black liquor gain decrease in these evaporators.

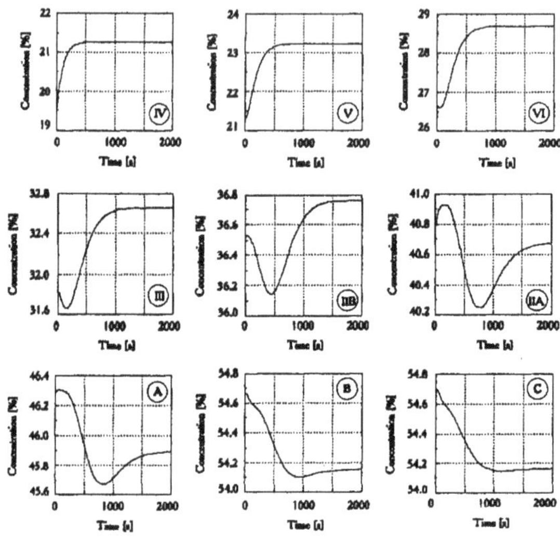

Figure 3 Evaporation system time response.

4 CONTROL STRATEGY

For eliminating the influence of measurable disturbances in weak black liquor upon solids content of black liquor leaving the evaporation plant, an optimal hierarchical infinite continuous time linear regulator is implemented. This regulator is synthesised according the interaction prediction method proposed by Takahara (Jamshidi, 1983) and taking fresh steam pressure in each concentrator section as manipulated variable.

4.1 Problem Formulation

The optimisation problem for N interconnected dynamical system, with state vectors $\boldsymbol{x}_1, \ldots, \boldsymbol{x}_N$, may be stated as follows: Find the admissible local control vectors $\boldsymbol{u}_1, \ldots, \boldsymbol{u}_N$, which minimises the quadratic cost function (13), under restrictions (14), Titli (1978).

$$J(\boldsymbol{u}(t)) = \frac{1}{2}\sum_{i=1}^{N}\left\{\|\boldsymbol{x}_i(t_f)\|_{P_i}^2 + \int_0^{\infty}\left[\|\boldsymbol{x}_i(t)\|_{Q_i}^2 + \|\boldsymbol{u}_i(t)\|_{R_i}^2\right]\mathrm{d}t\right\} \tag{13}$$

$$\dot{\boldsymbol{x}}_i(t) = \boldsymbol{A}_i\boldsymbol{x}_i(t) + \boldsymbol{B}_i\boldsymbol{u}_i(t) + \boldsymbol{F}_i\boldsymbol{w}_i(t) + \boldsymbol{C}_i\boldsymbol{z}_i(t);\ \ \boldsymbol{z}_i(t) = \sum_{\substack{j=1 \\ j\neq i}}^{N}\left(\boldsymbol{L}_{ij}\boldsymbol{x}_j(t) + \boldsymbol{M}_{ij}\boldsymbol{u}_j(t)\right) \tag{14}$$

where $\boldsymbol{P}_i$, $\boldsymbol{Q}_i$ are real symmetric positive semi-definite matrices, $\boldsymbol{R}_i$ is real symmetric positive definite matrix and $\|\boldsymbol{b}\|_L^2 = \boldsymbol{b}^{\mathrm{T}}\boldsymbol{L}\boldsymbol{b}$. $\boldsymbol{A}_i$, $\boldsymbol{B}_i$, $\boldsymbol{F}_i$ and $\boldsymbol{C}_i$ are, respectively, the state, input, disturbance and interaction matrices, whilst $\boldsymbol{L}_{ij}$ and $\boldsymbol{M}_{ij}$ are the state and control interaction matrices between the ith and jth subsystems. This problem can be solved by first introducing a set of Lagrange multipliers $\lambda_i(t)$ and costate vectors $\boldsymbol{p}_i(t)$ to the cost function integrand, forming the Lagrangian (15).

$$L(\cdot)=\sum_{i=1}^{N}\left\{\frac{1}{2}\left\|x_i(t_f)\right\|_{P_i}^2+\int_0^{\infty}\left[\frac{1}{2}\left(\left\|x_i(t)\right\|_{Q_i}^2+\left\|u_i(t)\right\|_{R_i}^2\right)+\lambda_i(t)\left(z_i(t)-\sum_{j=1}^{N}\left(L_{ij}x_{ij}(t)+M_{ij}u_{ij}(t)\right)\right)\right.\right.$$
$$\left.\left.+p_i(t)\left(-\dot{x}_i(t)+A_ix_i(t)+B_iu_i(t)+F_iw_i(t)+C_iz_i(t)\right)\right]dt\right\} \tag{15}$$

For a given coordination vector $[\lambda, z]$, equation (15) is additively separable and can be decomposed into N independent sub-Lagrangians. Each subsystem's Lagrangian is then minimised independently to the others. This procedure, which constitutes the first level of the hierarchical structure, is based on the necessary conditions for local optimality, Athans and Falb (1966). At the second level, where the solutions of all first-level subsystems are known, new coordenation vectors are iteratively evaluated according to (16).

$$\begin{bmatrix}\lambda_i(t)\\ z_i(t)\end{bmatrix}^{(k+1)}=\begin{bmatrix}-C_i^{\mathrm{T}}p_i(t)\\ \sum_{j=1}^{N}\left(L_{ij}x_{ij}+M_{ij}u_{ij}\right)\end{bmatrix}^{(k)} \tag{16}$$

For infinite time regulation, when $P_i = 0$ and A_i, B_i, R_i and Q_i, are constant matrices the solution of differential Riccati equation leads to a constant matrix. This means that the implementation of each local optimal controller consists of two parts: one of which is a local feedback fixed amplifier and the other is a pre-filter to determine the optimal driving function from disturbances and interactions.

4.2 Hierarchical Optimal Regulation of Evaporation Plant

In figure 4.a it is plotted the optimal path of black liquor solids content at the concentrator outlet, for a step disturbance of +10% in weak black concentration. As can be seen, the proposed control scheme reduces significantly the transient deviation from the set-point concentration (59.5%) and provides an adequate steady state solids content. Convergence to the optimal control took place in just 6 iterations of the second level (Fig. 4.b) which took 104 seconds to execute in a PC486 DX/2-66MHz, assuming the inicial values of coordination vector as $[0\ 0]^{\mathrm{T}}$ and an error adjoint vector of 10^{-12}.

5 CONCLUSIONS

In this paper it was proposed new dynamic nonlinear mathematical models for a long tube vertical evaporator and for a plate-type falling film concentrator. Evaluation of heat transfer rate takes into account flow regimes outside and inside the evaporator tubes and concentrator plates.

For controling a black liquor evaporation plant comprising six evaporators and one three-section falling film concentrator, it was designed an optimal two-level hierarchical infinite time linear regulator, using the interaction prediction method. Computed results show the effectiveness of this control strategy.

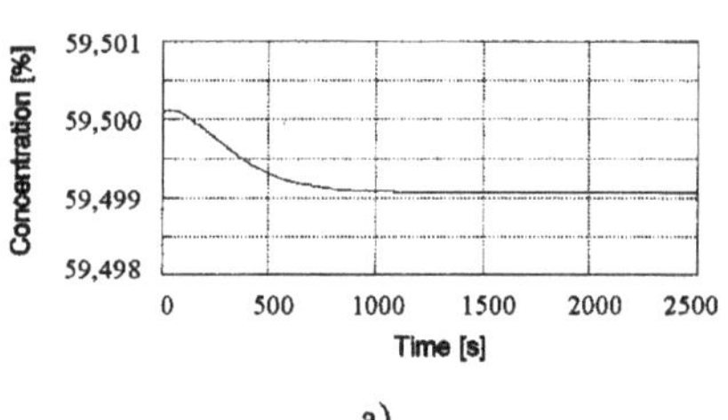

a)

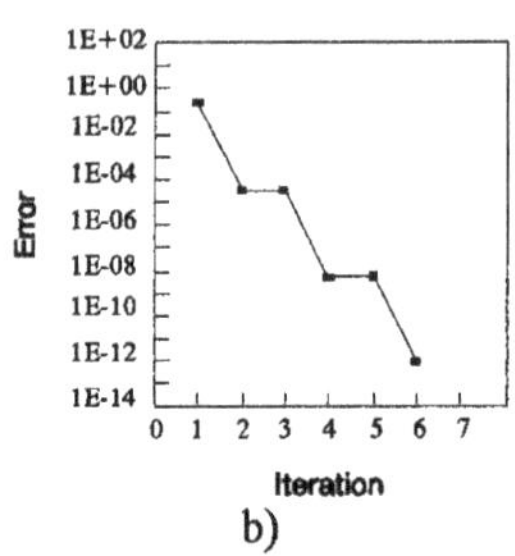

b)

Figure 4 a) Regulated evaporation plant time response for a step disturbance of +10% in weak black liquor concentration; b) Adjoint trajectory error vs iterations.

Symbols

h, enthalpy [J/kg]
$\dot{m}$, mass flux [kg/m^3]
K, thermal conductivity [W/m^2°C]
M, mass [kg]
N, number of elements
r, radius [m]
S, solids level [kg/kg]
T, temperature [°C]
t, time [s]
U, overall heat transfer coefficient [W/m^2°C]
V, volume [m^3]
ρ, density [kg/m^3]

Subscripts

Cond, condensation
Conv, convection
L, liquor
i, inlet
o, outlet
r, main liquor box
t, tube
Aft, afterheater
B, collecting box
I, incrustation

Aknowledgments: This work is carried on in cooperation with PORTUCEL-Viana (Ings. J.L. Amaral and D. Trancoso).

REFERENCES

Athans, M. and Falb, P (1966) Optimal Control: An Introduction to the Theory and Its Applications. McGraw Hill, New York.

Gil, P. (1995) Modelizaçăo Matemática e Controlo Hierárquico Contínuo de uma Cadeia de Evaporaçăo. M.S. Thesis, University of Coimbra, Coimbra, Portugal.

Jamshidi, M. (1983) Large Scale Systems: Modeling and Control. North-Holland, New York.

Niemi, A., Hartikainen, E. and Koistinen, R (1974) Dynamics and Control of Multiple-Effect Evaporator and Drum Washer Plants. Proc. 4th IFAC/IFIP International Conference on Digital Computer Applications to Process Control, 503-19.

Olauson, L. (1981) Heat Transfer in Climbing-Film Evaporators. *Svensk Papperstidning*, **84**, 61-71.

Titli, A. and Singh, M. (1978) Decomposition Optimization and Control. Pergamon Press, Oxford.

On the use of consistent approximations for the optimal design of beams*

C. Kirjner-Neto and E. Polak
Department of Electrical Engineering and Computer Sciences, University of California, Berkeley, CA 94720, USA.

Abstract
We present a strategy, based on the theory of consistent approximations, for solving a class of optimal beam design problems. The paper includes a numerical example.

Keywords
Optimal design, discretization theory, consistent approximations

1 INTRODUCTION

There is a considerable literature dealing with numerical methods for the solution of optimal design problems (see, e.g., Banichuk (1990), Pironneau (1984) and references therein). A major class of current methods constructs an infinite sequence of finite-dimensional approximating problems by means of numerical integration techniques. The theoretical treatment tends to be limited to showing that global minimizers of approximating problems converge to global minimizers of the original problem. The possibility of local minimizers of the approximating problems converging to nonstationary points of the original problem is ignored, and efficient diagonalization techniques (such as those in Polak (1993)) are not explored.

Recently, an abstract theory of consistent approximations for optimization problems has been proposed in Polak (1993). It is based on the concepts of epi-convergence and optimality functions, and provides conditions that ensure convergence of global minimizers, local minimizers and stationary points of the approximating problems to those of the original problem. It also provides diagonalization techniques that make it possible to use efficiently nonlinear programming libraries in solving infinite-dimensional optimization problems.

* The research reported herein was sponsored by the NSF grant ECS 9302926.

In this paper, we apply the theory of consistent approximations to the solution of a class of optimal Euler-Bernoulli beam design problems with continuum constraints, such as constraints on vertical deflection, and on stresses. We show that the original constraints must be relaxed to ensure that the approximating problems have solutions, and that norm preserving transformations must be used in solving the approximating problems, in order to preserve the conditioning of the original problem with respect to algorithms that are not scale invariant. For ease of exposition we restrict ourselves to beams with rectangular cross section, fixed width, and distributed loads. Extensions to beams of varying width and various cross sections are straightforward. A design example is included.

2 CONSISTENT APPROXIMATIONS

Let $\mathcal{B}$ be a topological vector space and consider the problem

$$\mathbf{P} \qquad \min_{z \in Z} f(z) \tag{2.1a}$$

where $f : \mathcal{B} \to \mathbb{R}$ is continuous and $Z \subset \mathcal{B}$ is the feasible set. Let $\{ \mathcal{B}_N \}_{N=1}^{\infty}$ be a family of finite-dimensional subspaces of $\mathcal{B}$ such that $\mathcal{B}_N \subset \mathcal{B}_{N+1}$, for all N, and consider the family of approximating problems

$$\mathbf{P}_N \qquad \min_{z \in Z_N} f_N(z), \quad N \in \mathbb{N}, \tag{2.1b}$$

where $f_N : \mathcal{B}_N \to \mathbb{R}$ is continuous, and $Z_N \subset \mathcal{B}_N$ (often $Z_N = \mathcal{B}_N \cap Z$).

Definition 2.1 [Attouch (1984)]. The problems in the family $\{ \mathbf{P}_N \}_{N=1}^{\infty}$ *converge epigraphically* to $\mathbf{P}$, ($\mathbf{P}_N \to^{Epi} \mathbf{P}$) if : *(a)* for every $z \in Z$, there exists a sequence $\{ z_N \}_{N=1}^{\infty}$, with $z_N \in Z_N$, such that $z_N \to z$ and $\overline{\lim} f_N(z_N) \le f(z)$; and *(b)* for every sequence $\{ z_{N_k} \}_{k=1}^{\infty}$, with $z_{N_k} \in Z_{N_k}$, such that $z_{N_k} \to z$ as $k \to \infty$, $z \in Z$ and $\underline{\lim} f_{N_k}(z_{N_k}) \ge f(z)$. □

Definition 2.2. A function $\theta : \mathcal{B} \to \mathbb{R}$ ($\theta_N : \mathcal{B}_N \to \mathbb{R}$) is an *optimality function* for $\mathbf{P}$ ($\mathbf{P}_N$) if *(i)* $\theta(\cdot)$ ($\theta_N(\cdot)$) is sequentially upper semicontinuous, *(ii)* $\theta(z) \le 0$ ($\theta_N(z) \le 0$)for all $z \in \mathcal{B}$ ($z \in \mathcal{B}_N$), and *(iii)* $\theta(\hat{z}) = 0$ ($\theta_N(\hat{z}) = 0$) for any $\hat{z} \in Z$ that is a local minimizer for $\mathbf{P}$ (for any $\hat{z} \in Z_N$ that is a local minimizer for $\mathbf{P}_N$). □

Definition 2.3. Let $\theta(\cdot)$, $\theta_N(\cdot)$, $N \in \mathbb{N}$, be optimality functions for $\mathbf{P}$, $\mathbf{P}_N$, respectively. The pairs $(\mathbf{P}_N, \theta_N)$, in the sequence $\{ (\mathbf{P}_N, \theta_N) \}_{N=1}^{\infty}$ are *consistent approximations* to the pair $(\mathbf{P}, \theta)$, *if (i)* $\mathbf{P}_N \to^{Epi} \mathbf{P}$, and *(ii)* for any sequence $\{ z_N \}_{N \in K}$, $K \subset \mathbb{N}$, *with* $z_N \in \mathcal{B}_N$ for all $N \in K$, such that $z_N \to z$, $\overline{\lim}\, \theta_N(z_N) \le \theta(z)$. □

Theorem 2.4. [Polak (1993)] Suppose that the pairs $(\mathbf{P}_N, \theta_N)$ in the sequence $\{ (\mathbf{P}_N, \theta_N) \}_{N=1}^{\infty}$ are consistent approximations to the pair $(\mathbf{P}, \theta)$, and that $\{ \hat{z}_N \}_{N=1}^{\infty}$ is a sequence such that $\hat{z}_N \in Z_N$ for all N and $\hat{z}_N \to \hat{z}$. *(a)* If the $\hat{z}_N$ are global minimizers for the $\mathbf{P}_N$, then $\hat{z}$ is a global minimizer of $\mathbf{P}$. *(b)* If $\hat{z}_N$ are strict local minimizers whose radii of attraction do not converge to zero, as $N \to \infty$, then $\hat{z}$ is a local minimizer of $\mathbf{P}$. *(c)* If $\overline{\lim}\, \theta_N(\hat{z}_N) = 0$, then $\theta(\hat{z}) = 0$.

3 CONSISTENT APPROXIMATIONS FOR A FIXED BEAM DESIGN

Beam model Consider a fixed Euler-Bernoulli beam with modulus of elasticity $E > 0$, length $L > 0$, rectangular cross-section, constant width $b > 0$, and variable depth defined by a positive, Lipschitz continuous function $h : [0, L] \to \mathbb{R}$. The beam is subjected to a vertical load with density $l(h, \cdot)$ of the form

$$l(h, x) = m(x) - Kh(x), \quad x \in [0, L], \tag{3.1}$$

where $K \geq 0$ is a constant, and $m(\cdot)$ is piecewise Lipschitz continuous, with finitely many points of discontinuity in $[0, L]$.

For a cantilever beam of length L, depth $h(x)$, and subject to the load density $l(h, \cdot)$, the bending moment, $M_c(h, \cdot)$, is the unique solution of the final value problem

$$M_c''(h, x) = l(h, x), \; x \in [0, L], \; M_c(h, L) = M_c'(h, L) = 0. \tag{3.2a}$$

The bending moment, $M(h, \cdot)$, in a fixed beam of length L, depth $h(x)$, and load density $l(h, \cdot)$, differs from $M_c(h, \cdot)$ only by an affine term in x, which is due to the reactions at the support points. Therefore

$$M(h, x) = M_c(h, x) + g_1(h)x + g_2(h), \quad x \in [0, L]. \tag{3.2b}$$

It follows from the dual formulation of the variational problem associated with the bending of the fixed beam (see Reedy (1984)), that $g(h) \triangleq [g_1(h)\; g_2(h)]^T$ satisfies the equation

$$\begin{bmatrix} \int_0^L \frac{x^2}{h(x)^3} dx & \int_0^L \frac{x}{h(x)^3} dx \\ \int_0^L \frac{x}{h(x)^3} dx & \int_0^L \frac{1}{h(x)^3} dx \end{bmatrix} \begin{bmatrix} g_1(h) \\ g_2(h) \end{bmatrix} = \begin{bmatrix} -\int_0^L \frac{M_c(h, x)x}{h(x)^3} dx \\ -\int_0^L \frac{M_c(h, x)}{h(x)^3} dx \end{bmatrix}. \tag{3.2c}$$

The shear force, $V(h, x)$, and the deflection of the beam, $y(h, \cdot)$, are given by

$$V(h, x) = -M'(h, x) = -M_c'(h, x) - g_1(h), \quad x \in [0, L], \tag{3.2d}$$

$$y''(h, x) = 12M(h, x)/Ebh(x)^3, \; x \in [0, L], \; y(h, 0) = y'(h, 0) = 0. \tag{3.2e}$$

For design purposes we assume that the depth function is an element of the set

$$H_{ad} \triangleq \{ h \in C[0, L] \mid 0 < \alpha \leq h(x) \leq \beta, \;\; |dh(x)/dx| \leq \gamma, \;\; \text{for a. e. } x \in [0, L] \}, \tag{3.3}$$

where $0 < \alpha < \beta < \infty$ and $\gamma \geq 0$ are given constants, and $C[0, L]$ is the space of continuous real-valued functions defined on $[0, L]$.

The "natural" norm on $C[0, L]$ for establishing continuity and differentiability of solutions of (3.2a-e) with respect to depth functions $h(\cdot)$ is the sup-norm, $\|\cdot\|_\infty$, while the "natural" norm for defining optimality functions is the $L_2[0, L]$ norm, $\|\cdot\|_2$. Hence, we will work in the "compromise" inner-product space $(C[0, L], \|\cdot\|_2, \langle\cdot,\cdot\rangle_2)$, where $\langle\cdot,\cdot\rangle_2$ denotes the usual inner-product on $L_2[0, L]$. As a consequence, we have to use the concepts of continuity and differentiability relative to H_{ad}.

Definition 3.1 Let $(V, \|\cdot\|_V)$ be a normed space and let ζ be a function from H_{ad} into V.

(a) $\zeta(\cdot)$ *is Lipschitz continuous relative to* H_{ad} if there exists an $C < \infty$, such that for all $h, h' \in H_{ad}$, $\|\zeta(h) - \zeta(h')\|_V \leq C\|h - h'\|_2$.

(b) $\zeta(\cdot)$ is *differentiable relative to* H_{ad} if for any $h \in H_{ad}$ there exists a continuous linear map $D\zeta(h\,;\cdot)$, called the H_{ad}-derivative of $\zeta(\cdot)$ at h, such that

$$\lim_{h' \in H_{ad},\, \|h'-h\|_2 \to 0} \|\zeta(h') - \zeta(h) - D\zeta(h\,;h'-h)\|_V / \|h'-h\|_2 = 0\,. \tag{3.4}$$

Theorem 3.2. The mappings $h \mapsto M(h,\cdot)$, $h \mapsto V(h,\cdot)$, and $h \mapsto y(h,\cdot)$, from H_{ad} into $(C[0,L], \|\cdot\|_\infty)$, are all Lipschitz continuously differentiable relative to H_{ad}. □

We will denote the H_{ad}-derivatives at h, of the mappings introduced above, by $D_1M(h,\cdot;\cdot)$, $D_1V(h,\cdot;\cdot)$, and $D_1y(h,\cdot;\cdot)$, respectively.

Problem formulation We will consider optimal beam design problems of the form

$$\mathbf{P} \qquad \min_{h \in H_{ad}} \{ f(h) \mid \psi(h) \triangleq \max_{j \in \mathbf{q}} \max_{x \in [0,L]} \phi^j(h,x) - r^j(x) \le 0 \}\,, \tag{3.5a}$$

where for any $q > 0$, $\mathbf{q} \triangleq \{1,\ldots,q\}$, and $f(\cdot)$ and $\phi^j(\cdot,\cdot)$, $j \in \mathbf{q}$, have the form

$$f(h) = \int_0^L \phi^0(h,x)\,dx\,, \tag{3.5b}$$

$$\phi^j(h,x) = \tilde{\phi}^j(h(x), M(h,x), V(h,x), y(h,x), x)\,,\; j \in \bar{\mathbf{q}}\,, \tag{3.5c}$$

with $M(h,\cdot)$, $V(h,\cdot)$, and $y(h,\cdot)$ determined by (3.2b-f), and for $j \in \bar{\mathbf{q}} \triangleq \{0,1,\ldots,q\}$, $\tilde{\phi}^j : [\alpha,\beta] \times \mathbb{R} \times \mathbb{R} \times \mathbb{R} \times [0,L] \to \mathbb{R}$.

Assumption 3.3. *(a)* The functions $r^j(\cdot)$, $j \in \mathbf{q}$, are Lipschitz continuously differentiable on $[0,L]$ and satisfy

$$\min_{j \in \mathbf{q}} \min_{x \in [0,L]} r^j(x) = \hat{r} > 0\,. \tag{3.6}$$

(b) The functions $\tilde{\phi}^j(\cdot,\cdot,\cdot,\cdot,\cdot)$, $j \in \bar{\mathbf{q}}$, are Lipschitz continuously differentiable.

(c) The feasible set for **P** is non-empty. □

It follows from (3.5b-c), the Lipschitz continuous differentiability of $\tilde{\phi}^j(\cdot,\cdot,\cdot,\cdot,\cdot)$, $j \in \bar{\mathbf{q}}$, and Theorem 3.2, that $h \mapsto \phi^j(h,\cdot)$, $j \in \bar{\mathbf{q}}$, and $h \mapsto f(h)$ are Lipschitz continuously differentiable functions relative to H_{ad}. We will denote by $D_1\phi^j(h,\cdot;\cdot)$, $j \in \bar{\mathbf{q}}$, and $Df(h\,;\cdot)$ their H_{ad}-derivatives.

Following Polak (1987), we we define the optimality function $\theta : H_{ad} \to \mathbb{R}_-$, for **P**, by

$$\theta(h) \triangleq \min_{h' \in H_{ad}} \tilde{F}(h,h')\,, \tag{3.7a}$$

where $\psi(h)_+ \triangleq \max\{\psi(h), 0\}$, and

$$\tilde{F}(h,h') \triangleq \max\{ Df(h\,;h'-h) - \psi(h)_+,\, \max_{j \in \mathbf{q}} \max_{x \in [0,L]} \phi^j(h,x) + D_1\phi^j(h,x\,;h'-h)$$

$$- \psi(h)_+ \} + \tfrac{1}{2}\|h'-h\|_2^2\,. \tag{3.7b}$$

Theorem 3.4. *(a)* $\theta : H_{ad} \to \mathbb{R}_-$ is upper semicontinuous; *(b)* If $\hat{h}$ is a local minimizer of **P**, then $\theta(\hat{h}) = 0$; *(c)* For any $\hat{h} \in H_{ad}$ such that $\psi(\hat{h}) \le 0$, $\theta(\hat{h}) = 0$ if and only if $dF(\hat{h}\,;h-\hat{h}) \ge 0$, for all $h \in H_{ad}$, where $F(h) \triangleq \max\{ f(h) - f(\hat{}), \psi(h) \}$.

Discretization Following the theory in Section 2, we select a family of finite-dimensional subspaces of $C[0,L]$, specified by basis sets. For every integer $N > 0$ we let $\Delta_N \triangleq L/N$ and define $x_{N,k} \triangleq (k-1)\Delta_N$, $k \in \mathbf{N+1}$. Let

$$\Pi_{N,k}(x) \triangleq \begin{cases} (x - x_{N,k-1})/\Delta_N , & \text{for all } x \in [x_{N,k-1}, x_{N,k}], \quad k \in \{2, \cdots, N+1\}, \\ (x_{N,k+1} - x)/\Delta_N , & \text{for all } x \in [x_{N,k}, x_{N,k+1}], \quad k \in \mathbf{N}, \\ 0, & \text{otherwise.} \end{cases} \tag{3.8}$$

Let by H_N the span of the basis set $\{ \Pi_{N,k} \}_{k=1}^{N+1}$ and let $H_{ad,N} \triangleq H_N \cap H_{ad}$.

Next, after transcription into first order form, we use Euler's method to discretize the initial value problem (3.2e) and the final value problem (3.2a), and the rectangle rule to approximate the integrals in (3.2c). It is not at all clear that using a higher order integration scheme would be more efficient in optimal design.

Let $M_N(h, x_{N,k})$, $V_N(h, x_{N,k})$, and $y_N(h, x_{N,k})$, $k \in \mathbf{N+1}$, denote the discretized bending moment, shear force, and deflection, respectively. The following result is a direct consequence of the Implicit Function Theorem.

Lemma 3.5. The maps $h \mapsto M_N(h, \cdot)$, $h \mapsto V_N(h, \cdot)$, and $h \mapsto y_N(h, \cdot)$, from $H_{ad,N}$ into $\mathbb{R}^{N+1}$, are Lipschitz continuous differentiable. □

Approximating problems. We now define the family of approximating problems $\mathbf{P}_N$, $N = 1, 2, \ldots$, as follows:

$$\mathbf{P}_N \qquad \min_{h \in H_{ad,N}} \{ f_N(h) \mid \psi_N(h) \triangleq \max_{j \in \mathbf{q}} \max_{k \in \mathbf{N+1}} \phi_N^j(h, x_{N,k}) - (1 + \Delta_N^{1/2}) r^j(x_{N,k}) \le 0 \}, \tag{3.9a}$$

where

$$f_N(h) \triangleq \textstyle\sum_{k=1}^{N} \phi_N^0(h, x_{N,k}) \Delta_N , \tag{3.9b}$$

$$\phi_N^j(h, x_{N,k}) = \bar{\phi}^j(h(x_{N,k}), M_N(h, x_{N,k}), V_N(h, x_{N,k}), y_N(h, x_{N,k}), x_{N,k}), \quad j \in \bar{\mathbf{q}}. \tag{3.9c}$$

The term $\Delta_N^{1/2}$ in (3.9a) is added to guarantee that for N large enough the feasible set for $\mathbf{P}_N$ is non-empty, a fact needed in the proof of Theorem 3.7 *(a)* below.

Equation (3.9c) defines the functions $\phi_N^j(h, \cdot)$, $j \in \bar{\mathbf{q}}$, only on the mesh points $x_{N,k}$, $k \in \mathbf{N+1}$. We define $\phi_N^j(h, \cdot) : [0, L] \to \mathbb{R}$ as the piecewise affine interpolation of the values $\phi_N^j(h, x_{N,k})$, $k \in \mathbf{N+1}$. It follows from (3.9c) and Assumption 3.3*(b)*, that the $\phi_N^j(\cdot, \cdot)$, $j \in \bar{\mathbf{q}}$, are Lipschitz continuously differentiable on $H_{ad,N}$. The derivatives of the mappings $h \mapsto \phi_N^j(h, \cdot)$, $j \in \bar{\mathbf{q}}$, and $h \mapsto f(h)$, denoted by $D_1\phi_N^j(h, \cdot; \cdot)$ and $Df_N(h; \cdot)$, can be obtained using (3.9b,c), Lemma 3.5 and the Chain Rule.

Next, we define the optimality functions $\theta_N : H_{ad,N} \to \mathbb{R}$ by

$$\theta_N(h) \triangleq \min_{h' \in H_{ad,N}} \tilde{F}_N(h, h'), \tag{3.10}$$

where $\tilde{F}_N(\cdot, \cdot)$ is defined as in (3.7b), with all the functions modified by the addition of the subscript N. Results analogous to Theorem 3.4 hold for $\theta_N(\cdot)$. Also note that one can evaluate $\theta_N(\cdot)$ by solving a positive definite quadratic program.

Consistency of Approximations The following is proved in Kirjner-Neto (1994).

Lemma 3.6. *(a)* For each $h \in H_{ad}$, and $N \ge 1$, there exists $h_N \in H_{ad,N}$ such that

$$\max_{x \in [0, L]} |h(x) - h_N(x)| \le \gamma \Delta_N . \tag{3.11a}$$

(b) There exists a $C \in \mathbb{R}$ such that for all $j \in \bar{\mathbf{q}}$, $N \ge 1$, $h \in H_{ad}$, and $h_N \in H_{ad,N}$,

$$\max_{x \in [0,L]} |\phi^j(h,x) - \phi^j_N(h_N,x)| \leq C[\Delta_N + \|h - h_N\|_2], \tag{3.11b}$$

$$|\psi(h) - \psi_N(h_N)| \leq C[\Delta_N^{1/2} + \|h - h_N\|_2], \tag{3.11c}$$

$$|f(h) - f_N(h_N)| \leq C[\Delta_N + \|h - h_N\|_2]. \tag{3.11d}$$

□

In view of the definitions of $\psi(h)$ and $\psi_N(h)$, it is clear that the feasible sets $\mathbf{Z}$ and $\mathbf{Z}_N$ for $\mathbf{P}$ and $\mathbf{P}_N$ satisfy: $\mathbf{Z} = \{ h \in H_{ad} \mid \psi(h) \leq 0 \}$, and $\mathbf{Z}_N = \{ h \in H_{ad,N} \mid \psi_N(h) \leq 0 \}$.

Theorem 3.7. (Epiconvergence) *(a)* For every $h \in \mathbf{Z}$, there exists a sequence $\{ h_N \}_{N=N_0}^{\infty}$, with $h_N \in \mathbf{Z}_N$, such that $f_N(h_N) \to f(h)$ as $N \to \infty$. *(b)* Let $\{ h_N \}_{N=N_0}^{\infty}$ be a sequence such that $h_N \in \mathbf{Z}_N$ and $h_N \to \hat{h}$ as $N \to \infty$, then $\hat{h} \in \mathbf{Z}$, and $f_N(h_N) \to f(\hat{h})$.

Proof. Suppose $h \in \mathbf{Z}$ is given. Then, by Lemma 3.9 *(a)*, for each integer N, there exists an $h_N \in H_{ad,N}$ such that (3.11a) holds. Clearly, $h_N \to h$ as $N \to \infty$. It follows from (3.11d) that $f_N(h_N) \to f(h)$ as $N \to \infty$. To complete the proof of part *(a)* it remains to show that there exists an N_0 such that $h_N \in \mathbf{Z}_N$ for all $N \geq N_0$. Indeed, since $h \in \mathbf{Z}$ by assumption we have $\psi(h) \leq 0$. Hence, using (3.11b) and (3.11a) we obtain

$$\psi_N(h_N) \leq \psi_N(h_N) - \psi(h)$$

$$= \max_{j \in \mathbf{q}} \max_{x \in [0,L]} [\phi^j_N(h_N,x) - r^j(x)(1 + \Delta_N^{1/2})] - \max_{j \in \mathbf{q}} \max_{x \in [0,L]} [\phi^j(h,x) - r^j(x)]$$

$$\leq \max_{j \in \mathbf{q}} \max_{x \in [0,L]} [|\phi^j_N(h_N,x) - \phi^j(h,x)| - r^j(x)\Delta_N^{1/2}]$$

$$\leq C[\Delta_N + \|h - h_N\|_2] - \hat{r}\Delta_N^{1/2} \leq C[\Delta_N + \|h - h_N\|_\infty] - \hat{r}\Delta_N^{1/2}$$

$$\leq C(1+\gamma)\Delta_N - \hat{r}\Delta_N^{1/2}, \tag{3.12}$$

where $\hat{r} > 0$ is as in (3.6). It follows from (3.12) that there exist an N_0 such that for all $N \geq N_0$, $\psi_N(h_N) \leq 0$, which proves *(a)*.

Let $\{ h_N \}_{N=N_0}^{\infty}$ be a sequence as in *(b)*. Since H_{ad} is closed and $\mathbf{Z}_N \subset H_{ad,N} \subset H_{ad}$ for all $N \in \mathbb{N}$, it follows that $\hat{h} \in H_{ad}$. The facts that $\hat{h} \in \mathbf{Z}$, that is, that $\psi(\hat{h}) \leq 0$, and that $f_N(h_N) \to f(\hat{h})$ follow directly from (3.11c) and (3.11d) respectively. □

Theorem 3.8. Suppose that $\{ h_N \}_{N=N_0}^{\infty}$, with $h_N \in H_{ad,N}$, is such that $h_N \to h$ as $N \to \infty$. Then $h \in H_{ad}$, and $\overline{\lim}_{N \to \infty} \theta_N(h_N) \leq \theta(h)$. □

Corollary 3.9. The sequence $\{ (\mathbf{P}_N, \theta_N) \}_{N=1}^{\infty}$ is a family of consistent approximations to the pair $(\mathbf{P}, \theta)$. □

4 TRANSCRIPTION INTO A NONLINEAR PROGRAMMING PROBLEM

The problems $\mathbf{P}_N$ are defined on the finite-dimensional subspaces $H_N \subset C[0,L]$. However, numerical computations must be carried out in a finite-dimensional space of coefficients, relative to some basis in $H_N \subset C[0,L]$. Note that the basis functions $\Pi_{N,k}(\cdot)$ are not orthonormal. Hence, if we define $W_N : H_N \to \mathbb{R}^{N+1}$ by $W_N(h) \triangleq (\eta_1, \eta_2, \ldots, \eta_{N+1})^T$, where η_k, $k=1,\ldots,N+1$ are the coefficients of h with respect to the basis functions $\Pi_{N,k}$, $k=1,\ldots,N+1$, then we find that $\|h\|_2^2 \neq \sum_{i=1}^{N+1} \eta_i^2 \triangleq \|W_N(h)\|^2$, i.e., $W_N(\cdot)$ is not an isometry. Hence it turns out that if we solve the problem $\mathbf{P}_N$, using a nonlinear programming algorithm in this coefficient space, we have inadvertently modified the metric on our original space, which can cause considerable deterioration of performance in non-scale invariant algorithms.

To compute in $\mathbb{R}^{N+1}$, using a metric that is equivalent to the one on H_N, one can either modify existing nonlinear programming software, something not easily undertaken when using a standard library of programs, or one can solve problems $\bar{\mathbf{P}}_N$, equivalent to the $\mathbf{P}_N$, and defined in terms of coordinates corresponding to an orthonormal basis.

To define the $\bar{\mathbf{P}}_N$ we define $Q_N \in \mathbb{R}^{(N+1)\times(N+1)}$ by $(Q_N)_{ij} \triangleq \int_0^L \Pi_{N,i}(x)\Pi_{N,j}(x)dx$, where for any matrix $A \in \mathbb{R}^{(N+1)\times(N+1)}$, A_{ij} denotes the i,j-th entry of A, and define $T_N : H_N \to \mathbb{R}^{N+1}$ by $T_N(h) \triangleq Q_N^{1/2} W_N(h)$, so that, for any $h \in H_N$, and $\bar{h} = T_N(h)$, we have

$$\|h\|_2^2 = \int_0^L \sum_{i,j=1}^{N+1} (W_N(h))_j \Pi_{N,j}(x)(W_N(h))_i \Pi_{N,i}(x)dx = (Q_N^{1/2} W_N(h))^T (Q_N^{1/2} W_N(h)) = \|\bar{h}\|^2. \quad (4.1)$$

Equation (4.1) implies that the distance between two elements h and h' of $H_N \subset (C[0,L], \|\cdot\|_2, \langle\cdot,\cdot\rangle)$ is equal to the Euclidean distance between $T_N(h)$ and $T_N(h')$.

Next, let $\bar{H}_N \triangleq T_N(H_N)$, and let $\bar{H}_{ad,N} \triangleq T_N(H_{ad,N}) \subset \mathbb{R}^{N+1}$. We define the mappings $\bar{f}_N : \bar{H}_{ad,N} \to \mathbb{R}$, and $\bar{\phi}_N^j(\cdot, x_{N,k}) : \bar{H}_{ad,N} \to \mathbb{R}$, $j \in \bar{\mathbf{q}}$, $k \in \mathbf{N+1}$ by

$$\bar{f}_N(\bar{h}) \triangleq \sum_{k=1}^{N} \phi_N^0(T_N^{-1}(\bar{h}), x_{N,k})\Delta_N, \quad \bar{\phi}_N^j(\bar{h}, x_{N,k}) \triangleq \phi_N^j(T_N^{-1}(\bar{h}), x_{N,k}), \quad (4.2)$$

which can be computed using (3.9c). Finally, we define the problems $\bar{\mathbf{P}}_N$, $N=1,2,\ldots$, by

$$\bar{\mathbf{P}}_N \qquad \min_{\bar{h} \in \bar{H}_{ad,N}} \{ \bar{f}_N(\bar{h}) \mid \max_{j \in \mathbf{q}} \max_{k \in \mathbf{N+1}} \bar{\phi}_N^j(\bar{h}, x_{N,k}) - (1 + \Delta_N^{1/2}) r^j(x_{N,k}) \leq 0 \}, \quad (4.3)$$

The following result, establishing equivalence between $\mathbf{P}_N$ and $\bar{\mathbf{P}}_N$, $N=1,2,\ldots$, follows directly from (3.9a-c),(4.2), and (4.3).

Proposition 4.1. *(a)* $h \in H_N$ is feasible for $\mathbf{P}_N$ if and only if $\bar{h} = T_N(h) \in \mathbb{R}^{N+1}$ is feasible for $\bar{\mathbf{P}}_N$, and *(b)* $h \in H_N$ is a global (local) minimizer for $\mathbf{P}_N$ if and only if $\bar{h} = T_N(h) \in \mathbb{R}^{N+1}$ is a global (local) minimizer for $\bar{\mathbf{P}}_N$. □

5 NUMERICAL RESULTS

To illustrate the use of consistent approximations together with a diagonalization strategy (see Polak (1993)), we solved a problem of the form **P**, with $E=10^7$ psi, $L=50$ in, $b=5$ in, $K=0$, $\alpha=1.0$ in, $\beta=5.0$ in, and $\gamma=0.15$, and continuum constraints on the maximum normal stress $\sigma_{max}(h, x)$, the maximum shear $\tau_{max}(h, x)$, and the deflection $y(h, x)$, as follows,

$$|\sigma_{max}(h, x)| \leq 30{,}000 \text{ psi}, \quad |\tau_{max}(h, x)| \leq 15{,}000 \text{ psi}, \quad |y(h, x)| \leq 0.1 \text{ in}, \quad (5.1a)$$

for all $x \in [0, 1]$. We used the cost function given by $f(h) = \int_0^L h(x)dx$. The load applied to the beam was $l(x) = -1500$ psi, $x \in [20, 30]$, $l(x) = 0$ otherwise.

The initial discretization was $N=8$ points, and the initial beam depth $h(x) \equiv 2.85$ in (see Figure 1.(a)). The initial cost was 142.5. Figure 1(b) shows the final design obtained in 16

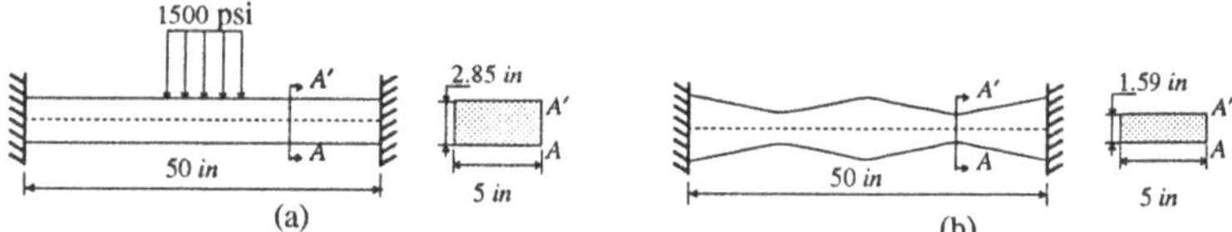

Fig. 1. *(a)* Initial Design; *(b)* Final Design.

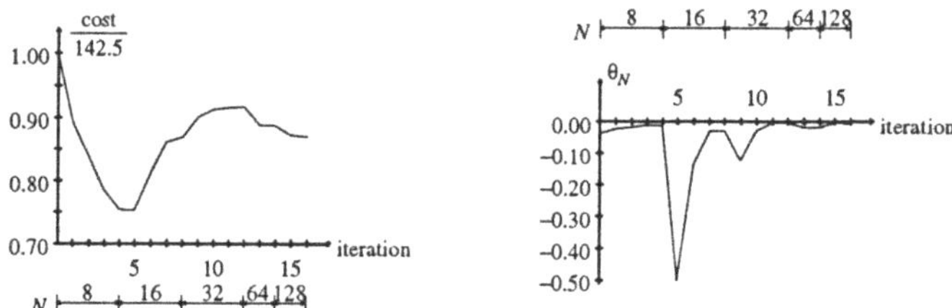

Fig. 2. Computed cost and computed optimality function

iterations of the algorithm in Polak (1991). The final discretization parameter was $N=128$ and cost was 124.05. In Figure 2, we find the cost as a percentage of the initial cost, the value of the optimality function θ_N, and the number of discretization points used at each iteration.

6 REFERENCES

Attouch, H. (1984) *Variational convergence for functions and operators*, Pitman.

Banichuk, N. V. (1990) *Introduction to optimization of structures*, Springer Verlag.

Kirjner-Neto, C. and Polak E. (1994) On the use of consistent approximations for the optimal design of beams,*University of California, Berkeley, Memo UCB/ERL M94/22.*

Pironneau, O. (1984) *Optimal shape design for elliptic systems*, Springer-Verlag.

Polak, E. (1987) On the mathematical foundations of nondifferentiable optimization in engineering design, *SIAM Review*, **29**, 21-91.

Polak, E. (1993) On the use of consistent approximations in the solution of semi-infinite optimization and optimal control problems, *Mathematical Programming*, **B-62,** 385-414.

Polak, E. and He, L. (1991) A Unified steerable phase I-Phase II method of feasible directions for semi-infinite optimization, *JOTA*, **69**, 83-107.

Reedy, J. N. (1984) *Energy and variational methods in applied mechanics*, Wiley & Sons.

A game-theoretical model for a controlled process of heat transfer

O.A. Malafeyev
Saint Petersburg University, Department of Applied Mathematics–Processes of Control
Frunze street 6, flat 225, St. Petersburg 196070, Russia.

M.S. Troeva
Yakut Research Institute for Applied Mathematics and Informatic
Petra Alekseeva 8/1, flat 16, Yakutsk 677000, Russia.

Abstract
The problem of finding an optimal technological mode for a controlled process of heat transfer is considered in this paper. This problem is formulated as a differential two-person zero-sum game of a technologist against 'nature'.
To solve the problem considered the numerical method based on the dynamic programming method and the finite difference method is proposed.

Keywords
Differential game, numerical method, piecewise-programmed strategies, dynamic programming, finite difference method

The problem of finding an optimal technological mode for a controlled process of heat transfer is considered in this paper.

A rectangular solid body in a medium with variable temperature is considered. The temperature field is controlled by heat (energy) supply in some way. The mathematical description of this action is described by a parameter w, controlled by the technologist. This parameter is an element of a set W of all control parameters. The temperature field of the body depends also on the state v of the medium. This is a "nature's" control parameter which belongs to a set V of all control parameters.

The quality of the process's control is estimated by a payoff function H.

The problem of choosing control function $w = w(t)$, $t \in [0, T]$, $w(t) \in W$, which guarantees the optimal value of the quality index H under any possible "nature's" function v arises. It is interesting for example to change the temperature field from initial state u_0 to given final state u_T with minimal energy expenses under uncertain "nature's" conditions by choosing control function $w(\cdot)$.

So, the problem can be formulated as a differential two-person zero-sum game $\Gamma(u_0, T)$

with initial position u_0 and duration T in the space $C_{\overline{\Omega}}$ of continuous functions with the domain $\overline{\Omega} = [0, l_1] \times [0, l_2]$.

The dynamics of the game $\Gamma(u_0, T)$ is described by the following boundary value problem for the heat transfer equation:

$$\begin{aligned}
c\frac{\partial u}{\partial t} &= \frac{\partial}{\partial x_1}\left(\lambda\frac{\partial u}{\partial x_1}\right) + \frac{\partial}{\partial x_2}\left(\lambda\frac{\partial u}{\partial x_2}\right), \; x_1 \in (0, l_1), \; x_2 \in (0, l_2), && (1)\\
\lambda\frac{\partial u}{\partial x_1} &= \alpha_1(u - v), \; x_1 = 0, \; x_2 \in [0, l_2], \; t > 0, && (2)\\
-\lambda\frac{\partial u}{\partial x_1} &= \alpha_1(u - v), \; x_1 = l_1, \; x_2 \in [0, l_2], \; t > 0, && (3)\\
\lambda\frac{\partial u}{\partial x_2} &= \alpha_1(u - v), \; x_1 \in [0, l_1], \; x_2 = 0, \; t > 0, && (4)\\
-\lambda\frac{\partial u}{\partial x_2} &= \alpha_2(u - w), \; x_1 \in [0, l_1], \; x_2 = l_2, \; t > 0, && (5)\\
u(x_1, x_2, 0) &= u_0(x_1, x_2), \; x_1 \in [0, l_1], \; x_2 \in [0, l_2], \; t = 0, && (6)
\end{aligned}$$

where $u(x, t)$ – the temperature at the point $x = (x_1, x_2) \in \overline{\Omega}$ at the moment t, $c = c(x, t)$ – volume heat capacity, $\lambda = \lambda(x, t)$ – heat conductivity coefficient, α_1, α_2 – heat exchange coefficients; w, v – the control's parameters of the players, P (technologist) and E (nature), respectively; $w \in W \subset R^p$, $v \in V \subset R^q$; W and V – are compact sets in Euclidean spaces R^p and R^q respectively.

The continuous function $w = w(t)$ ($v = v(t)$), satisfying the condition $w(t) \in W$ ($v(t) \in V$) for all $t \in [0, T]$ is called the admissible control of the player $P(E)$.

It is known (Tichonov and Samarsky, 1977), that for any finite $T < \infty$ for any admissible controls $w(t)$, $t \in [0, T]$ and $v(t)$, $t \in [0, T]$ there exists the unique solution $u(t) = u(u_0, t, w(t), v(t))$ of the problem (1)-(6) under any $u_0 \in C_{\overline{\Omega}}$ and $t \in [0, T]$.

At any moment t of the game the players are informed of the state of the game $u(t)$, initial moment $t_0 = 0$, terminal time T and the dynamics of the game.

For definiteness we will consider the above formulated problem. At the moment T the player E gets the payoff $H(\cdot)$ from the player P:

$$H(u(\cdot)) = \int_0^T h(u(t), w(t), v(t), t)\, dt, \qquad (7)$$

where $u(\cdot)$ is a trajectory of the process, corresponding to admissible controls $w(\cdot)$ and $v(\cdot)$ on the interval $[0, T]$, $h(u, w, v, t)$ is a continuous function bounded on the bounded sets. This function characterizes the energy expenses.

The aim of the player P is to minimize $H(\cdot)$ and the aim of the player E is contrary.

Let us cite (Malafeyev, 1993) the definition of strategies of players P and E in the game $\Gamma(u_0, T)$.

Definition 1 *A strategy φ (ψ) of player $P(E)$ in the game $\Gamma(u_0, T)$ is the pair (σ_1, K_{σ_1}) ((σ_2, K_{σ_2})), where σ_1 (σ_2) is an arbitrary finite partition of the interval $[0, T]$, while K_{σ_1} (K_{σ_2}) is a mapping which associates an admissible control $w_i(\tau)$, $\tau \in [t_i, t_{i+1})$ ($v_j(\tau)$, $\tau \in$*

$[t_j, t_{j+1})$) *with the information state of player* $P(E)$ *in the moment* $t_i \in \sigma_1, i=0,\ldots,N_{\sigma_1}-1$ ($t_j \in \sigma_2$, $j = 0,\ldots,N_{\sigma_2}-1$).

The set of strategies of the player $P(E)$ in the game $\Gamma(u_0, T)$ is denoted by $\Phi(\Psi)$.

The trajectory of the game $\chi(\varphi,\psi)$ is uniquely defined by the standard way for every strategy pair $(\varphi,\psi) \in \Phi \times \Psi$ under initial position u_0.

The payoff function of the player E for every strategy pair (φ,ψ) is defined as follows:

$$K(u_0,\varphi,\psi) = H(\chi(\varphi,\psi)(\cdot)), \tag{8}$$

where $\chi(\varphi,\psi)(\cdot)$ is a trajectory of the game $\Gamma(u_0,T)$, corresponding to the (φ,ψ).

Making use of the results of the dynamic games theory in complete metric spaces (Malafeyev, 1993), the existence of ε-equilibrium points in auxiliary approximate games of perfect information with discrimination of a player (upper and lower games $\overline{\Gamma}^{\sigma}(u_0,T)$ and $\underline{\Gamma}^{\sigma}(u_0,T)$ with discrimination of players P and E, respectively) is proved for the class of the piecewise-programmed strategies.

To find the optimal mode for the controlled process of heat transfer the numerical method based on the dynamic programming method (Bellman, 1960) and the finite difference method (Samarsky, 1989) is proposed.

On the domain $\overline{\Omega} = [0,l_1] \times [0,l_2]$ we construct the uniform net with steps h_1 on x_1 and h_2 on x_2

$$\begin{aligned}\overline{\omega}_h = \ &\Big\{x_1^{(i)} = ih_1,\ i=0,\ldots,N_1;\ x_1^{(0)} = 0,\ x_1^{(N_1)} = l_1;\\ &x_2^{(k)} = kh_2;\ k=0,\ldots,N_2;\ x_2^{(0)} = 0,\ x_2^{(N_2)} = l_2\Big\},\end{aligned} \tag{9}$$

where $h = (h_1,h_2)$.

On the interval $[t_s,t_{s+1}]$, $s=\overline{0,N_\sigma-1}$ we construct the uniform net with step δ

$$\overline{\omega}_\delta = \{\tau_j = j\delta,\ j=\overline{0,N_3};\ \tau_0 = t_s,\ \tau_{N_3} = t_{s+1}\}.$$

Here $t_s \in \sigma$, where σ is the time interval partition

$$\sigma = \{t_0 = 0, t_1, \ldots, t_{N_\sigma} = T\}.$$

Let us denote by y_{ik}^j the function defined on the net $\overline{\omega}_{h\delta} = \overline{\omega}_h \times \overline{\omega}_\delta$.

For approximate description of the attainability set at the moment t_{s+1} under any pair of admissible controls $w_s(t), v_s(t), t \in [t_s,t_{s+1})$ for the problem (1)-(6) we construct purely implicit locally one-dimensional difference scheme (Samarsky, 1989):

$$\begin{aligned}&c_{0,k}^{j+1/2}\frac{y_{0,k}^{j+1/2} - y_{0,k}^{j}}{\delta} =\\ &= 2\lambda_{1/2,k}^{j+1/2}\frac{y_{1,k}^{j+1/2} - y_{0,k}^{j+1/2}}{h_1^2} - 2\alpha_1\frac{y_{0,k}^{j+1/2} - v_s}{h_1},\quad i=0,\\ &c_{i,k}^{j+1/2}\frac{y_{i,k}^{j+1/2} - y_{i,k}^{j}}{\delta} =\end{aligned} \tag{10}$$

$$= \lambda_{i+1/2,k}^{j+1/2}\frac{y_{i+1,k}^{j+1/2}-y_{i,k}^{j+1/2}}{h_1^2}-\lambda_{i-1/2,k}^{j+1/2}\frac{y_{i,k}^{j+1/2}-y_{i-1,k}^{j+1/2}}{h_1^2},\ i=\overline{1,N_1-1}, \tag{11}$$

$$c_{N_1,k}^{j+1/2}\frac{y_{N_1,k}^{j+1/2}-y_{N_1,k}^{j}}{\delta}=$$

$$= -2\alpha_1\frac{y_{N_1,k}^{j+1/2}-v^s}{h_1}-2\lambda_{N_1-1/2,k}^{j+1/2}\frac{y_{N_1,k}^{j+1/2}-y_{N_1-1,k}^{j+1/2}}{h_1^2},\ i=N_1, \tag{12}$$

$$k=\overline{0,N_2},$$
$$j=\overline{0,N_3-1},$$

$$c_{i,0}^{j+1}\frac{y_{i,0}^{j+1}-y_{i,0}^{j+1/2}}{\delta}=$$

$$= 2\lambda_{i,1/2}^{j+1}\frac{y_{i,1}^{j+1}-y_{i,0}^{j+1}}{h_2^2}-2\alpha_1\frac{y_{i,0}^{j+1}-v^s}{h_2},\ k=0, \tag{13}$$

$$c_{i,k}^{j+1}\frac{y_{i,k}^{j+1}-y_{i,k}^{j+1/2}}{\delta}=$$

$$= \lambda_{i,k+1/2}^{j+1}\frac{y_{i,k+1}^{j+1}-y_{i,k}^{j+1}}{h_2^2}-\lambda_{i,k-1/2}^{j+1}\frac{y_{i,k}^{j+1}-y_{i,k-1}^{j+1}}{h_2^2},\ k=\overline{1,N_2-1}, \tag{14}$$

$$c_{i,N_2}^{j+1}\frac{y_{i,N_2}^{j+1}-y_{i,N_2}^{j+1/2}}{\delta}=$$

$$= -2\alpha_2\frac{y_{i,N_2}^{j+1}-w_s}{h_2}-2\lambda_{i,N_2-1/2}^{j+1}\frac{y_{i,N_2}^{j+1}-y_{i,N_2-1}^{j+1}}{h_2^2},\ k=N_2, \tag{15}$$

$$i=\overline{0,N_1},$$
$$j=\overline{0,N_3-1},$$

$$y_{i,k}^0=y_{i,k}^{N_3,s},\ i=\overline{0,N_1},\ k=\overline{0,N_2},\ j=0,\ s=\overline{0,N_\sigma-1}, \tag{16}$$

where

$$y_{i,k}^{N_3,0}=u_0(x_1^{(i)},x_2^{(k)}),\ i=\overline{0,N_1},\ k=\overline{0,N_2},\ j=0,\ s=0.$$

The constructed difference scheme (9)-(15) is solved by the sweep method (Samarsky, 1989).

The operator form of the difference scheme (9)-(15) is following:

$$\begin{aligned}&\mathcal{D}_\alpha^{j+1/2}y_{\delta\alpha}+A_\alpha^{j+\alpha/2}y^{j+\alpha/2}=\varphi_\alpha^{j+\alpha/2},\ \alpha=1,2;\ j=\overline{0,N_3-1},\\ &y_{i,k}^0=y_{i,k}^{N_3,s},\ i=\overline{0,N_1},\ k=\overline{0,N_2},\ j=0,\ s=\overline{0,N_\sigma-1},\end{aligned} \tag{17}$$

where

$$y_{\delta\alpha}=\frac{y^{j+\alpha/2}-y^{j+(\alpha-1)/2}}{\delta}$$

$$y_{i,k}^{N_3,0}=u_0(x_1^{(i)},x_2^{(k)}),\ i=\overline{0,N_1},\ k=\overline{0,N_2},\ j=0,\ s=0.$$

The following statements about the stability and the convergence of the difference scheme (9)-(15) are valid.

Theorem 1 *Let us consider the operators* $\mathcal{D}_{\alpha}^{j+\alpha/2} \geq c_0 E$ *and* $A_{\alpha}^{j+\alpha/2} = A_{\alpha}^{j+\alpha/2*} \geq c_1 \Lambda$ *that are positive defined and the operator* $A_{\alpha}^{j+\alpha/2}$ *that satisfies the Lipschitz condition*

$A_{\alpha}^{j+\alpha/2} \leq (1 + c_3\delta) A_{\alpha}^{j}, \ \alpha = 1, 2; \ j = \overline{0, N_{\sigma} - 1}.$

Then locally one-dimensional difference scheme (9)-(15) is absolutely stable and the a priori estimate is valid:

$$\|[y^{n+1}]\|_{A_2^{n+1}} \leq M_1 \left\{ \|[y^0]\|_{A_1^0} + \sum_{j=0}^{n} \sqrt{\frac{\delta}{4\varepsilon}} \sum_{\alpha=1}^{2} \|[\varphi_{\alpha}^{j+\alpha/2}]\| \right\}, \tag{18}$$

where $M_1 = e^{0.5 c_3 t_0}$.

Theorem 2 *For* $|h| \to 0$, $\delta \to 0$ *the solution of the difference scheme (9)-(15) converges to the solution of the problem (1)-(6) by the rate of order* $o(\delta + |h|^2)$ *and the following estimate is valid:*

$$\|[z^{n+1}]\|_{A_2^{n+1}} \leq M_2 \left\{ \sum_{j=0}^{n} \sqrt{\frac{\delta}{4\varepsilon}} \sum_{\alpha=1}^{2} \|[\psi_{\alpha}^{j+\alpha/2}]\| \right\}, \tag{19}$$

where $|h|^2 = h_1^2 + h_2^2, \ M_2 = const.$

Example. Let us consider the differential two-person zero-sum game $\Gamma(u_0, T)$ which describes the process of heating of the bar in a medium with variable temperature.

The temperature w on the left side of the bar is controlled by the technologist (the player P). The control parameter w is an element of a set $[W_1, W_2]$ of all control parameters. The heat exchange between the bar and the medium of temperature v (where v is a "nature's" (the player E) control parameter) takes place on the right side of the bar. The control parameter v is an element of a set $[V_1, V_2]$ of all control parameters.

The dynamics of the game $\Gamma(u_0, T)$ is described by the following boundary value problem for the heat transfer equation:

$$c\frac{\partial u}{\partial t} = \frac{\partial}{\partial x}\left(\lambda \frac{\partial u}{\partial x}\right), \ x \in (0, l), \ t > 0, \tag{20}$$

$$u(0, t) = w, \ x = 0, \ t > 0, \tag{21}$$

$$-\lambda \frac{\partial u}{\partial x} = \alpha(u - v), \ x = l, \ t > 0, \tag{22}$$

$$u(x, 0) = u_0(x), \ x \in [0, l], \ t = 0, \tag{23}$$

The problem is to support given temperature $u_{\xi}(t), t \in [0, T]$ at the fixed bar point $\xi \in (0, l)$ under any uncertain "nature's" conditions by means of choosing a control function $w = w(\cdot)$.

Thus, the player E gets the following payoff from the player P at the terminal moment T of the game:

$$H = \sqrt{\frac{1}{T}\int_0^T (u(\xi,t) - u_\xi(t))^2\,dt}, \tag{24}$$

where $u(\xi, t)$ is a temperature at the bar point $\xi \in (0, l)$ at the moment $t \in [0, T]$.

The aim of the player P is to minimize H and the aim of the player E is contrary.

The method suggested above is used for numerical solving for the described problem.

We assume the bar being uniform. Then one may consider the coefficients c, λ as constants. The numerical experiments were realized for the following input data: $W_1 = 10^0$C, $W_2 = 20^0$C; $V_1 = 1^0$C, $V_2 = 20^0$C; the length of bar $l = 1$m, $\xi = l/2$; heat conductivity coefficient λ=45.400 W/(m·^{0}C), density ρ=7900 kg/m^3, specific heat capacity c_p=462 J/(kg·^{0}C); heat exchange coefficient $\alpha = 50$ W/(m^2·^{0}C); the initial temperature distribution $u_0(x) = 12\,^0$C; the time of heating $T = 10$ hour.

The following function $u_\xi(t)$ has been considered at the fixed bar point ξ:

$$u_\xi(t) \quad = \quad 13 + \sin\left(2\pi t/T - \frac{\pi}{2}\right), \quad t \in [0, T]. \tag{25}$$

The value functions of the upper and lower games are:

$$Val(\overline{\Gamma}^\sigma(u_0, T)) = 0.149,$$

$$Val(\underline{\Gamma}^\sigma(u_0, T)) = 0.111.$$

The results of numerical experiments are given on the Figures 1-3.

REFERENCES

Tichonov, A.N., Samarsky, A.A. (1977) *Equations of Mathematical Physics.* Nauka, Moscow (Russian).

Malafeyev, O.A. (1993) *Dynamical control systems of conflict.* St. Petersburg (Russian).

Bellman, R. (1960) *Dynamic programming.* IL, Moscow (Russian).

Samarsky, A.A. (1989) *Difference schemes theory.* Nauka, Moscow (Russian).

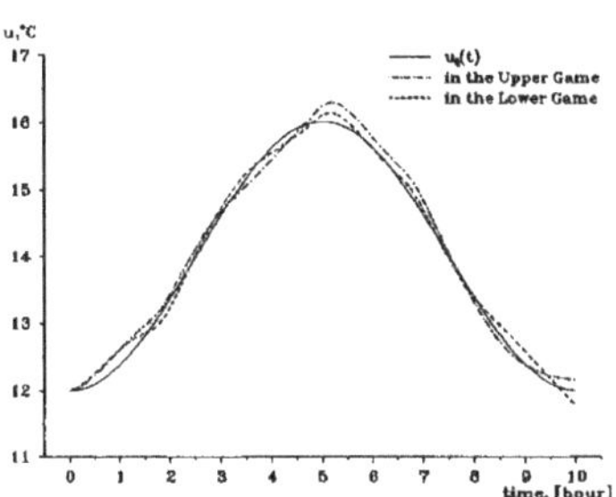

Figure 1 The Temperature at the Bar Point ξ.

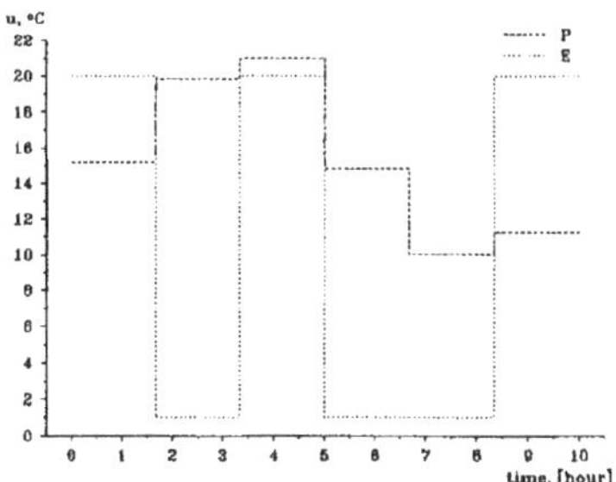

Figure 2 The Upper Game: The Optimal Control of the Players.

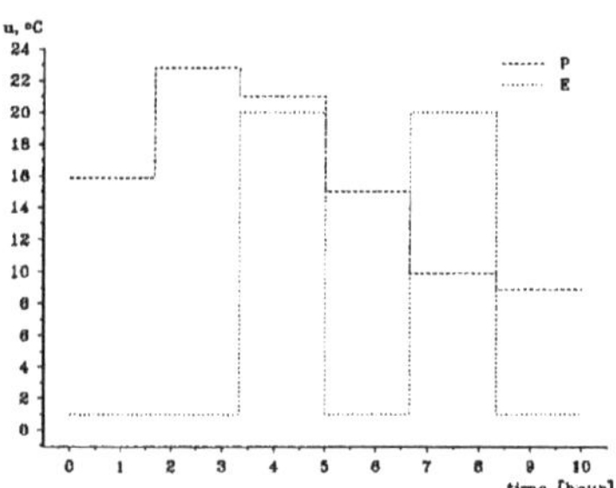

Figure 3 The Lower Game: The Optimal Control of the Players.

Constrained predictive control of a counter-current extractor

M. Mulholland, N.K. Narotam
Department of Chemical Engineering, University of Natal, King George V Avenue, Durban 4001, South Africa.
Tel: +27-31-2603123. Fax: +27-31-2601118.
e-mail: mulholla@che.und.ac.za

Abstract

Continuous counter-current liquid-liquid extraction of copper from an acidic aqueous solution is performed in a 6m pulsed packed column. Bubbles of an immiscible organic phase, containing the complexing agent LIX 64N, move upward as the continuous aqueous phase moves downward. It is desired to regulate the residual copper content of the departing aqueous phase, using as control actions the organic feed rate and the pulse frequency. In this distributed system, profiles of copper concentration become imprinted on the descending aqueous phase as a result of the history of controls and disturbances acting during the passage of a parcel. Initial work has involved the development of a dynamic model of the extractor, and the proving of a constrained model predictive controller on this model. The model is based on a spectral solution along the discretised column length, with special treatments of convection and the boundary conditions. The Linear Dynamic Matrix Controller (LDMC) minimises the summed magnitudes of deviation from setpoint, over a time horizon of up to 50 steps. The algorithm used is based on the method of Chang and Seborg (1983).

Keywords

Model predictive control, distributed system, liquid-liquid extraction, spectral model, linear dynamic matrix control.

1 INTRODUCTION

Acid leach liquors containing metallic ions result from metallurgical extraction processes. One means of concentrating the metals for recovery is by extraction into an organic phase (eg. kerosene) containing a metal-ion complexing agent such as LIX 64N (Forrest and Hughes, 1974; Kordosky, *et al*, 1987; Mickler and Uhlemann, 1992). This may be done by contacting the two phases in a series of mixer-settlers, or, as in this case (Figure 1), in a packed tower (Kumar and Hartland, 1994). Raschig ring packing in our tower breaks up the rising organic bubbles to offer more surface area for the inter-phase transfer to take place. This effect can be enhanced by pulsing at the base to superimpose an oscillatory vertical motion on the mean flow.

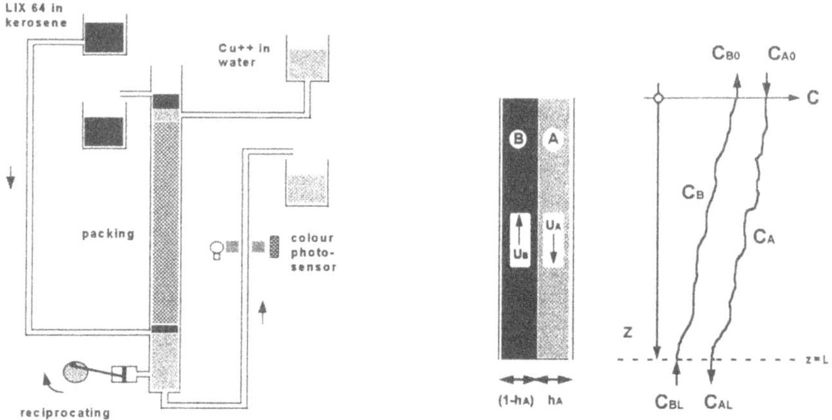

Figure 1 Arrangement of 6m pulsed packed column for the extraction of Cu^{++} from aqueous solution into kerosene containing LIX 64.

2 MODELLING

The system may be described by the pair of equations (Tsouris and Tavlarides, 1990; Hufnagel *et al*, 1991):

$$\frac{\partial C_A}{\partial t} = -u_A \frac{\partial C_A}{\partial z} + D_A \frac{\partial^2 C_A}{\partial z^2} - \frac{ka}{h_A}(C_A - \varepsilon C_B) \tag{1}$$

$$\frac{\partial C_B}{\partial t} = +u_B \frac{\partial C_B}{\partial z} + D_B \frac{\partial^2 C_B}{\partial z^2} + \frac{ka}{(1-h_A)}(C_A - \varepsilon C_B) \tag{2}$$

where C_A and C_B are the concentrations of copper in the aqueous and organic phases respectively. The term ka (transfer coefficient x surface area per unit volume) is expected to rise with more vigorous pulsing. Once the volumetric feeds f_A and f_B to the column are set, a characteristic volumetric holdup fraction is occupied by each phase, h_A and $h_B = (1-h_A)$, depending on which phase becomes continuous, and which is dispersed. The velocities are then determined by $u_A = f_A / [h_A A]$ and $u_B = f_B / [(1-h_A)A]$.

Notice that in this representation, the equilibrium concentration in the aqueous phase C_A^*, which would correspond to the present concentration C_B in the organic phase, is approximated as being proportional (ε) to C_B.

Various dynamic models of liquid-liquid extraction columns have been presented. Yoswathana *et al* (1985) used a simple lumped second-order discrete model for the aqueous exit concentration (conductivity). Bart, Bauer and Marr (1987) considered a detailed equilibrium mechanism, but only solved for the concentration profile in the column at equilibrium. The modelling approach of Najim *et al* (1987) is implicit, in that the algorithm treats the on-line process as a black box, and identifies those control levels which appear more successful, giving increasing weight to them. Najim *et al* (1988a, 1988b) use a lumped 3rd-order discrete model to represent the dependence of exit concentration (conductivity) on the pulse frequency. Najim (1988) reverts to the stochastic learning algorithm of Najim *et al* (1987), but adds a second control action, *viz.* the organic feed rate, to the pulse frequency , and includes a "feedforward" term in the form of a second conductivity measurement at some point in the column. Najim and Irving (1989) again used a lumped second-order SISO model for a liquid-liquid extractor. Najim and Youlal (1990) based a real-time identification on a lumped second-order discrete model , for pole-placement adaptive control. Tsouris and Tavlarides (1990,1991) stress that an increase in agitation (eg. pulsing) benefits extraction partly by increasing the organic phase holdup. They model this dependence of holdup on agitation as first order plus dead-time lag.

In the present work, it was recognised that the uniform velocity and axial diffusivity within each phase would lead to a simple solution by Fourier-transforming the dependence on z. Represent the two profiles of concentration by the single bivariate function

$$C_{jk} = C(y,z) \quad \text{with} \quad y = j\,\Delta y,\ j=1,2 \quad \text{and} \quad z = k\,\Delta z,\ k=1,N$$

such that $C_{1k} = C_A\,(k\,\Delta z)$ and $C_{2k} = C_B\,(k\,\Delta z)$.

If we have the 2-dimensional discrete Fourier transform of this function, it can be expressed in terms of the Fourier coefficients A_{jk} using the inverse Fourier transform:

$$C(y,z) = \frac{1}{2N}\sum_{k=1}^{N}\sum_{j=1}^{2} A_{jk}\, exp\{-i2\pi[\tfrac{(j-1)y}{2\Delta y} + \tfrac{(k-1)z}{2\Delta z}]\} = \frac{1}{2N}\sum_{k=1}^{N}\sum_{j=1}^{2} A_{jk}E_{jk} \tag{3}$$

Defining $f_k = -i2\pi k/(N\,\Delta z)$, note that

$$\frac{\partial C}{\partial z} = \frac{1}{2N}\sum_{k=1}^{N}\sum_{j=1}^{2} f_k A_{jk}E_{jk} \tag{4}$$

$$\frac{\partial^2 C}{\partial z^2} = \frac{1}{2N}\sum_{k=1}^{N}\sum_{j=1}^{2} f_k^2 A_{jk}E_{jk} \tag{5}$$

Substitute (3),(4) and (5) into (1) and (2) and equate the k harmonics to obtain

$$\frac{\partial A_{1k}}{\partial t} + \frac{\partial A_{2k}}{\partial t} = \left(-u_A f_k + D_A f_k^2 - ka/h_A\right)\left(A_{1k} + A_{2k}\right) + \left(ka\varepsilon/h_A\right)\left(A_{1k} - A_{2k}\right) = f_{A1k}A_{1k} + f_{A2k}A_{2k}$$

$$\frac{\partial A_{1k}}{\partial t} - \frac{\partial A_{2k}}{\partial t} = \left(+u_B f_k + D_B f_k^2 - ka\varepsilon/h_B\right)\left(A_{1k} - A_{2k}\right) + \left(ka/h_B\right)\left(A_{1k} + A_{2k}\right) = f_{B1k}A_{1k} + f_{B2k}A_{2k}$$

Thus

$$\frac{\partial A_k}{\partial t} = \boldsymbol{M}_k \boldsymbol{A}_k \quad \text{with} \quad \mathbf{A}_k = \begin{pmatrix} A_{1k} \\ A_{2k} \end{pmatrix} \quad \text{and} \quad \mathbf{M}_k = \tfrac{1}{2}\begin{bmatrix} (f_{A1k} - f_{B1k}) & (f_{A2k} - f_{B2k}) \\ (f_{A1k} + f_{B1k}) & (f_{A2k} + f_{B2k}) \end{bmatrix}$$

Integrate using the matrix exponential:

$$A_k(t+\Delta t) = exp(M_k \Delta t) A_k(t)$$

whence $C_A(z, t+\Delta t)$ and $C_B(z, t+\Delta t)$ are found using the inverse Fourier transform (3).

Since the Fourier transform is periodic, boundary conditions are handled by including in the solution, extensions of the z-range, at both ends of the column, which are longer than the convective jump on each time-step Δt. On each time-step, the concentration in this region is set to the feed condition of each phase, at the appropriate feed end. At the remote end, the "external" concentration is reset to the immediate internal concentration.

The above spectral solution is not accurate for convective jumps involving fractions of Δz. A minor approximation is made to overcome this problem for each phase. Records are kept of the "true" and "solution" distances of travel of each phase:

$$d_{Atrue} = \int_0^t u_A(t')dt' \qquad d_{Asoln} = \int_0^t u_{Aadj}(t')dt'$$

$$d_{Btrue} = \int_0^t u_B(t')dt' \qquad d_{Bsoln} = \int_0^t u_{Badj}(t')dt'$$

On each time-step, adjusted velocities u_{Aadj} and u_{Badj} are recalculated for use in the solution, such that they cause the solution travel distance to fall on the whole number of steps Δz which lies closest to the true travel distance for each phase.

3 CONTROL ALGORITHM

Several workers have considered the control problem for this distributed system. Yoswathana *et al* (1985) recursively estimated parameters for a lumped second-order discrete model on-line, for pole-placement adaptive control using a discrete PID controller. Najim *et al* (1987) and Najim (1988) used "black box" learning algorithm which automatically re-inforced controls which were detected to have a beneficial effect. Najim *et al* (1988a, 1988b) used an adaptive linear quadratic controller based on identified parameters for a 3rd-order discrete SISO model of the process. Najim and Irving (1989) used an unconstrained model predictive controller based on the discrete second-order lumped model relating exit conductivity to the pulse frequency. Najim and Youlal (1990) used pole-placement adaptive control with a regularization procedure to avoid the pole-zero cancellation problem.

In the present approach, it was recognised that an ideal controller should predict and counteract perturbations of the aqueous concentration profile moving towards the exit point. Both control actions will affect the entire concentration profile, so some minimisation of future deviations from setpoint over a time horizon is essential. A constrained multivariable Linear Dynamic Matrix Controller (LDMC), based on the linear programming solution of Chang and Seborg (1983), and the formulation of Morshedi *et al* (1985), has been formulated as follows:

Consider a vector of input control actions $\boldsymbol{m}$, which might contain both the organic phase feed rate and the pulse frequency. For a linear system, the response of the output vector x (eg. aqueous and organic concentrations) at the next N intervals can be constructed by superposition of the step responses for the sequence of control vector moves $\Delta \boldsymbol{m}_1, \Delta \boldsymbol{m}_2, \dots, \Delta \boldsymbol{m}_N$ as follows:

$$\begin{pmatrix} \mathbf{x}_1 \\ \mathbf{x}_2 \\ \mathbf{x}_3 \\ \vdots \\ \mathbf{x}_M \\ \mathbf{x}_{M+1} \\ \vdots \\ \mathbf{x}_N \end{pmatrix} = \begin{bmatrix} \mathbf{B}_1 & 0 & 0 & 0 & 0 & \cdots & 0 \\ \mathbf{B}_2 & \mathbf{B}_1 & 0 & 0 & 0 & \cdots & 0 \\ \mathbf{B}_3 & \mathbf{B}_2 & \mathbf{B}_1 & 0 & 0 & \cdots & 0 \\ \vdots & \vdots & \vdots & \vdots & \vdots & & \vdots \\ \mathbf{B}_M & \mathbf{B}_{M-1} & \cdots & \mathbf{B}_1 & 0 & \cdots & 0 \\ \mathbf{B}_M & \mathbf{B}_M & \mathbf{B}_{M-1} & \cdots & \mathbf{B}_1 & \cdots & 0 \\ \vdots & \vdots & \vdots & \vdots & \vdots & & \vdots \\ \mathbf{B}_M & \mathbf{B}_M & \mathbf{B}_M & \mathbf{B}_M & \mathbf{B}_{M-1} & \cdots & \mathbf{B}_1 \end{bmatrix} \begin{pmatrix} \Delta\mathbf{m}_1 \\ \Delta\mathbf{m}_2 \\ \Delta\mathbf{m}_3 \\ \vdots \\ \Delta\mathbf{m}_M \\ \Delta\mathbf{m}_{M+1} \\ \vdots \\ \Delta\mathbf{m}_N \end{pmatrix}$$

(Steady - state response achieved M steps ahead with $M < N$)

We represent this convolution model for future outputs as $\boldsymbol{x} = \boldsymbol{B\Delta m}$. Now, defining the $N \times M$ matrices:

$$\boldsymbol{B_{OL}} = \begin{bmatrix} \mathbf{B}_M & \mathbf{B}_M & \mathbf{B}_{M-1} & \mathbf{B}_{M-2} & \mathbf{B}_{M-3} & \cdots & \mathbf{B}_3 & \mathbf{B}_2 \\ \mathbf{B}_M & \mathbf{B}_M & \mathbf{B}_M & \mathbf{B}_{M-1} & \mathbf{B}_{M-2} & \cdots & \cdots & \mathbf{B}_3 \\ \mathbf{B}_M & \mathbf{B}_M & \mathbf{B}_M & \mathbf{B}_M & \mathbf{B}_{M-1} & \cdots & \cdots & \vdots \\ \mathbf{B}_M & \mathbf{B}_M & \mathbf{B}_M & \mathbf{B}_M & \mathbf{B}_M & \cdots & \cdots & \vdots \\ \vdots & \vdots & \vdots & \vdots & \vdots & & \vdots & \mathbf{B}_{M-1} \\ \mathbf{B}_M & \mathbf{B}_M & \mathbf{B}_M & \mathbf{B}_M & \mathbf{B}_M & \cdots & \mathbf{B}_M & \mathbf{B}_M \\ \vdots & \vdots & \vdots & \vdots & \vdots & & \vdots & \vdots \\ \mathbf{B}_M & \mathbf{B}_M & \mathbf{B}_M & \mathbf{B}_M & \mathbf{B}_M & \cdots & \mathbf{B}_M & \mathbf{B}_M \end{bmatrix}$$

$$\boldsymbol{B_0} = \begin{bmatrix} \mathbf{B}_M & \mathbf{B}_M & \mathbf{B}_{M-1} & \mathbf{B}_{M-2} & \mathbf{B}_{M-3} & \cdots & \mathbf{B}_3 & \mathbf{B}_2 \\ \mathbf{B}_M & \mathbf{B}_M & \mathbf{B}_{M-1} & \mathbf{B}_{M-2} & \mathbf{B}_{M-3} & \cdots & \mathbf{B}_3 & \mathbf{B}_2 \\ \mathbf{B}_M & \mathbf{B}_M & \mathbf{B}_{M-1} & \mathbf{B}_{M-2} & \mathbf{B}_{M-3} & \cdots & \mathbf{B}_3 & \mathbf{B}_2 \\ \mathbf{B}_M & \mathbf{B}_M & \mathbf{B}_{M-1} & \mathbf{B}_{M-2} & \mathbf{B}_{M-3} & \cdots & \mathbf{B}_3 & \mathbf{B}_2 \\ \vdots & \vdots & \vdots & \vdots & \vdots & & \vdots & \vdots \\ \mathbf{B}_M & \mathbf{B}_M & \mathbf{B}_{M-1} & \mathbf{B}_{M-2} & \mathbf{B}_{M-3} & \cdots & \mathbf{B}_3 & \mathbf{B}_2 \end{bmatrix}$$

and the present measurements (N) and past inputs (M):

$$\mathbf{x}_{0\,MEAS} = \begin{pmatrix} \mathbf{x}_{0_{meas}} \\ \mathbf{x}_{0_{meas}} \\ \mathbf{x}_{0_{meas}} \\ \mathbf{x}_{0_{meas}} \\ \vdots \\ \mathbf{x}_{0_{meas}} \end{pmatrix} \quad \text{and} \quad \Delta\mathbf{m}_{PAST} = \begin{pmatrix} \Delta m_{-M+1} \\ \Delta m_{-M+2} \\ \Delta m_{-M+3} \\ \Delta m_{-M+4} \\ \vdots \\ \Delta m_0 \end{pmatrix}$$

then the "open-loop" response, corrected for present model offset, is

$$x_{OL} = x_{0MEAS} + [B_{OL} - B_0]\, \Delta m_{PAST}$$

and the "closed loop" response up to the N-step horizon is obtained by including the contribution of the future control input steps Δm :

$$x_{CL} = x_{OL} + B\, \Delta m$$

On each time-step it is possible to compute the future open-loop response x_{OL} based on past inputs and the present output. Thus the control problem amounts to finding suitable Δm . Such an optimisation might solve for a limited sequence of the steps $\Delta m_1, \Delta m_2, \ldots, \Delta m_N$, but it is only the first step Δm_1 which is actually implemented, before the entire computation is repeated on the next time-step.

Define x_{SP} to contain a sequence of set-points for the outputs up to the time horizon, so that the open-loop error may be calculated in advance as $x_{OL} - x_{SP}$. Then the ideal control sequence Δm will be the one which minimises the magnitudes of the elements of the residual

$$r = x_{OL} - x_{SP} + B\, \Delta m$$

Transforming and adding a term for move suppression (Morshedi *et al*, 1985), we minimise rather the magnitude of

$$\rho = B^T(x_{OL} - x_{SP}) + [B^T B + \lambda \mathrm{I}]\Delta m$$

Minimisation of the size of each term in ρ , yet keeping Δm, m and x_{CL} within defined upper and lower constraints, is achieved using the linear programming technique of Chang and Seborg (1983). This requires that ρ is represented by two vectors of non-negative elements, *viz.* $\rho=\rho^+-\rho^-$, and that the constraints are defined in terms of

$$\Delta m = [B^T B+\lambda I]^{-1} [\rho^+ - \rho^- - B^T(x_{OL} - x_{SP})] = T\, [\rho^+ - \rho^- - \varepsilon_{OL}]$$

$$m = \begin{bmatrix} \mathrm{I} & 0 & 0 & 0 & 0 & \cdots & 0 \\ \mathrm{I} & \mathrm{I} & 0 & 0 & 0 & \cdots & 0 \\ \mathrm{I} & \mathrm{I} & \mathrm{I} & 0 & 0 & \cdots & 0 \\ \mathrm{I} & \mathrm{I} & \mathrm{I} & \mathrm{I} & 0 & \cdots & 0 \\ \vdots & \vdots & \vdots & \vdots & \vdots & & \vdots \\ \mathrm{I} & \mathrm{I} & \mathrm{I} & \mathrm{I} & \mathrm{I} & \cdots & \mathrm{I} \end{bmatrix} T\, [\rho^+ - \rho^- - \varepsilon_{OL}]$$

$$x_{CL} = x_{OL} + BT\, [\rho^+ - \rho^- - \varepsilon_{OL}]$$

Minimisation of $w^T(\rho^+ + \rho^-)$ within these constraints, where the vector w contains a sequence of specified weights up to the time horizon, then achieves the desired result.

4 CONTROL APPLICATION ON MODEL

A first application of the above LDMC algorithm has been in SISO control of the counter-current extractor model. The manipulated control action is the organic phase feed rate, with the aqueous exit Cu^{++} composition being the controlled process variable. The convolution model of the process was based on M=50 points of a step response near the normal operating point. Figure 2(a) shows the servo- and load-performance when only the first control step has been used to optimise the closed-loop response over a horizon of N=50 steps.. Notice the characteristic behaviour of the LDMC in which the control action moves directly to the compensation level following a step-change in set-point, or a step disturbance. For comparison, a proportional-integral controller was tuned using the recommended Ziegler-Nichols settings obtained from the reaction curve. The performance with the same set-point and disturbance sequence is shown in Figure 2(b). The system non-linearity results in differing behaviour of the PI controller over the organic phase flow range. Appreciable control action cycling is necessary to hold the exit Cu^{++} composition at set-point. Further advantages are expected from the LDMC approach for MIMO case, in which both organic phase feed rate (f_B) and the pulsing frequency (affecting ka) will be used as control actions.

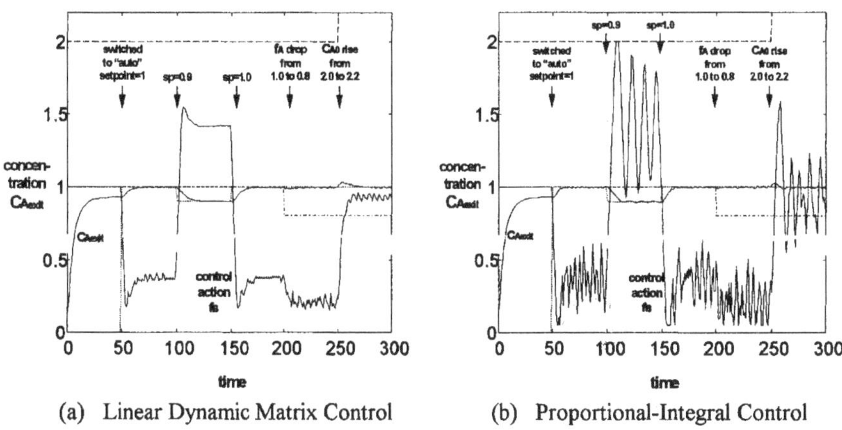

(a) Linear Dynamic Matrix Control (b) Proportional-Integral Control

Figure 2 Control of exit Cu^{++} ($C_{A\ exit}$) by manipulation of organic phase feed rate f_B .

6 ACKNOWLEDGEMENTS

The support of the Foundation for Research Development and the University of Natal Research Fund is gratefully acknowledged.

REFERENCES

Bart H.J., Bauer A. and Marr R. (1987) Modelling of mass transfer with chemical reaction in solvent extraction equipment, *Technische Universitat Graz*, Austria.

Chang T.S. and Seborg D.E. (1983) A linear programming approach for multivariable feedback control with inequality constraints, *Int.J.Control* **37**, 3, 583-597.

Forrest C. and Hughes M.A. (1974) The modelling of equilibrium data for the liquid-liquid extraction of metals: Part II: Models for the copper/LIX64 and chromate/ALIQUAT336 systems, *Hydrometallurgy* **I**, 139-154.

Hufnagel H., McIntyre M. and Blaβ E. (1991) Dynamic behaviour and simulation of a liquid-liquid extraction column, *Chem Eng. Technol* **14**, 301-306.

Kordosky G.A., Olafson S.M., Lewis R.G. and Deffner V.L. (1987) A state-of-the-art discussion on the solvent extraction reagents used for the recovery of copper from dilute sulfuric acid leach solutions, *Separation Science & Technology* **22**(2&3), 215-232.

Kumar A. and Hartland S. (1994) Prediction of drop size, dispersed-phase holdup, slip velocity, and limiting throughputs in packed extraction columns, *Trans IChemE* 72A, 89-104.

Mickler W. and Uhlemann E. (1992) Liquid-liquid extraction of copper from ammoniacal solution with β-diketones, *Separation Science and Technology* **27**(12), 1669-1674.

Morshedi A.M., Cutler C.R. and Skrovanek T.A. (1985) Optimal solution of Dynamic Matrix Control with Linear Programming techniques (LDMC), *American Control Conference*, Boston.

Najim K., Le Lann M.U. and Casamatta G. (1987) Learning control of a pulsed liquid-liquid extraction column, *Chemical Engineering Science* **42**, 7, 1619-1628.

Najim K., Irving E., Youlal H. and Najim M. (1988) Adaptive linear quadratic control of an extractor, *IFAC Symposium on Adaptive Control of Chemical Processes*, Copenhagen.

Najim K., Youlal H., Irving E. and Najim M. (1988) Linear quadratic self-tuning control of a liquid-liquid extraction column, *Optimal Control Applications & Methods* **9**, 273-284.

Najim K. (1988) Multivariable control of a liquid-liquid extraction column using a probabilistic automaton, *IEE Proceedings* **135D**, 6.

Najim K. and Irving E. (1989) Extended horizon self-tuning control of a pulsed liquid-liquid extraction column, *Optimal Control Applications and Methods* **10**, 375-383.

Najim K. and Youlal H. (1990) Regularized pole-placement adaptive control of a liquid-liquid extraction column, *Int.J.Systems Sci* **21**, 7, 1313-1323.

Tsouris C. and Tavlarides L.L. (1990) Application of the ultrasonic technique for real-time holdup monitoring and control of extraction columns, *Chemical Engineering Science* **45**, 10, 3055-3062.

Tsouris C. and Tavlarides L.L. (1991) Control of dispersed-phase volume fraction in multistage extraction columns, *Chemical Engineering Science* **46**, 11, 2857-2865.

Yoswathana N., Casamatta G. and Najim K. (1985) Adaptive control of a liquid-liquid extractor, *IFAC Symposium on Adaptive Control of Chemical Processes*, Frankfurt.

Optimal policies under different assumptions about target values: An optimum control analysis for Austria

R. Neck and S. Karbuz
Department of Economics, University of Bielefeld
P.O.Box 10 01 31, D - 33501 Bielefeld, Germany.
Tel: +49 521 106 48 59. Fax: +49 521 106 29 94.
e-mail: `neck@nw42.wiwi.uni-bielefeld.de`

Abstract

Optimal fiscal and monetary policies for the period 1993 to 2000 are calculated for Austria within the framework of a problem of quantitative economic policy. An intertemporal objective function is minimized subject to the constraints of a macroeconometric model. Exogenous variables of the model are forecast by time series methods. Using the optimum control algorithm OPTCON, approximately optimal policies are determined. The sensitivity of optimal policies with respect to the target values of the objective variables is investigated. It is shown that when designing intertemporal optimization problems, one must avoid to be either too ambitious or too cautious with respect to the postulated targets.

Keywords

Optimal control, economics, sensitivity, optimization, forecasting

1 INTRODUCTION

When planning policies to influence the development of an economy, it is usually not sufficient to rely on simple rules. Instead, trade-offs and side-effects of policies on variables other than those at which they are aimed have to be taken into account. Quantitative informations on these effects as well as on intertemporal preferences of policy-makers are required. Ideally, there should be decision support systems available to assist policy-makers in planning appropriate economic policies.

In this paper, we analyze these issues within a problem of quantitative economic policy, using an optimum control approach. We determine numerically optimal monetary and fiscal policies

---

Financial support from the "Jubiläumsfonds der Oesterreichischen Nationalbank" (project no. 5126) is gratefully acknowledged. Karbuz acknowledges support from the Institute for Advanced Studiens, Vienna.

for the nineties by minimizing an intertemporal objective function subject to the constraints given by an econometric model. The model, called FINPOL2, is a medium-size macroeconometric model for Austria. It relates policy and exogenous variables to objective variables of Austrian economic policies. Moreover, we postulate an objective function for Austrian policy-makers over the years 1993 to 2000, which penalizes deviations of objective variables from their desired values. The exogenous variables of the model are forecast over this planning horizon using time series methods. Next, we calculate optimal stabilization policies over this time horizon.

Here we are particularly interested in the sensitivity of optimal policies with respect to the target or "ideal" values of the objective variables from which deviations are penalized by the objective function. To answer this question, we perform several optimum control experiments under varying assumptions about the target paths of objective variables. The results show that there may be systematic dependences of optimal policies on these target paths. In particular, trade-offs between different policy goals may be sensitive with respect to postulated target values of the objective variables. This shows that when designing intertemporal optimization problems, and in particular when specifying an objective function numerically, one must avoid to be either too ambitious or too cautious with respect to the postulated targets.

2 THE ECONOMETRIC MODEL FINPOL2

The model FINPOL2 is based on traditional Keynesian macroeconomic theory in the sense of conventional IS-LM/aggregate demand-aggregate supply models. Stochastic behavioral equations for the demand side include a consumption function, an investment function, an import function and an interest-rate equation as a reduced-form money market model. Prices are largely determined by aggregate demand variables. Disequilibrium in the labor market, as measured by the excess of unemployed persons over vacancies, is modelled to depend on the real GDP growth rate and the rate of inflation, embodying both an Okun's law-type relation and a rudimentary Phillips curve. The main objective variables of Austrian economic policies, such as real GDP, the labor market disequilibrium variable (related to the rate of unemployment), the rate of inflation, the balance of payments and the ratio of the federal net budget deficit to GDP, are related directly or indirectly to those fiscal and monetary policy instruments which are used as control variables, namely to federal budget expenditures and revenues and to money supply.

The model, which is dynamic and nonlinear, was estimated first by OLS and then by simultaneous equations estimation methods using annual data over the period 1965 to 1992. Data have been obtained from the Austrian Institute of Economic Research (WIFO). All real data have dimension billions of 1983 Austrian Shillings. The estimates and test statistics together with ex-post simulation results suggest that the model provides a reasonable account of the development of economic variables in the recent past. Here we use the estimates of the parameters obtained by three-stage least squares. These estimation results together with the statistical characteristics of the regressions are given in Neck and Karbuz (1994).

3 THE OPTIMUM CONTROL APPROACH

In the theory of quantitative economic policy, optimum control theory has been used in several studies to determine optimal policies for econometric models (e.g., Chow 1975, 1981, Kendrick 1981). Here we use the algorithm OPTCON, developed by Matulka and Neck (1992); it determines approximate solutions of deterministic or stochastic optimum control problems with a quadratic objective function and a nonlinear multivariable dynamic model. The objective function is quadratic in the deviations of the state and control variables from their respective desired values. The dynamic system is required to be given in a state space representation.

For our simulation experiments, we choose the planning horizon as 1993 to 2000. Among the variables whose deviations from desired values are to be penalized, we distinguish two categories: First, there are five "main" objective variables which are of direct political relevance in assessing the performance of the Austrian economy. These are the rate of inflation ($PV\%_t$), the labor market excess supply variable (UN_t) as a measure for involuntary unemployment, the rate of growth of real GDP ($YR\%_t$), the balance of current account (LBR_t), and the federal net budget deficit as percentage of GDP ($DEF\%_t$). In all experiments, the desired levels for labor market excess supply (UN_t) and the balance of current account (LBR_t) are set equal to zero. Thus, we want to achieve equilibrium in the labor market and in the external relations. Assumptions about the desired (target) values for the rate of inflation ($PV\%_t$), the real growth rate ($YR\%_t$), and the deficit variable $DEF\%_t$ vary among different experiments.

Second, we introduce a category of "minor" objective variables. These include real private consumption (CR_t), real private investment (IR_t), real imports of goods and services (MR_t), the nominal rate of interest (R_t), real GDP (YR_t), real total aggregate demand (VR_t), the domestic price level (PY_t), the price level of public consumption (PG_t), nominal public consumption (G_t), and nominal public-sector net tax revenues (T_t), as well as the policy instrument (control) variables money supply ($M1_t$), federal budget net expenditures (NEX_t) and federal budget tax receipts (BIN_t). We take 1992 historical values of these "minor" objective variables (except for R_t) to be given and postulate desired growth rates for the planning horizon, which are varied in accordance with the target values for $YR\%_t$ and $PV\%_t$. The rate of interest R_t has a desired constant value of 7 for all periods.

In the weight matrix of the objective function, all off-diagonal elements are set equal to zero, and the main diagonal elements are given weights of 10 for the "main" objective variables and of 1 for the "minor" objective variables. The state variables that are not mentioned above get weights of zero, thus being regarded as irrelevant to the hypothetical policy-maker. The weight matrix is assumed to be constant over time.

For a simulation over a future planning horizon, projections (forecasts) of the exogenous (controlled and non-controlled) variables are needed. Here we use extrapolations of these variables for the years 1993 to 2000 calculated from linear stochastic time series models of the ARMA (mixed autoregressive-moving average process) type. After several trials and applying the usual diagnostic checking procedure for the time series under consideration, we decided to model money supply ($M1_t$) by an ARMA (2,1) process, federal budget expenditures (NEX_t) by an ARMA (2,1) process, federal budget revenues (BIN_t) by an ARMA (2,2) process, the import price level (PM_t) by an ARMA (1,1) process, real exports of goods and services (XR_t) by an ARMA (2,3) process, and the inventory change variable IIR_t by an AR (1) process.

The forecasts from these time series models imply moderate growth of the fiscal policy variables. The extrapolation implies that the federal budget deficit ($NDEF_t$) gets stabilized and falls gradually from 70.3 billions AS in 1993 to 60.5 billions AS in 2000, which seems to be a rather optimistic forecast. The development of the foreign sector variables PM_t and XR_t is even more optimistic: PM_t grows only by 1% p.a. or less, and XR_t grows by 5 to 7% p.a. IIR_t is positive but falling. Money supply $M1_t$ grows by 5 to 5.8% p.a.

As a first step, the model was simulated over the years 1993 to 2000 using the extrapolations of all exogenous variables from the time series models as input. This amounts to a dynamic forecast of the endogenous variables of the model. Next, we performed several optimization experiments. Here again the projections of the non-controlled exogenous variables from the time series models are used as inputs, being assumed to be known for certain, but the values of the policy instruments are determined endogenously as (approximately) optimal under the assumed objective function. In this paper, we are considering only the results of deterministic optimization experiments, although the algorithm OPTCON can solve stochastic control problems with additive and parameter errors as well. In several experiments, we found the results of deterministic and fully stochastic optimization runs to be very similar. Therefore we decided to concentrate on deterministic runs for a sensitivity analysis.

The projection scenario forecasts a recession for 1993 (or actually a continuation of the recession from 1992): growth of real GDP ($YR\%_t$) is negative, the rate of unemployment (related to UN_t) is high, the balance of trade (LBR_t) exhibits a deficit, the rate of inflation ($PV\%_t$) is moderate, and the ratio of the federal budget deficit to GDP ($DEF\%_t$) is high. Starting in 1994, however, a period of relatively high growth is projected, which is clearly above the one obtained on average during the eighties. The labor market excess supply variable (UN_t) falls gradually, with only slightly rising inflation ($PV\%_t$). High surpluses are obtained for the balance of trade (LBR_t), and the deficit variable $DEF\%_t$ falls gradually.

4 OPTIMAL POLICIES UNDER DIFFERENT ASSUMPTIONS ABOUT TARGET VALUES

In order to assess the scope for optimal stabilization policies, we conducted several optimization experiments with the model and the assumed intertemporal objective function. We are particularly interested in the sensitivity of optimal policies with respect to the target paths of the budget deficit on the one hand and of real and price variables on the other hand. As a benchmark, for experiment 1 we assume that the aim of the policy-maker consists in consolidating the federal budget deficit gradually such that the desired value of $DEF\%_t$ is reduced by 0.3 percentage points each year, from the historical value of 3.27% in 1992 down to 0.87% in 2000. Moreover, we consider 2% p.a. as the desired rate of inflation ($PV\%_t$) and 3.5% p.a. as the desired real growth rate ($YR\%_t$). In order to make target values consistent for the model variables, we also assume desired growth rates of 3.5% p.a. for all real variables, desired growth rates of 2% p.a. for the price level variables, and desired growth rates of 5.5% p.a. for the nominal variables among the "minor" objective variables.

Results for the instrument variables and the "main" objective variables from experiment 1 are given in Table 1. Optimal monetary and especially fiscal policies are more countercyclical than projected ones and imply smoother time paths of the endogenous variables of the model. The recession of 1993 is avoided by expansionary budgetary policies. Federal budget expenditures (NEX_t) are higher and federal budget revenues (BIN_t) are lower than both projected and desired levels of the variables. This results in a positive growth rate, lower unemployment, only slightly higher inflation, but distinctly higher deficits of the trade balance and the federal budget as compared to the projection. Also in 1994, optimal stabilization policies can be characterized as expansionary. The values of the instrument variables in 1995 and 1996 are close to those of the projection. From 1997 on, optimal monetary and fiscal policies are restrictive as compared to the projection. This results in lower growth, higher unemployment and lower inflation than in the reference scenario. The surplus of the trade balance starts increasing some years later than in the projection. The federal budget gets fully consolidated, with a balanced budget in 1999 and even a surplus in 2000. The optimal levels of aggregate demand and its components (private consumption, investment, government consumption, imports, tax revenues) all grow faster during the recession than in the projection. For the entire time horizon, however, optimal growth is slightly weaker on average than in the reference scenario.

Table 1 Optimal values of instruments and "main" objectives, experiment 1

year	$M1_t$	NEX_t	BIN_t	$PV\%_t$	UN_t	$YR\%_t$	LBR_t	$DEF\%_t$
1993	326.892	734.641	581.421	2.063	4.807	1.996	-22.144	7.196
1994	340.949	769.752	639.456	2.597	4.112	4.632	-7.253	5.671
1995	355.651	789.774	698.940	2.753	3.454	4.943	9.858	3.644
1996	373.033	833.577	746.124	2.658	3.428	3.365	12.669	3.282
1997	390.509	870.066	806.048	2.712	3.278	3.769	22.733	2.238
1998	409.455	907.822	870.236	2.681	3.279	3.390	32.310	1.229
1999	430.907	937.960	934.985	2.692	3.245	3.502	44.772	0.091
2000	457.717	959.630	974.370	2.838	2.920	4.423	55.274	-0.416

In order to investigate whether the consolidation of the federal budget is due to the ambitious goal of deficit reduction, we conducted an experiment where the target value of the deficit variable $DEF\%_t$ was set at a constant value of 2.5% of GDP for the entire planning horizon. The results are nearly identical to those of experiment 1. Monetary policy ($M1_t$) is slightly more restrictive after 1994, and fiscal policies (NEX_t and BIN_t) are slightly more expansionary than in experiment 1. The effects on the endogenous variables, including the federal budget deficit, are negligible as compared to experiment 1. The same holds true if we assume a constant desired value of 3.27% (the historical value of 1992) for $DEF\%_t$, and if we apply these modifications of the target values of the budget deficit to the experiments to be described below. From this, we conclude that in our model there is nearly no influence of the target values of the deficit variable $DEF\%_t$ on optimal policies, at least for "realistic" (not extreme) desired values of this variable.

The situation is quite different if we modify the target values of the real growth rate and of the inflation rate. In the following experiments, we retain the target values for $DEF\%_t$ from experiment 1. First, in experiment 2, we assume a desired growth rate of 2.5% for all real variables (and the same value for $YR\%_t$); the "ideal" value of the inflation rate (and of the growth rates of all price level variables in the model) is kept at 2%, and nominal variables get desired growth rates of 4.5%. This scenario, which can be interpreted as "pessimistic" with respect to the growth potential of the Austrian economy, gives results as shown in Table 2. Now optimal monetary and fiscal policies are considerably more restrictive than in experiment

1. In particular, money supply ($M1_t$) and federal budget expenditures (NEX_t) are always lower, and federal budget revenues (BIN_t), except for 1993 and 2000, are higher than in experiment 1. Especially in the later years of the planning horizon, all instruments follow a much more contractionary course than prescribed by their desired values. This implies a lower budget deficit than in experiment 1, with increasing budgetary surpluses starting already in 1997. Inflation ($PV\%_t$) and real GDP growth ($YR\%_t$) are lower, balance-of-payments surplus (LBR_t) and involuntary unemployment (UN_t) are higher than in experiment 1, especially during the last years of the planning horizon. Real, price, and nominal variables show lower levels than before, though they are still higher than their desired values.

Table 2 Optimal values of instruments and "main" objectives, experiment 2

year	$M1_t$	NEX_t	BIN_t	$PV\%_t$	UN_t	$YR\%_t$	LBR_t	$DEF\%_t$
1993	320.153	729.821	580.292	2.037	4.872	1.793	-21.549	7.039
1994	329.337	752.361	639.641	2.499	4.373	3.982	-4.279	4.953
1995	338.872	760.214	699.379	2.613	3.862	4.328	16.540	2.482
1996	350.561	790.330	747.983	2.479	3.958	2.744	24.135	1.630
1997	361.905	810.652	811.039	2.489	3.932	3.077	40.133	-0.014
1998	374.699	828.230	879.139	2.404	4.078	2.565	57.037	-1.744
1999	391.084	831.175	942.304	2.376	4.159	2.688	77.951	-3.597
2000	415.876	817.178	956.712	2.602	3.673	4.404	95.190	-4.192

These effects are intensified if we retain the specification of experiment 2, but assume more "optimistic" desired values for the price level variables and for the inflation rate by targeting zero inflation (implying "ideal" growth rates of 2.5% for nominal variables). Now all instrument variables exhibit lower values than in experiment 2. Money supply ($M1_t$) grows very slowly, federal budget expenditures (NEX_t) even decline over the entire period, and federal budget revenues (BIN_t) grow less than in experiment 2, resulting in a distinctly lower budget deficit (with a high surplus at the end of the planning horizon). All instrument variables show more contractionary behavior than their desired values. The rate of interest is higher, the rate of inflation ($PV\%_t$) only slightly lower than before; effects on growth ($YR\%_t$) and labor market excess supply (UN_t) differ between different years. Over the entire planning period, endogenous real variables mostly grow slightly more than in experiment 2, but less than in experiment 1. The behavior of nominal and price variables is similar as in experiment 2, in spite of the more ambitious goal regarding inflation. Nearly all endogenous variables are above their desired values, even with these very restrictive monetary and fiscal policies. From the point of view of actual policy-making, one can conclude that here (as already in experiment 2) the growth targets are too "pessimistic"; in addition, price targets are clearly too "optimistic", as prices do not react strongly upon these restrictive stabilization policies.

To explore the influence of growth and price targets further, we next assume more "optimistic" desired values for real economic variables. In experiment 3, we set the "ideal" growth rate of real variables at 4.5%, of price variables at 2% as in experiment 1, and of nominal variables at 6.5%. Results are given in Table 3. As compared to experiment 1, the behavior of the instruments from experiment 2 is now reversed: money supply ($M1_t$) and federal budget expenditures (NEX_t) are always higher, federal budget revenues (BIN_t), except for 1993 and 2000, are lower than in experiment 1, implying a more expansionary policy stance, also if compared with the desired values or the instrument variables. In this experiment, the federal budget does not get consolidated, with budget deficits being always higher than in experiment 1 and well above 100 billions Austrian Shillings until 2000. Prices and inflation rates, levels and growth rates of real variables, and levels of nominal variables are higher than in experiment 1; involuntary unemployment and especially the surplus of the trade balance are lower. The real GDP growth rate ($YR\%_t$) is close to its desired values. Among the "minor" objective variables, private consumption (after 1995), imports and prices are above, investment and real GDP are below their target values. Given the relatively "optimistic" forecast of the non-controlled exogenous variables, this scenario might seem to be attractive to policy-makers, although the budgetary goals are not fulfilled.

Table 3 Optimal values of instruments and "main" objectives, experiment 3

year	$M1_t$	NEX_t	BIN_t	$PV\%_t$	UN_t	$YR\%_t$	LBR_t	$DEF\%_t$
1993	333.789	739.023	582.922	2.084	4.754	2.162	-22.628	7.318
1994	352.870	786.984	639.412	2.695	3.860	5.287	-10.056	6.364
1995	372.988	819.558	698.488	2.895	3.047	5.572	3.316	4.776
1996	396.450	877.711	743.917	2.842	2.890	4.010	1.191	4.896
1997	420.590	931.508	800.041	2.946	2.605	4.497	5.003	4.438
1998	446.321	991.361	859.446	2.975	2.452	4.256	6.734	4.115
1999	473.424	1051.620	925.636	3.026	2.302	4.331	10.116	3.626
2000	502.602	1112.188	992.553	3.081	2.152	4.416	13.510	3.173

Finally, in experiment 4 we explore the consequences of "optimistic" target values for both real growth and price variables. Growth of real variables is targeted at 4.5%, growth of price variables at zero, and growth of nominal variables at 4.5%. The results are shown in Table 4.

Table 4 Optimal values of instruments and "main" objectives, experiment 4

year	$M1_t$	NEX_t	BIN_t	$PV\%_t$	UN_t	$YR\%_t$	LBR_t	$DEF\%_t$
1993	325.227	727.011	569.407	2.089	4.743	2.198	-22.734	7.386
1994	336.277	756.827	616.236	2.694	3.857	5.266	-10.180	6.062
1995	347.513	768.905	664.994	2.893	3.051	5.552	3.239	4.099
1996	361.245	804.129	699.356	2.836	2.905	3.973	1.268	3.836
1997	374.915	832.015	743.497	2.934	2.939	4.431	5.418	2.992
1998	389.787	862.027	788.520	2.957	2.500	4.190	7.640	2.297
1999	406.448	887.672	833.050	3.032	2.307	4.438	10.998	1.574
2000	426.678	912.031	863.474	3.192	1.929	5.118	11.684	1.279

All instruments have values well below those of experiments 1 and 3, and the course of stabilization policies is more expansionary than in experiment 2. As compared to their desired values, the instruments act in a countercyclical manner, being set at more expansionary values in the first years and at more restrictive values in the last years of the planning period. Comparing the results for the endogenous variables to those from experiment 3, it can be seen that except for public-sector variables, real economic variables grow more over the entire planning period, which is due to the more expansionary (lower) values of budget revenues (BIN_t), which are a rather effective instrument in our model. With respect to prices, this expansionary effect of the revenue side of the budget is compensated by more restrictive money supply ($M1_t$) and budget expenditures (NEX_t), and price variables (including inflation) have values very close to those of the previous experiment. Real GDP growth ($YR\%_t$) is first lower, then higher than in experiment 3, the balance of current account (LBR_t) is lower, and labor market excess supply (UN_t) is similar to that in experiment 3. The deficit of federal budget is reduced over the planning horizon, but it remains above target values until 2000. By and large, these results could also be acceptable to a more "conservative" policy-maker, especially because they imply a smaller public sector than those of experiment 3.

5 CONCLUSIONS

In this paper, we have examined the sensitivity of optimal stabilization policies for Austria during the nineties with respect to assumptions about target values of some objective variables in a framework of quantitative economic policy. The results show that it can be vital to avoid too "pessimistic" or too "optimistic" target values for some variables such as real quantities or prices; for the budget deficit, on the other hand, target values do not matter in our model. It also turns out that there is a trade-off between real growth and full employment on the one hand and price stability and budget consolidation on the other. Much more experiments are required to check whether these results hold true also for other econometric models. However, the present analysis can be seen as a first step towards developing a decision support system for policy-makers who, when confronted with results of alternative optimization experiments, should be assisted in revealing their preferences about macroeconomic policy objectives.

REFERENCES

Chow, G.C. (1975) *Analysis and Control of Dynamic Economic Systems*. Wiley, New York.

Chow, G.C. (1981) *Econometric Analysis by Control Methods*. Wiley, New York.

Kendrick, D. (1981) *Stochastic Control for Economic Models*. McGraw-Hill, New York.

Matulka, J. and Neck, R. (1992) OPTCON: An Algorithm for the Optimal Control of Nonlinear Stochastic Models. *Annals of Operations Research* **37**: 375 - 401.

Neck, R. and Karbuz, S. (1994) Optimal Stabilization Policies for the Nineties: A Simulation Study for Austria, in *Proceedings of the European Simulation Symposium 1994* (ed. A.R. Kaylan et al.), Vol. I, Istanbul.

30

Optimal usage of saline and non saline irrigation water; A policy tool

Arye Sadeh
Technology Management Department,
Center for Technological Education Holon
P.O.B. 305, Holon, Israel. e-mail: sadeh@barley.cteh.ac.il

1 INTRODUCTION

The shortage of good fresh water for agriculture has encouraged farmers to use water with differential quality. The usage of low quality water has three levels of consequences: immediate, intermediate and long run effects. The immediate consequences are on yield levels, intermediate impacts are reversible damages to soils, and the long run effects are on water aquifers and considerable damages to soils. Information about the economic value of using low quality water is of interest when the supply of water is limited and should be allocated among many farmers (Yaron and Voet, 1983).

This study focuses on a constrained optimization model to evaluate the immediate and the intermediate economic values of using differential water qualities for agricultural purposes, under conditions of water shortage.

2 BIO-ECONOMIC SYSTEM

The general problem is set of two sub-models, a biological model and an economic model for an individual decision maker. The biological model consists of two systems: a water-soil system and a soil-plant system.

2.1 The water-soil system

The soil in the root depth zone is the storage for water and salt. At any point of time there is a balance of water and balance of salt in the soil (Bresler et al., 1982). Water and salt are introduced to the soil via irrigation or rainfall. Water can be evaporated from the soil to the atmosphere directly, or to be transpirated via plants. Higher rates of transpiration are associated with better plants. On the other hand water can be percolated downward from the soil root zone, as the amount of water exceeds the soil ability to hold water.

Salt is introduced to soil via irrigation and can be leached downward via percolated water. Farmers may also reduce soil salinity rate by leaching salt below the root depth, using excess of low rate saline irrigation water.

It is assumed that mixture of saline and non saline water can be produced and applied to crops; therefore, infinite number of combinations of amount of water and salt concentrates are considered for irrigation.

2.2 The soil-plant system

Plants have to overcome a combination of matrix and osmotic pressures in their effort to absorb water from the soil. More water in the root depth results in a small matrix pressure. At the same time, a high soil salinity level results in a high osmotic pressure. That is, there is a room for substitution between water content and the rate of salinity in the soil during the growth period. The response of crop to available water is determined (Letey et al., 1986) as well as the response of crops to soil salinity (Maas and Hoffman, 1977). The existence of salt in soil water results in less transpiration. Consequently, more water stay in the soil, the matrix pressure decrease, and a feedback of the salt on crops' growth is revealed.

A seasonal simulation model is developed in order to describe the water-soil sub- system and the soil-plant sub-system. The model simulates the actual yield and soil salinity and soil's water content. The state variables are soil salinity and soil's water content, the output variables are yield, water and salt drainage below the root zone, and water transpirated to the atmosphere. The input variables are amount of water and the salt concentrate. The equations of motion are the responses of crops to water and salt and the dynamics of salt and water in the depth root zone. The initial values of soil salinity and soil water content should be provided along with other parameters of the equations of motions.

The results from the simulation model were incorporated into an economic decision model. A quadratic approximation of yield with respect to relevant ranges of initial values and inputs is used based on simulation's results.

3 ECONOMIC MODEL FOR AN INDIVIDUAL DECISION MAKER

Having the responses of different crops to soil salinity and water, decision maker has to allocate the lands by their initial salinity, and water by their salt content, among crops. The decision maker wants to maximize profits and cannot affect the markets of inputs or outputs. This is a mathematical programming problem with non linear objective and constraints.

3.1 Formulation of the model

A decision maker has soils of different types, irrigation water of different qualities, and a variety of crops as production alternatives.

L types of soils. They differ by their initial soil salinity at the beginning of a growing season.

N different crops (production technologies)

M sources of water differ by their salt concentration and unit cost per cube.

$k = 1, 2, \ldots, L$; k is index of soil type

$i = 1, 2, \ldots, N$; i is index of crop
$j = 1, 2, \ldots, M$; j is index of source of water.
W_{ijk} is the amount of water from source j for crop i on soil type k
Y_{ik} is the yield from one dunam of crop i on soil type k
C_j is the unit cost of water from source j.
P_i is the unit price of 1000 kg yield of crop i
X_{ik} is the number of dunams allocated for crop i of soil type k
TC_{ik} is the total Costs, except water costs, associated with growing crop i on soil type k. These are fixed costs or variable costs that are not functions of yield's levels (Y_{ik}).

Note the followings:
(1) The total amount of water applied to crop i on soil type k is

$$W_{ik} = \sum_{j=1}^{M} W_{ijk}.$$

(2) The concentration of salt is therefore

$$S_{ik} = \frac{\sum_{j=1}^{M} W_{ijk} \cdot S_j}{W_{ik}}$$

The decision variables are X_{ik} and W_{ijk} for all i, j and k.

The objective function is to maximize profits with respect to all X_{ik} and W_{ijk}:

$$\max_{X_{ik}, W_{ijk}} \left[\sum_i \sum_k \left(P_i \cdot X_{ik} \cdot Y_{ik} - \sum_j C_j W_{ijk} - TC_{ik} \right) \right]$$

subject to:

Production relationships:

$$Y_{ik} = Y_{ik}(W_{ik}, S_{ik}, SS_{ik})$$

$$W_{ik} = \sum_{j=1}^{M} W_{ijk}$$

$$S_{ik} = \frac{\sum_{j=1}^{M} W_{ijk} \cdot S_j}{W_{ik}}$$

$$SS_{ik} = f(SS_k, W_{ik}, S_{ik});$$

the relevant average soil salinity level, as a function of initial soil salinity crop characteristics and water and salt irrigated to the soil.

Sources constrains
Water constrains by type of water

$$\sum_i \sum_k X_{ik} \cdot W_{ijk} < Q_j$$

The summations are for a given source of water j.
Land constrains

$$\sum_i X_{ik} < \mathrm{LAND}_k$$

The summation is for a given land of type k.

Other managerial constraints *e.g.* labor constraints could be included.

Note the differences: SS_{ik} is the average soil salinity when producing crop i on soil type k
SS_k is the initial soil salinity of soil type k
S_{ik} is the salt concentrate of water applied to crop i on soil type k.

Number of dunams of each soil type k is an input, while SS_{ik} is solved internally in the simulation model and S_{ik} is a result of mixing (inputs) water of different quality.

4 APPLICATION AND RESULTS

The simulation model was run for cotton and tomatoes in southwest Israel. This is an area with limited rainfall and water is carried to this area from the northern part of Israel. The base scenario includes 4000 dunams with 1.4 million cube of good water. The soil's characteristics and crops' characteristics are adopted from other published experiments' results (Shalhevet, 1994), (Vinten et al., 1991). The production alternatives are cotton and tomatoes. Profits from one ton of cotton and tomatoes are \$650 and \$46, respectively. (excluding water's costs). Unit costs of one meter cube of good water and saline water are 15 cents and 4 cents, respectively.

4.1 Base run

All 4000 dunams and 1.4 million cube of good water are allocated for cropping tomatoes. The yield is 6.175 tons per dunam from 350 cube of good water. The dual costs (shadow price) of soil and water are \$36 and \$0.7, respectively. The objective value is \$1,091,018.

4.2 Second scenario

Half of the 1.4 Million water quote of the decision maker is substituted with saline water (EC = 5 dsm/m), with unit costs of 4 cents per one cube.

1796 dunams with 390 cube per dunam of good water are allocated for tomatoes. Cotton is grown on 2204 dunams with 327 cube/dunam saline water. The yield levels are 6.8 and 0.4 ton/dunam for tomatoes and cotton, respectively. The objective value is \$1,076,053 with \$0.63 and \$0.61 dual costs for good and saline water, respectively. The dual costs for soil is \$50.

4.3 Third scenario

The initial soil salinity in both base and second scenarios is 2 dsm/m. This is the electrical conductivity of the water content in the soil. For the third scenario the initial salinity is set to 3 dsm/m.

Tomatoes are grown on 1740 dunams with 402 cube of good water per dunam yielding 6.77 ton/dunam. On the other hand, cotton is grown on 2260 dunams with 310 cube of saline water, yielding 0.38 ton/dunam. The objective value is \$1,051,126 and the dual costs of soil is \$46.6. The dual costs of good and saline water are \$0.62 and \$0.61, respectively.

4.4 Fourth scenario

The soil salinity of 2000 dunams is set to 2 dsm/m and for the other 2000 dunams it is set to 3 dsm/m. All other parameters are the same as in scenario 3.

The 2000 saline dunams (3 dsm/m) are devoted for cotton along with 315 cube of saline water per dunam. the cotton yield is 0.4 ton/dunam. The 2000 dunams with 2 dsm/m salinity level are allocated to cotton (220 dunams) and tomatoes (1780). About 393 cub/dunam of good water are allocated to tomato, yielding 6.84 ton/ dunam. Only 321 of saline water are allocated for each dunam of cotton yielding 0.4 ton/dunam of cotton. The objective value is \$1,069,605. The dual costs of 2 dsm/m soil and of 3 dsm/m soil are \$51 and \$47.7 respectively. The dual costs of good and saline water are \$0.63 and \$0.61 respectively.

4.5 Policy consequences

The results of the four scenarios can be used to evaluate variables of interest. For example, the difference between the objective values of the base scenario and the second scenario (\$14,965) is a compensation for the farmer for using saline water rather than good water. The difference between the third and the second scenario (\$24,927) is the cost of having a salty soil. This implies that a farmer should be compensated for having salty soils due to a usage of saline water.

Farmers considering two management strategies, using saline water all over their soils or using saline water in particular fields over the years. The first alternative will yield low level saline soils, yet all soils will be somehow salty. The second alternative means that certain soils will be useful for crops that are tolerated to soil, *e.g.* cotton. The difference between the second and the fourth scenario is the costs of having half of the soils salty (3 dsm/m). That is, the previous year's irrigation policy came out with final salty soils.

The dual costs (shadow prices) of each resource constraint are very useful for marginal economic analysis. For example, the difference in shadow prices of saline and non saline, is much higher than those in the third scenario. This implies farmers cannot exercise the economic advantage of tomatoes over cotton. The advantage was revealed in the base run. Rather, the cotton, a tolerant crop toward saline water is a substitution for tomatoes. The high dual cost of water in the base run compared with other run's results, implies that farmers would like to have more good water even if the unit costs increase.

4.6 Recommendations for future research

A beneficial extension of the model will be by incorporating the short run model into a long run framework. This includes the feedback impact of having saline lands at the beginning of each growing season on a general long run objective function. There is a need to incorporate the impact of rainfall and the uncertainty consequences resulted from rainfall on decision making processes.

REFERENCES

Bresler, E., McNeal, B.L. and Carter, D.L. (1982) *Saline and Sodic Soils.* Springer-Verlag, Berlin.

Letey, J. and Dinar, A. (1986) Simulated Crop-Water Production Functions for Several Crops When Irrigated with Saline Water. *Hilgardia*, **54**, 1–32.

Maas, E.V. and Hoffman, G.J. (1977) Crop Salt Tolerance Current Assessment. *J. Irrig. and Drainage Div.*, ASCE, Vol 103 No. IR2, proc. Paper 12993, 115–34.

Shalhevet, J. (1994) Using Water of Marginal Quality for Crop Production: Major Issues. *Agricultural Water Management*, **25**, 233–69.

Vinten, A.J., Frenkel, A., Shalhevet, J. and Elston, D.A. (1991) Calibration and Validation of a Modified Steady State Model for Crop Response to Saline Water Irrigation Under Conditions of Transient Root Zone Salinity. *J. Contam. Hydrol.* **7**, 123–45.

Yaron, D. and Voet, H. (1983) Optimal Irrigation with Dual Quality (Salinity) Water Supply and the Value of Information, in *Agribusiness Planning and Decision Models* (eds. G. Schiefer and C.H. Hanf), Elsevier, Amsterdam.

The author wish to thank the office of the Israeli Water Commissioner, the Committee on Saline Water Evaluation and ARO for their help.

Fuzzy Systems

31

Fuzzy integer sharing problem with fuzzy capacity constraints

Hiroaki Ishii, Takeshi Itoh
Faculty of Engineering, Osaka University
2-1, Yamadagaoka, Suita, Osaka 565, Japan
TEL +81-6-879-7868, FAX +81-6-878-1385
ishiiha@ap.eng.osaka-u.ac.jp, takeshi@ap.eng.osaka-u.ac.jp

Abstract

The sharing problem is a method to find an equitable distribution of resources by maximizing the smallest value of all "tradeoff functions" where a tradeoff function is a function of the flux to a sink node. We generalize some researches about the problem and propose a fuzzy integer sharing problem with fuzzy capacity constraints. Our model has bicriteria, i.e., minimal satisfaction among all fuzzy capacity constraints and that among fluxes to all sink nodes, both of which are to be maximized.

Keywords

Sharing problem, integer flow, fuzzy capacity, bicriteria, nondominated flow pattern

1 INTRODUCTION

The sharing problem, originated by (Brown, 1979), is a method to find an equitable distribution of resources. Its objective is to maximize the smallest value of all "tradeoff functions" where a tradeoff function is a function of the flux to a sink node. In (Tada, Ishii, Nishida and Masuda, 1989), we considered a fuzzy version of the sharing problem by introducing a membership function designating a degree of satisfaction for the flux to each sink node and proposed an efficient algorithm for the problem. While, (Chanas and Kolodziejczyk, 1982) considered a fuzzy version of maximal flow problem by introducing the notion of fuzzy capacity constraint and derived an efficient algorithm for the problem by showing that the maximum flow minimum cut theorem also holds in this case. Further they extended this model to the integer flow case (Chanas and Kolodziejczyk, 1986). This paper combined both models, that is, our earlier model in (Tada, Ishii, Nishida and Masuda, 1989) and Chanas and Kolodziejczyk one and propose a fuzzy integer sharing problem with fuzzy capacity constraints. Our model is bicriteria one and this is another prominent feature of the model. Bicriteria are minimal satisfaction among all fuzzy capacity constraints and that among fluxes to all sink nodes, both of which are to be maximized.

Section 2 formulates our problem and defines nondominated flow patterns since generally speaking, there does not exist an optimal flow pattern maximizing both criteria. Section 3 proposes an algorithm for finding nondominated flow patterns and clarifies its validity. Section 4 discusses its complexity and improvement. Finally, section 5 summarize this paper and discusses further research problem.

2 PROBLEM FORMULATION

Let $G(N, A)$ be a distribution network where N is the set of nodes and A is the set of directed arcs connecting nodes. N includes special nodes, called source nodes and sink nodes. Let S, T be the set of source nodes and that of sink nodes, respectively. We attach to G a super source node s with directed arcs connecting from s to all source nodes and super sink node u with directed arcs connecting from all sink node to u. Let resulting network be $G'(N', A')$. Each arc $(i,j) \in A$ has a fuzzy capacity $\tilde{C}(i,j)$ with the membership function

$$\mu_{ij}(f_{ij}) = \begin{cases} 1 & (f_{ij} < c_{ij}) \\ \dfrac{\bar{c}_{ij} - f_{ij}}{\bar{c}_{ij} - c_{ij}} & (c_{ij} \leq f_{ij} \leq \bar{c}_{ij}) \end{cases} \tag{1}$$

where f_{ij} is a flow value in arc (i,j) and c_{ij}, $\bar{c}_{ij}$ are integers. We assume the capacity of each arc $(s, s_i), s_i \in S$ is ∞. While, the capacity of each arc $(t, u), t \in T$ is one of key points and it may be determined and updated as to maximize the second criterion under satisfaction function $\mu_t(f_t)$ of fluxes f_t of sink nodes $t \in T$ given as follows:

$$\mu_t(f_t) = \begin{cases} 1 & (f_t > b_t) \\ \dfrac{b_t - f_t}{b_t - a_t} & (a_t \leq f_t \leq b_t) \\ 0 & (f_t < a_t) \end{cases} \tag{2}$$

where a_t, b_t are integers. Further we restrict flow values f_{ij}, f_{ss_i}, f_t to be integer. Then we obtain the following bicriteria sharing problem **BSP**.

$$\begin{array}{lll} \textbf{BSP}: & & \\ \text{Maximize} & \min\limits_{(i,j)\in A} \mu_{ij}(f_{ij}), & \text{Maximize } \min\limits_{t\in T} \mu_t(f_t) \\ \text{subject to} & \sum\limits_{i\in N'-\{u\}} f_{ij} = \sum\limits_{k\in N'-\{s\}} f_{jk}, \ j \in N & \\ & f_{ij} : \textit{nonnegative integer}, \ (i,j) \in A & \end{array} \tag{3}$$

We define a flow pattern vector of flow pattern $\boldsymbol{f}$ to be vector

$$\boldsymbol{f} = (f^1, f^2) = (\min_{(i,j)\in A} \mu_{ij}(f_{ij}), \min_{t\in T} \mu_t(f_t)) \tag{4}$$

and nondominated flow patterns as follows.

Definition 1 *If $f_b^1 \leq f_a^1$, $f_b^2 \leq f_a^2$ and $(f_a^1, f_a^2) \neq (f_b^1, f_b^2)$ for $\boldsymbol{f}^a$ and $\boldsymbol{f}^b$, flow pattern $\boldsymbol{f}^a$ dominates flow pattern $\boldsymbol{f}^b$. And, if there exists no flow pattern vector $\boldsymbol{f}'$ that dominates $\boldsymbol{f}$, $\boldsymbol{f}$ is said to be nondominated flow pattern.*

Generally speaking, optimal flow pattern maximizing both criteria does not exist and so we seek nondominated flow patterns in the next section.

3 SOLUTION PROCEDURE FOR BSP

Since all flow values f_{ij}, f_{ss_i}, f_t are integer, we only need to consider integer capacity values. We first solve the following ordinary fuzzy sharing problem $\mathbf{BSP}(1)$ by the algorithm in (Tada, Ishii, Nishida and Masuda, 1989). That is, we first find the nondominated flow pattern $\boldsymbol{f}(1)$ whose corresponding flow pattern vector has the value 1 as a first component.

$$\begin{array}{ll} \mathbf{BSP}(1): & \\ \text{Maximize} & \min_{t \in T} \mu_t(f_t) \\ \text{subject to} & \sum_{i \in N'-\{u\}} f_{ij} = \sum_{k \in N'-\{s\}} f_{jk},\ j \in N \\ & 0 \leq f_{ij} \leq c_{ij},\ (i,j) \in A \end{array} \tag{5}$$

Of course, we must find the maximum flow value $v(1)$ sent from s to u in network $G'(N', A')$ wit the capacities c_{ij}, $(i,j) \in A$ before solving $\mathbf{BSP}(1)$. Then let the optimal value of $\mathbf{BSP}(1)$ be $f(1)^2$. Second, we solve the following fuzzy sharing problem $\mathbf{BSP}(0)$.

$$\begin{array}{ll} \mathbf{BSP}(0): & \\ \text{Maximize} & \min_{t \in T} \mu_t(f_t) \\ \text{subject to} & \sum_{i \in N'-\{u\}} f_{ij} = \sum_{k \in N'-\{s\}} f_{jk},\ j \in N \\ & 0 \leq f_{ij} \leq \bar{c}_{ij},\ (i,j) \in A \end{array} \tag{6}$$

Similarly, we must find the maximum flow value $v(0)$ from s to u in network $G'(N', A')$ with the capacities $\bar{c}_{ij}$, $(i,j) \in A$ before solving $\mathbf{BSP}(0)$. Let an optimal flow pattern and the optimal value of $\mathbf{BSP}(0)$ be $\boldsymbol{f}(0)$ and $f(0)^2$ respectively. Now sorting different $\mu_{ij}(k)$, $k \in (c_{ij}, \bar{c}_{ij})$, $(i,j) \in A$ and let result be $\mu^0 \equiv 1 > \mu^1 > \cdots > \mu^l > \mu^{l+1} \equiv 0$ (l is the number of different $\mu_{ij}(k)$). Then we obtain the following algorithm.

Algorithm

Step 0 : Set $q = 1$, $DS = \{\boldsymbol{f}(1)\}$ and $DV = \{(1, f(1)^2)\}$ and go to **Step 1**.

Step 1 : Solve the following fuzzy sharing problem $\mathbf{BSP}$ (μ^q) by the algorithm in (Tada, Ishii, Nishida and Masuda, 1989).

$$\begin{array}{ll} \mathbf{BSP}(\mu^q): & \\ \text{Maximize} & \min_{t \in T} \mu_t(f_t) \\ \text{subject to} & \sum_{i \in N'-\{u\}} f_{ij} = \sum_{k \in N'-\{s\}} f_{jk},\ j \in N \\ & 0 \leq f_{ij} \leq C_{ij}^q,\ (i,j) \in A \end{array} \tag{7}$$

where $C_{ij}^q = [(1-\mu^q)\bar{c}_{ij} + \mu^q c_{ij}]$, $(i,j) \in A$. Let an optimal flow pattern and the optimal value of $\mathbf{BSP}(\mu^q)$ be $\boldsymbol{f}(\mu^q)$ and $f(\mu^q)^2$ respectively. If flow pattern $\boldsymbol{f}(\mu^q)$ is dominated by some flow pattern of DS, then go to **Step 2**. Otherwise, set $DS = DS \cup \{\boldsymbol{f}(\mu^q)\}$ and $DV = DV \cup \{(\mu^q, f(\mu^q)^2\}$ and go to **Step 2**.

Step 2 : Set $q = q+1$. If $q \neq l+1$, then go to **Step 1**, else check whether $\boldsymbol{f}(0)$ is dominated by some flow pattern of DS. If dominated, terminate. Otherwise, set $DS = DS \cup \{\boldsymbol{f}(0)\}$ and $DV = DV \cup \{(0, f(0)^2)\}$.

Validity of the above algorithm is clear from the fact that it check all possibilities of the first component of nondominated flow pattern vectors and that the greater the flow value sent from s to u in network $G'(N', A')$ then so is $\min_{t\in T} \mu_t(f_t)$.

4 COMPLEXITY AND IMPROVEMENT OF THE ALGORITHM

The algorithm in the previous section is under the following theorem for the computation time.

Theorem 1 *The algorithm obtains nondominated flow patterns in at most*

$$O(L \times \max(\log L,\ |T|^2 cf(n,m))) \quad \textit{computational time,}$$

where $L = \sum_{(i,j)\in A}(\bar{c}_{ij} - c_{ij})$, $n = |N|$, $m = |A|$ *and* $cf(n,m)$ *is a computation time of the maximum flow problem for a graph* (N, A).

Proof

Solving $\mathbf{BSP}(1)$: $O(|T|^2 cf(n,m))$ (Tada, Ishii, Nishida and Masuda, 1989).

Solving $\mathbf{BSP}(0)$: $O(|T|^2 cf(n,m))$.

For an arc $(i,j) \in A$, the number of different $\mu_{ij}(k)$ is $(\bar{c}_{ij} - c_{ij})$ because of treating integer flow. Then it holds

$$l \leq L. \tag{8}$$

Therefore, Sorting $\mu^0, \mu^1, \ldots, \mu^{l+1}$ takes

$$O(L \log L). \tag{9}$$

On **Step 1**, Solving $\mathbf{BSP}(\mu^q)$: $O(|T|^2 cf(n,m))$.

On **Step 2**, without going to **Step 1**, judging whether there exists a flow pattern in DS which dominates $\boldsymbol{f}(\mu^q)$ or not : $O(L+1)$.

As **Step 1** $\sim$ **Step 2** is repeated at most L times by (8), solving $\mathbf{BSP}(\mu^q)$ for all μ takes

$$O(|T|^2 L cf(n,m)). \tag{10}$$

Therefore, the total complexity is

$$O(\max(L \log L,\ |T|^2 Lcf(n,m))) = O(L \times \max(\log L,\ |T|^2 cf(n,m))). \quad (11)$$

□

The computation time in above theorem is calculated under the condition that the maximum flow is obtained from scratch in the solution of **BSP**(μ^q) for every μ. However, current solution may be obtained from the previous solution by a little efforts, because capacities in the network do not decrease. Then the following improvement sophisticates the algorithm. That is, we insert the procedure

> **Step 0' :** Solve **BSP**(μ^1). Set $q = 2$. If $\boldsymbol{f}(\mu^1)$ is dominated by $\boldsymbol{f}(1)$, then go to **Step 1**. Otherwise, set $DS = DS \cup \{\boldsymbol{f}(\mu^1)\}$ and $DV = DV \cup \{(\mu^1, f(\mu^1)^2\}$ and go to **Step 1**.

between **Step 0** and **Step 1**. Of course, "**Step 1**" in **Step 0** should change into "**Step 0'** ".

Theorem 2 *The sophisticated algorithm obtains nondominated flow patterns in at most*

$$O(\max(L \log L,\ |T|^2 cf(n,m),\ |T|^2 Lmn^2)).$$

5 CONCLUSION

There may be many nondominated flow patterns with same corresponding flow pattern vector but our algorithm only find one of them. Further though in the worst case there exist $O(l)$ nondominated flow patterns, we should refine the algorithm by taking not such a case into account since the algorithm check all possibilities in this case also. These are further research problems. Another is extension to the more general case that f_{ij}, f_{ss_i}, f_t are restricted to multiple of some integer d. Besides, we are going to investigate a fuzzy sharing problem by the modal optimization. In the model, we take a fuzzy weight for each sink node into account and formulate the problem as a possibility programming problem.

REFERENCES

Brown, J.R. (1979) The Sharing Problem: *Operations Research*, **27**, 324-340.

Chanas, S. and Kolodziejczyk, W. (1982) Maximum Flow in a Network with Fuzzy Arc Capacities: *Fuzzy Sets and Systems*, **8**, 139-151.

Chanas, S. and Kolodziejczyk, W. (1986) Integer Flows in Network with Fuzzy Capacity Constraints: *Networks*, **16**, 17-31.

Tada, M., Ishii, H., Nishida, T. and Masuda, T. (1989) Fuzzy Sharing Problem: *Fuzzy Sets and Systems*, **33**, 303-313.

32

A fuzzy-PID-concept with minimal rule set

Dr. F. Roß

TU Ilmenau, Fakultät für Informatik und Automation

D-98684 Ilmenau, Germany

Tel (+49 3677) 691417, Fax (+49 3677) 840758

Dipl.-Inf. U. Döring

R3M Softwarebüro

Unterpörlitzer Straße 8, D-98693 Ilmenau, Germany

Tel (+49 3677) 670295, Fax (+49 3677) 840758

Abstract

In this paper an new fuzzy-PID-controller-concept is described. Such fuzzy-controllers consist of only two rules. By the development of such controllers only the membership functions of the inputs and the output must be defined. Also the inference method is reduced, because an operator for the action parts of the rules is not necessary.

Keywords

fuzzy-controller, membership function, inference

1 Introduction

The development of fuzzy systems, consisting of the steps fuzzification-inference-defuzzification, requires the definition of membership functions (MSF), of the inference method by choice of the operators for truth values, of the defuzzification method and last but not least of the rule set. The great number of existing Methods leads to a great number of ways to develop such fuzzy systems. Therefore special fuzzy solutions, e.g. fuzzy controllers, are developed by using of a reduced set of methods. In most cases trapezoids are used for MSF, MIN-MAX or MIN-PROD for inference and the modified center of gravity method for defuzzification. This reduces the development of fuzzy controllers to the definition of the MSF and the rule set.
In this paper we describe a new way for the design of fuzzy controllers, using only two rules. And so only the MSF must be defined.

2 Conventional fuzzy-PID-controllers

Like the classical PID controllers the fuzzy-PID-controllers also use control error e(t), its integral and its derivative as input information. With the usual constraints like trapezoids for the input-MSF, single tones for the output-MSF and symmetry for all MSFs we can define the following rule set for a fuzzy-PID-controller:

IF P=NEGATIVE THEN U=NEGATIVE
IF P=POSITIVE THEN U=POSITIVE
IF I=NEGATIVE THEN U=NEGATIVE
IF I=POSITIVE THEN U=POSITIVE
IF D=NEGATIVE THEN U=NEGATIVE
IF D=POSITIVE THEN U=POSITIVE
IF P=ZERO AND I=ZERO AND D=ZERO THEN U=ZERO

Here we use for the control error (P), its Integral (I) and its derivative (D) MSFs shown in Figure 1 and for the controller output (U) the MSF shown in Figure 2.

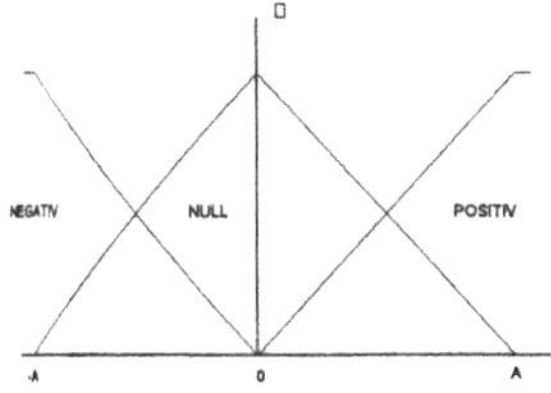

Figure 1 MSF-type for P, I and D.

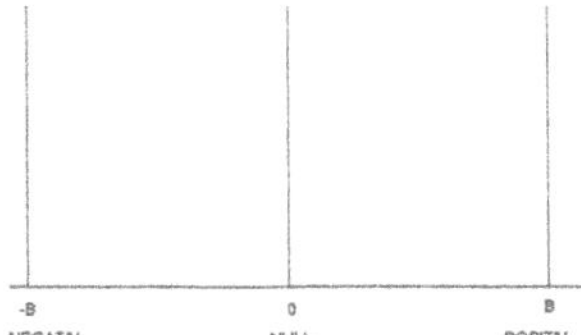

Figure 2 MSF-type for U.

This for the used MSFs minimal rule set is enlarged, if further attributes like VERY NEGATIVE or VERY POSITIVE are included. An essential disadvantage is caused by the used trapezoid form for the input MSF. In this case the transfere behavior is influenced by the break points of the MSFs (Figure 3).

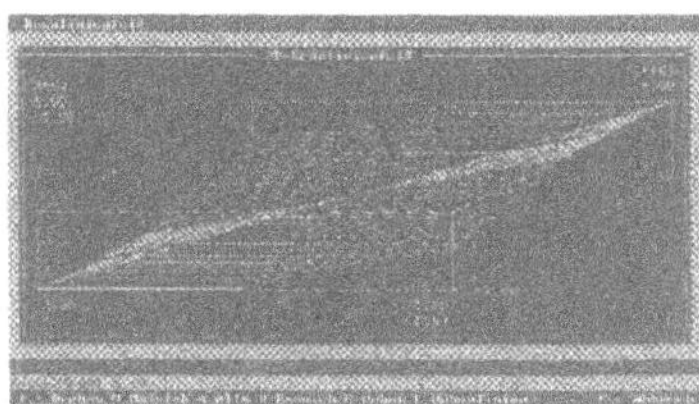

Figure 3 transfere behavior of the conventional fuzzy-PID-controller.

2 Fuzzy-PID-controller with minimal rule set

In the fuzzy-PID-concept shown above the MSFs contain attributes for the working point (P, I and D equal zero) explicitly (in the controller above this attributes are named ZERO). In contrast to this in the new fuzzy-controller we use MSFs shown in Figure 4 and Figure 5.

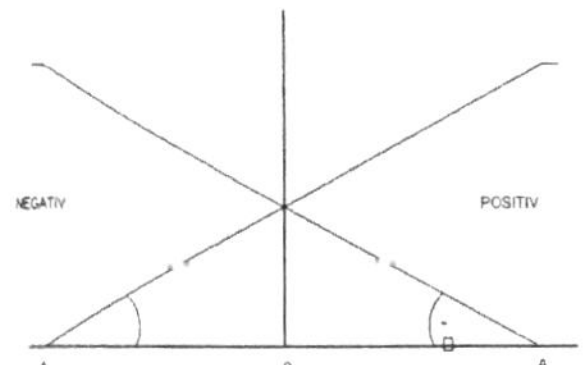

Figure 4 MSF-type for P, I and D.

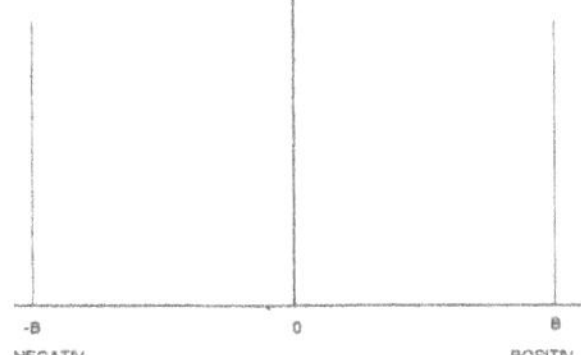

Figure 5 MSF-type for U.

The explicit consideration of the working point is not necessary, because in this point the values of POSITIVE and NEGATIVE are equal and so they compensate each other. After the defuzzification the controller output U gets value 0. Since for the used input MSFs both attributes (POSITIVE, NEGATIVE) in the working range [-A;A] get truth values different from 0 we can define the following minimal rule set:

IF P=NEGATIVE AND I=NEGATIVE AND D=NEGATIVE THEN U=NEGATIVE
IF P=POSITIVE AND I=POSITIVE AND D=POSITIVE THEN U=POSITIVE

Like classical PID-controllers the characteristic of this fuzzy-PID-controller is defined by the choise of the gains for P, I and D. Because we assume the MSFs to be symmetrical the choice of this gains is done by the choice of the angle α for P, I and D each. This way is not possible in available fuzzy development kits. Therefore the gains can also be set by choosing the parameter A. We recomment to use the operators ALGEBRAIC PRODUCT or ARITHMETIC SUM for the determination of the truth values of the conditional part. The offen used operator MINIMUM leads to a loss of information, because only the rule part with the minimal truth value is used. Since the contoller contains only two rules with different attributes in the action part, no combination of there truth values takes place. Therefore no additional operator is needed.
If we use trapezoids for the input MSFs, this fuzzy-PID-concept can be applied only if is ensured that the input is in the range of [-A;A]. To overcome this disadvantage we developed a new MSF-form called X-function. The equation is:

$\mu_{POSITIV} = 1/(2-X \cdot S)$ for $X \leq 0$

$\mu_{POSITIV} = 1-1/(2-X \cdot S)$ for $X > 0$

$\mu_{NEGATIV} = 1-\mu_{POSITIV}$

where S is the rise in the working point and X stands for control error, its integral or its derivativ respectivly (Figure 6).This MSF reaches the value 0 only for infinite values of X. So it is ensured, that really all conditional parts of the both rules ever yield a truth value greater zero an both rules are firing. If the X-function is used, further MSF-attributes like VERY NEGATIVE or VERY

POSITIVE do not increase the rule set. The following rule set can be defined for the MSFs shown in Figure 7.

IF P=NEGATIVE AND P=VERY NEGATIVE AND
I=NEGATIVE AND I=VERY NEGATIVE AND
D=NEGATIVE AND D=VERY NEGATIVE
THEN U=NEGATIVE
IF P=POSITIVE AND VERY POSITIVE AND
I=POSITIVE AND I=VERY POSITIVE AND
D=POSITIVE AND D=VERY POSITIVE
THEN U=POSITIVE

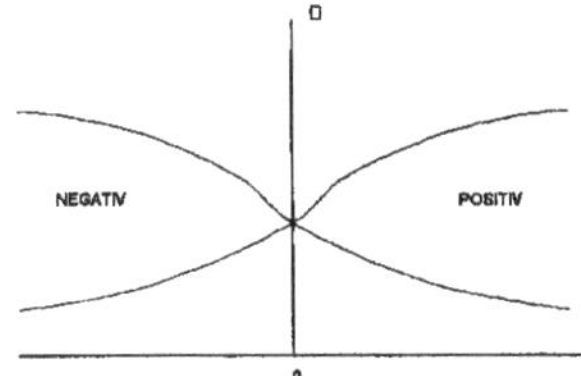

Figure 6 X-function.

Figure 7 enhanced X-function-MSF.

Systems with one output and multiple inputs (MISO) also do not increase the rule set. So an inverse pendulum can be stabilized with the following rules:

IF P1=NEGATIVE AND D1= NEGATIVE AND P2=NEGATIVE AND D2=NEGATIVE THEN U=NEGATIVE
IF P1=POSITIVE AND D1= POSITIVE AND P2=POSITIVE AND D2=POSITIVE THEN U=POSITIVE

Here is: P1 position of the car
D1 velocity of the car
P2 angle of the pendulum
D2 angular velocity of the pendulum

3 Conclusions

In this paper we described a new fuzzy-PID-controller with the following advantages over known fuzzy-PID-concepts:

1. The rule set is minimal and constant.
2. The behavior contains no break points (Figures 8 and 9).
3. More attributes do not increase the number of rules.
4. More system outputs do not increase the number of rules.
5. Only one operator for the determination of truth values is necessary.
6. The controller can be optimized well by known methods.

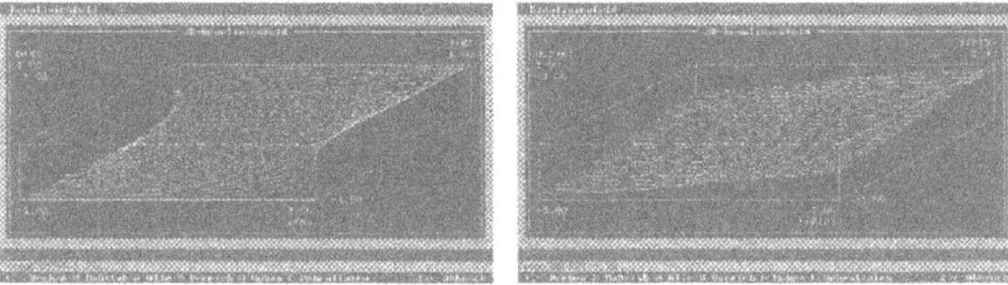

Figures 8 and 9 Behavior of the new fuzzy-PID-controller

The fuzzy-PID-concept described in this paper was used e.g. with an real invers pendulum, by the control of robots and by the on-line adaption of classical PID-controllers. Here we got very good results. The definition, the test and the optimization of such fuzzy-PID-controllers are possible with the fuzzy development tools FuzzyEXPERT and FuzzyEXPERT for Windows.

Game Problems

A numerical procedure for minimizing the maximum cost

Silvia Di Marco, Roberto L.V. González
Universidad Nacional de Rosario
Fac. Cs. Ex., Ing. y Agrim., Universidad Nacional de Rosario,
Avda. Pellegrini 250, (2000) Rosario, Argentina.
Fax: (54) 41 257164. e-mail: Dimarco@bibfei.edu.ar

Abstract
In this paper we consider the numerical solution of a minimax optimal control problem, where the cost to be minimized is the maximum of a function which depends on the state and the control. We present an approximation method which employs both discretization on time and on spatial variables. In this way, we obtain a computational implementable fully discrete problem. We give an optimal estimate for the error between the approximated solution and the optimal cost of the original problem.

Keywords
Minimax problems, numerical solution, quasi-variational inequalities

1 INTRODUCTION AND DESCRIPTION OF THE PROBLEM

We consider in the interval $[0,T]$ a dynamic system which evolves according to the ordinary differential equation

$$\left| \begin{array}{ll} \frac{dy}{ds}(s) = g(y(s),\alpha(s)) & 0 \leq t \leq s \leq T, \\ (t) = x \in \Omega \subseteq \mathrm{R}^r, & \Omega \text{ an open domain.} \end{array} \right. \tag{1}$$

The optimal control problem consists in minimizing the functional J

$$J : (t,x,\alpha(\cdot)) \in [0,T] \times \Omega \times \mathcal{U}(t,T) \mapsto \operatorname{ess\,sup} \{ f(y(s),\alpha(s)) : s \in [t,T) \} . \tag{2}$$

The set of controls is $\mathcal{U}(t,s) = \{\alpha : [t,s] \to A \subset \mathrm{R}^m : \alpha(\cdot) \text{ measurable}\}$.

This problem arises, for example, when we want to minimize the maximum deviation of the controlled trajectories with respect to a given special trajectory. This differs from those problems usually considered in the optimal control literature, where an accumulated cost is minimized. As considering an accumulated cost is not always the best method to qualify a controlled system with an unique scalar parameter, problems of this type has received considerable interest in recent publications (see *e.g.* (Barron and Ishii, 1989)).

The objective of this work is to approximate in a numerical way the value function u

$$u : (t,x) \in [0,T] \times \Omega \mapsto \inf \{J(t,x,\alpha(\cdot)) : \alpha \in \mathcal{U}(t,T)\}. \tag{3}$$

We assume that f and g are bounded and uniformly continuous functions on $\Omega \times A$, A is compact and the trajectory $y(\cdot)$ remains in Ω, for any control belonging to $\mathcal{U}(t,T)$. In that case, the function u has the following properties, which have been established by (Barron and Ishii, 1989):

- u is a bounded and uniformly continuous function and it is also Lipschitz continuous in its second variable.
- u satisfies the following dynamic programming principle $\forall\, t \in [0,T)$, $x \in \Omega$, $s < t$,

$$u(t,x) = \inf \left\{ \max \left\{ \operatorname*{ess\,sup}_{\tau \in [t,s]} f(y(\tau),\alpha(\tau))\,,\ u(s,y(s)) \right\} : \alpha \in \mathcal{U}(t,s), s \in (t,T] \right\},$$

with final condition

$$u(T,x) = \min \{f(x,a) : a \in A\}.$$

The principal results obtained in our work are the following ones:

1. We obtain a discrete time approximation using a finite differences scheme and we give an estimate of the error of this approximation.
2. By using linear finite elements, we obtain a fully discrete approximation that converges to the solution of the original problem with rate $\sqrt{k}$.
3. We show the optimality of the estimation $\sqrt{k}$.

2 A DISCRETE TIME SCHEME OF APPROXIMATION

Following the methodology used in (González and Tidball, 1992), we define an auxiliary problem which is a natural discretization of the optimal cost u defined in (3). We divide the interval $[0,T]$ into μ subintervals with common length $h = T/\mu$. We define recursively, $\forall\, n = 0,\ldots,\mu-1,\ x \in \Omega$.

$$u^h(n,x) = \min_{a \in A} \{\max \{f(x,a), u^h(n+1, x + hg(x,a))\}\}, \tag{4}$$

with the final condition

$$u^h(\mu,x) = \min_{a \in A} f(x,a). \tag{5}$$

The function u^h can be interpreted as the optimal cost function of a discrete time optimal control problem, which is a natural discretization of problem (3). From (4) and

(5), it is easy to prove that u^h is bounded and Lispschitz continuous. The function u^h approximates the function u with a rate of convergence given by the following theorem.

Theorem 1 *Let $u(t,x)$ be the optimal cost of the original problem and $u^h(n,x)$ the discrete time cost defined in (4) – (5), then the following estimate of the difference between u^h and u holds*

$$\left| u(n,x) - u^h(n,x) \right| \leq M\sqrt{h}. \tag{6}$$

3 FULLY DISCRETE SOLUTIONS

With the previous scheme we have approximated the function u with one obtained by discretizing the original problem in its time variable. This approximation scheme is not directly implementable to be computed numerically. To obtain a fully discrete approximation with this property, we discretize the space Ω, using the methodology described in (González and Rofman, 1985) or (González and Tidball, 1992)

3.1 A fully discrete scheme of approximation

We identify the discretization of the spatial variables with the parameter k, which also indicates the size of the discretization. We consider a family of quasi-uniform triangulations of Ω; i.e. a family of polyhedrons Ω_k given by a finite collection of closed simplices $\{S_j^k\}$, such that Ω_k converges to Ω as k goes to 0.

Let $V_k = \{x^i, i = 1, \ldots, N\}$ be the vertices of Ω_k, arbitrarily arranged and N its cardinal. Every $x \in \Omega_k$ is a convex combination of the vertices x^i of the simplex to which x belongs. Hence, $\forall\ a \in A$ there exists a matrix $N \times N$, with components $\gamma_j(x^i,a)$, such that:

$$\gamma_j(x^i,a) \geq 0, \ \sum_{j=1}^{N} \gamma_j(x^i,a) = 1, \ x^i + hg(x^i,a) = \sum_{j=1}^{N} \gamma_j(x^i,a)x^j. \tag{7}$$

We consider the set W_k of functions $w : \Omega_k \to \mathrm{R}$, $w \in W^{1,\infty}(\Omega_k)$, such that $\partial w/\partial x$ is constant in the interior of each simplex of Ω_k, i.e., the functions w are linear finite elements and they are characterized by their values on V_k. We denote $F_k = (W_k)^{\mu+1}$ and the elements of F_k will be denoted $w_k^h(n,x)$, $n = 0, \ldots, \mu$, $x \in V_k$.

Taking in mind the equation (4)-(5), we define the fully discrete solution to be the function $u_k^h \in F_k$ such that, $\forall\ x^i \in V_k$, $u_k^h(\mu, x^i) = \min_{a \in A} f(x^i,a)$ and, $\forall\ n = 0, \ldots, \mu - 1$, it verifies the relation

$$u_k^h(n,x^i) = \min_{a\in A} \left\{ \max \left\{ f(x^i,a), \sum_{j=1}^{N} \gamma_j(x^i,a) u_k^h(n+1,x^j) \right\} \right\}. \tag{8}$$

Obviously, the solutions of these equations are unique and can be computed recursively. This allows us to implement the computational procedure.

3.2 Central result: Rate of convergence

The central result of convergence is given by the following theorem which establishes an estimate of the difference between the optimal cost and the fully discrete solution. The proof is based on regularization techniques; essentially, it consists in obtaining estimates for the differences between subsolutions and supersolutions of problems introduced ad-hoc.

Theorem 2 *If there exist constants c_1 and c_2 such that $c_1 k \leq h \leq c_2 k$, then there exists C such that $\forall\, x \in V_k$, $\forall\, n = 0, \ldots, \mu$, it results*

$$\left|u(nh, x) - u_k^h(n, x)\right| \leq C\sqrt{k}. \tag{9}$$

Note 1 Some computational applications of our discretization procedure have been presented in detail in (Di Marco and González, 1995a).

4 OPTIMALITY OF THE ESTIMATE

In this problem, even though the data f and g are semiconcave in x, it is not possible to improve the estimate $\sqrt{h}$ which appears in (6), as it was done in the problem studied in (González and Tidball, 1992). There, under semiconcavity hypotheses on f and g, it was shown that the optimal cost function u also results semiconcave. In that case, the estimate for $\|u - u^h\|$ can be improved to order h, improvement that was crucial to prove an estimate of type $k^{2/3}$ for the fully discrete approximation. The following example shows that, for the minimax problem, an improvement of this type is not possible.

Let the dynamic of the system be

$$\begin{cases} \dfrac{dy}{dt}(t) &= \begin{pmatrix} 0 & 1 \\ -1 & 0 \end{pmatrix} y(t), \quad \forall\, t \in [0, 10], \\ y(0) &= x \in \mathrm{R}^2. \end{cases} \tag{10}$$

We define the instantaneous cost function as follows

$$f(y(t)) = \left(1 - (y_1^2(t) + y_2^2(t))\right) y_1(t), \tag{11}$$

where $y_i(\cdot)$ is the ith component of $y(\cdot)$.

The system moves freely in R^2 and the functions f and g verify the assumed hypotheses. Clearly, they are semiconcave.

Let $r = \sqrt{x_1^2 + x_2^2}$. It is easy to check, after elementary calculus that the optimal cost function is

$$u(0, x) = \max_{t \in [0,10]} f(y(t)) = \left|1 - r^2\right| r\,, \tag{12}$$

which is not semiconcave.
The discretization procedure introduced here coincides with the methodology studied by (González and Tidball, 1992). In that work, the authors proved that the estimate $\sqrt{k}$ is critical. In fact, here we can prove – using a special triangulation and calculations entirely similar to those employed in (González and Tidball, 1992) – that the error $|u(0,x) - u_k^h(0,x)|$ verifies

$$|u(0,x) - u_k^h(0,x)| \geq C\sqrt{k}. \tag{13}$$

In consequence, we have that for approximations of this type the better error that can be expected, is of order $\sqrt{k}$.

5 GENERALIZATION TO ACCUMULATED COSTS

Similarly to what we have done for function (2), we can deal with the general problem of minimizing a functional $\widehat{J}$ that includes an integral cost. Specifically, we consider the functional

$$\widehat{J}(t,x,\alpha(\cdot)) = \operatorname*{ess\,sup}_{s\in[t,T)} \left\{ f(y(s),\alpha(s)) + \int_t^s b(y(\theta),\alpha(\theta))\,d\theta \right\},$$

where $y(\cdot)$ is the solution of the dynamical system (1) and the function b is bounded and Lipschitz continuous in x. Here, the optimal
cost function u takes the form

$$u : (t,x) \in [0,T] \times \Omega \mapsto \inf\left\{ \widehat{J}(t,x,\alpha(\cdot)) : \alpha \in \mathcal{U}(t,T) \right\}. \tag{14}$$

Similarly to (8), we define recursively the fully discrete approximation of u as follows,

$$\forall\, x^i \in V_k,\ u_k^h(\mu, x^i) = \min_{a\in A} f(x^i, a) \text{ and, } \forall\, n = 0, \ldots, \mu - 1$$

$$u_k^h(n,x^i) = \min_{a\in A}\left\{ \max\left\{ f(x^i,a),\ hb(x^i,a) + \sum_{j=1}^{\mu} \gamma_j(x^i,a)\, u_k^h(n+1,x^j) \right\}\right\}. \tag{15}$$

where $\gamma_j(x^i,a)$ are the functions defined in (7).

Also for this problem, it holds the following convergence result:

Theorem 3 *If there exist constants c_1 and c_2 such that $c_1 k \leq h \leq c_2 k$, then there exists a constant C such that $\forall\, x \in V_k$, $\forall\, n = 0, \ldots, \mu$*

$$\left|u(nh,x) - u_k^h(n,x)\right| \leq C\sqrt{k}. \tag{16}$$

Hence, for the general problem we can get fully discrete approximations and an estimate of convergence of the same order as that obtained for the original problem.

6 FINAL COMMENTS

Here, we have developed a discretization procedure to obtain the numerical solution of the problem of minimizing the maximum cost, analyzed from the continuous point of view by Barron and Ishii (see (Barron and Ishii, 1989)).

The numerical procedure obtained is easily implementable and it converges to the solution of the original problem, with an error estimate of the form

$$|u - u_k^h| \leq C\sqrt{k}. \tag{17}$$

This estimate was shown to be optimal.

The optimality of the estimate (17) stems from the fact that this minimax problem is a disguised differential game problem. In that game, the controller is trying to minimize the cost

$$J(t, x, \alpha(\cdot), \tau) = f(y(\tau), \alpha(\tau)), \tag{18}$$

while a hidden opponent – using full information of actions of the other player – chooses at any instant the stopping time τ of the process. The pay–off (18) is given as out-come of the complete game. As a result of the action of the second privileged player, the first player must – in a strict way – minimize a functional that is not semiconcave with respect to the spatial variable y. In this way, once a fully discrete approximation – using finite differences or finite elements – is applied, the resultant fully discrete optimal control problem reflects this property in the validity of the estimate of type $\sqrt{k}$.

Except for very special trajectories, where some carefully chosen triangulations may be used, it seems not possible in general to get approximations with better convergence properties.

In fact, the error worsens as time grows and for the case $T = \infty$, the phenomenon of non–convergence arises. This undesirable effect is studied more detailed in (Di Marco and González, 1995b).

REFERENCES

Barron, E.N. and Ishii, H. (1989) The Bellman equation for minimizing the maximum cost. *Nonlinear Analysis, Theory, Methods & Applications*, **13**, 1067–90.

Di Marco, S.C. and González, R.L.V. (1995a) A numerical procedure for minimizing the maximum cost. *Rapport de Recherche N° 2454, INRIA*.

Di Marco, S.C. and González, R.L.V. (1995b) A numerical procedure for minimizing the maximum cost - The infinite time problem. *Work in progress.*

González, R.L.V. and Rofman, E. (1985) On deterministic control problems: An Approximation procedure for the optimal cost, Parts 1 and 2. *SIAM Journal on Control and Optimization*, **23**, 242–85.

González, R.L.V. and Tidball, M.M. (1992) Sur l'ordre de convergence des solutions discrétisées en temps et en espace de l'équation de Hamilton–Jacobi. *Comptes Rendus Acad. Sci. Paris*, Tome 314, Série I, 479–82.

34

Game of pursuit with zero stop probability

H. S. Kang
470 Carnegie Dr.
Milpitas, CA 95035, U.S.A.

Abstract

The problem of Hierarchical structure in communicating with distantly located objects is formulated as M player N player nonzero sum game. A index for smallest winning coalition is introduced. Behavior strategy for coalition formed for evasion, and mixed strategy for coalition formed for pursuit are obtained. Repeatability and Precesion are defined for their application in communicating with distant bio objects. Necessary conditions of reachability for distantly located object with zero stop probability is formulated as Bimatrix game. Solution of Stochastic Bimatrix game in finite countable space is obtained. It is shown that mixed strategy used by pursuer over infinitely large partition is also an optimum strategy for stochastic game.

Keywords

Stochastic Games, Decision Making, Hierarchical Structure, Space Technology.

1 INTRODUCTION

With recent advances in space technology there is increased effort to communicate with distantly located intelligent life in space. For its application in communicating with distantly located objects the terms Repeatability and Precision are defined in this writeup. Dynamic and Nondynamic components of system (Figure 1) are dependent on unknown factors (as current state of the object). Problem of Hierarchial structure in communicating with distantly located bio objects sharing common workspace is formulated as rectangular nonzero sum M x N player game. A index measure to induce players to participate in coalition is introduced. Index (Wmin) for smallest Winning coalition is also introduced. Essential games considered in this article may have infinitely many imputations. For reasons of its simplicity saddel point solution for class of games with restricted imputation core (Kang, 1972) is obtained. Necessary conditions for distantly located object may be reachable are obtained. Class of games (Friedman, 1971) with zero stop probability are discussed. Solution of stochastic game over a infinitely large partition is complex (Chris, 1971) and is not discussed in this article. Game of pursuit with zero stop probability is formulated as Bimatrix game over a finite countable space. It is shown that mixed strategy used by pursuer over infinitely large partition is also an optimum strategy for stochastic game. Results are illustrated with two numerical examples. When location of a intelligent object is restricted to a regional area Geosynchronous Satellite may be employed to track the bio object.

2 MATHEMATICAL FORMULATION

Given a dynamic system defined by subgroups of Borel sets (t, U_i, Y, U_o, X). State transition map for the system is given by $k\Delta t \otimes k\Delta t \otimes U_i \otimes Y \otimes U_o \otimes X \to X$, $T: Y \to X$ is given by $[(k+1)\Delta t, k\Delta t, U_i, Y, U_o, X] = T(Y)$, and inverse State transition map for the system is given by $k\Delta t \otimes k\Delta t \otimes U_o \otimes X \otimes U_i \otimes Y \to Y$, $T^{-1}: X \to Y$ is given by $[(k+1)\Delta t, k\Delta t, U_o, X, U_i, Y] = T^{-1}(x)$. Here elements of Borel sets (t, U_i, Y, U_o, X) are further defined as follows: t:$k\Delta t$ is a ordered abelian group for time Set. X : is a abelian group of vectors with $k\Delta t \to X$ a homomorphism of abelian group of vectors with T: $Y \to X$. Abelian group of X is regarded as reflexive over field R with x (-R^n . Y : is a abelian group of vectors with $k\Delta t \to Y$ of reflexive y (-Y, a m tuplet over field R^m with y(-R^m . U_i, U_o: are the abelian group of external stimulants.

3 DEFINITIONS

REPEATABILITY : Set S of reflexive field X is regarded as a Repeatable field if for all x (- S following holds ; $\alpha \neq 0$, $\alpha x \in S$.

For the purpose of application in communicating with distantly located intelligent life, Repeatability is further defined as Topological characterization of object field as it returns to given previously known initial state at the end every day.

PRECISION : Ability of system to track the distantly located living object as it moves to previously known state during assigned workhours.

Precision is of significance in its applications to the concept of Repeatability.

4 MULTI PLAYERS SHARING WORK SPACE

In the following it is assumed N players (Figure 1) in Hierarchical Structure share common work space and are integrated to form a evasion Coalition. It is also assumed M players of

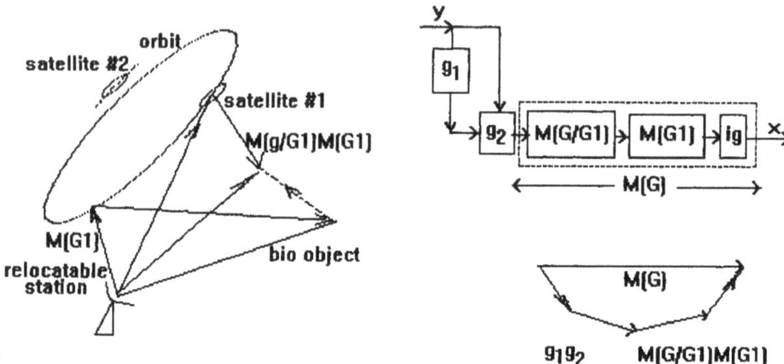

Figure 1. Compounded Transformation (satellite link)

pursuit coalition share the common work space (ref. Kang 1995). Further systems under consideration (ref. Kang 1995) may include relocatable control station and system interface board. Hierarchical structure payoff matrix may include Wmin, index for required smallest number of players combination to form a winning coalition. Modification of payoff matrix during game may serve the purpose of inducing evasion coalition players in pursuit coalition or for the purpose to exclude from evaders coalition. Hierarchical structure may consists of Upper Level the Coordinator, Intermediate level Organisors and Lower level Task Executors. Workspace for the game is topologically characterized as symetric transitive, further isomorphism of game is reflexive. Let probability distribution P(y) for pursuit Coalition Game function $K_j(p)$ in product space Y x X is given by $\int_{y,x} K_i\,(y,x)\,dP(y)\,dx$; j=1,2. In Hierarchical game structure multi level triplet of N players forming Coalition can be combined into larger game over a sum or product field.

4.1 Imputation Core

A game over Ring of field with product binary operation has payoff advantage over game with sum topology. It results in smaller value of Wmin the smallest number combination required to form a winning coalition. For the Hierarchical structure Essential game under consideration (Figure 2), class of core imputations are assumed as restricted.

THEOREM 4: Let η, μ , ξ be the solutions of game with product topology for Upper intermediate and Lower levels of Hierarchical Structure, and let x_j be the solution of K_2 $(y_i, \eta_{ji}) \otimes K_2\,(y_i, \mu_{ji}) \otimes K_2(y_i, \xi_{ji})$. Then Game function K_2 (y,x) for coalition and imputation for evasion are given by

$$K_2(y,x) = K_2(y,\eta_1) \otimes K_2(y,\eta_2) \otimes .. \otimes K_2(y,\mu_1) \otimes K_2(y,\mu_2) \otimes .. \otimes K_2(y,\xi_1) \otimes K_2(y,\xi_2) \qquad (1)$$

$$X = \bigcup_{0<\alpha<1} \{x1(\alpha 1) \oplus x2(\alpha 2) \oplus x3(\alpha 3)\}; \quad \text{with} \sum_{j=1}^{3} \alpha j = 1; \tag{2}$$

EXAMPLE: Consider N x N players game with evasion Coalition payoff (E_u, E_i, E_l) of the Upper Intermediate and Lower levels (Figure 2) of Hirarchical Structure respectively. Let P be the payoff for pursuit coalition. Using Theorem 4 $[E_u, E_i, E_l]_{max}$ and $[E_u, E_i, E_l]_{min}$ are the coalition imputation for evasion strategy. It may be verified that a payoff matrix modification may change imputations in core of Hierarchical structure to $[P, E_i]$ and $[P, E_l]$. Hyperplanes H_i and H_l are boundaries for permissible changes in Wmin index for evasion coalition imputation. In noncooperative game including games with possible defections, with multilevel Hierarchical structure, resulting index Wmin may not be within the permissible boundaries as specified by Hyperplanes. Further from Theorem 4 it can be verified (Figure 2) for a noncooperative games with possible defections imputation index may result in a lower payoff for evader coalition.

4.2 N x N Players Game With Mixed Extension

In Hierarchical structure information available to the Upper Intermediate and Lower levels is not directly available to other levels (Figure 2). The coordinator (dominant player) located at the Upper level acts as a generator for shared common workshape for other levels. Commands flow downwords and index for measure of performance flows upwords. Field of strategies available to Intermediate and Lower level players is dependent on strategy of dominant players.

The concept of Repeatability introduced in this paper considerably simplifies the analysis and design of systems for communicability. When Precision index of the game in the topology characterized by uncountable State space, is maximized over partitions, so formulated Game may have no solution . However, solution for the game exist on subgroup Ring of repeatable field. For Game so formulated Strategy for evader is characterized as Behavior strategy and Strategy for pursuer is mixed strategy. It can be shown mixed extension finite game may have a equilibrium saddle point solution.

5 COMPOUNDED TRANSFORMATION IN REPEATABLE SYSTEMS:

In the following a mathematical formulation of repeatable multistage process is given. Let group G_j be decomposed to sequence of subgroups G_{ji}; i=1, . ,n. Group can be represented by direct product of subgroup $G_j = X^n{}_{i=1} G_{ji}$. Further it is assumed abelian subgroup G_{jl} may be generated by generator $\langle g_{jl} \rangle$. For all g_{jl} (-G_{jl}, the group G_{jl}; may be represented by zone $G_{jl} = (g_{jl}, g_{jl}, . , g_{jl})$. Let m be the index and r be the period for the Repeatability Measure (m,r). Proposition 5.0 is stated in the following .

Proposition 5.0 : Given a repeatable subgroup of group Ring (R,+,.) over which evader employs the behavior strategy ($\eta_{ji}, \mu_{ji}, \xi_{ji}$); i=1, . ,n. There exists an optimal nonzero mixed strategy for the pursuer.

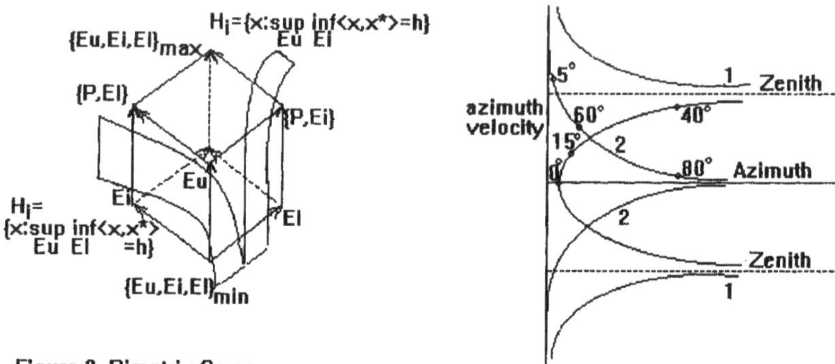

Figure 2. Bimatrix Game (Hierarchical Structure)

Figure 3.
1. zenith vs. azimuth vel.
2. azimuth vs. azimuth vel.

5.1 Design of the Repeatable System

In general system used for communicating with distantly located intelligent objects has space topology characterized by Repeatability measure (m,r) and generator $<g_{jl}>$, let ψ be the annihilator polynomial for the system with m= deg ψ . For repeatable system represented by countable field parameters system mapping matrix are given by

$$F = \begin{vmatrix} 0 & -\alpha m \\ I & B \end{vmatrix} \; ; \quad B = [-\alpha m\text{-}1, -\alpha m\text{-}2, \, .. \, , \alpha 1\,]^t \tag{3}$$

From proposition 5.0 the optimal solution for a repeatable system exists and it can be shown that optimal solution is obtained by minimizing precision index. Proposition 5.1 may be used to select the eigen vectors for the system . λ_0 is assumed to be known apriori.

Proposition 5.1 : Let v(t) and v(s) be the values of the payoff matrix at time t and s . Let the strategy used by the pursuer is completely mixed. Let λ_t and λ_s be the eigen values of the matrix, then

$$v_t / v_s = \lambda_t / \lambda_s \; ; \text{ where } \lambda_s = \lambda_0 , \; v_s = v_0 \; , \text{ at } s=0; \tag{4}$$

Precision index as given in equation (5) can be used to find self tuned controller for relocatable /mobil control station.

$$J[u_i(k)] = \sum_{k=1}^{r} E\,\{ Pii[x_i(k+1) - y_i(k+1)]^2 + Qii[e_i(k)]^2 + Dii[u_i(k)]^2 \} \tag{5}$$

here $e_i(k)$ is the white gaussian noise, P_{ii}, Q_{ii} and D_{ii} are the weighting matrices. Necessary condition for distantly located object with Repeatability Measure (m,r) be reachable is given by theorem 5.1.

THEOREM 5.1. : Let the decomposable group G_j consists of sequence of subgroups G_{ji};i=1,. ,n.Further define a group Ring (R(Gj),+,.). Let S_{ji} be the subset of subgroups G_{ji};i=1,. ,n. It is assumed following axioms are true: (i) $\phi \notin S_{ji}$, (ii) α_{1i} (- Sji, α_{2i} (-S_{ji}

implies $\alpha_{1i} \cap \alpha_{2i}$ (-S_{ji};i=1,. ,n,(iii) $\alpha_{1i} \supset \alpha_{2i}$, α_{2i} (-S_{ji} implies α_{1i} (-S_{ji}, (iv) For all x (-S_{ji} implies x+δx (-Sji, (v) Topological space (x,k Δt) be Hausdroff Space. Then using sequence x_{ji} (-S_{ji},i=1,.. ,n, the object in abelian Gji is reachable, i.e. for all x_l (-S_{ji}, ||x-S_{ji}|| =< ||x-S_{jl}||.

EXAMPLE: To illustrate results stated above consider a Satellite Tracker/Telescope used for studying cesestial bodies (Figure 1a) with system dynamics given by equation (6). It is desired to maximize the reachable zone. System may be configured using compounded transformation (Figure 1b).

$$(dZ/dH) = C1 \text{ Sin } (A) ; \quad (dA/dH) = C2 - C3 \text{ Cot } (Z) \text{ Cos } (A) \tag{6}$$

Here Z and A are the Zenith and Azimuth angles. H is the hour angle mark of planet. It is assumed Z(0) = Z0, A(0)=A0. For the purpose of analysis precision index in C^2 [0,T] space (ref. Gaughan 1968) may further be simplified for maximizing tracking velocity. Frechet differential of equation (6) for maximum azimuth velocity is given by (dA/dH) = 2*C1 (Cos A/Sin2Z), From the results as plotted in Figure 3 it is seen azimuth velocity is very large for low values of Zenith. Azimuth velocity of Tracker/ Telescope tends to be very low as azimuth becomes orthogonal to its axis horizon i.e. when location of a intelligent object is restricted to a regional area Geosynchronous Satellite may be employed to track the bio object. To establish reachability (Theorem 5.1) product space topology may be employed for the system under consideration. From equations (6) we obtain

$$\delta(dZ/dH) = 0 ; \; \delta(dA/dH) = C3 \, [\cot (z+\delta z) \cos (A + \delta A) - \text{Cot } Z \text{ Cos } A \tag{7}$$

For given constrains it can be verified δ(dZ/dH) = δ(dA/dH) = 0. Using convergence in product space as stated in Theorem 5.1, distantly located object in general may be continuously tracked.

6 COMPOUNDED TRANSFORMATION

Decomposition Lemma 6.0 for compounded transformation is stated below: Lemma 6.0 may be used to design and configure a relocatable Control Station/Satellite Link.

Lemma 6.0 : Given a group G_j, with a sequence of subgroups G_{ji};i=1,. ,n, M(G_{ji}) can be decomposed as cascade collection of components given by equation (8).

$$M(G_{ji}) = M(G_i / G_{i-1} .. \; G_1 , G_{i-1}/G_{i-2} .. \; G \; , G_{i-2}/G_{i-3} .. \; G_1 , \; .. \; , G_2/G_1) \tag{8}$$

When G_{ij}; i=1,. ,n are semigroups, equation (8) could further be simplified . Lemma 6.0 may be used for Figure 1 to obtain decomposition of Satellite Link frame system. Decomposed system can be used to design and configure a relocatable Control Station / Satellite Link frame (Figure 1).

7 STOCHASTIC GAME WITH ZERO STOP PROBABILITY :

Recently Kang (1995) has analyzed process for extending reachable zone and use of relocatable control station for the purpose. In this section a pursuit game with zero stop probability is formulated as a Bimatrix game.

7.1 Statement of Stochastic Bimatrix Game Problem

It is assumed generator for sequence of subgroup for Group G consists of Borel sets $<T \times U_i \times Y \times U_o \times X>$. We assume performance index for the system is given by $W(p,u_i,u_o)$ where u_i,u_o are given by axioms stated above. Given the current (i th) state, let p(j/i) be the conditional probability of transition to j th state at following stage. Assume C_j be the return for the j th state, β be the discount factor , S_p be the probability of stop, and (m,r) is the Repeatability measure. Limiting average payoff at j th state is given by equations (10) and conditional probability of exit from repeatable zone is given by equation (11)

$$W_{ij} = \mathop{Lt}_{n \to \infty} \sum_{n=1}^{N} [\beta^n C_j (1-S_p)\{p^j(i+1/i)\, p^j(i+2/i+1) \;..\; p^j(i+n/i+n-1)\}/n] \quad (9)$$

$$W_{ij} = \mathop{Lt}_{n \to \infty} \sum_{n=1}^{N} [\beta^n C_j (1-S_p)\{p(j/i)^{n/m}\} / n] \quad (10)$$

$$p(j/i) = p^j(i+1/i)\, p^j(i+2/i+1) \;..\; p^j(i+m/i+m-1) \quad (11)$$

7.2 Conditional Probability Distribution Matrices

Conditional probability matrices for the State of distant bio objects and for relocatable control station at i th stage are given in tables (7.1 - 7.2). It is assumed measurements for evader are available to pursuer. It is also assumed evader has no access to the measurements for the pursuer state. It is assumed due to restrictions placed by the system dynamics, relocatable control station may only move to neighboring grid node during a state transition. The probability p(y(i+1)/x( )) of pursuer state is given by Table 7.1.

Table 7.1

	y(i-1)	y(i)	y(i+1)
x(i-1)	p(i-1/i-1)	p(i-1/i)	0
x(i) .	p(i-1/i)	p(i/i)	p(i+1/i)
x(i+1)	0	p(i+1/i)	p(i+1/i+1)

Table 7.2

	x(i-1)	x(i)	x(i+1)	x(i-1)	x(i)	x(i+1)
x(i-1)	U(i-1)	U(i-1)	U(i-1)	p(i-1/i-1)	p(i/i-1)	p(i-1/i-1)
x(i) .	U(i)	U(i)	U(i)	p(i-1/i)	p(i/i)	p(i+1/i)
x(i+1)	U(i+1)	U(i+1)	U(i+1)	p(i-1/i+1)	p(i/i+1)	p(i+1/i+1)

$$p\{y(i+1)/x(i+1)\} = p\{y(i+1) / y_i\, x_i\}\, p\{x_i / x(i-1) \;..\; x(0)\}\, p\{y_i / y(i-1) \;..\; y_0\, x(i-1) \;..\; x(0)\} \quad (13)$$

The probability distribution of evader strategy and the probability of evader state as predicted by prsuer at i th state is given by table 7.2 and the probability p(x(i+1)) of evader state as estimated by pursuer is given by equation (14)

$$p\{ x(i+1) \} = p\{ x(i+1) / xi \} p\{xi/x(i-1) x(i-2) \ .. \ x(0) \} \quad (14)$$

Equations (10-14) can be used to compute mixed strategy for the pursuer and Behavior strategy for the evader. Payoff for game can be computed using equation (10).

8 CONCLUSIONS

This article addresses the problem of communicating with intelligent life in distant space. Problem of Hierarchical structure in communicating with distant bio objects is formulated in product space. Necessary conditions that distant bio object may be reachable are obtained. It is shown that given system may be decomposed as cascade collection of decomposed sequence of subgroups. Decomposed sequence so obtained is further used in compounded transformation for solution of process of communicating with distant bio object and implementing a relocatable control station / Satellite Link for system. The problem is formulated as Bimatrix game with zero stop probability in finitely countable space. It is shown saddle point equilibrium obtained over large partition is also the solution for the equivalent stochastic game.Results are illustrated with two numerical examples.

REFERENCES

Kang, H.S. (1972) Optimal Stochastic Control Via Differential Games. *Hawaii Systems Science Confererance, Hawaii, U.S.A.*

Tsokos, C.P., and Padsett, W.J. (1971) Random Integral Equation with Applications to Stochastic Systems. *Springer Verlag, Berlin.*

Gaughan, E. (1968) Introduction to Analysis. *Bokks/Cole Publication Cp, Div. of Wadisworth Publishing Co, Belmont.*

Friedman, A. (1971) Differential Games. *Wieley- Interscience,John Wieley & Sons*, London.

Kang, H.S.. (1970) On the Lyopunov Design of Systems with Zeros in Right Half Plane. *I.E.E.E.* Transactions *on Automatic Control, Vol 2, 1970, U.S.A.*

Kang, H.S. (1995) Canonical Transformation in Modelling Biosystems. *IFAC - 17th IFIP TC7. Conference on Systems Modelling and Otimization, July 1995, Prague, Czech Rep.*

Solution concepts in multicriteria cooperative games without side payments

Lech Kruś, Piotr Bronisz
Systems Research Institute, Polish Academy of Sciences
Newelska 6, 01-447 Warszawa, Poland.
e-mail: krus@ibspan.waw.pl, bronisz@ibspan.waw.pl

Abstract

The paper presents a study of n-person cooperative games without side payments in the case of vector payoffs of players. Recently, much attention has focused on problems with vector payoffs in the field of game theory, since multicriteria models can better apply to real-world situations. In the paper, solution concepts are formulated and analysed. A nucleolus concept is proposed which is invariant on affine transformation of criteria and coincides with the Schmeidler nucleolus in case of unicriteria games with payments and with Raiffa-Kalai-Smorodinsky solution in case of unicriteria bargaining problem.

Keywords

multicriteria decision making, theory of cooperative games, multicriteria optimization methods, decision support systems.

1 INTRODUCTION

The theory of cooperative games has been intensively investigated, especially in the case of games with side payments. The most interesting results can be found in the papers by Shapley, Schmeidler, Aumann, Maschler, to mention only some leading researchers. The theory of the games without side payments has not been so developed, however some important results have been obtained, among others, by Aumann (1961), Peleg (1963), Stearns (1964), Kalai (1975). The theory of cooperative games has been developed under a general assumption that the players outcomes are measured by explicitly given utility function satisfying some assumptions. In practical problems the outcomes are usually measured by some number of criteria, and the utility function are not explicitly given. Papers dealing with cooperative games with multicriteria outcomes of players are relatively rare. We would like to mention the paper by Bergstressen, Yu (1977), where Yu' domination structures were utilized to define and analyse the cooperative games with side payments. Some number of questions to be solved were formulated. However, we have not found in the literature any paper on multicriteria games without side payments.

In this paper we make an attempt to develop the theory of cooperative games without side payments in the case of multicriteria payoffs of the players. We do not assume any a priory given utility function of a player. However it is assumed that every player has his in mind preferences among the criteria. The preferences can be in general context dependent. Similarly as in aspiration function approach Wierzbicki (1982, 1986), we assume the reference points in the spaces of objectives of the players. The points define direction improving the criteria, which are used in an analogous way as in the goal attainment method Gembicki, Haimes (1975) to the sets describing payoffs of coalitions.

The study presented in the paper continues previous research of the authors on the multicriteria bargaining problem and decision support Kruś, Bronisz, Lopuch (1990), Kruś, Bronisz (1993) and on multicriteria noncooperative games Kruś, Bronisz (1994).

2 PROBLEM FORMULATION

Let $N = \{1, 2, \ldots, n\}$ be the finite set of the players, and let $\mathcal{N}$ be the set of all nonempty subsets of N. For any coalition $S \in \mathcal{N}$:
$E^S = \times_{i \in S} E_i$ be a decision space of players in S, where E_i is a decision space of the i-th player $i \in N$.
$G^S = \times_{i \in S} G_i$ be an objective space of players in S, where G_i is a k_i dimensional Euclidean space of outcomes of the i-th player $i \in N$,
$Q^S : E^S \rightarrow G^S$ be a vector valued function defining multicriteria payoffs of players in S, where $Q_i : E_i \rightarrow G_i$ is a vector valued function of the i-th player payoffs.

For simplicity of notation, we assume that each player tries to maximize all his criteria. For any $x = (x_i)_{i \in N} \in G^N$ let $x^S = (x_i)_{i \in S}$ denote payoffs of the players in S, where $x_i = (x_{i1}, x_{i2}, \ldots, x_{ik}) \in G_i = R^{k_i}$.

Let $E_0^S \subset E^S$ be a set of admissible decisions of the players in S. A cooperation of players can be defined by a collection $\{V^S\}_{S \in \mathcal{N}}$ of sets V^S, $V^S \subset G^S$, where $V^S = Q^S(E_0^S)$ denotes the set of attainable multicriteria payoffs of the players in S.

We employ a convention that for $x, y \in R^m$, and for any m:
$x \geq y$ implies $x_i \geq y_i$ for all $i = 1, 2, \ldots, m$,
$x > y$ implies $x_i \geq y_i$, $x \neq y$ for all $i = 1, 2, \ldots, m$,
$x >> y$ implies $x_i > y_i$ for all $i = 1, 2, \ldots, m$

Definition 1

A multicriteria n-person cooperative game without side payments (**n-person MCC game**) is described by a collection $V = \{V^S\}_{S \in \mathcal{N}}$ of sets V^S satisfying the following conditions:

1. V^S is closed and nonempty subset of G^S,
2. V^S is upper bounded, i.e. there exists $x^S \in G^S$ such that $V^S \subset \{y^S \in G^S : y^S \leq x^S\}$,
3. for any $x^S \in G^S$, $y^S \in V^S$, if there is $y^S < x^S$, then $y^S \in int(V^S)$,
4. for each two coalitions S, $T \in \mathcal{N}$, such that $S \cap T = \emptyset$, $V^S \times V^T \subset V^{S \cup T}$.

The formulation of MCC game is closely related to the formulation of cooperative game without side payments given by Aumann (1967).

3 GENERAL MULTICRITERIA SOLUTION CONCEPTS

We denote by Ω a class of all n-person MCC games. A **solution concept** is a function $F : \Omega \rightarrow G^N$, which associates to each game $V \in \Omega$ a set of payoffs $F(V) \subset V^N$.

Definition 2
A **core** of the game V is the set $core(V) = \{x \in V^N :$ for every coalition S there is no $y^S \in V^S$, such that $y_i > x_i$ for every $i \in S\}$.

A payoff belongs to a core if for every coalition, there is no payoff improving at least one criterion of each member.

Definition 3
A function $l_S : G^N \times \Omega \rightarrow R$ will be called an **excess function** for the coalition S if it satisfies the following conditions:

1. If $x, y \in G^N$ are such that $x_i = y_i$ for every $i \in S$, then for every game V,

   $l_S(x, V) = l_S(y, V)$

2. If $x, y \in G^N$ such, that $x_i > y_i$ for every $i \in S$, then for every game V,

   $l_S(x, V) < l_S(y, V)$

3. For any game V, if $x^S \in boundary(V^S)$ then $l_S(x, V) = 0$.
4. $l_S(x, V)$ is continuous jointly with respect to x and V.

The excess function $l_S(x, V)$ reflects the "attitude" of coalition S to the payoff x. The condition 1 means that the excess function does not depend on the remaining players in N. The condition 2 assures that if at least one criterion of some players in payoff increase, then an excess function of the coalition created by them decreases. Condition 3 divide payoffs onto 2 categories: attainable for any coalition S - when $l_S(x, V) \geq 0$ and such that coalition can not assure - when $l_S(x, V) < 0$. From the conditions 2 and 4 we have that if $x, y \in G^N$ are such that $x_i \geq y_i$ for every $i \in S$, then for every game V, $l_s(x, V) \leq l_S(y, V)$.
The proposed conditions are a generalization of the condition imposed on an excess function for classical cooperative game without side payment proposed in Kalai (1975) to MCC game.

Definition 4
A payoff $x \in V$ is called **individually rational** if it belongs to the set

$IR(V) = \{x \in V^N :$ for every $i \in N$ there is no $y \in V^{\{i\}}$ for which $y_i > x_i\}$.

A payoff $x \in V$ are called **group rational** if it belongs to the set

$GR(V) = \{x \in V^N :$ there is no $y \in V^N$ for which $y_i > x_i$ for every i $\in N\}$.

Individual rationality means that no player agrees on a payoff "worse" than he can get acting individually. Group rationality says that each player tries to maximize his payoff.

Theorem 1
For any collection of excess functions $\{l_S\}_{S \in \mathcal{N}}$, for every game V

$$core(V) = \{x \in GR(V) : \ l_S(x, V) \leq 0, \ \text{for every } S \in \mathcal{N} \backslash \{N\}\}.$$

The proof can be found in Kruś, Bronisz (1993).

Definition 5
Let $\Theta(x)$ be the vector in $R^{|\mathcal{N}|-1}$ obtained by arranging values of the excess functions $l_s(x, V)$ of all coalitions S in $\mathcal{N}$, $S \neq N$ in the nonincreasing order. The **nucleolus** is defined by

$$N(V) = \{x \in IR(V) : \ \Theta(x) \leq_{lex} \theta(y) \ \text{for any } y \in IR(V)\}.$$

For any vectors $x, y \in R^m$, $x \leq_{lex} y$ means that $x = y$, or that there is an integer k, $1 \leq k \leq m$, such that $x_i = y_i$, for $1 \leq i < k$ and that $x_k < y_k$.

Theorem 2
For any collection of excess functions $\{l_S\}_{S \in \mathcal{N}}$, for every game V, the nucleolus $N(V)$ is nonempty. Moreover, if the core is nonempty then

$$N(V) \subset core(V).$$

The proof can be found in Kruś, Bronisz (1993).

4 PROPOSED EXCESS FUNCTION, NUCLEOLUS AND THEIR PROPERTIES

Let $\underline{x} = (\underline{x}_1, \underline{x}_2, \ldots \underline{x}_n) \in G^N$ and $\overline{x} = (\overline{x}_1, \overline{x}_2, \ldots, \overline{x}_n) \in G^N$ be given points such that $\underline{x} \in wIR(V)$, $\underline{x}_i \in V^{\{i\}}$ and $\overline{x} >> \underline{x}$. A point $\underline{x}$ can be treated as a preferred payoff of players acting individually, and $\overline{x}$ as a payoff describing the players "aspiration levels". According to aspiration function approach Wierzbicki (1982, 1986), the payoffs $\underline{x}$ and $\overline{x}$ define a desirable improvement direction of the players payoffs. The improvement direction (assuming a normalization among the players) can be formulated by:

$$w(\underline{x}, \overline{x}) \in G^N, \ w(\underline{x}, \overline{x}) = (w_1(x, \overline{x}), \ldots, w_n(\underline{x}, \overline{x})),$$

$$w_i(\underline{x}, \overline{x}) \in G_i = R^{k_i}, \ w_i(\underline{x}, \overline{x}) = (w_{i1}(\underline{x}, \overline{x}), \ldots, w_{ik_i}(\underline{x}, \overline{x})), \ \text{for } i \in N,$$

$$w_{ij}(\underline{x}, \overline{x}) = \frac{\overline{x}_{ij} - \underline{x}_{ij}}{\sum_{j=1}^{k_i} (\overline{x}_{ij} - \underline{x}_{ij})}.$$

It can be easy noticed that for every $i \in N$,

$$\sum_{j=1}^{k_i} w_{ij}(\underline{x}, \overline{x}) = 1.$$

Let $w^S(\underline{x}, \overline{x}) = (w_i(\underline{x}, \overline{x}))_{i \in S} \in G^S$. In our approach, we are looking for solution concepts to the game which depend on preferences of every player. The composition of the players preferences is expressed by the points $\underline{x}$ and $\overline{x}$. The normalization of weights assures anonymity of the players.

For given points $\underline{x}$ and $\overline{x}$, we propose the following function l_S:

$$l_S(x, V) = h_S(x, V, \underline{x}, \overline{x}) = \sup\{t \in R : \ x^S + t \cdot \frac{t^S(\underline{x}, \overline{x})}{s} \circ w^S(\underline{x}, \overline{x}) \in V^S\},$$

where:
s denotes the number of players in S,
$t(\underline{x}, \overline{x}) = (t_1(\underline{x}, \overline{x}), \ldots t_n(\underline{x}, \overline{x})) \in R^n$, $t^S(\underline{x}, \overline{x}) = (t_i(\underline{x}, \overline{x}))_{i \in S}$,
$t_i(\underline{x}, \overline{x}) = \sup\{t \in R : \ (\underline{x}_i + t \cdot w_i(\underline{x}, \overline{x})) \in P^{\{i\}}(V^N)\}$,
$P^S : G^N \rightarrow G^S$ is the projection of G^N on G^S, i.e. $P^S(V^N) = \{x^S : \ x \in V^N\}$,
$t(\underline{x}, \overline{x}) \circ w(\underline{x}, \overline{x}) = (t_1 \cdot w_1(\underline{x}, \overline{x}), \ldots, t_n \cdot w_n(\underline{x}, \overline{x})) \in G^N$,
$t^S(\underline{x}, \overline{x}) \circ w^S(\underline{x}, \overline{x}) = (t_i \cdot w_i(\underline{x}, \overline{x}))_{i \in S} \in G^S$.

It has been shown Kruś, Bronisz (1993), that the following lemma fulfills:

Lemma 1

For given $\underline{x}$ and $\overline{x}$, the function $h_S(x, V, \underline{x}, \overline{x})$ is an excess function for MCC game.

Definition 6

A payoff $u(\underline{x}, \overline{x}) = (u_i(\underline{x}, \overline{x}))_{i \in N}$ is an **utopia payoff** relative to $\underline{x}$ and $\overline{x}$ if for each $i \in N$

$$u_i(\underline{x}, \overline{x}) = \sup\{x_i \in P^{\{i\}}(V^N) : \ x_i = \underline{x}_i + t \cdot (\overline{x}_i - \underline{x}_i) \text{ for some } t \in R\}.$$

It is easy to notice that the following equations are satisfied:

$$u(\underline{x}, \overline{x}) = \underline{x} + t(\underline{x}, \overline{x}) \circ w(\underline{x}, \overline{x}),$$

$$h_S(x, V, \underline{x}, \overline{x}) = h_S(x, V, \underline{x}, u(\underline{x}, \overline{x})),$$

and for any $z \in G^N$ such that $z = \underline{x} + t \cdot w(\underline{x}, \overline{x})$ for some $t \in R$, $t > 0$,

$$h_S(x, V, \underline{x}, \overline{x}) = h_S(x, V, \underline{x}, z).$$

From the equations it follows that the excess function $h_S(x, V, \underline{x}, z)$ depends only on $\underline{x}$ and on direction w generated by $\overline{x}$, but does not depend on value of $\overline{x}$. The players i in N are balanced with weights, which are proportional to the distances of $u_i - \underline{x}_i$. The distance for particular player describes his "bargaining power" in the MCC game.

Definition 7
Let $T = (T_1, \dots, T_n) : G^S \to G^S$ be an arbitrary affine transformation such that $T_{ij}x = (a_{ij} \cdot x_{ij} + b_{ij})$, where $a_{ij} > 0$, for $i \in N$, $j = 1, 2, \dots k_i$. We say that a solution concept $F(V)$ is **invariant under positive affine transformation of criteria** if

$$F(TV) = TF(V).$$

Proposition 1
For given $\underline{x}$ and $\overline{x}$, the nucleolus $N(V, \underline{x}, \overline{x})$ generated by the function $h_S(x, V, \underline{x}, \overline{x})$ is invariant under positive affine transformations of criteria, i.e.

$$N(TV, T\underline{x}, T\overline{x}) = TN(V, \underline{x}, \overline{x}).$$

Proof. Let T be as in the definition 7 and let $T^S = (T_i)_{i \in S} : G^S \to G^S$. To proof the proposition 1 we show that for any coalition S, any $x \in G^N$, and any $t > 0$,

$$T^S x^S + t \cdot \frac{t^S(T\underline{x}, T\overline{x})}{s} \circ w^S(T\underline{x}, T\overline{x}) = T^S \left(x^S + t \cdot \frac{t^S(\underline{x}, \overline{x})}{s} \circ w^S(\underline{x}, \overline{x}) \right).$$

For any $i \in N$ and j such that $1 \le j \le k_i$, from the above equations we have

$$w_{ij}(T\underline{x}, T\overline{x}) = \frac{a_{ij}(\overline{x}_{ij} - \underline{x}_{ij})}{\sum_{j=1}^{k_i} a_{ij}(\overline{x}_{ij} - \underline{x}_{ij})}$$

and

$$\frac{t_i(\underline{x}, \overline{x})}{\sum_{j=1}^{k_i} (\overline{x}_{ij} - \underline{x}_{ij})} = \frac{t_i(T\underline{x}, T\overline{x})}{\sum_{j=1}^{k_i} a_{ij}(\overline{x}_{ij} - \underline{x}_{ij})}.$$

We obtain

$$T_{ij}x_{ij} + t \cdot \frac{t_i(T\underline{x}, T\overline{x})}{s} \cdot w_{ij}(T\underline{x}, T\overline{x}) =$$

$$a_{ij}x_{ij} + b_{ij} + t \cdot \frac{t_i(\underline{x}, \overline{x}) \cdot \sum_{j=1}^{k_i} a_{ij}(\overline{x}_{ij} - \underline{x}_{ij})}{\sum_{j=1}^{k_i} (\overline{x}_{ij} - \underline{x}_{ij}) \cdot s} \cdot \frac{a_{ij}(\overline{x}_{ij} - \underline{x}_{ij})}{\sum_{j=1}^{k_i} a_{ij}(\overline{x}_{ij} - \underline{x}_{ij})} =$$

$$a_{ij} \left(x_{ij} + t \cdot \frac{t_i(\underline{x}, \overline{x})}{s} \cdot \frac{(\overline{x}_{ij} - \underline{x}_{ij})}{\sum_{j=1}^{k_i} (\overline{x}_{ij} - \underline{x}_{ij})} \right) + b_{ij} = T_{ij} \left(x_{ij} + t \cdot \frac{t_i(\underline{x}, \overline{x})}{s} \cdot w_{ij}(\underline{x}, \overline{x}) \right).$$

Therefore the excess function $h_s(x, V, \underline{x}, \overline{x})$ is independent on positive affine transformation of criteria, i.e. for any $S \in \mathcal{N}$ and an arbitrary affine transformation T such as in the definition 7,

$$h_S(Tx, TV, T\underline{x}, T\overline{x}) = h_S(x, V, \underline{x}, \overline{x}).$$

From the definition of the nucleolus we have immediately the result of the proposition.
□

It can be verified that the nucleolus generated by the functions h_S is a generalization to MCC games of the nucleolus originally defined in Schmeidler (1969) for unicriteria, cooperative games with side payments.

Let us assume that the game V is such that subcoalitions containing more then one player and less then n players are trivial, i.e. if $|S| \neq 1$, $|S| \neq N$ then

$$V^S = \times_{i \in S} V^{\{i\}}.$$

Let $(V^N, \underline{x})$ be an n-person multicriteria bargaining problem (MCB problem) defined as in Bronisz, Kruś (1988), Kruś, Bronisz (1993). If the solution concept $f^R(V^N, \underline{x}, u)$ proposed in Bronisz, Kruś (1988) for MCB problem $(V^N, \underline{x})$ is Pareto optimal in V^N then it can be shown that

$$N(V, \underline{x}, \overline{x}) = f^R(V^N, \underline{x}, u(\underline{x}, \overline{x})).$$

The solution $f^R(V^N, \underline{x}, u)$ is a generalization (see Bronisz, Kruś 1988) of the Raiffa-Kalai-Smorodinsky solution originally defined and analysed for classical bargaining problem (i.e. if $k_i = 1$ for every $i \in N$) in Raiffa (1953), Kalai, Smorodinsky (1975), Thomson (1980).

In the unicriteria case (i.e. if $k_i = 1$ for every $i \in N$), if the subcoalitions are trivial, the nucleolus proposed here coincides with the original Imai solution to bargaining problem Imai (1983). The Imai solution lexicographically improves the Raiffa-Kalai-Smorodinsky solution if the last one is not Pareto optimal.

5 CONCLUSIONS

In the paper, multicriteria cooperative games without side payment have been formulated. Concepts of core, excess function and nucleolus are formulated and analysed. In our approach we follow the way applied by Kalai (1975) in the classical case of unicriterial payoffs of players, trying to generalize it on the case of multicriteria games. Main result consists in the proposition of a new excess function which is independent on affine transformations of criteria. The nucleolus generated by the function is invariant on affine transformations of the players criteria. It is shown that the nucleolus coincides with the Schmeidler nucleolus in case of unicriteria games with side payments, with generalized Raiffa-Kalai- Smorodinsky solution proposed and discussed in Kruś, Bronisz (1993), Bronisz, Kruś (1988) in the case of multicriteria bargaining problem and with Raiffa-Kalai-Smorodinsky solution Raiffa (1953); Kalai, Smorodinsky (1975) in the case of classical unicriteria bargaining problem.

Further research will be devoted to construction of interactive procedures supporting players in analysis of the multicriteria game and in selection of an agreeable, consistent to the players preferences payoff. The proposed nucleolus seems to be candidate for such a payoff. Other forms of excess function and nucleolus concepts formulated with use of the functions will be analysed.

REFERENCES

Aumann, R.J. (1961) The Core of Cooperative Games without Side Payments. *Trans. Amer. Math. Soc.*, **98**, 539–52.

Aumann, R.J. and Maschler, M. (1964) The Bargaining Set for Cooperative Games, in *Advances in Game Theory* (eds. M. Dreshler, L.S. Shapley and A.W. Tucker), Annals of Mathematics Studies, No. 52, Princeton University Press, Princeton, New Jersey.

Bergstresser, K., Yu, P.L. (1977) Domination Structures and Multicriteria Problems in N-person Games. *Theory and Decision*, **8**, 5-48.

Bronisz, P., Kruś, L. (1988) Application of Generalized Raiffa Solution to Multicriteria Bargaining Support, in *System Modeling and Optimization* (eds. M. Iri, K. Yajima), *Lecture Notes in Control and Information Sciences*, Vol. 113, Springer-Verlag.

Gembicki, F., Haimes, Y.Y. (1975) Approach to Performance and Multiobjective Sensitive Optimization: the Goal Attainment Method. *IEEE Automatic Control* **AC-20**, No. 6.

Kalai, E. (1975) Excess Functions for Cooperative Games without Sidepayments. *SIAM J. Appl. Math.*, **29**, No. 1.

Kalai, E., Smorodinsky, M. (1975) Other Solutions to Nash's Bargaining Problem. *Econometrica*, **43**, 513–8.

Kruś L., Bronisz, P., Lopuch, B. (1990) MCBARG Enhanced. A System Supporting Multicriteria Bargaining, Collaborative Paper, CP-90-06, IIASA, Laxenburg, Austria.

Kruś L., Bronisz, P. (1993) Some New Results in Interactive Approach to Multicriteria Bargaining, in *User-Oriented Methodology and Techniques of Decision Analysis and Support* (eds. J. Wessels, A.P. Wierzbicki), Lecture Notes in Economics and Mathematical Systems, Vol. 397, Springer-Verlag.

Kruś, L., Bronisz, P. (1993) On Multicriteria Cooperative Games without Side-payments. Research Report, Systems Research Institute, Polish Academy of Sciences, Warsaw, Poland.

Kruś L., Bronisz, P. (1994) On n-person Noncooperative Games Describe in Strategic Form. *Annals of Operation Research.* **51**, J.C. Balzer AG, Sci. Publ., 83–97.

Peleg, B. (1963) Solutions to Cooperative Games without Side Payments. *Trans. Amer. Math. Soc.*, **106**, 280–92.

Raiffa, H. (1953) Arbitration Schemes for Generalized Two-Person Games. *Annals of Mathematics Studies*, **28**, Princeton University Press, Princeton, New Jersey, 361–87.

Stearns, R. (1964) On the Axioms for a Cooperative Game without Side Payments. *Proc. Amer. Math. Soc.*, **15**, 82–86.

Thomson, W. (1980) Two Characterization of the Raiffa Solution. *Economic Letters*, **6**, 225–31.

Wierzbicki, A.P. (1982) A Mathematical Basis for Satisficing Decision Making. *Mathematical Modelling*, **3**, 391–405.

Wierzbicki, A.P. (1986) On the Completeness and Constructiveness of Parametric Characterization to Vector Optimization Problems. *OR-Spectrum*, **8**, 73–87.

Immunology

36

Computer models for maximizing tumor cell kill and for minimizing side effects in radiation therapy

W. Düchting and T. Ginsberg
Department of Electrical Engineering and Computer Science
University of Siegen, 57068 Siegen, Germany.
Tel: 0271/740-4437. Fax: 0271/740-4382.
e-mail: duechting@hrz.uni-siegen.d400.de
W. Ulmer
Max-Planck-Institute of Biophysical, 37073 Göttingen, Germany.

Abstract

Previous studies have shown that systems analysis, control theory and computer science can stimulate new approaches to interpret cancer as an unstable closed-loop control circuit, to study tumor growth, and to optimize tumor treatment. The aim of this paper is: 1. modeling the growth of tumor spheroids; 2. simulating different clinical treatment schedules applied to irradiation of in-vitro tumor spheroids; 3. considering the side effects on normal tissue. A comparison of the simulation results with clinical experience demonstrates that the clinical reality can qualitatively be represented by the model. This method enables a reduction of time-consuming studies prior to clinical therapy.

Keywords

Cancer, computer simulation, fractionation, modeling, radiation therapy, side effects, tumor growth

1 INTRODUCTION

Our research group is intensively involved in the modeling of malignant and normal cell growth and applies methods of control theory. This approach and other aspects have brought about the hypothesis to associate tumor growth with a cellular division control circuit which has become structure-unstable. The scope of these investigations was to stepwise develop numerous models for the description of the chronological and spatial growth of tumors and of the time behavior of normal cell renewal systems (Düchting, 1990). The growth models were extended by a radiobiological dose-response model based on the linear-quadratic approach (Fowler, 1989). Thus, clinical irradiation schemes can be tested by computer experiments at in-vitro tumor spheroids and at normal cells providing a contribution to the optimization of therapy planning.

2 CELL GROWTH AND RENEWAL MODELS

Based on a control model which describes the cell division of a tumor cell (Figure 1) with cellkinetic data and on cell production and interaction rules, we constructed a model simulating the growth of a tumor spheroid (Düchting, 1990) with a steady state volume of about 1 mm^3. If we want to study the radiation side effects produced in normal cells of the target as well, we additionally have to model the renewal of rapidly proliferating cell systems (Figure 2) which are comprising a stem-cell compartment (Düchting et al., 1994), and the renewal of slowly proliferating parenchymal tissue (Figure 3) which mainly consists of a pool of indivisible resting G0-cells (Düchting et al., 1995).

3 IRRADIATION MODEL

In order to construct a model describing radiation treatment, it is necessary to know the number of cells hit by radiation. The computation of the percentage of the cells killed by irradiation is based on the „linear-quadratic model" (Fowler, 1989) with the survival function:

$$S(D) = e^{-\alpha D} * e^{-\beta D^2} \quad (1)$$

In equation 1, D stands for dose, and α, β are symbolizing parameters depending on the kind of cells and on the type of radiation. The number of the cells to be killed can be determined via equation 1. Subsequently, the killing is performed by means of pseudo-random-number generators in the computer model. To become more realistic, the simplified model was enriched by implementing the following additional considerations: repair mechanisms, reoxygenation, lysis, dose-dependent parameters (Düchting et al., 1992, 1994, 1995).

4 SIMULATION OF FRACTIONATED RADIATION THERAPY

It is well known in radiobiology and radiotherapy that, contrary to a single tumor irradiation, a fractionated application offers many advantages. Going into further detail, advanced results of radiobiology provide many rationales for modifications of the daily applied irradiation dose of 1 x 2 Gy (standard fractionation scheme, D_{TOTAL}= 60 Gy) usually found in clinical routine. Thus, the task was to develop a treatment scheme which leads to a maximum cell kill of tumor cells and to a minimum damage of normal cells. Because of the complexity and heterogenity of biological systems, we started with very simple simulation experiments to study clinical treatment schedules applied to in-vitro tumor growth and to normal tissue under different assumptions and restrictions. The simulation experiments were performed on a DEC 3000 Model 400 AXP workstation. The computing time takes about 40 minutes for simulating a

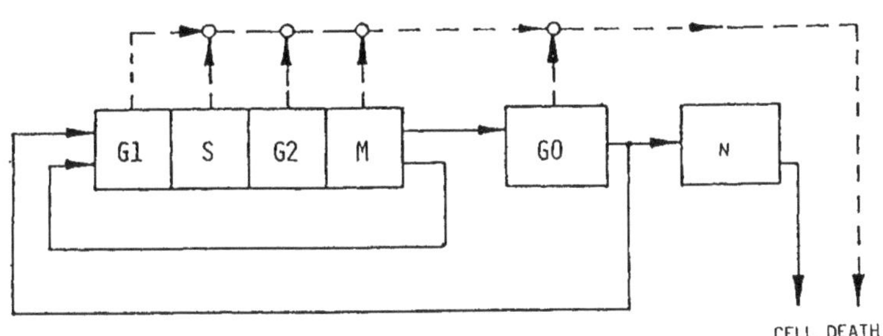

Figure 1 Simplified cytokinetic model of a tumor cell (G1, S, G2, M: Cell cycle phases; Go: Resting phase; N: Necrotic state).

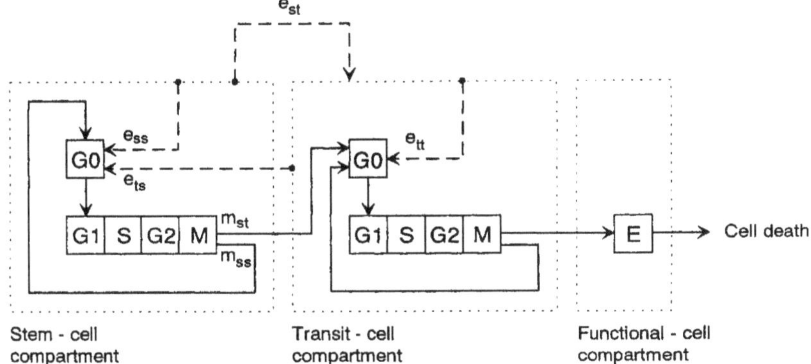

Figure 2 Cellkinetic model of radpidly renewing normal tissue (m_{ss}, m_{st}, e_{ss}, e_{ts}, e_{tt}, e_{st}: Control and regulation variables, see Düchting et al., 1994).

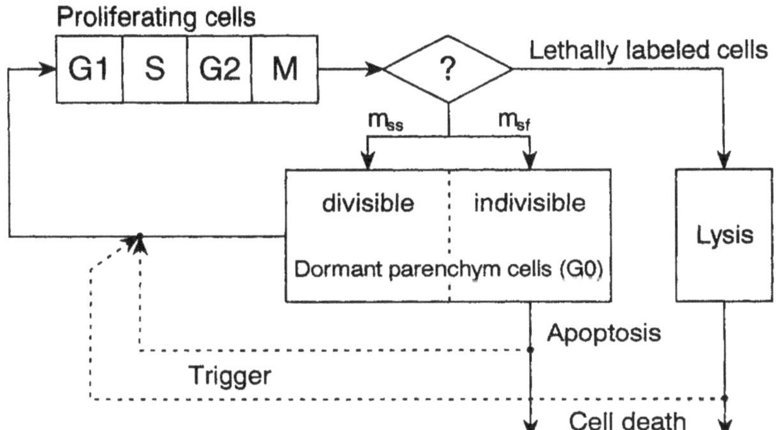

Figure 3 Cellkinetic model of slowly renewing normal tissue (parenchym; m_{ss}, m_{sf}: Control and regulation variables, see Düchting et al., 1995).

single therapy course of about 60 days. In the following sections five different fractionation schemes (Table 1) are simulated for different growing cell systems.

Table 1 Different fractionation schemes

Fractionation schemes		*Dose (Gy)*
Standardfractionation	1	1 x 2 Gy per day
Superfractionation	2	3 single doses per day in an interval of 4 h: 0.7/0.6/0.7 Gy
Hyperfractionation I	3	3 single doses per day in an interval of 4 h: 1/1/1 Gy
Hyperfractionation II	4	3 single doses per day in an interval of 4 h: 1.5/1.5/1.5 Gy
Weekly high single dose	5	1 x 6 Gy per week

4.1 Irradiation of a rapidly growing tumor spheroid

The constructed computer model (Düchting, 1990) based on Figure 1 allows the calculation and representation of the spatial configuration and of the time behavior of the irradiated tumor spheroid. To demonstrate the power of the model, Figure 4 shows the spatial configuration of the hyperfractionated irradiation (scheme 4, Table 1) of a moderately fast growing tumor spheroid (e.g. squamous carcinoma of the lung) at three different points of time starting at t = 500 hours after first dose application. The time course of the number of tumor cells of an irradiated small cell lung carcinoma spheroid is plotted with five different fractionation schemes in Figure 5 (data see Düchting, 1992).

4.2 Irradiation of rapidly renewing normal tissue

In radiation therapy of malignant tumors normal cells are inevitably exposed to ionizing rays, and therefore the effectivity of tumor destruction of a therapy regimen and the associated side-effects produced in normal cells of the target have to be balanced up. The clinical and radiobiological experiences have shown that the irradiation of normal cells has to imply both acute and late effects. Acute effects can already be observed a few days after the corresponding radiation exposition, whereas late responses may be registered after a latency period of 6-12 months. It was clinically verified that acute responses preferrably occur in rapidly renewing tissues (epidermis, mucosa, skin) and radiogenic late effects are preferrably observed in slowly renewing kinds of tissue (e.g. parenchym, connective tissue).

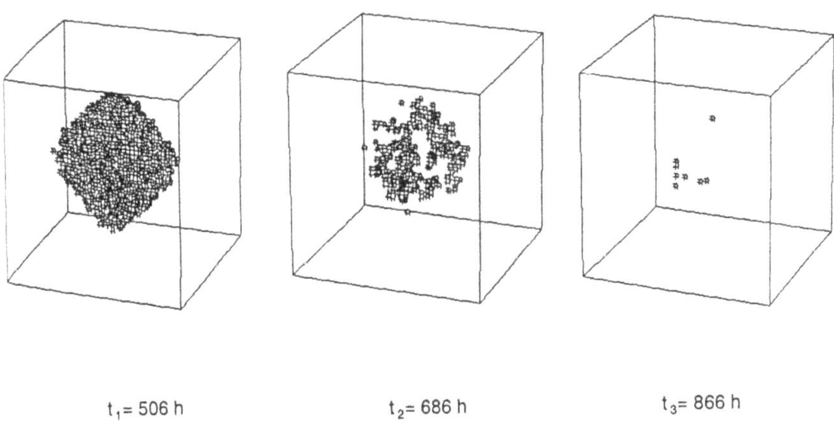

Figure 4 3D illustration of a hyperfractionated irradiation (scheme 3, see Table 1) of a moderately fast growing tumor spheroid (e.g. squamous carcinoma of the lung) in a nutrient medium.

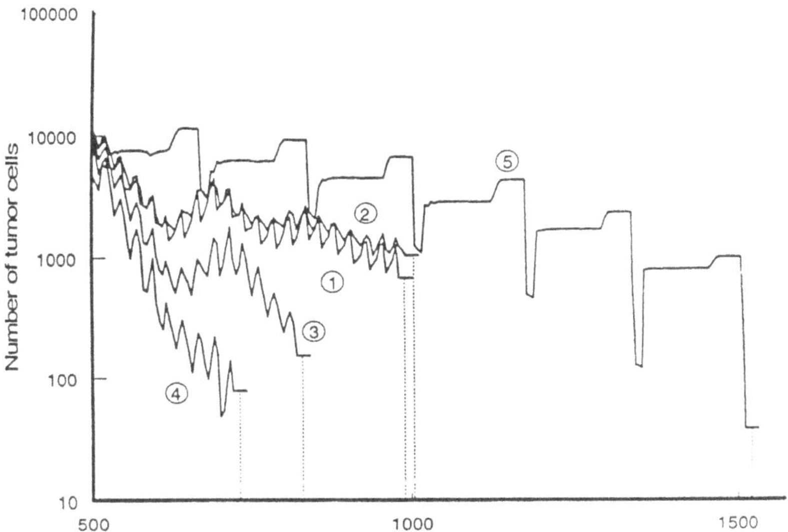

Figure 5 Irradiation of a rapidly growing tumor spheroid (e.g. small cell lung carcinoma) with 5 different fractionation schemes (Table 1; the dotted line indicates the upper limit of dose $D_{TOTAL} = 60$ Gy).

Based on the computer model (Figure 2) describing the dynamics of rapidly proliferating normal tissue (Düchting et al., 1994), Figure 6 represents the simulation run of the radiogenic response of the thin epidermis of the mouse induced by different fractionation schemes (Table 1).

4.3 Irradiation of slowly renewing normal tissue

Late effects appear after a typical latency period of approx. 3-6 months (or sometimes longer) and preferrably occur in slowly renewing normal tissue, e.g. parenchym (brain, lung), connective tissue. While rapidly proliferating cell systems consist of hierarchically structured compartments (stem -, transit -, functional cell compartment), parenchymal tissue contains highly differentiated functional cells residing in the resting phase G0 (Figure 3). The cell space of our computer model is only about 1 mm^3 which implies about 74.000 cells. Therefore, it is tolerated to assume a sufficient nutrient supply of all cells and to neglect an explicite representation of the capillary network. Based on the computer model (Figure 3) describing the dynamics of slowly renewing normal tissue (Düchting et al., 1995), the simulation result in Figure 7 represents the radiogenic response of the brain parenchym of the mouse induced by different fractionation schemes (Table 1).

5 DISCUSSION

Every fractionation scheme can only be evaluated by a juxtaposition of its tumor kill effectivity and severity of side effects in normal tissue. In Figure 5 we have presented a computer simulation result of fractionated radiation therapy applied to a tumor spheroid of a small cell lung carcinoma. A comparison with the results of Figure 6 referring to the same treatment schemes indicates that hyperfractionation II (scheme 4, Table 1) implies the highest tumor cell kill, but also severe acute responses of the thin epidermis of the mouse. Since repopulation in the irradiated target can hardly be expected, this scheme can only be taken into account, if the irradiated areas can be kept small or else the overall dose should be reduced.

Comparing the results of Figure 7 to those of Figures 5 and 6, hyperfractionation II (scheme 4, Table 1) tends in every case to produce severe radiation responses. On the other side, this schedule may be correlated with the most significant tumor kill effectivity. Therefore, the adequacy of this therapy modality is only ensured, if the tumor volume and the associated volume of normal tissue are small in order to prevent severe late and acute side-effects in an extended area. A further possibility is a reduction of the overall dose, e. g. $D_{TOTAL} = 50$ Gy instead of 60 Gy, as considered by some authors.

A difficult problem is the evaluation of scheme 5 (weekly high single dose, scheme 5, Table 1). Since the superior tumor kill effect of scheme 5 in Figure 5 could not be verified by some other authors, the more significant acute and late effects produced by scheme 5 (Figures 6 to 7) should be kept in view.

However, the presented computer simulation models may serve as additional tools for the therapists. On the basis of the experiences of the already published results, our group is paying much effort to develop improved simulation models.

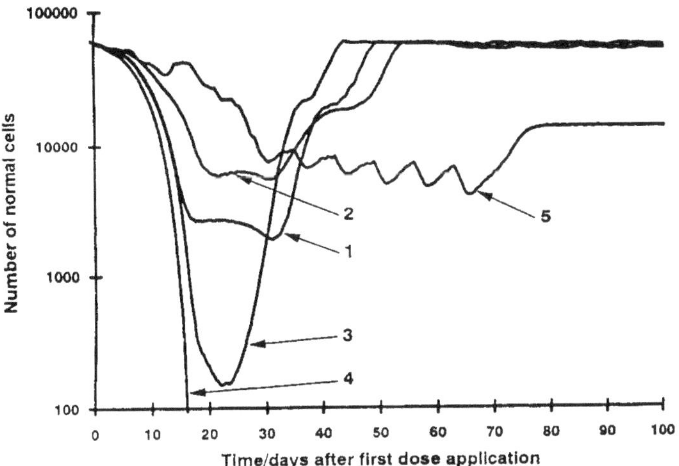

Figure 6 Irradiation of the thin epidermis of the mouse with 5 different fractionation schemes (see Table 1).

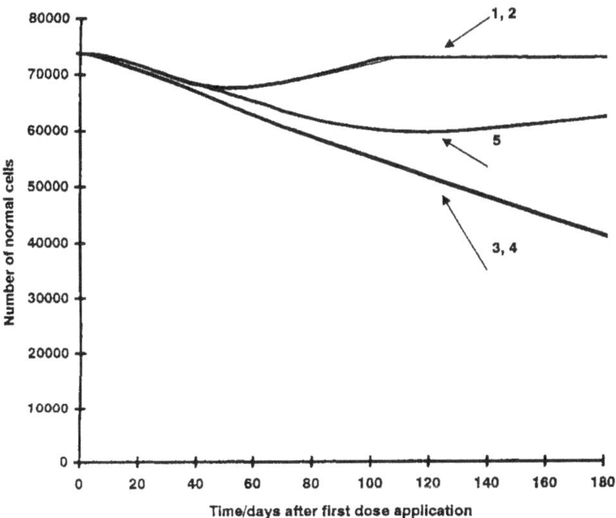

Figure 7 Irradiation of the brain parenchym of the mouse with 5 different fractionation schemes (see Table 1).

REFERENCES

Düchting, W. (1990) Computer simulation in cancer research, in *Advanced Simulation in Biomedicine* (ed. D.P.F. Möller), Springer, New York, 117-139.

Düchting, W., Ulmer, W., Lehrig, R., Ginsberg, T., and Dedeleit, E. (1992) Computer simulation and modelling of tumor spheroid growth and their relevance for optimization of fractionated radiotherapy. *Strahlenther. Onkol.*, **168**, 354-360.

Düchting, W., Ulmer, W., Ginsberg, T., and Saile, C. (1994) Radiogenic responses of normal tissue induced by fractionated irradiation - a simulation study. I Acute responses (submitted to publication).

Düchting, W., Ulmer, W., Ginsberg, T., Kikhounga-N'Got, O., and Saile, C. (1995) Radiogenic responses of normal tissue induced by fractionated irradiation - a simulation study. II Late responses. *Strahlenther. Onkol.* (in press).

Fowler, J.F. (1989) The linear-quadratic formula and progress in fractionated radiotherapy. *Brit. J. Radiol.*, **62**, 679-694.

37

Decision making problems: AIDS prevention and energy development

G. Dzemyda, V. Šaltenis and V. Tiešis
Institute of Mathematics and Informatics
Akademijos St., Vilnius 2600, Lithuania.
Tel.: +370 2 623724; Fax: 729209
e-mail: **dzemyda@osts.mii.lt**

Abstract

The problems of simulation and control of the HIV/AIDS infection spread and search for the optimal energy development strategy were investigated. The decisions in both applications are of great importance, because they have an influence on the main social and economic conditions of the society.

Keywords

HIV/AIDS infection, simulation, multicriteria decision support, energy development.

1 INTRODUCTION

Two problems of decision making are presented in this paper. In both cases optimal decisions were the objective of investigations. The first problem deals with the investigation of the HIV/AIDS infection spread. The goal is to determine the optimal level of preventive means allowing to slow down the infection spread and to plan the activities of health services. The second problem is to search for the optimal energy development strategy. It is necessary to select the best alternative from the finite set. In both applications the human decision is the final. A specialized computer software is developed for each multiple criteria application.

[1]The work was partly supported by the Lithuanian State Science and Education Foundation (grants No.141 and No.143, 1995.03.31).

2 AIDS PREVENTION

Several models are now available for studying AIDS epidemic spread within specific risk groups and among them (Dzemyda, Šaltenis and Tiešis, 1994; Kaplan 1990). These models characterize the dynamics and intercourse within and among the risk groups of individuals (e.g. homosexual men, highly promiscuous heterosexual men and women, drug users) and allow to analyze the means having the greatest influence on the spread rate. Our model describes any risk group by three differential equations, which simulate the dynamics of the amount of active susceptible x_i, active infected y_i, and passive (not spreading the infection) infected individuals z_i:

$$\frac{dx_i}{dt} = u_i^x - (\mu^i + \mu_2^i)x_i - \frac{x_i}{x_i + y_i}\sum_{j=1}^{n} b_{ij} p_{ij} y_j \tag{1}$$

$$\frac{dy_i}{dt} = u_i^y + \frac{x_i}{x_i + y_i}\sum_{j=1}^{n} b_{ij} p_{ij} y_j - (\mu^i + \mu_1 + \mu_2^i + k_i + k_a)y_i \tag{2}$$

$$\frac{dz_i}{dt} = u_i^z + (k_i + \mu_2^i + k_a)y_i - (\mu_i + \mu_1)z_i\,, \quad i = 1,2,\ldots,n, \tag{3}$$

where

n is the number of groups;

u_i^x, u_i^y, u_i^z are the numbers of susceptible, active infected and passive infected individuals, respectively, recruited into the group per time unit; u_i^x depends on the number of adolescents and on the immigration-emigration balance, u_i^y, u_i^z depends on the infection due to foreign relations

μ^i, μ_1 are the mortality rates not due to AIDS and due to AIDS, correspondingly;

μ_2^i is the rate at which an individual naturally becomes inactive (but does not die);

k_i is the rate at which an infected individual becomes inactive because of blood test for HIV;

k_a is the rate of transition from HIV infected to AIDS;

b_{ij} is the averaged number of contacts of an individual from group j with the persons from group i per unit time;

p_{ij} is the probability of infecting a susceptible person from group i by an infected one from group j per contact.

The model (1)-(3) of HIV/AIDS transmission requires to estimate both the initial extent of infection and various rates: mortality, partner-changing, prophylaxis and other ones. Some of these parameters may be evaluated directly from the statistical data. The estimates of transmission probabilities vary in a wide range. For example, the average probability for heterosexuals varies from 0.001 (Longini *et al.*, 1988) to 0.0018 (Laga *et al.*, 1988) or even to overestimated upper bound 0.011 (Goedert *et al.*, 1988), used by Stigum *et al.* (1991).

The balance of contacts must be satisfied:

$N_j \cdot b_{ij} = N_i \cdot b_{ji}$, where $N_j = x_j + y_j$.

The individuals from the same group can have various rates of contacts. Let the percentage of individuals N_j^k has rate b_{ij}^k. May and Anderson (1987) and Jacquez and Simon (1990) have considered weighted mean contact rate:

$$b_{ij} = \frac{\sum_k (b_{ij}^k)^2 N_i^k}{\sum_k b_{ij}^k N_i^k}$$

The main control parameters are such:

- K^i is the percentage of persons tested for the AIDS per year;
- α_i is the percentage of protected sexual contacts.

These parameters are related with the model (1)-(3) as follows:

$$k_i = v^i K^i / 10000,$$
$$p_{ij} = p_{ij}^* (1 + \alpha^j (r_{ij} - 1) / 100),$$

where

- v^i is the annual percentage of individuals terminating the infection spread after diagnosing it;
- p_{ij}^* is the transmission probability per unprotected contact;
- r_{ij} is the probability of protection failure.

In the beginning of the infection spread the parameters p_{ij}^*, k_a and μ_1 must be treated as increasing with time, because the percentage of individuals with later stages of HIV increases (Longini *et al.*, 1988). Their values are minimal at the beginning of infection spread and increase until some value. Dzemyda, Šaltenis and Tiešis (1993) proposed to use linear increasing function approximating such an increase.

The aim of investigations was to determine the optimal values of two control parameters (the percentage of HIV/AIDS blood tests K^i and the percentage of usage of preventive means α_i) when the HIV/AIDS infection may be stopped. Both computational experiments and analytical investigations were performed

Let us consider a case without an import of the infection, i.e. $u_i^y = 0$. Let $y = \sum_{i=1}^{n} y_i$, $\mu_s^i = \mu^i + \mu_1^i + \mu_2^i$. Then we have, that $dy/dt \le 0$ for any x_i and y_i, if and only if

$$\sum_{i=1}^{n} b_{ij} p_{ij}^{*} - \mu_s^j - k_a \leq k_j + (\alpha^j/100) \sum_{i=1}^{n} b_{ij} p_{ij}^{*} (1 - r_{ij}), \ j = \overline{1,n}. \quad (4)$$

That is, the spread of infection decreases in the case (4). In Figure 1 the points mark the area where (4) is satisfied for Lithuanian homosexuals. The values of transmission probabilities were assumed to be equal to 0.01 and 0.002 for single anal and heterosexual contact, respectively. On the left the rather optimistic case is shown assuming that the HIV positive person decreases his personal spread in 10 times (ν^1=90%) after diagnosis the infection. The case for ν^1=50% is shown on the right. The present level of AIDS tests is K^1=6%. Most likely this level will remain the same in the near future. From Figure 1 it follows that the necessary level of protected sexual contacts is 65% when v lies in the interval [50, 90].

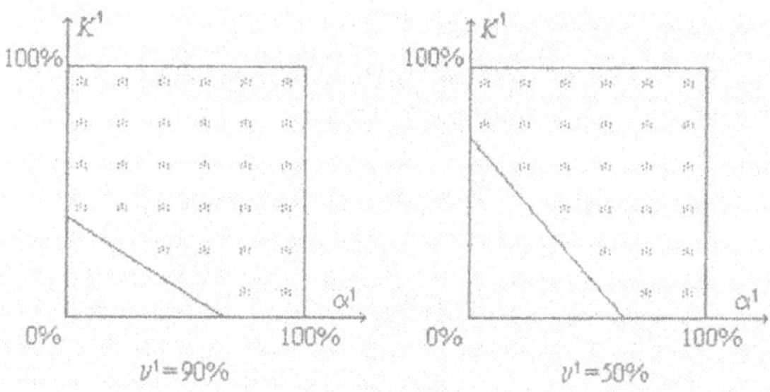

Figure 1 The regions were HIV infection decreases.

The economical and social prices of both control parameters (the percentage of HIV/AIDS blood tests K^i and the percentage of usage of preventive means α_i) grow with an increase of their values. Therefore, the boundary (4) dividing the definition area into the regions of increase and stabilization of infection, may be treated as the Pareto set of optimal solutions. The problem is to minimize the total cost of preventive means.

The infection spread in Lithuania was investigated by means of computational experiments for n=3 (homo/bisexual men, promiscuous heterosexual men and women). The values of α_i and K^i were varied. Assumptions were made that the usage of protective means during sexual contacts α ($\alpha = \alpha^i$, $i = 1,2,\ldots,n$) grows from the initial value α_1=10% to α_2 that is the forecasted value of α at the end of the simulation period. If the upper bounds of transmission probabilities are used, then the simulation shows that the infection spread may be stoped when α_2=85% and 20% of the every risk group undergo the HIV/AIDS blood tests (Dzemyda, Šaltenis and Tiešis, 1994). In the case of present level of HIV/AIDS tests (K^i=6%) the level of α_2 stabilizing the infection spread is equal to 65% if the medium probabilities are used in the model (the average transmission probability for heterosexuals is equal to 0.002). These results correspond with that obtained analysing (4).

3 MULTIPLE CRITERIA DECISION SUPPORT IN ENERGY DEVELOPMENT

The multiple criteria support system (Dzemyda and Šaltenis, 1994) solves the problem of selecting the best alternative (plan, item, option, candidate, and so on) from the finite set. Any alternative is characterized by several criteria. The criteria may be expressed as some values (e.g. capacity, power, weight), but abstract, verbal criteria (e.g. requirement, feasibility, social effect) are possible, too. The system is used to solve the problem with the help of judges (experts, voters). It allows to systematize the judge's knowledge and experience.

Three methods of increasing complexity are realized by the system:

- paired comparisons of alternatives (see Karpak and Zionts (1989)),
- Pareto (see Karpak and Zionts (1989)),
- Fuzzy (see Zhang Li Li and Chang Da Young (1992)).

The method of paired comparisons is the simplest one. The judge must only compare the alternatives, two at a time, and determine which of the two alternatives is better (or equivalent).

The Pareto method offers the Pareto subset of alternatives to the judge. He varies the weights of criteria and looks for the best alternative from this subset.

The Fuzzy method is similar to that of Pareto, only the judge has an opportunity to doubt as to his opinion. Naturally, such an approach requires for much more information from the judge and is most complicated. The reasons for which the Fuzzy approach may be more adequate are:

- uncertainty of a decision maker as to his preferences (hesitation),
- lack of information,
- existence of different opinions (in the group choice).

The procedure of normalizing the criteria values (within the interval 1-10) must be used by the judge prior to the usage of the methods (except the paired comparisons method). It is essential in the case of verbal criteria.

The decision support methods are linked by the common use of data. That provides some integration among these methods. Each part of the system uses the following information from the specialized data base:

- names of alternatives.
- names and values of criteria;
- names of judges;
- intermediate dialog actions of each judge;
- results of each judge;
- integral results.

The flow chart of the system is given in Figure 2.

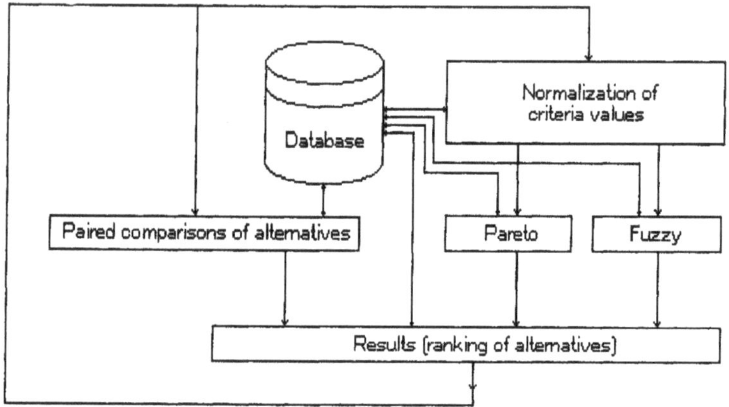

Figure 2 Flow chart of the system.

The main applications of the system were in the scope of decisions on the best energy development strategy for Lithuania. The urgent energy problems are related to the nuclear safety and night power consumption of Ignalina Nuclear Plant (2500 MW). The alternatives covered various scenarios of nuclear plant development, fuel import, electricity export, environmental impact, and so on.

Table 1 of alternatives illustrates the problem of reconstruction of the Kruonis Pumped Storage Hydro Power Plant (Hydro), Ignalina Nuclear Plant (Nuclear) and installation of the night electric heaters (Heaters).

Table 1. The description of alternatives.

Short name of alternative	Number of turbines in Hydro	Number of reactors in Nuclear	Power of Heaters (MW)
H-2, A-2	2	2	-
H-3, A-2	3	2	-
H-4, A-2	4	2	-
H-0, A-1	-	1	-
H-2, A-2, 400	2	2	400
H-0, A-2, 800	-	2	800

Ten criteria (export of energy, fuel saving, degree of nuclear risk, and so on) were used by judges to solve the problem of reconstruction of the Pumped Storage Hydro Power Plant, Ignalina Nuclear Power Plant, and installation of the night electric heaters.

The result was the ranking of alternatives by various methods. It is presented in Table 2.

Table 2. The ranking of alternatives: weights and order numbers.

Short name of alternative	Methods		
	Paired comparisons	Pareto	Fuzzy
H-2, A-2	1.00 (1)	0.96 (2)	0.94 (4)
H-3, A-2	0.83 (2)	0.96 (3)	0.95 (3)
H-4, A-2	0.67 (4)	0.96 (4)	0.95 (2)
H-0, A-1	0.50 (6)	0.90 (5)	0.78 (6)
H-2, A-2, 400	0.67 (5)	1.00 (1)	1.00 (1)
H-0, A-2, 800	0.83 (3)	0.87 (6)	0.79 (5)

The software for decision support is meant for general purposes. It may be successfully used to solve any other applied problem of similar type.

4 CONCLUSIONS

The decisions in both applications are of great importance, because they have an influence on the main future social and economic conditions of the society. Therefore the results of the epidemic spread simulation or the evaluation of the energy alternatives are not used directly. They may be treated as well founded advice for decision makers.

The software is available to PC AT of all modifications. It is suitable to the modifications of models and input data during the investigations.

REFERENCES

Dzemyda, G., Šaltenis V. (1994) Multiple criteria decision support system: methods, user's interface and applications. *Informatica*, **5**(1-2), 31-42, Vilnius.

Dzemyda, G., Šaltenis, V. and Tiešis, V. (1993) Control of the HIV/AIDS infection spread. *12th IFAC World Congress, Preprints of Papers*, Vol.4, Sydney.

Dzemyda, G., Šaltenis, V. and Tiešis, V. (1994) Means for the HIV/AIDS infection control. *Proceedings of the Asian Control Conference*, **1**, 13-16, Tokyo.

Goedert, J.J., Eyster, M.E., Ragni, M.V., Biggar, R.J. and Gail, M.H. (1988). Rate of heterosexual HIV transmission and associated risk with HIV-antigen. *IV-International Conference on AIDS*, Abstract 4019, Stockholm.

Jacquez, J.A. and Simon, C.P. (1990) AIDS: The epidemiological significance of two different mean rates of partner change. *IMA Journal of Mathematics Applied in Medicine and Biology*, **7**, 27-32.

Kaplan, E.H. (1990) An overview of AIDS modeling. *New Directions for Program Evaluation*, **46**, 23-36.

Laga, M., Taelman, H., Bonneux, L., Cornet, P., Vercauteren, G. and Piot, P. (1988) Risk factors for HIV infection in heterosexual partners of HIV infected Africans and Europeans. *IV-International Conference on AIDS*, Abstract 4004, Stockholm.

Longini, I.M., Clark, W.S., Haber, M. and Horsburgh, R. (1989). The stages of HIV infection waiting times and infection transmission probabilities. In *Lecture Notes in Biomathematics*, **83**, 112-137.

May, R.M. and Anderson, R.M. (1987) Transmission dynamics of HIV infection. *Nature,* **326**, 137-142.

Stigum, H., Gronnesby, J.K., Magnus, P., Sundet, J.M. and Bakketeig L.S (1991). The potential for spread of HIV in the heterosexual population in Norway: a model study. *Statistics in Medicine,* **10**, 1003-1023.

Zhang Li Li and Cheng De Yong (1992). Extent analysis and synthetic decision. In *Support Systems for Decision and Negotiation Processes, Preprints of the IFAC/IFORS/IIASA/TIMS Workshop*, Vol. 2. System Research Institute, Warsaw.

Karpak, B. and Zionts, S. (Eds., 1989) *Multiple Criteria Decision Making and Risk Analyzis Using Microcomputers.* NATO ASI Series, Series F: Computer and System Sciences, Vol. F56. Springer.

38

A mathematical model of HIV infection: The role of CD8$^+$ lymphocytes

Tomáš Hraba
Institute of Molecular Genetics
Academy of Sciences of the Czech Republic CZ-16637 Prague, Czech Republic. Tel: 42 2 3312282. Fax: 42 2 24310955.
e-mail: hraba@img.cas.cz

Jaroslav Doležal
Institute of Information Theory and Automation
Academy of Sciences of the Czech Republic, CZ-18208 Prague, Czech Republic. Tel: 42 2 66052062. Fax: 42 2 824755.
e-mail: dolezal@utia.cas.cz

Abstract

The previously developed mathematical model of CD4$^+$ lymphocyte dynamics in HIV infection simulated the respective lymphocyte dynamics well, however simulation of CD8$^+$ lymphocyte values was not satisfactory. Modifications of the model were attempted and satisfactory results were obtained, if the influx of immature CD4$^+$ and CD8$^+$ lymphocytes was assumed to be constrained by HIV infection.

Keywords

Mathematical model, HIV infection, CD4$^+$ lymphocyte depletion, CD8$^+$ lymphocyte dynamics

1 INTRODUCTION

Our mathematical model of CD4$^+$ lymphocyte dynamics in HIV infection (Hraba et al., 1990) assumes a helper T cell-dependent immune reaction limiting HIV proliferation. It simulates successfully the dynamics of CD4$^+$ lymphocyte depletion not only qualitatively, but fits the observed data quite well (Doležal and Hraba, 1991).

As CD4$^+$ lymphocyte numbers are a good prognostic indicator of HIV infection, the model is suitable for simulations of different therapeutic treatments. We have used the mathematical model for simulating the effects of zinovudine therapy on this infection (Hraba and Doležal, 1994), immunotherapy (Doležal and Hraba, 1994a) and of anti-CD8 antibody administration (Hraba and Doležal, 1995a). The latest treatment was suggested (Adleman and Wofsy, 1993), because it seemed highly probable that a homeostatic mechanism regulating T cell numbers did not discriminate between the loss of CD4$^+$ or CD8$^+$ lymphocytes. It was therefore assumed that activation of this mechanism by CD4$^+$ lymphocyte depletion might be inhibited in HIV-infected individuals by increased numbers of CD8$^+$ cells. In consequence, it was suggested to lower the CD8$^+$ cell numbers by administering anti-CD8 antibodies, and in this way, to increase CD4$^+$ lymphocyte level by a stronger influx of new T cells. The model could simulate the effect of anti-CD8 antibody administration, because it incorporated such homeostatic mechanism regulating T cell level (Hraba et al., 1990).

Simulations carried out with the originally described model showed that CD4$^+$ lymphocyte depletion was paralleled by CD8$^+$ cell number increase (Hraba et al., 1990). Detailed studies were carried out to compare simulated CD4$^+$ lymphocyte dynamics with clinical data (Doležal and Hraba, 1991). At that time no analogous attempts were made as far as CD8$^+$ lymphocytes were concerned. Quantitative comparisons of simulated and observed CD8$^+$ lymphocyte dynamics are carried out in this paper.

2 MODEL

The model considers immature and mature CD4$^+$ ($\bar{P}$ and P cells) and CD8$^+$ lymphocytes ($\bar{R}$ and R cells). As normal values of R cells equal approximately 2/3 of those of P cells, it is assumed that also normal $\bar{R}$ values correspond in a similar way to 2/3 of $\bar{P}$ cells. The sizes of these cell compartments at time t are described by Eqs (1)-(4). The amount of HIV products at time t is given by Eq. (5). Finally, Eq. (6) gives the number of cytotoxic T cells specific for HIV (C cells) at time t. In the model used in this paper, and in the preceding ones simulating therapeutic interventions in HIV infection, these cells both limit proliferation of HIV, as indicated in Eq. (5), and effect destruction of CD4$^+$ cells presenting HIV products according to Eqs (1)-(2).

$$\frac{d\bar{P}(t)}{dt} = \frac{I_P + f[(P_0 - P(t)) + (R_0 - R(t))]}{d(t)} - \bar{\tau}_P \bar{P}(t) - \bar{c}_P a(t) C(t) \bar{P}(t), \quad \bar{P}(0) = \bar{P}_0 \tag{1}$$

$$\frac{dP(t)}{dt} = \bar{\tau}_P \bar{P}(t) - \tau_P P(t) - c_P a(t) C(t) P(t), \quad P(0) = P_0 \tag{2}$$

$$\frac{d\bar{R}(t)}{dt} = \frac{2}{3}\, \frac{I_P + f[(P_0 - P(t)) + (R_0 - R(t))]}{d(t)} - \bar{\tau}_R \bar{R}(t), \quad \bar{R}(0) = \frac{2}{3}\bar{P}_0, \tag{3}$$

$$\frac{dR(t)}{dt} = \bar{\tau}_R \bar{R}(t) - \tau_R R(t), \quad R(0) = \frac{2}{3} P_0 \tag{4}$$

$$\frac{da(t)}{dt} = a(t)\,[\theta - \gamma\, C(t)], \quad a(0) = a_0 \tag{5}$$

$$\frac{dC(t)}{dt} = a(t)\,[\varepsilon\, I_C + \alpha\, C(t)] \left(\frac{P(t)}{P_0}\right)^{\nu} - \tau_C C(t), \quad C(0) = C_0, \tag{6}$$

where the influx constraining function was

$$d(t) = \begin{cases} 1 & \text{if } \ln\frac{a(t)}{a_0} < L \\ h \ln\frac{a(t)}{a_0} & \text{if } \ln\frac{a(t)}{a_0} \geq L \end{cases} \tag{7}$$

Here I_P is the influx of $\bar{P}$ cells, *i.e.* the rate (all rates are in days^{-1}) of differentiation of $\bar{P}$ cells from stem cells, $\bar{\tau}_P$ is the rate of maturation of $\bar{P}$ cells into P cells, and τ_P is the rate of natural death of P cells; the quantities $\bar{\tau}_R$ and τ_R are defined in a fully analogical way. Further, f is the amplifying coefficient of the linear feedback effect of P and/or R cell decrease on the influx of $\bar{P}$ and $\bar{R}$ cells at time t.

The quantity $\bar{c}_P a(t)C(t)$ is the rate of elimination of $\bar{P}$ cells due to the amount of HIV products $a(t)$ and the number of cytotoxic T cells $C(t)$ at time t. Analogously, $c_P a(t)C(t)$ is the rate of elimination of P cells. The value a_0 is the function of the infectious dose of HIV, θ characterizes the growth rate of HIV, and γ is the rate of inactivation of HIV products mediated by cytotoxic C cells. The maturation of these cells from their precursors is assumed to be dependent on the encounter with HIV products and the effect of HIV specific helper T cells. I_C is the influx of C cell precursors, ε their maturation rate, α the proliferation rate of C cells under the antigenic stimulation by HIV products and helper T cell influence, and τ_C their natural death rate. Helper T cell effect on maturation and proliferation of C cells is expressed by the ratio $P(t)/P_0$; the coefficient ν is introduced to characterize the intensity of this helper effect. The value h characterizes HIV-constraining intensity on the $\bar{P}$ and $\bar{R}$ cell influx. Value L defines the level, where such constraining (limiting) effect of $d(t)$ starts.

If not otherwise stated, the model parameters in simulation runs were selected as follows: $\bar{\tau}_P = 0.2$, $\tau_P = 0.01$, $\bar{\tau}_R = 0.2$, $\tau_R = 0.01$, $\tau_C = 0.01$, $I_P = 1.0$, $I_C = 0.2$, $\bar{P}_0 = 5.0$, $P_0 = 100.0$, $\bar{R}_0 = 3.33$, $R_0 = 66.7$, $C_0 = 0.0$, $a_0 = 0.0005$, $f = 0.01$, $\alpha = 0.3$, $\varepsilon = 0.144$, $\gamma = 0.7$, $\theta = 0.02$, $\nu = 2.0$. Only mature CD4$^+$ lymphocytes were assumed to be susceptible to HIV products, *i.e.* $\bar{c}_P = 0.0$, $c_P = 20.0$. As a rule, parameter ε was used for final adjustment of the respective simulation run.

In the subsequent figures numbers of mature CD4$^+$ lymphocytes (P cells – solid line) and CD8$^+$ lymphocytes (R cells – dashed line) are given as percentage of P_0. The value of R_0 is then 66.67 percent. Further, mean observed T cell values in HIV-infected individuals (Lang et al., 1989; Doležal and Hraba, 1991) are depicted: solid circles – CD4$^+$ lymphocytes, solid squares – CD8$^+$ lymphocytes.

3 RESULTS

Clinical data used for quantitative comparisons of simulated and observed CD8$^+$ lymphocyte dynamics are those of Lang et al. (1989) which were used for analogous studies with CD4$^+$ cells (Doležal and Hraba, 1991). In the above setting such configuration of the model is achieved having L large enough to imply $d(t) = 1$ in the model equations (1) and (3), *i.e.* no influx constraining effect was present.

Simulations of CD8$^+$ lymphocyte dynamics using such original model configuration were not satisfactory (Fig. 1). Numbers of these cells increased during the whole course of the infection, when no further increase in these cells was observed clinically, actually, it rather seemed that $CD8^+$ cell counts in the patients started to decline in later stages of the infection. No realistic parameter variation was able to change this character of the simulated curve, and it became evident that the model had to be modified to be able to achieve agreement between simulated and observed data. The attempt to improve simulation of $CD4^+$ lymphocyte dynamics by assuming that immature $CD8^+$ lymphocytes were susceptible to HIV products in a similar way as the $CD4^+$ cells was not successful (Hraba and Doležal, 1995b).

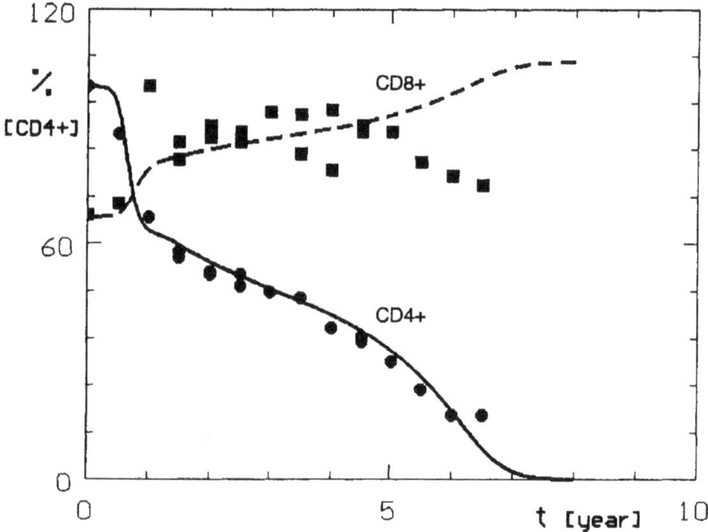

Figure 1 Mature $CD4^+$ and $CD8^+$ lymphocyte dynamics using the original version of the model

We therefore tried to achieve a satisfactory simulation by assuming that HIV infection affects common precursors of $CD4^+$ and $CD8^+$ cells. This situation could be obtained by decreasing the influx of immature $\bar{P}$ and $\bar{R}$ cells due to HIV infection. Simulation results giving best agreement with clinical data were obtained, when the influx was constrained by the logarithm value of HIV product concentration as also given in the definition of $d(t)$ function in Eq. (7). Roughly speaking, such construction allows to control both populations of T cells in intermediate and final phases of the infection.

In Fig. 2, it can be seen that satisfactory agreement between simulated and observed data was obtained under these assumptions with the following changed/additional parameter values: $\alpha = 0.7$, $\varepsilon = 0.535$, $\gamma = 0.3$, $\nu = 1.6, h = 0.3$, $L = 2.5$. The final abrupt drop of simulated $CD8^+$ lymphocyte numbers is only an extrapolation, occurring when patients are assumed to be dying, or are even dead, and the respective virus concentraions reach seemingly biologically not feasible values.

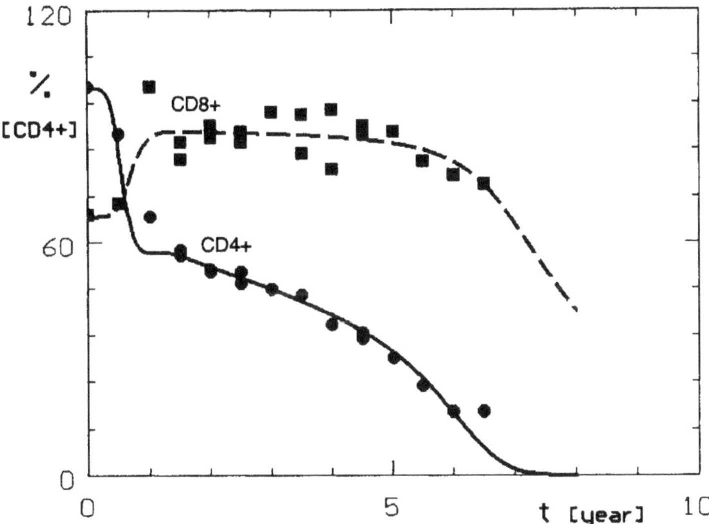

Figure 2 Mature $CD4^+$ and $CD8^+$ lymphocyte dynamics simulated by model modification with HIV-constrained T cell influx

4 CONCLUSIONS

The major advantage of our model is that it is able to simulate reasonably well $CD4^+$ lymphocyte dynamics in all stages of infection more appropriately than some alternative, more sophisticated models of HIV infection (Nowak et al., 1991; Perelson et al., 1993). The modified version simulates also $CD8^+$ lymphocyte dynamics satisfactorily. The model is rather versatile and easily adaptable to different assumptions. When the model was formulated originally, it was assumed that the cytopathogenic effect of HIV was not responsible for the $CD4^+$ lymphocyte depletion. It was postulated that HIV products induced this depletion by some other mechanism either directly, *e.g.* by apoptosis, or indirectly, *e.g.* by inducing an (auto)immune reaction.

The model version used in this, and most preceding papers, assumes that the specific immune response induced by HIV not only limits HIV proliferation but is also instrumental in $CD4^+$ lymphocyte elimination in a similar way as it was suggested by Zinkernagel et al. (1994) and recent findings (Ho et al., 1995; Wei et al., 1995). We preferred this version to that in which HIV products were assumed to induce $CD4^+$ cell elimination directly, because it gave a better fit to observed $CD4^+$ lymphocyte values. However, the difference was not substantial, and in absence of observation data permitting us to determine proper parameter values, we sticked to this arrangement just for its convenience. However, it is possible to adjust the model to any other assumption concerning pathogenic mechanisms, including even direct cytopathogenic effect of HIV on $CD4^+$ lymphocytes.

REFERENCES

Adleman, L.M. and Wofsy, D. (1993) T-cell homeostasis: implications in HIV infection. *J. AIDS*, **6**, 144–52.

Doležal, J. and Hraba, T. (1991) Model-based analysis of CD4$^+$ lymphocyte dynamics in HIV infected individuals. II. Evaluation of the model based on clinical observations. *Immunobiology*, **182**, 178–87.

Doležal, J. and Hraba, T. (1994) Mathematical modelling of immunotherapy in HIV infection. *Folia Biol. (Praha)*, **40**, 193–9.

Ho, D.D., Neumann, A.U., Perelson, A.S., Chen, W., Leonard, J.M. and Markowitz, M (1995) Rapid turnover of plasma virions and CD4 lymphocytes in HIV-1 infection. *Nature*, **373**, 123–6.

Hraba, T. and Doležal, J. (1994) Mathematical modelling of chemotherapy in HIV infection. *Folia Biol. (Praha)*, **40**, 103–11.

Hraba, T. and Doležal, J. (1995a) Mathematical modelling of HIV infection therapy. *Int. J. Immunopharmacol.* In press.

Hraba, T. and Doležal, J. (1995b) Mathematical modelling of CD8$^+$ lymphocyte dynamics in HIV infection. *Folia Biol. (Praha)*. In press.

Hraba, T., Doležal, J. and Čelikovský, S. (1990) Model-based analysis of CD4$^+$ lymphocyte dynamics in HIV infected individuals. *Immunobiology*, **181**, 108–18.

Lang, W., Perkins, H., Anderson, R.E., Royce, R., Jewell, N. and Winkelstein, W. (1989) Patterns of T lymphocyte changes with human immunodeficiency virus infection: from seroconversion to the development of AIDS. *J. AIDS*, **2**, 63–9.

Nowak, M.A., Anderson, R.M., McLean, A.R., Wolfs, T.F.W., Goudsmit, J. and May, R.M. (1991) Antigenic diversity thresholds and the development of AIDS. *Science*, **254**, 963–9.

Perelson, A.S., Kirschner, D.E. and de Boer, R. (1993) Dynamics of HIV infection of CD4$^+$ T cells. *Math. Biosci.* **114**, 81–125.

Wei, X., Ghosh, S.K., Taylor, M.E., Johnson, V.A., Emini, E.A., Deutsch, P., Lifson, J.D., Bonhoeffer, S., Nowak, M.A., Hahn, B.A., Saag, M.S. and Shaw, G.M (1995) Viral dynamics in human immunodeficiency virus type 1 infection. *Nature*, **373**, 117–22.

Zinkernagel, R.M. and Hengartner, H. (1994) T-cell-mediated immunopathology versus direct cytolysis by virus: implications for HIV and AIDS. *Immunol. Today*, **15**, 262–8.

39

Mathematical modelling of conjugate formation by cytotoxic lymphocytes and tumour cells

J. Waniewski [1], *A. K. Palucka* [2] *and A. Porwit* [3]

[1] *Baxter Novum, Department of Clinical Sciences, Huddinge University Hospital, Huddinge, Sweden and Institute of Biocybernetics and Biomedical Engineering, Warsaw, Poland.*
[2] *Department of Haematology, Karolinska Hospital, Hospital, Stockholm, Sweden.*
[3] *Department of Pathology, Karolinska Hospital, Stockholm, Sweden*

Jacek Waniewski, Department of Renal Medicine K56, Huddinge University Hospital, S-141 86 Huddinge, Sweden.
Tel: 46-8-746 3292. Fax: 46-8-746 3290.
e-mail: jacek@kfc.hs.sll.se

Abstract

Kinetics of conjugate formation between cytotoxic lymphocytes and tumour cells is described using the enzyme kinetics model and adapted for the evaluation of in vitro measurements by double fluorescence flow cytometry. The parameters of the model were estimated for two effector-target systems. A previously reported model of the first order kinetics is shown to be an approximation for the presented model. A parameter, derived from the model and called the strength of binding, is proposed for a simple and objective evaluation of the experiments.

Keywords

Conjugate formation, enzyme kinetics, first order kinetics, flow cytofluorometry

1 INTRODUCTION

The effect of cytotoxic lymphocytes on tumour cells includes usually three steps: 1) binding of the lymphocyte to a target and formation of a conjugate, 2) cytotoxic "hit" which causes a damage of the target membrane and invokes lethal processes inside the cell, and 3) dissolution of the conjugate and target's death. The formation of conjugates can be investigated in vitro using tumour cell lines or blasts from patients with leukaemia. It is possible to stimulate in vitro patient's cytotoxic lymphocytes with interleukin 2 (lymphokine activated killers, LAK, and adherent LAK, ALAK) and applied the stimulated cells in therapy of leukaemia. Therefore, a method for analysis of the lymphocyte cytotoxic effects at each stage of the interaction with targets is necessary. Whereas the overall cytotoxic effect is routinely evaluated by ^{51}Cr release assay (Thoma et al., 1978), various measures of the effectiveness of the conjugate formation have been previously proposed. Some of them are based on the dose-response curve analysis (Garcia-Penarrubia et al., 1989b). In contrast, we investigated in detail the kinetics of conjugate formation and applied a mathematical model based on an enzyme kinetics description (Waniewski et al., 1993).

2 EXPERIMENTAL METHODS

A short description of experimental methods applied in the investigations of conjugate formation between cells is necessary to understand the details of the presented model (Waniewski et al., 1993). Each population of the interacting cells is labelled with a specific fluorescent marker, e.g. effectors with a red fluorescence and targets with a green one. Some of cells are not labelled, as they do not bind specific markers (dark or negative cells). The labelled populations (usually with a small fraction of dark cells) are mixed so as to get a definite effector to target ratio (which should be about 1:1 for the experiments described by the model), spun for a short time in a centrifuge, so that they form a pellet, and incubated for various periods of time. After the incubation, the cells are resuspended and kept at 0°C to prevent further conjugation. The flow cytometric measurements (e.g. with FACScan) provide the numbers of characteristics counts of each population, e.g., single red fluorescence - free effectors, single green fluorescence - targets, double (red and green) fluorescence - conjugates, dark particles - not stained cells, per a definite number of total counts (usually 5 - 10 thousand).

2 MODEL

The model has been derived for the conditions in which the conjugates include only one effector cell (cytotoxic lymphocyte) and one target cell (tumour cell); this means that the initial ratio of target to effector cells should be close to 1 (Garcia-Penarrubia et al., 1989a, Waniewski et al., 1993). Three variables describe the interacting cells: the concentration of free, i.e. not bound in conjugates, effector cells, E_F, the concentration of free target cells, T_F, and the concentration of conjugates of effector and target cells, C. A number of other (dark) cells, with concentration D, is also present in the experiments but not involved in the cellular interactions.

The interactions between target and effector cells are described by the enzyme kinetics model (Murray, 1989), with additional possibility of the death of free effectors and free targets:

$$\dot{E}_F = -k_+ E_F T_F + k_- C - k_E E_F, \quad (1)$$
$$\dot{T}_F = -k_+ E_F T_F + k_- C - k_T T_F, \quad (2)$$
$$\dot{C} = k_+ E_F T_F - k_- C, \quad (3)$$

where k_+ is the association rate constant, k_- is the dissociation rate constant, k_E and k_T are the disappearance rate constants for free effectors and free targets, respectively. The population of dark cells may increase because of darkening of labelled T_F and E_F (which is interpreted as the death of these cells) and may decrease because of the disintegration of the dark cells:

$$\dot{D} = k_E E_F + k_T T_F - k_D D. \quad (4)$$

The total concentration of cells, N, given as $N = E_F + T_F + 2C + D$, changes with time according to the following law, which is the consequence of equations (1) - (4):

$$\dot{N} = -k_D D, \quad (5)$$

whereas the total amount of "particles" (or “events”) counted in flow cytometry, M, given as $M = E_F + T_F + C + D$, changes as:

$$\dot{M} = k_E E_F T_F - k_- C - k_D D, \quad (6)$$

If the time of the experiment is short enough to make the lysis of cells negligible (less than 30 min), then the total amount of cells, N, could be assumed to be approximately constant. In contrast, the amount of "particles", M, depends on the stage of the process of conjugate formation, see equation 6.

In the typical experimental conditions it is not possible to count whole involved cell populations, which usually include about 10^6 cells. Therefore, a sample from the whole population is investigated and the fraction of a particular subpopulation in the sample is considered to be an estimate of the fraction of this subpopulation in the whole population. The model has been adapted to this situation and new variables have been introduced: for free effectors $\varepsilon_F = E_F/N$, for free targets $\tau_F = T_F/N$, for conjugates $\gamma = C/N$, and for dark cells, $\delta = D/N$ (however, these fractions are not directly measured with FACScan and have to be calculated from the fractions of each population measured per number of counted “events”, see Perez et al. (1985), Segal et al. (1984), Waniewski et al. (1993)). Because $\varepsilon_F+\tau_F+2\gamma+\delta=1$, only three equations for subpopulation fractions may be considered:

$$\dot{\varepsilon}_F = -k_+ N \varepsilon_F \tau_F + k_- \gamma - k_E \varepsilon_F + k_D \delta \varepsilon_F, \quad (7)$$
$$\dot{\tau}_F = -k_+ N \varepsilon_F \tau_F + k_- \gamma - k_T \tau_F + k_D \delta \tau_F, \quad (8)$$
$$\dot{\gamma} = k_+ N \varepsilon_F \tau_F - k_- \gamma + k_D \delta \gamma. \quad (9)$$

The total concentration of cells, N, is usually relatively constant during short experiments aimed on the investigations of conjugate formation, and equal to its initial value, N_0. This means that $k_D\delta \ll 1$, c.f. equation (5), and therefore the terms with $k_D\delta$ in equations (7) - (9) may be considered negligible. The final form of the model is as follows:

$$\dot{\varepsilon}_F = -k_+N_0\varepsilon_F\tau_F + k_-\gamma - k_E\varepsilon_F, \tag{10}$$
$$\dot{\tau}_F = -k_+N_0\varepsilon_F\tau_F + k_-\gamma - k_T\tau_F, \tag{11}$$
$$\dot{\gamma}_F = k_+N_0\varepsilon_F\tau_F - k_-\gamma. \tag{12}$$

The concentration N_0 is the concentration of all cells in the pellet formed during centrifugation, and, as the volume of the pellet is not usually measured, N_0 is not known. Therefore, the lumped parameter k_+N_0, called the binding capacity constant, is estimated.

3 ANALYSIS OF THE MODEL

Strength of binding

Based on the model, a parameter Sb, defined as $Sb = k_+N_0/k_-$ and called the strength of binding, has been proposed for the evaluation of the experimental results (Palucka et al., 1995). Note that Sb is related to the equilibrium constant, $K = k_-/k_+$. This parameter may be evaluated using only one measurement of the cell population fractions, γ_S, ε_{FS} and τ_{FS}, within the period of the steady state of the system, as follows from equation (12) for $\dot{\gamma}_F = 0$:

$$Sb = \frac{\gamma_S}{\varepsilon_{FS}\tau_{FS}}. \tag{13}$$

The steady state values γ_S, ε_{FS} and τ_{FS} depend only on one parameter, Sb, which characterises the interaction between effectors and targets, and two additional parameters, the fraction of all effectors, $\varepsilon_W = \varepsilon_F + \gamma$, and the fraction of all targets, $\tau_W = \tau_F + \gamma$, which characterise experimental conditions and are relatively constant during the initial phase of conjugate formation (Waniewski et al., 1993, Palucka et al., 1995). Thus, the definition of Sb and equation (13) yield:

$$\gamma_S = \frac{1 + Sb(\varepsilon_W + \tau_W) - \sqrt{Sb^2(\varepsilon_W - \tau_W)^2 + 1 + 2Sb(\varepsilon_W + \tau_W)}}{2Sb} \tag{14}$$

The equations for ε_{FS} and τ_{FS} can be easily formulated as $\varepsilon_{FS} = \varepsilon_W - \gamma_S$ and $\tau_{FS} = \tau_W - \gamma_S$, respectively, together with equation (14) for γ_S. It should be noted however, that time which the system needs to reach the steady state depends on both kinetic parameters, k_+N_0 and k_-, in contrast to the values of variables at the steady state which depend only on the ratio of these kinetic parameters, i.e. on Sb.

Approximation with first order kinetics

Equations (10) - (12) may be written in another form using the relationships $\varepsilon_F = \varepsilon_W - \gamma$ and $\tau_F = \tau_W - \gamma$ as follows:

$$\dot{\varepsilon}_F = a_E - b_E \varepsilon_F, \tag{15}$$
$$\dot{\tau}_F = a_T - b_T \tau_F, \tag{16}$$
$$\dot{\gamma} = a_C - b_C \varepsilon_C, \tag{17}$$

where

$$a_E = k_- \varepsilon_W, \quad b_E = k_+ N_0 \tau_W + k_- - k_E - k_+ N_0 \gamma, \tag{18}$$
$$a_T = k_- \tau_W, \quad b_T = k_+ N_0 \varepsilon_W + k_- - k_T - k_+ N_0 \gamma, \tag{19}$$
$$a_C = k_- \varepsilon_W \tau_W, \quad b_C = k_+ N_0 (\varepsilon_W + \tau_W) + k_- + k_+ N_0 \gamma. \tag{20}$$

The parameters "a" are relatively constant if ε_W and τ_W are relatively constant, see Waniewski et al. (1993), and the parameters "b" may be considered approximately constant if γ (which is usually lower then ε_W and τ_W) is approximated by its average value. Therefore, the kinetic of the process of conjugate formation may be approximately described by the first order model, as proposed by Segal et al. (1984) and Perez et al. (1985). Note however, that this kind of kinetic description does not provide direct information about cellular interactions, and depends on the assumption that changes of γ, ε_W and τ_W over time may be neglected for the evaluation of the kinetic coefficients.

Parameter estimation

The parameters of the model k_+N_0, k_-, k_E, and k_T may be estimated with a simplified method of one- and two-parameter linear regressions using the following equations:

$$\varepsilon_W(t) = \varepsilon_W(0) - k_E \int_0^t \varepsilon_F(s)ds, \tag{21}$$

$$\tau_W(t) = \tau_W(0) - k_T \int_0^t \tau_F(s)ds \tag{22}$$

$$\gamma(t) = \gamma(0) + k_+ N_0 \int_0^t \varepsilon_F(s)\tau_F(s)ds - k_- \int_0^t \gamma(s)ds, \tag{23}$$

This method of parameter estimation has been found quite precise (Waniewski et al., 1993). Alternatively, the values of the transport parameters estimated with equations (21) - (23) may be used as the initial parameter values for a nonlinear regression applied to equations (10) - (12).

4 RESULTS

The model was applied for the evaluation of the experiments with lymphokine activated killers (LAK) interacting with K562 or Daudi cell lines (Waniewski et al., 1993). The fraction of the cell populations were examined at 0, 5, 10, 15, 20, 25 and 30 min from the beginning of the incubation using dual parameter flow cytometry (FACScan). The number of conjugates increased for the initial 15-20 min and then remained stable till the end of the experiment. The estimated parameters as well as the strength of binding, Sb, are presented in Table 1.

Table 1. Model parameters estimated for LAK cells incubated with K562 or Daudi targets

Parameters	K562-LAK mean±SD (n=8)	Daudi-LAK mean±SD (n=4)
$k_+ N_0$ [1/min]	0.196 ± 0.049	0.101 ± 0.020**
k_- [1/min]	0.053 ± 0.025	0.125 ± 0.045**
k_E [1/min]	0.010 ± 0.006	-0.000 ± 0.003*
k_T [1/min]	0.016 ± 0.012	-0.001 ± 0.005*
Sb[a]	3.8 ± 1.1	1.0 ± 0.5**

[a] calculated using equation (13)

*,** denote statistical significance (P<0.05 and P<0.01, respectively), according to Student's t test, for the difference between the K562-LAK and Daudi-LAK systems.

The model yielded the life times, calculated as the inverse of k_-, for the investigated conjugates, which were on average 8 min for Daudi - LAK and 19 min for K562 - LAK. A good correlation between the values of Sb calculated using its definition and the estimated values of k_+N_0 and k_- and the values of Sb calculated using equation (13) was found (Palucka et al., 1995).

For the evaluation of Sb using equation (13) an experiment may be simpler and less laborious than the study of kinetics of conjugate formation, and therefore Sb may be easily applied for investigation of different cell systems. In particular, interactions between peripheral blood lymphocytes, LAK, ALAK and various subpopulations of LAK as effectors and cell lines of lymphoid and myeloid origin as well as blast from patients with acute myeloid leukaemia as targets were described using Sb as a characteristic parameter (Palucka et al., 1995).

5 SUMMARY

The model provides a useful and precise description of the dynamic phenomenon of the conjugate formation by cytotoxic lymphocytes and tumour cells, and may be applied for evaluation and comparison of the interactions in various cell systems, including blasts from patients with leukaemia. An extension of the model for the description of formation of conjugates with more than two cells and the investigation of such conjugates with FACScan are possible, but have not been performed yet.

REFERENCES

Garcia-Penarrubia, P., Koster, F.T., Bankhurst, A.D. (1989a) Model for population distributions of lymphocyte-target cell conjugates. *J Theor Biol*, **138**, 77-92.

Garcia-Penarrubia, P., Koster, F.T., Bankhurst, A.D. (1989b) The maximum conjugate frequency (α_{max}) characterizes killer cell populations. *J Immunol Methods*, **118**, 199-208.

Murray, J.D. (1989) *Mathematical Biology*. Springer, Berlin

Palucka, A.K., Waniewski, J. and Porwit, A. (1995) Strength of binding between leukemic blasts and cytotoxic lymphocytes. Cytometry, submitted.

Perez, P., Bluestone, J.A., Stephany, D.A. and Segal DM. (1985) Quantitative measurements of the specificity and kinetics of conjugate formation between cloned cytotoxic T lymphocyte and splenic target cells by dual parameters flow cytometry. *J Immunol*, **134**, 478-85.

Segal, D.M. and Stephany, D.A. (1984) The measurement of specific cell: cell interactions by dual-parameter flow cytometry. *Cytometry*, **5**, 169-81.

Thoma, J.A., Touton, M.H. and Clark, W.R. (1978) Interpretation of ^{51}Cr -release data: a kinetic analysis. *J Immunol*, **120**, 991-7.

Waniewski, J., Palucka, A.K. and Porwit, A. (1993) Kinetic analysis of cytotoxic lymphocyte - target cell interaction as quantified by dual parameter flow cytometry. *Cytometry*, **14**, 393-400.

Information Systems

40

Reliability optimization of complex systems using sharp lower bounds

M. SOUISSI[1] *and Y. SMEERS*
CORE, Université Catholique de Louvain
34, voie du roman pays, 1348 Louvain-la-Neuve, Belgium.
Fax: (32)10-474301, e-mail: souissi@core.ucl.ac.be

Abstract

Reliability of production systems has become a prominent issue especially after the introduction of process oriented manufacturing techniques, (ex: Just-in-Time, M.R.P., O.P.T.). In general the reliable performance of a system is of the utmost importance in many other industrial, security and everyday life situations as well. This paper deals with reliability optimization of coherent systems by using redundancy. An algorithm based on some lower bounds of system reliability is presented and the numerical results are compared to the so far existing results.

Keywords

Reliability, optimization, nonlinear programming.

1 INTRODUCTION

The reliable performance of a system for a mission under various conditions is of the utmost importance in many industrial, security and everyday life situations. The increasing need for highly reliable systems and components of greater safety and less cost, has given rise to several developments in the past three decades. A good deal of effort has been centered in the field of optimal redundancy allocation. Most of the papers have considered the case of n-stage series models or series-parallel- series models, and developed different techniques to optimize their redundancy (dynamic programming, Linear integer programming, nonlinear programming with integer variables, geometric programming, ...). Some authors have given heuristics to optimize the reliability in complex systems. This paper deals with optimization of components redundancy into a coherent system. The algorithm can be used by a system designer faced with the problem of maximizing system reliability $R(S)$, subject to constraints (possibly nonlinear) on system resourses such as cost, weight, and power.

This problem was deeply studied in the case of series systems; the reader may refer to Bellman et al. (1958) and Tillman et al. (1977) In the case of complex systems, some

[1]Corresponding author: on leave from Ecole Mohammadia d'Ingénieurs, Rabat, Maroc

Heuristics have been developed, see Aggarwal (1976), Misra et al. (1979), Shi (1987), Yasutake-Baker et al. (1985).

The work presented here is based on the equivalent representation of coherent systems in terms of its minimal cuts or minimal paths; see Barlow and Prochan (1975). In this fashion we have a lower bound on system reliability which will be maximized instead of $R(S)$.

Before introducing our algorithm, here are some fundamental definitions as introduced by Barlow et al. (1975).

Coherent system: A system S of components is coherent if: *(a)* its structure function Φ is increasing, and *(b)* each component is relevant.

path : is a vector $x = (x_1, x_2, ..., x_n)$ such that $\Phi(x) = 1$

cut : is a vector $x = (x_1, x_2, ..., x_n)$ such that $\Phi(x) = 0$

minimal path : is a path x such that $\forall\, y < x,\ \Phi(y) = 0$. Let $P = \{i : x_i = 1\}$, P is called a minimal path set. It is a minimal set of components whose functioning insure the functioning of the system. For each minimal path set P_j we assign a series structure $\rho_j(x) = \prod_{i \in P_j} x_i$.

minimal cut : is a cut x such that $\forall\, y > x,\ \Phi(y) = 1$. Let $K = \{i : x_i = 0\}$ K is called a minimal cut set. It is a minimal set of components whose failure causes the system to fail. For each minimal cut set K_j we assign a parallel structure $\kappa_j(x) = \coprod_{i \in K_j} x_i$.

Proposition 1 *A coherent system (S, Φ) functions if and only if at least one of the minimal path structures is functioning. That is $\Phi(X) = \coprod_{j=1}^{p} \rho_j(x) = \coprod_{j=1}^{p} \prod_{j \in P_j} x_i$ where p is the number of minimal path sets.*

Similarly, (S, Φ) fails if and only if at least one of the minimal cut structures fails. That is $\Phi(X) = \prod_{j=1}^{k} \kappa_j(x) = \prod_{j=1}^{k} \coprod_{j \in K_j} x_i$ where k is the number of minimal cut sets.

2 ALGORITHM BASED ON MINIMAL CUT SETS

The main assumptions considered here are the statistical independency between the component states and the separability of constraints in terms of variables m_j. The problem of constrained redundancy optimization can be formulated as follows:

$$\begin{aligned} &Maximize \quad R(S) \\ &\text{s.t.} \quad \sum_{j=1}^{N} g_{ij}(m_j) \le b_i \quad i = 1, 2, ..., r \qquad (P1) \\ &\text{and} \quad m_j \in I\!N^* \quad j = 1, ..., N \end{aligned}$$

In general, it is quite difficult to compute the exact expression of the reliability function for complex systems. Ball (1979) has shown that all nontrivial reliability problems of general systems are $NP-hard$. Even for small systems, the mathematical expression of reliability function, as we will see in example 1, can be very complicated and heavy to manipulate. We have thus chosen to use an approach by lower bounds. In the first part, we consider the lower bound of system reliability. It is obtained from the representation of the initial system S in terms of its minimal cut sets given by $\underline{R}(S) = \prod_{l=1}^{k} R_l(S)$, where $R_l(S)$ is the reliability of the lth minimal cut. Instead of maximizing $R(S)$, we maximize $\underline{R}(S)$. The resulting problem is:

$$\begin{aligned} &Maximize \quad \underline{R}(S) = \prod_{l=1}^{k}(1 - \prod_{j\in K_l}(1-p_j)^{m_j}) \\ &\text{s.t.} \quad \sum_{j=1}^{N} g_{ij}(m_j) \le b_i \quad i = 1,2,...,r \qquad (P2) \\ &\text{and} \quad m_j \in I\!N^* \quad j = 1,...,N \end{aligned}$$

To solve this problem we need the following lemma:

Lemma 1 *Let f be a positive function defined on D and g be a function defined by $g(x) = \log f(x)$. If g is concave and has a unique maximum x^* on D, then f has the same unique maximum x^* on D.*

Proposition 2 *The function f defined on $D = [1, \infty[^n$ by $f(x_1, x_2, ..., x_n) = \sum_{l=1}^{k} \log(1 - \prod_{j\in K_l} q_j^{x_j})$ is concave.*

For proofs of Lemma 1 and Proposition 2 we can refer to Souissi et al. (1994).

After logarithmic transformation, the problem to be solved can be written as follows:

$$\begin{aligned} &Maximize \quad \sum_{l=1}^{k} ln[(1 - \prod_{j\in K_l}(1-p_j)^{m_j})] \\ &\text{s.t.} \quad \sum_{j=1}^{N} g_{ij}(m_j) \le b_i \quad i = 1,2,...,r \qquad (P3) \\ &\text{and} \quad m_j \in I\!N^* \quad j = 1,...,N \end{aligned}$$

By lemma 1, the objective function of (P2) has a unique maximum . Then the problem is to maximize a non-linear objective function having a unique maximum subject to linear or nonlinear convex constraints, with m_j integers. To this end we propose the following algorithm:

ALGORITM

This algorithm is based on the branch and bound principle

1 Relax the integrality constraints and solve the problem with the solver **MINOS5** for nonlinear programming.

2 - IF the solution is integer THEN **stop**, the solution is optimal. ELSE, GO TO 3.

3 - Put the upper bound UB as the value of the objective function of the nonlinear program of the first step.

Do a depth-first search to obtain an integer solution with cost C_0 which will constitute a lower bound LB on the objective function.

4 - Go up in the tree and explore the other branches updating lower and upper bounds.

We have developed a code of this algorithm in GAMS language and computed the optimal solution for different examples in the literature.
Recently we have purchased DICOPT, the MINLP solver, and we have written a GAMS code using this solver to test the same examples.

Example 1: The system considered in this example is the same as represented in Figure 0.1 and it was studied in Aggarwal (1976) and Shi (1987). The problem

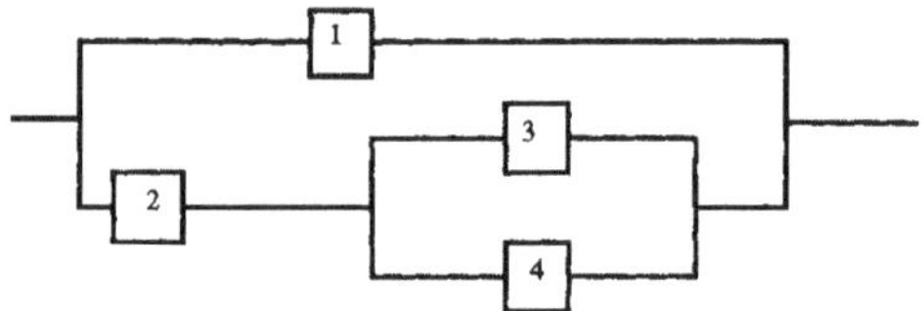

Figure 0.1: system S studied by Shi and Aggarwal

is to maximize system reliability under budget and weight constraints. Component relibilities, costs and weights are given in Table 1 and the total budget allocated is 40.

Table 1 Data of example 1.

component i	1	2	3	4
reliability p_i	0.80	0.75	0.70	0.65
costs c_i	6	4	3	2
weights w_i	9	4	4	3

The results of the different studies are drawn in Table 2 . We see that we have improved the results obtained by D.H. Shi (1986), which are better than those obtained by Aggarwal (1976).

Table 2 Comparison of computational results

Method	K.K. Aggarwal	D.H. Shi	Our Algorithm
Optimal solution	1 ,3 ,2 ,3	2 , 2, 1, 3	3, 1, 1, 1
Reliability value	0.9961	0.9970	0.99737

Further examples and comparisons

Recently we have got the DICOPT++ solver for mixed integer nonlinear programming and we have tested many examples drawn from literature using our algorithm. We have then compared the results obtained by the code using branch and bound and those obtained by the code using the DICOPT++ solver to those obtained by the authors of the different given examples. Some of the results of these computations are given in Table 3.

Table 3 Some computational results using the first lower bound.

author	known opt (1)	B. and B. (2)	DICOPT (2)	L. B. (3)
Misra(c=70)	2, 2, 1, 2, 1	1, 1, 2, 3, 2	3, 2, 2, 1, 1	
5 subs	0.987370	*0.990188	0.990188	0.9900046
Misra(c=90)	4, 3, 1, 2, 1	3, 2, 1, 3, 2	3, 2, 1, 3, 2	
5 subsy	0.998844	*0.998963	0.998963	0.9989553
Misra(c=110)	6, 3, 1, 2, 1	4, 2, 2, 4, 2	4, 2, 2, 4, 2	
5 subsy	0.999776	*0.999809	0.999809	0.9998085
Yasutake	1, 1, 2, 2, 1	1, 1, 2, 2, 1	1, 1, 2, 2, 1	
10 subsy	1, 1, 2, 3, 2	1, 1, 2, 3, 2	1, 1, 2, 3, 2	
	0.983708	0.983708	0.983708	0.9819554
Luus	3, 4, 6, 4, 3	3, 4, 5, 4, 3	3, 4, 6, 4, 3	
15 subsy	2, 4, 5, 4, 2	2, 4, 5, 4, 2	2, 4, 5, 4, 2	
	3, 4, 5, 4, 5	3, 4, 6, 4, 5	3, 4, 5, 4, 5	
	0.945613*	0.944860	*0.945613	0.9448599

3 ALGORITHM BASED ON MINIMAL PATH SETS

In this section we propose to use the bound of Prékopa to approximate the objective function of Problem (P1).We make also the assumption that the components are statisticaly independent.

3.1 Prékopa's bounds

Prékopa (1988) has presented a method, based on the so called Bool-Bonferroni inequalities, to obtain sharp lower and upper bounds for the probability that at least one-out-of a number of events in an arbitrary probability space will occur.

The knowledge of some of S_k's $k = 1, \ldots, m$, allows us to obtain a lower bound for the probability $P[\mu \geq 1]$ by solving the following linear programming

$$\begin{aligned} & Minimize \ \sum_{i=1}^{n} v_i \\ \text{s.t.} \quad & \sum_{i=1}^{n} \binom{k}{i} v_i = S_k \qquad k = 1, \ldots, m, \\ \text{and} \quad & v_i \geq 0 \qquad i = 1, \ldots, n, \end{aligned} \tag{P4}$$

Similarly, to obtain an upper bound for $P[\mu \geq 1]$, we maximize the objective function of the above linear program with the same constraints. In fact, the optimum value V_{max} of the objective function of the maximization problem can be greater than 1. If this is the case , $V_{max} > 1$, then we know that the sharp upper bound is equal to 1.

The optimum values V_{min} and V_{max} of problems mentioned above give sharp lower and upper bounds, respectively, of $P[\mu \geq 1]$

Remark

a) The more S_k's we know, the better are the bounds obtained by Prékopa's method.
b) To use Prékopa's algorithm for reliability problems, we have to know all minimal path sets and their probabilities, the probabilities of the intersection of all pairs of paths, and so on. These quantities are very difficult to determine.

3.2 Sharp bounds for $P[\mu \geq 1]$ in case of $m = 2$ and $m = 3$.

In this subsection we give the expressions of sharp lower and upper bounds on $P[\mu \geq 1]$ in term of S_i's for the case where $m = 3$. These bounds have been provided by Boros and Prékopa (1989) where we can find other expressions of these bounds.
a - case m=2: the lower bound is given by: $P(\mu \geq 1) \geq \frac{2}{l+1}S_1 - \frac{2}{l(l+1)}S_2$. This bound is sharp for $l-1 = \lfloor \frac{2S_2}{S_1} \rfloor$ where $\lfloor . \rfloor$ means the largest integer smaller than the number inside. This bound was obtained before by Dawson and Sankoff, Kwerel, Galambos.
b - case m=3: the lower bound is given by: $P(\mu \geq 1) \geq \frac{l+2n-1}{(l+1)n}S_1 - \frac{2(2l+n-2)}{l(l+1)n}S_2 + \frac{6}{l(l+1)n}S_3$. This bound is sharp for $l = 1 + \lfloor \frac{2((n-2)S_2-3S_3)}{(n-1)S_1-2S_2} \rfloor$

4 APPLICATION TO RELIABILITY FIELD

Let S be a coherent system as defined by Balow and Prochan (1975) and designate by P_i $(i = 1, 2, ..., p)$ the ith minimal path set of S. The events considered in this section are the minimal path sets occurence. Let E_r be the event that all components in the minimal path set P_r work. Then $P[E_r] = \prod_{i \in P_r} p_i$. *The system functions if and only if at least one of the minimal path sets operates.* Then the system operation corresponds to the event $\bigcup_{r=1}^{p} E_r$ (at least one path set is functioning), hence $P[S \text{ functions}] = P[\bigcup_{r=1}^{p} E_r] = P[\mu \geq 1] = R(S)$.

The exact value of $R(S)$ is given by $P[\mu \geq 1] = \sum_{k=1}^{n}(-1)^{(k-1)} S_k$. Its computation necessitates the knowledge of all the S_k's which is a difficult, even impossible task to perform. Hence we propose to approach this probability by the bounds obtained by Prékopa and Boros (1989).

4.1 Redundancy optimization using Prékopa and Boros bounds

We propose to maximize the system reliability using redundancy. But, as we mentioned above, the problem is that we haven't the exact expression of the reliability of the system. We shall use Prékopa and Boros bounds instead.
Let $\alpha_j = -\ln(1-p_j)$ and $z_j = \ln(1-e^{-\alpha_j x_j})$ and denote by $\mathcal{Z}_j$ $j = 1, ..., n$ the set

$\{z_j = \ln(1-\exp(-\alpha_j x_j))\ ,\ \ x_j \in I\!N\}$ and $\mathcal{Z} = \bigotimes_{j=1}^{n} \mathcal{Z}_j$

In the case of m=3 the problem can be stated as follows:

$$\begin{aligned} &Maximize\ \ \tfrac{l+2n-1}{(l+1)n} \sum_{i=1}^{p_1} \exp\left(\sum_{j\in P_i^1} zj\right) + \frac{6}{l(l+1)n} \sum_{i=1}^{p_3} \exp\left(\sum_{j\in P_i^3} zj\right) + \\ &\frac{2(2l+n-2)}{l(l+1)n} \sum_{i=1}^{p_2} \exp\left(\sum_{j\in P_i^2} zj\right) \\ &\text{s.t.}\ \ \sum_{j=1}^{n} -\beta_j\ \ln(1-e^{z_j})\ \leq c \\ &\text{and}\ \ z_j\ \in\ \mathcal{Z}_j \end{aligned} \qquad (P5)$$

where P_i^k is a union of k different minimal path sets (k different minimal path sets work simultaneously).

4.2 Convexity of the studied problems

In this subsection we will show that the problems mentioned above are of the type of minimisation of a difference of two convex functions over a convex domaine.

Lemma 2 *Let f be a numerical concave funcion from $I \subseteq I\!R$ to $I\!R$ and let φ be a concave function from $I\!R$ to $I\!R$ which is nondecreasing. Then $h(x) = (\varphi(f(x))$ is concave on I.*

Lemma 3 *Let A be a linear transformation from $I\!R^n$ to $I\!R^m$. Then for each convex function g on $I\!R^m$, the function gA defined by $gA(x) = g(A(x))$ is convex on $I\!R^n$.*

Proposition 3 *The functions S_k's are convex and the domain D of Problem (P5) is convex.*

Proof: immediate from Lemma 2 and Lemma 3. The reader can refer to Souissi et al. (1995).

Using Tuy (1987) formulation of the minimisation of a difference of two convex functions , we can write the Problem (P5) into the equivalent form:

$$\begin{aligned} &Maximize\ \ t - \left[\ \tfrac{l+2n-1}{(l+1)n} \sum_{i=1}^{p_1} \exp\left(\sum_{j\in P_i^1} zj\right) + \frac{6}{l(l+1)n} \sum_{i=1}^{p_3} \exp\left(\sum_{j\in P_i^3} zj\right) \right] \\ &\text{s.t.}\ \ \tfrac{2(2l+n-2)}{l(l+1)n} \sum_{i=1}^{p_2} \exp\left(\sum_{j\in P_i^2} zj\right) \leq t \\ &\qquad \sum_{j=1}^{n} -\beta_j\ \ln(1-e^{z_j}) \quad \leq c \\ &\text{and}\ \ z_j\ \in\ \mathcal{Z}_j \end{aligned} \qquad (P6)$$

The objective function of (P6) is now a concave function. If we relax the condition that z_j is discrete to that z_j is continuous, $z_j \in [\ln(1-e^{-\ \alpha_j}), \ln(1-e^{-10\ \alpha_j})]$, then the constraint set D (called domain) is a closed bounded and convex.

5 CONCLUSION

Using our approach we have, in most examples of literature, improved the known solutions. For example, the reliability of the bridge system, have been improved by comparison to Misra results, respectively for c=70, c=90 and c=110 . We have also improved the result of Shi (1986) for the system with 2 constraints. However for the Luus system we note that the solution obtained by the code of branch and bound is worse than the solution obtained by Luus, but we obtain the same optimal solution as Luus when using the DICOPT++ solver.

The computation of the absolute and relative errors shows that the lower bounds obtained with the minimal cuts in section 2 and with the minimal paths in section 3 are very tight to the exact system reliability values. These errors are comprised between $5 * 10^{-7}$ and $3.0377 * 10^{-3}$.

REFERENCES

Aggarwal, K.K. (1976) Redundancy optimization in general systems. *IEEE Transactions on Reliability*, **R-25 (5)**, pp 330-2.

Ball, M.O. (1980) Complexity of Network Reliability Computations. *Networks*, **10**, 153-65.

Barlow, R.E. and Prochan, F. (1975) *Statistical theory of reliability and life testing: probability models.* Holt, Rinehart and Winston, New York.

Bellman, R. and Dreyfus, S. (1958) Dynamic programming and the reliability of multicomponent devices. *Oper. Res.*, **Vol 6**, pp 200-6.

Boros, E. and Prékopa, A. (1989) Closed form tow-sided bounds for probabilities that at least r and exactly r out-of n events occur. *Mathematics of Operations Research*, **14**, 317-42.

Luus, R. (1975) Optimization of system reliability by a new nonlinear integer programming procedure. *IEEE Transactions on Reliability*, **R-24(1)**, 14-16.

Misra, R.B. and Agnihotri, G. (1979) Pecularities in optimal redundancy for a bridge network. *IEEE Transactions on Reliability*, **R-28(1)**, pp 70-2.

Prékopa, A. (1988) Boole-Bonferroni Inequalities and Linear Programming. *Operations Research*, **Vol 36**, pp 145-62.

Prékopa, A. (1990) Sharp Bounds on Probabilies Using Linear Programming. *Operations Research*, **Vol 38**, pp 227-39.

Shi, D.H. (1987) A New Heuristic Algorithm for Constrained Redundancy-Optimization in Complex Systems. *IEEE Transactions on Reliability*, **R-36(5)**, 621-3.

Souissi, M. and Smeers, Y. (1994) Constrained redundancy optimization of coherent systems. *Worksop On Reliability and Maintenance Modeling.*, Amsterdam.

Souissi, M. and Smeers, Y. (1995) Redundancy optimization of coherent systems. *Nantes Eurodays on Industrial Systems Design.*, Nantes, France.

Tillman, F.A. Hwang ,C.L. and Kuo, W. (1977) Optimization techniques for system reliability with redundancy - a review. *IEEE Transactions on Reliability*, **R-26(3)**, pp 148-55.

Tuy, H. (1987) Global Minimization of a Difference of two Convex Functions. *Mathematical Programming Study*, pp. 150-82,

Knowledge retrieval for autonomous agents

E. Szczerbicki
The University of Newcastle
Newcastle, NSW 2308, Australia.
Tel: (49) 21 6209. Fax: (49) 21 6946.
e-mail: mees@cc.newcastle.edu.au

Abstract
The paper addresses the problem of developing the domain knowledge base for autonomous manufacturing agents. In an organisational context, autonomous agents consist of elements (people, machines, robots, etc.) tied by the flow of information between an agent and its external environment as well as within an agent. Various tools can be used to develop and evaluate such an information flow. In the paper, the application of mathematical modelling, decision trees, and neural networks is investigated and illustrated with examples.

Keywords
Information flow, production rules, knowledge acquisition, autonomous manufacturing agents, decision trees, neural networks

1 AUTONOMOUS AGENTS MATHEMATICAL MODELLING AND REPRESENTATION

Structuring an information flow for a manufacturing agent belongs to the functional area of management and forms the basis for further agent integration. Mathematical models are designed to describe, understand, and finally support processes and activities that are primarily intellectual. In other words, models are developed mainly to create knowledge. The formal model is presented that can be used in creation of knowledge connected with an information flow evaluation in autonomous systems.

A general discussion of some principles of the functioning of autonomous systems and the role of information flow in an "information society" is included in Zeigler and Rozenblit (1990) and in Richards and Gupta (1985). The problem of information modelling for the integration of computerised manufacturing agents is discussed in Hsu and Rattner (1990). Mathematical representation of autonomous systems functioning in various decision situations described by the characteristics of external and internal environments is presented in Szczerbicki (1994). In this paper the approach developed earlier is used as the basis for quantitative and non-quantitative agent representation and the retrieval of knowledge concerning agent functioning.

1.1 Information flow for autonomous agents: principles and tools for modeling and evaluation

There are three components which are the key to understanding the functioning of an autonomous system. They are information structure, external environment, and internal environment. The growth of interest in these three components defining autonomous systems in all kinds of manufacturing organisations that strive for integration through the flow of information is apparent.

Formal representation of decision-making processes represents the core of the generalised model of autonomous agents functioning. Since most autonomous systems involve decision analysis with incomplete knowledge under uncertainty, the best alternative is very often defined, by assuming that the description of the environment is known statistically, as the one that maximises the expected utility. Let **A** represent the set of possible actions which can be undertaken by the element of an agent, **Z** the set of corresponding consequences, and **X** random variables describing the actual state of the external environment. It can be assumed that $z=f(a,x)$ as the particular consequence (z) depends usually on an action (a) undertaken in the particular state of the environment (x). On the other hand, the decision about particular action depends on information that is available about the state of the environment. If ß stands for the decision function, we have $a=\beta(d)$ where d represents information. It can be shown that, after considering certain correlation between information, action, and energy, we have:

$$f(a,x)=\mathbf{B}_0 - 2\mathbf{B}^T\mathbf{A} + \mathbf{A}^T\mathbf{Q}\mathbf{A}, \quad (1)$$

where $\mathbf{B}_0=b_0(x)$, $\mathbf{A}=[a_i]$, $\mathbf{B}=[b_i(x)]$, symmetric matrix $\mathbf{Q}=[q_{ij}]$ $(i,j=1, 2, ..., n)$ and n represents the number of elements of an agent. Minimum of (1) exists if $\mathbf{A}^T\mathbf{Q}\mathbf{A}$ is positively defined.

For an n-element agent we have:

$$\beta_i(d_i) + \sum_{j\neq i} q_{ij}E[\beta_j(d_j)|d_i] = E(b_i|d_i), \quad (2)$$

where $i, j = 1, 2, ..., n$.

Formalisation of agent decision making process expressed by (2) is a tool necessary for modeling and evaluation of information flow in an autonomous system. Information flow connects agent elements with the external environment described by random variables **X**. For our purposes we define information as knowledge about realisation **X** and the above connection is represented by information structure. This structure is modelled by matrix **C** in which $c_{ij}=1$ if the ith element has obtained (either by observation - sensoring, monitoring - or communication) information about the jth variable **X** realisation (if c_{ij}=0 it has not got it). The ith variable **X** realisation can be observed (sensored, monitored) only by the ith element of the agent. It can be informed about other realisations only when communication (information exchange) inside the agent is organised.

The value of information structure defined above is given as:

$$VC=\min E[f(\mathbf{A}, \mathbf{X})|\, \mathbf{C0}]-\min E[f(\mathbf{A}, \mathbf{X})|\mathbf{C}], \quad (3)$$

where min $E[f(\mathbf{A}, \mathbf{X})|\mathbf{C0}]$ represents the utility of information structure **C0** in which c_{ij}=0 for each i and j. Using (2) the VC can be represented by:

$$VC=E[\mathbf{b}^T\beta]. \quad (4)$$

1.2 Model-based generation of knowledge

With the mathematical tools presented shortly above it is possible to carry out various simulations of agent functioning in different decision situations. Such simulations help us to understand the role of various parameters describing agent external and internal environments in an information flow evaluation. The results of the model-based simulations can be formulated as IF...AND...THEN rules that provide useful knowledge about autonomous agents functioning in static and dynamic environments. Generation of such knowledge is an important factor in successful implementation of integration based on the flow of information. In static environment such factors as correlation, interaction, number of agent elements, and incompleteness of information were considered. In dynamic environment the corresponding rules consider the character of dynamics (stable, Brownian, and explosive), and delay. Examples of such rules are presented next.

Samples of production rules for static environment.

RULE 12

IF an external environment of an autonomous agent is static,

AND there is an interaction in the internal environment,

AND the relationship between variables describing the external environment is of statistical character,

THEN information structure should include observation (sensoring) and communication.

RULE 13

IF an external environment of an autonomous agent is static,

AND the relationship between variables describing the external environment is given by function dependence,

THEN communication between agent elements does not affect the value of information structure; information flow should be restricted to observation (sensoring).

Samples of production rules for dynamic environment.

RULE 14

IF an external environment of an autonomous group is described by stochastic process,

AND information is not delayed,

AND the external environment is described by Brownian movement,

THEN the value of information structures increases proportionally with the increase of time.

RULE 17
IF an external environment of an autonomous group is described by stochastic process,
AND information is delayed,
AND stochastic process is dependent,
AND the external environment is stable,
THEN the losses caused by delayed information stabilise with increasing value of delay.

2 NONQUANTITATIVE TOOLS AND TECHNIQUES

2.1 Decision tree classifiers

Decision tree classifiers are used successfully in many diverse areas. Their most important feature is the capability of capturing descriptive decisionmaking knowledge from the supplied data (Safavian and Landgrebe 1991). Decision tree can be generated from training sets. The procedure for such generation based on the set of objects (**S**), each belonging to one of the classes $\mathbf{C}_1$, $\mathbf{C}_2$, ..., $\mathbf{C}_k$ is as follows (Quinlan 1990):

Step 1. If all the objects in **S** belong to the same class, for example $\mathbf{C}_i$, the decision tree for **S** consists of a leaf labelled with this class.

Step 2. Otherwise, let T be some test with possible outcomes O_1, O_2, ..., O_n. Each object in **S** has one outcome for T so the test partitions **S** into subsets $\mathbf{S}_1$, $\mathbf{S}_2$, ... $\mathbf{S}_n$ where each object in $\mathbf{S}_i$ has outcome O_i for T. T becomes the root of the decision tree and for each outcome O_i we build a subsidiary decision tree by invoking the same procedure recursively on the set $\mathbf{S}_i$.

The above procedure is applied to the training set of objects in Table 1. The training sets are delivered from the analysis based on the quantitative model of agent functioning (Szczerbicki 1994). Each object is described by the relating attributes and belongs to one of the agent decision classes exchange_information ("yes" in the last column) or do_not_exchange_information ("no" in the last column).

Table 1 Training set for agent functioning

external environment	internal environment	type of dynamics	correlation	delay of information	decision
static	independent_actions	0	0	0	no
static	independent_actions	0	0.7	1	no
dynamic	dependent_actions	1.5	-0.5	1	yes
static	dependent_actions	0	0	0	yes
static	independent_actions	0	1	2	no
static	dependent_actions	0	0.5	2	yes
static	dependent_actions	0	-1	3	no
static	independent_actions	0	-1	0	no
static	dependent_actions	0	0.9	1	yes
static	dependent_actions	0	1	1	no

Suppose, for the sake of simplicity, that we are interested in decision making situations involving static environment only. When for this case the set is partitioned by testing on internal_environment and then on correlation, the resulting structure is equivalent to the decision tree shown in Figure 1.

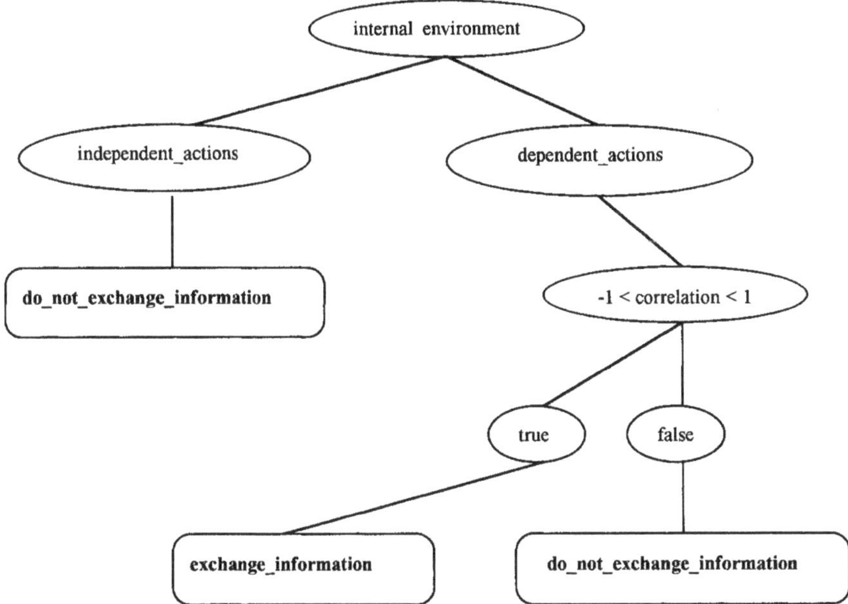

Figure 1 Decision tree classifier for agent decisionmaking.

The following rules can be delivered from Figure 1

RULE 1

IF an external environment of an agent is static
AND it is described by random variables
AND there is no interaction in the internal environment
THEN communication (exchange of information) between agent elements is not necessary

RULE 2

IF an external environment of an agent is static
AND it is described by random variables
AND there is interaction in the internal environment
AND the relationship between variables describing the external environment is of statistical character
THEN exchange of information between agent elements should be organised

RULE 3

IF an external environment of an agent is static
AND it is described by random variables

AND there is interaction in the internal environment
AND the relationship between variables describing the external environment is given by function dependence
THEN exchange of information between agent elements is not necessary

The simple tree in Figure 1 has been used to retrieve some of the knowledge concerning the functioning of an agent in static environment. The decision tree, once developed, can support decision situations that are not covered by the training set. That is why the production rules can be formulated as a generalised statements. As it has been illustrated the use of decision trees is simple and as effective as the analysis based on a rigorous mathematical model (the production rules formulated above are exactly the same as the Rules 4,5, and 6 given in Szczerbicki 1994). Another tool that can be of much help in the process of knowledge retrieval is the technique based on connectionist systems.

2.2. Connectionist systems

Neural networks which learn mappings between sets of patterns are called mapping neural networks (Maren et.al.). A key property of mapping networks is their ability to produce reasonable output vectors for input patterns outside of the set of training examples (please note the similarity to the decision tree classifiers). The above is especially important in areas such as discussed in this paper, i.e. areas for which it is possible to develop only a very limited number of IF...THEN rules and thus also to make inferences only for a very limited number of decision situations.

Problem solving tasks, such as information structure development for an autonomous agent, may be considered pattern classification tasks. The system analyst learns mappings between input patterns, consisting of characteristics of agent's external and internal environment, and output patterns, consisting of information structures to apply to these characteristics. Thus, neural networks (neural-based expert systems) offer a promising solution for automating the learning process of the analyst.

As we already know, systems analyst, while developing an information structure for an agent, transforms certain characteristics of an agent into recommendations concerning the flow of information. These characteristics represent the input for the system and their full description (for both static and dynamic environments) includes 5 parameters: correlation in the external environment (r), dynamics (t), interaction in the internal environment (q), delay (d), and type of the process describing the external environment (w). Output consists of the following decisions (recommendations): (i) observation (or sensoring) should be present, and (ii) exchange of information should be present. Please note that the decision concerning sensoring of information is added for this case (it was not considered in Section 2.1). An input portion together with an output portion of the data represents a training pair. The training pairs were used to train a 5-10-2 neural network.

The target values for each output node were normalised in such a way that the maximum target for each node received a value of 0.75 and the minimum target for each node received a value of 0.25. The training values for each input node were identically normalised. The learning rate and momentum term of 0.9 were used in the network. The network was trained using error back propagation procedure with a training tolerance of 5%. The network was considered trained if, for all training pairs and output nodes, |(desired output - actual output)/(desired output)| < tolerance.

After training, additional characteristics of an agent were generated for use of the network. Five sets of characteristics were submitted to the network. In response, the network suggested five information flow recommendations. As an example, Table 2 presents two sets of characteristics submitted and the obtained recommendations after the trained network has been used. For the first input set (no. 1 in Table 2) the network recommends decentralised information structure (only observation, no exchange of information). For the second, full information structure is recommended (observation and exchange). In both cases the recommendations agree with the IF ... AND ... THEN rules discussed in Szczerbicki (1994).

Table 2 The use of the trained network

no	value	description	observation	exchange
1	r=0.95	strong relationship between variables describing external environment	yes	no
	t=0	external environment is static		
	q=0.01	there is no interaction in internal environment		
	d=0	information is not delayed		
	w=0	process is independent		
2	r=0.2	weak relationship between variables describing external environment	yes	yes
	t=0	external environment is static		
	q=0.90	there is interaction in internal environment		
	d=0	information is not delayed		
	w=0	process is independent		

3 CONCLUSIONS

For the decision making process supporting agents functioning and information flow development, the knowledge domain includes many abstract concepts. The formal quantitative model can be used to generate some examples of the above domain knowledge. Quantitative models of an information flow evaluation, however, are often too complex to serve as the tools useful in the knowledge retrieval process. In the paper the preliminary results of two non-quantitative procedures applied in the process of knowledge acquisition have been presented and illustrated with examples. The procedures, decision trees and neural networks, show the potential for use in reasoning and retrieval of knowledge describing the flow of information between an agent and its external environment as well as within an agent. It was shown that the techniques applied are able to provide general knowledge about agent functioning in static and dynamic external environments. Both techniques, back propagation algorithm and decision tree classifier, illustrate the ease and appropriateness of such methods for dealing with implicit knowledge and also provide a model for extension into other expert domains.

REFERENCES

Hsu, C. and Rattner, L. (1990) Information modeling for computerised manufacturing. *IEEE Transactions on Systems, Man, and Cybernetics*, **20**, 758-776.

Maren, A.J., Harston, C.T. and Pap, R.M. (1990) *Handbook of neural computing applications*. Academic Press, New York.

Quinlan, J.R. (1990) Decision trees and decision making. *IEEE Transactions on Systems, Man, and Cybernetics*, **20**, 339-346.

Richards, L.D. and Gupta, S.K. (1985) The systems approach in an information society: a reconsideration. *Journal of Operational Research Society*, **36**, 833-843.

Safavian, S. and Landgrebe, D. (1991) A survey of decision tree classifier methodology. *IEEE Transactions on Systems, Man, and Cybernetics*, **21**, 660-674.

Szczerbicki, E. (1994) Model-based generation of knowledge for autonomous systems. *International Journal of Systems Science*, **25**, 453-472.

Zeigler, B. and Rozenblit, J. (ed.) (1990) *AI, Simulation and Planning in High Autonomy Systems*. IEEE Computer Society Press, Los Alamitos.

42

Simulation and optimization of complex systems reliability characteristics in grouped data structure

Ye. B. Tsoi, S. V. Tishkovskaya
Department of Applied Mathematics,
Novosibirsk State Technical University
20, K.Marx Prospekt, Novosibirsk, 630092, Russia.
Tel: (+7-3832) 46-03-01,46-05-11. Fax: +7-3832-46-02-09.
e-mail: ebcoi@nstu.nsk.su

Abstract

The present paper considers the problem of obtaining Bayesian estimates of the entity reliability characteristics. Bayesian estimates are calculated in the grouped data structure. Some properties and certain aspects of construction of Bayesian estimates in grouped data structure are studied.

Keywords

Complex systems, reliability characteristics, Bayesian estimation, grouped data.

1 INTRODUCTION

While studying or designing complex entities, there appear numerous problems requiring research of quantitative and qualitative laws of their functioning. To solve these problems, the entity considered is interpreted as a complex system, and mathematical simulation apparatus can be applied to its study. In constructing the mathematical model of the system studied, its major properties and laws of functioning are described by certain quantitative characteristics. Applying the mathematical simulation and optimization methods, these characteristics can be studied and, probably, improved.

The paper considers one of the characteristics of complex systems, its reliability. Reliability is the most important property of system's functioning quality, and the problem of its estimation reduces, in its turn, to the estimation problem of certain reliability quantitative properties such as average system lifetime, failure rate, etc. These properties have a probability character and are based on the assumption that system's lifetime duration is a random value. Note that everything concerning the system reliability and discussed above may be extended to its separate components.

2 PROBLEM STATEMENT

Let us denote by x the random value equal to the entity lifetime period, F(x) is its probability distribution function. The reliability function $\overline{F}(x) = 1 - F(x)$ is considered as the major property of the entity reliability. In practice, function F(x) and, consequently, $\overline{F}(x)$ is known as a rule to an accuracy of unknown parameter θ, i.e. $F(x) = F(x; \theta)$. (Values x and θ can be multivariate ones.) In this case the researcher faces a problem of unknown parameter estimation. The paper suggests to apply the Bayesian approach for obtaining estimates $\hat{\theta}$ of unknown parameters θ. Prior to considering directly the issues of Bayesian estimation, let us describe the data structure used for constructing estimates.

3 DATA STRUCTURE USED

According to the definition, the point estimate $\hat{\theta}$ is statistics, i.e. some measurable function of the sample used instead of the unknown parameter θ. Traditionally in statistics, almost everywhere it is assumed that the random sample, according to which the estimate $\hat{\theta}$ is constructed, consists of individually known observations. But in practice of statistical and experimental computations such a 'pure' data structure does not exist, as a rule. That is why development of methods for parameter estimation in the grouped data structure is of great practical value (Denisov, Lemeshko and Tsoi, 1993). Let us introduce the concept of grouped sample in the following way. Partition a set of $\mathbf{X}$ possible values of variable x into $k \geq 2$ non-intersecting intervals $\mathbf{R}_i = (x_{(i-1)}, x_{(i)}]$, $i = 1, \ldots, k$, $x_{(0)} = \inf(\mathbf{X})$, $x_{(k)} = \sup(\mathbf{X})$, $x_{(0)} < x_{(1)} < \ldots < x_{(k-1)} < x_{(k)}$. Then the grouped data are a population composed of non-random partition for random value x chahges' domain into k intervals $\mathbf{R}_i$ and discrete random values n_i, $i = 1, \ldots, k$, which represent a number of observations belonging to the ith grouping interval. Thus we obtain a grouped sample $\overline{n} = (n_1, \ldots, n_k)$.

A grouped sample is a special case of more general concept: partially grouped sample. Let $x_{i1}, \ldots, x_{in_i}$ are individual values of observations, belonging to the ith grouping interval. According to Kulldorf (Kulldorf,1961), let us introduce the definition of partially grouped sample.

Definition 1. The sample is called a partially grouped one if the information available is connected with the multitude of non-intersecting intervals which divide the range of random values such that each interval belongs to one of the two types:

$\mathbf{R}^*$, the ith interval belongs to the first type if the number n_i is known, but the individual values x_{ij}, $j = 1, \ldots, n_i$ are unknown;

$\mathbf{R}^{**}$, the ith interval belongs to the second type if not only the number n_i, but also all the individual values x_{ij} are known.

The concept of partially grouped sample combines non-grouped, grouped, and censored samples.

Let us note that some methods of estimation work only with data structures of a definite type. The Bayesian approach suggested in the paper do not impose any limitations on the data structure used. Thus, the Bayesian estimates can be constructed according to non-grouped, grouped, and partially grouped samples.

4 BAYESIAN METHOD OF PARAMETER ESTIMATION

Let us describe the scheme of constructing a Bayesian estimate for unknown parameter $\theta \in \Theta$. Let $f(x;\theta)$ be the density function corresponding to the distribution function $F(x;\theta)$. Let us consider the general case and assume that data **D** represent a partially grouped sample, i.e., the domain **X** is partitioned into $k \geq 2$ non-intersecting intervals, and part of these intervals belongs to the type $\mathbf{R}^*$, and the remaining part to the type $\mathbf{R}^{**}$. Then the likelihood function will have the form

$$l(\theta|\mathbf{D}) = \gamma \prod_{(1)} \left[\int_{x_{(i-1)}}^{x_{(i)}} f(x;\theta)\,dx \right]^{n_i} \prod_{(2)} \prod_{j=1}^{n_i} f(x_{ij};\theta)$$

where γ is a certain constant, indexes (1) and (2) indicate that multiplication is done correspondingly according to intervals of the first and second type, x_{ij} are individual values of observations belonging to the ith interval. Let $\pi(\theta)$ be an a priori density function θ. The a posteriori density function will have the following form

$$\pi(\theta|\mathbf{D}) = \frac{\pi(\theta)\,l(\theta|\mathbf{D})}{\int_{\Theta} \pi(\theta)\,l(\theta|\mathbf{D})\,d\theta}.$$

We take as the Bayesian estimation of parameter θ with respect to a priori distribution $\pi(\theta)$ the estimate minimizing the a posteriori risk $\int_{\Theta} \mathbf{L}(\theta, t(\mathbf{D}))\,\pi(\theta|\mathbf{D})\,d\theta$ where $\mathbf{L}(\theta, t(\mathbf{D}))$ is the given loss function, $t(\mathbf{D})$ is the estimation of parameter θ constructed according to the data **D** (Zacks, 1971). Let us denote by $\hat{\theta}(\mathbf{D}) = \hat{\theta}$ the Bayesian estimate according to partially grouped sample. Then

$$\int_{\Theta} \mathbf{L}(\theta, \hat{\theta}(\mathbf{D}))\,\pi(\theta|\mathbf{D})\,d\theta = \inf_{t \in T} \int_{\Theta} \mathbf{L}(\theta, t(\mathbf{D}))\,\pi(\theta|\mathbf{D})\,d\theta.$$

where **T** is set of all possible estimations of parameter θ.

5 CONVERGENCE OF BAYESIAN ESTIMATES IN THE STRUCTURE OF GROUPED DATA

As sample data **D**, let us consider a certain fixed non-grouped sample $\overline{x} = (x_1, \ldots, x_N)$ from distribution $f(x;\theta)$. Let us denote correspondingly by $\hat{\theta}^{ng}$ and $\pi^{ng}(\theta|\overline{x})$ the Bayesian estimate and a posteriori density function constructed from $\overline{x}$. Let $\overline{n} = (n_1, \ldots, n_k)$ be a grouped sample corresponding to $\overline{x}$ and connected with grouping intervals boundary points $x_{(0)} = \inf(\mathbf{X}) < x_1 < \ldots < x_{(k-1)} < x_{(k)} = \sup(\mathbf{X})$. Let us denote correspondingly by $\hat{\theta}^{gr}$ and $\pi^{gr}(\theta|\overline{n})$ the Bayesian estimate and the a posteriori density function constructed according to $\overline{n}$.

Let us note that there exists the following connection between the grouped data structure and non-grouped one. If $\Delta x_{\max} = \max_{i=2,\ldots,k-1}(x_{(i)} - x_{(i-1)}) \to 0$, $x_{(1)} \to x_{(0)}$, $x_{(k-1)} \to x_{(k)}$, then the number of grouping intervals $k \to \infty$ and grouped sample $\overline{n} = (n_1, \ldots, n_k)$ degenerates into non-grouped one $\overline{x} = (x_1, \ldots, x_N)$. In this connection a question arises: how do the function $\pi^{gr}(\theta|\overline{n})$ and estimate $\hat{\theta}^{gr}$ as $k \to \infty$ behave with respect to function $\pi^{ng}(\theta|\overline{x})$ and estimate $\hat{\theta}^{ng}$ respectively?

This paper shows that under performing a number of conditions functions $\pi^{gr}(\theta|\overline{n})$ converge uniformly on the set Θ to the function $\pi^{ng}(\theta|\overline{x})$. Before formulating these conditions let us introduce some designations and definitions.

We consider the set **A** of all possible partitions of the domain **X** into $k \geq 2$ non-intersecting intervals. Elements of set **A** are sets of points

$x_{(0)} = \inf(\mathbf{X}) < x_{(1)} < \ldots < x_{(k-1)} < x_{(k)} = \sup(\mathbf{X})$,

those define the partition of set **X** into grouping intervals. We will have to designate these partitions by one way from following ones

$$\alpha = \alpha^k = \{x_{(0)}, \ldots, x_{(k)}\} = \{x_{(0)}^{\alpha}, \ldots, x_{(k)}^{\alpha}\}.$$

Let us name the quantity $d(\alpha) = \max_{i=2,\ldots,k-1}(x_{(i)} - x_{(i-1)})$ as diameter of partition α. To every α we set in conformity the density $\pi_{\alpha}^{gr}(\theta|\overline{n}) = \pi_{\alpha}^{gr}(\theta)$ depending on partition α. Thus we obtained a sequence of functions $\{\pi_{\alpha}^{gr}(\theta)\}_{\alpha \in \mathbf{A}}$ where indexes α are ordered as follows:

$$\alpha_2 > \alpha_1 \Leftrightarrow d(\alpha_2) < d(\alpha_1) \wedge x_{(1)}^{\alpha_2} < x_{(1)}^{\alpha_1} \wedge x_{(k-1)}^{\alpha_2} > x_{(k-1)}^{\alpha_1}.$$

Then the definition of uniform convergence $\pi_{\alpha}^{gr}(\theta)$ to $\pi^{ng}(\theta|\overline{x}) = \pi^{ng}(\theta)$ can be formulated as follows.

Definition 2. The sequence of functions $\{\pi_{\alpha}^{gr}(\theta)\}_{\alpha \in \mathbf{A}}$ converges uniformly with respect to θ on the set $\boldsymbol{\Theta}$ towards the function $\pi^{ng}(\theta)$ as $d(\alpha) \to 0, x_{(1)} \to x_{(0)}, x_{(k-1)} \to x_{(k)}$ simultaneously if

$$\forall \varepsilon > 0 \quad \exists \delta_{\varepsilon} > 0 \quad \exists r_{\varepsilon} > x_{(0)}, s_{\varepsilon} < x_{(k)} \quad \forall \alpha \in \mathrm{A}: \quad d(\alpha) < \delta_{\varepsilon}, x_{(1)}^{\alpha} < r_{\varepsilon}, x_{(k-1)}^{\alpha} > s_{\varepsilon} \quad \forall \theta \in \Theta$$

$$|\pi_{\alpha}^{gr}(\theta) - \pi^{ng}(\theta)| < \varepsilon .$$

In practice the problem with $\mathbf{X} = (-\infty; +\infty)$ is often substituted by the problem in which $\inf(\mathbf{X}) < \infty$ and $\sup(\mathbf{X}) < \infty$. For this particular case the diameter of partition $d(\alpha)$ can be defined as $d(\alpha) = \max\limits_{i=1,\dots,k}(x_{(i)} - x_{(i-1)})$. Indexes of a considered sequence of functions $\{\pi_{\alpha}^{gr}(\theta)\}_{\alpha \in \mathbf{A}}$ will be ordered in accordance with the following rule: $\alpha_2 > \alpha_1 \Leftrightarrow d(\alpha_2) < d(\alpha_1)$.

We will formulate the following theorem about uniform convergence of the a posteriori density functions.

Theorem 1. Let $\pi(\theta)$ be a priori density function of θ; $f(x;\theta)$ be density function of distribution of $x \in \mathbf{X}$; θ be unknown parameter; $\overline{x} = (x_1, \dots, x_N)$ be a some fixed sample from distribution $f(x;\theta)$; $\pi_{\alpha}^{gr}(\theta)$ be a posteriori density function constructed according to the grouped sample connected with partition $\alpha \in \mathrm{A}$, where A is set of all possible partitions of the domain $\mathbf{X}$ into $k \geq 2$ non-intersecting intervals; $\pi^{ng}(\theta)$ be a posteriori density function constructed according to the non-grouped sample $\overline{x}$. Let the following conditions are performed:

1) $\exists \mathbf{M} < \infty \quad \forall \theta \in \Theta \quad \pi(\theta) < \mathbf{M}$;
2) $\exists \mathbf{K} < \infty \quad \forall x \in \mathbf{X} \quad \forall \theta \in \Theta \quad f(x;\theta) < \mathbf{K}$;
3) function $f(x;\theta)$ is continuous uniformly on the $\mathbf{X} \times \Theta$.

Then the sequence $\{\pi_{\alpha}^{gr}(\theta)\}_{\alpha \in \mathbf{A}}$ converges uniformly with respect to $\theta \in \Theta$ towards the function $\pi^{ng}(\theta)$ as $d(\alpha) \to 0$.

For the sequence of density functions $\{\pi_{\alpha}^{gr}(\theta)\}_{\alpha \in \mathbf{A}}$ there exists a sequence of independent random values $\{\theta_{\alpha}\}_{\alpha \in \mathbf{A}}$. Hence we can speak about a sequence of Bayesian estimations $\{\hat{\theta}_{\alpha}^{gr}\}_{\alpha \in \mathbf{A}}$ obtained as solution of the problem of minimization for corresponding a-posteriori risk.

Theorem 2. Let the conditions of Theorem 1 be performed. For squared-error loss function, if set Θ is compact, the convergence of Bayesian estimates $\hat{\theta}_{\alpha}^{gr}$ in the grouped data structure towards Bayesian estimate $\hat{\theta}^{ng}$ in the non-grouped data structure takes place as $d(\alpha) \to 0$.

6 COMPUTATIONAL DIFFICULTIES OF BAYESIAN ESTIMATION ACCORDING TO GROUPED DATA

As was mentioned above, the ideas of Bayesian approach are equally acceptable both for grouped and non-grouped data. Nevertheless, the Bayesian approach is mostly applied for constructing estimates from non-grouped and censored samples. For computing parameter estimates from completely grouped data, the Bayesian method practically is not applied. It may be due to the difficulties of computational character, the researcher faces in constructing Bayesian estimates with grouped data structure.

It is rather rare that one can succeed in obtaining an analytical expression for the Bayesian estimate from grouped data. That is why in the process of computing the estimate there is constantly a necessity to carry out numerical integration, and the integrals can have large dimension. The other practical difficulty consists in the fact that in the work with the grouped data it is necessary to compute the probabilities of the random value x falling in the ith grouping interval: $p_i(\theta) = \int_{R_i} f(x;\theta)dx, \quad i = 1, ..., k$. In the case when the dimension of the domain **X** is larger than 1, in computing $p_i(\theta)$, there appear some technical difficulties owing to the dimensions of the integrals and to the fact that the region of grouping R_i in the general case can have an arbitrary form.

One more problem, specific for the Bayesian estimation from the grouped data, consists in violating the property of closure of the conjugate distribution families. As is known, the conjugate distribution families which are widely applied to the Bayesian estimation, are characterized by the closure property in relation to the process of experimental data choice. In other words, if the a priori distribution $\pi(\theta)$ belongs to some conjugate family, then, with any sample $\overline{x}$, the a posteriori density function $\pi(\theta|\overline{x})$ will also belong to this family. Due to the violation of the closure property, the analytical or numerical computation of the Bayesian estimates from grouped data becomes even more complicated. In this paper, the following scheme of the Bayesian inference from grouped data, which allows to get round the problem of violating the property described.

Let us apply the binomial formula to the likelihood function from grouped data $l(\theta|\overline{n}) = \gamma \prod_{i=1}^{k} p_i^{n_i}(\theta) = \gamma \prod_{i=1}^{k} (F(x_{(i)};\theta) - F(x_{(i-1)};\theta))^{n_i}$. Then the a posteriori density function from the grouped data can be written down in the following way

$$\pi^{gr}(\theta|\overline{n}) = \sum_{i_k=0}^{n_k} \dots \sum_{i_2=0}^{n_2} \mathbf{K}_{i_2 \dots i_k}(\overline{n}) \pi_{i_2 \dots i_k}(\theta|\overline{n}),$$

where

$$\mathbf{K}_{i_2 \dots i_k}(\overline{n}) = \frac{\mathbf{C}_{n_2}^{i_2} \dots \mathbf{C}_{n_k}^{i_k} (-1)^{i_2+\dots+i_k} \int_{\Theta} \mathbf{G}_{i_2 \dots i_k}(\theta|\overline{n}) \pi(\theta) d(\theta)}{\sum_{i_k=0}^{n_k} \dots \sum_{i_2=0}^{n_2} \mathbf{C}_{n_2}^{i_2} \dots \mathbf{C}_{n_k}^{i_k} (-1)^{i_2+\dots+i_k} \int_{\Theta} \mathbf{G}_{i_2 \dots i_k}(\theta|\overline{n}) \pi(\theta) d(\theta)},$$

$$\pi_{i_2 \dots i_k}(\theta|\overline{n}) = \frac{\mathbf{G}_{i_2 \dots i_k}(\theta|\overline{n}) \pi(\theta)}{\int_{\Theta} \mathbf{G}_{i_2 \dots i_k}(\theta|\overline{n}) \pi(\theta) d\theta},$$

$$\mathbf{G}_{i_2 \dots i_k}(\theta|\overline{n}) = \mathrm{F}^{n_1+i_2}(x_{(1)})\, \mathrm{F}^{n_2-i_2+i_3}(x_{(2)}) \dots \mathrm{F}^{n_{k-1}-i_{k-1}+i_k}(x_{(k-1)}), \quad \mathbf{C}_n^i = \frac{n!}{i!(n-i)!}.$$

With this approach we can construct for the density functions $\pi(\theta)$ and $\pi_{i_2 \dots i_k}(\theta|\overline{n})$ the conjugate density families corresponding to the grouped data structure. Note that in some cases the conjugate families for grouped and non-grouped data will coincide. The scheme suggested allows, in a number of cases, to obtain an analytical form of Bayesian estimates from grouped data and avoid numerical integration. In the next section, an example of constructing a Bayesian estimate according to the scheme described is given.

7 EXAMPLE

Let the random value x belongs to the exponential distribution with distribution function $\mathrm{F}(x) = \mathrm{F}(x;\theta) = 1 - e^{-\theta x}$ and density function $f(x;\theta) = \theta e^{-\theta x}$, $x \geq 0$, $\theta \geq 0$, θ is unknown parameter. We will construct the Bayesian estimate of parameter θ from grouped data. As a priori distribution we take the gamma distribution with parameters α, β and with density function $\pi(\theta) = \alpha^{\beta} \theta^{\beta-1} e^{-\alpha\theta} / \Gamma(\beta)$. Consider the grouped sample $\overline{n} = (n_1, \dots, n_k)$. Likelihood function has the following form

$$l^{gr}(\theta|\overline{n}) = \gamma \prod_{i=1}^{k} (\mathrm{F}(x_{(i)}) - \mathrm{F}(x_{(i-1)}))^{n_i} = \gamma \prod_{i=1}^{k} (e^{-\theta x_{(i-1)}} - e^{-\theta x_{(i)}})^{n_i}.$$

In this example it is more convenient to apply the scheme from the section 6 to the likelihood function written as $l^{gr}(\theta|\bar{n})=\gamma\prod_{i=1}^{k}(\bar{F}(x_{(i-1)})-\bar{F}(x_{(i)}))^{n_i}$, where $\bar{F}(x)=1-F(x)$. Then we obtain

$$\pi^{gr}(\theta|\bar{n})=\sum_{i_{k-1}=0}^{n_{k-1}}\cdots\sum_{i_1=0}^{n_1}\mathbf{K}_{i_1\ldots i_{k-1}}(\bar{n})\,\pi_{i_1\ldots i_{k-1}}(\theta|\bar{n}),$$

$$\mathbf{K}_{i_1\ldots i_{k-1}}(\bar{n})=\frac{\mathbf{C}_{n_1}^{i_1}\ldots\mathbf{C}_{n_{k-1}}^{i_{k-1}}(-1)^{i_1+\ldots+i_{k-1}}(\varphi_{i_1\ldots i_{k-1}}(\bar{n}))^{-\beta}}{\sum_{i_{k-1}=0}^{n_{k-1}}\cdots\sum_{i_1=0}^{n_1}\mathbf{C}_{n_1}^{i_1}\ldots\mathbf{C}_{n_{k-1}}^{i_{k-1}}(-1)^{i_1+\ldots+i_{k-1}}(\varphi_{i_1\ldots i_{k-1}}(\bar{n}))^{-\beta}},$$

$$\pi_{i_1\ldots i_{k-1}}(\theta|\bar{n})=(\varphi_{i_1\ldots i_{k-1}}(\bar{n}))^{\beta}\,\theta^{\beta-1}\exp\{-\theta\varphi_{i_1\ldots i_{k-1}}(\bar{n})\}/\Gamma(\beta),$$

where $\varphi_{i_1\ldots i_{k-1}}(\bar{n})=\sum_{l=1}^{k-2}x_{(l)}(i_l-i_{l+1}+n_{l+1})+x_{(k-1)}(i_{k-1}+n_k)+\alpha$. We see that $\pi_{i_1\ldots i_{k-1}}(\theta|\bar{n})$ is density function of the gamma distribution with parameters $\varphi_{i_1\ldots i_{k-1}}(\bar{n})$, β. Thus we can write the analytical expression for the Bayesian estimate from grouped data with the squared-error loss function:

$$\hat{\theta}^{gr}(\bar{n})=\sum_{i_{k-1}=0}^{n_{k-1}}\cdots\sum_{i_1=0}^{n_1}\mathbf{K}_{i_1\ldots i_{k-1}}(\bar{n})\frac{\beta}{\varphi_{i_1\ldots i_{k-1}}(\bar{n})}.$$

REFERENCES

Denisov, V.I., Lemeshko, B.Yu., and Tsoi, Ye.B.(1993) Optimal grouping, parameter estimation, and design of regression experiments, (in Russian). Novosibirsk.

Kulldorf, G. (1961) Estimation from grouped data and partially grouped samples. John Wiley, New York.

Zacks,S. (1971) The Theory of Statistical Inference. John Wiley and Sons Inc., New York.

Multicriterial Problems

A modular system of software tools for multicriteria model analysis

J. Granat, T. Kręglewski, J. Paczyński, A. Stachurski, A.P. Wierzbicki
Institute of Control and Computation Engineering
Warsaw University of Technology
Nowowiejska 15/19, 00665 Warsaw, Poland.
Tel/Fax: +48 22 - 25 37 19. e-mail: J.Granat@ia.pw.edu.pl

Abstract
Applications of various analytical models in decision support systems motivate the development of efficient tools for the analysis of such models and their optimization. There are various approaches to the design of DSS based on analytical models. We focus here on developing well defined modular tools supporting selected phases of decision process, which can be used to build a DSS customized to a problem being solved.

Keywords
Modelling, decision support, multiobjective optimization

1 INTRODUCTION

Modular tools considered in this paper support selected phases of decision process. We assume that an analyst or an engineering designer uses substantive model expressing essential aspects of the decision situation.

The first set of tools supports the phase of model formulation and introductory model analysis. In this phase a user, usually an analyst, defines the substantive model and edits it on the computer. This phase was not explicitly supported in earlier versions of DIDAS-type systems (Lewandowski at al., 1983) and the user had to separately prepare (define and edit) his models, in nonlinear case together with user-supplied formulae for the derivatives of all outcome functions with respect to decision variables. It is a known fact that most mistakes in applying nonlinear programming methods are made when determining derivatives analytically; thus, this way of substantive model preparation required substantial experience in applications of nonlinear programming. The definition and edition of substantive models in the next generation of DIDAS-type systems, e.g. in IAC-DIDAS-N (Kręglewski at al., 1988) was supported by an easy standard format of a spreadsheet, together with an automatic support of calculations of all needed derivatives by an automatic differentiation program. However, IAC-DIDAS-N was a closed system, not easy to modify for specific classes of models being solved. This motivated the development of a modular system called IAC-DIDASN++, composed of modular tools written in the C++ language.

A model in IAC-DIDASN++ is prepared as an ASCII text in a *Model Description File* format. The model can be purely nonlinear or consist of a linear and nonlinear parts. At present, the linear part can be expressed in the form of MPS file. Equations appearing in the nonlinear part must be explicitly resolvable. In this system there is no spreadsheet. A separate module called *Model Compiler* reads the file with model description, verifies its correctness and completeness, and translates it into forms suitable for other modules. This includes a representation of model variables and texts of procedures for efficient calculation of model outputs and their Jacobians. The Model Compiler produces a binary file with the definitions of all model variables and parameters and a C++ code with procedures for the calculation of nonlinear functions and their Jacobians. This code can be compiled and linked with other modules. The binary file can be edited by *Model Editor*. The user can view inputs, outputs, parameters, lower and upper bounds, initial values as well as edit some of them.

The second set of tools provided in IAC-DIDASN++ and adaptable for other similar systems concerns various modelling tasks such as direct simulation, inverse simulation, analysis of ranges on input and output variables, parametric and sensitivity analysis, single criteria optimization and multiobjective analysis; such tools include also specialized optimization solvers adaptable for these modelling tasks.

Various tools are provided for data analysis in graphic form as well as for graphic interaction between an user and the model analysis system. Other application specific tools can be added due to open architecture of the system. Moreover, dynamic interaction between user and the system gives additional possibilities to the user. For example, when the aspiration led methodology is applied to solving multiobjective optimization problems, the specification of reference levels is usually imprecise. The imprecision of reference levels specification can be included the in mathematical formulation of the decision problem by using fuzzy sets. The membership functions might be defined and modified graphically in such a case.

Together, all these tools constitute a modular environment for model analysis, optimization and decision support. Such environment is represented by two more specific cases: the IAC-DIDASN++ system for nonlinear models (described in this paper) and a similar system developed in cooperation with IIASA and other group of Polish researches, based on MOMIP solver for mixed integer programming models (Makowski, 1994). However, other systems for analyzing various classes of models can be also developed when using such an environment.

2 STRUCTURE OF THE IAC-DIDASN++

The IAC-DIDASN++ system is an open one — it consists of a set of independent modules with well defined interfaces, see Figure 1. Each module can potentially be exchanged and replaced by a another one. The system consist of the following modules:

- the system management module;
- model compiler;

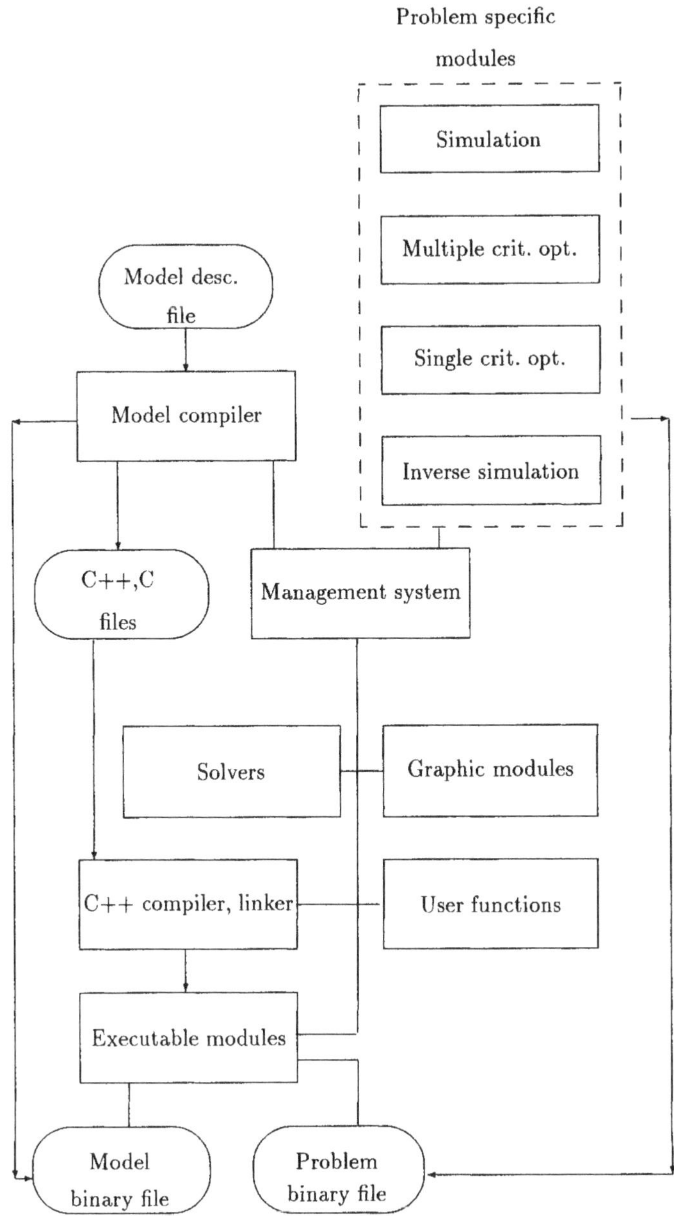

Figure 1 The structure of the IAC-DIDASN++

- model editor;
- problem editor;
- model debugger;
- modules to control the various optimization tasks;
- differentiable solver;
- nondifferentiable solver;
- graphical modules.

These modules can be divided into three groups:

- Executable modules: the system management, model compiler, model editor, problem editor.
- Modules which have to be linked with the models: model debugger, modules to control various optimization task.
- Modules, which are provided in the form of the libraries and could be used in the presented system as well as a tools in different systems: solvers, graphical modules.

The IAC-DIDASN++ system was written in C++ and developed on the Sun workstation under the operating system Solaris. The modules have been compiled with the gnu C++ compiler. The OpenLook widget set was used to develop graphical user interface . It was verified that the whole system can be used on IBM-PC computers working under Solaris x86 operating system. Some of the modules can also be run on an IBM-PC under MS-DOS. These modules were compiled with Borland C++. The C++ compiler must be installed on the user's computer since the C++ code of the model file must be compiled and linked with other modules.

3 THE MODEL FORMULATION AND INTRODUCTORY MODEL ANALYSIS

The models considered in IAC-DIDASN++ are nonlinear and can be expressed by the following format:

$$\begin{aligned} \mathbf{y}_1 &= \mathbf{N}(\mathbf{x}_1, \mathbf{z}, \mathbf{y}_1) + \mathbf{A}_{11}\mathbf{x}_1 + \mathbf{A}_{12}\mathbf{x}_2 + \mathbf{B}_1\mathbf{z} \\ \mathbf{y}_2 &= \mathbf{A}_{21}\mathbf{x}_1 + \mathbf{A}_{22}\mathbf{x}_2 + \mathbf{B}_2\mathbf{z} \\ & \mathbf{x}_l \le \mathbf{x} \le \mathbf{x}_u \\ & \mathbf{y}_l \le \mathbf{y} \le \mathbf{y}_u \\ & \mathbf{z}_l \le \mathbf{z} \le \mathbf{z}_u \end{aligned}$$

where

$\mathbf{x} \in R^n$	- are decision variables $\mathbf{x} = (\mathbf{x}_1, \mathbf{x}_2)$
$\mathbf{z} \in R^p$	- are parameters
$\mathbf{y} \in R^m$	- are outcomes, $\mathbf{y} = (\mathbf{y}_1, \mathbf{y}_2)$
$\mathbf{x}_l, \mathbf{y}_l, \mathbf{z}_l$	- are lower bounds
$\mathbf{x}_u, \mathbf{y}_u, \mathbf{z}_u$	- are upper bounds

A model is prepared in a *Model Description File (MDF)*. The model can be purely nonlinear or consist of a linear and of a nonlinear parts. At present the linear part can be expressed in the form of a MPS file, which consist a fragment of the MDF or is included from it. Equations appearing in the nonlinear part must be explicitly resolvable. However, they can be introduced in any order as they are automatically sorted.

The *Model Compiler* is a tool for convenient formulation of mathematical models, their verification and automatic transformation to the form accepted by the IAC-DIDASN++ solver. The Model Compiler verifies the correctness of both the linear and nonlinear parts and produces:

- a binary file with the definitions of all model variables and parameters;
- a text file with procedures for the calculation of nonlinear functions and their gradients (i.e. outcomes of a nonlinear part of the model and their Jacobians);
- an (auxiliary) text file with the summary of contents of the binary file;
- a log file containing documentation of the compilation process, plus additional information, warnings and error messages.

The first two files are used by other modules.

Only original Model Description File is platform-independent, i.e. it can be transferred both between different operating systems (MS-DOS and UNIX) and different architectures under UNIX (Intel 386 and SUN Sparc).

A model description is contained in a text file called the *Model Description File (MDF)*. Its general layout resembles slightly a classic MPS file format. Generally, a model can consist of a 'linear' and a 'nonlinear' part. The 'nonlinear part' means here — expressed in the form suitable for nonlinear models. In fact it can be linear, but the model compiler and the solver will treat it as a nonlinear one.

The syntax used presents a compromise between the orthodox MPS format and a 'free style' used in modern programming languages. The first parts of MDF, which are pertinent to the linear part inherit the MPS "by rows" structure. Its syntax is however relaxed slightly. There are no restrictions as to starting position of fields and their length. As a consequence of this the names used must be 'proper' in the typical programming sense, e.g. `Vit.A` is incorrect since it contains the dot.

The idea of defining nonlinear problems by extension of the MPS file was applied in the LANCELOT (Conn at al., 1992) system. Their SIF (Standard Input Format) adheres more closely to the MPS file including some of its rather archaic features.

The typical contents of a MPS file can be explicitly written into the MDF file — hence the restrictions forced by compatibility reasons. However there is another, perhaps more convenient form. An external MPS file can be simply included from the MDF file.

The nonlinear sections (if present) appear after linear section (if present). The "by rows" principle is abandoned and its contents can be edited in any reasonable fashion — subject to aesthetical limitations only.

Nearly all MDF key words have synonyms. They are introduced in order to allow a user to choose a form he/she is accustomed to. Key words in the linear and nonlinear parts are (with the common sense limitations) distinct, e.g. the identifier `ROWS` has no predefined meaning in the nonlinear part. Maximal freedom is assured as to the usage or omission of terminators or/and separators and the order of appearance of many language elements. However the use of a consistent style is encouraged.

A model file consists of the following sections:
NAME, ROWS, COLUMNS, RHS, RANGES, BOUNDS, PARAMETERS , VARIABLES, EQUATIONS, ENDATA
The simple model:
$\min y = a * (x_2 - x_1^2)^2 + (x_1 - 1)^2$
with bounds:
$-10 \leq x_1 \leq 10, \quad -10 \leq x_2 \leq 10, \quad -100 \leq y \leq 100$
can be expressed by the following MDF:

```
NAME  EXAMPLE
PARAMETERS
a 100
VARIABLES
x1; initial=0.0; lower=-10.0; upper=10.0;
x2; initial=0.0; lower=-10.0; upper=10.0;
EQUATIONS
y = a*(x2-x1^2)^2+(x1-1)^2;
lower=-100.0; upper=100.0;
MINIMIZE
y;
ENDATA
```

The *Model Debugger* can be used for investigation of the numerical properties of nonlinear models. The *Problem Editor* is a tool for the creation or modification of optimization problems. The user can view, define and modify optimization status of chosen outcomes.

4 THE VARIOUS MODELLING TASK

The following experiments are useful during the model analysis:

- direct simulation, including automatic calculation of the derivatives, step by step calculation of the model;
- analysis of ranges on the input and output variables as well as derivatives;
- single objective optimization and multiobjective analysis;
- parametric and sensitivity analysis;
- model inversion i.e.:
  $\min_{\mathbf{x} \in X} ||\tilde{\mathbf{y}} - \tilde{\mathbf{y}}_{\mathbf{ref}}|| + \rho||\tilde{\mathbf{x}} - \tilde{\mathbf{x}}_{\mathbf{ref}}||$
  subject to
  $\mathbf{y} = \mathbf{F}(\mathbf{x}, \mathbf{z}, \mathbf{y}), \quad \mathbf{F} : R^n \times R^p \times R^m \mapsto R^m$
  $\mathbf{x_l} \leq \mathbf{x} \leq \mathbf{x_u}, \quad \mathbf{y_l} \leq \mathbf{y} \leq \mathbf{y_u}, \quad \mathbf{z_l} \leq \mathbf{z} \leq \mathbf{z_u}$
  $x_{i,ref} : i \in I = \{1, \ldots, n\}, \quad y_{i,ref} : i \in J = \{1, \ldots, m\}$ - the reference values for the selected outcomes and the decision variables.

The optimization in IAC-DIDASN++ is performed by a specially developed nonlinear optimization algorithm called *solver.* Since this maximization is performed repetitively, at least once for each interaction with the user that changes the parameters , there are special requirements for the solver that distinguish this algorithm from typical nonlinear optimization algorithms. It should be robust, adaptable and efficient, that is, it should compute reasonably fast an optimal solution for optimization problems of a broad class without requiring of the user to adjust special parameters of the algorithm in order to obtain a solution. The experience in applying nonlinear optimization algorithms in decision support systems (see Kreglewski at al., 1988) has led to the choice of an algorithm based on penalty shifting technique and projected conjugate gradient method. The algorithm have been used in IAC-DIDAS-N system. In the present system there is a modified version implemented in C language. The experimental nondifferentiable optimization solver is also provided (Granat at al., 1994).

5 GRAPHICAL MODULES

There are three different aspects of graphical information presentation which are present in IAC-DIDASN++:

- Presentation of results in graphic form to improve understanding of these results as well as of relationships between components of the decision problem being solved (objectives, decision variables etc.).
- Providing support for dynamic interaction between user and Decision Support Systems during specification of reference levels. This function contributes to the learning process by allowing an analysis of the history of interaction which has resulted in achieving the current state of the decision making process.
- A possible use of graphic interaction during the specification of reference levels, together with an interpretation of multiple reference levels (aspiration and reservation, possibly other levels) in terms of fuzzy set theory.

The IAC-DIDASN++ provides two graphical modules in the form of libraries. These modules supports the phase of interactive analysis and comparison of results phase in the case of multiobjective optimization.

The first module is used for specifying preferences of the user in terms of fuzzy sets. For each of the objective functions the linear membership function is presented. The user can interactively change membership functions by moving characteristic points. The thick lines show the sensitivity of the solution to the changes of the membership function for the objective. The detailed description can be found in (Granat and Wierzbicki, 1994).

The second one is used for specifying aspiration and reservation levels by using bar charts. The interactive screen consists of two window. The upper one presents the history of the process of interaction. The second one presents the current state of the objectives. For each of the objectives the following information is displayed: the solution (in the form of bar, and the numerical value), the aspiration level (the green triangle and the numerical value), the reservation level (the red triangle and the numerical value), the upper and the lower bounds (the numerical values). The user can change aspiration and reservation levels by moving the red and green triangles or by editing the numerical field.

6 CONCLUSIONS

The IAC-DIDASN++ system have been applied for solving engineering problems in mechanics, automatic control and ship navigation (Wierzbicki at al., 1995) as well as for modelling environmental problems. The modular structure is especially useful for solving problems with real data. Usually in such cases the software should be customized for the specific problem being solved. Under development there is a new version of algebraic processing software and model compiler taking into account dynamic nonlinear models.

The development of the system was partly supported by the grant No. 8 8303 91 02 and No. 0958/P4/94/06 of the Committee for Scientific Research of Poland.

REFERENCES

Brooke, A., Kendrick, D. and Meeraus, A. (1988) *GAMS: a Users's Guide.* The Scientific Press. Redwood City. USA.

Conn, A.R., Gould, N. and Toint, Ph. (1992) *LANCELOT. A Fortran Package for Large-Scale Nonlinear Optimization.* Springer-Verlag. Berlin Heidelberg.

Granat, J. and Wierzbicki, A.P. (1994) *Interactive Specification of DSS User Preferences in Terms of Fuzzy Sets.* Working Paper of the International Institute for Applied Systems Analysis, WP-94-29, Laxenburg, Austria.

Granat, J., Kręglewski, T., Paczyński, J. and Stachurski, A. (1994) *IAC-DIDAS-N++ Modular Modeling and Optimization System. Part I: Theoretical Foundations, Part II: Users Guide.* Reports of the Institute of Automatic Control, Warsaw University of Technology, March 1994, Warsaw, Poland.

Kręglewski, T., Paczyński, J., Granat, J. and Wierzbicki, A.P. (1988) *IAC-DIDAS-N: A Dynamic Interactive Decision Support System for Multicriteria Analysis of Nonlinear Models with a Nonlinear Model Generator Supporting Model Analysis.* Working Paper of the International Institute for Applied Systems Analysis, WP-88-112, Laxenburg, Austria.

Lewandowski, A., Grauer, M., Wierzbicki, A.P. (1983). DIDAS: theory, implementation, in *Interactive Decision Analysis* (eds. M. Grauer, A.P. Wierzbicki), Proceedings Laxenburg 1983. Springer Verlag, Berlin Heidelberg (Lecture Notes in Economic and Mathematical Systems 229).

Makowski, M. (1994) *Methodology and a Modular Tool for Multiple Criteria Analysis of LP Models.* IIASA WP-94-102. Laxenburg. Austria.

Wierzbicki, A.P. and Granat, J. (1995) *Multi-Objective Modeling for Engineering Applications in Decision Support.* Paper presented at the Twelfth International Conference on Multiple Criteria Decision Making. Hagen. Germany.

44

Methodology and modular tools for aspiration-led analysis of LP models

Marek Makowski
International Institute for Applied Systems Analysis
A-2361 Laxenburg, Austria.
Tel: +43-2236-807561. Fax: +43-2236-71313.
e-mail: marek@iiasa.ac.at

Abstract
The paper presents a Multiple Criteria Model Analysis (MCMA) based approach to the analysis of Linear Programming (LP) models. A brief overview of different approaches based on aspiration-led MCDA is followed by an overview of the implemented methodology. The corresponding approach to design and implementation of model-based decision support systems is illustrated by its application to the regional water quality management problem of the Nitra River Basin (Slovakia).

Keywords
Decision making support, multiple-criteria optimization, aspiration-reservation-led decision support.

1 INTRODUCTION

Decision making often requires analysis of large amounts of data and complex relations. In such cases, an analysis of a mathematical model can support rational decision making. Computerized tools designed and implemented for such purposes are called Decision Support Systems (DSS). A DSS, which is typically a problem specific tool, helps in the evaluation of consequences of given decisions and advises what decision would be the best for achieving a given set of goals. In a traditional optimization approach, only one goal is used as an optimized performance index and constraints are set for other goals. Such an approach requires from the user both deep knowledge of and experience in using various Operations Research methods (including model building, optimization techniques, sensitivity analysis). Specialists in other fields and Decision Makers (DM) typically can not meet such requirements. Therefore, multiple-criteria model analyses (MCMA) are being more widely used. The advantages of using MCMA are their ability to handle several goals. This clearly corresponds to the needs of decision support because most of real-life problems are multiobjective. The main advantage of proper implementation of MCDA is due the way it is used. Namely, it helps to analyze the problem rather than providing a single optimal solution. In typical situations, the specification of a consistent set of attainable goals is practically impossible. Therefore a DM interactively changes goals upon analysis of solutions obtained for previously specified goals expressed in terms of aspiration and reservation values for each criterion. Such an approach is called aspiration-led decision support (ALDS). It corresponds well to the natural way of analysis of a broad range of different types of problems. Proper implementation of ALDS requires both deep knowledge of the underlying methodologies and technical skills. However, use of ALDS based DSS is not only easy. It's main advantage is due to the fact that it allows for analysis of the problem by non-specialists in Operations Research who can change his/her preferences during the analysis process and can easily generate and compare solutions (or alternatives) that correspond to those different preferences. This paper presents an extension of the ALDS aimed at allowing a user flexible specification of preferences, optionally in terms of Fuzzy Sets.

2 MULTIPLE CRITERIA OPTIMIZATION BASED DECISION SUPPORT

Any model-based DSS relies on mathematical programming models that can adequately represent decision situations. This means that the model can be used for predicting and evaluating consequences of decisions, which is a basic function of simulation based DSSs. In optimization based DSSs the model is also used to compute decisions that would result in attaining specified goals. A specification of a model to be used within a DSS differs from a specification of a traditional model used for simulation or for single-criterion optimization due to the way the model is used. In traditional approaches a number of constraints are added to the core of the model in order to implicitly define not only feasible but also acceptable solutions. This used to be a must for batch oriented optimization approaches but it should be avoided for the specification of a model that is to be used as a part of a DSS. Hence, it is practical to divide specification and generation of the model into two parts and the corresponding stages:

- First, a *core model* is specified and generated. This model contains only a set of constraints that correspond to logical and physical relations between variables.
- Second, during an interactive procedure a DM specifies goals and preferences, including values of objectives that he/she wants to achieve and to avoid. Such a specification usually results in the generation of additional constraints and variables, which are added to the *core model* thus forming an optimization problem.

Such an approach has several advantages over the traditional approach in which both a preferential structure of a user and logical and physical relations between variables are specified and implemented together. Some of the advantages of the two-stage approach are listed below:

- A core model defines implicitly a set of feasible solutions. Feasibility is understood in the sense of logical and physical relations that must always hold. Therefore this part of a model (once the model is verified) should not be modified during analysis of the model.

- A traditional model quite often has an unnecessarily narrow set of admissible solutions, which is caused by adding constraints aimed at making a solution not only feasible but also acceptable. Such additional constraints correspond to a preferential structure of a user and its implementation is done in a way similar to the constraints representing logical and physical relations. However, this often results in leaving out many interesting solutions beyond the analysis (because such solutions are not considered to be feasible in the strict sense of mathematical programming).
- Interactive analysis of the model is aimed at generation and analysis of rational solutions. Therefore a DM specifies interactively preferences that narrow the set of acceptable solutions. In other words, a DM examines solutions that fulfill both constraints specified by the core model and additional requirements specified by a DM. A DM typically changes those requirements substantially upon analysis of previously obtained solutions. Contrary to the constraints specified by a core model (which must not be violated) additional requirements are very often not attainable therefore they should not be represented as hard constraints.
- An interactive analysis of the model can be done with the help of modular tools that are not problem specific and can be used for a class of problems. Hence, software development is easier. Moreover, different methodologies and corresponding software modules for interactive analysis can be used without changing a core model formulation.

A more detailed discussion of specification of a core model and different traditional approaches to the model analysis for decision support is beyond the scope of this paper. A reader interested in those issues may want to consult (Makowski 1994b).

For the sake of brevity we have to skip the discussion of different multiple criteria based approaches. Those are discussed in detail *e.g.* by Gardiner and Steuer (1994) and by Makowski (1994b). Instead, we briefly summarize only the aspiration level based approach, originally proposed by Wierzbicki (1980). The essence of this method can be summarized as follows:

1. The DM selects, out of the potential objectives, a number (here denoted by n) of variables that will serve as criteria for evaluations of feasible solutions $x \in X_0$ defined by a core model. In typical applications there are 2–7 criteria.
2. The DM specifies (with a help of an interactive tool) an aspiration level composed of the values that he/she wants to achieve for each criterion. Therefore, $\bar{q} = \{\bar{q}_1, \ldots, \bar{q}_n\}$.
3. The problem is transformed by a DSS into an auxiliary parametric single-objective problem. Its solution gives a Pareto-optimal point. If a specified aspiration level $\bar{q}$ is not attainable, then the Pareto-optimal point is the nearest (in the sense of a Chebyshev weighted norm) to the aspiration level. If the aspiration level is attainable, then the Pareto-optimal point is uniformly better than $\bar{q}$.
4. The DM explores various Pareto-optimal points by changing the aspiration levels $\bar{q}$. The underlying formulation of the problem is minimization of an *achievement scalarizing function* that can be interpreted as an ad-hoc non-stationary approximation of the DM's value function, depending on the currently selected aspiration level.
5. The procedures described in points 2, 3 and 4 are repeated until a satisfactory solution is found.

The selection of the Pareto-optimal point depends on the definition of the achievement scalarizing function, which includes also a selected aspiration point. Most of the methods use the achievement scalarizing function in the form:

$$s(q, \bar{q}, w) = \max_{1 \le i \le n} \{w_i(q_i - \bar{q}_i)\} + \epsilon \sum_{i=1}^{n} w_i(q_i - \bar{q}_i) \tag{1}$$

where $q(x) \in R^n$ is a vector of criteria, $\bar{q} \in R^n$ is an aspiration point, $w_i > 0$ are scaling coefficients and ϵ is a given small positive number. Minimization of (1) for $x \in X_0$ generates a properly efficient solution with trade-off coefficients less then $(1 + 1/\epsilon)$. Setting a value of ϵ is itself a trade-off between getting a too restricted set of properly Pareto solutions or a too wide set practically equivalent to weakly Pareto optimal solutions. Assuming the ϵ parameter to be of a technical nature, the selection of efficient solutions is controlled by the two vector parameters: $\bar{q}$ and w.

There is a common agreement that the aspiration point is a very good controlling parameter for examining a Pareto set. Much less attention is given to the problem of defining the weighting* coefficients w. A detailed discussion on weights in a scalarizing function is beyond the scope of this paper. The four commonly used approaches are summarized by Makowski (1994b). In practical applications the most promising approach is based on calculation of weights (that are used in definition of Chebyshev norm mentioned above) with help of the aspiration level $\bar{q}$ and a reservation level $\underline{q}$ (the latter is composed of values of criteria that a user wants to avoid). Such approach has been introduced by the DIDAS family (described in (Lewandowski and Wierzbicki 1989)) of DSS.

3 ASPIRATION/RESERVATION LED DECISION SUPPORT

Following Ogryczak and Lahoda (1992) we will use for Aspiration-Reservation Based Decision Support techniques the acronym ARBDS. The ARBDS is an extension of the aspiration-based approach summarized in Section 2 and is based on the methodology proposed by Wierzbicki (1986), who also formulated general properties for the achievement scalarizing function. The achievement scalarizing function takes the form:

$$\mathcal{S}(q, \bar{q}, \underline{q}) = \min_{1 \le i \le n} u_i(q_i, \bar{q}_i, \underline{q}_i) + \epsilon \sum_{i=1}^{n} u_i(q_i, \bar{q}_i, \underline{q}_i) \tag{2}$$

where $\bar{q}, \underline{q}$ are aspiration and reservation levels, respectively. Maximization of the function (2) over the set of feasible solutions defined by the corresponding core model provides a properly Pareto-optimal solution with the trade-off coefficient smaller than $(1 + 1/\epsilon)$. Component achievement functions $u_i(\cdot)$ are strictly monotone (decreasing for minimized and increasing for maximized criteria, respectively) functions of the objective vector component q_i with values

$$u_i(q_i^U, \cdot) = 1 + \bar{\beta}, \qquad u_i(\bar{q}_i, \cdot) = 1, \qquad u_i(\underline{q}_i, \cdot) = 0, \qquad u_i(q_i^N, \cdot) = -\bar{\eta} \tag{3}$$

*Note that the weights w should not (see *e.g.* (Makowski 1994b, Nakayama 1994)) be used for a conversion of a multiple criteria problem into a into a single criterion problem with a weighted sum of criteria.

where $\bar{\beta}$ and $\bar{\eta}$, are given positive constants, typically equal to 0.1 and 10, respectively.

The piece-wise linear component achievement functions u_i proposed by Wierzbicki (1986) are defined by (4) and by (5) for minimized and maximized criteria, respectively.

$$u_i(q, \bar{q}, \underline{q}) = \begin{cases} \alpha_i w_i(\bar{q}_i - q_i) + 1, & \text{if } q_i < \bar{q}_i \\ w_i(\bar{q}_i - q_i) + 1, & \text{if } \bar{q}_i \leq q_i \leq \underline{q}_i \\ \beta_i w_i(\underline{q}_i - q_i) & \text{if } \underline{q}_i < q_i \end{cases} \tag{4}$$

$$u_i(q, \bar{q}, \underline{q}) = \begin{cases} \alpha_i w_i(\underline{q}_i - q_i) & \text{if } q_i < \underline{q}_i \\ w_i(\bar{q}_i - q_i) + 1, & \text{if } \underline{q}_i \leq q_i \leq \bar{q}_i \\ \beta w_i(\bar{q}_i - q_i) + 1, & \text{if } \bar{q}_i < q_i \end{cases} \tag{5}$$

where $w_i = 1/(\underline{q}_i - \bar{q}_i)$, and α_i, β_i $(i = 1, 2, \ldots, n)$ are given parameters. The parameters α_i and β_i are set in such a way that u_i takes the values defined by (3).

The ARBDS method outlined above can be also interpreted in terms of fuzzy sets as an extension of *interactive fuzzy multi-objective programming* as proposed by Seo and Sakawa (1988). In this approach the membership function is not elicited at an initial iteration but the user is allowed to interactively change it upon analysis of obtained solutions. This approach assumes the classical form of the membership function originally proposed by Zadeh (1965). However, in order to properly handle – within the framework of the component achievement function – the criteria's values worse than a reservation level, and better than an aspiration level, it is necessary to allow for values of a membership function that are negative or greater than one. Such an extension of the membership function has been proposed by Granat and Wierzbicki (1994), who also suggested a method of constructing order-consistent component achievement scalarizing functions based on membership functions describing the satisfaction of the user with the attainment of separate objectives.

The modular tool called LP-Multi (Makowski 1994b) handles the multicriteria analysis of LP and MIP models. It allows a user specification, in addition to aspiration and reservation levels for each criterion, and also interactive specification of preferences for criteria values between aspiration and reservation levels. Therefore, the piece-wise linear functions u_i (4,5) take a more general form. Namely they are defined by segments u_{ji}:

$$u_{ji} = \alpha_{ji} q_i + \beta_{ji}, \qquad q_{ji} \leq q_i \leq q_{j+1,i} \qquad j = 1, \ldots, p_i \tag{6}$$

where p_i is a number of segments for i-th criterion,

$$\alpha_{ji} = \frac{u_{j+1,i} - u_{ji}}{q_{j+1,i} - q_{ji}} \tag{7}$$

$$\beta_{ji} = u_{ji} - \alpha_{ji} q_{ji} \tag{8}$$

where points (u_{ji}, q_{ij}) are interactively defined with the help of FT-Tool due to Granat and Makowski (1995). Concavity of the piece-wise linear function $u(q)$ defined by (6) can be assured by a condition:

$$\alpha_{1i} > \alpha_{2i} > \ldots > \alpha_{p_i i} \tag{9}$$

This condition corresponds well to the nature of the problem since one accepts small changes of u_i when a criterion value is better or close to an aspiration level. The speed of such change should increase along with moving towards a reservation level and should increase even faster between reservation and nadir points. Such features are consistent with the commonly known properties of the membership function used in applications based on the fuzzy set approach. Therefore such an interpretation of the extended-valued membership function makes it possible to combine the extensions of the two methods: *Aspiration Reservation Based Multiple-Criteria Optimization* and *Fuzzy Multi-objective Linear Programming* into a uniform approach described in more detail by Makowski (1994b).

4 DESIGN AND IMPLEMENTATION

Although a DSS must be problem specific, there are methodologies and tools applicable to many different problems. Such methodologies and reusable modular tools have been developed by the Methodology of Decision Analysis Project and have been applied, in collaboration with other projects at IIASA, to several problems. For the sake of illustration we outline only one application, namely, the regional water quality management problem of the Nitra River Basin (Slovakia) documented by Makowski, Somlyódy and Watkins (1995). We consider a river basin or a larger region composed of several basins where the water quality is extremely poor. We also consider a set of waste water treatment plants (either existing or to be possibly constructed) and, at each plant, some technology (which may be composed of a set of technologies to be selected out of a bigger given set of possible technologies) that can be implemented in order to improve the water quality in a region. The traditional optimization based approach to solving such a problem consists of looking for a set of plants and technologies whose implementation would result in maintaining prescribed water quality standards at minimum cost. However, the application of such an approach would in this case, as in many other cases, result in an infeasible solution because of the costs involved. Therefore another approach to decision support has been applied for the Nitra River Basin. Namely, a system of models has been developed for supporting a decision making process. The system is composed of simulation and single criterion dynamic programming models by (Somlyódy, Masliev, Petrovic and Kularathna 1994) and of an aspiration-led multiple criteria optimization model due to Makowski et al. (1995) and it is envisaged to serve two purposes. First, as a decision-aid tool for analysts and high-level decision makers in establishing the effluent and/or ambient water quality standards and the associated appropriate economic instruments that can be enforced to control the waste water discharges. Second, to aid in evaluation of alternative treatment strategies (technologies in treatment plants) and/or in selecting the most appropriate strategy based on the water quality standards and on the costs (capital investment and operational).

The aspiration-led component of this DSS is composed of the following modular and portable software tools (see Figure 1):

1. A Graphical User Interface (GUI) for handling all the interaction with the user. It is linked with FT-Tool (Granat and Makowski 1995) that allows for interactive specification of aspiration and reservation levels (and optionally piece-wise linear membership function for criteria values between those levels), as well as the changing of a criterion's

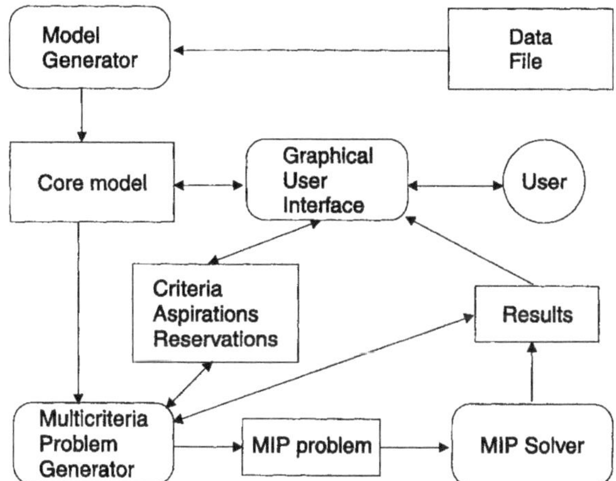

Figure 1 The structure of a Decision Support System for the water quality management in the Nitra River Basin.

status. Another modular tool, LP-Multi (see (Makowski 1994b) for details), processes the multiple criteria problem using the approach outlined in Section 3. The resulting MIP problem is based on the core model and the aspiration and reservation levels which represent current preferential structure of a DM.

2. A problem-specific model generator for generating the core model which relates waste water emissions, treatment decisions, and the resulting ambient water quality. It is important that the core model include only physical and logical relations, and not the preferential structure of the DM.
3. A modular solver for mixed integer programming problems MOMIP developed by Ogryczak and Zorychta (1994).
4. A data interchange tool LP-DIT described by Makowski (1994a). This tool provides an easy and efficient way for the definition and modification of MIP problems, as well as the interchange of data between a problem generator, a solver, and software modules which serve for problem modification and solution analysis.

Portability of the developed tools is achieved by using C++ programming language and a portable commercial tool for Graphical User Interface (GUI) documented in (Inm 1993b, Inm 1993a). Such an approach allows for reuse of most of the components needed for a DSS applied to other problems. It also facilitates experiments with different solvers and with modules providing problem specific interaction with a user. Note, that a new application only requires development of a model generator and, optionally, a problem specific module for a more detailed analysis of results. Hence, the approach presented is useful for fast development of DSS and for analysis of mathematical programming problems that result from the research activities at IIASA.

REFERENCES

Gardiner, L. and Steuer, R. (1994). Unified interactive multiple objective programming, *European Journal of Operational Research* **74**: 391–406.

Granat, J. and Makowski, M. (1995). FT - Modular tool for interactive specification of preferences in terms of fuzzy sets, *Working Paper WP-95-73*, International Institute for Applied Systems Analysis, Laxenburg, Austria.

Granat, J. and Wierzbicki, A. P. (1994). Interactive specification of DSS user preferences in terms of fuzzy sets, *Working Paper WP-94-29*, International Institute for Applied Systems Analysis, Laxenburg, Austria.

Inm (1993a). *zApp Interface Pack, Programmer's Guide & Reference.*

Inm (1993b). *zApp, The Portable C++ Application Framework, Programmer's Guide.*

Lewandowski, A. and Wierzbicki, A. (eds) (1989). *Aspiration Based Decision Support Systems: Theory, Software and Applications*, Vol. 331 of *Lecture Notes in Economics and Mathematical Systems*, Springer Verlag, Berlin, New York.

Makowski, M. (1994a). LP-DIT, Data Interchange Tool for Linear Programming Problems, (version 1.20), *Working Paper WP-94-36*, International Institute for Applied Systems Analysis, Laxenburg, Austria.

Makowski, M. (1994b). Methodology and a modular tool for multiple criteria analysis of LP models, *Working Paper WP-94-102*, International Institute for Applied Systems Analysis, Laxenburg, Austria.

Makowski, M., Somlyódy, L. and Watkins, D. (1995). Multiple criteria analysis for water quality management in the Nitra basin, *Water Resources Bulletin*. Submitted in May 1995.

Nakayama, H. (1994). Aspiration level approach to interactive multi-objective programming and its applications, *Working Paper WP-94-112*, International Institute for Applied Systems Analysis, Laxenburg, Austria.

Ogryczak, W. and Lahoda, S. (1992). Aspiration/reservation-based decision support — a step beyond goal programming, *Journal of Multi-Criteria Decision Analysis* **1**(2): 101–117.

Ogryczak, W. and Zorychta, K. (1994). Modular optimizer for mixed integer programming, MOMIP version 2.1, *Working Paper WP-94-35*, International Institute for Applied Systems Analysis, Laxenburg, Austria.

Seo, F. and Sakawa, M. (1988). *Multiple Criteria Decision Analysis in Regional Planning: Concepts, Methods and Applications*, D. Reidel Publishing Company, Dordrecht.

Somlyódy, L., Masliev, I., Petrovic, P. and Kularathna, M. (1994). Water quality management in the Nitra River Basin, *Collaborative Paper CP-94-02*, International Institute for Applied Systems Analysis, Laxenburg, Austria.

Wierzbicki, A. (1980). The use of reference objectives in multiobjective optimization, *in* G. Fandel and T. Gal (eds), *Multiple Criteria Decision Making, Theory and Applications*, Vol. 177 of *Lecture Notes in Economics and Mathematical Systems*, Springer Verlag, Berlin, New York, pp. 468–486.

Wierzbicki, A. (1986). On the completeness and constructiveness of parametric characterizations to vector optimization problems, *OR Spectrum* **8**: 73–87.

Zadeh, L. (1965). Fuzzy sets, *Information and Control* **8**: 338–353.

Interactive multiobjective optimization system NIMBUS applied to nonsmooth structural design problems

K. Miettinen
Department of Mathematics, University of Jyväskylä
P.O. Box 35, FIN-40351 Jyväskylä, Finland.
Tel: +358 41 602 743. Fax: +358 41 602 731.
e-mail: miettine@math.jyu.fi

M. M. Mäkelä
Department of Mathematics, University of Jyväskylä
P.O. Box 35, FIN-40351 Jyväskylä, Finland.
Tel: +358 41 602 764. Fax: +358 41 602 731.
e-mail: makela@math.jyu.fi

R. A. E. Mäkinen
Department of Mathematics, University of Jyväskylä
P.O. Box 35, FIN-40351 Jyväskylä, Finland.
Tel: +358 41 602 753. Fax: +358 41 602 731.
e-mail: rainom@math.jyu.fi

Abstract

We shortly describe an interactive method, called NIMBUS, for multiobjective optimization involving nondifferentiable and nonconvex functions. We illustrate the functioning of NIMBUS by numerical examples in the area of structural design. We consider a beam with varying thickness and our aim is to find a thickness distribution in such a way that the resulting structure is as good as possible.

Keywords

Nonsmooth optimization, multiobjective optimization, structural optimization

1 INTRODUCTION

In the solution process of optimal shape design and structural design problems we are often faced with optimization problems with several, not necessarily differentiable objective functions (see Mäkelä and Neittaanmäki, (1992), Mäkinen, (1989), and Miettinen and Mäkelä, (1993). However, methods applicable for solving nonsmooth multiobjective optimization problems appear rarely in the literature. Especially, the lack of numerically effective interactive solution methods is evident (as stressed in Miettinen, (1994)).

In this paper we shortly describe an interactive method, called NIMBUS, for multiobjective optimization involving nondifferentiable and nonconvex functions. It has been originally introduced in Miettinen, 1994, and Miettinen and Mäkelä, (1995). The motivations in the development have been numerical efficiency and easiness of use.

We illustrate the functioning of NIMBUS by numerical examples in the area of structural design. We consider a beam with a varying thickness and our aim is to find a thickness distribution in such a way that the resulting structure is as good as possible.

2 NIMBUS METHOD

Let us consider a multiobjective optimization problem of the form

$$\begin{array}{ll} \text{Minimize} & \left\{F_1(x), F_2(x), \ldots, F_k(x)\right\} \\ \text{subject to} & x \in S. \end{array} \tag{1}$$

Pareto optimality and weak Pareto optimality are used as optimality concepts. An essential part in the solution process of multiobjective optimization is played by a decision maker (DM).

The assumptions in the NIMBUS method are that

- All the objective functions are locally Lipschitz continuous.
- The feasible region S is convex.
- Less of any objective is preferred to more in the mind of the DM.

In the NIMBUS method, the idea is that the DM examines the values of the objective functions calculated at a current point x^h and divides the objective functions F_i, $i = 1, \ldots, k$, into up to five classes. Those classes are functions whose values

1. should be decreased ($i \in I^{<}$),
2. should be decreased down till some aspiration level ($i \in I^{\leq}$),
3. are satisfactory at the moment ($i \in I^{=}$),
4. are allowed to increase up till some upper bound ($i \in I^{>}$),
5. are allowed to change freely ($i \in I^{\diamond}$).

The DM is asked to specify the aspiration levels $\bar{F}_i$ for $i \in I^{\leq}$ and the upper bounds ε_i for $i \in I^{>}$. The difference between the classes $I^{<}$ and $I^{\leq}$ is that the functions in $I^{<}$ are to be minimized as far as possible but the functions in $I^{\leq}$ only till the aspiration level. Also weighting coefficients can be connected with the functions in the classes $I^{<}$ and $I^{\leq}$.

According to the classification and the connected information we form a new problem

$$\begin{array}{lll} \text{Minimize} & \left\{F_i(x)/w_i,\ \max_j \left[\max[F_j(x)/w_j - \bar{F}_j,\ 0]\right] \mid i \in I^{<},\ j \in I^{\leq}\right\} & \\ \text{subject to} & F_i(x) \leq F_i(x^h), \quad i \in I^{=} & \\ & F_i(x) \leq \varepsilon_i, \qquad i \in I^{>} & \\ & x \in S, & \end{array} \tag{2}$$

where $\sum_{i \in I^{<} \cup I^{\leq}} w_i = 1$ and $w_i > 0$ for $i \in I^{<} \cup I^{\leq}$, $\bar{F}_i < F_i(x^h)$ for $i \in I^{\leq}$ and $\varepsilon_i > F_i(x^h)$ for $i \in I^{>}$.

This problem is solved by a multiobjective proximal bundle (MPB) method, where the multiple objective functions are treated individually without employing any scalarization. The method is capable of handling several nonconvex, locally Lipschitzian objective functions subject to nonlinear (possibly nonsmooth) constraints and it produces weakly Pareto optimal solutions.

The detailed algorithm of the NIMBUS method is the following.

1. Choose a starting point $x^0 \in S$ and calculate its weakly Pareto optimal counterpart x^1 by setting $I^{<} = \{1, \ldots, k\}$ and employing MPB. Set the iteration counter $h = 1$.
2. Ask the DM to classify the objective functions into $I^{<}$, $I^{\leq}$, $I^{=}$, $I^{>}$, and $I^{\diamond}$ at x^h such that $I^{>} \cup I^{\diamond} \neq \emptyset$ and $I^{<} \cup I^{\leq} \neq \emptyset$. If either of the unions is empty, go to step 9. Ask the DM for the possible weighting coefficients w_i^h, aspiration levels $\bar{F}_i^h$ and the upper bounds ε_i^h.
3. Calculate $\hat{x}^h$ by solving the problem (2) by MPB. If $\hat{x}^h = x^h$, ask the DM whether (s)he wants to try another classification. If yes, set $x^{h+1} = x^h$, $h = h + 1$ and go to step 2. If no, go to step 9.
4. If the DM wants to see different alternatives between x^h and $\hat{x}^h$, set $d^h = \hat{x}^h - x^h$ and go to step 6. If the DM prefers x^h, set $x^{h+1} = x^h$, $h = h + 1$ and go to step 2.
5. Now the DM wants to continue from $\hat{x}^h$. Set $x^{h+1} = \hat{x}^h$, $h = h + 1$ and go to step 2.
6. Calculate P^h different vectors $x^h + t_j d^h$, where $t_j = \left(\frac{j-1}{P^h - 1}\right)$ and $j = 1, \ldots, P^h$.
7. Produce weakly Pareto optimal criterion vectors from the above vectors, employing MPB (with $I^{<} = \{1, \ldots, k\}$).
8. Present P^h alternatives to the DM and let her or him choose the most preferred one among them. Denote the corresponding variable by x^{h+1} and set $h = h + 1$. If the DM wants to continue, go to step 2.
9. Solve the problem (3). Let the solution be $(\tilde{x}, \tilde{\delta})$. Stop. The final solution is $\tilde{x}$.

In the last step, the Pareto optimality of the final solution is guaranteed by solving an additional problem

$$\begin{array}{ll} \text{Maximize} & \sum_{i=1}^{k} \delta_i \\ \text{subject to} & F_i(x) + \delta_i \leq F_i(x^h) \text{ for all } i = 1, \ldots, k, \\ & \delta_i \geq 0 \text{ for all } i = 1, \ldots, k, \\ & x \in S, \end{array} \tag{3}$$

with respect to (x, δ). If x^h is not Pareto optimal, then the solution $\tilde{x}$ is. For clarity of notations, it has not been mentioned in the algorithm that the DM may check the Pareto optimality at any time during the solution process.

The NIMBUS method has been successfully applied to several problems of optimal control, like continuous casting of steel and optimal shape design (see Miettinen, (194), and Miettinen and Mäkelä, (1995)). Next, we utilize our method in the solution processes of a structural design problem.

3 MULTIOBJECTIVE OPTIMIZATION OF ELASTIC BEAMS

We consider an Euler-Bernoulli beam with a varying thickness. The structure is analyzed under three different load cases, namely static loading, free vibration and linear stability. Our aim is to find a thickness distribution in such a way that the resulting structure would be as good as possible. The state problem is discretized with the finite element method. We assume that the beam has a circular cross-sectional shape and we approximate the cross-sectional area by a piecewise constant function. Therefore, we state the following multiobjective optimization problem

$$\text{Minimize} \quad \{F_1(x), F_2(x), F_3(x)\} \tag{4}$$

subject to the discretized state problems

$$\begin{cases} \mathbf{K}(x)\,\mathbf{u} &= \mathbf{f} \\ \mathbf{K}(x)\,\phi &= \lambda \mathbf{M}(x)\,\phi \\ \mathbf{K}(x)\,\psi &= \mu \mathbf{B}\,\psi, \end{cases} \tag{5}$$

the volume constraint

$$V(x) \leq V_{max} \tag{6}$$

and the simple bounds for variables

$$x_{min} \leq x \leq x_{max}. \tag{7}$$

Here x is the vector of design variables, $\mathbf{K}$ is the structural stiffness matrix, $\mathbf{M}$ is the mass matrix, $\mathbf{B}$ is the geometric stiffness matrix, and $\mathbf{u}$ is the nodal displacement vector. The eigenvalues λ and μ are the eigenfrequencies and the buckling load factors, respectively. The choice for objective functions F_i is the following:

$$\begin{aligned} F_1(x) &= \mathbf{f}^{\mathrm{T}}\mathbf{u}(\mathbf{x}) && \text{minimization of the compliance} \\ F_2(x) &= -\lambda_m(x) && \text{maximization of the } m\text{:th eigenfrequency} \\ F_3(x) &= -\mu_1(x) && \text{maximization of the buckling load.} \end{aligned} \tag{8}$$

We assume that the matrices in equations (5) are smooth functions of the design variable vector x. Then it is well-known that the displacement vector $\mathbf{u}$ depends smoothly on x. However, the situation is more complicated in the case of the eigenvalues. It can be shown that, in general, the eigenvalues are only directionally differentiable with respect to the design variables (Haug, Choi and Komkov, (1985)). Nonsmoothness of eigenvalues in structural optimization of beams has been recently studied by Rodrigues, Guedes and Bendsøe, (1995) and Seyranian, Lund and Olhoff, (1994).

Table 1 Intermediate alternatives of the third iteration

	F_1	F_2	F_3
1	0.36×10^{-2}	-4085.51	-47.21
2	0.36×10^{-2}	-4607.19	-46.81
[3]	0.39×10^{-2}	-5021.93	-43.51
4	0.43×10^{-2}	-5278.10	-38.33
5	0.50×10^{-2}	-5311.68	-31.99

4 NUMERICAL RESULTS

Next we will solve a design problem numerically to demonstrate the performance of the NIMBUS system in practical design problems.

The NIMBUS algorithm has been implemented in Fortran 77 and the test runs have been performed on an HP9000/735 (99MHz) computer. The implementation of the MPB routine (called MPBNGC) calls the quadratic solver QPDF4 derived in Kiwiel, (1986). After discretization the dimension of the problem is $n = 14$ and we have chosen $V_{max} = 1.0$, $x_{min} = 0.1$ and $x_{max} = 2.0$.

In the beginning the beam has a constant cross-sectional area, that is, the starting point is $x^0 = (1.0, 1.0, \ldots, 1, 0)$ and the corresponding criterion vector is $z^0 = (F_1(x^0), F_2(x^0), F_3(x^0)) = (0.52 \times 10^{-2}, -3804.06, -39.48)$. We start by minimizing all the objective functions simultaneously. This guarantees that we can begin the actual solution process from a (weakly) Pareto optimal solution. This time, z^0 cannot be improved and we begin our classification from $z^1 = z^0$.

At z^1, our primary goal is to decrease the value of the first objective function and thus $I^< = \{1\}$. Upper bounds are given to the others, $I^> = \{2,3\}$, by $\varepsilon_2^1 = -3000.0$ and $\varepsilon_3^1 = -39.0$. The solution obtained is $\hat{z}^1 = (0.36 \times 10^{-2}, -4085.51, -47.21)$. Notice that all of its components are better than those of z^1. This can be explained by the fact that we can only guarantee the local optimality of the solutions. The new solution is naturally selected for continuation without considering any intermediate alternatives, that is, $z^2 = \hat{z}^1$.

Next, we consider what happens when we decrease the value of the second objective. Thus, $I^{\leq} = \{2\}$ with $\bar{F}_2^2 = -5000.0$ and $I^\diamond = \{1,3\}$.

The result is $\hat{z}^2 = (0.50\times10^{-2}, -5311.68, -31.99)$. Because we did not restrict the values of the other objective functions in any way, we take a look at intermediate alternatives. A set of five candidates has been listed in Table 1.

From this table the third alternative is selected and we have $z^3 = (0.39\times10^{-2}, -5021.93, -43.51)$. The value of the first objective function is here satisfactory and we want to see whether we can further decrease the values of the second and the third function. The next classification is $I^< = \{2,3\}$ with $w_2^3 = 0.5$ and $w_3^3 = 0.5$. We let the first objective function change freely and $I^\diamond = \{1\}$. This setting produces $\hat{z}^3 = (0.38 \times 10^{-2}, -5023.04, -46.24)$ and we continue from it ($z^4 = \hat{z}^3$).

We would still like to see how the other objectives behave when we decrease the value of the second objective function. We classify $I^{\leq} = \{2\}$ with $\bar{F}_2^4 = -6000.0$ and $I^\diamond = \{1,3\}$.

Table 2 Intermediate alternatives of the fifth iteration

	F_1	F_2	F_3
1	0.38×10^{-2}	-5023.04	-46.24
2	0.39×10^{-2}	-5158.13	-45.09
3	0.40×10^{-2}	-5290.22	-43.70
4	0.41×10^{-2}	-5418.73	-42.09
5	0.43×10^{-2}	-5543.00	-40.28
6	0.44×10^{-2}	-5662.26	-38.30
7	0.47×10^{-2}	-5775.60	-36.18
8	0.49×10^{-2}	-5881.88	-33.92
9	0.52×10^{-2}	-5979.68	-31.57
10	0.56×10^{-2}	-6067.16	-29.13

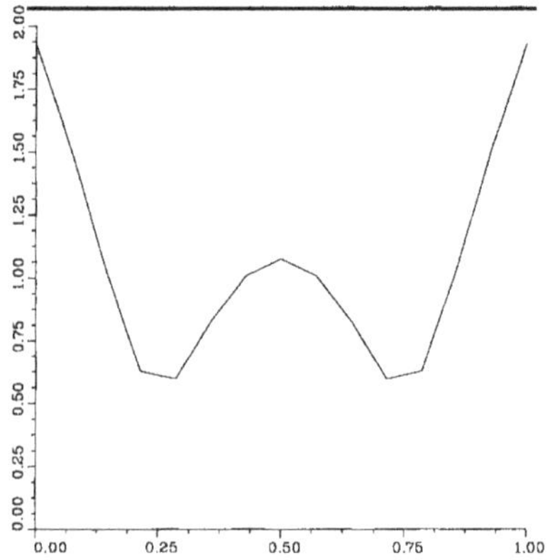

Figure 1 The cross-sectional area of the beam at the final solution.

This classification gives $\hat{z}^4 = (0.56 \times 10^{-2}, -6067.16, -29.13)$. Naturally, the value of the first objective function is too high and we consider ten alternatives listed in Table 2.

We select the second alternative as the final solution, that is, $z^5 = (0.39\times10^{-2}, -5158.13, -45.09)$. It is Pareto optimal. The cross-sectional area of the beam at the final solution is presented in Figure 1.

REFERENCES

Haug, E.J., Choi, K.K. and Komkov, V. (1985) *Design Sensitivity Analysis of Structural Systems*. Academic Press, Orlando.

Kiwiel, K.C. (1986) A Method for Solving Certain Quadratic Programming Problems Arising in Nonsmooth Optimization. *IMA Journal of Numerical Analysis*, **6**, 137–52.

Mäkelä, M.M. and Neittaanmäki, P. (1992) *Nonsmooth Optimization: Analysis and Algorithms with Applications to Optimal Control*, World Scientific Publishing Co.
Mäkinen, R.A.E. (1989) On Numerical Methods for State Constrained Optimal Shape Design Problems, in *International Series of Numerical Mathematics*, vol. 91, Birkhäuser Verlag, Basel.
Miettinen, K. (1994) *On the Methodology of Multiobjective Optimization with Applications.* Doctoral Thesis, University of Jyväskylä, Department of Mathematics, Report 60.
Miettinen, K. and Mäkelä, M.M. (1993) An Interactive Method for Nonsmooth Multiobjective Optimization with an Application to Optimal Control. *Optimization Methods and Software*, **2**, 31–44.
Miettinen, K. and Mäkelä, M.M. (to appear) Interactive Bundle-based Method for Nondifferentiable Multiobjective Optimization: NIMBUS, *Optimization.*
Rodrigues, H.C., Guedes, J.M. and Bendsøe, M.P. (1995) Necessary conditions for optimal design of structures with a nonsmooth eigenvalue based criterion. *Structural Optimization*, **9**, 52–6.
Seyranian, A.P., Lund, E. and Olhoff, N. (1994) Multiple eigenvalues in structural optimization problems. *Structural Optimization*, **8**, 207–27.

Nondifferentiable Optimization

Preliminary computational experience with a descent level method for convex nondifferentiable optimization

U. Brännlund
Royal Institute of Technology, Stockholm, Sweden.
e-mail: uffe@math.kth.se

K.C. Kiwiel
System Research Institute, Warsaw, Poland.
e-mail: kiwiel@ibspan.waw.pl

P.O. Lindberg
Royal Institute of Technology, Stockholm, Sweden.
e-mail: pol@math.kth.se

Abstract

We report on numerical tests with our recently introduced descent level bundle method for convex minimization. The test problems include standard nonsmooth problems, eigenvalue problems, and Lagrangian duals of traveling salesman, capacitated lotsizing, hierarchical production planning and unit commitment problems.

Keywords

Nondifferentiable optimization, proximal bundle methods, level methods.

1 INTRODUCTION

We have recently introduced a descent bundle method Brännlund *et al.* (1995) for minimizing a (possibly nondifferentiable) convex function $f : \mathbb{R}^N \to \mathbb{R}$ over a nonempty closed convex set $S \subset \mathbb{R}^N$. We assume that at each $x \in S$ we can compute $f(x)$ and an arbitrary subgradient $g(x) \in \partial f(x)$.

At the kth iteration, having generated *linearizations* $f^j(\cdot) = f(y^j) + \langle g(y^j), \cdot - y^j \rangle$ of f at *trial points* $y^j \in S$ for $j = 1\colon k$, we approximate f from below by the piecewise linear

Research supported by the Swedish Research Council for Engineering Sciences, the Göran Gustafsson Foundation and the Polish State Committee for Scientific Research under Grant 8S50502206.

cutting-plane *model* $\check{f}^k = \max_{j \in J^k} f^j$, where $J^k \subset \{1{:}\,k\}$ contains k and at most N other indices. We set

$$y^{k+1} = \arg\min\{\, |x - x^k|^2/2 : x \in S,\ \check{f}^k(x) \le f_{\text{lev}}^k \,\}, \tag{1}$$

where the *prox-center* x^k usually has the best f-value among the points $\{y^j\}_{j=1}^k$, and the *target level* $f_{\text{lev}}^k < f(x^k)$ is chosen to ensure $f_{\text{lev}}^k \to f^* := \inf_S f$ as $k \to \infty$. If a finite *lower bound* $f_{\text{low}}^k \le f^*$ is known, then usually $f_{\text{lev}}^k = f(x^k) - \kappa_l \Delta^k$, where $0 < \kappa_l < 1$ and the *optimality gap* $\Delta^k = f(x^k) - f_{\text{low}}^k$ provides the *optimality estimate* $f(x^k) - f^* \le \Delta^k$. Since $\check{f}^k \le f$, if the feasible set $S^k = \{x \in S : \check{f}^k(x) \le f_{\text{lev}}^k\}$ of (1) is empty then $f_{\text{lev}}^k < f^*$, so setting $f_{\text{low}}^k = f_{\text{lev}}^k$ reduces Δ^k by at least κ_l. Also f_{lev}^k is increased and y^{k+1} is recomputed if the *direction* $d^k = y^{k+1} - x^k$ is 'too large' relative to the *desired descent* $\delta^k = f(x^k) - f_{\text{lev}}^k$. A *descent* step to $x^{k+1} = y^{k+1}$ is taken if $f(y^{k+1}) \le f(x^k) - \kappa_d \delta^k$, where $0 < \kappa_d < 1$ (i.e., if the actual descent is at least a fraction of the desired one). Otherwise, a *null* step $x^{k+1} = x^k$ provides a new linearization f^{k+1}.

The original level methods of Lemaréchal *et al.* (1995) require S to be *compact*. They are hardly implementable because they employ $J^k = \{1{:}\,k\}$ and $f_{\text{low}}^k = \min_S \check{f}^k$. In contrast, our method needs bounded storage and is globally convergent without any compactness assumptions.

2 THE DESCENT PROXIMAL LEVEL ALGORITHM

The trial point finding subproblem (1) may be formulated as the QP problem

$$\begin{array}{lll} \text{minimize} & |x - x^k|^2/2 & \text{over all } x \in S \\ \text{satisfying} & f^j(x) \le f_{\text{lev}}^k & \text{for } j \in J^k. \end{array} \tag{2}$$

Algorithm 1

Step 0 (*Initiation*). Select an initial point $x^1 \in S$, a final optimality tolerance $\epsilon_{\text{opt}} \ge 0$, a multiplier bound $t_{\max} > 0$ and parameters $\kappa_d, \kappa_l, \kappa_\delta \in (0,1)$. Choose $f_{\text{low}}^1 \le f^*$ (*e.g.* , $f_{\text{low}}^1 = -\infty$). Set $\Delta^1 = f(x^1) - f_{\text{low}}^1$. Set $\delta^1 = \kappa_l \Delta^1$ if $\Delta^1 < \infty$; otherwise choose $\delta^1 > 0$. Set $J^1 = \{1\}$. Set the counters $k = 1$, $l = 0$ and $k(0) = 1$.

Step 1 (*Level feasibility check*). Set $f_{\text{lev}}^k = f(x^k) - \delta^k$. If (2) is feasible, go to Step 3.

Step 2 (*Update lower bound*). Choose $f_{\text{low}}^k \in [f_{\text{lev}}^k, f^*]$ (*e.g.* , $f_{\text{low}}^k = f_{\text{lev}}^k$ or $\inf_S \check{f}^k$). Set $\Delta^k = f(x^k) - f_{\text{low}}^k$, $\delta^k = \kappa_l \Delta^k$ and go to Step 1.

Step 3 (*Projection*). Find the solution y^{k+1} of (2) and its multipliers λ_j^k such that the set $\hat{J}^k = \{j \in J^k : \lambda_j^k > 0\}$ satisfies $|\hat{J}^k| \le N$. Set $t^k = \sum_{j \in J^k} \lambda_j^k$, $d^k = y^{k+1} - x^k$, $p^k = -d^k/t^k$, $\tilde{f}_S^k = \check{f}^k(y^{k+1}) + \langle p^k, \cdot - y^{k+1} \rangle$ and $\tilde{\alpha}_p^k = f(x^k) - \tilde{f}_S^k(x^k)$.

Step 4 (*Stopping criterion*). If $\sigma^k := \min(\Delta^k, \max\{|p^k|, \tilde{\alpha}_p^k\}) \le \epsilon_{\text{opt}}$, terminate.

Step 5 (*Multiplier check*). If $t^k > t_{\max}$, replace δ^k by $\kappa_\delta \delta^k$ and go to Step 1.

Step 6 (*Descent test*). If $f(y^{k+1}) \le f(x^k) - \kappa_d \delta^k$, set $t_L^k = 1$, $k(l+1) = k+1$ and increase the counter of descent steps l by 1; otherwise, set $t_L^k = 0$ (null step). Set $x^{k+1} = x^k + t_L^k d^k$.

Step 7 (*Selection*). Select $J_s^k \subset J^k$ such that $\hat{J}^k \subset J_s^k$. Set $J^{k+1} = J_s^k \cup \{k+1\}$.

Step 8 (*Gap update*). Set $f_{\text{low}}^{k+1} = f_{\text{low}}^k$ and $\Delta^{k+1} = f(x^{k+1}) - f_{\text{low}}^k$. If $t_L^k = 0$, set $\delta^{k+1} = \delta^k$; otherwise, choose $\delta^{k+1} \in [\min\{\delta^k, \kappa_l \Delta^{k+1}\}, \Delta^{k+1}]$. Increase k by 1 and go to Step 1.

3 MODIFIED LEVEL CONTROLS

The level control of Algorithm 1 can be modified in order to increase its efficiency without destroying convergence. We list a few (implemented) possibilities below.

Suppose Step 1 finds another lower bound $\hat{f}_{\text{low}}^k$ of f^* by computing $\inf_S \check{f}^k$, or from a feasible point to the primal problem in Lagrangian relaxation. If $\hat{f}_{\text{low}}^k \geq f(x^k) - \hat{\kappa}_l \Delta^k$ for some fixed $\hat{\kappa}_l \in [\kappa_l, 1)$ (i.e., $\hat{f}_{\text{low}}^k$ is significantly better than f_{low}^k), then Step 2 may be entered to set $f_{\text{low}}^k = \hat{f}_{\text{low}}^k$ and reduce Δ^k by $\hat{\kappa}_l$.

If $\inf_S \check{f}^k$ is not computed, then $f_{\text{lev}}^k \approx \inf_S \check{f}^k$ may be detected by the test $|p^k| \leq m_\alpha \tilde{\alpha}_p^k$ with a small $m_\alpha > 0$. Hence Step 5 may use this additional test for decreasing δ^k.

We may use the optimality measure $\tilde{\sigma}^k = \min_{j=1}^k\{|p^k| + \tilde{\alpha}_p^k\}$ for decreasing δ^k to $\min\{\kappa_l \Delta^k, \tilde{\sigma}^k\}$ and $\min\{\kappa_\delta \delta^k, \tilde{\sigma}^k\}$ at Steps 2 and 5 respectively, and for letting Step 8 choose $\delta^{k+1} \in [\min\{\delta^k, \kappa_l \Delta^{k+1}, \tilde{\sigma}^k\}, \Delta^{k+1}]$. This allows δ^k to decrease when $f(x^k)$ approaches f^*, as indicated by small $|p^k|$ and $\tilde{\alpha}_p^k$.

If the stepsize bound $t_{\max}$ is too small, the algorithm may crawl towards the solution. Hence our implementation sets $t_{\max} = 100t^1$, possibly increasing it to $\min\{10t_{\max}, 10^{10}\}$ at Step 5 if $t^k > t_{\max}$ and more than two consecutive descent steps occured.

4 SUBGRADIENT AGGREGATION

To trade off storage and work per iteration for speed of convergence, one may replace subgradient selection with aggregation as in Kiwiel (1995b). Alternatively, one may employ *selective aggregation* Kiwiel (1995c) as follows. If we pick $\hat{\imath}, \hat{\jmath} \in J^k$ with $\lambda_{\hat{\imath}}^k, \lambda_{\hat{\jmath}}^k > 0$, replace $f^{\hat{\jmath}}$ by $(\lambda_{\hat{\imath}}^k f^{\hat{\imath}} + \lambda_{\hat{\jmath}}^k f^{\hat{\jmath}})/(\lambda_{\hat{\imath}}^k + \lambda_{\hat{\jmath}}^k)$ and drop $\hat{\imath}$ from J^k, then the solution of (2) does not change.

5 SOLVING QP AND LP SUBPROBLEMS

Instead of using a separate LP solver for checking feasibility of (2) and possibly finding $\min_S \check{f}^k$, we employ the QP solver of Kiwiel (1989) within the exact penalty approach of Kiwiel (1995a). In this approach, given a penalty parameter $t > 0$, one solves

$$\text{minimize} \quad |x - x^k|^2/2 + t\max\{\check{f}^k(x), f_{\text{lev}}^k\} \quad \text{over all } x \in S. \tag{3}$$

Kiwiel (1995a) shows how to choose a sequence of penalty parameters so as to solve (2) or determine that (2) is infeasible and then deliver $\min_S \check{f}^k$. For handling large-scale problems, we intend to replace the QP solver of Kiwiel (1989) by that of Kiwiel (1994).

6 NUMERICAL EXPERIENCE

We present comparisons of Algorithm 1 with our implementation of the simplest level method of Lemaréchal *et al.* (1995). In our notation it can be stated as follows:

Algorithm 2
Step 0 (*Initiation*). Select an initial point $x^1 \in S$ and a final optimality tolerance $\epsilon_{\text{opt}} \geq 0$. If S is unbounded, choose $f^1_{\text{low}} \in (-\infty, f^*]$. Set $J^1 = \{1\}$.

Step 1 (*Call oracle*). Compute $f(x^k)$ and $g(x^k)$.

Step 2 (*Compute lower bound*). Compute $f^k_{\text{low}} = \inf_S \check{f}^k$.

Step 3 (*Optimality test*). Set $\Delta^k = f(x^k) - f^k_{\text{low}}$. If $\Delta^k < \epsilon_{\text{opt}}$ then stop.

Step 4 (*Projection*). Set $f^k_{\text{lev}} = f^k_{\text{low}} + \Delta^k/2$. Solve (2) to obtain x^{k+1}. Set $J^{k+1} = J^k \cup \{k+1\}$. Increase k by one and return to Step 1.

Our implementations of Algorithms 1 and 2, called DPLM and LNN respectively, were programmed in MATLAB, using mex-interfaces to Fortran QP routines. LNN calls CPLEX for its LP subproblems. Our results for LNN deviate slightly from those of Lemaréchal *et al.* (1995), perhaps because a different LP solver was employed.

DPLM stores at most $N+10$ subgradients, whereas LNN stores all of them. For problems where S is not compact we used a known lower bound on f^* for both methods. We tested two versions of DPLM: one with $f^k_{\text{low}} = \max\{\inf_S \check{f}^k, f^{k-1}_{\text{low}}\}$ calculated by the QP solver only when (1) is infeasible, and one in which f^k_{low} is calculated at every iteration using a separate LP solver. The latter version, which in the modified Step 1 used $\hat{\kappa}_l = 0.9999$, is denoted by DPLM(LP). We used the parameters $\kappa_d = 0.05$, $\kappa_l = 0.8$, $\kappa_\delta = 0.1$. The values of f^1_{low} are specified for each example. Step 4 of DPLM used the stopping criterion $\Delta^k \leq \epsilon_{\text{opt}}$, unless stated otherwise.

6.1 Standard test problems

Table 1 gives results for standard test problems Kiwiel (1990), reporting iteration numbers at which the methods find solutions optimal to about six digits.

6.2 Traveling salesman problems

Table 2 gives results for Lagrangian minimum spanning tree relaxations of traveling salesman problems with N cities, starting from the origin. Neither method solved the 442 node problem to the required accuracy; LNN had reached a value of -50434 after 600 iterations and DPLM had reached a value of -50499 after the same number of iterations.

6.3 Eigenvalue problems

Table 3 gives results for eigenvalue problems stated as $\min \lambda_1(M + P^T XQ + Q^T X^T P)$, where M, P and Q are constant matrices, X is a matrix variable, and $\lambda_1(A)$ is the largest eigenvalue of a symmetric matrix A. In these five examples X is 4×5 and M is 7×7, and we used $f^1_{\text{low}} = -0.1$, $S = [-1, 1]^N$ and $\epsilon_{\text{opt}} = 10^{-5}$. These problems are extremely

Table 1 Standard test problems

Problem	N	ϵ_{opt}	f^1_{low}	LNN	DPLM	DPLM(LP)
MXQUAD	10	10^{-5}	-10	74	60	60
GOFFIN	50	10^{-6}	-10	73	53	53
HILB	50	10^{-4}	-100	437	48	48
TR48	48	10^{-0}	-700000	209	183	183
SHUR	5	10^{-4}	0	40	33	36
L1HILN	10	10^{-6}	-10	32	22	22
MINSUM	6	10^{-4}	0	54	50	44
LOCATN	4	10^{-5}	0	24	31	32
POLAK2	8	10^{-5}	-10	110	86	86

Table 2 Travelling salesman problems

N	f^*	ϵ_{opt}	f^1_{low}	LNN	DPLM	DPLM(LP)
6	-617.000	10^{-3}	-1000	15	32	33
14	-3322.000	10^{-2}	-4000	38	42	42
29	-2013.500	10^{-2}	-3000	90	78	77
100	-20937.950	10^{-1}	-30000	312	125	130
120	-6911.250	10^{-2}	-8000	434	211	210
442	-50505.675	10^{-1}	-51000			

ill-conditionend. For termination we had to use the weaker criterion of Step 4 in DPLM and the corresponding one for LNN. DPLM(LP) performed exactly as DPLM.

Table 3 Eigenvalue problems

	LNN		DPLM	
Problem	$f(x^k)$	k	$f(x^k)$	k
EIG1	-2.03590866e-04	234	-2.035181426e-04	445
EIG2	-1.14572767e-05	204	-1.015554397e-05	366
EIG3	-3.89598493e-03	296	-3.898228667e-03	457
EIG4	-2.80409331e-03	326	-2.846003463e-03	520
EIG5	-1.51148612e-04	223	-1.806084067e-04	487
Mean		257		455

Table 4 Lotsizing problems

Test problem			LNN	DPLM	DPLM(LP)
Cap.	Cost	f^*			
Tight	High	27906	60	54	54
Tight	Med.	7997	52	49	58
Tight	Low	2893.3	20	15	15
Med. T	High	24363	54	36	39
Med. T	Med.	7722	27	28	35
Med. T	Low	2893.3	13	15	19
Med. L	High	20293	36	32	29
Med. L	Med.	7534	21	15	19
Med. L	Low	2865	1	1	1
Loose	High	18872	20	14	21
Loose	Med.	7464	13	13	12
Loose	Low	2865	1	1	1
		Mean	26.5	22.75	25.25

6.4 Lotsizing problems

The capacitated multi-item lotsizing problem is a scheduling model, which aims at scheduling production of several products over a planning horizon, while minimizing production costs, inventory holding costs and setup costs subject to demand and capacity constraints. The dual problem which is considered here arises from relaxing the capacity constraints using Lagragian multipliers. In each iteration we attempt to find a primal feasible solution in order to get a lower bound on the objective. The test problems have 8 variables which are constrained to the nonnegative orthant. For details, see Thizy and van Wassenhove (1985) and Brännlund (1993). We used $f^1_{\text{low}} = f^*$ and $\epsilon_{\text{opt}} = 10^{-6+\lceil \log_{10}(|f^*|+1) \rceil}$, i.e. 6 digits of accuracy.

6.5 Hierarchical Production Planning Problems

A variation of the multi-item lotsizing problem was introduced by Graves (1982). The test problems, which are specified in the appendix II of Graves (1982), are randomly generated. They fall into 3 different test sets. The test problems have different level of capacity limits (Tight, Medium, and Loose) and different levels of setup costs (High, Medium, Low). Each test problem has 36 dual variables. We generate one random problem for each test set, each capacity limit and each setup cost. In each iteration we attempt to find a primal feasible solution in order to get a lower bound on the objective. For details, see Graves (1982) and Brännlund (1993). We used $f^1_{\text{low}} = f^*$ and $\epsilon_{\text{opt}} = 10^{-6+\lceil \log_{10}(|f^*|+1) \rceil}$, i.e. 6 digits of accuracy.

Table 5 Hierarchical production planning problems

Test problem				LNN	DPLM	DPLM(LP)
Set	Cap.	Cost	f^*			
1	Tight	High	67435.0	143	130	143
1	Tight	Med.	27560.3	129	184	152
1	Tight	Low	13769.7	104	93	84
1	Med.	High	60163.2	133	130	100
1	Med.	Med.	21803.8	167	133	131
1	Med.	Low	5666.1	101	68	68
1	Loose	High	59118.5	121	112	124
1	Loose	Med.	21792.0	99	64	63
1	Loose	Low	5654.0	61	27	44
2	Tight	High	67334.0	128	146	178
2	Tight	Med.	28336.4	148	190	176
2	Tight	Low	14541.6	94	77	79
2	Med.	High	59698.9	138	116	150
2	Med.	Med.	22015.2	149	180	132
2	Med.	Low	7148.4	138	128	152
2	Loose	High	58116.9	126	129	169
2	Loose	Med.	21479.1	144	153	136
2	Loose	Low	5597.8	69	56	55
3	Tight	High	62346.1	88	110	129
3	Tight	Med.	32719.1	133	143	154
3	Tight	Low	22114.0	61	57	59
3	Med.	High	55385.9	122	109	123
3	Med.	Med.	26112.1	147	115	105
3	Med.	Low	15911.3	79	66	83
3	Loose	High	50000.5	102	99	98
3	Loose	Med.	20712.6	142	125	124
3	Loose	Low	10408.3	77	103	70
			Mean	116.4	112.7	114.1

6.6 Unit Commitment Problems

The thermal unit commitment problem is a large mixed-integer non-linear mathematical programming problem which arises in short-term power production planning. The problem consists of minimizing production costs over a planning period, satisfying system load and reserve constraints as well as technical constraints on each production unit.

We refer to Wood and Wollenberg (1984) for a description of the unit commitment problem. Thorough descriptions of the BARD test problem can be found in Bard (1988), and for the EPRI50 in Zeminger *et al.* (1977). BARD2 is a relaxed modification of BARD. ABB and ABB2 can be obtained from the authors. We used $f^1_{\text{low}} = f^*$.

Table 6 Unit commitment problems

Test problem	N	f^*	ϵ_{opt}	LNN	DPLM	DPLM(LP)
BARD	20	540952	1	399	709	386
BARD2	20	539923	1	334	751	260
ABB	40	105973	1	97	77	84
ABB2	40	143696	1	138	72	93
EPRI50	96	2843720	10	309	156	330
			Mean	255	353	231

REFERENCES

Bard, J.F. (1988) Short-term scheduling of thermal-electric generators using Lagrangian relaxation. *Operations Research*, **36**, 756–66.

Brännlund, U. (1993) *On relaxation methods for nonsmooth convex optimization.* Ph.D. thesis, Department of Mathematics, Royal Institute of Technology, Stockholm.

Brännlund, U., Kiwiel, K.C. and Lindberg, P.O. (1995) A descent proximal level bundle method for convex nondifferentiable optimization. *Operations Research Letters* (to appear).

Graves, S.C. (1982) Using Lagrangian techniques to solve hierarchical production planning problems. *Management Science*, **28**, 260–75.

Kiwiel, K.C. (1989) A dual method for certain positive semidefinite quadratic programming problems. *SIAM Journal on Scientific and Statistical Computing*, **10**, 175–86.

Kiwiel, K.C. (1990) Proximity control in bundle methods for convex nondifferentiable minimization. *Mathematical Programming*, **46**, 105–22.

Kiwiel, K.C. (1994) A Cholesky dual method for proximal piecewise linear programming. *Numerische Mathematik*, **68**, 325–40.

Kiwiel, K.C. (1995a) Finding normal solutions in piecewise linear programming. *Applied Mathematics and Optimization*, **32**, (to appear).

Kiwiel, K.C. (1995b) Proximal level bundle methods for convex nondifferentiable optimization, saddle-point problems and variational inequalities. *Mathematical Programming*, (to appear).

Kiwiel, K.C. (1995c) The efficiency of subgradient projection methods for convex optimization, part II: Implementations and extensions. *SIAM Journal on Control and Optimization*, (to appear).

Lemaréchal, C., Nemirovskii, A.S. and Nesterov, Yu.E. (1995) New variants of bundle methods. *Mathematical Programming*, (to appear).

Thizy, J.M. and van Wassenhove, L.N. (1985) Lagrangean relaxation for the multi-item capacitated lot-sizing problem: A heuristic implementation. *IIE Transactions*, **17**, 308–13.

Wood, A.J. and Wollenberg, B.F. (1984) *Power Generation Operation and Control.* John Wiley and Sons, New York.

Zeminger, H.W., Wood, A.J., Clark, H.K., Laskowski, T.F. and Burns, J.D. (1977) Synthetic electric utility systems for evaluating advanced technologies. **EM-285**. Electrical Power Research Institute (EPRI).

Bundle methods applied to the unit-commitment problem

C. Lemaréchal, C. Sagastizábal
INRIA, BP 105, 78153 Le Chesnay, France.
F. Pellegrino, A. Renaud
EdF, Dépt. Méthodes d'Optimisation et de Simulation, 92141 Clamart, France.

Abstract
On a daily-time scale, the power generation management poses a large-size and composite problem. For this reason, price decomposition has become one of the most commonly used strategies in this field. Classical spatial decomposition, obtained by dualizing the coupling constraints, is not the only way of decomposition. Some recent tests have shown that a s-pace/time decomposition is particularly well-suited to deal with transmission constraints. The resulting dual problem is large but tractable (more than ten thousand variables), while the spatial decomposition would yield exceedingly many dual variables. This paper is devoted to the application of bundle methods to such large-scale nondifferentiable problems. To assess the approach we present some numerical tests which show that, in practice, bundle methods can cope with such large problems.

1 THE POWER UNIT-COMMITMENT PROBLEM

1.1 General formulation

The optimal scheduling of many power units is a challenging problem, from both economical and mathematical points of view. In particular, to solve a short-term unit-commitment problem means to consider simultaneously:
- thermal power-plants with high start-up costs and discontinuous operation domain, introducing non-convex constraints;
- hydro-valleys of interconnected reservoirs, leading to tight dynamic constraints;
- various types of integral and logical constraints.

More precisely, we study a mix I of power-generation units (hydraulic valleys, nuclear plants and classical thermal units) for a discrete period of time $\{1, 2, \ldots, T\}$ (typically, 48 half-hour steps). Call p_i^t the production level of unit $i \in I$ at time $t \le T$. Througout this paper, we will denote by $p^t := (p_1^t, p_2^t, \cdots, p_{|I|}^t)$ the production vector of all the units at time t; likewise, $p_i := (p_i^1, p_i^2, \ldots, p_i^T)$ will be the production vector of unit i for the time period.

Altogether, our problem is

$$\begin{cases} \min \sum_{i \in I} c_i(p_i) \\ p_i \in \mathcal{D}_i\,, \quad \text{for all } i \in I\ , \\ p^t \in \mathcal{S}^t\,, \quad \text{for } t = 1, 2, \ldots, T\ , \end{cases} \tag{1}$$

where $\mathcal{D}_i$ describes the operating dynamic constraints of unit i and $\mathcal{S}^t$ stands for the static constraints such as satisfaction of demand and network transmission. As for the objective function, each c_i represents the whole production cost of each unit i over the time period.

1.2 The model

Dynamic constraints

Problem (1) can be considered as a general modelling of the unit-commitment problem. Yet, the operating constraints are not exactly the same from one utility to the other. Without entering into details, a general description of the $\mathcal{D}_i$'s for the French generation mix includes the following features:

Nuclear power plants: A minimum delay of two hours is necessary between two output variations; some output ranges are forbidden; the number of output variations is limited.

Classical thermal units: If the plant is online, then $p_i^t \in [\underline{p_i}, \overline{p_i}]$, with $\underline{p_i} > 0$; a plant must not have more than one start-up per day; during shutdown or start-up phases, fixed output curves must be followed.

Needless to say, the description of the operating constraints for these thermal plants involves 0-1 variables, to describe starts-up, number of output variations, etc.

Hydro-valleys: A valley consists of a set of interconnected reservoirs. During low-demand periods, water in a downstream reservoir can be pumped upstream, so that it can be discharged later, when generation costs become higher.

More formally, let $\mathcal{H}_v$ denote the set of water reservoirs of valley v, and let $\Gamma(h \leftarrow)$ be the set of plants upstream a particular reserve $h \in \mathcal{H}_v$.Then the following must hold for all $h \in \mathcal{H}_v$ and $t = 1, 2, \dots, T$:

$$V_h^{t+1} - V_h^t = a_h^t - T_h^t + Q_h^t - D_h^t + \sum_{g \in \Gamma(h \leftarrow)} \left(T_g^{t_{g \to h}} - Q_g^{t_{g \to h}} + D_g^{t_{g \to h}} \right),$$

where

- $V_h^t \in [\underline{V_h}, \overline{V_h}]$ and a_h^t are respectively the content and inflow of reservoir h at time t;
- $T_h^t \in [\underline{T_h^t}, \overline{T_h^t}]$ is the discharge, $D_h^t \in [\underline{D_h^t}, \overline{D_h^t}]$ the spillage and $Q_h^t \in [\underline{Q_h^t}, \overline{Q_h^t}]$ the quantity of water pumped by plant h at time t;
- $t_{g \to h} := t - d(g \to h)$, with $d(g \to h)$ denoting the discharge delay from plant g to reserve h.

Static constraints

Concerning the description of $\mathcal{S}^t$, we distinguish two types of constraints.

Demand: For each time-step $t = 1, 2, \dots, T$, the demand D^t is known and has to be satisfied: $\sum_{i \in I} p_i^t = D^t$.

"Spinning reserve" constraints could also be introduced, guaranteeing that there is enough power started up to quickly face most random events (unit outage, error in demand forecast, ...). These constraints are of the same type as above; for simplicity, we do not consider them here.

Transmission: The capacity of each arc in the network is limited. "Security constraints" can also be included, to ensure the satisfaction of the demand even after the outage of one line. In the flow or the DC approximation model, these constraints are linear; but, as opposed to demand or reserve, they are numerous: the network has several hundreds of nodes. With security constraints, there will be several hundred thousand linear constraints at each time step.

2 DECOMPOSITION SCHEMES

Abstractly, our problem can be written

$$\min c(p), \quad p \in \mathcal{D} \cap \mathcal{S}, \tag{2}$$

where $p = \{p_i^t\} \in R^I \times R^T$ represents the vector of productions, and the cost function c is a sum over $i \in I$ of "local costs" $c_i(p_i)$. As for the feasible sets $\mathcal{D}$ and $\mathcal{S}$, they both have the product form $\mathcal{D} = \prod_{i \in I} \mathcal{D}_i$ and $\mathcal{S} = \prod_{t=1}^{T} \mathcal{S}^t$; but $\mathcal{D}$ is nonconvex and has a rather complicated description, while $\mathcal{S}$ is described by linear constraints, say

$$\text{for each } t \in T, \quad p^t \in \mathcal{S}^t \Leftrightarrow A^t p^t = b^t. \tag{3}$$

Here, each matrix A^t is either a single row $(1, \cdots, 1)$ or an awfully large full matrix, depending on whether or not transmission constraints are considered.

Clearly, (2) is fairly difficult: it is a large-scale, nonlinear, mixed integer programming problem. However, it has a highly decomposable structure which can be exploited through *Lagrangian relaxation*. We study here two decomposition schemes, which lend themselves to a handy (approximate) numerical treatment of (2). For more details on dualization techniques see Chap. XII in (Hiriart-Urruty, Lemaréchal, 1993); see also (Grinold, 1970).

2.1 Spatial decomposition

Exploiting the linearity (hence decomposability) of the constraints (3), it is natural to write (2) as:

$$\min \sum_{i \in I} c_i(p_i) \quad \text{subject to} \sum_{i \in I} A_i p_i = b \text{ and } p_i \in \mathcal{D}_i \quad \text{for } i \in I, \tag{4}$$

where b denotes the compound vector $(b^1, \cdots, b^T)$ and similarly for the compound matrices A_i.

The writing in (4) exhibits the role of the production units p_i as "local agents", coupled by the constraints (3). For the present unit-commitment problem, the local agents are the power-plants, coupled by the "spatial" static constraints.

To achieve decomposition, associate a multiplier μ^t to the coupling constraints in (4) and form the Lagrangian

$$L_s(p, \mu) := \sum_{i \in I} [c_i(p_i) - \mu^\top A_i p_i] + \mu^\top b.$$

The associated dual function is then defined by

$$\Sigma(\mu) := \min_{p\in\mathcal{D}} L_s(p,\mu) = \mu^\top b + \min_{p\in\mathcal{D}} \sum_{i\in I} [c_i(p_i) - \mu^\top A_i p_i] \tag{5}$$

and can be computed in a decomposed way: each local agent $i \in I$ has to solve

$$\min\{c_i(p_i) - \mu^\top A_i p_i : p_i \in \mathcal{D}_i\}.$$

In other words, each power-plant has to optimize its own production, for a given "shadow price" μ. The whole issue is then to find appropriate values for μ; by duality theory, what we have to do is to maximize the function $\Sigma(\mu)$. This is the so-called *dual problem* associated with the formulation (4).

The above decomposition scheme has been used for rather long for the unit-commitment problem (Fisher, 1973), (Merlin, Sandrin, 1983).

2.2 Space-time decomposition

Our second scheme starts from a naive stratagem: artificially duplicating the variable p, we can write (2) in the "cross-form"*

$$\min c(p) \quad \text{subject to}\; p \in \mathcal{D},\; q \in \mathcal{S}, \quad p = q. \tag{6}$$

An obvious decomposition has thus appeared, which is actually enhanced by the particular form of $c, \mathcal{D}$ and $\mathcal{S}$. In fact, the variables $\{p_i^t\}$ and $\{q_i^t\}$ are "local agents", coupled by the constraints $p_i^t = q_i^t$.

To achieve decomposition, associate a multiplier λ_i^t to each coupling constraint and form the Lagrangian

$$L_\psi(p,q,\lambda) := c(p) - \lambda p + \lambda q.$$

The dual function is now defined by

$$\Psi(\lambda) := \min_{p\in\mathcal{D}, q\in\mathcal{S}} L_\psi(p,q,\lambda) = \min_{p\in\mathcal{D}} \sum_i [c_i(p_i) - \lambda_i p_i] + \min_{q\in\mathcal{S}} \sum_t \lambda^t q^t. \tag{7}$$

Here again, it can be computed in a decomposed way: the local agents have to solve respectively

$$\min\{c_i(p_i) - \lambda_i p_i : p_i \in \mathcal{D}_i\} \quad \text{and} \quad \min\{\lambda^t q^t : q^t \in \mathcal{S}^t\}. \tag{8}$$

The above p-problems are essentially the same as in spatial decomposition. As for the q-problems, they are just linear programs. When only demand-constraints are present, the feasible domain is a mere simplex. As for transmission constraints, they result in classical static network problems. For this scheme also, duality theory tells us that appropriate values of λ have to maximize $\Psi(\lambda)$.

---

*the objective function could also be $c(q)$.

For nonlinear unit-commitment problems, this decomposition scheme has been proposed in (Batut, Renaud, 1992). It somehow generalizes an earlier approach for mixed integer linear programming, given in (Guignard,Kim, 1987).

3 THE DUAL PROBLEMS

In the previous section, we have defined two approaches to solve (1)=(2), namely the two dual problems $\max_\mu \Sigma(\mu)$ of (5), and $\max_\lambda \Psi(\lambda)$ of (7). We now study these two new problems.

3.1 Theoretical assessment

A first question is: are the dual problems any good in terms of their respective primals (4) or (6)? Call $\bar{c}$ the optimal objective value in the primal problem (1)=(2)=(4)=(6). It is known that both dual optimal objective values $\bar{\Sigma}$ and $\bar{\Psi}$ are lower than $\bar{c}$, and a crucial property is whether any of them is equal to $\bar{c}$. However, this has no reason to be true, because (1) is a nonconvex optimization problem. In this sense, the dual approaches proposed in § 2 can only yield approximate solutions. Clearly, for such approaches to be performant, it is crucial that the (positive) *duality gaps* $\bar{c} - \bar{\Sigma}$ and $\bar{c} - \bar{\Psi}$ are sufficiently small.

Here comes an important result that we state here without proof. Because the static constraints $\mathcal{S}$ are linear,

there always holds $\bar{c} \geq \bar{\Psi} = \bar{\Sigma}$.

When (2) is an integer linear program, a proof can be found in (Guignard, Kim, 1987). For the general case it is a consequence of Proposition XII.5.1.3 of (Hiriart-Urruty, Lemaréchal, 1993). A detailed proof can be found in (Lemaréchal, Renaud, 1996), together with an interpretation of the optimal dual solutions $\bar{\mu}$ or $\bar{\lambda}$, as marginal prices associated with particular convexified forms of (1). As a result, if some nonconvex constraints are duplicated (taken into account in $\mathcal{S}$ *and* in $\mathcal{D}$), then the duality gap is reduced for the space-time decomposition.

Thus, for our problem, the space-time decomposition scheme can only give better dual solutions than the spatial one.

Another issue is the complexity of the resulting duals $\max \Sigma(\mu)$ and $\max \Psi(\lambda)$. We recall that a dual function such as Σ or Ψ, defined via a minimization problem such as (5) or (7), is concave; and that any optimal solution such as $\bar{p}$ in (5) or $(\bar{p}, \bar{q})$ in (7) gives a subgradient of the corresponding dual function:

$$\nabla_\mu L_s(\bar{p}, \mu) = b - \sum_{i \in I} A_i \bar{p}_i \in \partial\Sigma(\mu) \quad \text{and} \quad \nabla_\lambda L_\psi(\bar{p}, \bar{q}, \lambda) = \bar{q} - \bar{p} \in \partial\Psi(\lambda) . \tag{9}$$

Because, due to nonconvexity, such solutions $\bar{p}$ and $\bar{q}$ have no reason to be unique, $\nabla\Sigma$ and $\nabla\Psi$ have no reason to exist: the dual functions are nonsmooth. Now, the complexity of a nonsmooth optimization problem depends crucially on the number of variables.

In the case of space-time decomposition, the dual variables are indexed in $I \times \{1, \ldots, T\}$. For the French generation mix, we have $|I| \simeq 150$; with $T = 48$, this makes about 7000 dual variables λ_i^t.

For the spatial decomposition, according to our description of static constraints, there are two cases:

- If $\mathcal{D}$ involves only demand satisfaction, there is one linear constraint for each time-period t. The dual function Σ has then 48 variables μ^t, a reasonable number[†]
- If transmission constraints are taken into account, each righthand side b^t in (3) may have hundreds of thousands of coordinates; this results in a huge vector μ.

Thus, space-time decomposition becomes a must when transmission constraints are present; indeed, this has been the initial motivation of such an approach (Batut, Renaud, 1992).

3.2 Dual algorithms

Traditionally, subgradient optimization (Shor, 1985) has been used to maximize the non-smooth function Σ or Ψ; this is only conceivable for the simplest situation of spatial decomposition without transmission constraints. To handle the 10^4-10^5 variables appearing in the other cases, sophisticated methods must be applied: we choose bundle methods. They are a certain improvement of the classical cutting-plane algorithm, using ideas from the proximal regularization; for a general review, see Chapter XV in (Hiriart-Urruty, Lemaréchal,1993). We have conducted experiments with a particular implementation, namely Algorithm 19 of (Lemaréchal, Sagastizábal, 1993), using the objective regularization (g-f), diagonal quasi-Newton updates (dqN) and curved search. For the reader's convenience, we briefly sketch the algorithm here.

3.3 General Scheme

Consider the maximization of $\Psi(\lambda)$, with the help of local solvers able to compute $\Psi(\lambda)$ in (7), yielding $g(\lambda) \in \partial\Psi(\lambda)$ from (9). We choose a positive integer `MEMAX`. At the current iteration of the algorithm, we have on hand a "bundle" $\{\Psi(\lambda_k), g(\lambda_k)\}_{k\leq n}$ of $n \leq$`MEMAX` function- and subgradient-values. Therefore `MEMAX` is the maximum size of the bundle. We also have a current prox-center $\hat{\lambda}$. We define the polyhedral model $\hat{\Psi}_n(\lambda) := \min_{k\leq n}[\Psi(\lambda_k)+g(\lambda_k)(\lambda-\lambda_k)]$ and, for given prox-coefficient $t > 0$, we call λ^c the solution of the quadratic problem

$$\max_{\lambda}\,[\hat{\Psi}_n(\lambda) - \frac{1}{2t}|\lambda - \hat{\lambda}|^2]\,. \tag{10}$$

Step 1 *(t-adjustment)* Solve (10) and decide between the following actions:

- Declare "t too large": decrease t and solve (10) again.
- Declare "t too small": increase t and solve (10) again.
- Declare "descent-step": go to Step 2.
- Declare "null-step": go to Step 3.

Step 2 *(descent-step)* Update $\hat{\lambda} = \lambda^c$ and compute a new $t > 0$.

Step 3 *(bundle management)* If the size of the bundle has reached its maximum `MEMAX`, delete one element $\{\Psi(\lambda_k), g(\lambda_k)\}$. Append the new element $\{\Psi(\lambda^c), g(\lambda^c)\}$ to the bundle and loop to Step 1.

[†] if the "spinning" reserve constraints were taken into account, this number would be doubled.

We refer to (Lemaréchal, Sagastizábal, 1993) for details concerning the decision in Step 1 and the computation of t in Step 2. The quadratic code to solve (10) is due to K.C. Kiwiel, and is described in (Kiwiel, 1986).

3.4 Stopping test

The above description of the algorithm skips the question of the stopping criterion. This question is not straightforward in nonsmooth optimization; it becomes particularly crucial for the present problem, especially because of its large number of variables.

The resolution of (10) gives not only the candidate λ^c but also a certain $\varepsilon \geq 0$ and a certain "regularized subgradient" G, which is an ε-subgradient of Ψ at the prox-center $\hat{\lambda}$. Denoting by $\bar{\lambda}$ a maximizer of Ψ, we can write

$$\Psi(\bar{\lambda}) \leq \Psi(\hat{\lambda}) + \left\langle G, \bar{\lambda} - \hat{\lambda} \right\rangle + \varepsilon \leq \Psi(\hat{\lambda}) + \|G\| \, \|\bar{\lambda} - \hat{\lambda}\| + \varepsilon \,. \tag{11}$$

The rationale is then to request from the bundle algorithm a point $\hat{\lambda}$ which maximizes Ψ up to a tolerance $\underline{\varepsilon}$, chosen beforehand. Then we stop when both ε and $\|G\|$ are small, say $\varepsilon \leq \underline{\varepsilon}$ and $\|G\| \leq \mathtt{B}_G$. Ideally, (11) shows that we should have $\mathtt{B}_G = \underline{\varepsilon}/\|\hat{\lambda} - \bar{\lambda}\|$. Since this quantity is unknown, we need a value accounting for some "equanimity-in-the-adversity". For this, we assume that each coordinate $\|(\bar{\lambda} - \hat{\lambda})_i^t\|$ is smaller than some d; reasonable values of d are given by the nature of the problem. Then we obtain $\mathtt{B}_G = \underline{\varepsilon}/(\sqrt{n}d)$, where $n = |I|T$ is the number of (dual) variables.

4 NUMERICAL ILLUSTRATIONS

For numerical illustration, we have tested 6 datasets corresponding to 6 different days in the French mix. The set $\mathcal{S}$ is described by 48 demand-constraints in (3); the spatial dual problem has therefore 48 variables in all examples. Examples 1-4 have 207 plants in production, Examples 5 and 6 have 226 plants. For the space-time dual problems, this makes respectively $48 \times 207 = 9936$ and $48 \times 226 = 10848$ dual variables.

The table below compares the numbers of iterations with the two dualization schemes. The first column gives the number of resolutions of the decomposed problems (5) and (7) respectively. Between parentheses, we also give the number of descent-steps in the bundle algorithm of § 3.3. Then come the dual objective value after convergence, and the corresponding norm of regularized gradient. Note that $\mathtt{B}_G$ of § 3.4 corresponds to $d \simeq 0.1$FF/kW. In all our tests, the tolerance $\underline{\varepsilon}$ was set to 0.1%; this accuracy is good enough for a duality gap of 0.5%, typical for our unit-commitment problems. The aim of the column "partial" is to illustrate the difficulty to reach a reliable stopping criterion in nonsmooth optimization. For space-time decomposition (which has many variables), it gives the number of iterations and descent-steps necessary to obtain the required 3 digit accuracy. One sees how a bundle algorithm proceeds: first it improves objective values, and then it spends a rather long time reducing $\|G\|$ to appropriate values. Finally the last two columns give computing times: the proportion of CPU spent by the bundle algorithm itself (the rest being taken by Lagrangian minimizations) and the total CPU time on a SUN Sparc 20. Let us explain the drastic differences appearing in the last but one column. Most of the time used by a bundle algorithm is naturally due to the resolution of (10); but the complexity of the latter is far from what is commonly believed. Actually, one solves

Test	# iter	Dual value	$\|G\|$	Partial	Bundle	CPU(s)
1Σ	33(9)	5195575	150		0.02%	351
1Ψ	250(37)	5194556	100	85(22)	22%	1490
2Σ	19(5)	7842647	700		0.01%	327
2Ψ	98(16)	7841180	150	41(9)	11%	718
3Σ	25(70)	6499580	160		0.02%	237
3Ψ	280(33)	6499297	130	60(14)	27%	1480
4Σ	20(4)	8310985	690		0.02%	238
4Ψ	131(14)	8310235	160	37(8)	14%	590
5Σ	58(15)	4153291	110		0.04%	480
5Ψ	561(56)	4152531	70	209(49)	31%	3880
6Σ	34(10)	5679565	160		0.03%	326
6Ψ	216(28)	5678200	110	83(20)	22%	1480

the dual of (10), which is of the type $\min_\alpha \left[\sum_{k,\ell=1}^{n} \alpha_k \alpha_\ell \langle g(\lambda_k), g(\lambda_\ell)\rangle + \sum_{k=1}^{n} p_k \alpha_k\right]$. This requires first the computation of the scalar products $\langle g(\lambda_k), g(\lambda_\ell)\rangle$, which become fairly expensive when the dual space has 10^4 variables. Afterwards, the minimization problem is quickly solved: its number of variables does not exceed the number of iterations, which is itself reasonable; besides, it differs very little from the quadratic problem of the previous iteration.

REFERENCES

Batut, J., Renaud, A. (1992) Daily Generation Scheduling with Transmission Constraints: A New Class of Algorithms. *IEEE Transactions on Power Systems*, **7**, 982–9.

Fisher, M.L. (1973) Optimal Solution of Scheduling Problems Using Lagrangian Multipliers: Part I. *Operation Research*, **21**, 1114–27.

Grinold, R.C. (1970) Lagrangian subgradients. *Management Science*, **17**, 185–8.

Guignard, M., Kim, S. (1987) Lagrangean decomposition: a model yielding stronger lagrangean bounds. *Mathematical Programming*, **39**, 215–28.

Hiriart-Urruty, J.-B., Lemaréchal, C. (1993) *Convex Analysis and Minimization Algorithms*. Springer-Verlag.

Kiwiel, K.C. (1986) A Method for Solving Certain Quadratic Programming Problems Arising in Nonsmooth Optimization. *IMA Journal of Numerical Analysis*, **6**, 137–52.

Lemaréchal, C., Renaud, A. forthcoming paper.

Lemaréchal, C., Sagastizábal, C. (1994) An Approach to Variable Metric Bundle Methods in *Lecture Notes in Control and Information Sciences*, 197: System Modelling and Optimization.

Merlin, A., Sandrin, P. (1983) A New Method for Unit Commitment at Electricité de France. *IEEE Transactions on Power Apparatus and Systems*, **5**, 1218–25.

Shor, N. (1985) *Optimization methods for non-differentiable functions*, Springer-Verlag.

Nondifferentiable optimization solver: basic theoretical assumptions

Andrzej Stachurski
Institute of Automatic Control, Warsaw University of Technology
Nowowiejska 16/19, Warszawa, Poland. Tel: & Fax: +48 22-253719.
e-mail: stachurski@ia.pw.edu.pl

Abstract
The paper presents algorithms realized in NDOPT - the code designed for solving nondifferentiable convex optimization problems. The main algorithm is based on the use of the exact penalty function and linearizations of the nonlinear constraints. Its novelty lies in the original development of the solution method of the QP direction search problem. A strictly convex reduced primal direction search problem is solved, not dual as usual.

Keywords
nondifferentiable convex optimization, exact penalty function, linearization, primal direction search problem

1 INTRODUCTION

NDOPT is a part of the DIDASN++ package designed for multicriteria optimization and modelling. It is one of the optimization solvers included in DIDASN++. The aim of this paper is to present the algorithms realized in this code.

NDOPT is a collection of standard ANSI C functions designed to solve small-scale nondifferentiable optimization problems of the following standard form

$$\text{minimize} \qquad f(x) = \max\{\ f_j(x): \quad j = 1, \ldots, m_O\ \}, \tag{1}$$

subject to

$$y_j^L \leq F_j(x) \leq y_j^U, \quad for \quad j = 1, \ldots, m, \tag{2}$$

$$b^L \leq Ax \leq b^U, \tag{3}$$

$$x_i^L \leq x_i \leq x_i^U \qquad \text{for} \qquad i = 1, \ldots, n, \tag{4}$$

Sponsored by KBN, grant No. 3 P40301806 "Parallel computations in linear and nonlinear optimization" where the vector $x = (x_1, \ldots, x_n)^T$ has n components, f_j and F_j are locally Lipschitz continuous functions, A is an $m_L \times n$ matrix, y^L and y^U are constant m - vectors, b^L and b^U are constant m_L -vectors and x^L and x^U are constant n-vectors. Matrix A is assumed to be dense.

The nonlinear functions f_j and F_j have to be Lipschitz continuous, however they need not be continuously differentiable (have continuous gradients, i.e. vectors of partial derivatives). In particular, they may be convex. The user has to provide a C function for evaluating the problem functions and their single subgradients (called generalized gradients) at each x satisfying the linear constraints. For example, if F_j is continuously differentiable then its subgradient $g_{F_j}(x)$ is equal to the gradient $\nabla F_j(x)$ and if F_j is a max function

$$F_j(x) = \max\{\varphi(x, z); \quad z \in Z\} \tag{5}$$

i.e. $F_j(x)$ is a pointwise maximum of smooth functions and Z is compact, then $g_{F_j}(x)$ may be calculated as the gradient $\max_{z \in Z(x)} \partial_x \varphi(x, z)$ (with respect to x), where $Z(x)$ is an arbitrary solution to the maximization problem in (5). (Surveys of subgradient calculus, which generalizes rules like $\partial(F_1 + F_2)(x) = \partial F_1(x) + \partial F_2(x)$, may be found in Clarke (1983).

NDOPT implements the descent methods applied to the exact penalty function for problem (1–4). That approach have been recently intensively studied by Kiwiel in numerous papers, *e.g.* Kiwiel (1985,1988).

2 EXACT PENALTY FUNCTION

We assume the use of the exact penalty function with respect to the nonlinear constraints. Problem (1-4) is therefore internally transformed to the following condensed form:

$$\text{minimize} \quad f(x) \quad \text{over all} \quad x \in R \tag{6}$$

subject to

$$F(x) \leq 0, \tag{7}$$

$$b^L \leq Ax \leq b^U, \tag{8}$$

$$x^L \leq x \leq x^U, \tag{9}$$

where f is the objective function, F is the aggregate constraint function

$$F(x) \quad = \quad \max(F^L, F^U), \qquad \text{where}$$

$F^L = \max\{y_j^L - F_j(x) : j = 1, \ldots, m\}$ - aggregate lower inequality constraint function,
$F^U = \max\{F_j(x) - y_j^U : j = 1, \ldots, m\}$ - aggregate upper inequality constraint function.

The m_L inequalities (8) are called the general linear constraints, whereas the box constraints (9) specify upper and lower simple bounds on all variables.

The standard form (1-4) is more convenient to the user than (6-9), since the user does not have to program additional operations for evaluating the functions F^L and F^U and their subgradients. On the other hand, the condensed form simplifies the description of algorithms.

Let S_L denotes the feasible set defined by the linear constraints

$$S_L = \{x \in R^n;\ b_L \leq Ax \leq b^U,\ x^L \leq x^U\}$$

and let's define the exact penalty function with the penalty coefficient $c > 0$

$$e(x; e) = f(x) + c\max\{F(x), 0\}.$$

Given a fixed $c > 0$, any solution x_e to the problem

$$\min e(x; c) \qquad \text{over all} \qquad x \in S_L \tag{10}$$

solves problem (6 -9), if x_e is feasible, i.e. $(F(x_e) \leq 0)$. This holds, if the penalty parameter is sufficiently large, problem (6 - 9) has a solution and its constraints are sufficiently regular.

3 THE LINEARIZATION ALGORITHM

We assume that NDOPT algorithms has to generate points feasible with respect to the linear constraints, i.e. consecutive approximate solutions to problem (6 - 9) should lie within S_L. The user has to specify an initial estimate x^0 of the optimal solution and its orthogonal projection on S_L is taken as the algorithm's starting point x^1.

The algorithms in NDOPT are based on the concept of descent methods for nondifferentiable minimization. Starting from a given approximation to a solution of (1- 4), an iterative method of descent generates a sequence of points, which should converge to a solution. The property of descent means that successive points have lower objective (or exact penalty) function values. To generate a descent direction from the current iterate, the method replaces the problem functions with their piecewise linear (polyhedral) approximations. Each linear piece of such approximation is a linearization of the given function, obtained by evaluating the function and its subgradient at a trial point of an earlier iteration. (This generalizes the classical concept of using gradients to linearize smooth functions.) The polyhedral approximations and quadratic regularization are used to derive a local approximation to the original optimization problem, whose solution (found by quadratic programming) yields the search direction. Next, a line search along this direction produces the next approximation to a solution and the next trial point, detecting the possible gradient discontinuities. The successive approximations are formed to ensure convergence to a solution without storing too many linearizations. Rejecting the oldest subgradients from the bundle of those collected from the beginning of the computational process one avoids the potential problems with the lack of computer memory.

Each step of the algorithm consists of the following three main parts:

- generation a consecutive piecewise linear approximation of all nonlinear functions

- solution of the direction finding subproblem
- directional minimization to either find a new approximate solution with a smaller objective function value or deduce that the current search direction is not the descent direction. In both cases we get a new trial point which gives information necessary to generate a new linear approximation.

The first part is obvious. In the second an original primal approach is proposed. This is in contrary to the usual dual approach. In the third we shall investigate an approach based on the use of two linear supporting functions with opposite derivative signs. The last property ensures the existence of the intersection point inside the interval of the two trial points. These topics are discussed in next sections. Formulation of the main algorithm follows below:

Step 0 (Initialization). Choose a starting point $x^1 \in S_L$, final accuracy tolerance $\epsilon_s \geq 0$, direction search parameter $m \in (0,1)$, maximal number m_g of stored subgradients of $e(x;c)$, starting value $c^1 > 0$ of the penalty parameter and the starting value of the accuracy tolerance $\delta^1 > 0$ of the penalty function minimization. Set $y^1 = x^1$, $J_e^1 = \{1\}$, $e_1^1 = e(y^1;c^1)$, $g_e^1(y^1;c^1)$, $k = 1$.

Step 1 (Direction finding) Find the optimal solution (d^k, v^k) of the QP direction search problem:

$$\min \frac{1}{2}|d|^2 + v \qquad \text{with respect to} \quad (d,v) \in R^{N+1}$$

$$e_j^k + < g_e^{k,j}, d > \leq v \qquad j \in J_e^K,$$

$$x^k + d \in S_L$$

and the Lagrange multipliers λ - corresponding to the inequalities following from the hyperplanes supporting the epigraph of the penalty function from below, ν - corresponding to general linear constraints and μ - corresponding to box constraints respectively. Let $\hat{J}_e^k \in J_e^k$ denotes the subset of the indices of the first group of constraints with positive multipliers ($\lambda_j > 0$, $j \in \hat{J}_j^k$). Compute $V^k = v^k - e)x^k; c^k)$.

Step 2 (Stopping criterion) If $V^k \geq -\epsilon^s$ and $F(x^k) \leq \epsilon_s$ then STOP else continue.

Step 3 (Penalty coefficient modification) If $|V^k| \leq \delta^k$ and $F(x^k) > |V^k|$ then set $c^{k+1} = 2 * c^k$ and $\delta^{k+1} = \delta^k/2$ and recalculate respectively values of stored subgradients of the penalty function g_j^k; $j \in J_e^k$; otherwise set $c^{k+1} = c^k$ and $\delta^{k+1} = \delta^k$.

Step 4 (Direction search) Set $y^{k+1} = x^k + d^k$. If

$$e(y^{k+1}; c^{k+1}) \leq e(x^k; c^{k+1}) + m * V^k,$$

then set $x^{k+1} = y^{k+1}$ (serious step); otherwise set $x^{k+1} = x^k$ (zero step). Compute g_e^{k+1}.

Step 5 (Modification of linearizations) Choose a subset of $\tilde{J}_e^k \subset J_e^k$ satisfying $\tilde{J}_e^k \supset \hat{J}_e^k$ and $|\tilde{J}_e^k| \leq M_g - 1$. Augment $\tilde{J}_e^k$ with the newest linear approximation of the penalty function $e(y^{k+1}; c^{k+1})$ setting $J_e^{k+1} = \tilde{J}_e^k + k + 1$ and recalculate values of linearizations e_j^{k+1}, $j \in J_e^{k+1}$, if necessary. Increase k by 1 and return to Step 1.

4 AUXILIARY DIRECTION SEARCH QP PROBLEM

The direction finding subproblem has got the form of a following convex (unfortunately not strictly convex) QP programming problem:

$$\min_{(d,v)\in R^n\times R^1} \frac{1}{2}|d|^2 + v \tag{11}$$

subject to

$$-\alpha_i + P_i^T d \le v, \quad \text{where } i = 1, 2, \ldots, m, \tag{12}$$

$$y^L \le A^T(x^k + d) \le y^U \tag{13}$$

$$x^L \le x^k + d \le x^U \tag{14}$$

where $P = [P_1, \ldots, P_m]$ is an $n \times m$ matrix, $\alpha_1^T = (\alpha_1, \ldots, \alpha_m)$ - m-vector.

- α^T denotes the transposition of the column vector α.
- $|d|$ is the euclidean norm of vector $d \in R^n$.
- m denotes the current number of trial points stored by the algorithm (equal to number of preserved linear approximations to the exact penalty functions).

The left-hand sides of inequalities (12) represents the approximations to the exact penalty function $e(x; c)$ recalculated so that all of them are represented at the same point x^k (the current approximate solution of the main penalty problem).

Formula $-\alpha_i + P_i^T d$ in (12) appears as the result of the construction of linear approximation of the minimized penalty function at the i-th trial point y_i. All formulae (12) have been reformulated to describe the supporting hyperplanes directly by means of the current point x^k and vector perpendicular to it (see Lemarechal (1977), Mifflin (1982), Kiwiel (1985, 1988)).

Usually the dual problem to (11-14) is solved, *e.g.* Lemarechal (1977), Mifflin (1982), Kiwiel (1989). The reason is that one can easy generate the staring feasible point taking the optimal solution of the previous dual problem and assuming components corresponding to the newly added constraints equal to zero.

We have decided to use the primal one. The reasons are:

- We solve a transformed problem, that has a form of finding orthogonal projection onto a linearly constrained feasible set.
- The transformed problem is strictly convex
- There exists an easy way to construct a starting feasible point for a new transformed direction finding problem on the basis of the optimal solution of the previous one (see sections 4 and 5).
- Q, R factors of the matrix of active constraints at the optimal point of the previous problem after simple recalculation give the Q, R factors of the matrix of constraints active at the starting point of the new transformed problem.

The only difference between the consecutive problems (11- 14) is that one constraint is added, eventually one deleted (it is always a constraint inactive at the optimal solution (d^k, v^k) of the previous one and hence it may be neglected) and the linearizations are recalculated (it impacts exclusively the α_i^k scalars). The latter operation takes place only when the step is successfull. Then $x^{k+1} = x^k + d^k$ and we may take $d_0 = 0$ as the starting point for the QP problem number $(k+1)$. If the step is not successfull then the old linearizations do not vary and we take $d_0 = d^k$ as the starting feasible point. The rules accepted when the stepsize differs from 1 are easily deduced in the same manner.

The only problem left is with the newly added constraint. Let the pair (d^k, v^k) be the optimal solution of the k-th QP problem. Now two distinct cases are possible:

$$1) \quad \tilde{\alpha}_p + P_p^T d_2 \leq v^k, \tag{15}$$

$$2) \quad \tilde{\alpha}_p + P_p^T d_2 > v^k. \tag{16}$$

In the first case, pair (d_0, v^k) is the optimal solution of the problem number $(k+1)$. In the second one, much more interesting case, the newly added constraint will be surely active at the optimal solution of the new problem. This fact we use to eliminate variable v and form the reduced primal direction search problem which is strictly convex while the original is merely convex.

5 REDUCED DIRECTION SEARCH QP PROBLEM

As a result of elimination of variable v we obtain the following strictly convex QP program (least distance problem):

$$\min_d \frac{1}{2}|d + P_p|^2 \tag{17}$$

$$\tilde{\alpha}_i - \alpha_p + (P_i - P_p)^T d \leq 0, \qquad i = 1, \ldots, m \tag{18}$$

$$x^L \leq y^k + d \leq x^k, \tag{19}$$

$$y^L \leq A(y^k + d) \leq y^U. \tag{20}$$

There exists many effective methods for solving such problems. We have chosen the Gill and Murray primal solution method Gill and Murray (1978). The details of the algorithm are omitted due to the lack of space. Gill and Murray's method requires a feasible starting point and maintains this property in all iterations.

In the second case discussed at the end of the previous section $\alpha_p + P_p^T d_0 > v^k = \alpha_q + P_q^T d_0$ hence in the new reduced problem we are substracting the value greater than v^k which was substracted in the previous problem. Therefore d_0 shall be feasible for the new direction search problem.

Furthermore, point d_0 possesses some special properties as follows:

i) set of active general linear and box constraints remains intact, i.e. is the same as it was in the optimal point d^k of the previous reduced direction search subproblem.
ii) the set of active constraints reflecting the linear functions approximating from below the epigraph of the minimized nonsmooth function is not inherited. In fact, it may happen that none of these constraints shall be active at point d_0.

REFERENCES

Clarke, F.H. (1983) *Optimization and nonsmooth Analysis.* Wiley-Interscience, New York.
Gill, P.E. and Murray, W. (1978) Numerically Stable Methods for Quadratic Programming. *Mathematical Programming*, **14**, 349–72.
Kiwiel, K.C. (1985) *Methods of descent for nondifferentiable optimization.* Springer Verlag, Berlin, Heidelberg, New York.
Kiwiel, K.C. (1988) *Some Computational Methods of Nondifferentiable Optimization.* (in polish), Ossolineum, Wrocław.
Kiwiel, K.C. (1989) A dual method for certain positive semidefinite quadratic programming problem. *SIAM Journal on Scientific and Statistical Computing*, **10**, 175–86.
Lemarechal, C. (1977) *Nonsmooth Optimization and Descent Methods.* Technical Report RR-78-4, International Institute for Applied Systems Analysis, Laxenburg, Austria.
Mifflin, R. (1982) A Modification and an Extension of Lemarechal's Algorithm for Nonsmooth Minimization, in: *Nondifferentiable and Variational Techniques in Optimization* (eds. D.C. Sorensen and R.J.-B. Wets), Mathematical Programming Study 17, North Holland, Amsterdam.

Optimal Control

49

Discrete approximation of nonlinear control problems

Riho Lepp
Institute of Cybernetics, Estonian Academy of Sciences
Akadeemia 21, EE-0026 Tallinn, Estonia.
Tel: +3722-527902. Fax: +3722-527901.
e-mail: lprh@ioc.ee

Abstract
The basic optimal control problem with bounded measurable controls and absolutely continuous trajectories is approximated by a sequence of finite-dimensional problems. Using the notion of the discrete convergence of elements and operators, conditions are presented that quarantee the discrete convergence of trajectories and the weak* discrete convergence of controls.

Keywords
Nonlinear control equation, bounded measurable controls, discrete convergence

1 INTRODUCTION

Consider the following basic optimal control problem (Daniel, 1973): Minimize

$$f(x,u) = \int_0^T f(t, x(t), u(t))dt, \tag{1}$$

over the set of functions $\{(x(t), u(t))\}$ satisfying the differential equation

$$\dot{x}(t) = g(t, x(t), u(t)) \;\; for\ almost\ all \;\; (a.a.) \;\; t \in [0,T], \;\; x(0) = x_0, \tag{2}$$

and the additional control constraints

$$u(t) \in U_{ad} \;\; for \;\; a.a. \;\; t \in [0,T], \tag{3}$$

where functions $f(t,x,u)$ and $g(t,x,u)$ are nonlinear in all variables (t,x,u), $f : R \times R^r \times R^m \to R$, $g : R \times R^r \times R^m \to R^r$.

In order to to write the problem as an optimization problem in a Banach space, we have to choose controls $u(t)$ as elements of the space of essentially bounded measurable functions, $u \in L^\infty([0,T])$, otherwise the functional (1) and operator (2) are not differentiable and thus, certain difficulties arise with formulation of the second order optimality conditions. The trajectories $x(t)$, corresponding to controls $u(t)$, are assumed to be either absolutely continuous (Daniel, 1973), $x \in W^{1,1}([0,T])$, or Lipschitz continuous, $x \in W^{1,\infty}([0,T])$ (Hager, 1990).

In practice, the minimization of control problems is carried out in finite–dimensional subspaces, replacing integrals by sums and infinite number of constraints by finite ones. Correctness of this replacement is usually guaranteed by systems of projectors $\{P_n\}$ such that

$$\| P_n u - u \| \to 0, \quad n \in N, \ \forall u \in U_{ad} \tag{4}$$

(*e.g.* by orthoprojectors in the Ritz–Galerkin method (Daniel, 1973)).

This approximation scheme works well in L^p-spaces, $1 \leq p < \infty$, and results, generally speaking, from the fact that in Lebesgue spaces we can approximate functions by polynomials with rational coefficients.

The situation changes drastically if controls belong to the space of essentially bounded measurable functions $L^\infty([0,T])$ since the space $C([0,T])$ of continuous functions is not dense there and thus, it is not clear, how to construct systems of subspaces and projectors which could satisfy (4). At the same time it is more convenient to formulate nonlinear control problems in L^∞ since in L^p-spaces, $1 \leq p < \infty$, functions $f(t,x,u)$ and $g(t,x.u)$ should satisfy certain growth conditions and this in turn, together with the weak convergence of controls brings along linearity in u of functions $f(t,x,u)$ and $g(t,x,u)$, and, as it was mentioned earlier, functional in (1) and operator in (2) what define the problem, are not differentiable in L^p–spaces.

Still, there exists at least one possibility to approximate nonlinear control problems in L^∞. This possibility is based on the concept of the discrete convergence of elements and operators, and was formulated and developed in the 70 –ies by many authors in numerical analysis (see, *e.g.* , (Stummel, 1971), (Vainikko, 1978)). The main differences between the classical projection method and the new discretization method approaches are:

1) in the discretization approach P_n–s could not to be projectors since approximate problems need not to be formulated in subspaces;

2) in the discretization approach instead of strong convergence of projected elements (4) only consistency of norms between initial and approximate spaces is needed. This norm consistency condition guarantees the nondegeneracy of norms of the discretized elements and thus, uniqueness of the limit process.

The norm consistency condition is weaker than the convergence condition (4). For instance, in reflexive L^p –spaces, $1 < p < \infty$, from the weak convergence of elements and from the convergence of their norms it follows the strong (norm) convergence of elements (the Radon – Riesz property).

The norm consistency condition between the space of essentially bounded functions L^∞ and the sequence of Euclidean spaces with increasing dimension was formulated and proved in (Lepp, 1988) for a system of linear piecewise integral connection operators, where the partition of the integration domain was realized using only the sets with measure zero of their boundary. Note that for projectors P_n, defined in such way, it can happen that there exist L^∞ –functions for which the limit (4) does not exist (see, *e.g.* , (Atkinson, 1983)).

In this paper we will follow the next scheme. Relying on the Banach fixed point theorem, we can show for fixed controls u and u_n that there exist trajectories x and x_n as solutions of initial and approximate equations. Since the set of admissible controls U_{ad} is bounded, we can separate from a minimizing control sequence $\{u_n\}$ a weakly* discretely convergent subsequence. Then, showing that the approximate control equations are collectively compact (Anselone, 1971), we can guarantee the discrete convergence of trajectories. Finally, the convergence of approximate optimal controls and trajectories follows from the properties of the cost function.

Schemes of analysis of the stability of discrete approximations of infinite dimensional extremum problems rely on discrete analogues of the existence theorems (see, *e.g.* , (Daniel, 1969), (Vasin, 1982)). Hence, we at first formulate a theorem what gives us the existence of an optimal solution in $C([0,T]) \times L^\infty([0,T])$. Due to technical reasons we can not guarantee compactness in x of the integral operator $g(x,u) = \int_0^t g(s,x(s),u(s))ds$ in $W^{1,1}([0,T])$, but only in $C([0,T])$ for a fixed $u \in U_{ad}$. Consequently, convergence analysis of trajectories is carried out in a space with weaker topology than the topology of the space of absolute continuous functions.

Since the existence problems for differential equations are more convenient to handle in their integral form, we will reformulate the equation (2) as an integral equation:

$$x(t) = \int_0^t g(s,x(s),u(s))ds + x_0, \quad t \in [0,T].$$

Assume that

f.1) the function $f(t,x,u)$ is Riemann integrable in t for all $(x.u)$, continuously differentiable in (x,u) for a.a. t and for all $h > 0$ and all $|x|, \ |y|, \ |u| \leq h$ there exist Riemann integrable functions $a_{1h}(t), \ a_{2h}(t)$, such that $|f(t,x,u)| \leq a_{1h}(t)$ and $|f(t,x,,u) - f(t,y,u)| \leq a_{2h}(t)\,|x-y|$;

f.2) the function $f'_u(t,x,u)$ is Riemann integrable in t for all (x,u), and continuous in (x,u) for a.a. $t \in [0,T]$, there exists a Riemann integrable function $a_{3h}(t)$ such that $|f'_u(t,x,u)| \leq a_{3h}(t)$ for all $h > 0, \ |x|, |u| \leq h$;

g.1) the function $g(s,x,u)$ is Riemann integrable in s for all (x,u), continuously differentiable in (x,u) for a.a. t and for all $h > 0$ and all $|x|, |u|$ there exists a Riemann integrable function $b_{1h}(s)$ such that $|g(s,x,u)| \leq b_{1h}(s)$;

g.2) the function $g'_x(s,x,u)$ is Riemann integrable in s for all (x,u), continuous in (x,u) for a.a. $s \in [0,T]$ and continuous in x uniformly in u, for all $h > 0$ and all $|x|, |u| \leq h$ there exists a Riemann integrable function $b_{2h}(s)$ such that $|g'_x(s,x,u)| \leq b_{2h}(s)$.

Let us present now conditions what guarantee the existence of an optimal solution of the problem (1) - (3).

Theorem 1 *Let functions $f(t,x,u)$ and $g(s,x,u)$ satisfy conditions f.1), f.2) and g.1), g.2). Let the function $f(t,x,u)$ be convex in u, the set U_{ad} be bounded, convex and weakly* closed. Let in a certain ball $S(x_u,r)$, $r > 0$, the linearized homogeneous equation $z(t) = \int_0^t g'_x(s,x_u(s),u(s))z(s)ds$ have only the trivial solution $z(t) = 0$, $t \in [0,T]$. Then optimal control and trajectory (u,x) exist in $L^\infty([0,T]) \times C([0,T])$.*

The proof of the theorem relies on the Banach fixed point theorem. Using this theorem, we can show that the differential equation (2) (more concretely, its integral equivalent $x(t) = \int_0^t g(s, x(s), u(s))ds + x_0$) has for a fixed admissible control $u(t) \in U_{ad}$ the solution in $C([0,T])$. Then, relying on the Weierstrass existence theorem, it is not difficult to present conditions, what guarantee the existence of optimal control and trajectories.

2 DISCRETIZATION OF THE PROBLEM

In order to approximate the control problem (1) – (3) by a sequence of finite dimensional problems and to analyze stability of such approximations, we should define the discrete convergence in $L^\infty([0,T])$. We start from a convergent quadrature process. Let

$$\sum_{i=1}^{n} h(t_{in}) \triangle t_{in} \rightarrow \int_0^T h(t)dt, \quad n \in N, \tag{5}$$

for any continuous on $[0,T]$ function $h(t)$, $h \in C([0,T])$.
Convergent quadrature process (5) defines us the simpliest system of connection operators $\mathcal{P} = \{p_n\}$ from $C([0,T])$ to R^n:

$$(p_n x)_{in} = x(t_{in}), \; i = 1, ..., n, \; n \in N. \tag{6}$$

Note that the connection system (6) is also applicable for the Riemann integrable functions. For Lebesgue L^p–spaces, $1 \leq p \leq \infty$, we are forced to define the system of linear connection operators $\mathcal{Q} = \{q_n\}$, $q_n : L^p([0,T]) \rightarrow R^n$, only in the integral form:

$$(q_n u)_{in} = \triangle t_{in}^{-1} \int_{t_{i-1,n}}^{t_{in}} u(t)dt \tag{7}$$

with $0 = t_{0n} < t_{1n} < ... < t_{nn} = T$ and $\triangle t_{in} = t_{in} - t_{i-1,n}$, $max_i \triangle t_{in} \rightarrow 0$, $n \in N$.

Defined in such a way connection system satisfies the norm consistency property:

$$\| q_n u \|_n \rightarrow \| u \|, \; n \in N, \forall u \in L^p([0,T]), \; 1 \leq p \leq \infty \tag{8}$$

(here $\| \cdot \|_n$ denotes the Euclidean norm of a vector). For the L^p-spaces, $1 \leq p < \infty$, the proof of the convergence (8) relies on the fact that $C([0.T])$ is dense in these spaces. For the space $L^\infty([0,T])$ the proof of the convergence (8) is presented in (Lepp, 1988).

Define the $\mathcal{Q}$–convergence (or in other words – the discrete convergence) of the sequence of vectors $\{u_n\}$ to the function $u(t)$, $u \in L^\infty([0,T])$, in the following way.

Definition 1 *The sequence* $\{u_n\}$ $\mathcal{Q}$*–converges to the function* $u \in L^\infty$ *if*

$$\| u_n - q_n u \|_n \rightarrow 0, \; n \in N. \tag{9}$$

Remark 1 *The consistency property (8) guarantees the nondegeneracy of norms of discretized in R^n, $n \in N$, elements of L^∞ and thus, uniqueness of the limit process (9).*

In this paper we together with the Q–convergence of vectors need also their discrete weak* convergence. Let $z(t)$ be an integrable function, $z \in L^1([0,T])$.

Definition 2 *The sequence $\{u_n\}$ w^*Q–converges to the function $u \in L^\infty$ if*

$$\sum_{i=1}^{n}((q_n z)_{in}, u_{in}) \triangle t_{in} \rightarrow \int_0^T (z(t), u(t))dt, \; n \in N \;\; \forall z \in L^1([0,T]).$$

Remark 2 *The consistency property (8) holds also for any $z \in L^1([0,T])$.*

Denote for brevity sums $f_n(x_n, u_n) = \sum_{i=1}^{n} f(t_{in}, x_{in}, u_{in}) \triangle t_{in}$ and $g_n(x_n, u_n) = \sum_{j=1}^{i} g(s_{jn}, x_{jn}, u_{jn}) \triangle t_{jn}$. Formulate the discretized problem: Minimize

$$\sum_{i=1}^{n} f(t_{in}, x_{in}, u_{in}) \triangle t_{in} \tag{10}$$

over the set of vectors $\{(x_n, u_n)\}$, satisfying constraints

$$x_{in} = \sum_{j=1}^{i} g(s_{jn}, x_{jn}, u_{jn}) \triangle t_{jn} + x_{0n}, \; i = 1, ..., n, \tag{11}$$

and

$$u_{in} \in U_{nad}, \; i = 1, ..., n. \tag{12}$$

In order to apply in approximation the existence theorem scheme, we need conditions what guarantee the discrete approximation of the integral equation (2) by the sequence of sums (11).

Proposition 1 *Let the function $g(s, x, u)$ satisfy conditions g.1) and let quadrature process (5) converge. Then*

$$\| g_n(p_n x, q_n u) - p_n g(x, u) \|_n \rightarrow 0, \; n \in N.$$

Proposition 2 *Let the function $g(s, x, u)$ satisfy conditions g.1), g.2). Let quadrature process (5) converge. Then from the weak* discrete convergence of discrete controls $\{u_n\}$ to control $u \in L^\infty$ it follows strong discrete convergence of a subsequence of operators $g_n(x_n, u_n)$, $n \in N'' \in N' \subset N$, to a continuous function $y(t)$.*

Remark 3 *Note that from the last convergence it does not follow the norm (strong) convergence of $\{u_n\}$ to u, $n \in N''$ (if, e.g. , the operator g is linear in u, then its inverse is not bounded – (2) is the first kind equation relative to u).*

Approximate now the integral cost functional $f(x,u)$ by the sequence $\{f_n(x_n,u_n)\}$ of finite dimensional sums from (10).

Proposition 3 *Let function $f(t,x,u)$ satisfy conditions f.1), f.2) and be convex in u. Let quadrature process (5) converge. Then*

$$lim\ sup\ f_n(x_n,u_n)\ \leq f(x,u)\ as\ \mathcal{P}-lim\ x_n\ =\ x,\ \mathcal{Q}-lim\ u_n\ =\ u,\ n\in N,$$

$$lim\ inf\ f_n(x_n,u_n)\ \geq f(x,u)\ as\ \mathcal{P}-lim\ x_n\ =\ x,\ w^*\mathcal{Q}-lim\ u_n\ = u,\ n\in N.$$

Remark 4 *Due to limited average of the paper, we can not present proofs of propositions. Technically they are quite lengthy and sophisticated.*

Assume, relying on the Theorem 1, that the initial and approximate problems have optimal solutions (x^*,u^*) and (x_n^*,u_n^*), respectively.

Relying on Propositions 1 - 3, we can now together with Theorem 1 formulate the main result of the paper on the approximate solution of nonlinear optimal control problems.

Theorem 2 *Let functions f and g satisfy conditions $f_1), f_2), g_1), g_2)$. Let nonempty set of admissible controls U_{ad} be convex, bounded and weakly* closed in L^∞, and function $f(t,x,u)$ be convex in u. Let quadrature process (5) converge and let $U_{nad}\ \subseteq\ q_nU_{ad}$. Let the linearized homogeneous equation $z(t)=\int_0^t g_x'(s,x^*(s),u^*(s))z(s)ds$ have only the trivial solution $z(t)=0$. Then*
1) $f_n^*\ \rightarrow\ f^*,\ n\in N$;
2) a subsequence of discrete optimal trajectories $\{x_n^*\}$ $\mathcal{P}$-converges and a subsequence of discrete optimal controls $\{u_n^*\}$ $w^*\mathcal{Q}$-converges, $n\in N'$, to the optimal trajectory and control of the initial problem (1) - (3).

Proof. By Theorem 1 optimal solutions (x^*,u^*) and (x_n^*,u_n^*) of both, initial and approximate problems, exist. By Proposition 2 the sequence of operators $\{g_n(x_n^*,u_n^*)\}$ is $\mathcal{P}$-compact. Thus, both subsequences $\{x_n^*\}$ and $\{g_n(x_n^*,u_n^*)+x_{0n}\}$ $\mathcal{P}$-converge in uniform norm sense to the same limit. Take from the sequence $\{u_n^*\}$, $n\in N'$, a $w^*\mathcal{Q}$-convergent subsequence $\{u_n^*\}$, $n\in N''\in N'\in N$. Let $w^*\mathcal{Q}-lim\ u_n=v.\ n\in N''$, $v\in U_{ad}$, and let corresponding to the control $v(t)$ trajectory be $y(t)$. Then by Proposition 3

$$f^*\ \leq\ f(y,v)\ \leq\ lim\ inf\ f_n(x_n^*,u_n^*)\ =\ lim\ inf\ f_n^*.$$

Prove the opposite inequality. Consider the difference $f_n(x_n^*,u_n^*)\ -\ f^*$. Take the sequence $\{(y_n,q_nu^*)\}$, where y_n is the solution of the equation $x_n=g_n(x_n,q_nu^*)+x_{0n}$. Then

$$f_n^*\ -\ f^*\ =\ f_n(x_n^*,u_n^*)\ -\ f^*\ \leq\ f_n(y_n,q_nu^*)\ -\ f^*.$$

By the definition of the $\mathcal{Q}$-convergence we have $\mathcal{Q}-lim\ q_nu^*\ =\ u^*,\ n\in N$. By $\mathcal{P}$-compactness of the sequence $\{g_n(y_n,q_nu)\}$ $\mathcal{P}-lim\ y_n=x^*$, where x^* is the solution of the equation (2) with the control u^*. Hence, $f_n^*\ -\ f^*\ \leq\ \varepsilon$ for a small $\varepsilon>0$ and all $n\geq n_\varepsilon$. Consequently,

$$f_n^*\ \rightarrow\ f^*,\ n\in N.$$

The second statement of the theorem follows assuming in contrary. □

Corollary 1 *If the function $f(t,x,u)$ is strictly convex in u then the whole sequence of discrete optimal solutions $\{(x_n^*, u_n^*)\}$ of problems (10) - (12) $\mathcal{PQ}$-converges to the optimal solution (x^*, u^*) of the initial problem (1) - (3).*

Let us compare the approach presented with papers published earlier.

1. We do not assume that for each $u \in U_{ad}$ there exist a unique solution x_u of the equation (2), as it was done in (Daniel, 1973). Relying on the Banach fixed point theorem, we present conditions what guarantee its existence.

2. We do not use the implicit function theorem approach. The implicit function theorem and its modifications need in stability analysis of an equation $F(x) = y$ the boundedness of the inverse of the differential $F'(x_0)$. From the boundedness of $F'(x_0)^{-1}$ it follows non-compactness of the linear operator $F'(x_0)$. But, assuming only pointwise convergence of operators, one can say nothing about the convergence of solutions. Thus, the strongly regular generalized equations approach (Robinson, 1980), applied in many times in nonlinear control theory (see, *e.g.* , (Alt and Malanowski, 1993)), is not applicable in our case. Here we use the Banach fixed point theorem, assuming that the linearized homogeneous control equation has only the trivial solution zero.

3. We do not use the subspace approximation approach, presented in (Hager, 1990) since we do not know, how to construct a system of approximating subspaces for the space of essentially bounded functions L^∞.

4. We do not use the approach, presented in (Mordukhovich, 1978), where the measurable control was approximated by a continuous function, using the Luzin's theorem. This theorem is not applicable in numerical analysis – there is no algorithmic way, how to realize the Luzin's C-property for measurable functions.

5. We do not need convexity of $g(t,x,u)$ in u, usually assumed in nonlinear control theory (see, *e.g.* , (Fattorini, 1994)). Also, as it was written in comment 2., we assume that the linearized homogeneous control equation has only the trivial solution zero.

REFERENCES

Alt, W. and Malanowski, K. (1993) The Lagrange-Newton method for nonlinear optimal control problems. *Computational Optimization and Applications*, **2**, 77–100.

Anselone, P.M. (1971) *Collectively compact approximation theory.* Prentice-Hall, New Jersey.

Atkinson, K., Graham, I. and Sloan, I. (1983) Piecewise continuous collocation for integral equations. *SIAM Journal on Numerical Analysis*, **20**, 172–86.

Daniel, J.W. (1969) On the approximate minimization of functionals. *Mathematics of Computation*, **23**, 573–81.

Daniel, J.W. (1973) The Ritz-Galerkin method for abstract optimal control problems. *SIAM Journal on Control*, **11**, 53–63.

Fattorini, H.O. (1994) Existence theory and the maximum principle for relaxed infinite-dimensional optimal control problems. *SIAM Journal on Control and Optimization*, **32**, 311–31.

Hager, W.W. (1990) Multiplier methods for nonlinear optimal control. *SIAM Journal on Numerical Analysis*, **27**, 1061–80.

Lepp, R. (1988) Discrete approximation conditions of the space of essentially bounded functions. *Proceedings of the Academy of Sciences of Estonian SSR. Physics - Mathematics*, **37**, 204–8 (in Russian).

Mordukhovich, B.Sh. (1978) On the difference approximations of systems of optimal control. *Prikladnaya matematika i mekhanika*, **42**, 431–40 (in Russian).

Robinson, S. (1980) Strongly regular generalized equations. *Mathematics of Operations Research*, **5**, 43–62.

Stummel, F. (1971) Diskrete Konvergenz linearer Operatoren, I. *Mathematische Annalen*, **190**, 45–92.

Vainikko, G. (1978) Approximative methods for nonlinear equations. *Nonlinear Analysis. Theory, Methods and Applications*, **2**, 647–87.

Vasin, V.V. (1982) Discrete approximation and stability in extremal problems. *U.S.S.R. Computational Mathematics and Mathematical Physics*, **22**, 57–74.

Convergence of Lagrange-Newton method for control-state and pure state constrained optimal control problems

Kazimierz Malanowski
System Research Institute, Polish Academy of Sciences
ul.Newelska 6, 01-447 Warszawa, Poland.
e-mail: kmalan@ibspan.waw.pl

Abstract
The Lagrange-Newton method (LNM) is applied to a class of optimal control problems for nonlinear ODEs subject to pointwise mixed control-state and pure state constraints. Using the so called **two-norm approach**, developed in stability analysis of solutions to this class of problems, it is shown that (LNM) is locally superlinearly convergent and the rate of convergence is estimated.

Keywords
Lagrange-Newton method, sequential quadratic programming, optimal control, nonlinear ordinary differential equations, control-state and pure state constraints.

1 INTRODUCTION

The Lagrange-Newton method (LNM) is a highly effective sequential quadratic programming method, which is obtained by applying Newton's procedure to Kuhn-Tucker points of optimization problems.

In studying the local convergence of (LNM), Robinson's stability results for the solutions to generalized equations, or related concepts, have been used (see Robinson (1974), Robinson (1980), Alt (1990), Alt (1991), Alt, Malanowski (1993), Dontchev et al. (1995)). In this approach the crucial role is played by the concept of **strong regularity** (SRC). It can be shown that if (SRC) is satisfied, then (LNM) is locally quadratically convergent.

In case of optimal control problems subject to state constraints, due to the phenomenon of **two-norm discrepancy**, we are not able to assure that (SRC) is satisfied. To overcome this difficulty the so called **two-norm approach** can be used. This method is based on stability results for solutions of parametric optimization and optimal control problems developed in Malanowski (1993) and Malanowski (1995), where additional information

Supported by grant No.3 P403 002 05 from the State Committee for Scietific Research (KBN).

on regularity of the solutions is exploited. Using this method the conditions were derived in Alt, Malanowski (1995) under which (LNM), applied to control and state constrained optimal control problems is locally superlinearly convergent.

In the present paper the same approach is applied to more general class of optimal control problems with mixed initial-terminal boundary conditions and subject to pointwise mixed control-state and first order pure state constraints. Moreover, second order sufficient optimality conditions weaker than that in Alt, Malanowski (1995) are assumed. Using stability results for the solutions of this class of problems derived in Malanowski (1994), the local superlinear convergence of (LNM) is obtained and the rate of convergence is estimated.

Comprehensive numerical studies of application of (LNM) to state constrained optimal control problems can be found in Machielsen (1987).

2 LNM FOR ABSTRACT OPTIMIZATION PROBLEMS

In this section we recall convergence results for (LNM) applied to optimization problems with **two-norm discrepancy** in abstract Banach spaces. These results were obtained in Alt, Malanowski (1995).

Let Z and Y be two Banach spaces, the space of arguments and constraints, respectively. Moreover, two Hilbert spaces $\widehat{Z}$ and $\widehat{Y}$ are given, such that $Z \subset \widehat{Z}$ and $Y \subset \widehat{Y}$, with dense and continuous embeddings. The norms in Banach and Hilbert spaces are denoted by $[\![\cdot]\!]$ and $\| \cdot \|$, respectively, with the appropriate subscript. By $(\cdot,\cdot)_{\widehat{Z}}$ there is denoted a duality pairing between $\widehat{Z}^*$ and $\widehat{Z}$ or Z^* and Z. We put $\widehat{Y}^* = \widehat{Y}$ and by $(\cdot,\cdot)_{\widehat{Y}}$ we denote the inner product in $\widehat{Y}$, extended by continuity to $Y^* \times Y$. We define $X = Z \times Y$, and $\widehat{X} = \widehat{Z} \times \widehat{Y}$. In Y there is given a closed convex cone K with the vertex at the origin. By $\widehat{K}$ we denote the closure of K in $\widehat{Y}$ and we put $K^+ = \{\lambda \in Y^* \mid (\lambda, y)_{\widehat{Y}} \geq 0 \quad \text{for all } y \in K\}$.

We consider the following optimization problem

$$\text{(P)} \qquad \min_{z \in Z} F(z) \quad \text{subject to } \varphi(z) \in K.$$

Assume:

- (I.1) There exists a possibly local solution $\widetilde{z}$ of (P).
- (I.2) The functions $F : Z \mapsto \mathbf{R}^1$ and $\varphi : Z \mapsto Y$ are two times Fréchet differentiable in a neighborhood $\mathcal{O} \subset Z$ of $\widetilde{z}$ and some compatibility conditions are satisfied, which assure that F_x, φ_x are well defined on $\widehat{X}$. These conditions are always satisfied for optimal control problems (cf. Alt, Malanowski (1995)).
- (I.3) The local solution $\widetilde{z}$ is regular in the sense of Robinson, i.e., $\varphi_z(\widetilde{z})Z - K + [\varphi(\widetilde{z})] = Y$, where $[y] = \{\alpha y \mid \alpha \in \mathbf{R}^1\}$.

Let us introduce the following Lagrangian associated with (P):

$$\mathcal{L}: Z \times Y^* \mapsto \mathbf{R}^1, \qquad \mathcal{L}(z, \lambda) = F(z) - (\lambda, \varphi(z))_Y .$$

By (I.3) there exists a Lagrange multiplier $\widetilde{\lambda} \in Y^*$ associated with $\widetilde{z}$ such that the Kuhn-Tucker conditions

$$\mathcal{L}_z(\widetilde{z}, \widetilde{\lambda}) := F_z(\widetilde{z}) - \varphi_z(\widetilde{z})^* \widetilde{\lambda} = 0, \qquad (\widetilde{\lambda}, \varphi(\widetilde{z}))_Y = 0, \qquad \widetilde{\lambda} \in K^+ \tag{1}$$

hold. We will assume that $\widetilde{\lambda}$ is more regular:

(I.4) $\widetilde{\lambda} \in Y$.

In the Lagrange-Newton method applied to (P) the following iterative procedure is constructed:

(LNM) *Having* $w_k = (z_k, \lambda_k)$, *compute a Kuhn-Tucker point* $w_{k+1} = (z_{k+1}, \lambda_{k+1})$ *of the following linear-quadratic problem:*

$$(\mathrm{QP}_{w_k}) \qquad \begin{array}{l} \min_{z\in Z} \widetilde{F}(z, w_k) := \frac{1}{2}(\mathcal{L}^2_{zz}(z_k, \lambda_k)(z - z_k), z - z_k)_{\widehat{Z}} + (F_z(z_k), z - cz_k)_{\widehat{Z}} \\ \text{subject to } \ \varphi_z(z_k)(z - z_k) + \varphi(z_k) \in K. \end{array}$$

If $w_k \to \bar{w} = (\bar{z}, \bar{\lambda})$, then $\bar{w}$ is a Kuhn-Tucker point of (P). If, in addition a sufficient optimality condition is satisfied at $\bar{w}$, then $\bar{z}$ becomes a solution of (P) and $\bar{\lambda}$ - an associated Lagrange multiplier.

(QP_{w_k}) can be treated as a linear-quadratic optimization problem depending on the parameter w_k. As in Alt, Malanowski (1995), to analyse convergence of (LNM) we will use stability results for parametric optimization problems with norm discrepancy, obtained in Malanowski (1993). To this end we will need the family of the following **accessory linear-quadratic problems**, depending on the perturbation $\delta = (a, b) \in \widehat{X}^*$:

$$(\mathrm{LP}_\delta) \qquad \begin{array}{l} \min_{z\in Z}\{\frac{1}{2}(\mathcal{L}^2_{zz}(\widetilde{z}, \widetilde{\lambda})(z - \widetilde{z}), z - \widetilde{z})_{\widehat{Z}} + (F_z(\widetilde{z}) - a, z - \widetilde{z})_{\widehat{Z}}\} \\ \text{subject to } \ \varphi_z(\widetilde{z})(z - \widetilde{z}) + \varphi(\widetilde{z}) - b \in \widehat{K}. \end{array}$$

The following assumptions constitute a weakened version of the so called **strong regularity condition** introduced in Robinson (1980).

(I.5) There exist a closed convex set $\Delta \subset \widehat{X}^*$ of regular perturbations and a convex set Γ compact in X, such that $0 \in \Delta$ and for each $\delta \in \Delta$ there exists, a unique in Γ, Kuhn-Tucker point $\zeta_\delta = (x_\delta, \mu_\delta)$ of (LP_δ). Moreover, there exists $c > 0$ such that

$$\|x_{\delta_1} - x_{\delta_2}\|_{\widehat{Z}}, \ \|\mu_{\delta_1} - \mu_{\delta_2}\|_{\widehat{Y}} \le c\|\delta_1 - \delta_2\|_{\widehat{X}^*} \quad \text{for all } \delta_1, \delta_2 \in \Delta. \tag{2}$$

As in Robinson (1980), Alt (1990), Alt (1991), Alt, Malanowski (1993), Alt, Malanowski (1995), for $w = (z_w, \lambda_w)$, $\xi = (z, \lambda) \in X$ we define the following functions:

$$\ell: X \times X \mapsto X^*, \quad \ell(\xi, w) = \begin{pmatrix} \ell_1(\xi, w) \\ \ell_2(\xi, w) \end{pmatrix}, \quad \text{where}$$

$$\begin{aligned} &\ell_1(\xi, w) = F_z(\widetilde{z}) + \mathcal{L}^2_{zz}(\widetilde{z}, \widetilde{\lambda})(z - \widetilde{z}) - \varphi_z(\widetilde{z})^*\lambda - \\ &- F_z(z_w) - \mathcal{L}^2_{zz}(z_w, \lambda_w)(z - z_w) + \varphi_z(z_w)^*\lambda, \\ &\ell_2(\xi, w) = \varphi(\widetilde{z}) + \varphi_z(\widetilde{z})(z - \widetilde{z}) - \varphi(z_w) - \varphi_z(z_w)(z - z_w). \end{aligned}$$

(I.6) $\ell(\xi, w) \in \Delta$ for all $\xi, w \in \Gamma$.

REMARK **21** In the original paper Robinson (1980), as well as in the other papers dealing with the convergence of the Lagrange-Newton method (cf. e.g., Alt (1990), Alt (1991), Alt, Malanowski (1993), Dontchev et al. (1995)) the so called **strong regularity condition** was used instead of (I.5). In that condition it is required that (z_δ, μ_δ) are locally Lipschitz continuous functions of δ in the stronger norm of the original space X, for all perturbations from a ball in X^*. However, in applications to nonlinear optimal control problems subject to pure state constraints we can expect at most that the weaker condition (2) is satisfied (cf. Section 4 in Malanowski (1993)). Since Γ is a compact set, (I.5) imposes some **regularity conditions** on the Kuhn-Tucker points of (LP_δ), whereas assumption (I.6), assures that the set of perturbations Δ is **sufficiently rich**. These assumptions were not needed in Robinson (1980).

It can be shown (cf. Lemma 2.1 in Malanowski (1993)) that, by compactness of Γ, there exists a continuous non-decreasing function $\zeta: \mathbf{R}^+ \mapsto \mathbf{R}^+$, $\zeta(0) = 0$, such that

$$[\![w - \widetilde{w}]\!]_X \leq \zeta(\|w - \widetilde{w}\|_{\widehat{X}}) \qquad \text{for all } w \in \Gamma. \tag{3}$$

The following result is proved in Alt, Malanowski(1995) (cf. Theorem 2.5):

THEOREM **22** *If assumptions (I.1)-(I.6) are satisfied then there exists a ball $\mathcal{B}_{\widehat{X}}(\widetilde{w}, \rho) := \{w \in \widehat{X} \mid \|w - \widetilde{w}\|_{\widehat{X}} \leq \rho\}$ of radius ρ about $\widetilde{w}$ such that for each starting point $w_0 \in \Gamma \cap \mathcal{B}_{\widehat{X}}(\widetilde{w}, \rho)$ the Lagrange-Newton method generates a unique sequence $\{w_k\}$, $w_k = (z_k, \lambda_k)$ convergent to $\widetilde{w}$ and*

$$\|w_{k+1} - \widetilde{w}\|_{\widehat{X}} / \|w_k - \widetilde{w}\|_{\widehat{X}} \leq c\, \zeta(\|w_k - \widetilde{w}\|_{\widehat{X}}). \tag{4}$$

REMARK **23** Since $\zeta(\cdot)$ is a non-decreasing function and $\lim_{k \downarrow 0} \zeta(k) = 0$, (4) shows that the Lagrange-Newton method converges superlinearly for any starting point $w_0 \in \Gamma \cap \mathcal{B}_{\widehat{X}}(\widetilde{w}, \rho)$. In case without two-norm discrepancy, $\zeta(\|w_k - \widetilde{w}\|_{\widehat{X}})$ in (4) is substituted by $\|w_k - \widetilde{w}\|_{\widehat{X}}$, and we obtain quadratic convergence (cf., e.g., Theorem 3.3 in Alt(1990)).

3 OPTIMAL CONTROL PROBLEM

In this section we introduce optimal control problems subject to mixed control-state and pure state constraints. We recall stability results for the solutions of these problems derived in Malanowski (1994). These results are used in the convergence analysis of (LNM).

Let us consider the following optimal control problem:

(O) Find $(\widetilde{x}, \widetilde{u}) \in W^{1,\infty}(0, T; \mathbf{R}^n) \times L^\infty(0, T; \mathbf{R}^m)$ such that
$F(\widetilde{x}, \widetilde{u}) = \min\{F(x, u) := \int_0^T f^0(x(t), u(t))dt + g(x(T))\}$
subject to

$$\begin{aligned}
&\dot{x}(t) - f(x(t), u(t)) = 0, && \text{for a.a. } t \in [0, T],\\
&\xi(x(0), x(T)) = 0, &&\\
&\theta(x(t), u(t)) \leq 0, && \text{for a.a. } t \in [0, T],\\
&v(x(t)) \leq 0, && \text{for all } t \in [0, T],
\end{aligned}$$

where $\xi: \mathbf{R}^n \times \mathbf{R}^n \mapsto \mathbf{R}^d$, $\theta: \mathbf{R}^n \times \mathbf{R}^m \mapsto \mathbf{R}^k$, $v: \mathbf{R}^n \mapsto \mathbf{R}^l$.
Denote by $I = \{1, ..., k\}$ and $J = \{1, ..., l\}$ the sets of indices of constraints.

We are going to apply (LNM) to find the solution of (O) and we will use Theorem 22 to analyse convergence of the method. To do that we will use stability results obtained in Malanowski (1994). Let us recall assumptions from that paper. It is assumed that:

(II.1) There exists a possibly local solution $(\widetilde{x}, \widetilde{u}) \in C^1(0, T; \mathbf{R}^m) \times C(0, T; \mathbf{R}^m)$ of (O).

(II.2) Functions $f^0(\cdot,\cdot), g(\cdot), f(\cdot,\cdot), \xi(\cdot,\cdot), \theta(\cdot,\cdot), v(\cdot)$ and $v_x(\cdot)$ are two times Fréchet differentiable in all arguments, and the respective derivatives are locally Lipschitz continuous.

In order to represent (O) in the form (P) we put

$$Z = W^{1,\infty}(0,T;\mathbf{R}^n) \times L^\infty(0,T;\mathbf{R}^m), \qquad \widehat{Z} = W^{1,2}(0,T;\mathbf{R}^n) \times L^2(0,T;\mathbf{R}^m).$$

Moreover we put $F(z) = F(x,u)$ and $\varphi(z) = (\dot{x} - f(x,u), \xi(x(0),x(T)), -\theta(x,u), -v(x))$. Hence it is natural to choose

$$\begin{aligned} Y &= L^\infty(0,T;\mathbf{R}^n) \times \mathbf{R}^d \times L^\infty(0,T;\mathbf{R}^k) \times W^{1,\infty}(0,T;\mathbf{R}^l), \\ \widehat{Y} &= L^2(0,T;\mathbf{R}^n) \times \mathbf{R}^d \times L^2(0,T;\mathbf{R}^k) \times W^{1,2}(0,T;\mathbf{R}^l), \end{aligned}$$

with $K = K_1 \times K_2 \times K_3 \times K_4$, where

$$\begin{aligned} &K_1 = \{0\} \qquad K_2 = \{0\}, \\ &K_3 = \{u \in L^\infty(0,T;\mathbf{R}^k) \mid u^i(t) \geq 0, \quad i = 1,\dots,k \quad \text{for a.a. } t \in [0,T]\}, \\ &K_4 = \{x \in W^{1,\infty}(0,T;\mathbf{R}^l) \mid x^j(t) \geq 0, \quad j = 1,\dots,l \quad \text{for all } t \in [0,T]\}. \end{aligned}$$

The cone $\widehat{K}$ is defined as above but with Y substituted by $\widehat{Y}$.

Note that by (II.2) conditions (I.2) are satisfied. Now we will introduce constraint qualifications. To simplify notation let us put for $w(t) = (x(t), u(t))$:

$$\begin{aligned} &A(w(t)) := f_x(x(t),u(t)), && B(w(t)) := f_u(x(t),u(t)), \\ &\Xi_0(w) := \xi_{x(0)}(x(0),x(T)), && \Xi_T(w) := \xi_{x(T)}(x(0),x(T)), \\ &\Theta_x(w(t)) := \theta_x(x(t),u(t)), && \Theta_u(w(t)) := \theta_u(x(t),u(t)), \\ &\Upsilon(w(t)) := v_x(x(t)). \end{aligned}$$

Let us introduce the sets of indices of active constraints $\widehat{I}(t) = \{i \in I \mid \theta^i(\widetilde{x}(t),\widetilde{u}(t)) = 0\}$, $\widehat{J}(t) = \{j \in J \mid v^j(\widetilde{x}(t)) = 0\}$, and denote by $\widehat{\Theta}_x(t)$, $\widehat{\Theta}_u(t)$ and $\widehat{\Upsilon}(t)$ the matrices, whose rows are the functions $\theta^i_u(\widetilde{x}(t),\widetilde{u}(t))$, $\theta^i_x(\widetilde{x}(t),\widetilde{u}(t))$ for $i \in \widehat{I}(t)$ and $v_x(\widetilde{x}(t))$ for $j \in \widehat{J}(t)$, respectively. We assume that the following constraint qualifications hold:

(II.3) There exists $\beta > 0$ such that $\left| \left[B(\widetilde{w}(t))^*\widehat{\Upsilon}(\widetilde{w}(t))^*, \widehat{\Theta}_u(\widetilde{w}(t))^*\right] \chi\right| \geq \beta|\chi|$ for all χ of appropriate dimension and all $t \in [0,T]$.

(II.4) For any $\vartheta \in \mathbf{R}^d$ there exists a solution (y,v) to the following system of equations (complete controllability condition is satisfied):

$$\begin{aligned} &\dot{y}(t) - A(\widetilde{w}(t))y(t) - B(\widetilde{w}(t))v(t) = 0, \\ &\Xi_0(\widetilde{w})y(0) + \Xi_T(\widetilde{w})y(T) = \vartheta, \\ &\text{subject to} \\ &\widehat{\Theta}_x(\widetilde{w}(t))y(t) + \widehat{\Theta}_u(\widetilde{w}(t))v(t) = 0, \\ &\widehat{\Upsilon}(\widetilde{w}(t))[A(\widetilde{w}(t))y(t) + B(\widetilde{w}(t))v(t)] = 0. \end{aligned}$$

REMARK **31** It is shown in Malanowski (1994) that (II.4) can be expressed as a rank condition for a certain controllability matrix.

It follows from Lemma 3.3 in Malanowski (1994) that, by (II.3) and (II.4) constraints qualification of the type of **stable linear independence of gradients of active constraints** are satisfied (cf. condition (A3) in Malanowski (1992)). In particular it implies that (I.3) holds.

Let us introduce the following Lagrangian associated with (O):

$$\begin{array}{l}\mathcal{L} : W^{1,\infty}(0,T;\mathbf{R}^n) \times L^\infty(0,T;\mathbf{R}^m) \times (L^\infty(0,T;\mathbf{R}^n))^* \times \mathbf{R}^d \times \\ \times \left(L^\infty(0,T;\mathbf{R}^k)\right)^* \times \left(W^{1,\infty}(0,T;\mathbf{R}^l)\right)^*, \\ \mathcal{L}(x,u,q,\rho,\kappa,\mu) = F(x,u) + (q,\dot{x} - f(x,u)) + \langle \rho, \xi(x(0),x(T)\rangle + (\kappa,\theta(x,u)) + \\ +\langle \mu(0), v(x(0))\rangle + (\dot{\mu}, v_x(x)f(x,u)).\end{array} \tag{5}$$

By Lemma 3.3 in Malanowski (1994) there exist unique Lagrange multipliers $(\widetilde{q},\widetilde{\rho},\widetilde{\kappa},\widetilde{\mu}) \in \widehat{X}^*$ such that the first order necessary optimality conditions in the Kuhn-Tucker form are satisfied, and in the same way as in Hager (1979), it can be shown that $\dot{\widetilde{q}},\widetilde{\kappa},\dot{\widetilde{\mu}}$ are continuous on $[0,T]$.

For any $\alpha \geq 0$ define the sets $I_\alpha^+(t) = \{i \in \widehat{I}(t) \mid \widetilde{\kappa}^i(t) > \alpha\}$, and denote by $\Theta_x^\alpha(t)$ and $\Theta_u^\alpha(t)$ the matrices, whose rows are the functions $\theta_x^i(\widetilde{x}(t),\widetilde{u}(t))$ and $\theta_u^i(\widetilde{x}(t),\widetilde{u}(t))$, respectively, for $i \in I_\alpha^+(t)$. Define the following subspaces $\mathcal{D}_\alpha(t) = \{v \in \mathbf{R}^m \mid \Theta_u^\alpha(t)v = 0\}$ of $\mathbf{R}^m$ and introduce the augmented Hamiltonian

$$\mathcal{H}(t) = f^0(\widetilde{x}(t),\widetilde{u}(t)) - \langle\widetilde{q}(t), f(\widetilde{x}(t),\widetilde{u}(t))\rangle + \langle\widetilde{\kappa}(t),\theta(\widetilde{x}(t),\widetilde{u}(t))\rangle + \langle\dot{\widetilde{\mu}}(t), v_x(\widetilde{x}(t))f(\widetilde{x}(t),\widetilde{u}(t))\rangle.$$

In addition to (II.1)-(II.4), we assume :

(II.5) There exists a constant $\gamma > 0$ such that

$$\langle u, \mathcal{H}_{uu}^2(t)u\rangle \geq \gamma|u|^2, \quad \gamma > 0, \quad \text{for all } u \in \mathcal{D}_0(t) \text{ and all } t \in [0,T].$$

(II.6) There exist $\bar{\alpha} > 0$ and $\bar{\gamma} > 0$ such that the following coercivity condition holds:

$$\left((y,v), \begin{pmatrix} \widetilde{\mathcal{L}}_{xx}^2 & \widetilde{\mathcal{L}}_{xu}^2 \\ \widetilde{\mathcal{L}}_{ux}^2 & \widetilde{\mathcal{L}}_{uu}^2 \end{pmatrix} (y,v)\right) \geq \bar{\gamma}(\|y\|_2^2 + \|v\|_2^2) \tag{6}$$

for all $(y,v) \in W^{1,2}(0,T;\mathbf{R}^n) \times L^2(0,T;\mathbf{R}^n)$ such that

$$\begin{array}{l}\dot{y}(t) - A(\widetilde{w}(t))y(t) - B(\widetilde{(t)})v(t) = 0, \\ \Xi_0(\widetilde{w})y(0) + \Xi_T(\widetilde{w})y(T) = 0, \\ \Theta_x^{\bar{\alpha}}(t)(t)y(t) + \Theta_u^{\bar{\alpha}}(t)(t)v(t) = 0.\end{array} \tag{7}$$

REMARK **32** Note that in (6) coercivity holds not in the norm of the space Z in which problem (O) is well defined and differentiable, but in the weaker norm of the space $\widehat{Z}$ in which, in general, (O) is not well defined. This phenomenon is called **two-norm discrepancy.**

In Malanowski, Maurer (1994) condition (II.6) is expressed in terms of existence of a bounded solution to an auxiliary Riccati equation, satisfying certain boundary conditions.

In Malanowski (1994) it is shown that if assumptions (II.1)-(II.6) hold then $(\widetilde{x},\widetilde{u})$ is a locally isolated, local minimizer of (O).

In Alt, Malanowski (1995) a coercivity condition stronger than (6) was used. Namely, the last equation in (7) was not present.

We are going to analyse convergence of (LNM) for optimal control problem (O). We will use abstract results presented in Section 2 in the same way as in Alt, Malanowski (1995), where less general optimal control problem was considered. The basis of our analysis will be the stability results for parametric optimal control problems derived in Malanowski (1994). Let us denote $w_k = (x_k, u_k, q_k, \rho_k, \kappa_k, \mu_k)$. For optimal control problem (O), the quadratic optimization problem (QP_{w_k}) used in (LNM) takes on the form:

$$
(\mathrm{QO}_{w_k}) \quad \begin{array}{l}
\min_{(y,v)\in Z} \widetilde{F}(y,v,w_k) := \int_0^T \widetilde{f}^0(y(t),v(t),w_k(t))dt + \widetilde{g}(y(0),y(T),w_k) \\
\text{subject to} \\
\dot{y}(t) - f(x_k(t),u_k(t)) - A(w_k(t))(y(t)-x_k(t)) - B(w_k(t))(v(t)-u_k(t)) = 0, \\
\xi(x_k(0),x_k(T)) + \Xi_0(w_k)(y(0)-x_k(0)) + \Xi_T(w_k)(y(T)-x_k(T)) = 0, \\
\theta(x_k(t),u_k(t)) + \Theta_x(w_k(t))(y(t)-x_k(t)) + \Theta_u(w_k(t))(v(t)-u_k(t)) \leq 0, \\
v(x_k(t)) + \Upsilon(w_k(t))(y(t)-x_k(t)) \leq 0,
\end{array}
$$

where $\widetilde{F}(y,v,w_k)$ is defined as in (QP_{w_k}), using Lagrangian (5).

In view of the form of the mapping given by the derivatives of the Lagrangian (5), we choose as the space of perturbations the space

$$
\begin{array}{l}
\widehat{X}^* = L^2(0,T;\mathbf{R}^n) \times L^2(0,T;\mathbf{R}^m) \times \mathbf{R}^n \times \mathbf{R}^n \times \\
\times L^2(0,T;\mathbf{R}^n) \times \mathbf{R}^d \times L^2(0,T;\mathbf{R}^k) \times W^{1,2}(0,T;\mathbf{R}^l).
\end{array}
$$

For $\delta = (a_1,a_2,a_3,a_4,b_1,b_2,b_3,b_4) \in \widehat{X}^*$, the problem (LO_δ), corresponding to (LP_δ) is defined as follows:

$$
(\mathrm{LO}_\delta) \quad \begin{array}{l}
\min_{(y,v)\in Z} \widetilde{F}(y,v,\delta) := \\
:= \int_0^T \Big(\widetilde{f}^0(y(t),v(t),\widetilde{w}(t)) - \langle a_1, y(t)-\widetilde{x}(t)\rangle - \langle a_2, v(t)-\widetilde{u}(t)\rangle \Big) dt + \\
+\widetilde{g}(y(0),y(T),\widetilde{w}) - \langle a_3, y(0)-\widetilde{x}(0)\rangle - \langle a_4, y(T)-\widetilde{x}(T)\rangle \\
\text{subject to} \\
\dot{y}(t) - f(\widetilde{x}(t),\widetilde{u}(t)) - A(\widetilde{w}(t))(y(t)-\widetilde{x}(t)) - B(\widetilde{w}(t))(v(t)-\widetilde{u}(t)) - b_1(t) = 0, \\
\xi(\widetilde{x}(0),\widetilde{x}(T)) + \Xi_0(\widetilde{w})(y(0)-\widetilde{x}(0)) + \Xi_T(\widetilde{w})(y(T)-\widetilde{x}(T)) - b_2 = 0, \\
\theta(\widetilde{x}(t),\widetilde{u}(t)) + \Theta_x(\widetilde{w}(t))(y(t)-\widetilde{x}(t)) + \Theta_u(\widetilde{w}(t))(v(t)-\widetilde{u}(t)) - b_3(t) \leq 0, \\
v(\widetilde{x}(t)) + \Upsilon(\widetilde{w}(t))(y(t)-\widetilde{x}(t)) - b_4(t) \leq 0,
\end{array}
$$

We have to verify assumptions (I.5) and (I.6) of Theorem 22.

In the same way as in Malanowski (1994) (cf. also Malanowski (1995) and Alt, Malanowski (1995)), as the set $\Gamma \in X$ of Kuhn-Tucker points of (LO_δ) we choose the set of functions $\mathbf{x} = (y,v,r,\sigma,\lambda,\nu)$ having the same regularity as $\widetilde{\mathbf{x}} = (\widetilde{x},\widetilde{u},\widetilde{\rho},\widetilde{\kappa},\widetilde{\mu})$ i.e., uniformly bounded and uniformly Lipschitz continuous, with the same bound and Lipschitz modulus. By the Arzela-Ascoli theorem the set Γ is compact in X.

It follows from Lemmas 6.6 and 6.7 in Malanowski (1995) that if the set $\Delta \subset X^*$ of regular perturbations is chosen also as the set of uniformly bounded and uniformly Lipschitz continuous functions, with sufficiently small Lipschitz modulus, then, for each $\delta \in \Delta$ there exists a unique in Γ Kuhn-Tucker point $\mathbf{x}_\delta := (z_\delta, w_\delta, r_\delta, \sigma_\delta, \pi_\delta, \nu_\delta)$ of (LO_δ) and

$$
\|y_{\delta_1} - y_{\delta_2}\|_{1,2}, \|v_{\delta_1} - v_{\delta_2}\|_2, \|r_{\delta_1} - r_{\delta_2}\|_{1,2}, |\sigma_{\delta_1} - \sigma_{\delta_2}|, \|\pi_{\delta_1} - \pi_{\delta}\|_2, \|\nu_{\delta_1} - \nu_{\delta_2}\|_{1,2} \leq c\, \|\delta_1 - \delta_2\|_{\widehat{X}}.
$$

for all $\delta_1, \delta_2 \in \Delta$.

Thus all assumptions of Theorem 22 are satisfied.

Moreover, it was shown in Section 7 of Alt, Malanowski (1995) that the function ζ in (3) has the form $\zeta(r) = c\, r^{\frac{2}{3}}$. Hence we obtain the following convergence result for (LNM):

THEOREM **33** *If assumptions (II.1)-(II.6) are satisfied, then there exists $\rho > 0$ such that for each starting point $w_1 \in \Gamma \cap \mathcal{B}_{\hat{X}}(\widetilde{w}, \rho)$ the Lagrange-Newton method applied to (O) generates a unique sequence $\{w_k\}$ convergent to $\widetilde{w}$ with*

$$\|w_{k+1} - \widetilde{w}\|_{\hat{X}} \leq c\|w_k - \widetilde{w}\|_{\hat{X}}^{\frac{5}{3}} \quad \text{and} \quad [\![w_{k+1} - \widetilde{w}]\!]_X \leq c[\![w_k - \widetilde{w}]\!]_X^{\frac{4}{3}}.$$

REFERENCES

Alt, W. (1990) The Lagrange-Newton method for infinite-dimensional optimization problems. *Numerical Functional Analysis and Optimization,* **11**, 201–24.

Alt, W. (1991) Parametric programming with applications to optimal control and sequential quadratic programming. *Bayreuther Mathematische Schriften,* **35**, 1–37.

Alt, W. and Malanowski, K. (1993) The Lagrange-Newton method for nonlinear optimal control problems. *Computational Optimization and Applications,* **2** , 77–110.

Alt, W. and Malanowski, K. (1995) The Lagrange-Newton method for state constrained optimal control problems. *Computational Optimization and Applications,* **4**, 217–39.

Dontchev, A.L., Hager, W.W., Poore, A.B. and Yang, B. (1995) Optimality, stability and covergence in nonlinear control. Applied Mathematics and Optimization, **31**, 297–326.

Hager, W.W. (1979) Lipschitz continuity for constrained processes. *SIAM Journal on Control and Optimization,* **17**, 321–38.

Machielsen, K.C.P. (1987) Numerical solution of optimal control problems with state constraints by sequential quadratic programming in function space *CWI Tract,* **53**, Amsterdam.

Malanowski, K. (1992) Second order conditions and constraint qualifications in stability and sensitivity analysis of solutions in optimization problems in Hilbert spaces. *Applied Mathemathics and Optimization,* **21**, 98–110.

Malanowski, K. (1993) Two-norm approach in stability and sensitivity analysis of optimization and optimal control problems. *Advanes in Mathemathical Sciences and Applications,* **2**, 397–443.

Malanowski, K. (1994) Sufficient optimality conditions in optimal control. *System Research Institute,* **WP-1-1994**, Warszawa.

Malanowski, K. (1995) Stability and sensitivity of solutions to nonlinear optimal control problems (to appear).

Malanowski, K. and Maurer, H. (1994) Sensitivity analysis for parametric optimal control problems with control-state constraints. *Universität Münster, Angewandte Mathematik und Informatik,* **4/94-N**, Münster.

Robinson, S.M. (1974) Perturbed Kuhn-Tucker points and rates of convergence for a class of nonlinear-programming algorithms. *Mathematical Programming,* **7**, 1-16.

Robinson, S.M. (1980) Strongly regular generalized equations. *Mathematics of Operations Research,* **5**, 43–62.

Descent methods for optimal periodic hereditary control problems

Krystyna Nitka-Styczeń
Institute of Engineering Cybernetics, Technical University of Wrocław
Janiszewskiego 11/17, 50-372 Wrocław, Poland.
Tel: 48 71 202745. Fax: 48 71 203408.

Abstract
Continuously operated autonomous processes governed by hereditary state equations are considered. The problem of improving of the optimal steady-state solution by periodic variations of control and state variables is discussed. A new formulation of the periodic hereditary control problem exploiting the heredity shift operator is presented, and its advantages are pointed out. This problem is solved iteratively by descent methods using variable metric ideas and implemented with the help of the Mathematica system. The proposed approach is illustrated by a simple example of ecological hereditary system.

Keywords
Hereditary autonomous process, periodic control problem, heredity shift operator, descent method, Mathematica system.

1 INTRODUCTION

Long horizon continuously operated autonomous processes are encountered in many chemical, biochemical, biotechnological, and ecological systems (May, 1973), (Reber, 1979), (Ray, 1981), (Smith and Waltman, 1995). The optimization of such processes is connected with the question if the optimal steady-state solutions can be improved by periodic variations of the state and control variables. Adequate models of many processes under discussion should mirror miscellaneous hereditary phenomena (Abulesz and Lyberatos, 1987), (Colonius, 1988), (Smith and Waltman, 1995). This fact complicates the determination of the optimal periodic control, because necessary optimality conditions formulated as the maximum principle (Colonius, 1988) or as the Banach space optimality conditions (Colonius, 1988) may take a quite complicated form. In particular, the periodic delayed-advanced Hamiltonian system arising from the maximum principle may be unstable and sensitive to parameter changes similarly to the case of undelayed problems, where the appropriate Hamiltonian systems cannot be asymptotically stable according to the Poincare-Lyapunov theorem (Demidovich, 1967).

For the above reasons descent optimization methods may be helpful in searching of nearly optimal solutions to periodic hereditary control problems. To depict such methods a new formulation of the problem discussed is presented. It uses the process dynamics description in the form of an integral equation equivalent to the basic periodic differential hereditary equation. The equivalent integral equation exploits the so-called heredity shift operator, and acts in the space of continuous functions defined on the process operation period. The simple form of the heredity shift operator (for a simple discrete delay) has been introduced by Krasnosel'skii (1966) in the context of the theory of periodic nonlinear equations.

We generalize the definition of this operator to the case of many lumped and distributed delays, and we adopt it to periodic hereditary control problems. The main advantage of our formulation is connected with the interpretation of periodic admissible trajectories as fixed points of a completely continuous integral operator. Such operators possess high accuracy finite dimensional approximations (Krasnoselskii, 1969). Therefore, the formulation proposed is especially suitable for computational purposes, and in particular for the construction of iterative descent optimization methods.

In the approach considered the control space may be chosen as the Banach space of essentially bounded measurable functions or as the Hilbert space of square-integrable functions. While the first choice is useful for the analysis of optimality conditions, the second one may be advantageously exploited to design enhanced descent optimization directions (Nitka-Styczeń, 1995).

We use the following spaces of n-dimensional functions defined on the interval (α, β): $C^{0,n}(\alpha, \beta)$ the space of continuous functions, $L_p^n(\alpha, \beta)$ the space of pth power integrable functions, $W_p^{1,n}(\alpha, \beta)$ the space of differentiable functions with derivative in $L_p^n(\alpha, \beta)$.

2 OPTIMAL PERIODIC HEREDITARY CONTROL PROBLEM

Consider the following optimal periodic hereditary control (OPHC) problem: minimize the time-averaged objective function

$$J(\tau, x, u) = \frac{1}{\tau} \int_0^\tau g(x(t), u(t))dt \tag{1}$$

subject to

$$\dot{x}(t) = f(x_t, u(t)), \qquad t \in [0, \tau], \tag{2}$$

$$x_0 = x_\tau, \tag{3}$$

$$\frac{1}{\tau} \int_0^\tau \phi(x(t), u(t))dt = 0, \tag{4}$$

$$\tau \in \mathcal{T}, \quad u(t) \in U, \quad t \in [0, \tau], \tag{5}$$

where τ is the operation period, $\mathcal{T} = [\tau^0, \tau^1] \subset R_+$, $r \ll \tau_0$, $x \in \mathcal{X} = W_p^{1,n}(0, \tau)$ is the momentary state trajectory, $x_t \in \mathcal{X}_0 = W_p^{1,n}(-r, 0)$ is the complete state at the time t, $u \in \mathcal{U} = L_p^m(0, \tau)$ is the control, U is a (bounded and closed) parallelepiped in R^m, and

$$g : R^n \times R^m \to R^1, \qquad \phi : R^n \times R^m \to R^\nu,$$

$f : \mathcal{X}_0 \times R^m \to R^n,$

$f(x_t, u(t)) = \tilde{f}(\xi_0, ..., \xi_q, \xi_{q+1}, ..., \xi_{q+\tilde{q}}, u(t)), \quad \tilde{f} : R^{n\times(1+q+\tilde{q})} \to R^n,$

$\xi_h = x(t - r_h), \quad h = 0, 1, ..., q; \quad 0 = r_0 < r_1 < ... < r_q \leq r,$

$$\xi_{q+h} = \int_{-\tilde{r}_h}^{0} \Gamma_h(\theta)x(t+\theta)d\theta, \quad \Gamma_h : R \to R^{n\times n}, \quad h = 1, ..., \tilde{q}, \; 0 < \tilde{r}_1 < ... < \tilde{r}_{\tilde{q}} \leq r.$$

Thus we deal with the case of many lumped and distributed delays.

Assumption 1 *The functions $g, \tilde{f}, \phi$ are continuously differentiable, and the functions Γ_h are continuous.*

Together with the OPHC problem we consider the following optimal integral hereditary control (OIHC) problem: minimize

$$J(\tau, x, u) = \frac{1}{\tau}\int_0^{\tau} g(x(t), u(t))dt \tag{6}$$

subject to

$$x(t) = x(\tau) + \int_0^t \tilde{f}(S_\theta\, x, u(\theta))d\theta, \quad t \in [0, \tau], \tag{7}$$

$$\frac{1}{\tau}\int_0^{\tau} \phi(x(t), u(t))dt = 0, \tag{8}$$

$$\tau \in \mathcal{T}, \quad u(t) \in U, \quad t \in [0, \tau], \tag{9}$$

where $x \in \tilde{\mathcal{X}} = C^{0,n}(0, \tau), \; u \in \mathcal{U} = L_p^m(0, \tau)$, and the heredity shift operator

$$S_t : \tilde{\mathcal{X}} \to R^{n\times(1+q+\tilde{q})}$$

is defined as follows:

$S_t\, x = (S_{t,h}\, x)_{h=0,1,...,q+\tilde{q}},$

$$S_{t,h}\, x = \begin{cases} x(t - r_h) & \text{if } \; r_h \leq t \leq \tau, \\ x(\tau + t - r_h) & \text{if } \; 0 \leq t < r_h, \end{cases} \quad h = 0, 1, ..., q,$$

$$S_{t,h}\, x = \int_{-t}^{0} \Gamma_h(\theta)x(t+\theta)d\theta + \int_{-r_h}^{-t} \Gamma_h(\theta)x(\tau + t + \theta)d\theta, \quad p = q+1, ..., q+\tilde{q}.$$

Lemma 1 *The problems OPHC and OIHC are equivalent.*

The proof follows from the fact that the set of solutions (x, u) of the equations (2) and (3) and the set of solutions (x, u) of the equation (7) coincide.

Lemma 2 *The operator $A : \tilde{\mathcal{X}} \to \tilde{\mathcal{X}}$ defined as*

$$(Ax)(t) = x(\tau) + \int_0^t \tilde{f}(S_\theta\, x, u(\theta))d\theta, \quad t \in [0, \tau], \tag{10}$$

is completely continuous for every admissible control u.

The proof follows from the fact that the heredity shift operator transforms every equibounded set X in $\tilde{\mathcal{X}}$ into an equibounded one in the same space. This allows one to show under Assumption 1 equiboundedness and equicontinuity of the functions (10) on the set X.

Thus the convertion of the OPHC problem into the OIHC problem is justified by the simplicity of the latter problem taking the form of controlled fixed points of the complete continuous operator A.

3 DESCENT METHOD IN THE PERIOD AND CONTROL SPACE

We assume that the additional constraints (4) an (5) are handled by penalty or/and projection methods. We consider at first the possibility of the local reduction of the OIHC problem to the control space for fixed period τ.

This reduction allows one to construct a descent sequence, for which a high accuracy resolution of the periodic hereditary process equation is preserved.

We write the process equation (7) as the equality $P(\tau, x, u) = 0$, where the nonlinear operator $P : R \times \tilde{\mathcal{X}} \times \mathcal{U} \to \tilde{\mathcal{X}}$ is defined as

$$P(\tau, x, u)(t) = x(t) - x(\tau) - \int_0^t \tilde{f}(S_\theta\, x, u(\theta))d\theta, \quad t \in [0, \tau]. \tag{11}$$

Assumption 2 *The unity does not belong to the spectrum of the operator $A : \tilde{\mathcal{X}} \to \tilde{\mathcal{X}}$.*

This assumption guarantees that the F-derivative $P_x : \tilde{\mathcal{X}} \to \tilde{\mathcal{X}}$ defined by the formula

$$(P_x \delta x)(t) = \delta x(t) - \delta x(\tau) - \int_0^t \tilde{f}_x(S_\theta\, x, u(\theta))\delta x(\theta)d\theta, \quad t \in [0, \tau], \tag{12}$$

where $\delta x \in \tilde{\mathcal{X}}$, is continuously invertible.

Lemma 3 *The OIHC problem is under Assumption 2 locally reducible for fixed period τ to the control space $\mathcal{U}$, i.e. there exists a uniquely determined function $x(\tau, u)$ solving the equation (7) in the vicinity of the control u with fixed parameter τ.*

The proof is a consequence of the standard implicit function theorem.

Let us denote the objective function reduced to the period and control space as

$$\tilde{J}(\tau, u) = J(\tau, x(\tau, u), u).$$

Then the differential of the functional $\tilde{J}$ can be written in the form

$$\tilde{J}_u \delta u = \frac{1}{\tau} \int_0^\tau (g_x(x(t), u(t))(x_u(\tau, u)\delta u)(t) + g_u(x(t), u(t))\delta u(t))\, dt, \tag{13}$$

where $x_u(\tau,u)\delta u = P_x^{-1}P_u\delta u$. The formula (13) can be used to develop the following principal descent optimization algorithm for the OIHC problem:

Algorithm 1

1. *Choose a triple* (τ, x, u).
2. *Solve the process equation (7) for a given* (τ, u) *with the initial approximations* x *by means of the Newton algorithm*

$$x_+ = x - \delta x, \quad P_x(\tau,x,u)\delta x = P(\tau,x,u). \tag{14}$$

3. *Compute an improved control*

$$u_+ = u - \alpha_+ D\tilde{J}_u^*,$$

where $D\tilde{J}_u^*$ *is a descent direction determined by a positive operator* D *and gradient* $\tilde{J}_u$*, while* α_+ *is a line search coefficient.*

4. *Repeat 3. to satisfy the stop criterion*

$$\| \tilde{J}_u \|_{\mathcal{U}} \leq \epsilon.$$

5. *Go to 1.*

4 COMPUTATIONAL IMPLEMENTATION

To make the principal algorithm computationally implementable we approximate the control by step functions and the state by roof functions

$$u(t) = \sum_{k=1}^{K} e_k(t)u_k, \qquad x(t) = \sum_{k=0}^{K} \psi_k(t)x_k.$$

The choice of the roof basis (Daniel 1971, Wierzbicki 1984) for the state is justified by its compatibility with the control approximation.

We denote by the upper index K operators and functions discretized on the time grid $t_k = \frac{\tau}{K}k$, $k = 1, \ldots, K$

$$x^K = \mathrm{vec}_{k=1}^K x(t_k) = \mathrm{vec}_{k=1}^K x_k.$$

The discretized version of the shift operator may be written as follows.

$$\tilde{S}_k : R^{n\times K} \rightarrow R^{n\times(1+q+\tilde{q})}$$

is defined as follows

$$\tilde{S}_k x^K = (\tilde{S}_{k,h}x^K)_{h=0,\ldots,q+\tilde{q}}$$

$$\tilde{S}_{k,h}x^K = \begin{cases} x_{k-[r_h\frac{T}{K}]}, & for \quad k-[r_h\frac{T}{K}] \geq 0 \\ x_{K+k-[r_h\frac{T}{K}]}, & for \quad k-[r_h\frac{T}{K}] < 0 \end{cases} \quad h = 0,...,q,$$

$$\tilde{S}_{k,h}x^K = \begin{cases} \sum_{j=k-[r_h\frac{T}{K}]+1}^{k} \Gamma_{j-k}x_j & for \quad k-[r_h\frac{T}{K}] \geq 0 \\ \sum_{j=1}^{k} \Gamma_j x_j & for \quad k-[r_h\frac{T}{K}] < 0 \quad h = q+1,...,q+\tilde{q}, \\ +\sum_{j=K+k-[r_h\frac{T}{K}]+1}^{K} \Gamma_{j-K-k}x_j & \end{cases}$$

where $[z]$ denotes integer of z closest to z.

The solution of the integral hereditary state equation (7) by the Newton method given by the formulas (14) can be written in the discretized version

$$x_{k+} = x_k - \delta x_k, \qquad k = 1,...,K,$$

$$\delta x_k - \delta x_K - \sum_{h=0}^{q+\tilde{q}} \sum_{j=1}^{k} \frac{T}{K} \tilde{f}_{\xi_h}(\tilde{S}_j x^K, u_j) S_{j,h} \delta x^K = x_k - x_K - \sum_{j=1}^{k} \frac{T}{K} \tilde{f}(S_j x^K, u_j).$$

We compose these equations in the matrix form encompassing all discrete time moments, and we write the second step of the principal descent optimization algorithm as

$$x_+^K = x^K - \delta x^K, \quad \delta x^K = [I - J - \sum_{h=0}^{q+\tilde{q}} \tilde{F}_{\xi_h}^K M_h^K]^{-1} \mathcal{P}^K, \tag{15}$$

where I is the identity matrix, J is the matrix with the elements equal to 1 in the last column and others equal to 0, $\tilde{F}_{\xi_h}^K$ is lower diagonal matrix with elements $\frac{T}{K}\tilde{f}_{\xi_h}(S_j x^K, u_j)$ below the diagonal in the jth column,

$$M_h^K \delta x^K = \mathrm{vec}_{j=1}^K S_{j,h} \delta x^K, \quad \mathcal{P}^K = \mathrm{vec}_{k=1}^K [x_k - x_K - \sum_{j=1}^{k} \frac{T}{K} \tilde{f}(S_j x^K, u_j)].$$

We also write the gradient (13) in the discretized version

$$\tilde{J}_{u^K} = \frac{1}{K}(\mathcal{G}_x^K [I - J - \sum_{h=0}^{q+\tilde{q}} \tilde{F}_{\xi_h}^K M_h^K]^{-1} \mathcal{P}_u^K + \mathcal{G}_u^K) \tag{16}$$

where $\mathcal{P}_u^K$ is the lower diagonal matrix with the elements $\frac{T}{K}\tilde{f}_u(\tilde{S}_j x^K, u_j)$ in the jth column, $\mathcal{G}_x^K = [g_x(x_1, u_1), ..., g_x(x_K, u_K)]$, $\mathcal{G}_u^K = [g_u(x_1, u_1, ..., g_u(x_K, u_K)]$.

The convergence of the discretized optimization method is connected with the convergence of the process equation approximation, the approximation of its derivative with respect to the control, and with the convergence of the approximate integration formula for the objective function. The two first question can be analysed, due to the complete continuity of the operator A, by applying the Banach-Steinhaus theorem in the form (Krasnosel'skii, 1969)

$$\| \mathcal{A}^K \| \leq c, \quad K = K_0, K_0 + 1, ...,$$

$$lim_{K\to\infty} \parallel E_K Ax - \mathcal{A}^K E_K x \parallel = 0, \quad E_K x = \text{vec}_{k=0}^{K} x(t_k).$$

Next the convergence of nearly optimal solutions for discretized problems can be analysed, for example, along the lines of Daniel (1971).

The computational algorithm (15) and (16) with the variable metric matrix has been applied with the help of the Mathematica system to the following example of the OIHC problem.

Example 1 *Consider the following OPHC control problem for a single-species population model with time lag: minimize*

$$J(\tau, x, u) = -\frac{1}{\tau}\int_0^{\tau} u(t)x(t)dt,$$

subject to

$$\dot{x}(t) = cx(t)\left(1 - \frac{x(t-r)}{a}\right) - u(t)x(t),$$

$$u(t) \in [0, u^{max}].$$

This problem corresponds to the maximization of the harvesting effect with the help of the periodic operation (optimal periodic harvesting problem).

We assume the following parameter values $c = 1.3797$, $a = 3.8187$, $r = 2.2764$. *Then the optimal steady-state process has been computed* $\overset{o}{x} = 1.9093$, $\overset{o}{u} = 0.6898$, $J(\overset{o}{x}, \overset{o}{u}) = -1.3172$.

The Π *form (Colonius, 1988) has the following expression*

$$\Pi(\omega) = \frac{ac\cos(\omega r)}{c^2/2 + 2\omega^2 - 2c\omega\sin(\omega r)}.$$

It follows from the shape of the curve $\Pi(\omega)$ *that the choice of the operation frequency has considerable influence on the effectiveness of periodic control. The periodic operation should be especially advantageous for the* $\omega = 0.6900$ *i.e. for the period* $\tau = 9.1059$.

Starting from the steady control with harmonic variation, and from the steady state, after three steps of the algorithm we have obtained suboptimal periodic control process shown in Figure 1, with the cost function $J = -2.700$.

The main advantage of the proposed approach consists in the simplicity of the computation of periodic solutions to the hereditary state equation. A fast convergence of the Newton method applied to this end has been observed. The disadvantage of the method consists in the necessity of applying of small control steps to guarantee the stability of the Newton method.

The suboptimal control computed may be used for final optimization performed with the help of the maximum principle.

The paper carried out within the research project Nr 8 S505 011 07 of Polish State Committee for Scientific Research

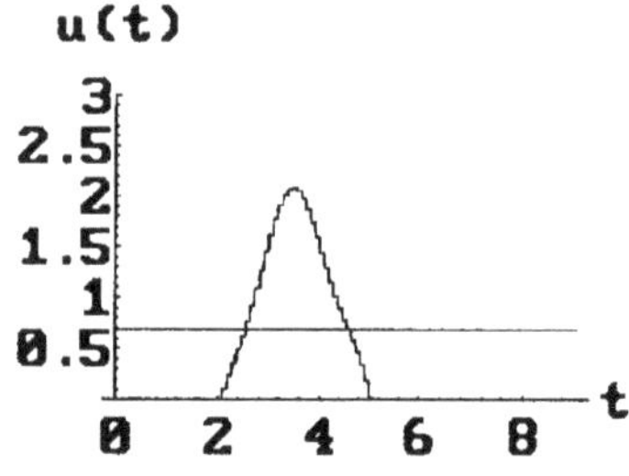

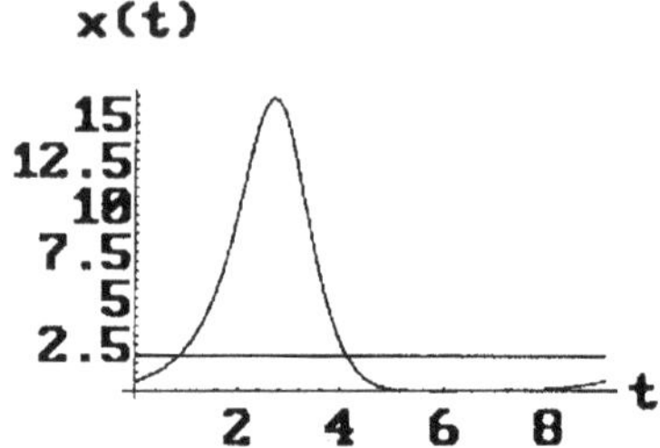

Figure 1 The steady-state process and suboptimal periodic process for the Example 1

REFERENCES

Abulesz, E.M. and Lyberatos, G. (1987) Periodic optimization of continuous microbial growth processes. *Biotechnology and Bioengineering*, **29**, 1059–65.

Colonius, F. (1988) *Optimal Periodic Control.* Lecture Notes in Mathematics Nr 1313. Springer-Verlag, Berlin.

Daniel, J.W. (1971) *The Approximate Minimization of Functionals.* Prentice Hall, Englewood Cliffs, New Jersey.

Demidovich, B.P. (1967) Lectures on Mathematical Stability Theory. Nauka, Moscow.

Krasnosel'skii, M. (1966) *The Shift Operator Along the Trajectories of Differential Equations.* Nauka, Moscow (in Russian).

Krasnosel'skii, M. (1969) *The Approximate Solving of Operator Equations.* Nauka, Moscow (in Russian).

May, R.M. (1973) *Stability and Complexity in Model Ecosystems.* Princeton, University Press, Princeton, N.J.

Nitka-Styczeń, K. (1995) Optimal Retarded Control Problems with Free Initial States: A Descent Method. *System Analysis Modelling Simulation* (to appear).

Ray, W.H. (1981) *Advanced Process Control.* Mc Grow Hill, New York.

Reber, D.C. (1979) A finite difference technique for solving optimization problems governed by linear functional differential equations. *Journal of Differential Equations*, **32**, 193–232.

Smith, H.L. and Waltman, P. (1995) *The Theory of the Chemostat. Dynamics of Microbial Competition.* Cambridge University Press, Cambrigde.

Wierzbicki, A. (1984) *Models and Sensitivity of Control Systems.* Elsevier, Amsterdam.

Wolfram Research, Inc. (1994) *Mathematica. A System for Doing Mathematica by Computer.* Wolfram Research, Inc., Champaign, Illinois.

Aircraft trajectory optimization using nonlinear programming

T. Raivio, H. Ehtamo, R. P. Hämäläinen
Systems Analysis Laboratory, Helsinki University of Technology
Otakaari 1 M, 02150 Espoo, Finland. Tel. +358-0-451 3059.
Fax: +358-0-451 3096. e-mail: tuomas.raivio@hut.fi

Abstract

We describe two discretization methods, direct collocation and a scheme based on differential inclusion, that enable the solution of optimal control problems by nonlinear programming. We apply the methods in calculating optimal trajectories for a modern fighter aircraft. Unlike collocation, the differential inclusion scheme converges robustly even in the presence of singular controls.

Keywords

Trajectory optimization, direct methods, discretization, nonlinear programming, singular controls.

1 INTRODUCTION

Aircraft trajectory optimization problems constitute a challenging class of optimal control problems. A six degrees of freedom aircraft model consists of 12 first order differential equations. In trajectory optimization the equations of rotation can be neglected, but the remaining model still includes six nonlinear first order equations of motion (see, *e.g.* , Miele (1962)).

The optimal control is usually solved either by indirect or direct methods. Indirect methods solve the multipoint boundary value problem arising from the necessary conditions for optimality. The methods provide accurate results but require a good initial guess. This stems mainly from the nonlinear and unstable nature of the boundary value problem and Newton-type solution methods.

Infinite dimensional direct methods attempt to overcome these difficulties. They improve a given nominal solution by a gradient search in a function space. A larger convergence domain is often achieved, but other difficulties, like the treatment of constraints, arise.

In many cases the accuracy of the solution is not as important as is the convergence of the solution method. The convergence could be improved by replacing the original infinite-dimensional problem with a finite-dimensional approximation, in which the differential equation constraint is satisfied only pointwise. Restricting to a finite dimension also allows the use of ordinary nonlinear optimization.

The discretization of the problem can be carried out in a number of ways. Hargraves et al. (1981) present the state trajectories with high-order patched polynomials, whose coefficients are the decision variables. Betts and Huffman (1993) discuss trapetzoidal, Hermite-Simpson and Runge-Kutta discretization. In the following we describe two schemes, direct collocation and a recently proposed method based on differential inclusions Seywald (1994). In the former approach the states and controls are approximated by piecewisely defined low-order polynomials. The latter scheme makes use of differential inclusion (see, *e.g.* , Aubin and Frankowska (1989)) and the concept of attainability. We demonstrate and compare the performance of these methods and the accuracy of the results. The numerical examples involve also singular controls.

2 AIRCRAFT MODEL

The dynamics of a point mass aircraft can be described by the following system of equations (see, *e.g.* , Miele (1962)):

$$\begin{aligned}
\dot{x} &= v\cos\gamma\cos\chi \\
\dot{y} &= v\cos\gamma\sin\chi \\
\dot{h} &= v\sin\gamma \\
\dot{\gamma} &= \frac{g}{v}(n\cos\mu - \cos\gamma) \\
\dot{\chi} &= \frac{g\,n\sin\mu}{v\,\cos\gamma} \\
\dot{v} &= \frac{1}{m}[uT_{max}(h, M(h,v)) - D(h,v,M(h,v),n)] - g\sin\gamma.
\end{aligned}$$

The state variables x, y, h, v, γ and χ are the x and the y coordinates, altitude, velocity, flight path angle and heading angle of the aircraft, respectively. The acceleration due to gravity, g, is assumed constant. $T_{max}(h, M(h,v))$ denotes the maximum available thrust force, u the throttle setting, $D(\cdot)$ the drag force and $M(\cdot)$ the Mach number. For short flight times the mass of the aircraft can be assumed constant.

The normal acceleration of the aircraft is controlled with the normal load factor n and the tangential acceleration with the throttle setting $u \in [0,1]$. To produce a horizontal turn, the normal load factor can be directed away from the vertical plane with the bank angle $\mu \in [-\pi, \pi[$.

The load factor n cannot be chosen freely. At low velocities, a large load factor requires a large angle of attack which results in loss of lift force and stall. At higher velocities, the magnitude of the load factor is constrained by the largest acceleration that the pilot and the aircraft withstand. Here the smallest allowed load factor is set to zero to ensure the uniqueness of the control variable combinations.

The aircraft drag is assumed to obey a shifted quadratic polar emerging from the actively controlled aircraft wing Ehtamo et al. (1994):

$$C_D(M(h,v),n) = C_{D_0}(M(h,v)) + K(M(h,v))(C_L(\cdot) - a)^2, \quad a > 0.$$

Here $C_D(\cdot)$, $C_{D_0}(\cdot)$ and $K(\cdot)$ denote the total, zero-lift and induced drag coefficients, respectively. The drag force becomes

$$D(h, v, M(h, v), n) = C_D(M(h, v), n)Sq(h, v) =$$

$$(C_{D_0}(M(h, v)) + K(M(h, v))(\frac{nmg}{Sq(h, v)} - a)^2)Sq(h, v), \quad q(h, v) = \frac{1}{2}\varrho(h)v^2.$$

S and $q(h, v)$ stand for the reference wing area and the dynamic pressure. The coefficients $C_{D_0}(M(\cdot))$ and $K(M(\cdot))$ are approximated by rational polynomials on the basis of realistic tabular data. The maximum thrust data is approximated by a two-dimensional polynomial. The air density and the speed of the sound are taken from the standard ISA atmosphere.

3 THE DISCRETIZATION METHODS

Direct collocation
In direct collocation the state trajectories and admissible controls are constrained to lie in the space of piecewise polynomials of time with given degree. The polynomials must satisfy the state equation in a finite number of points at each interval. In this paper the state trajectories are interpolated by Hermite interpolation and cubic polynomials. The controls are approximated piecewise linearly. The slope of the approximating polynomial must coincide with the state equation value at the middle of each discretization interval. The optimal values of the control and state variables in the discretization points are selected through nonlinear optimization to minimize the cost function. The method or its variants have been applied to various trajectory optimization tasks (see, *e.g.* , Hargraves and Paris (1987)) but also to facilitate the solving of complex pursuit-evasion games by providing an initial guess for the solution Lachner et al. (1994).

Differential inclusion scheme
Another way to discretize the problem is to require that each subsequent state can be attained from the preceding state. Given t_0, an initial state vector $x(t_0) = x_0$ and t_1, the *set of attainability* $K(x_0, t_0, t_1)$ is defined as the collection of the states that can be reached from x_0 in $[t_0, t_1]$ using admissible controls $u(t)$, $t \in [t_0, t_1]$ (Lee and Markus, 1986). In general, this set cannot be expressed explicitly. To approximate it we use the *set of attainable state rates* at state $x(t)$ defined by

$$\mathcal{H}(x(t)) = \{\dot{x}(t) \in R^n \mid \dot{x}(t) = f(x(t), u(t)), \ u(t) \text{ admissible}\}.$$

Here $f(\cdot)$ refers to the RHS of the state equations. In the following, we drop the argument t for clarity. The set $\mathcal{H}(x)$ is sometimes called the *hodograph* of the system (*e.g.* Seywald (1994)). We next assume that u can be eliminated from the definition of $\mathcal{H}(x)$. That is, there exist smooth functions $p : R^n \times R^n \mapsto R^p$ and $q : R^n \times R^n \mapsto R^q$ such that $\mathcal{H}(x)$ can be expressed as

$$\mathcal{H}(x) = \{\dot{x} \in R^n \mid p(\dot{x}, x) = 0, q(\dot{x}, x) \leq 0\}.$$

The existence of p and q depends on the problem considered. In practice, they are derived by eliminating the controls from the state equations and then using the control constraints.

The first order approximation of the set of attainability is

$$\hat{K}(x,t,t+\Delta t) = \{y \in R^n \mid y = x + \Delta t \cdot \mathcal{H}(x)\}$$

where $\Delta t \cdot \mathcal{H}(x) = \{\Delta t \cdot \dot{x} \mid \dot{x} \in \mathcal{H}(x)\}$. The condition that takes the system dynamics into account is that each subsequent discretized state must lie in the approximated set of attainability of its predecessor.

Two remarks are to be made. First, in this form the differential inclusion scheme is merely a first order discretization of the system. Nevertheless, the elimination of the control variables is expected to be computationally beneficial, especially in optimization problems that include singular controls. Second, the system equations must be invertible with respect to the control variables.

4 NUMERICAL EXAMPLES

In the first numerical example the task is to find a minimum time trajectory from the initial conditions

$$\begin{array}{lll} x(0) = 0 \text{ m}, & y(0) = 0 \text{ m}, & h(0) = 2,000 \text{ m}, \\ v(0) = 200 \text{ m/s}, & \gamma(0) = 0 \text{ rad}, & \chi(0) = 0 \text{ rad} \end{array}$$

to the point ($x(T) = 10,000$ m, $y(T) = 15,000$ m, $h(T) = 7,000$ m), final heading $\chi(T) = 0.35$ rad and to level flight, $\gamma(T) = 0$ rad. The mass of the aircraft was set to 10,000 kg. The problem was solved using Sequential Quadratic Programming and 5, 10, 15 and 20 equidistant discretization points. A representative solution of both methods, together with a reference solution obtained by multiple shooting, is presented in fig. 1. The iteration results are summarized in table 1.

With differential inclusion the errors decrease rapidly when the amount of discretization points is increased. In collocation, the situation is opposite: in our case five nodes provided the best result. The reason is probably that with short discretization intervals the best approximation approaches a line segment, which turns the collocation constraints more linearly dependent.

The dynamic pressure constraint $q(h,v) - q_{max} \leq 0$ in minimum time problems is known to lead to first order singularity on the controls (see, *e.g.* , Seywald et al. (1994)). To compare the convergence properties of the methods when singular controls arise, a minimum time descent was computed from the initial conditions

$$\begin{array}{lll} x(0) = 0 \text{ m}, & y(0) = 0 \text{ m}, & h(0) = 5,000 \text{ m}, \\ v(0) = 300 \text{ m/s}, & \gamma(0) = 0 \text{ rad}, & \chi(0) = 0 \text{ rad} \end{array}$$

to the point $x(T) = 20,000$ m, $y(T) = 10,000$ m and $h(T) = 1,000$ m, with free final flight path angle and heading with the condition

$$q(h,v) \leq 80,000 N/m^2.$$

No. of nodes	*Direct collocation* 5	10	15	20	*Differential inclusion* 5	10	15	20
Final time ($T_{ref} = 60.2$ s)	59.6	61.6	60.3	61.2	60.0	59.2	59.3	59.4
Dimension	41	81	121	161	31	61	91	121
Simple bounds	15	30	45	60	5	10	15	20
Constraints	35	65	95	125	45	85	125	165
Major iterations	30	19	18	21	18	31	45	55
Minor iterations	83	74	51	60	53	158	296	1099

Table 1 Summary of the iteration characteristics. 'Simple bounds' means the amount of decision variable bounds, whereas 'constraints' refers to the number of true constraints. 'Major iterations' means the amount of QP problems solved and 'minor iterations' the cumulative amount of iterations needed to solve them.

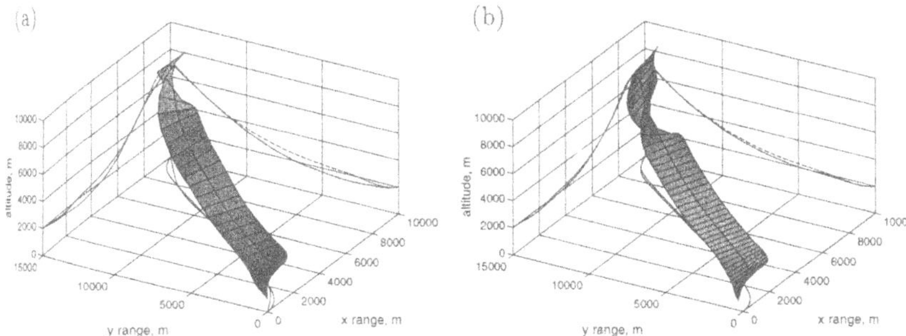

Figure 1 The solution obtained with (a) direct collocation and 5 nodes and (b) differential inclusion and 15 nodes. The stripe describes the bank angle of the aircraft. The dashed projections represent the reference solution obtained with multiple shooting.

In an unconstrained solution the maximum pressure would be over $120,000\ N/m^2$, which is untolerable for most aircraft.

Six nodes were employed in the collocation and 15 nodes in the differential inclusion. The solution trajectories are presented in fig. 2. The state and control variables are presented in fig. 3. The pressure constraint becomes active at $t \cong 27$s. After that the controls turn singular. The constraint remains active for the rest of the trajectory. Note that the singularity of the controls cannot be deduced without some additional knowledge on the problem.

Singular controls shrunk the convergence domain of collocation for convergence was obtained only after numerous attempts with different initial guesses. The differential inclusion scheme did not exhibit such convergence problems, as expected.

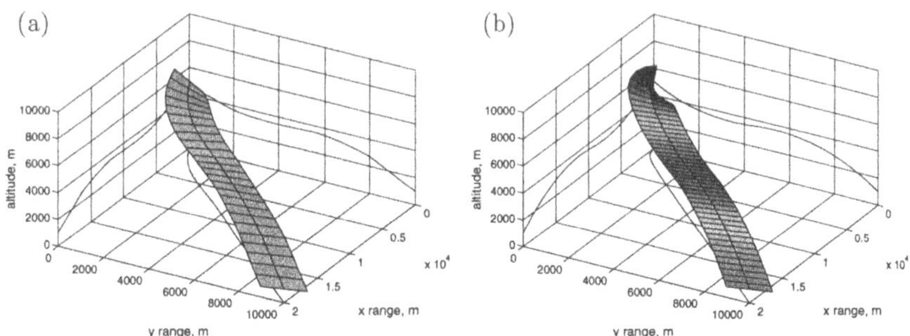

Figure 2 The descent trajectories obtained with (a) collocation and 6 nodes and (b) differential inclusion and 15 nodes. Both methods predicted a final time of approximately 58 s.

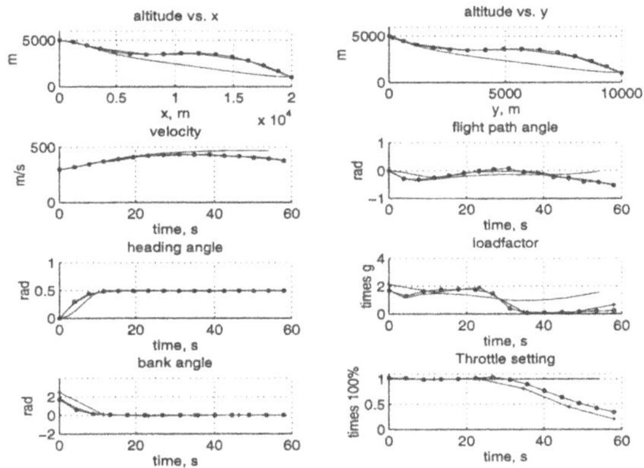

Figure 3 The state and control variable histories of the descent. '+' corresponds to collocation and 'o' to differential inclusion. The dashed line indicates the unconstrained solution. The control variables corresponding to the differential inclusion solutions have been calculated afterwards.

5 CONCLUSIONS

We have described two discretization schemes for direct trajectory optimization. The presented numerical examples suggest that both methods are capable of approximating optimal trajectories. The equidistant discretization scheme used in the examples is simple to implement but by no means the most efficient one. The accuracy can be further improved by adaptively redistributing the discretization nodes.

Both schemes converge robustly in the case of regular controls. Singular controls affect the convergence of the collocation method, but do not considerably influence the performance of the differential inclusion scheme. This is consistent with previous results (see

Seywald (1994)). In addition, the differential inclusion scheme describes the system with smaller number of decision variables and constraints than collocation.

It should be noted that the methods do not provide any explicit information on the structure of the solution. The activity of state and control constraints can be deduced from the solution data, but singular control intervals cannot be identified without deriving the necessary conditions. On the other hand, these approaches do not require the optimal switching structure in advance, which is the case with indirect methods.

REFERENCES

Aubin, J-P. and Frankowska, H. (1989) *Set-valued Analysis.* Birkhäuser.

Betts, J. and Huffman, W. (1993) Path-Constrained Trajectory Optimization Using Sparse Sequential Quadratic Programming. *Journal of Guidance, Control and Dynamics*, **16**, 1.

Ehtamo, H.K., Raivio, T. and Hämäläinen, R.P. (1994) A method to generate trajectories for minimum time climb, in *Preprints of the 6th international symposium on differential games and applications*, St. Jovite, Canada.

Hargraves, C., Johnson, F., Paris, S. and Rettie, I. (1981) Numerical Computation of Optimal Athmospheric Trajectories. *Journal of Guidance, Control and Dynamics*, **4**, 4.

Hargraves, C. and Paris, S. (1987) Direct Trajectory Optimization Using Nonlinear Programming and Collocation. *Journal of Guidance, Control and Dynamics*, **10**, 4.

Lachner, R., Breitner, M., and Pesch, H.J. (1994) Three-Dimensional Air Combat: Numerical Solution of Complex Differential Games, in *Preprints of the 6th international symposium on differential games and applications*, St. Jovite, Canada.

Lee, E.B. and Markus, L. (1986) *Foundations of optimal control theory.* Wiley, New York.

Miele, A. (1962) *Flight Mechanics.* Addison-Wesley, Massachusetts.

Seywald, H. (1994) Trajectory Optimization Based on Differential Inclusion. *Journal of Guidance, Control and Dynamics*, **17**, 3.

Seywald, H. Cliff, E. and Well, K. (1994) Range Optimal Trajectories for an Aircraft Flying in the Vertical Plane. *Journal of Guidance, Control and Dynamics*, **17**, 2.

Feedback control of state constrained optimal control problems

D.A. Redfern and C.J. Goh
Department of Mathematics, University of Western Australia
Nedlands, Western Australia 6907. e-mail: goh@maths.uwa.edu.au

Abstract
Solutions to nonlinear optimal control problems are usually computed in open-loop form, especially if the problem is subjected to state constraints. Open-loop controls, however, are not very useful from a control perspective due to their lack of robustness. We discuss a couple of useful techniques which reduce a state-constrained problem into an unconstrained problem, and subsequently apply a technique to synthesize an optimal state feedback controller for the reduced problem. The controller is able to recover the open-loop optimal state and control for arbitrary initial conditions arising from a nontrivial subset of the state space, while ensuring that the state constraint is satisfied at all time.

Keywords
Feedback control, state constraints, optimal control, neural networks.

1 INTRODUCTION

The linear-quadratic regulator is one of the most popular design methodologies in modern control theory. It is also one of the rare optimal control problems where the optimal control is furnished in the form of a feedback controller which is independent of the system's initial condition. For most nonlinear optimal control problems, one is usually concerned with the computation of an open-loop optimal control which is dependent on the system's initial condition. In general, optimal feedback control law for nonlinear optimal control problems with general cost functionals are extremely difficult to compute. For some simple problems such as the linear quadratic regulator, the optimal feedback controller can be obtained by solving the corresponding Hamilton-Jacobi-Bellman (HJB) dynamic programming equation. In practice, the curse of dimensionality associated with dynamic programming incurs serious computational difficulties for other problems. Sometimes local perturbation methods can be used for constructing the feedback controller around a nominally optimal trajectory, see for example Bryson and Ho (1975). However, these perturbation methods are unlikely to be effective in the event of a large shift in initial conditions.

In Edwards and Goh (1995), a direct training method is proposed for the synthesis of an optimal feedback controller in the form of a feedforward neural network for continuous-time nonlinear dynamical systems. The controller is independent of the system's initial condition provided that it arises from some bounded domain of the state space, and is not constricted by any fixed model structure for the dynamical system. The underlying idea is motivated by a paper of Nguyen and Widrow (1990) where a trailer-truck is trained to back up to a loading dock from arbitrary initial conditions. Although it was not explicitly mentioned in Nguyen and Widrow (1990), the problem addressed therein was really a discrete-time optimal control problem. The controller in this case is also a neural network, but it is possible that other versatile function emulators such as radial basis functions can also be used for the controller.

It is the goal of this paper to extend the result in Edwards and Goh (1995) to nonlinear feedback optimal control problems subjected to further continuous state constraints. It is well-known that, even for computing open-loop optimal solutions, the state constrained optimal control problem is notoriously difficult. While there is no shortage of theoretical results in the form of necessary conditions for the constrained problem, practical methods for constructing the constrained solutions have been few. In this paper, we discuss two methods to transform a state-constrained problem into an unconstrained problem, and subsequently apply the method in Edwards and Goh (1995) to synthesize the optimal feedback controller for the resulting unconstrained problem.

2 FEEDBACK CONTROL OF UNCONSTRAINED OPTIMAL CONTROL PROBLEMS

We introduce the underlying notion of a feedback controller by formulating the unconstrained problem first. Consider the continuous-time dynamical system defined in the finite interval $[0, T]$:

$$\dot{\mathbf{x}} = \mathbf{f}(\mathbf{x}, \mathbf{u}, t), \tag{1}$$

$$\mathbf{x}(0) = \boldsymbol{\xi}, \tag{2}$$

where $\mathbf{x} \in \mathbb{R}^n$ is the state; $\mathbf{u} \in \mathbb{R}^m$ is the control; and $\mathbf{f} = (f_1, f_2, \cdots, f_n)^\top : \mathbb{R}^n \times \mathbb{R}^m \times [0, T] \to \mathbb{R}^n$ is assumed known, smooth and Lipschitz. The initial state $\boldsymbol{\xi}$ is distributed randomly in some bounded subset Γ of the state space according to some density distribution $\rho(\boldsymbol{\xi})$ where ρ vanishes identically outside Γ. The control is subjected to the simple bound constraint:

$$\alpha_i \leq u_i(t) \leq \beta_i, \quad \forall i = 1, \cdots, m, \quad \text{and} \quad \forall t \in [0, T]. \tag{3}$$

A control $\mathbf{u}$ is said to be 'admissible' if it is a measurable function which satisfies the bound (3). Let $\mathcal{U}$ be the class of all such admissible controls. The (open-loop) optimal control problem can be stated as:

Problem P1: Given a fixed initial condition $\mathbf{x}(0) = \boldsymbol{\xi}$, find a $\mathbf{u}(\cdot; \boldsymbol{\xi})$ from $\mathcal{U}$ such that the cost functional:

$$J_1(\mathbf{u}(\cdot); \boldsymbol{\xi}) = \Phi(\mathbf{x}(T)) + \int_0^T L(\mathbf{x}(t), \mathbf{u}(t), t)dt \tag{4}$$

is minimized, where $L : \mathbb{R}^n \times \mathbb{R}^m \times [0, T] \to \mathbb{R}$ and $\Phi : \mathbb{R}^n \to \mathbb{R}$ are known smooth functions. Note that the (open-loop) optimal control $\mathbf{u}^*$ which solves P_1 is dependent on

the initial condition $\boldsymbol{\xi}$. Assuming that full state measurement is possible, the feedback problem is defined by:
Problem P2: Construct a state feedback controller $\mathbf{u} = \mathbf{g}(\mathbf{x}, t)$, such that for any given initial condition in Γ, the feedback control will give rise to the same minimum cost functional, optimal control and optimal state trajectory as that of the open-loop optimal solution to problem P1.

Such a feedback controller is then independent of any shift in initial condition (provided that it remains in Γ), and is almost surely more robust than the open-loop controller in the presence of noise and parameter variation. An exact analytical solution for $\mathbf{g}(\mathbf{x}, t)$ is only possible in trivial cases such as the linear quadratic regulator. We shall return in section 4 to discuss how we can synthesize an approximate optimal feedback controller.

3 TWO TECHNIQUES FOR REDUCING CONSTRAINED TO UNCONSTRAINED PROBLEMS

In some practical situations such as obstacle avoidance in robot control, the system is subjected to one or more of the following state constraint:

$$S(\mathbf{x}(t), t) \leq 0 \tag{5}$$

where $S : \mathbb{R}^n \times [0, T] \to \mathbb{R}$ is assumed to be smooth. We say that the constraint (5) is *of order* K if $dS/dt, d^2S/dt^2, \cdots, d^{K-1}S/dt^{K-1}$ do not contain $\mathbf{u}$ explicitly, and $d^K S/dt^K$ is the first term that contains $\mathbf{u}$ explicitly. We shall discuss two techniques to transform a state constrained problems into an unconstrained problem, and subsequently the direct training method of Edwards and Goh (1995) will be used in section 4 to construct an optimal feedback controller which, due to the prior transformation, will ensure that these state constraints are satisfied.

3.1 The penalty function approach

The first technique is based on the commonly used concept of exact penalty function. In particular, we use a special constraint transcription technique as presented in Teo et. al. (1992) which, apart from satisfying the usual constraint qualification, has several nice properties. While the penalty function method is entirely general, and may account for vector control with multiple state constraints, it is often slow in convergence, especially when one tries to train the feedback controller for the transformed problem. Furthermore, it is almost impossible to achieve perfect constraint satisfaction. We summarize the technique as follows and refer the readers to Teo et. al. (1992) for further details. As with all penalty function techniques, we append the constraint into the cost functional as follows:

$$J(\mathbf{u}(\cdot)) = \Phi(\mathbf{x}(T), T) + \int_0^T (L(\mathbf{x}(t), \mathbf{u}(t), t) + \mu L_\varepsilon(\mathbf{x}(t), t))\, dt, \tag{6}$$

$$\text{where } L_\varepsilon(\mathbf{x}(t), t) = \begin{cases} 0, & \text{if } S(\mathbf{x}(t), t) < -\varepsilon, \\ \frac{(S(\mathbf{x}(t),t)+\varepsilon)^2}{4\varepsilon}, & \text{if } -\varepsilon < S(\mathbf{x}(t), t) < \varepsilon, \\ S(\mathbf{x}(t), t), & \text{if } S(\mathbf{x}(t), t) > \varepsilon, \end{cases} \tag{7}$$

and μ is some large penalty weight. To determine the solution to the constrained problem, ε is initially set to some moderately small positive value of, say 10^{-2}, and μ is set to some moderately large value, say 10^2. Note that if $\varepsilon > 0$ and if the penalty term is identically zero, then the constraint will be satisfied more than is necessary. In fact $S(t) \leq -\varepsilon < 0$. The optimal control problem is then solved and the feasibility of the constraint checked. If there is any constraint violation, increase μ, otherwise decrease ε, and the problem is solved again, until ε is less than a certain prescribed threshold. The rate of increasing μ and decreasing ε is based on some heuristic which comes about as the result of much trial and error. Multiple state constraints can be treated similarly by appending one penalty term for each constraint to the cost functional.

3.2 The Valentine transformation

The second technique is based on a result due to Valentine (1937) from the early calculus of variation literature. This result has subsequently been applied to compute the open-loop solution for state constrained optimal control problems in Jacobson and Lele (1969) and Miele et. al. (1979). We further extend the open-loop result to the synthesis of an optimal feedback controller. Unlike the penalty function approach, the Valentine transformation increases the order of the state space (by up uo the order of the constraint) using a new control. For some problems, this method works extremely well and the constraint is always satisfied exactly. The feedback controller can be trained in a fraction of the time taken by the penalty function approach. Nevertheless for the past two decades, very little attention has been paid to this seemingly elegant technique, but no reasons have been given to the best of our knowledge. We shall discuss in detail, the application of the Valentine transformation to the synthesis of feedback controllers, and point out the limitation which prevents this technique from becoming popular.

We consider firstly the case involving a single control and a single constraint. Generalization of the method to problems involving multiple constraints and multiple controls is slightly more tricky but not difficult. Firstly the inequality constraint is converted to an equality constraint using a slack variable:

$$S(\mathbf{x}(t), t) + \frac{1}{2}(\alpha(t))^2 = 0. \tag{8}$$

Repeated differentiation of equation (8) leads to the following set of equations, where $S^{(k)}$ and $\alpha^{(k)}$ are the kth derivative of S and α with respect to t, respectively:

$$\begin{aligned} &S^{(1)}(\mathbf{x}(t), t) + \alpha(t)\alpha^{(1)}(t) = 0, \\ &S^{(2)}(\mathbf{x}(t), t) + \left(\alpha^{(1)}(t)\right)^2 + \alpha(t)\alpha^{(2)}(t) = 0, \\ &S^{(3)}(\mathbf{x}(t), t) + 3\alpha^{(1)}(t)\alpha^{(2)}(t) + \alpha\alpha^{(3)}(t) = 0, \\ &\quad\vdots \\ &S^{(K)}(\mathbf{x}(t), u(t), t) + \{\text{terms involving } \alpha^{(1)}(\mathrm{t}), ..., \alpha^{(\mathrm{K}-1)}(\mathrm{t})\} + \alpha(\mathrm{t})\alpha^{(\mathrm{K})}(\mathrm{t}) = 0. \end{aligned} \tag{9}$$

If the constraint (9) is of order K, then $\partial S^{(K)}/\partial u \neq 0$. The implicit function theorem asserts that the control u can be expressed as a function of the other variables:

$$u = \phi(\mathbf{x}, \alpha, \alpha^{(1)}, ..., \alpha^{(K-1)}, \alpha^{(K)}, t). \tag{10}$$

If this cannot be done analytically then an approximate implicit function can be emulated using a function emulator, typically either a neural network or a radial basis function. Treating $\alpha, \alpha^{(1)}, ..., \alpha^{(K-1)}$ as additional state variables and $v = \alpha^{(K)}$ as the new control variable, and using (10), the following unconstrained problem is obtained.

$$\min J(v(\cdot)) = \Phi(\mathbf{x}(T),T) + \int_0^T L(\mathbf{x}(t), \phi(\mathbf{x},\alpha,\alpha^{(1)},...,\alpha^{(K-1)},v(t),t),t)dt \tag{11}$$

$$\begin{aligned} \dot{\mathbf{x}} &= f(\mathbf{x}(t), \phi(\mathbf{x},\alpha,\alpha^{(1)},...,\alpha^{(K-1)},v(t),t),t) \\ \dot{\alpha} &= \alpha^{(1)} \\ &\vdots \\ \dot{\alpha}^{(K-1)} &= v(t) \end{aligned} \tag{12}$$

with the initial conditions $\mathbf{x}(0) = \xi, \alpha(0), \alpha^{(1)}(0), ..., \alpha^{(K-1)}(0)$; where $\alpha(0), \alpha^{(1)}(0), ...,$ $\alpha^{(K-1)}(0)$ are obtained directly from equation (8) and the first $K-1$ equations from (9):

$$\begin{aligned} &\alpha(0) = \sqrt{-2S(\mathbf{x}(0),0)} \\ &\alpha^{(1)}(0) = -\frac{S^{(1)}(\mathbf{x}(0),0)}{\alpha(0)} \\ &\alpha^{(2)}(0) = -\frac{(\alpha^{(1)}(0))^2 + S^{(2)}(\mathbf{x}(0),0)}{\alpha(0)} \\ &\vdots \end{aligned} \tag{13}$$

Despite the guarantee of feasibility, this transformation technique does have a major drawback which severely limits the problems for which is appropriate. To be specific, for a certain class of problem this transformation results in a control which is unbounded at various points in the state space. As this singularity problem has not been reported elsewhere to the best of our knowledge, we shall illustrate it using a simple example, which is modified from a problem discussed in Bryson and Ho (1975).

Consider the following modified Zermelo problem. The problem represents the control of a ship traveling in a region of variable current strength. Instead of fixing the speed and controling the orientation of the ship as in the original Zermelo's problem, we fix the orientation (θ) and control the speed instead. The state constraint in this case is given by a parabolic boundary which the ship is to stay clear of.

$$\begin{aligned} \text{Min } J(u(\cdot)) &= \int_0^1 u^2 dt \\ \text{with } \dot{x}_1 &= u\cos\theta + \beta x_2 \\ \dot{x}_2 &= u\sin\theta \end{aligned}$$

subject to the terminal constraints $x_1(1) = 0,\ x_2(1) = 0$; and the first order constraint:

$$x_2 \le a(x_1 - b)^2 + c. \tag{14}$$

Addition of an appropriate slack variable α leads to

$$x_2 + \frac{1}{2}(\alpha(t))^2 = a(x_1 - b)^2 + c. \tag{15}$$

Differentiating the above once and introducing the new control $v(t) = \alpha^{(1)}(t)$ to obtain:

$$u\sin\theta + \alpha v = 2a(x_1 - b)(u\cos\theta + \beta x_2), \tag{16}$$

and solving for $u(t)$ we get

$$u = \frac{2a\beta x_2(x_1 - b) - \alpha v}{\sin\theta - 2a(x_1 - b)\cos\theta}. \tag{17}$$

Note that u becomes unbounded whenever the denominator vanishes, i.e., when

$$x_1 = \frac{\tan\theta}{2a} + b. \tag{18}$$

For realistic values of the parameters a, b and θ this often causes problems. For example if we have $a = 5$, $b = -0.5$ and $\theta = \frac{\pi}{4}$ then the control u becomes unbounded whenever x_1 approaches -0.4. Thus the method will fail if we wish to find the optimal trajectory from an initial condition with $x_1(0)$ less than -0.4. In general, if $x_1(0)$ is less than $\tan(\theta)/2a + b$, this singularity problem will surely arise. This difficulty was first observed when we attempt to control the movement of an inverted pendulum which results in consistent numerical singularity. The above example was then constructed to illustrate this singular effect. Despite this limitation, the Valentine transformation technique has been found to be generally effective for other nonsingular problems.

4 SYNTHESIS OF OPTIMAL FEEDBACK CONTROLLER

Once the appropriate transformation has been applied to reduce the constrained problem into an unconstrained problem, the technique of Edwards and Goh (1995) can be applied to construct a feedback controller that solves problem P2. The basic idea is to approximate the optimal controller by some parameterized model $\mathbf{u}(t) = \hat{\mathbf{g}}(\mathbf{x}(t), t; \mathcal{W})$, where $\mathcal{W}$ is the set of parameters for the controller, which is determined optimally by solving the following optimal parameter selection problem:

Problem P3: Find the optimal set of parameters $\mathcal{W}$ such that the cost functional

$$J_2(\mathcal{W}) = E(J_2'(\mathcal{W}; \boldsymbol{\xi})) = \int_{\boldsymbol{\xi} \in \Gamma} J_2'(\mathcal{W}; \boldsymbol{\xi}) \rho(\boldsymbol{\xi}) d\boldsymbol{\xi} \tag{19}$$

is minimized with respect to $\mathcal{W}$, where

$$J_2'(\mathcal{W}; \boldsymbol{\xi}) = \Phi(\mathbf{x}(T)) + \int_0^T L(\mathbf{x}, \hat{\mathbf{g}}(\mathbf{x}, t; \mathcal{W}), t) dt, \tag{20}$$

and $\mathbf{x}(\cdot)$ is uniquely determined by solving the (homogeneous) state differential equation:

$$\dot{\mathbf{x}} = \mathbf{f}(\mathbf{x}, \hat{\mathbf{g}}(\mathbf{x}(t), t; \mathcal{W}), t), \tag{21}$$

$$\mathbf{x}(0) = \boldsymbol{\xi}. \tag{22}$$

There are several possibilities for choosing the parameterized controller, the key requirement is that it must be flexible enough to emulate the true optimal controller. Theoretically, any function approximation schemes can be used, although in practice, our experience suggests that only radial basis functions or feedforward neural networks have any chance of success. In particular, when the controller has a large input dimension (greater than 5 say), as is usually the case after the application of Valentine transformation, neural networks appear to be the only feasible model without requiring an unwieldily large number of parameters. Furthermore, a neural network with a bounded sigmoidal output has the natural advantage of modeling the bounded control output as given by (3). It is also versatile enough to emulate discontinuous control as is often encountered in bang-bang type control. In what follows, we shall assume that a neural network will be used for the controller, although the idea can also be used for any other parameterized model.

Problem P3 is a nonlinear programming problem in disguise, although the determination of the optimal set of weight parameters have to be computed in a somewhat roundabout way. The direct training algorithm for the feedback controller in the form of a neural network is based on the gradient formulae for optimal parameter selection problems (see Teo et. al. (1992)), in conjunction with a steepest descent algorithm using instantaneous gradient similar to, although not quite the same, as the backpropagation training of neural networks. We shall summarize the important steps required in the training algorithm, and refer the readers to Edwards and Goh (1995) for details.

Modified Backpropagation Algorithm.

Initialization: generate a set of initial conditions $\Delta = \{\boldsymbol{\xi}_i, i = 1, \cdots, N\}$ from the distribution ρ which fills the set Γ in a sufficiently dense manner. Select an appropriate neural network and randomize all the weights $w \in \mathcal{W}$ to small values. While the average performance index can be reduced further, (more specifically, if the average performance index in subsequent iteration reduces by more than some pre-assigned threshold) do:

i. For each $\boldsymbol{\xi}_i$ solve the differential equation (21) and (22) forward in time from $t = 0$ to $t = T$.

ii. Solve the corresponding costate differential equation

$$\dot{\boldsymbol{\lambda}}^{\top} = -\frac{\partial L}{\partial \mathbf{x}} - \frac{\partial L}{\partial \mathbf{u}}\frac{\partial \hat{\mathbf{g}}}{\partial \mathbf{x}} - \boldsymbol{\lambda}^{\top}\left(\frac{\partial \mathbf{f}}{\partial \mathbf{x}} + \frac{\partial \mathbf{f}}{\partial \mathbf{u}}\frac{\partial \hat{\mathbf{g}}}{\partial \mathbf{x}}\right) \tag{23}$$

$$\boldsymbol{\lambda}^{\top}(T) = \frac{\partial \Phi}{\partial \mathbf{x}(T)}, \tag{24}$$

backward in time from $t = T$ to $t = 0$, where $\partial \hat{\mathbf{g}}/\partial \mathbf{x}$ is the Jacobian of the controller to be computed by some special formulae derived in much the same way as the backpropagation formulae (see equations (13)-(15) of Edwards and Goh (1995)).

iii. Compute, for each weight parameter $w \in \mathcal{W}$

$$\frac{\partial J_2'}{\partial w} = \int_0^T \left(\frac{\partial L}{\partial \mathbf{u}}\frac{\partial \hat{\mathbf{g}}}{\partial w} + \boldsymbol{\lambda}^{\top}\frac{\partial \mathbf{f}}{\partial \mathbf{u}}\frac{\partial \hat{\mathbf{g}}}{\partial w}\right) dt \tag{25}$$

where $\partial \hat{\mathbf{g}}/\partial w$ is computed by the usual backpropagation formulae.

iv. Update the weight $w, \forall w \in \mathcal{W}$, by $w \leftarrow w - \eta \partial J_2'/\partial w$, where η is some small learning rate.

The algorithm described above is only conceptual. For successful implementation of the algorithm, there are a number of fine details that warrant critical attention. Firstly, the choice of network's structure and parameterization (number of layers and nodes) for implementing the modified backpropagation algorithm is problem dependent. At this stage of research, there are no clever ways of choosing the network apart from systematic trial and error. Unfortunately, this has remained a major criticism of the application of neural networks, despite the fact that they have been used successfully to solve many difficult problems. Secondly, when the system is inherently unstable, difficulty will arise during the integration of the state and costate equations in the initial training stage before the weights are appropriately initialized. An ad hoc way to overcome this is to start integrating with a small terminal time T, so that the system does not have sufficient time to escape to infinity, and gradually increase T until the desired T is reached. Nevertheless this has not been an entirely satisfactory way for highly unstable systems. Other issues such as the choice of training rate and the sequence in which the training initial conditions are used are also yet to be completely resolved.

5 CONCLUDING REMARKS

The proposed method has been tested on several non-trivial test problems, using both the penalty function method and the Valentine's transformation, with varying degree of success. In a test problem considered by Jacobson and Lele (1969), the trained feedback controller is able to achieve within 1% of the performance of the open-loop controller for all initial conditions arising from a fairly large subset. Furthermore, the feedback controller is also capable of generating the optimal solution for initial conditions not included in the training set, and even sometime for initial conditions *outside* the training set. Due to space constraint, these simulation results are not included but are readily available from the second author upon request.

REFERENCES

Bryson, A.E. and Ho, Y.C. (1975) *Applied Optimal Control.* Hemisphere publishing, Washington D.C..

Edwards, N.J. and Goh, C.J. (1995) A direct training method for continuous-time nonlinear optimal feedback controller. *Journal of Optimization Theory and Applications* **84(3)**.

Jacobson, D.H. and Lele, M.M. (1969) A transformation technique for optimal control problems with a state variable inequality constraint. *IEEE Trans. on Automatic Control,* **AC-14(5)**, 457–64.

Miele, A., Wu, A.K. and Liu, C.T. (1979) A transformation technique for optimal control problems with partially linear state inequality constraints. *Journal of Optimization Theory and Applications*, **28(2)**, 185-212.

Nguyen, D.H. and Widrow, B. (1990) Neural networks for self learning control systems. *IEEE Control Systems Magazine*, April, 18–23.

Teo, K.L., Goh, C.J. and Wong, K.H. (1991) *A Unified Computational Approach to Optimal Control Problems.* Pitman Series in Pure and Applied Mathematics, Longman.

Valentine, F.A. (1937) The problem of lagrange with differential inequalities as added side conditions. *Contribution to the Calculus of Variation*, University of Chicago press.

Optimization Algorithms and Methods

Primal-dual interior point method for multicommodity network flows with side constraints and comparison with alternative methods

J. Castro, N. Nabona
Statistics and Operations Research Dept., Univ. Politècnica de Catalunya
Pau Gargallo 5, 08071 Barcelona, Spain.
Tel: 34-3-4017335. Fax: 34-3-4017040. e-mail: jcastrop@eio.upc.es

Abstract
This document presents a primal-dual interior point algorithm for the solution of large multicommodity network flow problems with or without side constraints. The method exploits the structure of the problem and uses a preconditioned conjugate gradient solver. The algorithm has been implemented for the case of pure multicommodity problems (without side constraints), and some computational results are presented, comparing the performance of the code developed with alternative ones.

Keywords
Computational Benchmarks, Interior Point Methods, Linear Programming, Multicommodity Network Flows, Primal-Dual Algorithm, Side Constraints

1 INTRODUCTION

Multicommodity network flows (Kennington and Helgasson (1980)) are used as a modelling tool in many applications in routing, telecommunication networks, allocation and transportation problems and in electrical power systems. It is thus important to have efficient tools to optimize this kind of problems. Interior point methods have recently gained wide recognition as an optimization procedure for applications with general linear constraints and both their general primal-dual and dual-affine-scaling formulations have been extended to the linear multicommodity network flow problem (Choi and Goldfarb (1990), Kamath et al. (1993)).

In the work described here the primal-dual interior point algorithm has been specialized for solving multicommodity network flow problems, considering additional side constraints. The method is heavily based on the use of a preconditioned conjugate gradient algorithm for solving repeatedly a part of the systems of the type $ASA^t dy = \bar{b}$, ASA^t being symmetric and positive definite. Two specialized multicommodity network flow codes (Kennington (1979), Castro and Nabona (1995)), both based on the primal partitioning algorithm (Kennington and Helgasson (1980)), have been run on the same test problems solved with the interior point code in order to compare the performance of the interior point solution with that of the specialized network codes. The performance of the multicommodity primal-dual interior point code developed is also compared with that of a general primal-dual interior point code (Vanderbei (1993)) so that it is possible to appreciate the computational advantages of using multicommodity specialization within the interior point scheme. Randomly generated multicommodity test problems of sizes ranging from 100 to 10000 arcs and numbers of commodities ranging from 1 to 200 were solved and their results are reported.

2 OUTLINE OF THE PRIMAL-DUAL INTERIOR POINT FOR UPPER-BOUNDED LINEAR PROGRAMMING

Let us consider the following minimization problem with upper bounds in some variables

$$\min \quad c^t x \tag{1}$$
$$\text{subj. to} \quad Ax = b \tag{2}$$
$$\underline{0} \leq x_u \leq \overline{x}_u \tag{3}$$
$$\underline{0} \leq x_l \tag{4}$$

where $x_u \in \mathbb{R}^{n_u}$, $x_l \in \mathbb{R}^{n_l}$, $x = (x_u^t, x_l^t)^t$, $x \in \mathbb{R}^n$ (thus $n = n_u + n_l$), $c \in \mathbb{R}^n$, $b \in \mathbb{R}^m$ and $A \in \mathbb{R}^{m \times n}$. Considering an appropriate partitioning of c and A, equations (1) and (2) can be rewritten as:

$$c_u^t x_u + c_l^t x_l \tag{5}$$
$$A_u x_u + A_l x_l = b \tag{6}$$

where $c_u \in \mathbb{R}^{n_u}$, $c_l \in \mathbb{R}^{n_l}$, $A_u \in \mathbb{R}^{m \times n_u}$ and $A_l \in \mathbb{R}^{m \times n_l}$. The dual of the minimization problem stated can be cast as:

$$\begin{aligned} \max \quad & b^t y - \overline{x}_u^t w \\ \text{subj. to} \quad & A_u^t y + z_u - w = c_u \\ & A_l^t y + z_l \qquad = c_l \\ & z = (z_u^t, z_l^t)^t \geq 0 \quad w \geq 0 \end{aligned}$$

where $z_u \in \mathbb{R}^{n_u}$, $z_l \in \mathbb{R}^{n_l}$ (thus $z \in \mathbb{R}^n$) and $w \in \mathbb{R}^{n_u}$.

Considering a logarithmic barrier function (μ being its penalty term) for the nonnegativity constraints of the variables, and adding slacks $f \in \mathbb{R}^{n_u}$ for the upper bounds ($x_u + f = \overline{x}_u$), the Kuhn-Tucker optimality conditions for both the dual and the primal can be written as:

$$b_{1_l} \equiv \mu e_{n_l} - X_l Z_l e_{n_l} = 0 \tag{7}$$
$$b_{1_u} \equiv \mu e_{n_u} - X_u Z_u e_{n_u} = 0 \tag{8}$$
$$b_2 \equiv \mu e_{n_u} - F W e_{n_u} = 0 \tag{9}$$
$$b_3 \equiv b - (A_u x_u + A_l x_l) = 0 \tag{10}$$
$$b_{4_l} \equiv c_l - (A_l^t y + z_l) = 0 \tag{11}$$
$$b_{4_u} \equiv c_u - (A_u^t y + z_u - w) = 0 \tag{12}$$

e_l being the l-dimensional vector of 1's, and where matrices X_u, X_l, Z_l, Z_u, F and W are diagonal and defined as $M \in \mathbb{R}^{l\times l} = \text{diag}(m_1, \ldots, m_l)$. It is clear than when $n_u = 0$ ($n = n_l$) only equations (7, 10 and 11) hold, thus having the optimality conditions of the standard primal-dual algorithm with no upper bounds.

When using Newton's method to find a point satisfying (7–12), linear systems of the type $J_i d_i = -f_i$ must be solved at each iteration i. These solutions amount to finding dy and then computing dx, dw, dz_u, dz_l, in:

$$(ASA^t)dy = b_3 + ASr \tag{13}$$

$$dx = S(A^t dy - r) \tag{14}$$

$$dw = F^{-1}(b_2 + W dx_u) \tag{15}$$

$$dz_u = b_{4_u} + dw - A_u^t dy \tag{16}$$

$$dz_l = b_{4_l} - A_l^t dy \tag{17}$$

where

$$\begin{array}{c} r = (r_u^t, r_l^t)^t \quad r \in \mathbb{R}^n \quad r_u \in \mathbb{R}^{n_u} \quad r_l \in \mathbb{R}^{n_l} \\ r_u = F^{-1}b_2 + b_{4_u} - X_u^{-1}b_{1_u} \quad r_l = b_{4_l} - X_l^{-1}b_{1_l} \end{array} \tag{18}$$

and

$$\begin{array}{c} S = \begin{pmatrix} S_u & \mathbf{0} \\ \mathbf{0} & S_l \end{pmatrix} \quad S \in \mathbb{R}^{n\times n}, \quad S_u \in \mathbb{R}^{n_u \times n_u}, \quad S_l \in \mathbb{R}^{n_l \times n_l u} \\ S_u = FX_u(Z_uF + X_uW)^{-1} \qquad S_l = Z_l^{-1}X_l \end{array} \tag{19}$$

(where S_u and S_l can be directly computed, since they are made of products and sums of diagonal matrices). A justification of this process can be found in Castro (1995a).

It is quite clear that the main computational burden in solving system (7–12) is the repeated solution of the linear system (13).

3 FORMULATION OF THE LINEAR MULTICOMMODITY NETWORK FLOWS WITH SIDE CONSTRAINTS

The multicommodity network flow problem corresponds to the minimization problem (1–4). Let us consider a network with m^* nodes, n^* arcs (where the last one is a rooted arc added to avoid the singularity of the network matrix $A^* \in \mathbb{R}^{m^* \times n^*}$) and k commodities. Adding slacks to the mutual capacity constraints ($s_{mc} \in \mathbb{R}^{n^*}$) and the side constraints ($s_{sc} \in \mathbb{R}^t$, $t \geq 0$), and denoting by $x_i = (x_{i_a}^t x_{i_r})^t \in \mathbb{R}^{n^*}$ i=1,...,k the flows for each commodity ($x_{i_r} \in \mathbb{R}$ being the rooted arc and $x_{i_a} \in \mathbb{R}^{n^*-1}$ the remaining ones, for commodity i) with capacities $\overline{x}_i \in \mathbb{R}^{n^*-1}$ (thus considering the rooted arcs as uncapacitated ones), by $b_{mc} \in \mathbb{R}^{n^*}$ the mutual capacities (for the rooted arcs an arbitrary mutual capacity can be considered), by $b_i \in \mathbb{R}^{m^*}$ i=1,...,k the node supplies/demands of each commodity i, by $T_i \in \mathbb{R}^{t\times n^*}$ the matrices defining the side constraints structure, and by $\overline{b}_{sc}, \underline{b}_{sc} \in \mathbb{R}^t$ the upper and lower bounds of the side constraints, then the multicommodity network problem can be stated as:

$$\min \quad \sum_{i=1}^{k} c_i^t x_i \tag{20}$$

subj. to

$$\begin{array}{|c|c|c|c|c|c|}\hline A^* & 0 & \dots & 0 & 0 & 0 \\ \hline 0 & A^* & \dots & 0 & 0 & 0 \\ \hline \vdots & \vdots & \ddots & \vdots & \vdots & \vdots \\ \hline 0 & 0 & \dots & A^* & 0 & 0 \\ \hline \mathbb{1} & \mathbb{1} & \dots & \mathbb{1} & \mathbb{1} & 0 \\ \hline T_1 & T_2 & \dots & T_k & 0 & \mathbb{1} \\ \hline \end{array} \begin{array}{|c|}\hline x_1 \\ \hline x_2 \\ \hline \vdots \\ \hline x_k \\ \hline s_{mc} \\ \hline s_{sc} \\ \hline \end{array} = \begin{array}{|c|}\hline b_1 \\ \hline b_2 \\ \hline \vdots \\ \hline b_k \\ \hline b_{mc} \\ \hline \overline{b}_{sc} \\ \hline \end{array} \tag{21}$$

$$\underline{0} \leq x_{i_a} \leq \overline{x}_i \quad 0 \leq x_{i_r} \quad i = 1, \dots, k \tag{22}$$

$$\underline{0} \leq s_{sc} \leq \overline{b}_{sc} - \underline{b}_{sc} \tag{23}$$

$$\underline{0} \leq s_{mc} \leq b_{cm} \tag{24}$$

In this case the total number of variables and constraints is given by $n = (k+1)n^* + t$ and $m = km^* + n^* + t$, and the partitioning of the variables $x = (x_u^t, x_l^t)^t$ is $x_u^t = (x_{1_a}^t, \dots, x_{k_a}^t, s_{mc}^t, s_{sc}^t)$ and $x_l^t = (x_{1_r}, \dots, x_{k_r})$.

In the multicommodity problem, matrix S defined in (19) can be partitioned as:

$$S = \begin{bmatrix} S_1 & & & & \\ & \ddots & & & \\ & & S_k & & \\ & & & S_{mc} & \\ & & & & S_{sc} \end{bmatrix} \tag{25}$$

Applying equations (13–17) to the multicommodity problem, one can take advantage of the special structure of matrix A, especially when solving $(ASA^t)dy = b_3 + ASr$. In this case, and considering (21) and (25), the structure of matrix ASA^t is as follows:

$$ASA^t = \begin{array}{|c|c|c|c|c|c|}\hline A^*S_1A^{*^t} & 0 & \dots & 0 & A^*S_1 & A^*S_1T_1^t \\ \hline 0 & A^*S_2A^{*^t} & \dots & 0 & A^*S_2 & A^*S_2T_2^t \\ \hline \vdots & \vdots & \ddots & \vdots & \vdots & \vdots \\ \hline 0 & 0 & \dots & A^*S_kA^{*^t} & A^*S_k & A^*S_kT_k^t \\ \hline S_1A^{*^t} & S_2A^{*^t} & \dots & S_kA^{*^t} & \sum_{i=1}^k S_i + S_{mc} & \sum_{i=1}^k S_iT_i^t \\ \hline T_1S_1A^{*^t} & T_2S_2A^{*^t} & \dots & T_kS_kA^{*^t} & \sum_{i=1}^k T_iS_i & S_{sc} + \sum_{i=1}^k T_iS_iT_i^t \\ \hline \end{array}$$

$$= \begin{array}{|c|c|c|}\hline B & C_1 & C_2 \\ \hline C_1^t & D_1 & D_2 \\ \hline C_2^t & D_3 & D_4 \\ \hline \end{array} = \begin{array}{|c|c|}\hline B & C \\ \hline C^t & D \\ \hline \end{array} \tag{26}$$

4 NUMERICAL SOLUTION SCHEME

Because of the structure of $A^* S_i A^{*^t}$ and that of $A^* S_i$, when a solution for system (13) is attempted directly using the Cholesky decomposition, submatrix D_1 become completely dense. Since the dimension of D_1 is n^* this would mean having to store and process $n^*(n^*+1)/2$ values. For large networks this amount of memory can become prohibitive. This is stated in Choi and Goldfarb (1990), but no procedure is given there to circumvent this difficulty. The algorithm developed considers the solution of the linear system (13) ($ASA^t dy = \bar{b}$) taking into account the partition indicated in (26); thus the system to be solved can be written as:

$$\begin{bmatrix} B & C \\ C^t & D \end{bmatrix} \begin{bmatrix} dy_1 \\ dy_2 \end{bmatrix} = \begin{bmatrix} \bar{b}_1 \\ \bar{b}_2 \end{bmatrix}$$

whose solution is directly obtained by block multiplication:

$$\begin{aligned} (D - C^t B^{-1} C) dy_2 &= (\bar{b}_2 - C^t B^{-1} \bar{b}_1) \\ B dy_1 &= (\bar{b}_1 - C dy_2) \end{aligned} \tag{27}$$

B is made of k diagonal blocks $A^* S_i A^{*^t}$. Each block has the same very sparse topological structure of nonzero elements. If $\mathcal{A}$ is the set of arcs of the network, and $\mathcal{I}_v$ the set of incident arcs to node v, $A^* S_i A^{*^t}$ can be computed as follows:

$$A^* S_i A^{*^t} = \underset{\substack{v=1,\ldots,m^* \\ w=1,\ldots,m^*}}{(a_{vw})} = \begin{cases} \sum\limits_{\forall a} - S_{i_{(a)}} & \text{if } a \equiv (v,w) \in \mathcal{A} \text{ and } (w,v) \notin \mathcal{A} \\ \sum\limits_{\forall a,b} (- S_{i_{(a)}} - S_{i_{(b)}}) & \text{if } a \equiv (v,w) \in \mathcal{A} \text{ and } b \equiv (w,v) \in \mathcal{A} \\ \sum\limits_{\forall a \in \mathcal{I}_v} S_{i_{(a)}} & \text{if } (v = w) \\ 0 & \text{otherwise} \end{cases} \tag{28}$$

and any solution having B as the system matrix can be decomposed in k systems of equations (thus the process could be parallelized). The minimum order degree algorithm was used to reorder the nodes of the network to avoid fill-in when making the Cholesky decomposition of the k blocks of B. Neither the calculation of $B^{-1}\bar{b}_1$ nor the solution of $Bdy_1 = (\bar{b}_1 - Cdy_2)$ involves too much work. However, when computing dy_2, matrix $\widehat{D} = D - C^t B^{-1} C$ should be formed, which would mean solving $n^* + t$ systems of equations to obtain $B^{-1}C$ and, afterwards, a Cholesky decomposition of $\widehat{D}$.

A better choice is to use a preconditioned conjugate gradient (PCG) algorithm to obtain dy_2, and then to compute dy_1 directly. In the PCG algorithm the only operation directly made with the system matrix is a product of it by a vector v. But we have: $(D - C^t B^{-1} C)v = Dv - C^t B^{-1} w$, with $w = Cv$. Thus one can take advantage of the sparsity of D and C, and the fact that the computation of $B^{-1}w$ is numerically efficient.

In order to speed up the PCG algorithm, a positive definite matrix M must be determined such that $M^{-1}\widehat{D}$ becomes less ill-conditioned than $\widehat{D}$, and that the computation of $Mz = r$ does not involve too much work. For pure multicommodity network problems (without side constraints) the system matrix is $\widehat{D}_1 = D_1 - C_1^t B^{-1} C_1$. In this case, and considering a splitting of matrix $\widehat{D}_1$ such that $\widehat{D}_1 = P - Q$, where $P = D_1$ and $Q = C_1^t B^{-1} C_1$ (both positive definite), it can be proved (Castro (1995b)) that:

$$\widehat{D}_1^{-1} = (\sum_{i=0}^{\infty}(P^{-1}Q)^i)P^{-1} \tag{29}$$

The preconditioner M^{-1} to be used will be an approximation of $\widehat{D}_1^{-1}$, and can be obtained by truncating (29) at some term ϕ:

$$M^{-1} = (\mathbb{1} + (P^{-1}Q) + (P^{-1}Q)^2 + \ldots + (P^{-1}Q)^{\phi})P^{-1} \tag{30}$$

The higher ϕ is, the better the preconditioning, and the fewer iterations of the PCG will be required. However, it must be noted that the product of Q by a vector r implies the solution of $B^{-1}(C_1 r)$, and this should be performed at each iteration of the PCG algorithm, increasing the execution time considerably. Thus ϕ must be chosen in order to balance both objectives: to decrease the PCG iterations and to improve the time per PCG iteration. Various tests have shown than, in general, the best results are obtained with $\phi = 1$. In this case $M^{-1} = P^{-1} = D_1^{-1}$, which means that $z = M^{-1}r$ can be obtained in $O(n)$ operations (since D_1 is a diagonal matrix).

5 COMPUTATIONAL RESULTS

The multicommodity primal-dual interior point algorithm outlined in the above sections was implemented for the case of problems without side constraints, using the preconditioning previously stated. The code was written in ANSI-C, and to test its performance four types of problems, obtained from different network generators, were used: Rmfgen, Grid-on-torus, Gridgraph and Gridgen (Dimacs (1991)). These generators do not consider the case of multicommodity flows, and the output networks had to be converted to a multicommodity one. The conversion algorithm is described in Castro (1995b). Five particular instances were created with each of these generators. The first two are problems with few commodities and medium-sized networks, whereas the last three correspond to small-sized networks with many commodities. Each problem will be denoted by $L_j^{i)}$, i=1,...,4, j=1,...,4, i denoting the generator employed (1 for Rmfgen, 2 for Grid-on-torus, 3 for Gridgraph and 4 for Gridgen). Table 1 presents the characteristics of each problem, showing for each test problem the number of commodities, nodes and arcs of the network, and the total number of constraints and variables of the linear program to be solved (columns *Rows A* and *Columns A*).

The interior point multicommodity code developed (denoted by IPM) was compared with MINOS 5.3 (Murtagh and Saunders (1983)), a general-purpose package, PPRN (Castro and Nabona (1995)) and MCNF85 (Kennington (1979)), two specialized multicommodity network flow codes, and LoQo (Vanderbei (1993)), a state-of-the-art primal-dual interior point code. Table 2 shows the CPU seconds required by each code. The fastest execution for each test is marked with an asterisk (*). All runs were carried out on a SunSparc 10/41 (one CPU), with a 40MHz clock, ≈100Mips and ≈20Mflops CPU, and 64Mbytes of main memory. From Table 2 it can be concluded that the performance of IPM increases with the size of the problem, this code thus being a good choice

Table 1 Linear test problems

Test	*Commodities*	*Nodes*	*Arcs*	*Rows A*	*Columns A*
$L_1^{1)}$	8	2048	9472	25856	85248
$L_2^{1)}$	16	2048	9472	42240	161024
$L_3^{1)}$	50	128	496	6896	25296
$L_4^{1)}$	150	128	496	19696	74896
$L_5^{1)}$	200	128	496	26096	99696
$L_1^{2)}$	8	1500	9000	21000	81000
$L_2^{2)}$	16	1500	9000	33000	153000
$L_3^{2)}$	50	100	600	5600	30600
$L_4^{2)}$	150	100	600	15600	90600
$L_5^{2)}$	200	100	600	20600	120600
$L_1^{3)}$	8	2502	5000	25016	45000
$L_2^{3)}$	16	2502	5000	45032	85000
$L_3^{3)}$	50	227	450	11800	22950
$L_4^{3)}$	150	227	450	34500	67950
$L_5^{3)}$	200	227	450	45850	90450
$L_1^{4)}$	8	976	7808	15616	70272
$L_2^{4)}$	16	976	7808	23424	132736
$L_3^{4)}$	50	101	606	5656	30906
$L_4^{4)}$	150	101	606	15756	91506
$L_5^{4)}$	200	101	606	20806	121806

for large multicommodity network flow problems, especially for the case of small-sized networks with many commodities.

6 REFERENCES

Castro, J. (1995a) An implementation of a primal-dual interior point algorithm with upper bounded variables. *Qüestiió*, **19**, to appear. (written in Catalan)

Castro, J. (1995b) *Efficient methods for the solution of multicommodity network flow problems.* Ph.D. dissertation, Statistics and Operations Reseach Dept., Universitat Politècnica de Catalunya, Barcelona, Spain. (written in Catalan)

Castro, J. and Nabona, N (1995) An implementation of linear and nonlinar multicommodity network flows. Accepted for publication in the *European Journal of Operational Research.*

Choi, I.C. and Goldfarb, D. (1990) Solving multicommodity network flow problems by an interior point method. *SIAM Proceedings in Applied Mathematics*, **46**, 58–69.

Table 2 CPU seconds of each code for the linear test problems

Test	IPM	MINOS	PPRN	MCNF85	LoQo
$L_1^{1)}$	7095.4	15147.7	737.9*	1778.2	$^{(d)}$
$L_2^{1)}$	16737.9	$^{(a)}$	6838.7	5651.3*	$^{(d)}$
$L_3^{1)}$	178.6*	2639.4	275.1	398.6	3402.8
$L_4^{1)}$	1839.9*	$^{(b)}$	8069.0	11319.0	$^{(d)}$
$L_5^{1)}$	1710.0*	$^{(c)}$	15415.3	26479.9	$^{(d)}$
$L_1^{2)}$	12296.3	$^{(b)}$	4962.2	4833.0*	$^{(d)}$
$L_2^{2)}$	$^{(d)}$	$^{(c)}$	37470.5	34383.0*	$^{(d)}$
$L_3^{2)}$	287.0	1402.8	169.2*	466.9	5211.1
$L_4^{2)}$	4352.4*	105082.5	7605.5	15836.2	$^{(d)}$
$L_5^{2)}$	11974.4*	$^{(c)}$	22218.4	81903.5	$^{(d)}$
$L_1^{3)}$	2818.3	$^{(b)}$	1409.2*	2134.1	$^{(d)}$
$L_2^{3)}$	16485.5	$^{(c)}$	14139.8*	14709.9	$^{(d)}$
$L_3^{3)}$	236.3*	3172.8	364.9	533.9	3180.4
$L_4^{3)}$	962.9 *	$^{(a)}$	4480.5	6142.8	$^{(d)}$
$L_5^{3)}$	2083.1*	$^{(a)}$	11736.8	19458.0	$^{(d)}$
$L_1^{4)}$	12216.2	205836.0	3424.7*	$^{(e)}$	$^{(d)}$
$L_2^{4)}$	$^{(d)}$	$^{(c)}$	40974.1*	$^{(e)}$	$^{(d)}$
$L_3^{4)}$	114.1	918.2	39.4*	$^{(e)}$	$^{(d)}$
$L_4^{4)}$	584.8	15236.6	415.8*	$^{(e)}$	$^{(d)}$
$L_5^{4)}$	915.2*	$^{(c)}$	1273.4	$^{(e)}$	$^{(d)}$

$^{(a)}$ Too many constraints. $^{(b)}$ Error during execution. $^{(c)}$ Execution too long.
$^{(d)}$ Not enough memory. $^{(e)}$ Feasibility error.

DIMACS. (1991) The first DIMACS international algorithm implementation challenge: The bench-mark experiments. Technical Report, DIMACS, New Brunswick, NJ.

Kamath, A.P, Karmarkar, N.K. and Ramakrishnan, K.G. (1993) Computational and Complexity Results for an Interior Point Algorithm on Multicommodity Flow Problems. TR-21/93, Dipartimento di Informatica, Università di Pisa, Italy.

Kennington, J.L. and Helgasson, R.V. (1980) *Algorithms for Network Programming.* John Wiley & Sons, Inc., New York, NY.

Kennington, J.L. (1979) A primal partitioning code for solving multicommodity flow problems (version 1). Technical Report 79008. Dept. of Industrial Engineering and Operations Research, Southern Methodist University, Dallas, USA.

Murtagh, B.A. and Saunders, M.A. (1983) MINOS 5.0. User's guide. Dept. of Operations Research, Stanford University, CA, USA.

Vanderbei, R.J. and Carpenter, T.J. (1993) Symmetric indefinite systems for interior point methods. *Mathematical Programming*, **58**, 1-32.

Dual Bregman proximal methods for large-scale 0–1 problems

K.C. Kiwiel
System Research Institute, Warsaw, Poland.
e-mail: kiwiel@ibspan.waw.pl

P.O. Lindberg, Andreas Nõu
Royal Institute of Technology, Stockholm, Sweden.
e-mail: pol@math.kth.se, andreasn@math.kth.se

Abstract

We describe an extension of the Bregman proximal method for convex programming, employing B-functions as generalizations of Bregman functions that cover more applications. We allow inexact subproblem solutions, increasing their accuracy successively to retain global convergence. Our framework is applied to Lagrangian relaxations of large-scale set covering problems that arise in airline crew scheduling.

Keywords

Proximal point methods, Bregman functions, set covering problems.

1 INTRODUCTION

We describe an extension of the Bregman proximal method for convex programming (Censor and Zenios, 1992; Chen and Teboulle, 1993; Eckstein, 1993). First, we consider B-functions of (Kiwiel, 1994) as generalizations of Bregman functions (Censor and Lent, 1981) that cover more applications. Second, we allow inexact subproblem solutions, increasing their accuracy successively to retain global convergence. Our framework is applied to Lagrangian relaxations of large-scale set covering problems that arise in airline crew scheduling (Wedelin, 1993). We report encouraging preliminary computational experience.

Our notation is fairly standard. $\langle\cdot,\cdot\rangle$ and $|\cdot|$ are the Euclidean inner product and norm respectively. $\mathbb{R}_+$ and $\mathbb{R}_>$ are the nonnegative and positive reals respectively. $\operatorname{cl} C$ and $\operatorname{ri} C$ denote the closure and relative interior of $C \subset \mathbb{R}^n$. We let $\imath_C(x) = 0$ if $x \in C$, ∞ otherwise, and $\imath_C^* = \sup_{x \in C} \langle\cdot, x\rangle$. For a convex function f, $\mathcal{D}_f = \{x : f(x) < \infty\}$, $\partial f(\cdot) = \{g : f(x) \geq f(\cdot) + \langle g, \cdot - x\rangle \ \forall x\}$ and $\mathcal{D}_{\partial f} = \{x : \partial f(x) \neq \emptyset\}$.

Research supported by the Swedish Transport and Communications Research Board, the Göran Gustafsson Foundation and the Polish State Committee for Scientific Research under Grant 8S50502206.

2 B-FUNCTIONS

For any convex function h on $\mathbb{R}^n$, we define its *difference functions*

$$D_h^\flat(x,y) = h(x) - h(y) - \imath^*_{\partial h(y)}(x-y) \qquad \forall x,y \in \mathcal{D}_h,$$
$$D_h^\sharp(x,y) = h(x) - h(y) + \imath^*_{\partial h(y)}(y-x) \qquad \forall x,y \in \mathcal{D}_h.$$

By convexity, $h(x) \geq h(y) + \imath^*_{\partial h(y)}(x-y)$ and

$$0 \leq D_h^\flat(x,y) \leq h(x) - h(y) - \langle g, x-y\rangle \leq D_h^\sharp(x,y) \quad \forall x,y \in \mathcal{D}_h, g \in \partial h(y). \tag{1}$$

$D_h^\flat$ and $D_h^\sharp$ generalize the usual *D-function* of h (Censor and Lent, 1981), defined by

$$D_h(x,y) = h(x) - h(y) - \langle \nabla h(y), x-y\rangle \quad \forall x \in \mathcal{D}_h, y \in \mathcal{D}_{\nabla h}, \tag{2}$$

since

$$D_h(x,y) = D_h^\flat(x,y) = D_h^\sharp(x,y) \quad \forall x \in \mathcal{D}_h, y \in \mathcal{D}_{\nabla h}. \tag{3}$$

(Readers only interested in our applications may assume h is s.t. $\partial h(y) = \{\nabla h(y)\}$ for all points y of interest.)

Definition 1 (Kiwiel, 1994) A closed proper (possibly nondifferentiable) convex function h is called a B-function (*generalized Bregman function*) if
(a) h is strictly convex on $\mathcal{D}_h$.
(b) h is continuous on $\mathcal{D}_h$.
(c) For every $\alpha \in \mathbb{R}$ and $x \in \mathcal{D}_h$, the set $\mathcal{L}_h^\flat(x,\alpha) = \{y \in \mathcal{D}_{\partial h} : D_h^\flat(x,y) \leq \alpha\}$ is bounded.
(d) For every $\alpha \in \mathbb{R}$ and $x \in \mathcal{D}_h$, if $\{y^k\} \subset \mathcal{L}_h^\flat(x,\alpha)$ is a convergent sequence with limit $y^* \in \mathcal{D}_h \setminus \{x\}$, then $D_h^\sharp(y^*, y^k) \to 0$.

General properties of B-functions are discussed by (Kiwiel, 1994). From lack of space, we only quote the following simple consequence of Lemmas 2.8–2.10 of (Kiwiel, 1994).

Lemma 2 *Let ψ be a proper convex function on $\mathbb{R}$. Then $h(x) = \sum_{i=1}^n \psi(x_i)$ is a B-function iff ψ is closed, essentially strictly convex and $\mathcal{D}_{\psi^*} = \operatorname{ri}\mathcal{D}_{\psi^*}$, where $\psi^*(\cdot) = \sup_t \langle \cdot, t\rangle - \psi(t)$ is the conjugate of ψ. In particular $\mathcal{D}_{\psi^*} = \operatorname{ri}\mathcal{D}_{\psi^*} = \mathbb{R}$ if $\mathcal{D}_\psi$ is bounded.*

Examples 3 Let $\psi : \mathbb{R} \to (-\infty, \infty]$ and $h(x) = \sum_{i=1}^n \psi(x_i)$; then $h^*(y) = \sum_i \psi^*(y_i)$. In each example, h is an essentially smooth B-function with $\partial h = \{\nabla h\}$ and $\mathcal{D}_{h^*} = \mathbb{R}^n$.
1 (Fermi/Dirac). $\psi(t) = t\ln t + (1-t)\ln(1-t)$ on $\mathcal{D}_\psi = [0,1]$ $(0 \ln 0 = 0)$. Then $\psi^*(t) = \ln(1+\exp t)$ on $\mathcal{D}_{\psi^*} = \mathbb{R}$.
2 (Burg's entropy). $\psi(t) = -\ln t - \ln(1-t)$ on $\mathcal{D}_\psi = [0,1]$. Then $\psi^*(t) = \frac{1}{2}[(t^2+4)^{1/2} + t - 2] + \ln[(t^2+4)^{1/2} - 2] - \ln t^2$ for $t \neq 0$, $\psi^*(0) = -\ln 4$.
3 (Hellinger) . $\psi(t) = -\sqrt{t(1-t)}$ on $\mathcal{D}_\psi = [0,1]$. Then $\psi^*(t) = \frac{1}{2}(t + \sqrt{1+t^2})$.

3 AN INEXACT B-PROXIMAL METHOD

Consider the convex minimization problem $f_* = \inf_X f$, where $f : \mathbb{R}^n \to (-\infty, \infty]$ is closed proper convex and X is a nonempty closed convex set in $\mathbb{R}^n$. Suppose h is a B-function s.t. $\mathcal{D}_{f_X} \cap \mathcal{D}_h \neq \emptyset$, where $f_X = f + \imath_X$ is the essential objective. Let $\{t_k\} \subset \mathbb{R}_>$ and $\{\epsilon_k\} \subset \mathbb{R}_+$ satisfy $\sum_{k=1}^{\infty} t_k = \infty$ and $\lim_{l\to\infty} \sum_{k=1}^{l} t_k\epsilon_k / \sum_{k=1}^{l} t_k = 0$.

The *inexact Bregman-prox method* of (Kiwiel, 1995) runs as follows. At iteration $k \geq 1$, having $x^k \in \mathcal{D}_{f_X} \cap \mathcal{D}_{\partial h}$, $\gamma^k \in \partial h(x^k)$ and $D_h^k(\cdot, x^k) = h(\cdot) - h(x^k) - \langle \gamma^k, \cdot - x^k \rangle$, minimize $f_X + D_h^k(\cdot, x^k)/t_k$ approximately to find x^{k+1}, γ^{k+1} and p^{k+1} satisfying

$$\gamma^{k+1} \in \partial h(x^{k+1}), \tag{4}$$

$$t_k p^{k+1} + \gamma^{k+1} - \gamma^k = 0, \tag{5}$$

$$p^{k+1} \in \partial_{\epsilon_k} f_X(x^{k+1}), \tag{6}$$

$$f_X(x^{k+1}) \leq f_X(x^k). \tag{7}$$

We note that (4)–(6) imply

$$x^{k+1} \approx \arg\min\{ f(x) + D_h^k(x, x^k)/t_k : x \in X \} \tag{8}$$

with $0 \leq D_h^\flat(\cdot, x^k) \leq D_h^k(\cdot, x^k) \leq D_h^\sharp(\cdot, x^k)$; in fact x^{k+1} is an ϵ_k-minimizer of $f_X + D_h^k(\cdot, x^k)/t_k$ (Kiwiel, 1995). The *exact* Bregman-prox method (Censor and Zenios, 1992; Chen and Teboulle, 1993; Eckstein, 1993) corresponds to assuming that $\epsilon_k \equiv 0$ and $\mathcal{D}_{\partial h} = \mathcal{D}_{\nabla h}$, so that $p^{k+1} \in \partial f_X(x^{k+1})$, $\gamma^k = \nabla h(x^k)$ and $D_h^k = D_h$ for all k (cf. (3)). A particular implementation of our inexact method will be described in the next section.

The following global convergence results are established in (Kiwiel, 1995).

Theorem 4 (a) $f_X(x^k) \downarrow \inf_{\mathcal{D}_h} f_X = \inf_{\mathrm{cl}(\mathcal{D}_h \cap \mathcal{D}_{f_X})} f$. *Hence* $f_X(x^k) \downarrow \inf_X f$ *if* $\mathcal{D}_{f_X} \subset \mathcal{D}_h$. *If* $\mathrm{ri}\,\mathcal{D}_h \cap \mathrm{ri}\,\mathcal{D}_{f_X} \neq \emptyset$ *(e.g. ,* $\mathrm{int}\,\mathcal{D}_h \cap \mathcal{D}_{f_X} \neq \emptyset$*) then* $\inf_{\mathcal{D}_h} f_X = \inf_{\mathrm{cl}\,\mathcal{D}_h} f_X = \inf_{(\mathrm{cl}\,\mathcal{D}_h)\cap(\mathrm{cl}\,\mathcal{D}_{f_X})} f$. *If* $\mathrm{ri}\,\mathcal{D}_{f_X} \subset \mathrm{cl}\,\mathcal{D}_h$ *(e.g. ,* $\mathcal{D}_{\partial f_X} \subset \mathrm{cl}\,\mathcal{D}_h$*) then* $\mathrm{cl}\,\mathcal{D}_h \supset \mathrm{cl}\,\mathcal{D}_{f_X}$ *and* $\mathrm{Arg\,min}_X f \subset \mathrm{cl}\,\mathcal{D}_h$.

(b) *If* $\sum_{k=1}^{\infty} t_k\epsilon_k < \infty$ *and* $X_* = \mathrm{Arg\,min}_{\mathcal{D}_h} f_X$ *is nonempty then* $\{x^k\}$ *converges to some* $x^\infty \in X_*$, *and* $x^\infty \in \mathrm{Arg\,min}_X f$ *if* $\mathcal{D}_{f_X} \subset \mathcal{D}_h$.

(c) *If* $\mathcal{D}_{f_X} \subset \mathcal{D}_h$ *and* $X_* = \emptyset$ *then* $|x^k| \to \infty$.

In certain applications it may be difficult to satisfy the descent requirement (7). Fortunately, it may be replaced by another condition on the tolerances $\{\epsilon_k\}$ (Kiwiel, 1995).

Theorem 5 *Let* $s_k = \sum_{j=1}^{k} t_j$ *for all* k. *The assertions of Theorem 4 hold if the descent condition* (7) *is replaced by the accuracy condition* $\sum_{k=1}^{l} s_k\epsilon_k/s_l \to 0$.

For choosing $\{\epsilon_k\}$, one may use the following result of (Kiwiel, 1995).

Lemma 6 (i) *If* $\epsilon_k \to 0$ *then* $\sum_{k=1}^{l} t_k\epsilon_k/s_l \to 0$ *as* $l \to \infty$.

(ii) *If* $\sum_{k=1}^{\infty} \epsilon_k < \infty$ *and* $\{t_k\} \subset (0, t_{\max}]$ *for some* $t_{\max} < \infty$ *then* $\sum_{k=1}^{\infty} t_k \epsilon_k < \infty$.
(iii) *If* $\sum_{l=1}^{\infty} t_l / s_l = \infty$ *and* $\lim_{k \to \infty} \epsilon_k s_k / t_k = 0$ *then* $\lim_{l \to \infty} \sum_{k=1}^{l} s_k \epsilon_k / s_l = 0$.
(iv) *If* $\{t_k\} \subset [t_{\min}, t_{\max}]$ *for some* $0 < t_{\min} \leq t_{\max}$ *and* $k\epsilon_k \to 0$ *then* $\sum_{k=1}^{l} s_k \epsilon_k / s_l \to 0$.

4 LAGRANGIAN RELAXATIONS OF SET COVERING PROBLEMS

Let $A \in \{0,1\}^{m \times n}$, $e = (1, \ldots, 1)^T \in \mathbb{R}^m$ and $c \in \mathbb{R}^n$. Consider the *set covering problem*

$$\min \left\{ c^T x : Ax \geq e, \ x \in \{0,1\}^n \right\}. \tag{9}$$

To solve its LP *relaxation*

$$\min \left\{ c^T x : Ax \geq e, \ x \in [0,1]^n \right\} \tag{10}$$

via the Bregman-prox method, let $f(x) = c^T x$, $X = \{x : Ax \geq e\} \cap [0,1]^n$ and choose a B-function h as in Ex. 3, so that $\operatorname{cl} \mathcal{D}_h = [0,1]^n$ and $\partial h = \{\nabla h\}$ on $\mathcal{D}_{\partial h} = (0,1)^n$. Then $\gamma^k = \nabla h(x^k)$ and (8) reads

$$x^{k+1} \approx \arg\min\{ c^T x + [h(x) - \nabla h(x^k)^T x]/t_k : Ax \geq e \}. \tag{11}$$

The *dual* problem, obtained by assigning a multiplier λ to the constraints $Ax \geq e$, is

$$\max\{ q(\lambda) : \lambda \in \mathbb{R}_+^m \}, \tag{12}$$

where q is the concave *continuously differentiable* dual functional (Kiwiel, 1994)

$$\begin{aligned} q(\lambda) &= \min_x \{ c^T x + [h(x) - \nabla h(x^k)^T x]/t_k + \lambda^T (e - Ax) \} \\ &= -h^*(t_k(A^T \lambda - c) + \nabla h(x^k))/t_k + e^T \lambda \end{aligned} \tag{13}$$

with $\nabla q(\lambda) = e - Ax(\lambda)$ for all $\lambda \in \mathbb{R}^m$, where

$$\begin{aligned} x(\lambda) &= \arg\min_x \{ c^T x + [h(x) - \nabla h(x^k)^T x]/t_k + \lambda^T (e - Ax) \} \\ &= \nabla h^*(t_k(A^T \lambda - c) + \nabla h(x^k)). \end{aligned} \tag{14}$$

Note that $q(\lambda)$ and $x(\lambda)$ are easily computed for h as in Ex. 3 even in the large-scale case.

In effect, (12) may be solved approximately in the large-scale case by coordinate or block coordinate ascent (Kiwiel, 1994). However, having obtained an approximate solution λ^k of (12), we cannot simply set x^{k+1} to $x(\lambda^k)$, because $x(\lambda^k)$ will not, in general, be feasible in (11). The following result helps in constructing x^{k+1} from $x(\lambda^k)$ to satisfy (4)–(6).

Theorem 7 *Let* λ^k *be an approximate solution of* (12). *Suppose* $x^{k+1} \in \mathcal{D}_{\nabla h}$ *satisfies* $Ax^{k+1} \geq e$ *and* $(Ax^{k+1} - e)^T \lambda^k + e^T \delta_+^k - \langle \delta^k, x^{k+1} \rangle \leq \epsilon_k$, *where* $\delta^k = [\nabla h(x(\lambda^k)) - \nabla h(x^{k+1})]/t_k$ *and* $[\delta_+^k]_i = \max\{\delta_i^k, 0\}$, $i = 1{:}n$. *Then* (4)–(6) *hold with* $\gamma^{k+1} = \nabla h(x^{k+1})$.

The algorithm of (Kiwiel, 1994) applied to (12) generates a sequence $\{\lambda^{(l)}\} \in \mathbb{R}^m_+$ s.t. $(Ax(\lambda^{(l)}) - e)^T\lambda^{(l)} \to 0$, $x(\lambda^{(l)}) \to \hat{x}^{k+1} \in (0,1)^n$ and $\nabla h(x(\lambda^{(l)})) \to \nabla h(\hat{x}^{k+1})$ as $l \to \infty$, where $\hat{x}^{k+1}$ solves (11), so that $A\hat{x}^{k+1} \geq e$. Hence, for l sufficiently large, one may take $\lambda^k = \lambda^{(l)}$ and obtain x^{k+1} by increasing slightly certain components of $x(\lambda^k)$ (using $A \in \{0,1\}^{m\times n}$) to satisfy the conditions of Theorem 7 for any $\epsilon_k > 0$. In effect, Theorem 5 applies, i.e., $\{c^T x^k\}$ converges to the optimal value of the LP relaxation (10), and $\{x^k\}$ converges to a solution of (10) if $\sum_k t_k\epsilon_k < \infty$.

Integer programming (IP) problems of the form (9) are usually solved by branch and bound, and their LP relaxations are often highly degenerate, which makes them difficult for simplex-based methods (Lindberg and Nöu, 1994). Our Bregman-prox method may be used not only for solving LP relaxations, but also for finding approximate solutions to the IP problem (9). Specifically, while solving the current dual subproblem (12), we may interpret $x_j(\lambda^{(l)}) \in (0,1)$ as the probability of having $x_j = 1$ in a solution of (9). Hence we may employ the randomized heuristic of (Lindberg and Nöu, 1994) for generating primal feasible solutions (PFS's).

Broadly speaking, our randomized PFS heuristic runs as follows.

- Start with $\check{x} = 0$, i.e., all rows uncovered.
- While the set of uncovered rows $\mathcal{V} = \{i : (A\check{x} - e)_i < 0\}$ is nonempty:
  - Choose an uncovered row $i \in \mathcal{V}$.
  - Choose $\hat{\jmath} \in \{j : A_{ij} = 1\}$ at random, using the probabilities $x_j(\lambda^{(l)})$. Set $\check{x}_{\hat{\jmath}} = 1$.
- If possible, set $\check{x}_j = 0$ for one or more js s.t. the final $\check{x}$ satisfies $A\check{x} \geq e$.

Of course, this heuristic may be implemented in various ways. The next section shows that it may produce points $\check{x}$ close to the primal integer optimum.

5 NUMERICAL RESULTS

We now report on preliminary numerical experience obtained with an implementation of our method programmed in Matlab. To save space, we only mention some of its features.

We chose h as the Fermi/Dirac '$x \log x$' entropy of Example 3.1, leaving other possibilities for future work.

The dual subproblems (12) were solved to relatively high accuracy by the coordinate ascent method of (Kiwiel, 1994). Each coordinate ascent employed a safeguarded Newton's scheme to reduce the directional derivative of q by at least a quarter, i.e., the relaxation factor stayed in $[0.75, 1]$, unless a bound became active. For simplicity, the coordinates (rows) were chosen cyclically. The randomized PFS heuristic was called once every three cycles (minor iterations), i.e., complete sweeps over the constraint set. We used constant proximal stepsizes $t_k \equiv 100/\max_i c_i$ (with $c > 0$ for our problems) and the initial accuracy tolerance $\epsilon_1 = v_{\mathrm{LP}}/10$, where v_{LP} was the (known) optimal value of the LP relaxation (10). The accuracy tolerances were decreased relatively slowly by setting $\epsilon_{k+1} = \kappa_\epsilon\epsilon_k$, where $\kappa_\epsilon = 0.8$.

We present numerical results for a set of real-world set covering problems, emanating from airline crew scheduling problems (Wedelin, 1993). Table 1 highlights main features

Table 1 Airline scheduling test problems

Name	Rows	Columns	Density (%)	NNZ col	NNZ row	IP-gap (%)
B727	29	157	8.2	2.4	12.8	0.372
ALITALIA	116	1163	3.1	3.6	35.8	0
A320	195	6927	2.4	4.6	163.1	0
LH_A320	234	18752	1.9	4.4	351.4	0.003
SASjump	725	10353	0.57	4.1	58.9	0.025
SASD9_2	1360	25026	0.33	4.4	81.5	≤ 0.018

Table 2 Convergence to the LP relaxation value

Name	Relative accuracy (%) at minor iteration			
	10	25	50	100
B727	7.13	3.09	0.74	0.11
ALITALIA	1.12	0.54	0.13	0.03
A320	9.91	0.98	0.37	0.23
LH_A320	10.36	0.83	0.40	0.28
SASjump	7.85	2.15	0.66	0.41
SASD9_2	6.77	1.72	0.70	0.16

of these problems (NNZ stands for the mean number of nonzeros, and IP-gap for the difference in the optimal values of (9) and (10)). We may add that preprocessing has considerably reduced the size of these problems, and that CPLEX (version 2.1beta) could not solve the largest problem (Wedelin, 1993).

Table 2 shows the accuracy in the objective value relative to the (known) LP-value v_{LP} obtained after a given number of minor iterations (cycles). The numbers of major prox-iterations were fairly small, *e.g.* , 15 for the ALITALIA problem. Anyway, most effort goes into minor iterations.

Table 3 shows the accuracy of the best primal solution obtained by the PFS heuristic relative to the (known) IP-value v_{IP} after a given number of sweeps (v_{IP} is not known for SASD9_2). We note that the PFS heuristic was not called every cycle.

REFERENCES

Censor, Y. and Lent, A. (1981) An iterative row action method for interval convex programming. *Journal of Optimization Theory and Applications*, **34**, 321–53.

Censor, Y. and Zenios, S.A. (1992) Proximal minimization algorithm with D-functions. *Journal of Optimization Theory and Applications*, **73**, 451–64.

Table 3 Convergence to the IP value

Name	Relative accuracy (%) at minor iteration			
	10	25	50	100
B727	10.54	6.67	3.50	0.64
ALITALIA	1.39	1.39	1.39	1.39
A320	1.35	0.86	0.86	0.75
LH_A320	2.13	1.64	1.64	1.64
SASjump	22.86	8.71	7.51	6.41
SASD9_2	15.78	8.19	6.98	5.69

Chen G. and Teboulle, M. (1993) Convergence analysis of a proximal-like minimization algorithm using Bregman functions. *SIAM Journal on Optimization*, **3**, 538–43.

Eckstein, J. (1993) Nonlinear proximal point algorithms using Bregman functions, with applications to convex programming. *Mathematics of Operations Research*, **18**, 202–26.

Kiwiel, K.C. (1994) *Free-steering relaxation methods for problems with strictly convex costs and linear constraints.* WP-94-89, International Institute for Applied Systems Analysis, Laxenburg, Austria, September. Revised version: July 1995.

Kiwiel, K.C. (1995) *Proximal minimization methods with generalized Bregman functions.* WP-95-24, International Institute for Applied Systems Analysis, Laxenburg, Austria, March.

Lindberg, P.O. and Nõu, A. (1994) *Dual methods for large scale* 0-1 *problems.* Technical report, Department of Mathematics, Royal Institute of Technology, Stockholm, April.

Rockafellar, R.T. (1970) *Convex Analysis.* Princeton University Press, Princeton, NJ.

Wedelin, D. (1993) *Efficient algorithms for probabilistic inference, combinatorial optimization and the discovery of causal structure from data.* Ph.D. thesis, Department of Computer Sciences, Chalmers University of Technology, Göteborg.

On long-step surrogate projection methods for solving convex feasibility problems

Krzysztof C. Kiwiel, Bożena Łopuch
Systems Research Institute
Newelska 6, 01-447 Warsaw, Poland.
e-mail: kiwiel@ibspan.waw.pl *and* lopuch@ibspan.waw.pl

Abstract
We present methods for finding common points of finitely many closed convex sets in a Hilbert space. Every iteration approximates each set by a halfspace given by an approximate projection of the current iterate. The next iterate is found by projecting the current one on a surrogate halfspace formed by taking a convex combination of the halfspace inequalities. The resulting methods are block-iterative and hence lend themselves to parallel implementation. We extend to accelerated methods some recent results of Bauschke and Borwein on convergence of projection methods.

Keywords
Successive projections, relaxation methods, convex feasibility problems, surrogate inequalities.

1 INTRODUCTION

Let $\{C_i\}_{i=1}^N$ be a collection of closed convex subsets of a real Hilbert space $\mathcal{H}$ with inner product $\langle\cdot,\cdot\rangle$ and norm $\|\cdot\|$. Suppose $\cap_{i=1}^N C_i \neq \emptyset$. Consider the *convex feasibility problem* (CFP)

$$\text{find any point } x \text{ in } C := \cap_{i=1}^N C_i. \tag{1}$$

Kiwiel (1995a) has recently studied long-step surrogate projection methods for CFP's in finite-dimensional spaces, motivated mainly by their possible applications in methods for convex nondifferentiable optimization (Kiwiel, 1995b). We now give convergence and rate of convergence results in the infinite-dimensional setting. Such results indicate that our methods should work well even when the dimension of the space becomes very large. Our results extend to long-step methods some results of Bauschke and Borwein (1995) concerning short-step methods, which in turn generalize those in (Aharoni and Censor,

Research supported by the Polish State Committee for Scientific Research under Grants 8S50502206 and 8T11A00409.

1989; Flåm and Zowe, 1990); see (Bauschke and Borwein, 1995) for more references and an extensive list of applications of CFP's.

A general surrogate projection method is described in §2. In §§3–4 we give a sample of results on convergence and rate of convergence that supplement those in Kiwiel (1995a). For brevity, we omit the proofs given in (Kiwiel and Lopuch, 1994).

A set of the form $H = \{x : \langle a, x\rangle \le b\}$, where $a \in \mathcal{H}$ and $b \in \mathbb{R}$, is called a halfspace, including $H = \mathcal{H}$ or $\emptyset$. Given a closed convex set D, $P_D x = \arg\min_{y\in D} \|x - y\|$ is the projector on D and $d_D = \inf_{y\in D} \|\cdot - y\|$ is the *distance function* of D. $\operatorname{int} C$ denotes the interior of C. We let $\{1{:}\,N\} = \{1, 2, \ldots, N\}$ and $\mathbb{R}^N_+ = \{\lambda \in \mathbb{R}^N : \lambda \ge 0\}$. We need

Lemma 1 *If D is a closed convex set and $\alpha \ge 0$ then $R_{D,\alpha} = (1-\alpha)I + \alpha P_D$ satisfies $\|c - R_{D,\alpha}x\|^2 \le \|c - x\|^2 - \alpha(2-\alpha)\|x - P_D x\|^2$ $\forall c \in D$, $x \in \mathcal{H}$.*

2 A GENERAL ALGORITHM AND ITS BASIC PROPERTIES

We first state a general algorithm for (1).

Algorithm 2

Step 0 (*Initiation*). Select an initial $x^0 \in \mathcal{H}$ and a weight threshold $\lambda_{\min} \in (0, \frac{1}{N}]$. Set $n = 0$.

Step 1 (*Working set selection*). Choose a nonempty set $\hat{I}^{(n)} \subset \{1{:}\,N\}$, so that for each $i \in \{1{:}\,N\}$, $i \in \hat{I}^{(n)}$ for infinitely many n.

Step 2 (*Halfspace selection*). For each $i \in \hat{I}^{(n)}$, choose a projection operator $P_i^{(n)} : \mathcal{H} \to \mathcal{H}$ with $P_i^{(n)}\mathcal{H} \supset C_i$, let $x^{in} = P_i^{(n)} x^n$ and $H_i^n = \{x : \langle a^{in}, x\rangle \le b_{in}\}$ with

$$(a^{in}, b_{in}) = (x^n - x^{in}, \langle x^n - x^{in}, x^{in}\rangle), \quad i \in \hat{I}^{(n)}. \tag{2}$$

Step 3 (*Surrogate construction*). Find a *weight* vector $\lambda^n \in \mathbb{R}^N_+$, $\lambda_i^n = 0$, $i \notin \hat{I}^{(n)}$, $\sum_{i=1}^N \lambda_i^n = 1$, for the *surrogate inequality* $\langle \hat{a}^n, x\rangle \le \hat{b}_n$ with

$$(\hat{a}^n, \hat{b}_n) = \sum_{i\in\hat{I}^{(n)}} \lambda_i^n (a^{in}, b_{in}) = \sum_{i\in\hat{I}^{(n)}} \lambda_i^n (x^n - x^{in}, \langle x^n - x^{in}, x^{in}\rangle) \tag{3}$$

such that the *surrogate halfspace* $\hat{H}^n = \{x : \langle \hat{a}^n, x\rangle \le \hat{b}_n\}$ satisfies

$$d_{\hat{H}^n}(x^n) \ge \lambda_{\min} \max_{i\in\hat{I}^{(n)}} d_{H_i^n}(x^n). \tag{4}$$

Step 4 (*Relaxation*). Select a relaxation parameter $\alpha^{(n)} \in (0, 2]$ and set (cf. Lem. 1)

$$x^{n+1} = R_{\hat{H}^n,\alpha^{(n)}} x^n = x^n + \alpha^{(n)}(P_{\hat{H}^n} x^n - x^n), \tag{5}$$

$$\sigma_n = \alpha^{(n)}(2 - \alpha^{(n)}) d^2_{\hat{H}^n}(x^n). \tag{6}$$

Step 5. Increase n by 1 and go to Step 1.

At Step 2, by Lem. 1, $C_i \subset P_i^{(n)}\mathcal{H} \subset H_i^n$, $\forall i \in \hat{I}^{(n)}$. Let $I^{(n)} = \{i : \lambda_i^n > 0\}$. By (2)–(3),

$$\langle a^{in}, x^n \rangle - b_{in} = \|a^{in}\|^2 = \|x^n - x^{in}\|^2 \quad \forall i \in \hat{I}^{(n)}, \tag{7}$$

$$\langle \hat{a}^n, x^n \rangle - \hat{b}_n = \sum_{i \in I^{(n)}} \lambda_i^n \|x^n - x^{in}\|^2 \quad \text{and} \quad \|\hat{a}^n\| = \|\sum_{i \in I^{(n)}} \lambda_i^n (x^n - x^{in})\|, \tag{8}$$

so $x^{in} = P_{H_i^n}(x^n)$, $d_{H_i^n}(x^n) = \|x^n - x^{in}\| = \|a^{in}\|$, $i \in \hat{I}^{(n)}$. At Step 3, (4) holds if $\lambda_{\hat{\imath}}^n \geq \lambda_{\min}$ for some $\hat{\imath} \in \operatorname{Arg\,max}_{i \in \hat{I}^{(n)}} \|x^n - x^{in}\|$, since

$$d_{\hat{H}^n}(x^n) = \frac{\langle \hat{a}^n, x^n \rangle - \hat{b}_n}{\|\hat{a}^n\|} \geq \lambda_{\min} \frac{\max_{i \in \hat{I}^{(n)}}(\langle a^{in}, x^n \rangle - b_{in})}{\max_{i \in \hat{I}^{(n)}} \|a^{in}\|} = \lambda_{\min} \max_{i \in \hat{I}^n} d_{H_i^n}(x^n)$$

from $\langle \hat{a}^n, x^n \rangle - \hat{b}_n \geq \lambda_{\min} \|a^{\hat{\imath}n}\|^2$ and $\|\hat{a}^n\| \leq \sum_i \lambda_i^n \max_i \|a^{in}\|$ by the convexity of $\|\cdot\|$. Since $\cap_{i \in I^{(n)}} C_i \subset \cap_{i \in I^{(n)}} H_i^n \subset \hat{H}^n$ from $\lambda^n \geq 0$, by Lem. 1 and (6),

$$\|c - x^{n+1}\|^2 \leq \|c - x^n\|^2 - \alpha^{(n)}(2 - \alpha^{(n)}) d_{\hat{H}^n}^2(x^n) = \|c - x^n\|^2 - \sigma_n \quad \forall c \in \cap_{i \in I^{(n)}} C_i. \tag{9}$$

Thus progress towards the solution is measured by $d_{\hat{H}^n}(x^n)$, and one is interested in *deep* cuts that have $d_{\hat{H}^n}(x^n)$ as large as possible (Kiwiel, 1995a).

Remark 3 Other ways of finding cuts are given in (Kiwiel, 1995a). First, the weights $\lambda_i^n = \|a^{in}\|^\gamma / \sum_j \|a^{jn}\|^\gamma$, $i \in \hat{I}^{(n)}$, are admissible if $\gamma \geq 0$, since $\max_{i \in I_{\max}^{(n)}} \lambda_i^n \geq 1/|\hat{I}^{(n)}|$; e.g., $\gamma = 0$ gives $\lambda_i^n = 1/|I_+^{(n)}|$, $i \in I_+^{(n)} := \{i : a^{in} \neq 0\}$. Second, the *deepest* surrogate cut that maximizes $d_{\hat{H}^n}(x^n)$ is obtained for weights that solve (cf. (7)–(8))

$$\max \left\{ \frac{\sum_{i \in \hat{I}^{(n)}} \lambda_i \|a^{in}\|^2}{\|\sum_{i \in \hat{I}^{(n)}} \lambda_i a^{in}\|} : \lambda_i \geq 0, i \in \hat{I}^{(n)}, \sum_{i \in \hat{I}^{(n)}} \lambda_i = 1 \right\}, \tag{10}$$

in which case $P_{\hat{H}^n} x^n = P_{\cap_{i \in I^{(n)}} H_i^n} x^n$. Approximate solutions to (10) (that satisfy (4) with $\lambda_{\min} = 1$) are discussed in (Kiwiel, 1995a, §8).

Remark 4 To compare Algorithm 2 with that of Bauschke and Borwein (1995), let

$$\hat{c}_n = (\langle \hat{a}^n, x^n \rangle - \hat{b}_n) / \|\hat{a}^n\|^2. \tag{11}$$

Then

$$P_{\hat{H}^n}(x^n) - x^n = -\hat{c}_n \hat{a}^n = \frac{\sum_{i \in I^{(n)}} \lambda_i^n \|x^n - x^{in}\|^2}{\|x^n - \sum_{i \in I^{(n)}} \lambda_i^n x^{in}\|^2} (\sum_{i \in I^{(n)}} \lambda_i^n x^{in} - x^n), \tag{12}$$

$$d_{\hat{H}^n}(x^n) = \frac{\langle \hat{a}^n, x^n \rangle - \hat{b}_n}{\|\hat{a}^n\|} = \frac{\sum_{i \in I^{(n)}} \lambda_i^n \|x^n - x^{in}\|^2}{\|\sum_{i \in I^{(n)}} \lambda_i^n (x^n - x^{in})\|} \tag{13}$$

and $x^{n+1} = x^n - \alpha^{(n)}\hat{c}_n\hat{a}^n$ if $\hat{a}^n \neq 0$ (cf. (3), (5), (8)). The method of Bauschke and Borwein (1995) would produce

$$\tilde{x}^{n+1} = \sum_{i\in I^n} \lambda_i^n R_{H_i^n,\alpha^{(n)}}(x^n) = x^n + \alpha^{(n)}(\sum_{i\in I^n} \lambda_i^n x^{in} - x^n) \tag{14}$$

instead of x^{n+1}. To see why this method works, apply Lem. 1 to obtain

$$\|c - R_{H_i^n,\alpha^{(n)}}(x^n)\|^2 \leq \|c - x^n\|^2 - \alpha^{(n)}(2 - \alpha^{(n)})\|x^n - x^{in}\|^2 \quad \forall c \in H_i^n \supset C_i;$$

multiply this by λ_i^n, sum over i and use $\sum_i \lambda_i^n = 1$ and the convexity of $\|\cdot\|^2$ to get

$$\|c - \tilde{x}^{n+1}\|^2 \leq \|c - x^k\|^2 - \alpha^{(n)}(2 - \alpha^{(n)}) \sum_{i\in I^{(n)}} \lambda_i^n \|x^n - x^{in}\|^2 \quad \forall c \in \bigcap_{i\in I^{(n)}} C_i. \tag{15}$$

Thus to compare the methods (5) and (14) via their Fejér estimates (9) and (15), we may use the ratio of σ_n (given by (6) and (13)) and (cf. (8))

$$\tilde{\sigma}_n = \alpha^{(n)}(2 - \alpha^{(n)}) \sum_{i\in I^{(n)}} \lambda_i^n \|x^n - x^{in}\|^2 = \alpha^{(n)}(2 - \alpha^{(n)})(\langle \hat{a}^n, x^n\rangle - \hat{b}_n). \tag{16}$$

Lemma 5 *Let* $\tilde{x}^{n+1} = x^n - \alpha^{(n)}\hat{a}^n$ *denote the point produced by the iteration* (14) (cf. (3)), *and let* x^{n+1} *be given by* (5). *Then:*

(i) $x^n \in \hat{H}^n \iff \langle \hat{a}^n, x^n\rangle = \hat{b}_n \iff \sum_{i\in I^{(n)}} \lambda_i^n \|x^n - x^{in}\|^2 = 0 \iff x^n \in H_i^n$ $\forall i \in I^{(n)} \iff \tilde{\sigma}_n = 0 \iff \sigma_n = 0$. *Moreover, each of these conditions implies* $\hat{a}^n = 0$ *and* $x^{n+1} = \tilde{x}^{n+1} = x^n$.

(ii) *If* $\cap_{i\in I^{(n)}} H_i^n \neq \emptyset$ *then* $\hat{a}^n = 0 \iff \hat{H}^n = \mathcal{H} \iff$ *all the conditions of* (i) *hold.*

(iii) *If* $\hat{a}^n \neq 0$ *then* $x^{n+1} = x^n - \alpha^{(n)}(\sigma_n/\tilde{\sigma}_n)\hat{a}^n$, *where*

$$\sigma_n/\tilde{\sigma}_n = \hat{c}_n = \frac{\langle \hat{a}^n, x^n\rangle - \hat{b}_n}{\|\hat{a}^n\|^2} = \frac{\sum_{i\in I^{(n)}} \lambda_i^n \|x^n - x^{in}\|^2}{\|\sum_{i\in I^{(n)}} \lambda_i^n (x^n - x^{in})\|^2} \geq 1, \tag{17}$$

and $\sigma_n = \tilde{\sigma}_n \iff x^{in} = x^{jn}$ $\forall i, j \in I^{(n)}$.

(iv) *If* $\alpha^{(n)} = 1$ *then* $x^{n+1} \in \hat{H}^n$, *whereas* $\tilde{x}^{n+1} \in \hat{H}^n \iff \tilde{\sigma}_n = \sigma_n$.

We note that both the *long-step* method (5) and the *short-step* method (14) *use the same direction negative to the normal of the surrogate inequality.* Lemma 5 says that the long-step method can take a much longer step in this direction, thus providing a much larger guaranteed progress when $\sigma_n \gg \tilde{\sigma}_n$ in (9) and (15) (i.e., some tangent cone to $\cap_{i\in I^k} H_i^k$ is 'flat').

3 BASIC CONVERGENCE RESULTS

Let $\underline{\alpha}^{(\infty)} = \underline{\lim}_n \alpha^{(n)}$ and $\overline{\alpha}^{(\infty)} = \overline{\lim}_n \alpha^{(n)}$.

Lemma 6 (i) $\|c - x^n\|^2 - \|c - x^{n+1}\|^2 \geq \alpha^{(n)}(2 - \alpha^{(n)})d_{\hat{H}^n}^2(x^n)$ *if* $c \in \cap_{i\in I^{(n)}} C_i$.

(ii) $\|c - x^n\|^2 - \|c - x^m\|^2 \geq \sum_{l=n}^{m-1} \alpha^{(l)}(2 - \alpha^{(l)})d_{\hat{H}^l}^2(x^l)$ *if* $c \in \cap_{l=n}^{m-1} \cap_{i\in I^{(l)}} C_i$, $m \geq n \geq 0$.

(iii) $\{x^n\}$ *is Fejér monotone w.r.t.* C *(hence bounded) and* $\sum_{l=0}^{\infty} \alpha^{(l)}(2 - \alpha^{(l)})d_{\hat{H}^l}^2(x^l) < \infty$.

Corollary 7 (i) *If* $\operatorname{int} C \neq \emptyset$ *then* $\{x^n\}$ *converges in norm to some* $x \in \mathcal{H}$.
(ii) *If* $\underline{\lim}_n d_C(x^n) \to 0$, *then* $\{x^n\}$ *converges in norm to some* $x \in C$.

Corollary 8 *If* $\operatorname{int} C \neq \emptyset$ *then* $\sum_n \|x^{n+1} - x^n\| < \infty$ *and* $\sum_n \alpha^{(n)} d_{\hat{H}^n}(x^n) < \infty$.

Corollary 9 *If* $\overline{\alpha}^{(\infty)} < 2$ *then the algorithm is* regular, *i.e.*, $\|x^{n+1} - x^n\| \to 0$.

Further assumptions are necessary to ensure that the algorithm converges.

Definition 10 We say the algorithm is *focusing* if for each index i and subsequence $\{x^{n_k}\}$, the conditions $x^{n_k} \rightharpoonup x$, $d_{H_i^{n_k}}(x^{n_k}) \to 0$ (i.e., $x^{n_k} - P_i^{(n_k)} x^{n_k} \to 0$) and $i \in I^{(n_k)}$ $\forall k$ imply $x \in C_i$, where $\to$ and $\rightharpoonup$ stand for norm and weak convergence respectively.

Remark 11 The above definition of Bauschke and Borwein (1995) is equivalent to the concept of cut map introduced by Kiwiel (1995a) when $\dim \mathcal{H} < \infty$.

Theorem 12 *Suppose the algorithm is focusing. If* $0 < \underline{\alpha}^{(\infty)}, \overline{\alpha}^{(\infty)} < 2$ *then* $\{x^n\}$ *either converges in norm to some point in* C *or has no norm cluster points at all.*

Remark 13 If there exists $\epsilon > 0$ s.t. $\{\alpha^{(n)}\} \subset [\epsilon, 2-\epsilon]$ then $0 < \underline{\alpha}^{(\infty)}, \overline{\alpha}^{(\infty)} < 2$.

For motivation, we quote the following result of Bauschke and Borwein (1995).

Proposition 14 *Suppose for every index* i, $\{P_i^{(n)}\}$ converges actively pointwise *to* P_{C_i}, *i.e.*, $\lim_{n:i \in I^{(n)}} P_i^{(n)} x = P_{C_i} x$ *for every* $x \in \mathcal{H}$. *Then the algorithm is focusing.*

Definition 15 If there is a positive integer p s.t. $i \in \hat{I}^{(n)} \cup \hat{I}^{(n+1)} \cup \cdots \hat{I}^{(n+p-1)}$ for every index i and all n, then we speak of an *intermittent* or p-*intermittent* algorithm.

Theorem 16 (i) *Suppose the algorithm is focusing and intermittent, and* $0 < \underline{\alpha}^{(\infty)}$ *and* $\overline{\alpha}^{(\infty)} < 2$. *Then* $\{x^n\}$ *is regular and converges weakly to some point in* C.
(ii) *Suppose the algorithm is focusing, p-intermittent, regular (cf. Cor. 9) and* $\hat{\nu}_n = \min\{\alpha^{(l)}(2-\alpha^{(l)}) : np \le l < (n+1)p\}$ $\forall n \ge 0$. *If* $\sum_n \hat{\nu}_n = \infty$ *then* $\{x^n\}$ *has a unique weak cluster point in* C. *In particular, this happens whenever* $0 < \underline{\alpha}^{(\infty)}, \overline{\alpha}^{(\infty)} < 2$.
(iii) *Suppose the algorithm is focusing and* $x^n \rightharpoonup x$. *If* $\sum_{n:i \in \hat{I}^{(n)}} \alpha^{(n)}(2 - \alpha^{(n)}) = \infty$ *for some index* i, *then* $x \in C_i$. *Consequently,* $x \in C$ *if* $\sum_{n:i \in \hat{I}^{(n)}} \alpha^{(n)}(2 - \alpha^{(n)}) = \infty$ *for* every *index* i.

Corollary 17 *Suppose the algorithm is focusing,* $\sum_{n:i \in \hat{I}^{(n)}} \alpha^{(n)}(2 - \alpha^{(n)}) = \infty$ *for every index* i *and* $\operatorname{int} C \neq \emptyset$. *Then* $\{x^n\}$ *converges in norm to some point in* C.

Remark 18 If $0 < \underline{\alpha}^{(\infty)}, \overline{\alpha}^{(\infty)} < 2$ then $\sum_{n:i \in \hat{I}^{(n)}} \alpha^{(n)}(2 - \alpha^{(n)}) = \infty$ for all i.

Theorem 19 *Suppose the algorithm is focusing.*
(i) *If* $\sum_{n:\hat{I}^{(n)} = \{1:N\}} \alpha^{(n)}(2 - \alpha^{(n)}) = \infty$ *then* $\{x^n\}$ *converges weakly to some point in* C.

(ii) *If there exist* $\epsilon > 0$ *and a subsequence* $\{x^{n_k}\}$ *s.t.* $\epsilon \le \alpha^{(n_k)} \le 2-\epsilon$ *and* $\hat{I}^{(n_k)} = \{1:N\}$ *for all* k *then* $x^n \rightharpoonup x$ *for some* $x \in C$.

Definition 20 We say the algorithm is *linearly focusing* if there is $\beta > 0$ s.t. $\beta d_{C_i}(x^n) \le d_{H_i^n}(x^n)$ for all large n and every $i \in \hat{I}^{(n)}$; it is *strongly focusing* if for every index i and every subsequence $\{x^{n_k}\}$, the conditions $x^{n_k} \rightharpoonup x$, $d_{H_i^n}(x^{n_k}) \to 0$ and $i \in \hat{I}^{(n_k)}$ imply $d_{C_i}(x^{n_k}) \to 0$. Thus (cf. Def. 10): linearly focusing $\Rightarrow$ strongly focusing $\Rightarrow$ focusing.

Corollary 21 *Suppose the algorithm is linearly focusing,* $0 < \underline{\alpha}^{(\infty)}, \overline{\alpha}^{(\infty)} < 2$ *and either* $\mathcal{H}$ *is finite-dimensional or* $\operatorname{int} C \ne \emptyset$. *Then* $\{x^n\}$ *converges in norm to some point in* C.

Theorem 22 *Suppose the algorithm is focusing and* $\operatorname{int} C \ne \emptyset$, *so that* $\{x^n\}$ *converges to some* x. *If* $\sum_{n:i\in\hat{I}^{(n)}} \alpha^{(n)} = \infty$ *for some index* i, *then* $x \in C_i$. *Consequently,* $x \in C$ *if* $\sum_{n:i\in\hat{I}^{(n)}} \alpha^{(n)} = \infty$ *for* every *index* i.

Definition 23 We say the algorithm *considers remotest sets* if $\hat{I}^{(n)}$ contains some *active remotest index* $i_n \in \hat{I}^{(n)}_{\text{rem}} := \{i : d_{C_i}(x^n) = \max_{j=1}^N d_{C_j}(x^n)\}$, for all n.

Theorem 24 *Suppose the algorithm is strongly focusing and considers remotest sets.*
(i) *If* $\sum_n \alpha^{(n)}(2-\alpha^{(n)}) = \infty$ *then* $\{x^n\}$ *converges weakly to some point in* C.
(ii) *If* $0 < \underline{\alpha}^{(\infty)}, \overline{\alpha}^{(\infty)} < 2$ *then* $x^n \rightharpoonup x$ *for some* $x \in C$ *and* $\max_{j=1}^N d_{C_j}(x^n) \to 0$.

4 NORM CONVERGENCE AND BOUNDED REGULARITY

Guaranteeing norm convergence requires further assumptions.

Definition 25 The N-tuple $(C_1,\ldots,C_N)$ is *regular* if $\forall\epsilon > 0\ \exists\delta > 0\ \forall x \in \{x \in \mathcal{H} : \max_{j=1}^N d_{C_j}(x) \le \delta\}$: $d_C(x) \le \epsilon$. It is *boundedly regular* if this holds only on bounded sets, i.e., $\forall$ bounded $S \subset \mathcal{H}\ \forall\epsilon > 0\ \exists\delta > 0\ \forall x \in \{x \in S : \max_{j=1}^N d_{C_j}(x) \le \delta\}$: $d_C(x) \le \epsilon$.

Theorem 26 *Suppose the algorithm is strongly focusing, p-intermittent, regular (cf. Cor. 9) and* $\hat{\nu}_n = \min\{\alpha^{(l)}(2-\alpha^{(l)}) : np \le l < (n+1)p\}\ \forall n \ge 0$. *If* $\sum_n \hat{\nu}_n = \infty$ *and* $(C_1,\ldots,C_N)$ *is boundedly regular then* $\{x^n\}$ *converges in norm to some point in* C. *In particular, this happens whenever* $0 < \underline{\alpha}^{(\infty)}, \overline{\alpha}^{(\infty)} < 2$.

Theorem 27 *Suppose the algorithm is strongly focusing and considers remotest sets, and* $(C_1,\ldots,C_N)$ *is boundedly regular. If* $\sum_n \alpha^{(n)}(2-\alpha^{(n)}) = \infty$ *then* $\{x^n\}$ *converges in norm to some point in* C. *In particular, this happens whenever* $0 < \underline{\alpha}^{(\infty)}, \overline{\alpha}^{(\infty)} < 2$.

As in (Bauschke and Borwein, 1995), guaranteeing *linear* convergence requires *linear* regularity.

Definition 28 The N-tuple $(C_1,\ldots,C_N)$ is called *linearly regular* if $\exists\kappa > 0\ \forall x \in \mathcal{H}$: $d_C(x) \le \kappa \max_{j=1}^N d_{C_j}(x)$. It is called *boundedly linearly regular* if $\forall$ bounded $S \subset \mathcal{H}$ $\exists\kappa_S > 0\ \forall x \in S$: $d_C(x) \le \kappa_S \max_{j=1}^N d_{C_j}(x)$.

We say $\{x^n\}$ *converges linearly with rate* $\beta \in [0,1)$ if there are $x \in \mathcal{H}$ and $\alpha \geq 0$ s.t. $\|x^n - x\| \leq \alpha\beta^n$ for all n.

Theorem 29 *Suppose the algorithm is linearly focusing and p-intermittent,* $\{\alpha^{(n)}\} \subset [\epsilon, 2-\epsilon]$ *for some* $\epsilon > 0$ *and* $(C_1, \ldots, C_N)$ *is boundedly linearly regular. Then* $\{x^n\}$ *converges linearly to some point in* C*; the rate of convergence is independent of the starting point whenever* $(C_1, \ldots, C_N)$ *is linearly regular.*

Theorem 30 *Suppose the algorithm is linearly focusing, considers remotest sets,* $0 < \underline{\alpha}^{(\infty)}, \overline{\alpha}^{(\infty)} < 2$ *and* $(C_1, \ldots, C_N)$ *is boundedly linearly regular. Then* $\{x^n\}$ *converges linearly to some point in* C*; the rate of convergence is independent of the starting point whenever* $(C_1, \ldots, C_N)$ *is linearly regular.*

Theorem 31 *Suppose the algorithm is strongly focusing and* $(C_1, \ldots, C_N)$ *is boundedly regular. If* $\sum_{n: \hat{I}^{(n)} = \{1:N\}} \alpha^{(n)}(2 - \alpha^{(n)}) = \infty$ *then* $\{x^n\}$ *converges in norm to some point in* C*. In particular, this happens whenever there exist* $\epsilon > 0$ *and a subsequence* $\{n_k\}$ *s.t.* $\epsilon \leq \alpha^{(n_k)} \leq 2 - \epsilon$ *and* $\hat{I}^{(n_k)} = \{1:N\}$ *for all* k.

Theorem 32 *Suppose the algorithm is linearly focusing,* $\hat{I}^{(n)} = \{1:N\}$ *for all* n, $0 < \underline{\alpha}^{(\infty)}, \overline{\alpha}^{(\infty)} < 2$ *and* $(C_1, \ldots, C_N)$ *is boundedly linearly regular. Then* $\{x^n\}$ *converges linearly to some point in* C*; the rate of convergence is independent of the starting point whenever* $(C_1, \ldots, C_N)$ *is linearly regular.*

To illustrate, we only quote one fact from (Bauschke and Borwein, 1995).

Fact 33 *If* $C_{\hat{\imath}} \cap \operatorname{int} \bigcap_{i \neq \hat{\imath}} C_i \neq \emptyset$ *for some* $\hat{\imath}$ *then* $(C_1, \ldots, C_N)$ *is boundedly linearly regular.*

REFERENCES

Aharoni, R. and Censor, Y. (1989) Block-iterative projection methods for parallel computation of solutions to convex feasibility problems. *Linear Algebra and Applications*, **120**, 165–75.

Bauschke, H.H. and Borwein, J.M. (1995) On projection algorithms for solving convex feasibility problems. *SIAM Review*, (to appear).

Flåm, S.D. and Zowe, J. (1990) Relaxed outer projections, weighted averages and convex feasibility. *BIT*, **30**, 289–300.

Kiwiel, K.C. (1995a) Block-iterative surrogate projection methods for convex feasibility problems. *Linear Algebra and Applications*, **215**, 225–59.

Kiwiel, K.C. (1995b) The efficiency of subgradient projection methods for convex optimization, part II: Implementations and extensions. *SIAM Journal on Control and Optimization*, (to appear).

Kiwiel, K.C. and Lopuch, B. (1994) *Surrogate projection methods for finding fixed points of firmly nonexpansive mappings.* Systems Research Institute, Warsaw.

Theoretical and experimental analysis of random linkage algorithms for global optimization

Marco Locatelli
Università di Milano
Dip. di Scienze della Informazione, via Comelico 39/41, 20139 Milano, Italy. Tel: +39 - 2 - 55006 300.
e-mail: locatelli@hermes.mc.dsi.unimi.it

Fabio Schoen
Università di Firenze
Dip. di Sistemi e Informatica, via di Santa Marta 3, 50139 Firenze, Italy. Tel: +39 - 55 - 4796 358. Fax: +39 - 55 - 4796 363.
e-mail: schoen@ingfi1.ing.unifi.it
www: www-dsi.ing.unifi.it/~schoen/home.html

Abstract

In this paper a class of randomized algorithms for the optimization of multimodal functions is introduced and some theoretical results briefly recalled. A comparison, both theoretical and experimental, is also performed between some members of this class of algorithms and a well known method of global optimization - Multi Level Single Linkage. It is shown that it is possible to obtain similar, and even better, results with the proposed algorithms, with significantly lower computational effort.

Keywords

Global Optimization, Stochastic Algorithms, Clustering, Multi Level Single Linkage

1 PROBLEM STATEMENT

We consider the following problem: find

$$f^\star := \max_{x \in \Omega} f(x) \tag{1}$$

where $\Omega \subseteq \mathbb{R}^d$ is a compact set, $d \geq 1$, $f \in \mathcal{C}(\Omega)$; in this paper it will be assumed that $\Omega = [0,1]^d$. The function f is usually non concave, so that classical "local" methods cannot be employed to solve the problem. However, if f is sufficiently regular, good and reliable local optimization algorithms usually do exist, so that it is quite natural to design global optimization methods which mix in an appropriate way local and global procedures. As problem (1) is NP-hard (even very special sub-classes of this problem have been proven to be NP-hard - see, for example, (Pardalos and Vavasis, 1991)), randomized algorithm often are considered as a good choice; one quite attractive such an approach, and one which has been succesfully implemented in recent years, consists of repeatedly sampling points from a uniform distribution over the feasible set, coupled with local searches started from selected points in the sample. At two opposite extrema of this class of algorithms we find Multistart, in which a local search is started from each sampled point, and Single Start, where a single local search is started from the best (record) point at the end of the sampling process.

In this paper we introduce and analyze a class of algorithms which in some sense is intermediate between Multistart and Single Start, where the decision whether to start or not a local search from a sampled point is made on-line, depending both on the relative merit (function value) of the current point and on the distance between the current point and other points in the sample. This way it is hoped that unnecessary local searches, possibly leading to already discovered or to non interesting local optima, can be avoided, while retaining some good convergence property.

2 MULTI LEVEL SINGLE LINKAGE

Multi Level Single Linkage (MLSL for short) is a popular stochastic global optimization algorithm introduced in (Rinnooy Kan and Timmer, 1987 a-b). Let $N > 0$, $\sigma > 0$, $\gamma \in (0,1]$ be parameters chosen by the user and let $k := 0$; MLSL can be defined as follows:

1. let $k := k+1$;
2. let

$$r_{k;N;\sigma} := \pi^{-1/2}\left(\Gamma(1+d/2)\,\sigma\,\frac{\log kN}{kN}\right)^{1/d}; \tag{2}$$

3. generate a uniform random sample of size N in Ω;
4. sort the whole sample of Nk points in order of decreasing function value, and select the γNk best of them; the resulting sample is called *reduced sample*;
5. apply a local search algorithm from each point in the reduced sample except if there exists another point with higher function value at a distance which is less than or equal to the threshold $r_{k;N;\sigma}$;
6. check if some stopping condition is satisfied, otherwise repeat from 1.

The rationale behind MLSL is that if a sampled point is "near" to a better one, either a local search was started from the latter, or that point was in turn "near" to another one, with even better function value, and so on. Eventually, a non decreasing chain of sampled points will be detected, the last point being a point from which a local search has been started. The true meaning of "near" obviously depend on the sample size, and is reflected in (2).

Among the main drawbacks, in the authors' opinion, of MLSL, there are both the necessity of sampling in batches of N points (with a related difficult decision about which value to assign to N) and the necessity of reconsidering, after each sample, each point in the reduced sample (except those points from which a local search was already started) in order to judge whether, with the updated threshold (2), a local search should be started. Moreover the original definition of MLSL does not allow starting a local search from a point which is too near to the boundary of Ω (this fact is crucial in the proofs of convergence). On the positive side, MLSL enjoys strong theoretical properties, among which most notably are:

1. convergence of the record value to $f^\star$ with probability 1;
2. the probability of starting a local search decreases to 0 if $\sigma > 2$;
3. the total expected number of local searches performed, even if the algorithm is run forever, is finite if $\sigma > 4$.

3 RANDOM LINKAGE

Starting from MLSL, the authors reconsidered the basic idea of the algorithm, and generalized it, while introducing significant simplifications. The theoretical analysis of Random Linkage is reported in full details in (Locatelli and Schoen, 1995). Here we simply recall that, differently from MLSL, a Random Linkage algorithm

1. samples sequentially (*i.e.* $N = 1$);
2. does not reconsider previous decisions about whether to start or not a local search; in other words, at iteration k a local search may possibly be started only from X_k, the most recently sampled point;
3. a local search may be started also from points near or over the boundary of the feasible region;
4. the decision whether to start or not a local search may be randomized, with the probability of starting a local search being a non increasing function of the distance between the current point and the nearest better point in the sample (the name "Random Linkage" came from this particular acceptance/rejection mechanism).

In this paper we will report only results relative to a sub-class of Random Linkage, named Threshold Random Linkage (TRL) which is characterized by the fact that the decision whether to start or not a local search is not randomized, but instead it is given by the following deterministic rule: start a local search from X_k, having observed f at $X_1, \ldots, X_{k-1}$ if and only if

$$\min\{\|X_k - X_j\|, j = 1, \ldots, k-1 : f(X_j) \geq f(X_k)\} \geq \alpha_k \tag{3}$$

where $\{\alpha_k\}$ is a non-increasing positive sequence. For this special class of algorithms, and the more general one with randomized decisions, a complete characterization was given in (Locatelli and Schoen, 1995). In particular, if the criterion for deciding whether to start a local search is slightly relaxed as follows:

$$\min\{\|X_k - X_j\|, j : f(X_j) \geq f(X_k) - \epsilon\} \geq \alpha_k \tag{4}$$

where ϵ is a small positive constant, then it was proven that

Theorem 1 *The probability of starting a local search tends to 0 with $k \to \infty$ if and only if*

$$\alpha_k k^{1/d} \to \infty \tag{5}$$

If α_k is chosen in a way similar to (2), that is letting

$$\alpha_k = r_{k;1;\sigma}, \tag{6}$$

then the probability of starting a local search will decrease to 0 for any choice of $\sigma > 0$; moreover the expected number of local searches started even if the algorithm is run forever is finite if $\sigma > 2^d/(2d)$. If no local search is allowed to start from within a prefixed distance from the boundary, then finiteness of the total number of local searches is achieved for all $\sigma > 1$. In any case, the global optimum will be found in finite time with probability 1.

4 COMPARISON BETWEEN MLSL AND TRL

A first numerical comparison between different Random Linkage implementations and MLSL with different values for σ is reported in (Locatelli and Schoen, 1995). There the results of running these algorithms over a set of test functions with dimension d ranging from 2 to 30 and with a number of local optima between 2 and 10^{30} were compared. The difference in standard statistics, like, *e.g.* , number of function evaluations, number of local searches, total sample size at stopping, were not very high. One of the difficulties in making a fair comparison was the very high computed variance in all of these statistics. Numerical experiments have thus been repeated using, for TRL and MLSL, the same sequence of randomly generated sample points: this way the only visible difference between the algorithms is caused by the different criterion for the choice of starting points for local searches. In table 1 we report the median statistics computed over 100 independent runs of TRL with threshold defined as in (6) and $\sigma = 1$ and of MLSL with $N = 50$ and $\sigma = 4$. Moreover in MLSL it was assumed that $\gamma = 1$ (*i.e.*, no reduction in the sample).

It is quite evident even from these numerical results that the performance of the two algorithm is very similar; it seems that, for difficult test functions, TRL is quite superior to MLSL. CPU times were not recorded, mainly because they strongly depend upon the implementation details and, in particular, the data structure used to store sampled points and to compute nearest neighboor distances. It should however be quite evident that, being the number of local searches quite comparable in the two algorithms, the main difference in performance should be caused by the number of nearest neighboor distance computed; a very naive implementation of both MLSL and TRL, using the simplest data structure, confirmed this analysis: for example, 100 runs of TRL on function "pen. shubert

function	method	funct.ev.	grad.ev.	loc. srch.	sample size
quartic	MLSL	73	22	1	50
$d = 2$	TRL	34	31.5	1	1
6 hump	MLSL	68	17	1	50
$d = 2$	TRL	33	31	1	1
shubert	MLSL	102	34	2	50
$d = 2$	TRL	175	110	6	48
pen.shubert 0.5	MLSL	2765	623.5	35	1975
$d = 2$	TRL	2174	505	25.5	1589.5
pen.shubert 1.	MLSL	2017.5	433	24	1625
$d = 2$	TRL	1725.5	404	20.5	1209
picc. 1	MLSL	69.5	18.5	1	50
$d = 2$	TRL	63	49.5	3	4
picc. 1	MLSL	87.5	36.5	1	50
$d = 3$	TRL	93.5	84.5	3	4
picc. 1	MLSL	107.5	56	1	50
$d = 4$	TRL	117.5	109	3	4
picc. 2	MLSL	361.5	308	4	50
$d = 5$	TRL	435	409	6	11.5
picc. 2	MLSL	908.5	855	6	50
$d = 8$	TRL	1126	1095.5	8	18.5
picc. 2	MLSL	1241.5	1189.5	7	50
$d = 10$	TRL	1420	1395	7	15
picc. 2	MLSL	4149	4096	6	50
$d = 20$	TRL	3924.5	3904.5	6	13
picc. 2	MLSL	7005	6952.5	6	50
$d = 30$	TRL	8844	8810.5	8	18

Table 1 Comparison between MLSL and TRL

0.5" (the most difficult of the test functions) required roughly 24′ on a SUN Sparc Station, while MLSL on the same test required 7 hours and 20′.

5 THEORETICAL ANALYSIS OF MLSL AND TRL

One of the strongest features of MLSL is that, thanks to the fact that the starting points for local searches may be found also in previously sampled points, the overall process consists in building chains of sampled points, with better and better function value, up to a point from which a local search is started; this way the algorithm can adapt very well the chain of sampled points to the shape of the region of attraction of local (and global) optima. This mechanism is absent in TRL, so it is quite natural to ask whater TRL makes sensible choices; the theoretical analysis in this section is devoted to proof that, with high probability, MLSL and TRL make the same choices, that is, they both decide to start local searches from the same points.

For the comparison we assume that in MLSL the choice $N = 1$ has been made. The analysis of the general case is actually being investigated and will be published elsewhere.

Let us define $M_k(h)$ (resp. T_k) to be the events that MLSL (TRL) started a local search from the k–th sampled point X_k; the index h in $M_k(h)$ indicates that the local search starting from X_k was decided at iteration $h \geq k$ by MLSL.

We start with a straightforward observation: if the same threshold is used in MLSL and TRL, then

$$P(M_k(k) \mid T_k) = 1. \tag{7}$$

For what concerns the reverse, *i.e.* the probability that TRL starts a local search from X_k given that *at some iteration* MLSL will decide the same, the situation is a little bit more complex. However the following result is valid, which states that, when the threshold parameter used in MLSL is strictly greater than that in TRL (like in our experiments, where $\sigma = 4$ was chosen for MLSL and $\sigma = 1$ for TRL), then TRL will surely start a local search if MLSL will start a local search from the same point *not too late*. It should be recalled that one of the main distinguishing features of MLSL versus TRL is the fact that MLSL might decide to start a local search from X_k at a later iteration $h \geq k$, when the threshold for acceptance/rejection has been decreased enough. The next theorem says that, if the decision of starting a local search from X_k with MLSL does not occur "too late", then surely also TRL will start a local search as soon as it will sample X_k, *i.e.* at iteration k. This is, in the authors' opinion, a strong property, which enables to use TRL as a much simpler and computationally efficient implementation of MLSL.

Theorem 2 *Let the threshold parameter for TRL be $\sigma > 0$ and let the analogous parameter for MLSL be $\beta\sigma$ with $\beta > 1$. Then, there exists $\bar{k}$ such that, for all $k \geq \bar{k}$,*

$$P(T_k \mid \bigcup_{h=k}^{\beta k} M_k(h)) = 1 \tag{8}$$

Proof. As f is continuous and Ω is compact, f is uniformly continuous; then it is easy to see that, thanks to the relaxation (4), there exists $\bar{k}$ sucht that, for all $k \geq \bar{k}$, a local search is started in TRL if and only if

$$\min\{\|X_k - X_j\|, j = 1, \ldots, k-1\} \geq \alpha_k \tag{9}$$

and similarly for MLSL (for a proof of this statement, see the cited paper (Locatelli and Schoen, 1995), lemma 3.2). This means that, at least for sufficiently large k, the dependance of the acceptance criterion on f can be neglected and all of the analysis can be carried out with explicit reference only to the relative positions of the sampled points.

If we denote by $\alpha_k = r_{k;1;\sigma}$ the threshold used in TRL, then the threshold for MLSL is $\beta^{1/d}\alpha_k$. The event that a local search is started from X_k in MLSL at iteration h with $k \le h \le \beta k$ is equivalent to the event

$$\min\{\|X_j - X_k\|, j = 1, \ldots, h, j \neq k\} \ge \beta^{1/d}\alpha_h \tag{10}$$

We have also

$$\min\{\|X_j - X_k\|, j = 1, \ldots, k, j \neq k\} \ge \min\{\|X_j - X_k\|, j = 1, \ldots, h, j \neq k\}$$

so that a local search in TRL will surely be started from X_k if

$$\beta^{1/d}\alpha_h \ge \alpha_k$$

(notice that $k \le h$). But, recalling the definition of the threshold, we have that $\beta^{1/d}\alpha_h \ge \alpha_k$ if and only if

$$(\beta\frac{\log h}{h})^{1/d} \ge (\frac{\log k}{k})^{1/d}$$

and, using the fact that $k \le h \le \beta k$, we obtain

$$\beta\frac{\log h}{h} \ge \beta\frac{\log k}{\beta k} = \frac{\log k}{k}$$

which completes the proof. □

6 CONCLUSIONS AND DIRECTIONS FOR FUTURE RESEARCH

After the preceding theorem, we observe that, while in both MLSL and TRL the probability of starting a local search decreases to zero, as the iterations proceed it is more and more likely that both algorithms will perform very similar decisions regarding the points from which to start a local search. Obviously, being the analysis based on some sort of asymptotic result (being in particular strongly dependent upon the consequence of the uniform continuity of f), MLSL and TRL might significantly differ in the first iterations. It seems that an accurate analysis of this "transient" phase of both algorithms is very difficult and no deep result is expected, unless very strong assumptions on f are assumed. On the other end, it seems plausible that a more complete characterization of the relative behaviour of MLSL and TRL can be effectively obtained, allowing to state results more general then that of the last theorem and comprising the event that MLSL starts a local search at any iteration later than k (and also later then βk). In this case it seems plausible that the probability of starting a local search from TRL will be strictly positive (although not necessarily equal to 1), thus confirming our impression on the similarity of the two algorithms. The details of this more elaborate analysis are a subject of current research and will appear elsewhere.

REFERENCES

Locatelli, M. and Schoen, F. (1995) Random Linkage: a family of acceptance/rejection algorithms for global optimisation. Submitted. (Available for anonymous `ftp` at `ftp-dsi.ing.unifi.it`, directory `pub/schoen/papers/randlink`).

Pardalos, P.M. and Vavasis, S.A. (1991) Quadratic programming with one negative eigenvalue is NP-Hard. *Journal of Global Optimization*, **1**, 15–22.

Rinnooy Kan, A.H.G. and Timmer, G.T. (1987a) Stochastic global optimization methods. Part I: Clustering methods. *Mathematical Programming*, **39**, 27–56.

Rinnooy Kan, A.H.G. and Timmer, G.T. (1987b) Stochastic global optimization methods. Part II: Multi Level methods. *Mathematical Programming*, **39**, 57–78.

ACKNOWLEDGEMENT

This research has been partially supported by MURST research project "Metodi di Ottimizzazione nella Ricerca Operativa".

58

A dynamic list heuristic for 2D-cutting

Luiz Antonio N. Lorena, Fábio Belo Lopes

INPE - Instituto de Pesquisas Espaciais
LAC - Laboratório Associado de Computação e Matemática
Aplicada, Caixa Postal 515
12201 - São José dos Campos - SP - Brazil

Abstract

We present a new heuristic for the 2D-cutting problem, called Dynamic List Heuristic (DLH). The objective is to eliminate the combinatorial explosion of the all rectangle combination's type heuristics, maintaining their good results of reduced waste. Algorithm DLH uses a dynamic list of constructed patterns, included and deleted at convenient times. DLH is based on the A* approach, a best-first tree of candidate solutions, and the use of an auxiliary tabu list for alternate best patterns to avoid excessive repetition. The computational tests for some problems of the literature and many random generated problems, confirm good results using microcomputers.

Keywords

Cutting Stock, Heuristics, Best-first search, Tabu search.

1 INTRODUCTION

The 2D-cutting problem received great attention in the cutting stock problem context. It is a problem of combining a limited demand of given rectangles to fit in a plate with smallest total waste. Dowsland and Dowsland (1992) and Israni and Sanders (1985) give surveys including 2D-cutting problems. For an annotated bibliography see Sweeney and Paternoster (1992).

Wang (1983) proposed two heuristics for the problem. The good results obtained are consequence of a combinatorial explosive sequence of horizontal or vertical combinations of rectangles. Its indication is only for small problems with medium rectangles. Trying to improve Wang algorithms, Oliveira and Ferreira (1990), used bounds of the unlimited demand heuristic to improve the Wang's pruning criteria (algorithm one). The paper accelerates Wang algorithm one, maintaining all their original characteristics.

We present in this paper a new heuristic. We not guarantee optimal solutions, but a fast approximate algorithm for large scale problems, using microcomputer. It appears definitively indicated to use in microcomputers. In section two we present the motivation and describe the algorithm DLH. Section three presents computational tests for problems of the literature and randomly generated test problems. Finally we conclude with suggestions in section four.

2 HEURISTIC DLH

We begin defining rectangle constructions and the type of wastes resulted. Given a plate with dimensions W (width) and H (height), and primary rectangle's $r_1, r_2, ..., r_m$, with bound limits $b_1, b_2, ..., b_m$. L is a list of rectangles, initially composed by primary rectangles. A rectangle $R_i \in L$, fitting in the plate, is constructed by horizontal or vertical build of two other rectangles, primary and/or another rectangle $R_j \in L$ already constructed. Two constructed rectangles are equivalent when formed by the same primary rectangles and have the same waste.

Denote the internal waste in R_i by $I_w(R_i)$. The difference $(WH - R_i)$ is the external waste of R_i. The final pattern will be of guillotined type, and the objective is construct rectangles R_i minimizing total waste (internal + external).

Wang (1983) proposes two algorithms based in the following increasing power of pruning criteria for R_i:

$$I_w(R_i) > \beta_1 WH, \text{ for a given } \beta_1, \ 0 \le \beta_1 \le 1, \text{ and} \tag{1}$$

$$I_w(R_i) > \beta_2 R_i, \text{ for a given } \beta_2, \ 0 \le \beta_2 \le 1. \tag{2}$$

The list L of R_i's increases combining recursively each two existing rectangle, that not satisfies pruning criteria (1) (algorithm one) or pruning criteria (2) (algorithm two), satisfies the bound limits for the primary rectangles, and avoids equivalent rectangles. For small primary rectangles, the number of combinations (patterns) fitting in the plate can be combinatorially explosive. The indications of Wang algorithms are only for small problems with medium/large primary rectangles.

Heuristic DLH is an A* like algorithm (1984). For a given R_i, defines a function $f(\alpha, R_i) = I_w(R_i) + \alpha(WH - R_i)$, $\alpha \ge 0$. The parameter α is a percentage of external waste that, along which the internal waste $I_w(R_i)$, constitute the expected total waste. That is, for a given β, $0 \le \beta \le 1$, the expected total waste is βWH and it is desirable that $f(\alpha, R_i) \le \beta WH$. Interpreting in the A* context, the function f is a node-selection function, consisting of two parts, g and h, where $g(R_i) = I_w(R_i)$, is the current waste, and $h(\alpha, R_i) = \alpha(WH - R_i)$ is an estimate of the total waste.

The new pruning criteria for R_i is

$$I_w(R_i) + \alpha(WH - R_i) > \beta WH. \tag{3}$$

For $\alpha = 0$, $I_w(R_i) > \beta WH$, reproduces pruning criteria (1) for $\beta = \beta_1$, and for $\alpha = \beta$, $I_w(R_i) > \beta R_i$, reproduces pruning criteria (2) for $\beta = \beta_2$.

Heuristic DLH makes a tree with a new R_i for each node. Build the nodes by the rules governing the construction of rectangles, that is, avoiding the pruning criteria (3) and equivalent rectangles, and satisfying the bound limits for primary rectangles. For the new list L, instead of combines each two existing rectangles to construct new rectangles, the rectangle corresponding to the minimum $f(\alpha, R_i)$ can be combined with the other existing rectangles, characterizing a best-first search. The parameter α must be (smoothly) updated at each iteration, beginning small, $\alpha < \beta$, and increasing as the external waste decreases. When α increases, the pruning criteria (3) increase in power.

These are basic lines for the algorithm design, but some points must be carefully explained, such that:

- When α increases some candidate patterns of L are deleted characterizing a dynamic list of patterns;
- The "good" candidate patterns must remains in L to continue the combinations, at least for a "convenient" time;
- The stopping criteria can be when the list L is empty, where the best pattern (minimum total waste) was retained;
- The primary rectangles are also pruned for some values of α, so these values must only be reached after some time to improve the list L;
- For all $R_i \in L$ the function $f(\alpha, R_i)$ changes with α, and must be recalculated, what seems to be inefficient for implementation purposes.

For given β, $0 \le \beta \le 1$ and $R_i \in L$, the auxiliary parameters γ_i, δ_i, and ε_i are defined as

$$\gamma_i = I_W(R_i)/\beta WH,$$

$$\delta_i = R_i/WH, \text{ and}$$

$$\varepsilon_i = (1 - \gamma_i)/(1 - \delta_i).$$

Parameter γ_i, $0 \le \gamma_i \le 1$, is the proportion of total waste reached by R_i, and δ_i, $0 \le \delta_i < 1$, the proportion of total area occupied by R_i. The pruning criteria (3) is then equivalent to

$$\alpha/\beta > \varepsilon_i . \tag{4}$$

If $\delta_i = 1$ or $R_i = WH$, $\varepsilon_i \to \infty$, and we arrive in a possible optimal solution for that β. To continue the algorithm and search better solutions for updated β it is necessary set $\delta_i = 0.99$ in this case.

Let $\varepsilon_{sp} = \min\{\varepsilon_i\}$, for ε_i corresponding to primary rectangles, that is $\varepsilon_i = 1/(1-\delta_i)$. It is immediate that $\varepsilon_{sp} > 1$. For all $\varepsilon_i \in [0,1]$, $\gamma_i \ge 1 - \varepsilon_i$ and for all $\varepsilon_i \in (1,2]$, $\delta_i \ge 1/\varepsilon_i$.

Given α and β, DLH maintain two kind of lists. An ordered list L of possible candidate solutions R_i, and a list F for temporary constructed solutions, subsequently incorporated in L. The list L is ordered considering a non-decreasing order of $|\varepsilon_{sp} - \varepsilon_i|$. At each step, criteria (4) possibly delete some candidate solutions of L, the rectangle R_s with minimum total waste $[I_W(R_s) + (WH - R_s)]$ is retained, and if $\beta WH > [I_W(R_s) + (WH - R_s)]$, then β can be updated as $\beta = [I_W(R_s) + (WH - R_s)]/WH$, and R_s is a possible optimal solution.

The best-first part of DLH is the selection of the index p corresponding to the minimum $\{|\varepsilon_{sp} - \varepsilon_i|\}$, and combining R_p with the existing rectangles in L for construction the new rectangles of F. The original proposition was to use the index of minimum $f(\alpha, R_i)$, but as ε_i is independent of α, its use is more efficient for implementation, because α changes at each iteration, and the ε_i's change only with β changes. The first F is formed combining the smallest primary rectangle with the other ones. Note that $\varepsilon_{sp} > 1$, and primary rectangles can be pruned only for $\alpha \geq \beta$. Using $|\varepsilon_{sp} - \varepsilon_i|$ primary rectangles are R_p even for $\alpha < \beta$, a characteristic that accelerates the method. When $\alpha = 2\beta$, all rectangles with $\varepsilon_i < 2$ are pruned or equivalently all rectangles with area < 50% of the total area are pruned, and the rectangles remaining can not be candidate for new combinations. Heuristic DLH stops for $\alpha \geq 2\beta$ or for an empty list L, with $\{R_s\}$ and a possibly updated β. To permit a more generalized selection of index p and avoid excessive repetition, a tabu list of the last five candidates is maintained. If an index is choosed to be the index p at current iteration, only after five iterations it can be reconsidered to be the index p again.

Parameter α is in the kernel of DLH. When it increases, some patterns are deleted of L, and good solutions can be lost in the process. Then, parameter α must be carefully updated. Parameter ε_i also gives some insight in controlling α. For primary rectangles, $I_w(r_i) = 0$, $i = 1,\ldots,m$, and consequently $\varepsilon_i = 1/(1 - \delta_i)$. Using criteria (4), primary rectangles up to 50% of the total area, are pruned for $\beta < \alpha < 2\beta$. Then it is interesting a smoothly increasing of α between β and 2β, providing some more time for combining primary rectangles.

For primary rectangles we have used the smallest ε_i (ε_{sp}) for selection of the candidate rectangle for combinations. Parameter ε_{sp} and the medium ε_i ($\varepsilon_{mp} = \Sigma\, \varepsilon_i /m$) and the greater ε_i ($\varepsilon_{gp} = \max\{\varepsilon_i\}$) can be of value to determine the step size for α. As a suggestion, for a given β, α can be a function of β, setting

$$\alpha := \alpha + (0.01)\beta, \text{ for } \alpha < [\varepsilon_{sp} - 0.35(\varepsilon_{mp} - \varepsilon_{sp})]\beta,$$

$$\alpha := \alpha + [(\varepsilon_{mp} - \varepsilon_{sp})/300]\beta, \text{ for } [\varepsilon_{sp} - 0.35(\varepsilon_{mp} - \varepsilon_{sp})]\beta < \alpha \leq \varepsilon_{mp}\beta,$$

$$\alpha := \alpha + [(\varepsilon_{gp} - \varepsilon_{mp})/300]\beta, \text{ for } \varepsilon_{mp}\beta < \alpha \leq [\varepsilon_{gp} + 0.35(\varepsilon_{gp} - \varepsilon_{mp})]\beta,$$

$$\alpha := \alpha + (0.01)\beta, \text{ for } \alpha > [\varepsilon_{gp} + 0.35(\varepsilon_{gp} - \varepsilon_{mp})]\beta.$$

For an initial $\alpha = 0$, DLH perform $[\varepsilon_{sp} - 0.35(\varepsilon_{mp} - \varepsilon_{sp})]/(0.01)$ iterations with $\alpha < [\varepsilon_{sp} - 0.35(\varepsilon_{mp} - \varepsilon_{sp})]\beta$, and no primary rectangles are pruned. For the next iterations, until $\varepsilon_{mp}\beta$, the tolerance factor $0.35(\varepsilon_{mp} - \varepsilon_{sp})$ avoid immediate pruning of the smallest primary rectangle, and the step size $(\varepsilon_{mp} - \varepsilon_{sp})/300$ is considerably decreased, because $(\varepsilon_{mp} - \varepsilon_{sp}) < 1$ for a "normal" ($\varepsilon_{mp} < 2$) set of problems. As the zone $\varepsilon_{sp} \pm \Delta$ is of desirable rectangles, the smoothly variation is beginning. After $\varepsilon_{mp}\beta$, the new step size $(\varepsilon_{gp} - \varepsilon_{mp})/300$ determine the variation. In general, for "normal" set of primary rectangles we also have $(\varepsilon_{gp} - \varepsilon_{mp}) < 1$ and the smoothly variation continues. The final iterations are conducted with the first step size (0.01).

We examine in the following a formal description of DLH:

Heuristic DLH

```
-----------------------------------------------
Begin
        Choose a value for β, 0 ≤ β ≤ 1 and
for α, 0 ≤ α ≤ β;
        Given W, H;  r1,r2,...,rm;
                b1,b2,...,bm;
                total_waste := WH;
                mult1 := 0.01;
        For i = 1 to m do
                Ri := ri ;
                δi := Ri /WH ;
                εi := 1/(1-δi ) ;
                waste := WR-Ri ;
                if waste < total_waste then
                        total_waste := waste;
                        s := i;
                end_if
        end_for
        Compute εsp := min {εi};
                εmp := Σ εi /m;
                εgp := max {εi};
                mult2 := (εmp - εsp)/300;
                mult3 := (εgp - εmp)/300;
            factor1 := εsp - 0.35(εmp - εsp);
          factor2 := εgp + 0.35(εgp - εmp);
        end_compute
        Define initial list L(0) = {R1 ,R2
                                ,...,Rm} ;
                k := 1; l := 0; n :=m;
While L(k-1) is non empty and α < 2β do
        t := 0;  n := n + l;  l := 0;
        for i = 1 to n do
                if α/β > εi  then
                  L(k-1) := L(k-1) - {Ri};
                  t := t + 1;
                end_if
        end_for
        n := n - t;
        Compute
                p := index of tabu list for
                        min {|εsp - εi|} ;
                F(k) which is the set of l rectangles
        Rj, j = 1,...,l, satisfying
                (i)  each Rj is formed by an
        horizontal or vertical build of rectangle Rp
        and a rectangle from L(k-1);
                (ii)  α/β ≤ εj = (1 - γj)/(1 - δj), where
        γj := Iw(Rj)/βWH and δj :=Rj/WH,    for
        Rj ≠ WH and δj = 0.99 for Rj = WH,
                                        j = 1,...,l;
                (iii) those rectangles ri, i = 1,...,m,
        appearing in Rj do not violate the bound
        constraints b1, b2,...,bm;
        end_compute
        for i = 1 to l do
                waste := (WH - Ri) + Iw(Ri);
                if waste < total_waste then
                  total_waste := waste;
                  s := i ;
                end_if
        end_for
        If βεs >1 then
                β := [Iw(Rs) + (WH -
                                Rs)]/WH;
                for i = 1 to n do
                        γi := Iw(Ri)/[Iw(Rs) +
                                (WH - Rs)];
                        εi := (1 - γi)/(1 - δi);
                end_for
        end_if
        L(k) = L(k-1) ∪ F(k) and remove
any equivalent rectangle from L(k);
        k := k + 1;
        If α < (factor1)β then α := α +
                                (mult1)β
        else    if α < εmpβ then α := α +
                                (mult2)β
                else    if α < (factor2)β
                then α := α + (mult3)β
                else α := α + (mult1)β
                end_if
        end_while
        Write {Rs}, total_waste and β;
end
-----------------------------------------------
```

3 COMPUTATIONAL TESTS

The Algorithm DHL was tested on some problems of the literature and a number of randomly generated test problems. The problems of the literature are from Wang (1983), three from Christofides and Whitlock (1977) (with no values assigned to rectangles), and two from Oliveira and Ferreira (1990). The generated problems can be seem from Table 1.

Table 1: Random generated problems

Problem Set	Area		Aspect Ratio		Bound Limits		m	No. of Problems
	mean	σ	mean	σ	mean	σ		
P10	9%	20%	1	80%	2	60%	10	3
P20	9%	20%	1	80%	2	60%	20	3
P50	9%	20%	1	80%	2	60%	50	3
P100	9%	20%	1	80%	2	60%	100	3
P200	9%	20%	1	80%	2	60%	200	3
P200.2	2%	20%	1	80%	2	60%	200	1

The method used to generate the random problem is given by the following expressions:

$$\text{Area}(r_i) = N(A, sdtA)WH$$

$$\text{Width}(r_i) = \sqrt{N(Asp, stdAsp) \,.\, \text{Area}(r_i)}$$

$$\text{Height}(r_i) = \text{Area}(r_i) \,/\, \text{Width}(r_i)$$

$$b_i = N(BL, stdBL)$$

where

$\text{Area}(r_i)$, $\text{Width}(r_i)$ and $\text{Height}(r_i)$: area width and height of primary rectangle i, respectively.

$N(m,\sigma)$: normal distribution with mean m and standard deviation σ.

A, stdA: average area and the standard deviation of the rectangles, respectively.

Asp, stdAsp: average aspect ratio (width / height) of the rectangles and it's standard deviation, respectively.

BL, stdBL: average bound limits and it's standard deviation, respectively.

Some cares were taken to assure that the rectangles do not exceed the plate width or height. Finally, the plate size was 70 x 40 for all problems; the number of different rectangles ranged from 10 to 200, and we assume that the rectangles were of fixed orientation, although it is not a restriction to the algorithm.

The results were obtained by coding the algorithm DLH in C (Turbo C 2.0), and executing on an IBM-PC 486 computer working at 33 MHz, under MS-DOS 6.0.

Since DLH forms a new rectangle by a horizontal *or* vertical build, it is somewhat random. Therefore, the program was run five times for each test problem. The results are shown in Table 2. For all test problems, $\alpha = 0$ and $\beta = 0.08$ at the start of DLH.

Some comments about Table 2:

- All wastes are smaller than 10% of plate size, except in one case, the problem P2200. For this problem we reached good results halfing the plate. The cut can be horizontal (P2200(1)) or vertical (P2200(2)), and the best wastes are reduced to 7% and 10%, repectively.
- The number of partial solutions generated by DLH is smaller than one generated by the algorithms of Wang (1983) and Oliveira and Ferreira (1990), for problems of similar size.

Table 2: Computational results.

Problem	Waste		Average Time (sec)	Maximum Size of L	No. of partial solutions
	Min.	Max.			
Wang	100	100	16.21	228	911.8
Chris1	9	10	12.50	147	372.8
Chris2	163	278	47.57	574	1469.8
Chris3	100	127	15.49	228	723.8
Olive1	87	140	17.14	218	808.2
Olive2	110	278	12.75	165	540.6
P10-1	80	106	21.15	409	1255.4
P10-2	60	144	21.34	405	812.8
P10-3	124	136	11.62	265	444.8
P20-1	112	149	63.01	1571	2066.8
P20-2	163	184	138.44	1723	4837.0
P20-3	154	169	123.56	1629	4712.4
P50-1	126	172	201.30	1611	6950.2
P50-2	138	180	201.81	1640	7213.0
P50-3	166	198	219.53	1651	8369.4
P100-1	159	200	305.11	1654	11197.6
P100-2	126	184	349.33	1888	13209.8
P100-3	133	160	337.47	1783	12398.8
P200-1	117	165	375.15	2003	12327.0
P200-2	117	171	334.07	1669	10570.0
P200-3	141	189	393.83	1882	13574.4
P2200	1063	1349	629.08	2479	21216.4
P2200(1)	99	122	575.13	2075	22625.2
P2200(2)	136	207	482.26	2014	22851.6

(1) Plate of 70x20
(2) Plate of 35x40

The tests in table 2 was obtained with a fixed value of β. Varying this value better results can be obtained. For example, the best results for the problems of the literature are presented in table 3, with the refered β values. All but one optimal solutions are obtained.

Table 3: Computational results varying β.

Problems	Optimal solution	β	Waste	Times (sec.)	Max. size of L	Number of partial solutions
Wang	79	0.25	79	128	1314	3400
Chris1	6	0.07	6	65	129	480
Chris2	22	0.05	60	76	346	630
Chris3	79	0.25	79	128	1314	3400
Olive1	63	0.06	63	60	83	235
Olive2	110	0.06	110	64	349	121

4 CONCLUSION

We proposed in this paper a new heuristic for the 2D-cutting problem. Heuristic DLH performs a best-first search in a tree of candidate solutions, builded by horizontal or vertical combinations of rectangles. The main characteristic is the use of a dynamic list of candidate patterns, pruned by dynamic pruning criteria, getting the essence of other Wang's based heuristics, while eliminate combinatorial explosion.

The good results obtained in section three can be improved if we admit rotations for rectangles. For the case of small rectangles (2% of plate), we can improve the criteria for the choice of the first cut. The case of selection of plates can be directly extended solving a knapsack area based, considering a tolerance for waste, at least for the case of greater demand.

REFERENCES

Christofides, N., and Whitlock, C. (1977), An algorithm for two-dimensional cutting problems, *Operations Research* **25**, 31-44.

Dowsland, K.A., and Dowsland, W.B. (1992), Packing problems, *European Journal of Operational Research* **56/2**, 2-14.

Israni, S.S., and Sanders, J.L. (1985), Two-dimensional cutting stock problem research: a review and a new rectangular layout algorithm, *J. Manufacturing Sys.* **1**, 169-182.

Oliveira, J.F., and Ferreira, J.S. (1990), An improved version of Wang's algorithm for two-dimensional cutting problems, *European Journal of Operational Research* **44**, 256-266.

Pearl, J. (1984), *Heuristics: intelligent search strategies for computer problem solving*, Addison-Wesley Publishing Co., London.

Sweeney, P.E., and Paternoster, E.R. (1992), Cutting and packing problems: a categorized, application-orientated research bibliography, *Journal of the Operational Research Society* **43/7**, 691-706.

Wang, P.Y. (1983), Two algorithms for constrained two-dimensional cutting stock problems, *Operations Research* **31**, 573-586.

About solving linear integer programs through Hermite Normal Form decomposition

J.MAUBLANC, A.QUILLIOT
Université BLAISE PASCAL,
ISIMA, BP 125, Campus des cézeaux, 63173 CLERMONT-FD, FRANCE, Tel 73 40 50 04, Fax 73 40 50 01

Abstract
This paper discusses the ways Hermite Normal Form decompositions may be used in order to solve integer linear programs.

Keywords
Hermite Normal Form, Integer Linear Programming

1 INTRODUCTION

Integer linear programming (ILP) is a central combinatorial optimization problem. More than the boolean satisfiability problem 3-SAT, which is used as the reference problem in complexity theory (Garey and Johnsson, 1979), it provides a simple way of expressing problems belonging to the class NP-Time. In many cases, this translation process can be automatized. Thus, ILP plays the role of some kind of universal problem inside NP-Time, in a more practical way than the other NP-Complete problems.

The ILP (or MLP: Mixed linear programming) formalism is mostly applied to (Schrijver, 1986) combinatorial problems involving decision variables, planning and scheduling problems, and diagnostic problems for fault localization. Its universal status provides a justification to the very huge amount of work which has been developped about it, as well as the large variety of software built in order to handle it (CPLEX, XMP, OSL, ...). Recently, research concerning constraint programming languages (CHIP, CLP, PECOS, PROLOG III) have focused on integrating efficient ILP libraries inside logic programming formalisms.

The most classical methods for dealing with ILP are derivations of the Simplex algorithm. Some stem from specificities of the underlying problem (total unimodularity, facet generation). Others are designed for general purposes and involve cutting plane generation mechanisms (Gomory, 1963; Glover, 1966; Chvatal, 1985), branch and bound combined with some kind of relaxation, or Benders decomposition (Dakin, 1965; Schrijver, 1986; Srinivasan, 1965).

Something somewhat disappointing is that few among these methods can exploit the geometrical properties of the domains they consider or the arithmetical properties of the integral numbers which describe the problem. While Lenstra result about the polynomiality of ILP in fixed dimension is based upon considerations about the thickness of a polyedron, it doesn't yield any efficient algorithm. In the same way, Hermite and Smith decomposition processes applied to the constraint matrix allow to solve linear diophantine equation systems in a polynomial time, but are scarcely used to solve more general ILP problems.

For this reason, we investigate here what kind of geometrical or arithmetical arguments may be introduced in order to help in dealing with ILP.

First, we notice that if a given square matrix defines a fractional vertex of some polyedron, then setting this matrix in Hermite Normal Form means performing some change of variables which preserves the integrality constraint, provides some local regularization of the polyedron in the neighbourhood of this vertex, and makes appear an integral "candidate" vector. It induces the possible implementation of cutting plane methods and of branch and bound methods.

Next we combine these methods with a stochastic control allowing the current vertex to jump throughout the polyedron, in search of the best areas for cutting it.

We end by discussing the impact of the thickness of the constraint polyedron on the behavior of descent algorithms for ILP, and tried to provide methods for increasing this thickness through well-choosen variable changes.

2 HERMITE NORMAL FORM AND DIOPHANTINE SYSTEMS

Recall: Normal Hermite Form.
A n*m ($n \geq m$) integral matrix B is said to be in Hermite Normal Form (HNF), if it is lower triangular on its m first columns, null on its other columns, if all the $B_{i,i}$, i = 1..m, are positive and if on any row i = 1..m, its non diagonal elements are negative or null and with absolute value strictly less than the corresponding $B_{i,i}$.

It is known (Schrijver, 1986) that for any integral matrix n*m matrix C, with range $m \leq n$, there exists an integral n*n matrix U, with determinant equal to 1 or -1, such that B = CU be in HNF. Then (Damich, Kannan and Trotter, 1987), B is unique and computing U and B can be done in polynomial time through modulo determinant arithmetic. Practically, one needs to be sure that the largest common divisor of all the n*n subdeterminants of A is not too large in order to avoid handling very large integers while performing this process.

It follows that a linear diophantine equalities system Cx = b, x in Z^n (b integral with length m, C as above) can be solved in polynomial time (Schrijver, 1986), through the change of variable x = UX, which preserves the integrality constraint.

Let us consider now the following satisfiability problem:

Problem (P): { Find x in Z^n, such that $Ax \leq b$, where A is an integral n*m matrix with rank n and b is an integral vector with length $m \geq n$};

and let us denote by P* the polyedron defined by: P* = { x in Q^n such that $Ax \leq b$}.

Let us suppose that I included into {1..m} defines a vertex of P*: that means that I is such that the submatrix A^I induced by the rows of A which are indexed on I is inversible and satisfies

$$A.(A^I)^{-1}.b^I \leq b.$$

We may apply to P the following change of variable:

x = UX, U being an integral matrix with determinant 1 or -1 such that A^I. U is in HNF.

This change of variable has several consequences:

- It induces a local regularization of the polyedron P* at the neighbourhood of the vertex defined by I, that means it increases the conical angulus defined by I, making more efficient descent methods initialized on this vertex.
- It yields cutting planes, called Hermite Cutting Planes associated with I, and some peculiar integral vector $\beta(I)$, which we call "Candidate vector associated with I". We only have to set B = AU and to label the row indices of B in such a way that I = 1..n, and to write:

$\beta_1 = [b_1 / B_{1,1}]$ = Integral lower bound of b1 / B1,1;

.....

$\beta_n = [(b_n - B_{n,1}.\beta_1 - B_{n,n-1}.\beta_{n-1})/ B_{n,n}]$;

and deduce the following constraints on : $X_i \leq \beta_i$, i dans 1..n, as well as an integral vector $\beta(I) = (\beta_1 .. \beta_n)$.

These constraints are called "Hermite cuts associated with I" and the vector $\beta(I)$is called "Candidate vector associated with I". One check that these Hermite cuts are not Gomory cuts.

3. AN ALGORITHM BASED ON HERMITE CUTS

From the above remarks, we deduce the following algorithm:

HERMI1
Input: the system $Ax \leq b$, x in Z^n;
Output: a solution of this system if there exists some;

{Not Stop; While Not Stop do
{Determine I included into {1..m}, which defines a vertex of P* and which is minimal for the lexicographic order; (*)
If I doesn't exist then Stop (Failure) else
{Perform from I the Hermite variable change, compute the Hermite cuts, insert them inside the definition of P, at the end of the array A;
Let α be the first element in I; For $i < \alpha$, replace b_i by $b_i - 1$; (**)
Compute the Candidate vector $\beta(I)$ associated with I; If $\beta(I)$ is solution of (P) then Stop (Success)}}}

Remark 1
Making reference to the lexicographic order in the instructions (*) and (**) strengthens the system with what we call Lexicographical Cuts. In fact, for $i < \alpha$ above, it will not be possible to find any integral vector x (writen in relation with the current basis), such that $A^i. x \geq b_i$.

Remark 2
Using the candidate vector $\beta(I)$ speeds the convergence of the method, since it is not anymore necessary to wait (as it was the case for the Gomory cut mechanism) for the vertex associated with I to be integral. It avoids the utilisation of fractionnal number during the execution of the algorithm.

Remark 3
In order to make HERMI1 efficient, we need to keep the system under the form of a family of diophantine linear inequalities. For this, we work through the dual version of the Simplex algorithm on the dual version of the program which defines P*. In order to avoid the appearance of very large numbers during the execution of HERMI1, we compute the HNF of a given square matrix through an algorithm (Damich, Kannan and Trotter, 1987) which uses modulo determinant arithmetic.

Theorem 1
If P* is bounded, then HERMI1 solves (P) in a finite number of iterations.

Proof
We proceed here the same way as in the proof of the convergence of the Gomory cuts algorithm. Let us suppose that some execution of HERMI1 doesn't end and proceed by induction on n. At some time during the process, the value of the index α used in the

description of HERMI1 becomes constant and equal to 1. If it were false, the first coordinate of b would indefinitely decrease (because of the instruction (*)) and the polyedron P* wouldn't be bounded. Thus, from some time during the process, the first constraint in (P) remains writen in an unmodified way: $A_{\alpha,1}.X_1 \leq b_\alpha$, b_α being divisible by $A_{\alpha,1}$.
From this time, a short reasoning makes appear that the process behaves as if it were working only on the variables $X_2,..,X_n$, with X_1 remaining constrained to be equal to $b_\alpha/A_{\alpha,1}$. Then it becomes possible to conclude by induction. END.

Introduction of a stochastic control
HERMI1 may converge relatively slowly, as it is the case for the general Gomory algorithm. In such a situation, the vertex associated with I is almost not affected by the introduction of the new cuts, and the candidate vector $\beta(I)$ remains far from P*. Its is possible to counter this drawback by using a linear control, whose parameters are modified in a stochastic way, and which makes the current vertex associated with I jump throughout the polyedron P*. This very simple trick prove itself to improve the algorithm in a surprinsing way.

HERMI1 can be then rewriten as an algorithm HERMI2 by replacing the (*) instruction of HERMI1 with the following instruction (*bis) and by suppressing the instruction (**):

Instruction (*bis):
Randomly generate a vector c in Z^n ;
Determine I included into {1..m} and defining a vertex of P* which maximizes c.x;

4. A TRY AND TEST ALGORITHM BASED ON HERMITE CUTS

An other way to take profit of the Hermite variable change techniques consists in performing a try/test tree search pruned by the insertion of the Hermite cuts mechanism and by a success test on the candidate vector. in order to describe it, let us consider the following problem (P):

{ Find x in Z^n such that $Ax \leq b$, A being a n*m integral matrix with rank n , b being an integral vector with length m};

The change of variable x = UX associated with the FNH decomposition of the submatrix A^I defined by some vertex of the polyedron P* makes appear a first Hermite cut $X_1 \leq \beta_1$ (cf previous notations). Then this cut yields in a natural way a binary branching process, defined by the two following options:

$X_1 = \beta_1$ and $X_1 \leq \beta_1 - 1$;

The first one will be tried first, as the most promizing and since it decreases the dimension of the problem.

This branching process will be summarized as follows:

BRANCH-HERMI
(1). Randomly generate c in Z^n; (*)
Solve the program: { $Ay \leq b$, y in Q^n, zmin = c.y };
If this program admits no solution then Failure and Stop
else continue after denoting by I the subset{1..m} which defines a vertex of the polyedron {$Ax \leq b$, y in Q^n } which is a solution of this program;

(2). Perform the change of variable x = UX associated with the FNH decomosition of A^I;
On the basis of this reformulation of the system, generate the Hermite cuts $X_i \leq \beta_i$, i =

1..n, and the candidate vertex $\beta(I)$ associated with I and defined by : $\beta(I) = \beta_1 .. \beta_n$;
If $A.\beta \leq b$ then Success and Stop else continue;

(3). Recursively solve the system (P1), obtained from (P) after reformulation through the change of variable x = UX and insertion of the Hermite cuts $X_i \leq \beta_i$, i = 2..n, and of the constraint $X_1 = \beta_1$;
If some solution appears, then Success, else recursively solve the system (P2) obtained from (P) after reformulation through the change of variable x = UX and insertion of the cuts $X_i \leq \beta_i$, i = 2.. n and of the constraint $X_1 \leq \beta_1 - 1$.

Explanations
The above process is doubly filtered: on one side, the test on the integral relaxation of the program (P) allows an anticipation on failure and on an other side, the test on the candidate vertex helps in detecting the success situations. It also helps in handling rounding: in fact, the main part of the above algorithm may be writen while using only integers.

5. NUMERICAL TESTS AND EXTENSIONS

We tested the above methods by randomly generating linear systems $Ax \leq b$ which admits at least one solution and by adding them some constraint $cx \leq k + 1$, k being the integral upper bound of the minimal fractional value of cx, taken for all the vectors x such that $Ax \leq b$.Thus, the systems generated this way were "semantic free", which means without any underlying structure.

For every example generated this way, we tested both HERMI1 and HERMI2, while focalizing our attention on the number executions of the main loop and comparing with the behaviour of the classical Gomory algorithm (Gomory, 1963).

We also tested BRANCH-HERMI, as well as its deterministic version DET-BRANCH-HERMI (obtained after replacing the first instruction (*) by an instruction "c := 0") while counting the number of nodes of the search tree which are visited during the execution of the process, and comparing with the behaviour of a simple branch and bound scheme RELAX-SIMPLE, combined a simple integral relaxation (Dakin, 1965; Srinivasan, 1965).

These experiments, realized for examples which involve from 10 to 40 variables and between 20 and 100 constraints, make appear:

- an iteration number almost 10 times less for HERMI2 than for HERMI1, and almost 3 times less for HERMI1 than for GOMORY.
- a node number almost 3 times less for BRANCH-HERMI than for RELAX-SIMPLE.

Still, one must take into account, while evaluating these results, the additionnal costs due to the FNH decomposition processes involved in the execution of HERMI1, HERMI2 and BRANCH-HERMI and the global tendancy of the method to generate very large numbers.

Comparizon between GOMORY, HERMI1 and HERMI2

	GOMORY	HERMI1	HERMI2
10	27	11	3.4
20	52	26	8.2
30	97	38	16
40	201	80	32

Comparizon between SIMPLE-RELAX, BRANCH-HERMI, DET-BRANCH-HERMI

	SIMPLE-RELAX	BRANCH-HERMI	DET-BRANCH-HERMI
10	13	3.5	5.2
20	32	8	13
30	58	17	25
40	105	33	45

6. MORE ON GEOMETRY: THICKNESS OF A POLYEDRON AND RELATED VARIABLE CHANGES

We just saw the way a Hermite variable change may help in generating cuts or pruning a tree search in order to solve an ILP. We are going to see now that other kind of improvement may come from some well choosen change of variable.
Its is known (LENSTRA, 1983) that if a n dimensional polyedron P is bounded and doesn't contain any integral vector, then there exists an integral non null vector c such that for any pair x, x' of vectors in P:

$$c.(x-x') \leq n(n+1).2^{1/2.n.(n+1)} + 1 \quad .$$

This result yields the famous Lenstra Theorem about the complexity of ILP in fixed dimension.

We define here, what we call the thickness Th(P) of P:

$$Th(P) = \inf_{c \text{ dans } R^n / \|c\| = 1} \ \sup_{x,x' \text{ dans } P} c.(x-x').$$

Intuitively, we feel that this quantity Th(P) must be related with the probability of existence of an integral vector in P. As a matter of fact, we may assert:

Theorem 2
If $Th(P) \geq 2n$, then P contains at least one integral vector.

Proof
Let us first prove that if $Th(P) \geq 2n$ and if $u_1..u_n$ are vectors in R^n which define an euclidian basis of R^n, then for any i in 1..n, there exists x_i in P such that $x_i + 2n.u_i$ is also in P. We clearly may suppose that P is bounded.
For any u such that $\| u \| = 1$, we may set:

$$k_u = \sup_{x \text{ in } C} \lambda \text{ such that } x + \lambda.u \text{ is in } C;$$

For such an u, let us separate the product P*P and $Q_u = \{ x,y \text{ in } R^{2n} \ / \ y - x = k_u.u \}$ with a hyperplan $a.x + b.y = c$ such that for any x,y in Q_u : $a.x + b.y \geq c$. (E1)
From (E1) it comes that $b = - a$ and $c = k_u b.u$. By choosing x_o in P such that :

$$a.x_o = \gamma_o = \text{Sup } a.x \ , \ x \text{ in } P,$$

we see that for any y in P: $a.y \leq \gamma_o$ and $a.y \geq \gamma_o - k_u.b.u$; (E2)
If we suppose $\| a \| = 1$, then : $k_u.b.u \leq k_u.\|a\| . \| u \| \leq k_u$;

which means, because of (E2), that $Th(P) \leq k_u$.
Then we only have to compute k_u for any $u := u_i$, $i = 1..n$, in order to get our initial assertion.
Let us suppose now that $x_1..x_n$ exist in P such that for any i in 1..n, $x'_i = x_i + 2n.u_i$ is also in P, $u_1..u_n$ defining an euclidian basis of R^n. We may set:

- for any i in 1..n, $x_i = n.a_i + p_i$ with a_i in Z^n and $0 \leq p_i < n.1$;
and also :
- x = gravity center of the x_i, x'_i ,i in1..n $= A + C/n + U$,
with $A = \sum_{i \text{ in } 1..n} a_i$, $C = \sum_{i \text{ in } 1..n} p_i$, $U = \sum_{i \text{ in } 1..n} u_i$;
- $x^* = A + U$.

If x^* is in P then it is over, else there exists λ in R^n,μ in R, such that: (Séparation)

$\sum_{i \text{ in } 1..n} |\lambda_i| = 1$; $\lambda.x^* < \mu$; for any i in 1..n, $\lambda.x_i$ and $\lambda.x'_i \geq \mu$.

Let us then set: $I = \{ i \text{ in } 1..n / \lambda_i \geq 0\}$,$J = \{1..n\} - I$, and also, for any i in 1..n:
$y_i = x_i$ if i in I and $y_i = x'_i$ else ;
$y'_i = x'_i$ if i in I and $y'_i = x_i$ else .
In any case, we may write: $y_i = n.b_i + p_i$ with b_i in Z^n ;
$y'_i = y_i + 2n.v_i$ (with $v_i = u_i$ if i in I and $v_i = -u_i$ else);
Then x^* may be rewriten $B + V$ (with $B = \sum_{i \text{ in } 1..n} b_i$ and $V = \sum_{i \text{ in } 1..n} v_i$)

while λ and μ satisfy:
for any i in1..n, $\lambda.y_i \geq \mu$ which implies: $\lambda.B + \lambda.C/n \geq \mu$ and $\lambda.x^* = \lambda.B + \lambda.V < \mu$
and yields : $1 = \lambda.V < \lambda.C/n$ which means a contradiction. END.

Practical interpretation
We want now try to find some practical interpretation of the above considerations and thus ask the following question:
In which way does a variable change which increases the thickness of a polyedron while conserving the integrality constraint makes easier solving the related system of inequations ?

Remark
The variable change associated with Hermite Normal decomposition doesn't always have this property.

In order to answer this question, we proceed by first writing the following descent heuristic, which work on any system $Ax \leq b$, which admits some integral solution and defines a bounded polyedron.

Algorithm TEST-THICK
Input . A ,b defining some polyedron P of R^n ;
Output : x such that $Ax \leq b$,x integral, or Failure ;
Initialize x in Z^n ; Counter := 0;
While x not in P and Counter $\leq$ 100 do
Randomly generate p in N^m ; Possible ;
While not Possible do
Try to find x', which differs from x by exactly one coordinate and such

that $V(p,x') < V(p,x)$ where $V(p,x) = V(p,x) = \sum_{i=1..m} p_i \cdot Sup(A_i(x) - b_i, 0)$;

If x' doesn't exist then Not Possible else x := x' ;

Counter := Counter + 1;

Then we compare, for a wide sample of such systems $Ax \leq b$, the behaviour of this heuristic before and after some variable change $x = UX$, which induces the multiplication of the thickness of P by some coefficient $\alpha > 1$. Then we get an evolution of the ratio:

$$R = \frac{\text{Number of failures for TEST-THICK working on the initial system}}{\text{Number ...on the system rewriten through the variable change } x = UX}$$

as follows:

	α in [1, 1.5]	α in [1.5, 2]	α in [2,3]	α in [3,4]
R	1.4	2.2	3.1	3.9

Remark

In order to systematically take profit from this kind of analysis, we shall need to be able to characterize (and compute) the variable change, which, for a given system $Ax \leq b$, will make increase the thickness of the related polyedron. At this time, we can't answer this question.

7. CONCLUSION

We saw that a conveniently designed variable change may modify the arithmetical or geometrical characteristics of an ILP, and thus make easier its resolution. Still, several questions remain open. While a Hermite change of variable may speed the execution of some kind of cutting process, it may induce trouble with very large subdeterminants. On an other side, using an other kind of variable change in order to "smooth" the polyedron (to increase its thickness) may be usefull before applying some very simple descent processes, but it raises the question of finding a way to automatically determine such a well designed variable change.

8. REFERENCES

CHVATAL.V. (1985) Cutting planes and combinatorics. *European Journal of combinatorics, 6,* 217-26.

DAKIN.R.J. (1965) A tree search algorithm for mixed integer programming problems. *The Computer Journal* , **8,** 250-255.

DAMICH.P.D, KANNAN.R and TROTTER.L. (1987) Hermite normal form computation using modulo determinant arithmetic. *Math operat Research* , **12.1,** 50-59.

GAREY.M and JOHNSSON.D. (1979) *Computer and intractability.* W.Freeman and Co, N.Y.

GLOVER.F. (1966-7) "Generalized cuts in diophantine programming. *Management Sciences, 13, 254-68.*

GOMORY.R.E. (1963) An algorithm for integer solutions to linear programs, in *Recent Advances in Math Programming*, (R.L.Graves and P.Wolfe eds), Mac Graw Hill, N.Y, 269-302.

LENSTRA.H. (1983) Integer Programming with a fixed number of variables. *Maths of Operat Research* , **8,** 538-48.

SCHRIJVER.A. (1986) *Theory of linear and integer programming.* Wiley, Chichester.

SRINIVASAN.A.V. (1965) An investigation of some computational aspects of integer programming. *JACM* , **12,** 525-35.

60

Software system for solving multi-scale optimization problems

E. Semenkin, K. Abramovich
Siberian Aerospace Academy
P.O.Box 486, Krasnoyarsk, 660014, Russia.
Tel: +7-391-2-33-34-20. Fax: +7-391-2-33-47-09.
e-mail: semenkin@stu.krasnoyarsk.su

Abstract

Optimization problems with different type of variables are described. Methods of decomposition and unification are suggested for solving these problems. Software system, that realizes mentioned methods, is described. Results of testing and ways of further development are discussed.

Keywords

Multi-scale optimization, decomposition, unification, software system

1 EXAMPLES OF PROBLEM

Optimization problems with the variables of different type (discrete, boolean, continuous, combinatorial, etc.) are called multi-scale or mixed problems. They arise in various design and planning problems when mathematical formulations and modelling are possible. Consider two examples of problems of that type.

1) CAD of some large scale system consisting of several interconnected units (subsystems).

The objective $\mathbf{F}(\mathbf{B}, \mathbf{D}, \mathbf{X})$ is some quality index, representing the ability of the system to perform its functions. $\mathbf{B} \in \mathbf{B}_n$ is the vector of Boolean variables representing the system structure. If $b_i = 1$ then i-th subsystem is included in the system and vice verse. $\mathbf{D} \in \mathbf{Z}_m$ is the vector of integer variables, representing the number of backup copies of subsystems. $\mathbf{X} \in \mathbf{R}_k$ is the vector of continuous variables representing the physical characteristics of the subsystems (mass, power consumption, etc.).

2) Planning of optimal sequence of an industrial projects' implementation.

Let some enterprise have several industrial investment projects. The problems are to choose the best ones, to define the sequence of their realization and to estimate the optimal values of project parameters. In this case, the objective $F(\mathbf{Y}) = \{f_1(\mathbf{Y}), f_2(\mathbf{Y}), \ldots, f_k(\mathbf{Y})\}$ is a multicriterial quality index, where, for example, $f_1(\mathbf{Y})$ - the value of necessary credit resources, $f_2(\mathbf{Y})$ - pay-back period, $f_k(\mathbf{Y})$ - maximal cash outflow, etc. $\mathbf{Y} = \{\mathbf{B}, \mathbf{C}, \mathbf{X}\}$ is the multi-scale vector of controllable variables. **B** is the vector of Boolean variables representing the totality of projects to be realized, **C** is the combinatorial variable defining the sequence of the projects, **X** is the vector of continuous variables describing some features of projects (production volume, cost structure, etc.).

Note that the quality index of such problems is usually multicriterial one and some of the criteria are given in implicit form, i.e., they have not evident analytical form but can be calculated algorithmically.

2 METHODS OF DECOMPOSITION AND UNIFICATION

Let's consider two basic approaches for solving the mixed optimization problems (Semenkin, 1992).

First approach, called decomposition method, has the following essence: the original multi-scale problem is replaced by several problems with variables of unique type. For this purpose the priority levels for each variable type are introduced, for instance: first (highest) level - Boolean variables; second level - discrete variables, etc. The multilevel optimization problem is then solved, a process, during which the solution of a lower-level problem corresponds to the evaluation of the objective function value in one current point for higher-level problem. Optimization on each level is implemented by the algorithm of according type, independently on the other levels.

The advantages of this method are the following:

- it is possible to choose for each level the most suitable algorithm regarding the problem features;
- there is the possibility to introduce the levels' priority according to that of the real problem.

The disadvantage is the great laboriousness of the method, i.e., the necessity of the large number of objective function evaluations. The effectiveness of the method depends entirely on the effectiveness of involved algorithms.

Second approach for solving of mixed optimization problems is called method of variables' unification and has the following essence: variables of different type are replaced by the variables of unique type, namely by continuous (relaxation), integer (discretization) or by boolean (binarization) variables.

The replacing of all variables by continuous ones is the most suitable way from optimization point of view, because the mathematical programming methods are well-known and relatively powerful. However, as it was mentioned above, mixed optimization problems are described often by implicit objective function and, in this case, it is impossible to calculate the objective function value beyond the discrete lattice. That is why the relaxation was not used by the development of the software described in the next section.

Transformation of continuous variables into discrete (integer) ones is performed easily when an error margin of the minimum point location in continuous space is known and when left and right bounds are given. Boolean variables can be treated as an integer ones bounded by 0 and 1. Thus one can use integer variables as the unification base.

In turn, discrete variables can be transformed to boolean ones using binary code of integers. Hence unification to boolean variables is possible too.

The advantage of the unification method is essentially lower number of objective function evaluations in comparison with decomposition method. However, there are some disadvantages that cannot be neglected.

The first of them is that the treatment of boolean variables in form of integers, bounded by 0 and 1, with discrete optimization algorithms leads, in some cases, to the bad work of these algorithms, that are not destined to this situation.

Other disadvantages, related to the binarization method, are the following:

- difference of topological properties in original space and in binarized one (near points become far ones and vice versa), which leads to the loss of useful properties of the objective function (monotonicity, convexity, etc.),
- considerable growth of problem dimension in comparison with the dimension of an initial problem,
- existence of relatively large number of points in binarized space that have no corresponding points in initial space.

3 OPTIMIZATION SUPPORT SYSTEM

Approaches described above (decomposition and unification using integer and boolean variables) were realized in the software system for solving mixed optimization problems (Abramovich, 1995). Aim of this software is to supply the convenient tool and to guide the user by solving problems of such type. This software can be also used for solving usual (not mixed) optimization problems, both continuous, and discrete and boolean ones. The system was developed to be flexible in sense of modifications.

Optimization algorithms included in the system, are of 3 type: continuous, boolean and discrete. The origin of them is different. Some algorithms were taken from other optimization systems (Dolezal and Fidler, 1992). Rest of them was developed in Siberian Aerospace Academy (Semenkin and Semenkina (1994), Semenkina (1995)). All of these algorithms are direct search methods of different kinds. This method was chosen because of its universality and applicability to implicit objective functions. The algorithm library is not reach at present, but there is a possibility to add easily any necessary algorithm.

Besides the algorithm library there are the following elements of the software system:

- C compiler, which is aimed at the creation of an executable program code after the preparation of initial data,
- interface program which interacts with the user of the system and manages the process of work,
- interface files used for introduction of initial data and for interacting of optimization algorithms with the main program,
- objective function prototype that should be prepared by user according to the problem solved.

Let's consider the sequence of steps during the work with software system.

The first step is the choice of the solving method (decomposition, discretization or binarization). If decomposition method was chosen, the user has to introduce priority levels for each type of variables presenting in the problem.

The second step is the choice of an optimization algorithm from the list of available ones. After that, some initial data should be introduced to the interface files.

The third step is the introduction of the objective function. This function can be of any complexity. The programming language is Microsoft C, version 6.0.

The fourth step is the executable program generation and implementation. After that, the user can change either optimization algorithm or objective function or method of problem solving and repeat the described steps, or the work with the system can be finished.

On-line help is available during all the steps of a problem solving.

4 RESULTS OF TESTING AND FURTHER DEVELOPMENTS

After the software was developed some test problems were solved using this system. Objective functions with variables of different type were taken and treated by different methods and algorithms. The results were different within the same method, depending on the algorithms applied, but in all cases a local minimum point was found.

Testing of optimization methods allows to draw some useful conclusions:

- unification method operates essentially faster and needs less number of objective function evaluation in comparison to decomposition,
- direct treatment of boolean variables in form of integers, bounded by 0 and 1, leads, in some cases, to the bad work of discrete algorithms. The most evident reason for that is the impossibility to define search direction in boolean space.
- binarization method leads to the steep increase of the problem dimension and hence to the very slow work of boolean optimization algorithm. To overcome this effect the adaptive approach for the choice of the search accuracy for continuous variables was realized, the essence of which is that this accuracy is not fixed but changes step by step to keep the number of points relatively small.

Some real problems of design of spacecraft's' systems were solved with developed software system. These problems are:

- load planning for subsystems of the control complex of the orbital group of communication satellites (Abramovich, 1995),
- maximization of expected profit when exploitation of communication satellites (Semenkina, 1995),
- optimal choice of the variant of technological contour of the control system of orbital group of spacecrafts (Semenkina, 1995).

When solving mentioned real problems, developed software system has showed good workability and allowed to obtain the results, that were interesting for developers of space systems.

There are several directions of the software system development.

The first (and the most evident) direction is the addition of new optimization algorithms to the library. It is necessary to add searching algorithms with global properties (multistart of local search) and to introduce one more variable type (combinatorial one) after development of according algorithms. It seems reasonable also to include such promising and powerful algorithms of adaptive random search as Genetic and Simulating Annealing ones, which are currently under development. The last direction of the expansion of algorithm library is the development of the algorithms that realize main approaches for multicriterial decision making.

Other direction of the system improving is connected with the realization of newly appeared ideas about using decomposition and unification methods.

The first idea is to apply the decomposition approach not only for variables of different type but for variables of the same type. For example vector X of continuous variables contains some components on which the objective function is differentiable or unconstrained. Then we can place these components on the lower level and use more effective algorithm for minimum point search (for example gradient algorithm or unconstrained one).

The second idea is the combination of both approaches: decomposition and unification. As it was mentioned above, the presence of boolean variables leads to the bad work of discrete algorithm if we use the unification to integer variables. The outcome is to solve two-level problem instead of one-level one, first level containing the boolean variables and second one containing unified discrete and continuous variables.

We plan also to develop the service possibilities of the software system to ensure a support of decision making in choice of suitable way of optimization and appropriate algorithmic tool.

REFERENCES

Abramovich, K. (1995) *Load Planning for Ground-Based Control Complex of Orbital Group of Communication Satellites*. Diploma work. Siberian Aerospace Academy, Krasnoyarsk.

Dolezal, J. and Fidler, J. (1992) *Dialogue System OP^{ti}A for Minimization of Functions of Several Variables*. UTIA, Prague.

Semenkin, E. (1992) Search Discrete Optimization for CAD of Spacecrafts, in *Optimization-Based Computer-Aided Modelling and Design* (eds. A.J.M. Beulens, J. Dolezal and H.-J. Sebastian), IFIP Working Group 7.6, Dagstuhl.

Semenkin, E. and Semenkina, O. (1994) OP^{ti}A Software Extension by Discrete Optimization Algorithms, in *Optimization-Based Computer-Aided Modelling and Design* (eds. J. Dolezal and J. Fidler), IFIP Working Group 7.6, Prague.

Semenkina, O. (1995) *Search Techniques for Synthesis of Spacecrafts' Control Systems*. Ph.D. thesis. Siberian Aerospace Academy, Krasnoyarsk. (In Russian).

Dual barrier-projection and barrier-Newton methods in linear programming

Vitali G. Zhadan
Computing Center of the Russian Academy of Sciences
40 Vavilov Str., 117967 Moscow GSP-1, Russia. Tel: (095)-135-25-39.
Fax: (095)-135-61-59. e-mail: zhadan@ccas.rus

Abstract
In this paper the stable version of dual barrier-projection method and the dual barrier-Newton method for solving linear programming problems are proposed. These methods are obtained by using space transformation techniques which are applied for releasing from nonnegativity of dual slack variables. The continuous and discrete versions of both methods are considered and their local convergence is investigated. A modification of the dual barrier-projection method is proposed in which a steepest descent is utilized.

Keywords
Linear programming, dual problem, space transformation, gradient projection method, Newton's method, steepest descent

1 INTRODUCTION

The paper deals with dual algorithms for solving linear programming problems. These algorithms are similar to primal ones which were proposed by Evtusenko and Zhadan (1978, 1994) and which are based on space transformation techniques. Using a surjective transformation we convert the original linear programming problem with nonnegative variables to a new problem where variables belong to the whole space. We can apply various numerical methods for solving this new problem, in particular, the stable version of the gradient projection method (Tanabe, 1980) and the Newton method. After an inverse transformation to the original space a family of barrier-projection and barrier-Newton methods are obtained. As a result of a space transformation additional diagonal matrices appear in the right-hand sides of formulas describing methods. These matrices prevent trajectories from leaving the positive orthant of the space. Different numerical methods are obtained by different choices of the space transformation.

This paper is devoted to dual barrier-projection and barrier-Newton methods. The space transformation technique is applied there for dual slack variables. In section 2 we outline the basic approach and propose a family of dual barrier-projection methods. These methods are described by systems of ordinary differential equations. Numerical algorithms are obtained as discretization of dynamical systems. In the case of exponential space transformation these methods are similar to the dual affine scaling algorithm proposed by Adler, Karmarkar, Resende and Veiga (1989). We show that if the quadratic space transformation is used then we obtain an exponential rate of convergence for continuous methods and a linear rate of convergence for discrete versions.

In section 3 we consider a variant of the dual barrier-projection method with a steepest descent which converges to the solution of the dual problem in a finite number of iterations. In section 4 we propose the barrier-Newton method.

2 DUAL BARRIER-PROJECTION METHOD

We consider a linear programming problem

$$\begin{array}{lrcl} \text{minimize} & c^T x & & \\ \text{subject to} & Ax & = & b \\ & x & \geq & 0_n, \end{array} \tag{1}$$

and its dual

$$\begin{array}{lrcl} \text{maximize} & b^T u & & \\ \text{subject to} & A^T u & \leq & c, \end{array} \tag{2}$$

where A is a full rank $m \times n$ matrix where $m < n$, x and c are n-dimensional vectors, and b and u are m-dimensional vectors.

We define the set of feasible solutions of dual problem (2) and its interior

$$U = \{u \in R^m : v = c - A^T u \geq 0_n\}, \quad U_0 = \{u \in R^m : v = c - A^T u > 0_n\}.$$

Throughout the paper we assume that U_0 is nonempty and that the primal and dual nondegeneracy holds.

Let $v = \varphi(w)$ be a componentwise space transformation from R^n to the nonnegative orthant R^n_+, i.e. $v^i = \varphi^i(w^i), 1 \leq i \leq n$. Suppose also that this transformation is differentiable and is such that the closure of $\varphi(R^n)$ coincides with R^n_+. In this case every vector in R^n_+ is the image of at least one vector in R^n or it is the image of a limit point of a sequence in R^n.

Let $w^i = \psi^i(v^i)$ denote the inverse transformation of $\varphi^i(w^i)$. This transformation exists at least in a neighborhood of a point $v^i_0 = \varphi^i(w^i_0)$, as long as $\dot{\varphi}^i(w^i_0) \neq 0$. If $\tilde{J}(w)$ is the Jacobian matrix of the transformation $\varphi(w)$, then the Gram matrix $G(v) = \tilde{J}(\psi(v))\tilde{J}^\top(\psi(v))$ is diagonal, i.e.

$$G(v) = D(\theta(v)), \quad \theta(v) = [\theta^1(v^1), \theta^2(v^2), \dots, \theta^n(v^n)],$$

where $\theta^i(v^i) = (\gamma^i(v^i))^2, \gamma^i(v^i) = \dot{\varphi}^i(\psi^i(v^i)), 1 \leq i \leq n$.

We impose the following conditions on the space transformation $\varphi(w)$:

C_1. The functions $\theta^i(v^i)$ are defined and continuous in some neighborhood of R^1_+ and $\theta^i(v^i) = 0$ if and only if $v^i = 0$, where $1 \leq i \leq n$.

C_2. The functions $\theta^i(v^i)$ are continuously differentiable in some neighborhood of R^1_+ and $\dot{\theta}^i(0) > 0$, $1 \leq i \leq n$.

Here we consider only two simple surjective transformations:

$$v = \frac{1}{4}D(w)(w), \quad v = e^{-w}, \tag{3}$$

where the i-th component of the n-vector e^{-w} is e^{-w^i}. For these transformations we obtain, respectively

$$\theta(v) = v, \quad G(v) = D(v); \qquad \theta(v) = D(v)v, \quad G(v) = D^2(v).$$

In both cases the Gram matrix is singular on the boundary of R^n_+. These transformations satisfy C_1. Condition C_2 holds only for the first quadratic transformation (3).

By extension of the space and by converting the inequality constraints to equalities, we transform the original dual problem (2) into the following equivalent problem in the extended space

$$\begin{array}{ll} \text{maximize} & b^T u \\ \text{subject to} & \varphi(w) + A^T u - c = 0_n. \end{array} \tag{4}$$

The Lagrangian associated with this problem is defined by

$$\tilde{L}(u, w, x) = b^T u + x^T[\varphi(w) + A^T u - c].$$

For solving problem (4) we use the stable version of the gradient projection method which is proposed by Tanabe (1980). The method is stated as an initial-value problem involving the following system of ordinary differential equations

$$\frac{du}{dt} = \tilde{L}_u(u, w, x(u, w)), \qquad \frac{dw}{dt} = \tilde{L}_w(u, w, x(u, w)). \tag{5}$$

The function $x(u, w)$ is chosen to satisfy the following condition

$$\tilde{L}_{xu}(u, w, x)\dot{u} + \tilde{L}_{xw}(u, w, x)\dot{w} = -\tau \tilde{L}_x(u, w, x). \tag{6}$$

Since $\dot{v} = \varphi_w \dot{w}$, (5) can be rewritten in terms of u and v as follows

$$\frac{du}{dt} = b - Ax(u, v), \quad \frac{dv}{dt} = -G(v)x(u, v), \tag{7}$$

where $\Phi(v)x(u, v) = A^T b + \tau(v + A^T u - c)$ and $\Phi(v) = G(v) + A^T A$.

Denote $v(u) = c - A^T u$.

Proposition 1 *Let the space transformation $\varphi(w)$ satisfy C_1. Then the matrix $\Phi(v(u))$ is positive definite for all $u \in U_0$.*

Proposition 2 *If all extreme points of the bounded set U are nondegenerate, then $\Phi(v(u))$ is positive definite for any $u \in U$.*

Let $[u(t,z_0), v(t,z_0)]$ denote the solution of the Cauchy problem (7) with initial condition $u(0,z_0) = u_0$, $v(0,z_0) = v_0$, $z_0^T = [u_0^T, v_0^T]$. This system of differential equations has the first integral

$$v(t,z_0) + A^T u(t,z_0) - c = (v_0 + A^T u_0 - c)e^{-\tau t}.$$

Thus, if the initial point z_0 is such that $v_0 = v(u_0)$, then $v(t,z_0) \equiv c - A^T u(t,u_0)$ for all $t \geq 0$. In this case the system (7) can be simplified

$$\frac{du}{dt} = b - Ax(u), \quad (G(v) + A^T A)x(u) = A^T b, \tag{8}$$

where $u(0,u_0) = u_0 \in U$. For this system we obtain the following inequality

$$b^T \frac{du}{dt} = \|b - Ax(u)\|^2 + x^T(u)G(v(u))x(u) \geq 0. \tag{9}$$

Hence the objective function of the dual problem monotonically increases on a feasible set.

Numerical algorithms are obtained as discretization of dynamical systems (7), (8). By applying the Euler numerical integration method we obtain the following iterative algorithm

$$u_{k+1} = u_k + \alpha_k(b - Ax_k), \quad v_{k+1} = v_k - \alpha_k G(v_k)x_k, \tag{10}$$

$$\left(G(v_k) + A^T A\right) x_k = A^T b + \tau \left(v_k + A^T u_k - c\right).$$

Similarly for the system (8) we have

$$u_{k+1} = u_k + \alpha_k \left(b - Ax_k\right), \quad \left(G(v_k) + A^T A\right) x_k = A^T b, \tag{11}$$

where $v_k = v(u_k)$. Both variants solve the primal and dual problems simultaneously.

Theorem 1 *Let x_* and u_* be unique nondegenerate solutions of Problems (1) and (2), respectively, and let $v_* = c - A^T u_*$. Assume that the space transformation $\varphi(w)$ satisfies conditions C_1, C_2 and $\tau > 0$. Then the following statements are true:*

1. *The pair $[u_*, v_*]$ is an asymptotically stable equilibrium state of system (7).*
2. *The solutions $u(t,z_0), v(t,z_0)$ of system (7) converge locally to the pair $[u_*, v_*]$. The corresponding function $x(u(t,z_0), v(t,z_0))$ converges to the optimal solution x_* of the primal problem (1).*
3. *The point u_* is an asymptotically stable equilibrium state of system (8)*
4. *The solutions $u(t,u_0)$ of system (8) converge locally to the optimal solution u_* of the dual problem (2). The corresponding function $x(u(t,u_0))$ converges to the optimal solution x_* of the primal problem (1).*
5. *There exists an $\alpha_* > 0$ such that for any fixed $0 < \alpha_k < \alpha_*$ the sequence $\{u_k, v_k\}$ generated by (10) converges locally with a linear rate to $[u_*, v_*]$ while the corresponding sequence $\{x_k\}$ converges to x_*.*

6. *There exists an $\alpha_* > 0$ such that for any fixed $0 < \alpha_k < \alpha_*$ the sequence $\{u_k\}$ generated by (11) converges locally with a linear rate to u_* while the corresponding sequence $\{x_k\}$ converges to x_*.*

If $u \in U_0$, then we can transform (8) to the equivalent system

$$\frac{du}{dt} = (I_m + AG^{-1}(v)A^T)^{-1}b,$$

where I_m is the $m \times m$ identity matrix. In the case where $G(v) = D^2(v)$ this method is similar to the dual affine scaling algorithm which was proposed and investigated by Adler, Karmarkar, Resende and Veiga (1989).

3 STEEPEST DESCENT

It follows from (9) that in order to maximize the value of the objective function at each iteration in the method (11) we must choose the step length α_k as large as possible provided that the point u_{k+1} belongs to the feasible set. This way of stepsize choice leads to a variant of the dual barrier-projection method (11) in which a steepest descent is utilized.

Let $u \in U$ and assume that

$$v(u) = \begin{bmatrix} v^B(u) \\ v^N(u) \end{bmatrix}, \quad v^B(u) \in R^s, \quad v^N(u) \in R^d,$$

where $v^B(u) = 0_s, v^N(u) > 0_d, d = n - s, s \leq m$. Denote the $m \times s$ matrix determined by the first s columns as B and denote the $m \times d$ matrix determined by the rest columns as N. Supposing for simplicity that $G(v) = D(v)$, we can write

$$\Phi(v) = \begin{bmatrix} B^TB & B^TN \\ N^TB & D(v^N) + N^TN \end{bmatrix}. \tag{12}$$

If B is a full rank matrix then, using the Frobenious formula, we obtain

$$\Phi^{-1} = \begin{bmatrix} Q & -QB^TWND_N^{-1} \\ -D_N^{-1}N^TWBQ & D_N^{-1} - D_N^{-1}N^T[W - WBQB^TW]ND_N^{-1} \end{bmatrix}, \tag{13}$$

where $Q = (B^TWB)^{-1}, \quad W = \left(I_m + ND_N^{-1}N^T\right)^{-1}, \quad D_N = D(v^N)$

Substituting (13) in (8) yields

$$\frac{du}{dt} = \left[W - WBQB^TW\right] b. \tag{14}$$

Moreover, we have

$$x(u) = \begin{bmatrix} x^B(u) \\ x^N(u) \end{bmatrix}, \quad x^B(u) = QB^TWb, \quad x^N(u) = D_N^{-1}N^T\left[W - WBQB^TW\right] b.$$

Thus

$$a_i^T \frac{du}{dt} = 0_s, \quad i \in J^B(u), \tag{15}$$

where a_i is the i-th column of the matrix A, and $J^B(u) = \{1 \le i \le n : v^i(u) = 0\}$. We supposed previously that $J^B(u) = \{1, 2, \ldots, s\}$.

The equality (15) shows that an adhesion property holds and that the trajectory can't leave the boundary of the constraint $a_i^T u \le c^i$ even if $x^i(u) < 0$. In order to overcome this drawback we will perturb the righthand side of the method (14). Denote

$$J_-^B(u) = \{i \in J^B(u) : x^i(u) < 0\}, \quad J_+^B(u) = \{i \in J^B(u) : x^i(u) \ge 0\}.$$

Let $\varepsilon > 0$ and let

$$y \in S^B(u) = \{z \in R_+^s : \sum_{i \in J^B(u)} z^i = 1, \ z^i = 0, i \in J_+^B(u)\}.$$

We introduce a diagonal matrix Y of order s such that $Y^2 = D(y)$ and modify the matrix $\Phi(v)$ changing only the left upper submatrix

$$\Phi(v) = \begin{bmatrix} B^T B + \varepsilon Y^2 & B^T N \\ N^T B & D(v^N) + N^T N \end{bmatrix}.$$

Then instead of (14) we obtain

$$\frac{du}{dt} = \{W - WB\,[Q - \varepsilon P]\, B^T W\}\, b, \tag{16}$$

where $P = QY\,[I_s + \varepsilon YQY]^{-1}\, YQ$. Now equality (15) takes place only for $i \in J^B(u)$ such that $y^i = 0$. Moreover, if ε is sufficiently small and $y^i > 0$ for $i \in J_-^B(u)$, then

$$a_i^T \frac{du}{dt} > 0, \quad i \in J_-^B(u). \tag{17}$$

There are several approaches to choosing the vector y from the set $S^B(u)$. The simplest one is the approach where a single index $i_k \in J_-^B(u_k)$ is selected and y is set to be equal to the vector $e_{i_k} \in S^B(u_k)$ with the components

$$e_{i_k}^i = 0, \ i \ne i_k, \qquad e_{i_k}^{i_k} = 1.$$

In this case the inequality (17) holds for any $\varepsilon > 0$.

We denote by q_i the i-th column of the matrix Q and by q_i^j the j-th element of q_i. We also introduce a vector

$$\Delta = A^T W\left[I_m - BQB^T W + (1 + q_{i_k}^{i_k})^{-1} B q_{i_k} q_{i_k}^T B^T W\right] b$$

and index sets

$$J^N(u) = \{1 \le i \le n : v^i(u) > 0\}, \quad J_+^N(u) = \{i \in J^N(u) : \Delta^i > 0\}.$$

Let, for simplicity, $\varepsilon = 1$. Then there is a variant of the method (11) with a steepest descent which can be written as

$$u_{k+1} = u_k + \alpha_k W \left[I_m - BQB^T W + (1 + q_{i_k}^{i_k})^{-1} B q_{i_k} q_{i_k}^T B^T W \right] b, \tag{18}$$

where

$$\alpha_k = \min_{i \in J_+^N(u_k)} \frac{v_k^i}{\Delta_k^i}.$$

Theorem 2 *Let the set U be bounded and let the function $b^T u$ have different values at extreme points of U. Then the method (18) solves the problem (2) in a finite number of iterations.*

Geometrically, the adding of the matrix $D(y)$ means the shifting of active constraints in dual problem (2). In method (18) we shift only a single active constraint. This simplest variant of the dual method possesses the following property, namely, if the point u_k is an extreme point of the set U, then the further behavior of the method is the same as a dual simplex method.

4 DUAL BARRIER-NEWTON METHOD

The method (11) can be interpreted as a contraction mapping for solving the system of nonlinear equations

$$b - Ax(u) = 0_m,$$

where the function $x(u)$ is defined from the condition

$$(D(\theta(v)) + A^T A)x(u) = A^T b. \tag{19}$$

We can apply the Newton method for this purpose. The continuous version of the Newton method leads to the following system of ordinary differential equations

$$\Lambda(u) \frac{du}{dt} = Ax(u) - b, \quad \Lambda(u) = -A \frac{dx}{du}. \tag{20}$$

By differentiating the relation (19) with respect to u, we obtain

$$-D(\dot{\theta}(v))D(x)A^T + \left(D(\theta(v)) + A^T A\right) \frac{dx}{du} = 0_{nm}.$$

Therefore

$$\Lambda(u) = -A \left(D(\theta(v)) + A^T A\right)^{-1} D(\dot{\theta}(v))D(x(u))A^T.$$

Proposition 3 *Let all conditions of Theorem 1 be fulfilled. Then $\Lambda(u_*)$ is a nonsingular matrix.*

Theorem 3 *Let all conditions of Theorem 1 hold. Then u_* is an asymptotically stable equilibrium point of system (20). If the matrix $\Lambda(u)$ satisfies a Lipschitz condition in a neighborhood of u_*, then the discrete version*

$$u_{k+1} = u_k + \alpha_k \Lambda^{-1}(u_k)(Ax_k - b), \quad x_k = x(u_k), \tag{21}$$

locally converges with at least linear rate to the point u_* if the stepsize α_k is fixed and $0 < \alpha_k < 2$. If the matrix $\Lambda(u)$ satisfies a Lipschitz condition in a neighborhood of u_* and $\alpha_k = 1$, then the sequence $\{u_k\}$ converges quadratically to u_*.

Let us suppose now that we use the quadratic space transformation (3). Then the method (20) can be rewritten as follows

$$\frac{du}{dt} = \left[A\left(D(v(u)) + A^T A\right)^{-1} D(x(u))A^T\right]^{-1} (b - Ax(u)). \tag{22}$$

On the set U_0 the right-hand side of this equation can be simplified

$$\frac{du}{dt} = \left[AD(x(u))D^{-1}(v(u))A^T\right]^{-1} b, \quad x(u) = \left[I_m - \left(I_m + AD^{-1}(v)A^T\right)^{-1}\right] b. \tag{23}$$

Thus the objective function of the dual problem increases along the Newton direction.

According to Theorem 3 the trajectories of the system (22) converge locally to u_*. The trajectories of the system (23) converge locally to u_* on the set U_0. The discrete version of the method (22) is similar to (21). It is possible to consider the discrete version of the method (23) using a steepest descent.

5 ACKNOWLEDGEMENT

Research supported by the Russian Scientific Fund, grant number 94-01-01379, and partially by the Czech Research Grant Agency, grant number 201/93/0429.

REFERENCES

Adler, I., Karmarkar, N., Resende, M. and Veiga, G. (1989) An implementation of Karmarkar's algorithm for linear programming. *Mathematical programming*, **44**, 297-335.

Evtushenko, Yu. and Zhadan, V. (1978) A relaxation method for solving problems of nonlinear programming. *U.S.S.R. Computational Mathematics and Mathematical Physics*, **17**, 73-87.

Evtushenko, Yu. and Zhadan, V. (1994) Stable Barrier-projection and Barrier-Newton Methods in Linear Programming. *Computational Optimization and Applications*, **3**, 289-303.

Tanabe K. (1980) A geometric method in nonlinear programming. *Journal of Optimization Theory and Applications*, **30**, 181-210.

Production Systems

Flow and release optimization in manufacturing systems represented as timed event graphs

A. Di Febbraro, R. Minciardi, M. Profumo, and S. Sacone
Department of Communication, Computer, and System Sciences, University of Genova, Via Opera Pia 13, I-16145 Genova, Italy.
Tel: +39-10-3532748. Fax: +39-10-3532948.
e-mail: angela@dist.unige.it, riccardo@dist.unige.it

Abstract

Timed Event Graphs can be applied successfully to the modelling, the analysis and the optimization of manufacturing systems. The choice of using such a particular class of Petri Nets allows to apply a set of related algebraic results to determine some performance measures as functions of a set of decision parameters. On this basis, performance optimization problems can be posed and solved as mathematical programming problems. In this paper, the problem consisting in the maximization of the weighted system throughput, while keeping the steady-state flow time of each product limited by an a-priori upper bound, is considered. This problem turns out to be a linear fractional programming one, which can be solved by means of standard techniques and software tools.

Keywords

Manufacturing systems, Petri Nets, optimization.

1 INTRODUCTION

A special class of Petri Nets, the Timed Event Graphs (or Timed Marked Graphs), has been investigated in the last years, as regards the development and the application of mathematical tools for the performance analysis. In particular, in (Cohen et al., 1985; 1989; Baccelli et al., 1992) a powerful set of algebraic results which allow a complete mathematical analysis of Timed Event Graphs have been provided. The main purpose of this paper is that of showing how such results can be applied to evaluate performance indices of interest, for a manufacturing system, as functions of a set of decision parameters. In this way, it is made possible to state and solve performance optimization problems through the application of proper mathematical programming techniques.

To this end, it is necessary to represent the manufacturing system model as a Timed Event Graph, that is, as a Timed Petri Net where each place has exactly one input and one output arc. This somehow limits the generality of the manufacturing system model to be considered, yet it is possible to treat significant models and optimization problems.

2 THE MODEL OF THE MANUFACTURING SYSTEM AND ITS REPRESENTATION AS A TIMED EVENT GRAPH

2.1 General characteristics of the model

A model of a manufacturing system is considered with p different classes of products, $P_1,..,P_p$. Starting from basic components, each product of class P_i, i=1,..,p, is obtained through a given set of operations o(i,1) ... o(i,n_i), structured in an in-tree, where o(i,n_i) is the final operation, and each join corresponds to an assembly operation. Operations can be simple, i.e., requiring one part as an input and giving one part as an output, or assembly ones. It is assumed that: a) no assembly operation requires as input two or more *identical* parts; b) when a basic component enters the system it is immediately assigned to a certain class of products; c) there may be a minimum waiting time to be observed between any pair of subsequent operations.

A (physical) function (e.g., painting, welding) ϕ=f(o(i,j)) is associated with each operation o(i,j), i=1,..,p, j=1,..,n_i. Each operation o(i,j), i=1,..,p, j=1,..,n_i, can be executed by any machine M_k, $k \in \mathcal{M}(i,j)$, where $\mathcal{M}(i,j)$ is defined as the set of indices of machines that can accomplish the function f(o(i,j)). Machines are never off and no preemption is allowed.

It is assumed that the set of m machines, $M_1,..,M_m$, is partitioned into a set BM of *batch machines,* and a set $\overline{BM}$ of ordinary machines. A batch machine can execute a *batch operation,* i.e., an operation resulting from the simultaneous execution of (simple) operations requiring the same function (e.g., heating, washing). Each batch machine can perform only one function. No function, hence no operation, implemented by a batch machine can be executed by any machine in $\overline{BM}$. Each machine $M_k \in \overline{BM}$ includes s_k independent equivalent servers, $s_k \geq 1$, each of which can perform a single operation at a time. A fixed *set-up time* may be necessary between the executions, on the same server of a machine in $\overline{BM}$, of two operations requiring different functions.

The presence of set-up times, as well as the fact that the elementary operations needed to manufacture products could not be so relevant to be considered individually, makes it convenient to group the single operations performed by machines in $\overline{BM}$ in *macro-operations.* If t(i,j,k) is the time (deterministic and known) required by machine M_k to perform o(i,j), then T(i,j,k) = t(i,j,k)·N(i,j,k) is the time required to perform the corresponding macro-operation O(i,j,k) on a lot of dimension N(i,j,k). In the case of an assembly macro-operation O(i,j,k), N(i,j,k) not only represents the size of the lots of assembled parts, but also that of the lots of the parts to be assembled.

As regards batch operations, the lot sizes do not affect the execution times, but are to be taken into account in order to ensure the fulfilment of 'capacity' constraints affecting such operations. In general, there is a number of batch operations implemented by a machine $M_h \in$ BM, say BO(h,1),...,BO(h,b_h). Each one of such operations arises from the simultaneous execution of a set of operations on a set of corresponding lots of parts. Let Λ[BO(h,r)] the set of operations corresponding to the batch operation BO(h,r). Then, the following constraint must be satisfied

$$\nu^{h,r}_{min} \leq \sum_{o(i,j) \in \Lambda[BO(h,r)]} N(i,j,h) \leq \nu^{h,r}_{max} \qquad \forall M_h \in BM,\ j=1,..,n_i,\ r=1,..,b_h$$

where $v_{min}^{h,r}$ ($v_{max}^{h,r}$) is an a-priori given lower (upper) bound. The time required to implement any batch operation BO(h,r) is TB(h), independent of r and of the lot sizes N(i,j,k) of the parts involved. The sets Λ(BO(h,r)), $M_h \in$ BM, r=1,..,b_h, defining the batch operations for machine M_h, are supposed to be a-priori given.

It is assumed that for each kind of product i, the lots of all the relevant basic components are released synchronously and at a constant rate $1/\pi(i)$. The time instant in which the first lot(s) of basic component(s) is (are) released is denoted by $\tau(i)$, with $\tau(i) \geq 0 \ \forall$ i=1,...,p. The number of lots to be released is infinite.

The complete definition of the system model requires to specify the system behaviour whenever alternatives are possible as regards: i) how to assign operations to machines; ii) how to sequence macro-operations, or batch operations, assigned to the same machine. As regards the first issue, the execution of operation o(i,j) is shared out among machines M_k, $k \in \mathcal{M}(i,j)$, by specifying the sizes of the lots N(i,j,k), $k \in \mathcal{M}(i,j)$, which are among the decision variables of the problem. As for the operation sequencing, it is assumed that, for each server of machine M_k in $\overline{BM}$, and for each machine M_k in BM, a fixed (known) service sequence $\mathcal{S}_k$ of macro- or batch operations is applied. The set of decision variables affecting the considered model can be represented as the collection of the components of vector ξ = col [N(i,j,k), π(i), τ(i), i=1,...,p; j=1,..,n_i; $k \in \mathcal{M}(i,j)$]. Instead, the initial marking of the net is considered to be fixed.

2.2 An example

In order to show that the above general model admits a representation as a Timed Event Graph, it is worth considering a specific example, namely a manufacturing system with two classes of products, P_1 and P_2, and five machines M_1,..,M_5, where M_5 is a batch machine, performing only one batch operation. Machines M_1 and M_2 have two servers, whereas M_3 and M_4 are monoserver. Products of class P_1 and P_2 are obtained through a processing sequence of four simple operations [o(i,1), o(i,2), o(i,3), o(i,4)], i=1,2. The batch operation BO(5,1) is made up of operations o(1,3) and o(2,3). The set of indices of machines that can execute operations not involved in the batch operation are:

$$\mathcal{M}(1,1)=\{1,2\} \quad ; \quad \mathcal{M}(1,2)=\{1\} \ ; \quad \mathcal{M}(1,4)=\{4\} \ ;$$
$$\mathcal{M}(2,1)=\{2\} \quad ; \quad \mathcal{M}(2,2)=\{3\} \ ; \quad \mathcal{M}(2,4)=\{3,4\} \ .$$

The structure of the Timed Event Graph representing this manufacturing system is given in Figure 1. In this example, it is supposed that, for any service sequence, changing from an operation to another one implies also a change in the function implemented, so that a set-up time is required. Moreover, some nonzero minimum waiting times have to be fulfilled, namely between o(1,2) and o(1,3), between o(2,1) and o(2,2), and between o(2,2) and o(2,3).

Transitions are immediate, whereas nonzero token holding times are associated with some places. Their values are represented by the numbers of vertical bars in the places. Of course, the system is assumed to work under the Early Operational Mode (i. e., transitions fire as early as allowed). The notation x_i/x_k is used to denote the place which has x_k as input transition and x_i as output transition. In the figure, places represented by thick black circles stand for macro- or batch operation executions. Only the holding times of places corresponding to macro-operations are affected by the choice of the lot sizes. A second class of places, the grey circles, has fixed holding times and represent minimum waiting times (x_9/x_8, x_{11}/x_6, x_{11}/x_{10}), set-up times (x_1/x_6, x_2/x_8, x_7/x_4, x_{18}/x_{10}, x_9/x_{20}, x_{15}/x_{19}, x_{17}/x_{16}), and lot decomposition/recomposition times (x_{14}/x_{13}, x_{12}/x_{11}). The remaining places model conditions representing the availability of machines and lots, and have zero holding times, too.

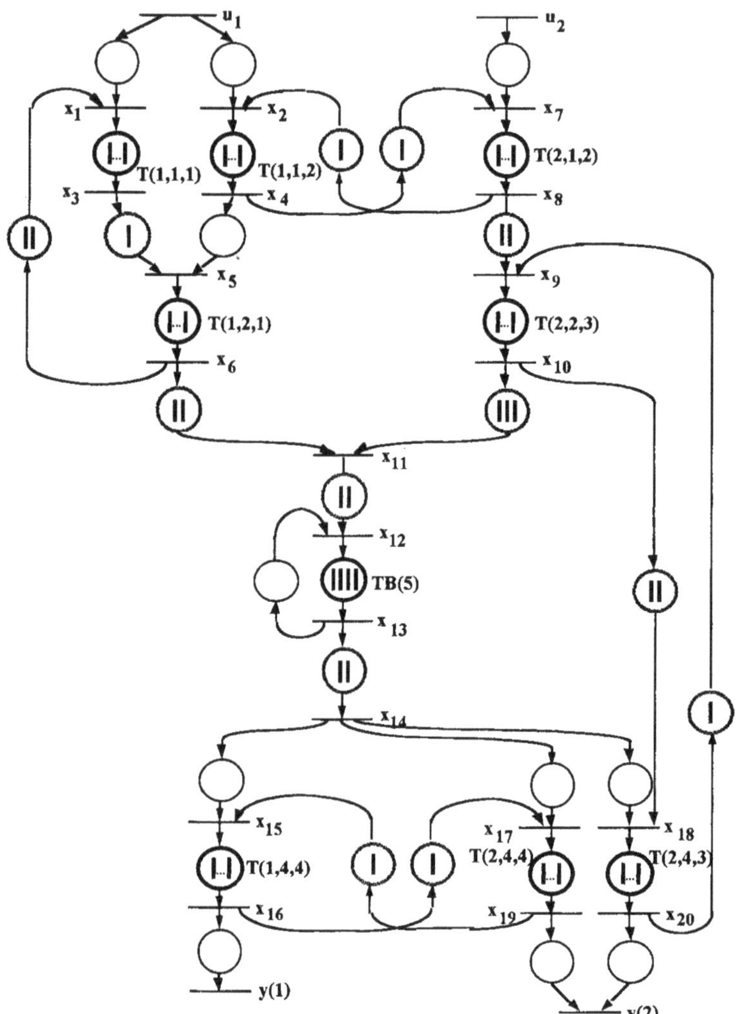

Figure 1

3 ALGEBRAIC COMPUTATION OF THE PERFORMANCE INDICES

3.1 Representation of Timed Event Graphs as linear systems

A major feature of Timed Event Graphs is that they can be represented through equations formally similar to those of linear time-invariant systems, provided that some tools peculiar to the so-called Max-Plus algebra are used.

The system representation is obtained in a two-dimensional domain resulting from the juxtaposition of an *event domain* and a *time domain* (Cohen et al., 1989). To obtain this kind of representation, the introduction of two classes of variables is required: the first class refers to the dates of event occurrences (*daters*), whereas the second class refers to the number of events occurred (*counters*). The use of such an algebraic structure, known as MinMax<<γ,δ>> allows to get, for any event graph, the following equations:

$$\mathbf{x} = \mathbf{A}\otimes\mathbf{x} \oplus \mathbf{B}\otimes\mathbf{u}$$
$$\mathbf{y} = \mathbf{C}\otimes\mathbf{x} \qquad (1)$$

where:

- $\mathbf{x}$ = col[x_i, i=1,..,ns], ns=number of intermediate transitions (i.e., transitions which have both input and output places), is the state vector: x_i represents all the information about the i-th transition;
- $\mathbf{u}$ = col[u_i, i=1,..,ni], ni=number of input transitions (transitions which have only output places), is the input vector: u_i represents all the information about the i-th input transition;
- $\mathbf{y}$ = col[y_i, i=1,..,no], no=number of output transitions (transitions which have only input places), is the output vector: y_i represents all the information about the i-th output transition;
- matrices $\mathbf{A}$ (dim($\mathbf{A}$) = ns × ns), $\mathbf{B}$ (dim($\mathbf{B}$) = ns × ni), and $\mathbf{C}$ (dim($\mathbf{C}$) = no × ns) represents the relations between intermediate transitions input transitions, and output transitions;
- $\otimes$ and $\oplus$ are the two fundamental operators of the considered algebraic structure.

The elements of matrices **A**, **B**, and **C** are written in terms of two shift operators, the first one (γ) acting in the event domain, and the second (δ) acting in the time domain.

It is important to remark that equations (1) depend not only on the TEG topological structure, but also on the numbers of bars representing the holding times of places, and on the initial positions of tokens. Once defined the system matrices, it is possible to determine a transfer matrix for the system, describing the relations between vectors **u** and **y**. The general expression of a transfer matrix is given by

$$\mathbf{H}=\mathbf{C}\mathbf{A}^{*}\mathbf{B} \qquad (2)$$

with $\mathbf{A}^{*}= \mathbf{E} \oplus \bigoplus_{i=1}^{+\infty} \mathbf{A}^{i}$ where **E** is the identity matrix in the Max-Plus algebra (Baccelli et al., 1992). Once obtained **H**, **y** can be computed as **y=Hu**.

3.2 Conditions for system stability

In the following, u(i) will denote the arrival process of the lot(s) relevant to the i-th product. Then, one can write **u**=col[u(i), i=1,..,p] where $u(i) = \delta^{\tau(i)}(\gamma\delta^{\pi(i)})^{*}$, which represents a sequence of events occurring at a constant rate 1/π(i), starting at instant τ(i). For instance, in the example introduced in Section 2, the input vector is given by

$$\mathbf{u}=\mathrm{col}[(\gamma\delta^{\pi(1)})^{*}, \delta^{\tau(2)}(\gamma\delta^{\pi(2)})^{*}]$$

implicitly assuming τ(1) = 0.

Since it is assumed that the arrival processes last over an infinite horizon, it is necessary to investigate under which conditions the stability of the system is ensured (the system is stable if the number of tokens in any place of the net remains always limited, i. e., if the net is k-bounded, for some finite k). To this end, let us first give the following definition: an *irreducible* TEG modelling a manufacturing system is a system which cannot be partitioned into *completely separate* subsystems (i.e., subsystems which do not share any machine). Then, define the *cycle time* $C(\chi)$ of the elementary circuit χ as the sum of the holding times of the places making up the elementary circuit χ, divided by the number of tokens in the circuit. It is apparent that $C(\chi)$ is an affine function of the decision variables (more specifically, the lot sizes) for any elementary circuit χ.

Result. An irreducible TEG modelling a manufacturing system is *stable* if and only if the two following conditions are satisfied:

1. $\pi(i) = \bar{\pi}$ $\qquad$ $i=1,..,p$

2. $\bar{\pi} \geq C(\chi)$ $\qquad$ $\forall$ elementary circuit χ in the network

Δ

Proof. Let us first demonstrate the necessity. As regards the first condition, let us recall that the irreducibility assumption imposes that no production route is completely disjoint from the rest of the system. Thus, consider product i, and suppose that it shares a machine M_h with another product j. If we admit $\pi(i) > \pi(j)$, then there will be an accumulation of tokens in the buffer preceding M_h, representing items of product P_j. This proves the necessity of the first condition.

As regards the second condition, suppose that an elementary circuit χ such as $\bar{\pi} < C(\chi)$ exists. Then, since $1/C(\chi)$ is the maximum rate at which any transition in that circuit can fire, and owing to the irreducibility assumption, the number of tokens representing parts waiting for entering circuit χ is increasing. Thus, also the second condition is necessary.

Then, let us come to the sufficiency part. It can be shown (Cohen et al., 1989; Baccelli et al., 1992; Parodi et al., 1994) that the only terms of the type $(\gamma^{\alpha}\delta^{\beta})^*$ appearing in matrix **H** are those associated with the elementary circuits, and are characterized by a value of α equal the number of tokens and of β equal to the cycle time of the considered circuit. On the counterpart, the only * term appearing in vector **u** is $(\gamma\delta^{\bar{\pi}})^*$, if the first condition holds. Due to the simplification rules in the Max-Plus algebra, and to the second condition, the * terms appearing in vector **y** are all of the same type, that is, $\gamma\delta^{\bar{\pi}}$. This implies that the output rate for any product i in the steady-state equals the input rate, thus proving the stability of the overall system.

Δ

As a consequence of the above first stability condition, all decision variables $\pi(i)$ reduce to a single variable $\bar{\pi}$.

3.3 Evaluation of the performance indices

The computation of matrix **A*** can be performed following the rules reported in (Baccelli et al., 1992). It is worth remarking that several additional simplification rules have to be applied in the presented model, since matrix **A*** has an additional dependence on the components of ξ (besides to γ and δ). These rules are reported in detail in (Parodi et al., 1994), and are omitted here for brevity. Similar considerations hold for the determination of **H**.

If the above stability conditions hold, the system is stable, and thus it makes sense to consider its steady-state behaviour. More specifically, the indices of interest will be the throughputs and the steady-state flow times.

In general, the overall throughput of the i-th product, namely X(i) is given by

$$X_i = \sum_{k \in \mathcal{M}(i,n_i)} N(i,n_i,k) / \bar{\pi} \qquad (3)$$

The flow time ft(i,r) of token r for the i-th class of products (this is the (r+1)-th token of product P_i that flows in the system, since the first is token 0) can be determined as the difference between the exit time instant and the entrance time instant. This requires the determination of the output sequence for the considered product. In the proposed model, it turns out (Parodi et al., 1994) that all output sequences (note that each product has only one output sequence) have the structure

$$y(i) = Q(i) \otimes (\gamma \delta^{\bar{\pi}})^* \qquad (4)$$

where Q(i) is a polynomial in the operators γ and δ, having the structure

$$Q(i) = \bigoplus_{q=0}^{n(i)} \gamma^q \delta^{f_q(i)} \qquad (5)$$

where the coefficients $f_q(i)$, i=1,...,p, are affine functions (in the conventional algebra) of parameters N(l,j,k) and $\tau(l)$, $\forall$ l,j,k.

On this basis, it is possible to evaluate the steady-state flow time relevant to product P_i (Di Febbraro et al., 1994), which turns out to be

$$\overline{ft}(i) = \max\{f_0(i)+n(i)\bar{\pi}, f_1(i)+(n(i)-1)\bar{\pi}, \ldots, f_{n(i)}(i)\} - (n(i)\bar{\pi} + \tau(i)) \qquad (6)$$

4 THE OPTIMIZATION PROBLEM

Having so determined the performance indices of interest, several optimization problems can be stated. In this paper, the problem consisting in the maximization of the weighted system throughput, while keeping the steady-state flow time of each product limited by an a-priori upper bound, is considered.

Problem 1: Maximize the objective function

$$\max_{\xi} \sum_{i=1}^{p} w_i \sum_{k \in \mathcal{M}(i,n_i)} N(i,n_i,k) / \bar{\pi} \qquad (7)$$

subject to

$$N_{min}(i,j,k) \,.\, N(i,j,k) \,.\, N_{max}(i,j,k) \qquad i=1,..,p,\ j=1,..,n_i,\ k \in \mathcal{M}(i,j) \qquad (8)$$

$$\bar{\pi} \geq C(\chi) \qquad \forall \text{ elementary circuit } \chi \text{ in the network} \qquad (9)$$

$$\overline{ft}(i) \leq \bar{F}_i \qquad i=1,\ldots,p \qquad (10)$$

$$\sum_{k \in M(i,1)} N(i,1,k) = \sum_{k \in M(i,j)} N(i,j,k) \qquad i=1,..,p,\ j=2,..,n_i \qquad (11)$$

$$v_{min}^{h,r} \cdot \sum_{o(i,j) \in \Lambda[BO(h,r)]} N(i,j,h) \,.\, v_{max}^{h,r} \qquad \forall M_h \in BM,\ j=1,..,n_i,\ r=1,..,b_h \qquad (12)$$

where

- coefficients w_i are a-priori given constants;
- constraints (8) impose a-priori lower and upper bounds on the lot sizes of components of product class P_i;
- constraints (9) guarantee the fulfilment of the second stability condition;
- in constraints (10), $\bar{F}_i$ is the a-priori given upper bound for the steady-state flow time of the class of product P_i;
- constraints (11) guarantee the 'flow conservation' of the parts.

Δ

Owing to the structure of the terms $\bar{ft}(i)$, as given by (6), to the fact that the terms $f_q(i)$, as well as the cycle time $C(\chi)$, are affine functions of the decision variables, Problem 1 turns out to be a linear fractional programming problem which can be solved by means of standard techniques (Bazaraa and Shetty, 1980) and software tools.

The solution of the above optimization problem, with reference to a numerical example corresponding to the net structure depicted in Figure 1, has required a computation time of about 2 hours on a Workstation HP 9000/720, by use of a purposely developed software package based on the commercial code MAPLE V.

Finally, it is worth remarking that assuming to have fixed the initial marking of the net is not too restrictive. In fact, the number of tokens in a machine circuit is constrained to be equal to the number of servers of that machine. Moreover, the tokens in the initial marking representing the parts already present in the system are actually not relevant as regards the steady-state behaviour of the system, provided that the stability conditions are fulfilled.

REFERENCES

Baccelli, F.L., Cohen, G., Olsder, G.J. and Quadrat, J.P. (1992) *Synchronization and Linearity*. Wiley, New York.

Bazaraa, M. and Shetty, C. (1980) *Nonlinear Programming*. Wiley, New York.

Cohen, G., Dubois, D., Quadrat, J. P. and Viot, M. (1985) A linear-system-theoretic view of discrete-event processes and its use for performance evaluation in manufacturing. *IEEE Transactions on Automatic Control*, **30**, 210-220.

Cohen, G., Moller, P., Quadrat, J. P. and Viot, M. (1989) Algebraic tools for the performance evaluation of discrete event systems. *Proc. IEEE*, 77, 39-58.

Di Febbraro A., Minciardi, R. and Sacone S. (1994) Performance optimization in manufacturing systems by use of Max-Plus algebraic techniques. *Proc. IEEE Int. Conference on Systems, Man, and Cybernetics*, **2**, 1995-2001.

Parodi, G., Prati, A. and Profumo, M. (1994) Metodi algebrici per l'analisi e l'ottimizzazione di sistemi di produzione automatizzata rappresentati tramite Reti di Petri. Tesi di Laurea, Universita di Genova (in Italian).

A control model for assembly manufacturing systems

A. B. Dolgui
Belarussian State University of Informatics and Radio Electronics
6, P. Brovki Street 220027 Minsk, Belarus
CRIN - Ecole des Mines de Nancy, 54042 Nancy Cedex, France
Phone 83.58.41.77 fax 83.57.97.94 e-mail dolguia@loria.fr

M. C. Portmann
CRIN - Ecole des Mines de Nancy, 54042 Nancy Cedex, France
Phone 83.58.41.85 fax 83.57.97.94 e-mail portmann@loria.fr

J. M. Proth
INRIA - Lorraine, Project SAGEP
CESCOM, Technopole Metz 2000, 57070 Metz, France
Phone 87.20.35.00 fax 87.76.39.77 e-mail proth@ilm.loria.fr

Abstract
In this paper, medium term planning problems related to multi-product assembly manufacturing systems are considered. The assembly lines are fed by external suppliers. Their delivery delays, not known in advance, are crucial for the assembly process. If some required components are not available at assembly time, then the decision system must decide what to produce instead. This decision system uses a sliding horizon and tries to optimize costs on any given horizon. A model and some approaches for its resolution are proposed. The optimisation criterion is the average production cost (component inventory cost and backlogging cost).

Keywords
Assembly manufacturing systems, random supply, inventory control, medium term planning, linear programming, simulation

1 INTRODUCTION

We consider multi-product assembly systems, which use external ordered components to assemble products. We study the medium term planning and inventory control problems when delivery delay of components is critical. To control such systems, the following problems arise:

- How to maintain constant levels of component inventories on the average in order to limit the consequences of random delivery delay deviations ? A trade-off must be made between component inventory cost and product backlogging cost (product inventories are useless with unlimited capacity hypothesis since components not used at previous periods are used only just in time when the corresponding product is required).
- In case of insufficient component inventory level, how to manage conflicts between the products requiring simultaneously the same set of components ? The corresponding criterion value depends strongly on the maximum deviations of both component inventory cost values and backlogging cost values.

In the literature, reported works essentially deal with supplier selection strategies (Cohen and Lee, 1989; Hendrik and Rush, 1988; Trevelen and Schweikhar, 1988). In (Gurnani et al., 1990) the authors address the supplier diversification when the quantity delivered is uncertain. Ramasesh et al. (1991) have considered the randomness of the lead time but only in the case of one product and with uniform or exponential laws.

We suppose that we know the distribution laws for component delivery delays, as well as the inventory cost for each component and the backlogging cost for each product. We search a solution approach considering only steady-states for this system. So the cost is minimized only in average. If the reader is interested to know how to stabilize a transient system into a given steady-state, some approaches can be found in (Chu et al., 1994). We do not consider this problem in this paper and assume that it is possible to reach the optimal steady-state, which can be obtained by solving the model proposed.

The problem is formulated in section 2. Then, the global solution approach is presented in section 3. Finally, some numerical examples are given in section 4.

2 PROBLEM FORMULATION

In the following, we consider a steady-state of the system on an infinite horizon. Time axis is divided into a number of equal time intervals called periods. Notations used are:

- **N** the matrix which describes the product structure, where N_z^j is the number of components j required to assemble one unit of product z , $1 \leq z \leq n$, $1 \leq j \leq d$.
- **y** the vector $(y_1, y_2, ..., y_n)$, where y_z is the number of products z to be manufactured at the end of each period.
- **r(i)** the vector $(r_1(i), r_2(i), ..., r_n(i))$, where $r_z(i)$ is the number of backloggings of product z at the beginning of period i.
- **s(i)** the vector $(s_1(i), s_2(i), ..., s_d(i))$, where $s_j(i)$ is the inventory level of component j at the beginning of period i.
- α the vector $(\alpha_1, \alpha_2, ..., \alpha_d)$, where α_j is the period inventory cost of component j.
- β the vector $(\beta_1, \beta_2, ..., \beta_n)$, where β_z is the period backlogging cost of product z .
- **x(i)** the vector $(x_1(i), x_2(i), ..., x_n(i))$, where $x_z(i)$ is the number of products z which are assembled during the period i.
- **q** the vector $(q_1, q_2, ..., q_d)$, where q_j is the number of components j ordered at the beginning of each period.
- **f(t)** the vector $(f_1(t), f_2(t), ..., f_d(t))$, where $f_j(t)$ is the distribution law for delivery delay of component j.

Let $CR_1(h)$ and $CR_2(h)$ be respectively the total inventory cost and the total backlogging cost from period 0 to period h -1 included.

$$CR_1(h) = \sum_{k=0}^{h-1} \sum_{j=1}^{d} \alpha_j \left| s_j(k) - \sum_{z=1}^{n} \left(x_z(k)\, N_z^j \right) \right|, \tag{1}$$

$$CR_2(h) = \sum_{k=0}^{h-1} \sum_{z=1}^{n} \beta_z \left(y_z + r_z(k) - x_z(k) \right), \tag{2}$$

$$s_j(k) = s_j(k-1) - \sum_{z=1}^{n} x_z(k-1)\, N_z^j + \sum_{u=0}^{k-1} q_j\, d_{u,k}^j\ , \tag{3}$$

$$d_{u,k}^j = \begin{cases} 1 & \text{if a delivery of type } j \text{ ordered at period } u \text{ occurs at the end of period } k-1, \\ 0 & \text{otherwise,} \end{cases} \tag{4}$$

$$q_j = \sum_{z=1}^{n} N_z^j\, y_z. \tag{5}$$

The goal is to minimize, $CR = \lim_{h->\infty} \left\{ \frac{1}{h} \left[CR_1(h) + CR_2(h) \right] \right\}$. (6)

3 SOLUTION APPROACH

3.1 Planning model with sliding horizon

Working on an infinite horizon is not easy, thus we decide to optimize decisions on a sliding horizon of h periods. When taking h equal to 1 we obtain a simplified model called "the basic model". The following example illustrates the fact that using only the basic model does not give an optimal approach. This is due to the fact that solving optimally and successively conflicts between products for each period may be worse than solving globally conflicts between products for a series of periods even if some data are not known in advance but only forecast.

Example, $\mathbf{N}=\begin{bmatrix} 0 & 1 \\ 1 & 1 \end{bmatrix}$, $\mathbf{y}=(2,2)$, $\mathbf{s(1)}=(3,4)$, $\alpha=(10,20)$, $\beta=(30,1000)$, $\mathbf{f_j(t)}=(0.5,0.5)$, $\forall j=1, 2$, where 0.5 is the probability of regular component delivery. We consider two periods.

Basic model: **x(1) = (2,2)** => C = 10+0 = 10.
1.1 With probability 0.5, s(2) = (1,0), **x(2) = (0,0)** => C = 10+2060 = 2070.
1.2 With probability 0.5, s(2) = (3,4), **x(2) = (2,2)** => C = 10+0 = 10.
The average total cost E[C/h]=(10+2080×0.5)/2=**525**.

Optimal solution: **x(1) = (1,2)** => C = 30 + 30 = 60.
2.1 With probability 0.5, s(2) = (1,1), **x(2) = (0,1)** => C = 0+1060 = 1060.
2.2 With probability 0.5, s(2) = (3,5), **x(2) = (2,2)** => C = 30+0 = 30.
The average total cost E[C/h] = (60+1090×0.5)/2 = **302.5**.

Due to the reasons explained earlier, we decide to work with a sliding horizon which is modelled by the following model (7)-(10). The periods of a given horizon are indexed by t_0, t_1, ..., t_{h-1}. For the first sliding horizon, t_0 is equal to 0 and increases by 1 at each iteration of the general solving method. For any $i = 0, 1, 2, ...$, when we have to decide the number of products to be assembled, we solve this model by using integer linear programming (exact or approximation) methods with $t_0 = i$ and we only use the solutions $\mathbf{x}(\mathbf{t_0})$ associated with current period i.

Note that the criterion (7) is not the cost sum but a simplified form obtained by removing a constant part and by transforming a minimization into a maximization.

$$\sum_{k=0}^{h-1} \sum_{z=1}^{n} (h-k)\, x_z(t_k) \left(\beta_z + \sum_{j=1}^{d} N_z^j \alpha_j \right) \quad \rightarrow \quad max, \tag{7}$$

$$\sum_{k=0}^{v} \sum_{z=1}^{n} x_z(t_k)\, N_z^j \leq s_j(t_0) + L_j^*[t_0, t_v], \quad \forall v = 0, 1, 2, ..., h-1, \tag{8}$$

$$\sum_{k=0}^{v} x_z(t_k) \leq (v+1)\, y_z + r_z(t_0), \quad \forall z = 1, 2, ..., n, \quad \forall v = 0, 1, 2, ..., h-1, \tag{9}$$

$$x_z(t_k) \geq 0, \quad \forall z = 1, 2, ..., n, \quad \forall k = 0, 1, 2, ..., h-1, \tag{10}$$

where $L^*_j[t_0,t_v]$ is a forecast value of the components delivered within the interval $[t_0,t_v]$, $L^*_j[t_0,t_0] = 0$.

It is possible to compute an upper bound of h denoted by h' with the following equation:

$$h^* = max_j [\, max_z(\beta_z / N_z^j) / \alpha_j], \quad \forall z = 1, 2, ..., n, \quad \forall j = 1, 2, ..., d, \quad N_z^j \neq 0. \tag{11}$$

The best experimental value of h obtained by simulation (section 4) is strictly less than h' and varies with different other factors such as delivery components rate, forecasting methods, etc.

3.2 Inventory level determination

When some components are missing, the model (7)-(10) allows to optimally assign existing components to a limited part of the required products. Although we do not consider the transient phase, it provides us with average component inventory levels. We search for the average component inventory levels which minimize the total cost by simulation.

To obtain a good average inventory level ω_j^* for any component j in the steady-state, we build an iterative algorithm. At the beginning of step k, the component inventory levels are given by the vector $\omega^k = (\omega_1^k, \omega_2^k, ..., \omega_d^k)$. We assume that for a small variation ε_j^k of one of the component levels ω_j^k, the corresponding cost variation is linear. The coefficient of the cost variation obtained with $\varepsilon_j^k = 1$ is given after one simulation of M periods by the equation:

$$\gamma_j^k = 1/M \left(\sum_{i=0}^{M-1} \left(\alpha_j\, \varsigma_j^i \right) - \sum_{i=0}^{M-1} \sum_{z=1}^{n} \left(\beta_z\, \psi_{z,j}^i / N_z^j \right) \right), \ \forall k, \tag{12}$$

where $\varsigma^i_j = 1$ if there remains at least one component j in the inventory at the end of period i of the simulation and 0 otherwise, $\psi^i_{zj} = 1$ if at least one product z is not assembled during period i of the simulation by lack of component j and 0 otherwise.

For all j, ω_j^0 is obtained by uniform random generation between a lower and an upper bound (minimum and maximum delivery delay for component $\times q_j$).

At step k, we first simulate and compute the values $\gamma_j^k(\omega^{k-1})$. Then we search a good value for ε_j^k by an algorithm (similar to dichotomy) using the results of the above simulation and we set $\omega_j^k=\omega_j^{k-1}+\varepsilon_j^k$ ($\varepsilon_j^k=0$ when $\gamma_j^k=0$). The iterative algorithm stops when for all j, $\varepsilon_j^k=0$.

3.3 Forecasting methods

To compute forecast component future delivery $L^*_j[t_0,t_k]$ for the model (7)-(10), we experiment the following methods.

Average inventory forecasting (method M_1)

$$L_j^*[t_i, t_k] = \sum_{m=i+1}^{k} \sum_{v\in\theta_j(t_i)} q_j \operatorname{Prob}\{del_j(t_v|t_i) = t_m - t_v\} \tag{13}$$

where $\theta_j(t_i)$ is the set of indices of periods before t_i where orders of components j have been done, $del_j(t_v|t_i)$ is the conditional delivery delay of components ordered at the beginning of period t_v, knowing that this components are not yet delivered at the beginning of period t_i.

Random sample forecasting (method M_2)

$$L_j^*[t_i, t_k] = \sum_{m=i+1}^{k} \sum_{v\in\theta_j(t_i)} q_j\, \delta_j(t_m|t_v,t_i), \tag{14}$$

$$\delta_j(t_m|t_v,t_i) = \begin{cases} 1 & \text{if } del_j^*(t_v|t_i) = t_m - t_v, \\ 0 & \text{otherwise,} \end{cases} \tag{15}$$

$del_j^*(t_v\,|\,t_i)$ is the value obtained randomly by using the distribution function laws of delivery delay of component j.

3.4 Simulation model

In order to adjust the parameters of the proposed model and control the assembly system, we designed a simulation program running under UNIX on Sun workstations. To date, the program allows to treat examples with 100 products, 300 components and 100 forecasting periods. This simulation uses the integer linear programming part of OSL software. One step of the simulation *algorithm* works as follows:

1. Simulate the arrival of the components.
2. Compute the conditional distribution laws for the orders whose deliveries are not arrived at the current period.
3. Compute the forecasts for future deliveries.
4. Calculate the optimum values of production using OSL.
5. Compute the new inventory levels of components.
6. Calculate the production cost.

To obtain an ideal reference value for the average period cost, we made a series of experiments where the simulated delivery delay are taken equal to the randomly generated forecast delays. This forecasting method is called "100% forecast" and denoted by M_0.

The following *algorithm* explains how to work with our model:

1. Search the best inventory level ω_j, for all j.
2. Compute h' using equation (11).
3. For $h = h_1, h_2, \ldots$ $(h_1 = h', h_i > h_{i+1}, h_i \geq 1)$, use simulation with method M_0 and $s_j(t_0) = \omega_j$ for all j.
4. Keep the period length h which gives the minimal total cost.
5. If $h=1$, go to 8.
6. Apply the forecasting methods M_1 and M_2 with h and keep the best one M_{i*} whose cost is $C(h, \mathbf{s}, M_{i*})$.
7. If $C(h, \mathbf{s}, M_{i*}) \geq C(1, \mathbf{s})$ then $h=1$.
8. Use sliding plan with a time h, the forecasting method M_i* (if $h \neq 1$) and the quantity of components to be required q_j, for all j.

4 NUMERICAL EXAMPLES

To test the model efficiency, we have generated 5 examples. For each example, the values of matrix **N** were generated uniformly between 1 and 10. This provided us with different associations between the products and the components used. We took also different generated values for the other parameters of our model:

- Backlogging costs: the same for all the products or very different.
- Distribution laws for delivery delays: uniform on a small interval or with one peek in the distribution law or with several peeks in the distribution law or with large standard deviation.
- Inventory levels of components: small or "optimal".

Tables 1 and 2 present the results computed for 800 periods of simulation with the following parameters, α=(10,100,1000), β=(10,10,10,10), for all j, 3 delay values and $f_j(t)$=(0.333, 0.333, 0.334) and inventory component levels equal to their lower bounds. In table 1:

- %1 is the maximal cost improvement (in percentage) that may be obtained by using h periods instead of using the basic model. It is computed by using the ideal forecasting method M_0 (here we may see that it varies from 21% to 43%).
- %2 (resp. %4) is the cost improvement obtained by using h periods and the forecasting method M_1 (resp. M_2) instead of using the basic model (these percentages vary from 3% to 17% for M_2 and from 0,3% to 3% for M_1 which is apparently worse).
- %3 (resp. %5) is equal to %2 (resp. %4) divided by %1 and multiplied by 100. It represents the "true" performances obtained by method M_1 or M_2 which may be used to control on line the assembly system. Notice in table 1 that the performances vary considerably with the examples.

Table 1 Cost comparison between forecasting methods

	Basic	*100% forecast(M_0)*		*Average inventory forecasting (M_1)*			*Random sample forecasting (M_2)*		
	model	*CR*	*%1*	*CR*	*%2*	*%3*	*CR*	*%4*	*%5*
example 1	8874	6173	30.4	8588	3.2	10.6	8589	3.2	10.6
example 2	7198	4091	43.2	7017	2.5	5.9	5966	17.1	39.7
example 3	7815	5762	26.3	7792	0.3	1.1	7336	6.1	23.3
example 4	10 009	7925	20.8	9794	2.2	10.3	9718	2.9	14.0
example 5	7408	4743	36.0	7354	0.7	2.0	6883	7.1	19.7

Table 2 illustrates the search and interest of inventory level optimization. These results were obtained with $h=1$ to limit the computation time and to obtain only one cost for each set of inventory values (if optimal inventory levels are computed with M_1 or M_2 and $h>1$, the results are quite similar to the ones obtained with $h=1$).

Table 2 Cost obtained with optimization of component inventory levels

	Results of optimization				*Basic model*	
	"Optimal" inventory levels	*CR*	*NI*	*%6*	*Lower bounds of inventory levels*	*CR*
example 1	360, 404, 180, 250	4193	264	52.8	160, 160, 80, 120	8874
example 2	190, 100, 310, 228	2761	75	61.6	80, 50, 150, 100	7198
example 3	280, 407, 230, 230	4028	143	48.5	130, 170, 100, 110	7815
example 4	442, 150, 356, 342	6288	59	37.2	220, 150, 180, 160	10 009
example 5	261, 269, 360, 258	3660	126	50.6	110, 110, 140, 110	7408

Where *NI* is the number of iterations and %6 shows the percentage of gain obtained by using "optimal inventories" (from 37% to 61%). The "optimal" inventories are nearly twice bigger than the lower bound inventories.

5 CONCLUSION

A model has been proposed for the control of some multi-product assembly systems where component delivery delays are only known by distribution laws, but some other hypotheses are rather restrictive (constant product requirement and unlimited assembly capacity). A resolution approach is given which uses linear programming and simulation to improve some parameter choices. Experimentation shows the validing both of the model and the solving approach.

Introducing non constant product requirements in the model is quite obvious, for example with the hypothesis of cyclic requirements when working on an infinite horizon or non cyclic requirements on a finite horizon. In this case, a procedure must be added in the solving approach to compute at each period the component order quantities.

Introducing limited assembly capacity in the model and in the solving approach is quite obvious if anticipated production is forbidden which increases backlogging cost in average.

Further research work must be developed to introduce anticipated production and corresponding product inventory costs in the model and in the solution method.

6 REFERENCES

Chu, C., Proth, J.M. and Xie, X. (1992) Supply management in assembly systems: the basic problem. *Naval Research Logistic,* **39**, 265-83.

Chu, C., Proth, J.M., Wardi, Y. and Xie X. (1994) Supply management in assembly systems: the Case of Random Lead Times, in *Proc. 11th International Conference on Analysis and Optimisation of Systems*, Sophia-Antipolis, June 15th, 1994, Springer-Verlag.

Cohen, M. and Lee H. (1989) Resource development analysis of global manufacturing and distribution network. *Journal of Manufacturing and Operations Management,* **2.2**, 81-104.

Gurnani, H.B., Akella, R. and Lehoczky J. (1990) Supply management in assembly systems - General yield model, *Working Paper,* Graduate School of Industrial Administration, Carnegie Mellon University.

Hendrick, T.E. and Ruch, W.A. (1988) Determining performance appraisal criteria for the buyer. *Journal of Purchasing and Materials Management,* **24.2**, 18-26.

Ramasesh, R.V., Ord, J.K., Hayya, J.C. and Pan, A. (1991) Sole versus dual sourcing i stochastic lead-time (s,Q) inventory models. *Management Science,* **37.4**, 428-443.

Treleven, M. and Schweikhar, S.B. (1988) A risk/benefit analysis of sourcing strategies: single vs. multiple sourcing. *Journal of Operations Management,* **7.4**, 93-114.

Numerical experiment on the 2D cutting-stock algorithms based on local optimization

T. Sakamoto
Okayama Prefectural University
Soja, Okayama 719-11 Japan
TEL +81-866-94-2089
FAX +81-866-94-2199
E-mail sakamoto@c. oka-pu. ac. jp

Abstract

A numerical experiment on the computational complexity and the efficiency of layout is reported for the 2D cutting-stock algorithms based on local optimization. The time and the space complexity are respectively found to be $O(n^2)$ and better than $O(e_0)$ by the regression analysis, where n is the number of rectangular parts and e_0 is the edge length of square sheets. The efficiency of layout is also discussed with respect to the aspect ratio of rectangular sheets.

Keywords

Cutting-stock algorithms, computational complexity, local optimization, one-to-one mapping, merge operation

1 INTRODUCTION

In the first paper (Sakamoto, 1993), we have proposed the six kinds of 2D cutting-stock algorithms based on local optimization. These approximate algorithms require no combinatorial process for the rectangular parts in the traditional ones, but can realize the efficiency of layout comparable to that by the dynamic programming (Adamowicz, 1976).

In this paper, we report the result of the numerical experiment on the computational complexity and the efficiency of layout of these six algorithms.

2 OUTLINE OF THE ALGORITHMS

Figure 1 illustrates the flow of the algorithms. The algorithms are constructed recursively using the four operations, search, divide, merge and stack. Both the shapes of parts and regions are assumed rectangular. The search operation is defined as the one-to-one mapping $f: P \mapsto W$, where P and W are the sets of parts p and partial regions w generated by the divide operation, respectively. A stack is used to store the data of w. The merge operation is used to generate a larger region using those in the stack. Figure 2 illustrates an example of the merge operation in the Cartesian coordinates. In this example, a part p is allocated in the region w after the search operation. The residual area of w is divided into the three partial regions w_A, w_B and w_C. These regions are stored in the stack and used in the subsequent search operation. The order of search for the partial regions is assumed as w_A, w_B, $M(w_A, w_C)$ and $M(w_B, w_C)$, where $M(w_A, w_C)$(or $M(w_B, w_C)$) denotes the region generated by the merge operation with w_A(or w_B) and w_C. The types of merge operation is classified into four due to the geometrical condition of p and w(Sakamoto, 1993). The local optimization used in this paper means the determination of minimizing the cost function defined by $C(p,w) = e_x e_y - c_x c_y$. By modifying the cost function, other approximate solutions can be obtained.

The significance of merge operation is explained as follows: suppose the result of search for the region w_A(or w_B) is successful. Without the merge operation, the regions w_A(or w_B) and w_C are separately used in the later search operation, so that a part p' larger than these regions would not be allocated. However, since w_C shares an edge with w_A(or w_B), the region $M(w_A, w_C)$(or $M(w_B, w_C)$) generated by the merge operation becomes obviously larger as long as the area of w_C is not zero, so that the possibility for allocating p' would increase.

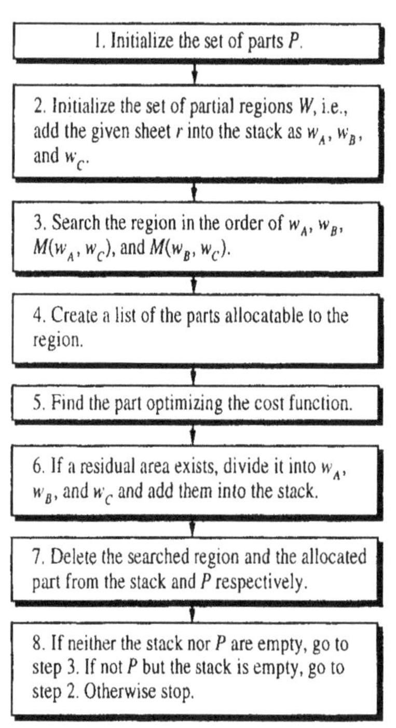

Figure 1 Flow of the algorithms.

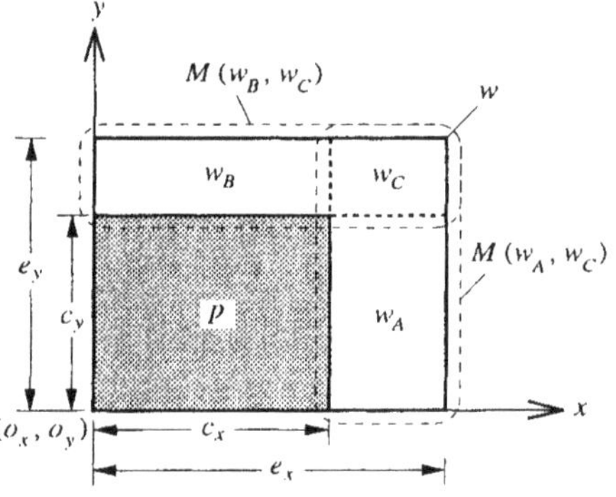

Figure 2 Conceptual diagram of merge operation in Cartesian coordinate.

Table 1 The six algorithms considered in this paper

Algorithms	*Orientation of the parts*	*Search order of regions*
A1	region-matched	horizontal
A2	region-matched	vertical
A3	horizontal	horizontal
A4	horizontal	vertical
A5	vertical	horizontal
A6	vertical	vertical

In the stack operation, it suffices in the decision of the possibility for the merge operation to check the three consecutive addresses from the memory pointer immediately after the divide operation, not the entire data in the stack. When the search and the divide operations are executed recursively, a larger region, in general, has a larger possibility to be stacked in the lower part and the search operation is applied with a priority for the smaller region. In other words, the small region, which has a smaller possibility of being allocated, has a shorter-period stay in the memory. Thus, using a stack as a memory of partial regions is effective in improving the efficiency of the data management.

In the search operation, the following two ambiguities exist;

(1) orientation of parts,

(2) order of search for partial regions.

Therefore, several kinds of algorithms can be constructed. Table 1 shows the six algorithms considered in this paper.

Figure 3 illustrates a layout example by the algorithm A2. The degree of freedom of parts in allocation is assumed two, i.e. either vertical or horizontal placement is possible. (a) shows the initial state. A square with unit area (1×1) is used as the reference. The set of sheets R is composed of a single rectangular region $r(5\times 4)$, i.e., $R=W$. Similarly the set of parts P is composed of the five rectangular parts $p_1(2\times 3)$, $p_2(1\times 3)$, $p_3(1\times 4)$, $p_4(1\times 1)$, and $p_5(2\times 1)$. The initial state of the stack is set at w_C=(5, 0, 0, 4), w_B=(5, 0, 0, 0), and w_A=(0, 0, 5, 4), where the content of the stack denotes (o_x, o_y, e_x, e_y). (b) shows the placement of p_1 into w_A. First, the set $\Pi=\{p_1, p_2, p_3, p_4, p_5\}$ is created as the list of parts allocatable to the region w_A. Then $f(p_1)$=(0, 0, 3, 2) is determined as the mapping minimizing the cost function $C(p, w_A)$ for $p\in\Pi$, where the mapping f denotes (o_x, o_y, e_x, e_y) and p_1 is rotated. The excess region is divided into w_{A_1}, w_{B_1}, and w_{C_1}. After removing p_1 and w_A from P and W respectively, the data of these three subregions are stored into the stack. (c) shows the result where the same operation as in (b) is iterated for three times. Namely, $f(p_2)$=(0, 2, 3, 1) is first determined for w_{A_1}. Then

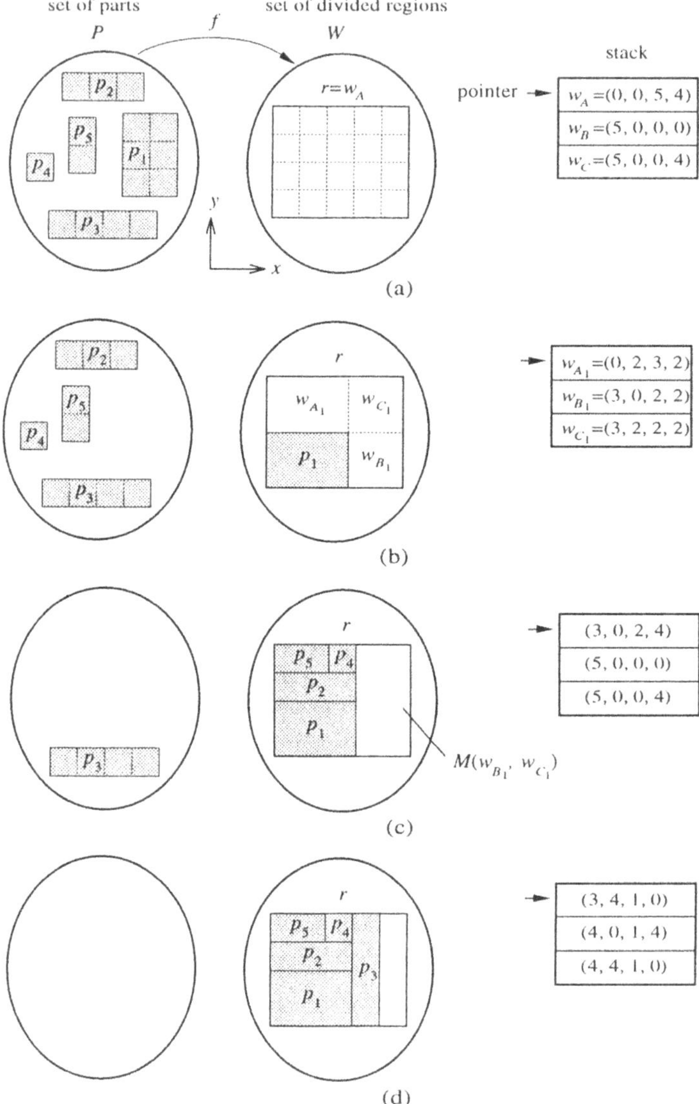

Figure 3 A layout example by the algorithm A2. (a) The initial state, (b) The result of search operation for w_A, (c) The result of the series of search operations for w_{A_1}, (d) The result of search operation for $M\left(w_{B_1}, w_{C_1}\right)$.

$f(p_5)$=(0, 3, 2, 1) and $f(p_4)$=(2, 3, 1, 1) are determined for w_{A_2} and w_{A_3} respectively. When $f(p_2)$ is determined, w_{B_1} and w_{C_1} are stored in the stack as $M(w_{B_1}, w_{C_1})$ by the merge operation for the later search operation. (d) shows the result where $f(p_3)$=(3, 0, 1, 4) is determined for $M(w_{B_1}, w_{C_1})$. When p_3 is removed from P, $P=\phi$ holds and the algorithm stops, where ϕ denotes an empty set.

3 NUMERICAL EXPERIMENT AND DISCUSSION

Table 2 shows the data of parts used in the numerical experiment (Sakamoto, 1993), (Adamowicz, 1976). The degree of freedom of parts in allocation is assumed two. When the total number of parts is changed, the proportion of each number of parts is kept constant. The size of sheet, i.e., the initial region is $12\,030\,mm \times 2\,550\,mm$. When the aspect ratio of sheet γ is changed, the area is also kept constant.

Figure 4 shows the time complexity. By the regression analysis, the time complexity is found to be $O(n^2)$, where n is the number of part. From the flow of the algorithms, this result can be explained as follows: in the search operation, all the parts not allocated yet are tested if they can be allocated in a given region. The amount of those parts is $c_1 n$ with $0 \le c_1 \le 1$. When several parts are found as a candidate to be allocated, the optimal part is decided among those parts be allocated in a given region. The amount of those parts is $c_1 n$ with $0 \le c_1 \le 1$. When several parts are found as a candidate to be allocated, the optimal part is decided among those parts based on the criterion of local optimization. The amount of those parts is $c_2 n$ with $0 \le c_2 \le c_1$. Therefore, the total time complexity is $c_1 n \times c_2 n = O(n^2)$.

Table 2 The data of parts used in the numerical experiment (Sakamoto, 1993), (Adamowicz, 1976)

Part No.	*No.*	*Edge length*	
		c_x *(mm)*	c_y *(mm)*
1	96	450	270
2	48	400	270
3	16	1 140	170
4	8	2 500	300
5	8	2 500	139
6	40	300	270
7	32	1 500	270
8	32	270	110
9	8	1 993	165
10	16	710	675
11	16	500	500
12	32	800	500
13	30	300	74
14	8	1 360	160
15	8	1 760	785

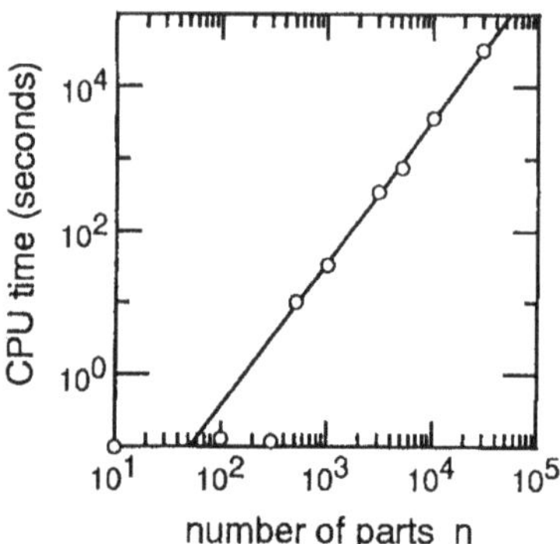

Figure 4 Time complexity.

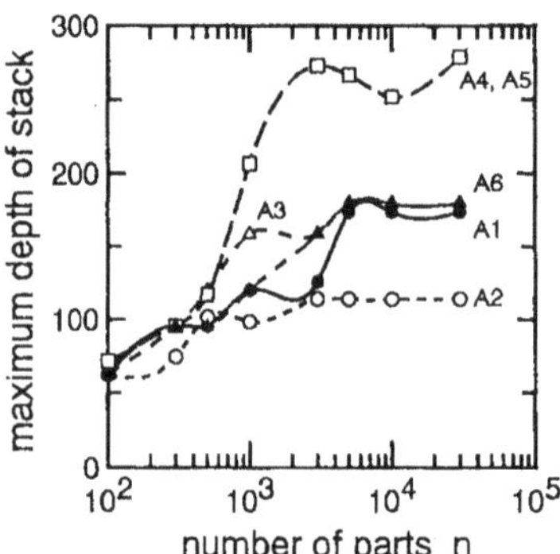

Figure 5 Dependence of the maximum depth of the stack on the number of parts n.

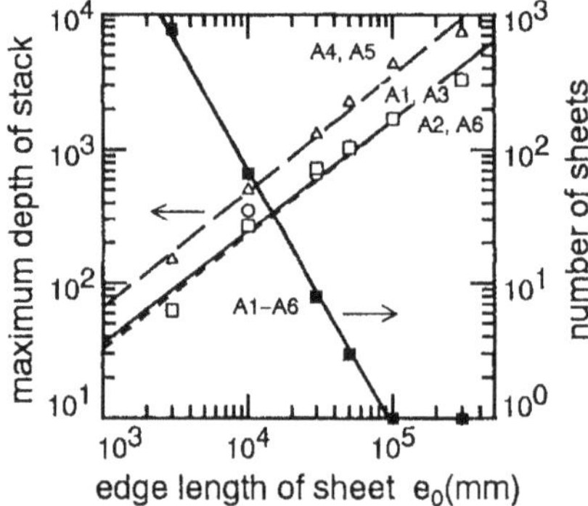

Figure 6 Space complexity at n=30 000 and γ=1.

Since the size of stack cannot be predicted, the maximum depth of stack is used as a measure for the space complexity. Figure 5 shows thedependence of the maximum depth of the stack on the number of parts n. It appears that the maximum depth tends to saturate as increasing n and is almost constant at $n = 30\ 000$. Thus, to analyze the space complexity, n and γ are respectively set at 30 000 and one. Figure 6 shows the result. The space complexity is found to be better than $O(e_0)$ by the regression analysis, where e_0 is the edge length of the sheet. Also the required number of sheets is $O(e_0^{-2})$ as is expected because the total area of parts is kept constant. Figure 7 shows the efficiency of layout η, where η is defined as the ratio of the total area of parts and that of sheets. (a) shows the influence of n with γ=1. η varies seriously among the algorithms for smaller n, but converges as increasing n. The required number of sheet is nearly proportional to n. (b) shows the influence of γ with $n = 30\ 000$. Also η varies seriously among the algorithms as γ departs from one, but there exists a pair of algorithms whose η's are symmetric around γ=1. At this aspect ratio, every algorithm generates a similar highest η.

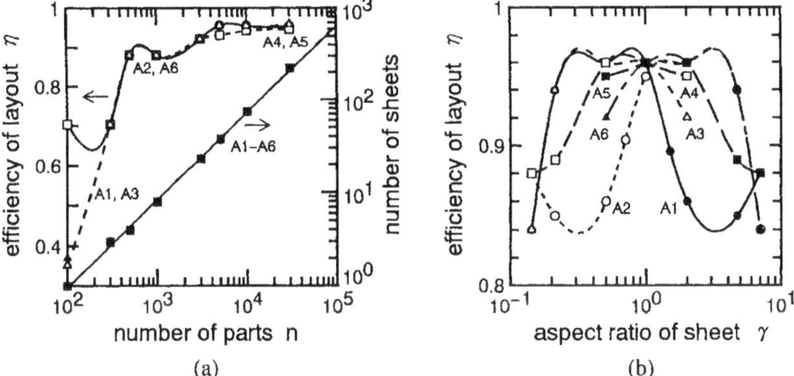

Figure 7 Efficiency of layout. (a) dependence on the number of parts, $\gamma = 1$, (b) dependence on the aspect ratio of sheet, $n = 30\ 000$.

4 SUMMARY

A numerical experiment on the computational complexity and the efficiency of layout has been reported for the six kinds of 2D cutting-stock algorithms based on local optimization. The time and the space complexity are respectively found to be $O(n^2)$ and better than $O(e_0)$ by the regression analysis, where n is the number of part and e_0 is the edge length of the rectangular sheet.

REFERENCES

Sakamoto, T. (1993) 2D cutting-stock algorithms based on local optimization, *Trans. IEICE Japan*, **76-A**, 605-11.

Adamowicz, M. and Albano, A. (1976) A solution of the rectangular cutting-stock problem, *IEEE Trans. Syst. Man Cyber.*, **SMC-6**, 302-10.

Scheduling Problems

65

An algorithm for the transportation problem with given frequencies

L. Bertazzi, M. G. Speranza
Dept. of Quantitative Methods – University of Brescia
Corso G. Mameli, 27 – I–25122 Brescia, Italy. Tel: 39–30–2988.1.
Fax: 39–30–2400925. e-mail: speranza@master.cci.unibs.it

W. Ukovich
D.E.E.I.-University of Trieste
Via A. Valerio, 10 – I – 34127 Trieste, Italy. Tel: 39–40–6767135.
Fax: 39–40–6763460. e-mail: ukovich@univ.trieste.it

Abstract

In this paper we consider the problem of shipping several products from an origin to a destination, when a discrete set of shipping frequencies is available only, in such a way that the sum of the transportation and inventory cost is minimized. This problem, which is known to be NP-hard, has applications in transportation planning and in location analysis. We derive some dominance rules for the problem solutions which allow to tighten the bounds on the problem variables. Moreover, we present a branch-and-bound algorithm and we evaluate its performance on randomly generated problem instances.

Keywords

Transportation, inventory, dominance rules, branch-and-bound.

1 INTRODUCTION

We consider the situation in which each product of a given set is made available at an origin and needed at a destination at a constant rate. A discrete set of available shipping frequencies is given and the shipment is carried out by trucks with given capacity. The problem is to decide how much of each product to ship at each frequency in such a way that the sum of the transportation and inventory cost is minimized.

This problem was introduced in Speranza and Ukovich (1994) together with other models for shipping products from an origin to a destination. Similar models, where the shipping frequencies are allowed to take any continuous value, are considered in Burns et al (1984), Blumenfeld et al (1985), Anily and Federgruen (1990). The problem was shown to be NP-hard in Speranza and Ukovich (1991), where an optimal branch-and-bound algorithm was presented. The computational results showed that the algorithm was efficient on problem instances of small/medium size. The more complex problem of shipping products from several origins to a destination, when a given set of shipping frequencies is available, was considered in Bertazzi, Speranza and Ukovich (1994), where different heuristics were presented, based on solving in a first phase a one origin – one destination problem for each of the given destinations, and then on improving the solution through local search techniques.

In this paper, we derive some dominance rules for the solution of the one-origin-one-destination problem. Such rules allow to tighten the bounds on the problem variables. Then we present a new branch-and-bound algorithm in which we apply the dominance rules in order to improve its efficiency and we compare it with the branch-and-bound proposed in Speranza and Ukovich (1991).

The paper is organized as follows. In Section 2 the problem is described and its mathematical programming formulation is given. The dominance rules and their implications are presented in Section 3. In Section 4 the new branch-and-bound is described and the computational results are given in Section 5.

2 PROBLEM DESCRIPTION AND FORMULATION

A set of products is made available at an origin and needed at a destination at a constant rate. Trucks, with given capacity, are available for shipping products, and each product can be partly shipped by trucks travelling at different frequencies. Each product can be continuously divided. Two cost factors are considered, namely the transportation and the inventory cost. Shipping products at a high frequency generates low inventory costs but high transportation costs, while shipping at a low frequency generates high inventory costs and low transportation costs. The problem is to decide the fraction of each product which is shipped at each frequency in such a way that the sum of the transportation and the inventory costs is minimized.
The following notation is used.

I: set of products;
J: set of frequency indices;
n: number of frequencies ($=|J|$);
i: products;
j: index of frequency;
f_j: jth frequency, with $j \in J$;
t_j: period of frequency f_j ($= 1/f_j$);
H: time horizon (equal to the minimum common multiple of the periods);
q_i: rate at which product i is produced at the origin and needed at the destination;
v_i: unit volume of product i;
h_i: inventory cost of one unit of product i per unit time;
r_j: capacity of each truck travelling at frequency f_j;
c_j: cost of a single trip of a truck travelling at frequency f_j;

x_{ij}: fraction of product i shipped at the frequency f_j;
y_j: number of trucks which travel at the frequency f_j.

The problem can be formulated as follows.

$$\min \sum_{i \in I} \sum_{j \in J} h_i t_j q_i H x_{ij} + \sum_{j \in J} c_j (H/t_j) y_j \tag{1}$$

$$t_j \sum_{i \in I} v_i q_i x_{ij} \leq r_j y_j \qquad j \in J \tag{2}$$

$$\sum_{j \in J} x_{ij} = 1 \qquad i \in I \tag{3}$$

$$0 \leq x_{ij} \leq 1 \qquad i \in I, j \in J \tag{4}$$

$$y_j \geq 0 \qquad \text{integer} \qquad j \in J. \tag{5}$$

The objective function (1) represents the sum of the inventory cost and the transportation cost on the time horizon H. Constraints (2) are capacity constraints, which state that the number of trucks y_j must be sufficient to load all the products assigned to frequency f_j. Constraints (3) impose that all the quantity of product i produced at the origin must be shipped to the destination at some of the given frequencies. The mixed integer programming problem defined through (1)–(5) will be referred to as Problem $\mathcal{P}$.

In the following, we assume that the unit time is chosen in such a way that all t_j are integer and that the frequencies are ordered in the decreasing order, that is $f_1 > f_2 > \ldots > f_n$. Moreover, we assume that all trucks have the same capacity, independently of the frequency, that is that $r_j = r$, $\forall j$, and, without loss of generality, we normalize the capacity to $r = 1$. As all trucks have the same capacity, we assume $c_j = c$, $\forall j$.

3 DOMINANCE CONDITIONS

In this section we present the most important theoretical results we have obtained for the Problem $\mathcal{P}$. For their formal proofs, we refer to Bertazzi, Speranza and Ukovich (1995).

Dominance is a concept, common in multiobjective optimization, that turns out to be convenient for our problem. In fact, Problem $\mathcal{P}$ can be considered as stemming from an optimization problem with two conflicting objectives: minimize the inventory cost and minimize the transportation cost.

In general terms, a feasible solution for a multiobjective optimization problems is dominated if another feasible solution exists with not worse performance with respect to each objective, and a strictly better performance with respect to at least one objective (cf. for instance Keeney and Raiffa, 1976).

The following result states a general condition guaranteeing that a solution is dominated.

Lemma 1 A solution $(\bar{x}, \bar{y})$ for Problem $\mathcal{P}$ is dominated if $\exists\ \alpha \in \Re_+$ (the set of positive real numbers), $j, k \in J$ such that:

$$t_k < t_j \tag{6}$$

$$\alpha \leq \bar{V}_j / t_j \tag{7}$$

$$\lceil \bar{V}_j - \alpha t_j \rceil / t_j + \lceil \bar{V}_k + \alpha t_k \rceil / t_k \leq \bar{y}_j / t_j + \bar{y}_k / t_k \tag{8}$$

where $\bar{V}_s = t_s \sum_i v_i q_i x_{is}$ for $s = j, k$.

It is worth to point out that the above result generalizes the saturation property for Problem $\mathcal{P}$ presented in Speranza and Ukovich (1994):

Corollary 1 Consider an optimal solution $(\bar{x}, \bar{y})$ for Problem $\mathcal{P}$. Suppose $\exists j, k \in J$ such that $j > k$, $\bar{V}_j > 0$. Then $\bar{y}_k = \bar{V}_k$.

As a consequence of the last result, in the sequel we always consider solutions satisfying the saturation condition. Then Lemma 1 yields a more convenient condition for domination:

Theorem 1 A feasible solution $(\bar{x}, \bar{y})$ for Problem $\mathcal{P}$ is dominated if $\exists \alpha \in \Re_+$, $j, k \in J$ such that conditions (6) and (7) of Lemma 1 are satisfied, and

$$\hat{c}_T = \lceil \bar{V}_j - \alpha t_j \rceil + \lceil \alpha t_k \rceil t_j / t_k \leq \bar{y}_j \tag{9}$$

Corollary 2 A solution $(\bar{x}, \bar{y})$ for Problem $\mathcal{P}$ is dominated if $\bar{y}_j t_k / t_j \geq 1$ for some $j > k$ with t_j / t_k integer.

Corollary 2 allows to tighten the bounds for the variables y_j, $j > 1$, whenever t_j admits a proper submultiple in the given set of periods $\{t_j\}$. Let us denote by $J' \subseteq J$ the set of indices of the periods which admit a proper submultiple. Then

$$y_j \leq UB1_j = \min_k \{t_j / t_k - 1 : t_j / t_k \text{ is integer}\} \qquad j \in J'. \tag{10}$$

In the case $J' = J$, then $y_j \leq UB1_j$, $\forall j$, and the range for y_1 can be made tighter. In fact,

$$y_1 \geq t_1 (\sum_i v_i q_i - \sum_{j=2}^{n} y_j / t_j) \geq t_1 (\sum_i v_i q_i - \sum_{j=2}^{n} \min(UB0_j, UB1_j) / t_j) = LB_1 \tag{11}$$

where $UBO_j = \lceil t_j \sum_{i \in I} v_i q_i \rceil$.

Observe that, in this case ($J' = J$), the width of the range of the variables does not depend on the volume to be shipped, but on the periods only. Let us emphasize this feature. Since $UB1_j \leq t_j - 1$, then $y_1 \geq \lceil t_1 (\sum_i v_i q_i - n + 1 + \sum_{j=2}^{n} f_j) \rceil \geq UB0_1 - t_1(n-1)$.

A complete enumeration of the solutions generates a number of solutions not larger than $(t_1(n-1)+1)t_2 \dots t_n$. This quantity depends on the number and on the values of the periods only, which in practice are quite moderate. Thus, for fixed n, such a "brute force" enumeration turns out to be just pseudo–polynomial.

A different kind of bounds for the variables y_j, $j > 1$ can be inferred from a new domination result, which in turn is another consequence of Theorem 1:

Corollary 3 A solution $(\bar{x}, \bar{y})$ for Problem $\mathcal{P}$ is dominated if $\bar{V}_j \geq t_j$ for some $j > 1$.

As a consequence of the last result, then $\bar{V}_j < t_j$ in any optimal solution, hence $\lceil \bar{V}_j \rceil = y_j \leq t_j$, $\forall j > 1$. Furthermore, note that if $y_j = t_j$, then the above Corollary still applies (i.e. the corresponding solution is still dominated), thus yielding a tighter upper bound for y_j, $j > 1$, equal to $t_j - 1$.

Another different dominance rule is given by the following theorem. which states that a solution is dominated if all products travelling at a certain frequency can be shipped by full load trucks at a higher frequency.

Corollary 4 A solution $(\bar{x}, \bar{y})$ for Problem $\mathcal{P}$ is dominated if $\bar{V}_j t_k / t_j$ is an integer positive number for some $j > k$.

Corollary 4 can be further generalized as follows.

Theorem 2 Given a solution $(\bar{x}, \bar{y})$, if a subset $\bar{J}$ of the frequencies and a frequency f_k exist such that $t_j > t_k$, $\forall j \in \bar{J}$, and $\sum_{j \in J} \bar{V}_j t_k / t_j$ is an integer positive number, then the solution is dominated.

4 A BRANCH-AND-BOUND ALGORITHM

In this section we present a branch-and-bound algorithm for the solution of the Problem $\mathcal{P}$. In this algorithm we apply the dominance rules we have derived in Section 3; this algorithm will be compared in Section 5 with the branch-and-bound proposed in Speranza and Ukovich (1991) in order to evaluate its performance.

The branch-and-bound works on a search tree where each level corresponds to a frequency f_j and each node ν represents the vehicles y_j used at this frequency. This number of vehicles belongs to a set of values defined in the following way.

We recall that, by the Corollary 3, $y_j < t_j$ $\forall j > 1$ and that, in the particular case in which the period t_j admits a submultiple in the given set of the periods $\{t_j\}$, by the Corollary 2, $y_j \leq UB1_j = \min_k \{t_j/t_k - 1 : t_j/t_k \text{ is integer}\}$. Furthermore, given a partial solution $(\bar{y}_1, \dots, \bar{y}_{j-1})$, $y_j \leq UB2_j$ with $UB2_j = \left\lceil (\sum_i v_i q_i - \sum_{k=1}^{j-1} \bar{y}_k / t_k) t_j \right\rceil$. Therefore, the upper bound for y_j, $j > 1$, is, in general, $y_j \leq \min(t_j - 1, UB2_j)$ and, in the particular case in which t_j has a submultiple in $\{t_j\}$, $y_j \leq \min(t_j - 1, UB1_j, UB2_j)$. The lower bound for y_j is $LB1$ for $j = 1$, as defined in (11) and $y_j = 0$ for all $j > 1$.

The tree is explored in the depth first way.

The fathoming criteria are the following. An upper bound U is initially set to the minimum value obtained by solving the heuristics presented in Bertazzi, Speranza and

Ukovich (1995) on the instance of the problem; this upper bound mantains the best current upper bound during the algorithm.

In each node ν, a lower bound L_ν is calculated by solving the relaxed problem by the Procedure Ext-Greedy presented in Speranza and Ukovich (1991). If the solution of the relaxed problem is integer, this node is fathomed. If $L_\nu > U$, the node ν is fathomed. Furthermore, in each node ν the dominance rules presented in Section 3 are applied; if the obtained solution is dominated, the node ν is fathomed. When the number of vehicles y_j are fixed in all the level of the tree, if the solution is feasible, the optimal value is calculated and the node ν is fathomed. If the optimal value is lower than the current upper bound U, then U is modified.

5 COMPUTATIONAL RESULTS

The branch-and-bound proposed in Speranza and Ukovich (1991) and the new branch-and-bound have been implemented in FORTRAN on a Personal computer with an Intel 486dx/66 processor.

The computational results have been obtained by generating random instances with the following characteristics:

- set of frequencies: 1, 1/2, 1/3, 1/4, 1/5, 1/6, 1/7, 1/8, 1/9, 1/10;
- number of products: 1000, 5000, 10 000;
- quantity per product per unit time (q_i): randomly generated between 0.1 and 5, between 5 and 100 and between 0.1 and 100;
- unit volume per product (v_i): randomly generated between 0.001 and 0.01;
- unit inventory cost per product (h_i): randomly generated between 0.001 and 0.05, between 0.05 and 1 and between 0.001 and 1;
- transportation cost per trip (c): 300.

The results are shown in Table I. Each row corresponds to 5 problem instances with the characteristics described in columns 1-3. Column 1 gives the number of products; Column 2 gives the range of the quantity per product and Column 3 the range of the inventory cost h_i.

Columns 4-7 show the computational results obtained by the branch-and-bound proposed in Speranza and Ukovich (1991), referred as Old branch-and-bound; Column 4 gives the average number of visited nodes (the maximum number of possible nodes has been fixed in 30 000); Column 5 gives the number of times we have obtained the optimal solution on 5 instances; Column 6 the node in which the optimal solution has been found (average on the 5 instances); finally, Column 7 gives the average CPU time (minutes and seconds). In columns 8-11 the computational results obtained by the new branch-and-bound, referred as New branch-and-bound, are shown as in columns 4-7.

The computational results show that the new branch-and-bound is more efficient than the one proposed in Speranza and Ukovich (1991), both in terms of number of visited nodes and CPU time.

Table I Comparison of the branch-and-bound algorithms

Parameters			Old branch-and-bound				New branch-and-bound			
Prod.	Quant.	Inventory	Nodes	N	Opt.	Time	Nodes	N	Opt.	Time
1000	0.1-5	0.001-0.05	13 784	3	1064	05.09	3431	5	1031	00.53
1000	0.1-5	0.05-1	148	5	125	00.02	51	5	37	00.00
1000	0.1-5	0.001-1	192	5	160	00.03	136	5	114	00.02
1000	5-100	0.001-0.05	22 290	3	4676	06.44	8411	5	2021	01.41
1000	5-100	0.05-1	1363	5	431	00.26	410	5	185	00.06
1000	5-100	0.001-1	3996	5	1899	01.23	1100	5	728	00.18
1000	0.1-100	0.001-0.5	9274	4	1387	02.47	1697	5	589	00.22
1000	0.1-100	0.05-1	949	5	221	00.18	227	5	47	00.03
1000	0.1-100	0.001-1	1970	5	1123	00.39	644	5	392	00.10
5000	0.1-5	0.001-0.05	15 255	3	1175	16.37	6251	5	1344	03.55
5000	0.1-5	0.05-1	3341	5	944	03.36	681	5	300	00.31
5000	0.1-5	0.001-1	1321	5	1090	01.26	610	5	507	00.30
5000	5-100	0.001-0.05	30 000	0	0	34.28	19070	5	690	12.15
5000	5-100	0.05-1	2070	5	434	01.58	249	5	117	00.14
5000	5-100	0.001-1	24 608	1	1223	26.56	4776	5	666	03.06
5000	0.1-100	0.001-0.05	30 000	0	0	34.41	13 097	5	645	08.33
5000	0.1-100	0.05-1	2212	5	305	02.08	364	5	122	00.18
5000	0.1-100	0.001-1	24 798	1	92	27.08	6588	5	368	04.68
10 000	0.1-5	0.001-0.05	30 000	0	0	59.02	10454	5	1730	12.01
10 000	0.1-5	0.05-1	3097	5	683	06.17	854	5	211	01.13
10 000	0.1-5	0.001-1	2562	5	689	05.13	850	5	345	01.14
10 000	5-100	0.001-0.05	4079	5	1306	07.18	584	5	515	00.52
10 000	5-100	0.05-1	3388	5	217	05.52	244	5	75	00.24
10 000	5-100	0.001-1	3615	5	794	06.21	418	5	348	00.39
10 000	0.1-100	0.001-0.05	4722	5	1954	08.42	803	5	577	01.10
10 000	0.1-100	0.05-1	3917	5	377	07.00	494	5	157	00.45
10 000	0.1-100	0.001-1	4725	5	1995	08.43	804	5	649	01.10

REFERENCES

Anily, S. and Federgruen, A. (1990) One Warehouse Multiple Retailer Systems with Vehicle Routing Costs. *Management Science*, **36**, 92–114.

Bertazzi, L., Speranza, M.G. and Ukovich, W. (1994) *Minimization of logistic costs with given frequencies.* Technical Report 79, Department of Quantitative Methods, University of Brescia (submitted).

Bertazzi, L., Speranza, M.G. and Ukovich, W. (1995) Dominance Rules and Approximate Solutions for the Transportation Problem with Discrete Available Frequencies, (in preparation).

Blumenfeld, D.E., Burns, L.D., Diltz, J.D. and Daganzo, C.F. (1985) Analyzing trade-offs between transportation, inventory and production costs on freight networks. *Transportation Research, 19B* 361–80.

Burns, L.D., Hall, R.W., Blumenfeld, D.E. and Daganzo, C.F. (1984) Distribution Strategies that Minimize Transportation and Inventory Cost. *Operations Research*, **33**, 469–90.

Keeney, R. L., and Raiffa, H. (1976) *Decisions with multiple objectives.* Wiley, New York.

Speranza, M.G. and Ukovich, W. (1991) *A capacitated transportation problem with factoring costs.* Technical Report 26, Department of Quantitative Methods, University of Brescia (submitted).

Speranza, M.G. and Ukovich, W. (1994) Minimizing transportation and inventory costs for several products on a single link. *Operations Research*, **42**, 879–94.

The Traveling Salesman Problem with Precedence Constraints and Binary Costs

L. Bianco[‡], *P. Dell'Olmo*[‡], *S. Giordani*[§]
[‡]*Electronic Eng. Dept.*, [§]*"V. Volterra" C.–Univ. of Rome "Tor Vergata"*
Viale della Ricerca Scientifica, I-00133 Rome, Italy. Ph.: +39-6-77161.
Fax: +39-6-7716461. email: giordani@iasi.rm.cnr.it

Abstract

Given a set of n elements $V = \{1, 2, \ldots, n\}$ and a precedence relation $<_P \subseteq V \times V$ defined on V, where $i <_P j$ or $(i,j) \in <_P$ means that i *precedes* j, let us consider the complete digraph $G = (V, A)$ with no loops and with $(j,i) \notin A$ if and only if $i <_P j$. Let be assigned a cost $c_{ij} \in \{0,1\}$ to each arc $(i,j) \in A$ with $c_{ij} = 0$ if and only if $(j,i) \notin A$. Let G' be the digraph obtained from G adding a dummy vertex s, the Traveling Salesman Problem with Precedence Constraints and Binary Costs (TSP-PC-BC) is to find a hamiltonian tour of minimum cost in G', starting from vertex s, visiting each vertex $i \in V$ after having visited every vertex $j \in V$ that must precede i and returning to vertex s. In this paper we investigate the TSP-PC-BC using a different approach from those used in previous works on TSP: we replace the graph G' with a partially ordered set $P = (V, <_P)$, defined by $<_P$ on V and reformulate the problem as the Jump Number Problem of P. We also propose exact and heuristic solution algorithms. Computational results on a set of randomly generated test problem are discussed.

Keywords

TSP, binary costs, posets, linear extensions, heuristic, dynamic programming.

1 INTRODUCTION

Given a set of n elements $V = \{1, 2, \ldots, n\}$ and a precedence relation $<_P \subseteq V \times V$ defined on V, where $i <_P j$ or $(i,j) \in <_P$ means that i *precedes* j, let us consider the complete digraph $G = (V, A)$ with no loops and with $(j,i) \notin A$ if and only if $i <_P j$. Let be assigned a cost $c_{ij} \in \{0,1\}$ to each arc $(i,j) \in A$ with $c_{ij} = 0$ if and only if $(j,i) \notin A$. Let G' be the digraph obtained from G adding a dummy vertex s, the Traveling Salesman Problem with Precedence Constraints and Binary Costs (TSP-PC-BC) is to find a hamiltonian tour of minimum cost in G', starting from vertex s, visiting each vertex $i \in V$ after having visited every vertex $j \in V$ that must precede i and returning to vertex s.

The Traveling Salesman Problem with Precedence Constraints is known to be $\mathcal{NP}$-

hard and has several practical applications in sequencing and routing problems, like crew scheduling and vehicle routing. It has been investigated by Bianco Mingozzi and Ricciardelli (1994) and Savelsberg (1990) for the general case and in the *dial-a-ride problem* version by Psaraftis (1980, 1983). At the best of our knowledge the particular case TSP-PC-BC has not yet been studied, and questions about this problem, e.g. computational complexity, are not yet explicitly answered.

In this paper we investigate the TSP-PC-BC using a different approach from those used in previous works: we replace the graph with a partially ordered set $P = (V, <_P)$, defined by the precedence relations between the elements of V and reformulate the problem as the jump number problem of P as explained in the sequel.

Given the precedence relation $<_P$, let us consider the relational set $(V, <_P)$, that is a *partially ordered* set (*poset*) $P = (V, <_P)$, since the relation $<_P$ is irreflexive, antisymmetric and transitive. Notationally, $u <_P v$ means that u *precedes* v, while $u \parallel_P v$ means that u and v are incomparable and $u \not<_P v$ simply means that u does not precede v, regardless if v precedes or not u; moreover, two elements $u, v \in V$ are comparable ($u \sim_P v$) if $u <_P v$ or $v <_P u$. Given a partially ordered set $P = (V, <_P)$, a linear extension $L = (V, <_L)$ of P is a total order on the same ground set V of P, such that each couple of elements $u, v \in V$ for which $u <_P v$ implies $u <_L v$. Given two consecutive elements $u, v \in L$ the couple (u, v) is a *jump* or *setup* of L if u, v are not comparable in P; the jump number of L, denoted by $s(L, P)$ is the number of jumps in L, while the jump number of P, denoted by $s(P)$, is the minimum of $s(L, P)$ taken over all linear extensions of P. A linear extension L of P is called *jump optimal* if $s(L, P) = s(P)$. The *jump number problem* of P consists in determining $s(P)$ and a jump optimal linear extension L of P.

For the above formulation some complexity results are known. In fact, the jump number problem is known to be $\mathcal{NP}$-hard for general posets (Pulleyblank, 1981), even if there are some classes of posets for which the problem may be optimally solved in polynomial time. In the non polynomial cases, an exact algorithm for general posets has been developed by Bianco, Dell'Olmo and Giordani (1994a) and approximation algorithms exist only for interval orders (Mitas, 1991).

In this work, we show that the TSP-PC-BC is equivalent to the Jump Number problem of $P = (V, <_P)$, proving that TSP-PC-BC remains $\mathcal{NP}$-hard. We propose both heuristic and exact algorithms, which solve TSP-PC-BC finding a jump optimal linear extension of P. Computational results show that the algorithms presented in this paper overcome the performances of the algorithm proposed by Bianco, Dell'Olmo and Giordani (1994a).

2 THE TSP-PC-BC AND THE JUMP NUMBER PROBLEM

Let $G = (V, A)$ be a complete digraph (i.e. a directed graph with at least one arc joining every pair of vertices) without loops and with the following cost assignment:

(a) $c_{ij} \in \{0, 1\}$;
(b) $c_{ij} = 0$ if and only if $(i, j) \in A$ and $(j, i) \notin A$;
(c) $c_{ik} + c_{kj} \geq c_{ij}$.

Furthermore, let $<_P$ be a binary relation on the elements of V defined as:

$$i <_P j \text{ for } c_{ij} = 0. \tag{1}$$

Proposition 1 *The relation $<_P$ stated as above on the graph $G = (V, A)$ defines a poset $P = (V, <_P)$ if and only if the costs of the arcs of G fulfill properties (a), (b) and (c).*

Proof. Let G_P be the digraph obtained from G considering only the zero cost arcs. The binary relation $<_P$ defined on the set V by the arc set of G_P is transitive by property (c), antisymmetric by property (b) and irreflexive since G_P is with no loops. Therefore $P = (V, <_P)$ is a poset and, furthermore, G_P is a transitive orientation of the edges of a comparability graph (Möring, 1985).

Vice-versa, given a poset $P = (V, <_P)$, let denote with $G'' = (V, A'')$ the complete symmetric digraph with the following arc costs: $c_{ij} = 0$ and $c_{ji} = \infty$ for $i <_P j$; $c_{ij} = 1$ and $c_{ji} = 1$ for $i \parallel_P j$; $c_{ii} = \infty$. Removing from G'' the infinity cost arcs we obtain a complete digraph $G = (V, A)$ without loops. By definition the costs of the arcs in A respect properties (a) and (b). Moreover, from the transitive property of the binary relation $<_P$ the property (c) follows. □

Then, we may say that the poset $P = (V, <_P)$ and the complete digraph $G = (V, A)$ defined above with the relation $<_P$ as in (1), are equivalent. See, for example, the digraph $G = (V, A)$ and the equivalent poset $P = (V, <_P)$ depicted in Figure 1a. Moreover, it is easily proved that:

Corollary 1 *Let be given $P = (V, <_P)$ and the equivalent representation $G = (V, A)$. It results $i \parallel_P j$ if and only if $c_{ij} = c_{ji} = 1$ in G.*

Corollary 2 *Let us consider $P = (V, <_P)$ and the equivalent representation $G = (V, A)$. It results $i \not<_P j$ if and only if $(j, i) \in A$.*

Proposition 2 *Given $G = (V, A)$ and the equivalent poset $P = (V, <_P)$, a total order $\Pi = (a_1, a_2, \ldots, a_n)$ of the elements of V is a linear extension L of P if and only if Π corresponds to a hamiltonian path H_L in G respecting the precedence relation $<_P$.*

Proof. Let us consider a total order Π of the elements of V and suppose that it represents a linear extension $L = a_1, a_2, \ldots, a_n$ of P. L defines a path in G since for each couple of consecutive elements (a_i, a_{i+1}) in L it results $a_i <_P a_{i+1}$ or $a_i \parallel_P a_{i+1}$, that is $(a_i, a_{i+1}) \in A$ from Corollary 1. Actually, $H_L = (a_1, a_2, \ldots, a_n)$ is a hamiltonian path of G since L contains every element of P exactly once. Moreover, by definition of linear extension, H_L respects the precedence relations defined by P.

Vice-versa, let us consider a hamiltonian path $H_L = (a_1, a_2, \ldots, a_n)$ in $G = (V, A)$. It represents a total order of the elements of V; to represent a linear extension $L = a_1, a_2, \ldots, a_n$ of P it is necessary that for each couple (a_i, a_{i+1}) of H_L $a_{i+1} \not<_P a_i$; but this is the case, since each arc $(a_i, a_{i+1}) \in A$ implies that $a_{i+1} \not<_P a_i$, by Corollary 2. Moreover, H_L has to respect the precedence relation on the elements of V, otherwise in $H_L = (a_1, \ldots, a_i, \ldots, a_{i+k}, \ldots, a_n)$ there will be at least a couple of elements (say a_i and a_{i+k}) for which $a_{i+k} <_P a_i$, that is L is not a linear extension. □

Corollary 3 *The jump number $s(L, P)$ of a linear extension L of a poset $P = (V, <_P)$ is equal to the cost of the corresponding hamiltonian path H_L in the equivalent $G = (V, A)$.*

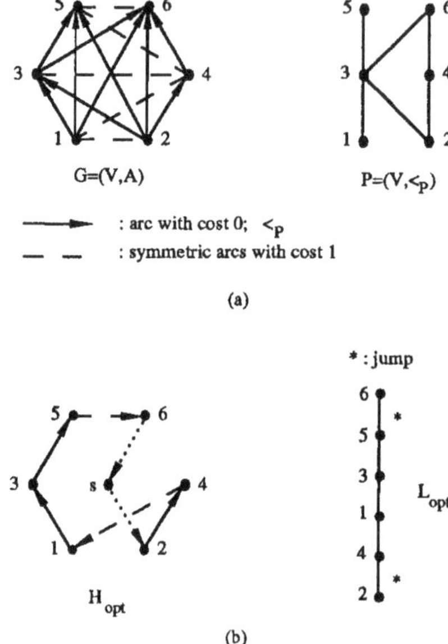

Figure 1 (a): A digraph $G = (V, A)$ with relation $<_P$ and the *Hasse diagram* of the equivalent poset P; (b): an optimal solution H_{opt} for the TSP-PC-BC on $G' = (V \cup \{s\}, A \cup B)$ and an optimal solution L_{opt} for the jump number problem on P.

Proof. It follows directly from the costs of the arcs in A. In fact, by Corollary 1 the arcs in H_L whose cost is not equal to zero but one are only those arcs corresponding to incomparable couples of elements in P, that the jumps in L. Therefore the cost of H_L is equal to the jump number of L. □

Now, we may assert that finding the jump number of a poset P is equivalent to find a minimum cost hamiltonian path, respecting the precedence relations defined by P, in a complete digraph $G = (V, A)$ without loops and binary costs for the arcs. Moreover, let $G' = (V \cup \{s\}, A \cup B)$ be the digraph obtained adding to G a dummy vertex s and the zero cost arcs in $B = \bigcup_{v \in V}\{(s, v), (v, s)\}$. The jump number problem on $P = (V, <_P)$ is equivalent to the Traveling Salesman Problem with Precedence Constraints and Binary Costs (TSP-PC-BC) on G', that is to find a minimum cost hamiltonian tour starting from vertex s, visiting every vertex i for which $i <_P j$ before j and returning to s. An example is given in Figure 1.

Theorem 1 *The TSP-PC-BC is $\mathcal{NP}$-hard.*

Proof. It follows from the fact that the jump number problem, which is $\mathcal{NP}$-hard (Pulleyblank, 1981), *polynomially transforms* to TSP-PC-BC as explained before. □

3 HEURISTIC ALGORITHM

First of all, we will report some further notations and definitions. Given a poset $P = (V <_P)$, a *subposet* $P|X$ of P is a subset of P with the induced order. A *chain* C is a set of pairwise comparable elements of P, that is a subposet of P which is a total order of P; moreover, let $Min(P)$ and $Max(P)$ be the sets of minimal and maximal elements of P, respectively. Given an element $v \in P$, let us consider the sets $Pred(v)$ and $Suc(v)$: we say that v is *down-accessible* (*up-accessible*) if $Pred(v)$ ($Suc(v)$) is a chain in P. Furthermore, a chain C in P will be called *greedy* (**greedy*) if the top element of the chain C, denoted with $sup_P(C)$ (the bottom element of the chain C, denoted with $inf_P(C)$) is a maximal down-accessible (up-accessible) element of P. We will consider greedy chains C_{u_k} for which $u \in C_{u_k}$ and $u \in Min(P)$, and *greedy chains $C^*_{v_l}$ for which $v \in C^*_{v_l}$ and $v \in Max(P)$, and we will call them, respectively, *greedy starting chains* and **greedy ending chains* of P. Moreover, let us say that $v \in P$ is *S-maximal* if $\forall u \in P, Suc(u) \subseteq Suc(v)$. It is clear that each S-maximal element of P is also a member of $Min(P)$. Likewise, let us define $v \in P$ an *S*-maximal* element of P if $\forall u \in P, Pred(u) \subseteq Pred(v)$ (obviously $v \in Max(P)$ too).

It is known (see Faigle and Schrader 1985) that given a poset P, if the set $S_{max}(P)$ of S-maximal elements of P exists then there also exists a jump optimal linear extension whose starting subset is composed by $S_{max}(P)$, regardless the order in which the S-maximal elements are chosen. A similar result may be obtained considering the S*-maximal elements of P and choosing them as the ending subset of a linear extension. Consider the following situation. Given a poset $P = (V, <_P)$, suppose that:

the S-maximal $S_{max}(P)$ and S*-maximal $S^*_{max}(P)$ sets of P exist and do not intersect, (i)

and, considering the poset $P_1 = P|(V - (S_{max}(P) \cup S^*_{max}(P)))$, suppose that:

the S-maximal $S_{max}(P_1)$ and S*-maximal $S^*_{max}(P_1)$ sets of P_1 exist and do not intersect. (ii)

Proposition 3 *If hypotheses (i) and (ii) hold, there exists a jump optimal linear extension $L = L^s_0 \oplus L_1 \oplus L^e_0$ of P, where L^s_0 is any linear order of $S_{max}(P)$ and L^e_0 is any linear order of $S^*_{max}(P)$, and L_1 is a jump optimal linear extension of P_1, whose starting subset and ending subset are, respectively, a particular linear order of $S_{max}(P_1)$ where the starting element is $u_1 \in S_{max}(P_1)$, such that $sup_L(L^s_0) <_P u_1$ if such u_1 exists, and a particular linear order of $S^*_{max}(P_1)$, where the ending element is $z_1 \in S^*_{max}(P_1)$, such that $z_1 <_P inf_L(L^e_0)$ if such z_1 exists.*

In the cases in which the set $S_{max}(P_1)$ ($S^*_{max}(P_1)$) is empty or there is no elements $u_1 \in S_{max}(P_1)$ such that $sup_L(L^s_0) <_P u_1$ ($z_1 \in S^*_{max}(P_1)$ such that $z_1 <_P inf_L(L^e_0)$), it is possible to consider the set of greedy starting chains $\{C_{u_k}\}$ of P_1 for each $u \in Min(P_1)$ (the set of *greedy ending chains $\{C^*_{v_l}\}$ of P_1 for each $v \in Max(P_1)$); from this set of chains it will be convenient to choose the greedy starting chain C_q of maximum size such that $sup_L(L^s_0) <_P inf_P(C_q)$, if it exists (the *greedy ending chain C^*_r of maximum size such that $sup_P(C^*_r) <_P inf_L(L^e_0)$, if it exists) and such that $C_q \cap C^*_r = \emptyset$.

All these considerations lead to the definition of a heuristic algorithm, in which a linear extension of P is constructed from the bottom and from the top, adding at each iteration two disjointed subsets of the elements of P, chosen, respectively, from the S-maximal set and the greedy starting chains, and from the S*-maximal set and the *greedy ending chains of the residual poset. For a complete description of the proposed heuristic algorithm see (Bianco, Dell'Olmo and Giordani, 1994b).

4 EXACT ALGORITHM

In the following, we design an exact algorithm based on dynamic programming. To this end, we illustrate some further properties of a poset.

Proposition 4 *Given a poset $P = (V, <_P)$, for any subposet $P|X$, where X is a subset of the ground set V of P, the jump number of $P|X$ is $s(P|X) \leq s(P)$.*

Theorem 2 *Given a poset $P = (V, <_P)$ for each greedy starting chain C of P it results $s(P|(V-C)) \leq s(P) \leq s(P|(V-C)) + 1$.*

Corollary 4 *The minimum jump number in some linear extension L of P with C as starting subset is equal to $s(P|(V-C)) + 1$.*

Theorem 3 *Given a poset $P = (V, <_P)$ there exists a jump optimal linear extension L of P, whose starting subset is composed by a greedy starting chain C.*

On the basis of Theorem 3, a jump optimal linear extension may be found generating only linear extensions which have a greedy starting chain as a starting subset. Let us consider the couple (X, C_i) representing the subposet $P|X$ and an its greedy starting chain C_i. Let $f(X, C_i)$ be the minimum jump number in some linear extensions of $P|X$ with C_i as starting subset. Let $\mathcal{G} = (\mathcal{S}, \mathcal{A})$ be a digraph (called *state space graph*) whose vertex set $\mathcal{S}$ contains the *states* (X, C_i), and whose arc set $\mathcal{A} = \{((X, C_i), (X', C_j)) : X = X' \cup C_i\}$ represents all feasible transitions from one state $(X, C_i) \in \mathcal{S}$ to another $(X', C_j) \in \mathcal{S}$. The cost of each arc $((X, C_i), (X', C_j)) \in \mathcal{A}$ is equal to 1, as for Corollary 4. Then, the dynamic programming recursion that defines $f(X, C_i)$ may be stated as follows:

$$f(X, C_i) = \min_{\{(X', C_j) \in \mathcal{S} : ((X, C_i), (X', C_j)) \in \mathcal{A}\}} \{f(X', C_j) + 1\},$$

initialized by $f(Y, Y) = 0$, since $P|Y$ is a chain. The optimal solution to the problem is then given by $z^* = \min_{(V, C_v) \in \mathcal{S}} f(V, C_v)$.

Moreover, let us consider the following two dominance rules. We refer to a poset P that, without loss of generality, is not a chain. The dominance rules may be respectively established on the basis of the following lemmas, which may be easily proved:

Lemma 1 *Given a poset $P = (V, <_P)$ and a greedy starting chain C of P, consider the element $a = sup_P(C)$. If $a \in Max(P)$ then $s(P) = s(P|(V-C)) + 1$, that is there exist an optimal linear extension of P starting with C.*

Lemma 2 *Given a poset $P = (V, <_P)$ and a greedy starting chain C of P, consider the element $a = sup_P(C)$. If $a \in S_{max}(P|((V-C) \cup \{a\}))$ then $s(P) = s(P|(V-C)) + 1$, that is there exist an optimal linear extension of P starting with C.*

The previous considerations allow to design a dynamic programming algorithm which recursively generates the states, computes their costs and marks an outgoing arc in $\mathcal{A}$

Table 1 Computational results ($n = 50 \div 250$, $p = 0.1 \div 0.8$)

n	p	*States*	*CPU Secs.*	$\frac{UB-LB}{LB}$	$\frac{UB-z^*}{z^*}$	LB	z^*	UB
50	0.1	168.8	4.69	0.26	0.08	15.2	17.8	19.2
	0.2	705.4	11.36	0.83	0.12	10.2	16.2	18.0
	0.4	62.4	2.06	0.55	0.07	9.8	13.8	14.8
	0.6	10.6	0.95	0.25	0.03	9.0	10.6	11.0
	0.8	1.2	0.28	0.04	0.00	6.2	6.6	6.6
100	0.1	2017.2	159.94	1.05	0.18	18.6	32.0	37.8
	0.2	791.6	107.75	1.19	0.11	14.8	29.2	32.4
	0.4	257.4	66.32	0.79	0.11	17.0	27.0	30.0
	0.6	36.0	24.47	0.26	0.01	19.8	24.4	24.6
	0.8	1.6	3.66	0.03	0.00	15.2	15.6	15.6
150	0.1	4178.2	1239.19	1.62	0.22	20.8	44.8	54.4
	0.2	1339.6	633.75	2.02	0.21	17.2	41.2	49.8
	0.4	338.0	233.82	0.77	0.12	25.6	40.2	45.0
	0.6	80.8	169.03	0.28	0.04	29.4	36.2	37.6
	0.8	6.8	32.55	0.05	0.00	20.6	21.6	21.6
200	0.1	9118.0	6300.20	2.39	0.20	19.8	55.8	67.0
	0.2	1987.8	1788.52	1.74	0.14	22.6	54.4	61.8
	0.4	446.2	1478.02	0.71	0.11	35.0	53.8	59.6
	0.6	104.8	747.17	0.23	0.03	40.2	47.8	49.2
	0.8	15.6	154.35	0.04	0.00	27.2	28.2	28.2
250	0.1	8137.4	12031.00	2.87	0.21	21.6	68.8	83.4
	0.2	2642.0	8027.60	1.76	0.16	29.4	68.6	79.6
	0.4	362.6	2992.76	0.68	0.09	42.4	64.8	70.8
	0.6	204.2	4038.94	0.27	0.03	47.2	58.4	60.2
	0.8	20.4	512.64	0.04	0.00	39.0	40.4	40.4

of the states. The algorithm, looks ahead in the decision process and, in solving the subproblem related to the states (X, C_i), marks the arc $((X, C_i), (X', C_j)) \in \mathcal{A}$ for which $f(X', C_j)$ is minimum. Finally, when every state (V, C_u) is solved, the optimal solution can be obtained going along the path on $\mathcal{G}$, starting from the minimum cost state (V, C_u^*) and following the marked arcs of $\mathcal{G}$, up to a state without outgoing arcs.

An upper bound $UB(X, C_i)$, issued by the heuristic algorithm, and a lower bound $LB(X, C_i)$, calculated by means of the the defect of $P|(X - C_i)$ (i.e. $LB(P|(X - C_i)) = def_K(P|(X - C_i)) - 1$) (Gierz and Poguntke, 1983), of the jump number of a minimum jump linear extension of $P|X$ starting with C_i, are used in order to reduce the number of generated states. For the proofs of the theorems and a complete description of the exact algorithm see (Bianco, Dell'Olmo and Giordani, 1994b).

5 COMPUTATIONAL RESULTS

The heuristic and exact algorithms have been implemented in C programming language and run on a IBM RISC System/6000. They have been experimentally evaluated on a set of randomly generated test problems, for which, the set of precedence relations, defining a poset P, are generated as the transitive closure of a randomly generated direct acyclic graph (dag) without loops, in which it is possible to choose the number n of vertices and the probability threshold p of presence of an arc. That is, for each pair of vertices (i, j)

with $i < j$ the presence of the arc (i,j) in the dag will be considered if $RN \leq p$, where the RN's are pseudorandom numbers from a uniform distribution on $[0,1]$.

We have considered test problems of size $n = 50, 100, \ldots, 250$, and for each size we have randomly generated posets with $p = 0.1, 0.2, 0.4, 0.6, 0.8$. For each (n,p) combination the results reported are averages over 5 randomly generated problems. The experimental results are summarized in Table 1, where we have reported the lower bound LB, the optimal solution values z^* and the heuristic solution values UB. We have also measured the heuristic performance as the relative gaps $(UB - z^*)/z^*$, $(UB - LB)/LB$, respectively related to the optimal and lower bound values.

Heuristic algorithm often finds optimal solutions: this happens for problems with $p = 0.8$, and in general for problem of small size; the worst case is obtained for $n = 150$, $p = 0.1$ in which the average relative error is $(UB - z^*)/z^* = 0.22$.

As for the exact algorithm, it arises that problem complexity increases with n and decreases with p. Nevertheless, the number of states generated by the algorithm are also strictly related to the relative difference $(UB - LB)/LB$. The worst case occurs for posets for $n = 250$, $p = 0.1$ for which the average gap value $(UB - LB)/LB$ is 2.87 and the average number of generated states is 8137.4; this case is also the worst one in terms of CPU time for which the time spent is about 12000 seconds, while the worst case in terms of number states occurs for $n = 200$, $p = 0.1$ with 9118 generated states.

6 REFERENCES

Bianco, L., Dell'Olmo, P. and Giordani, S. (1994a), Exact and Heuristic Algorithms for the Jump Number Problem, in *Operations Research Proceedings 1994* (ed. U. Derigs), Springer Verlag, 145–150; Tech. Rep. IASI R.376.

Bianco, L., Dell'Olmo, P. and Giordani, S. (1994b), The Traveling Salesman Problem with Precedence Constraints and Binary Costs. Tech. Rep. IASI R.397.

Bianco, L., Mingozzi, A. and Ricciardelli, S. (1994), Dynamic Programming Strategies for the Traveling Salesman Problem with Time Windows and Precedence Constraints. *Opns. Res.*, to appear.

Faigle, U. and Schrader, R (1985), A setup heuristic for interval orders. *Oper. Res. Letters*, **4**, 185–188.

Gierz, G. and Poguntke, W. (1983), Minimizing Setups for Ordered Sets: a Linear Algebraic Approach. *SIAM J. Alg. Disc. Meth.*, **4**, 132–144.

Mitas, J. (1991), Tackling the Jump Number Problem for Interval Orders. *Order*, **8**, 115–132.

Möhring, R.H. (1985), Algorithmic Aspects of Comparability Graphs and Interval Graphs, in *Graphs and Order* (ed. I. Rival), NATO ASI Series C, vol. 147, D. Reidel Publisher Company, 41–101.

Psaraftis, H. (1980), A dynamic programming solution to the single vehicle many-to-many immediate request dial-a-ride problem. *Trans. Sci.*, **14**, 130–154.

Psaraftis, H. (1983), k-Interchange procedures for local search in a precedence-constrained routing problem. *E.J.O.R.*, **13**, 391–402.

Pulleyblank, W.R. (1981), Preliminary manuscript.

Savelsberg, M.W.P. (1990), An efficient implementation of local search algorithms for constrained routing problems. *E.J.O.R.*, **47**, 75–85.

Cost oriented competing processes - a new handling of assignment problems

Jens Starke
University of Stuttgart, II. Institute of Theoretical Physics
Pfaffenwaldring 57/III, D-70550 Stuttgart, Germany.
Tel: +49 711 685-4932. Fax: +49 711 685-4902.
e-mail: jens@theo2.physik.uni-stuttgart.de

Abstract
The method of *Cost Oriented Competing Processes (COCP)* is a new approach to handle *assignment problems* and *combinatorial optimization problems.* The COCP method works with a system of nonlinear coupled ordinary differential equations with suitable stable points. In contrast to many other techniques, this method can easily be generalized and adapted to different assignment problems, even to those of higher order and with additional constraints. In assignment problems of higher order, the single costs which have to be considered for the total costs depend on several decisions instead of only one decision as in linear assignment problems. The handling of those higher order assignment problems is important for several scheduling problems such as manufacturing planning and time table planning.

Keywords
Combinatorial optimization, assignment problems, quadratic assignment problems (QAP), assignment problems of higher order, scheduling, selforganization, manufacturing planning, neural networks

1 LINEAR ASSIGNMENT PROBLEMS

In an assignment problem (see *e.g.* (Eiselt et al., 1987), (Zimmermann, 1990)) there are decision variables $x_{ij} \in \{0, 1\}$. The aim of linear assignment problems is to minimize

$$\sum_{i,j} c_{ij} \cdot x_{ij} \tag{1}$$

with the costs c_{ij} and with respect to the constraints

$$\sum_i x_{ij} = 1 \quad \forall j \in \{1,...,N\} \quad \text{and} \tag{2}$$

$$\sum_j x_{ij} = 1 \quad \forall i \in \{1,...,N\}. \tag{3}$$

The matrix (x_{ij}) is called the permutation matrix. The row index i indicates *e.g.* the number of a particular worker and the column index j relates to a particular job. The costs $c_{ij} \in \mathbb{R}$ have to be incurred if the i-th worker is assigned to the j-th job, i. e. if and only if $x_{ij} = 1$.

1.1 Handling of assignment problems using COCP

The Cost Oriented Competing Processes handle assignment problems and combinatorial optimization problems by using a system of nonlinear coupled ordinary differential equations with suitable stable points. In other words: the discrete problem is handled with a method for continuous variables. This is possible because the positions of the stable points of the differential equations of the COCP are discrete.

First, a linear transformation of the cost array is made to change the minimizing problem to a maximizing problem:

$$\xi_{ij}(t_0) = a - b \cdot c_{ij} \quad \text{with} \quad a, b > 0 \quad \text{and} \quad \xi_{ij}(t_0) \geq 0 \quad \forall i, j. \tag{4}$$

The condition $\xi_{ij}(t_0) \geq 0 \quad \forall i, j$ ensures a nonnegative output of the method. Because of numerical reasons a and b are chosen in a way that $\xi_{ij} \leq 1 \quad \forall i, j$. If this condition is not fulfilled, the values of the following differential equations (7) will become too large for a numerical treatment. Second, the variables ξ_{ij} are submitted to a dynamic competing process. The stable points ξ^*_{ij} of the dynamics of the COCP fulfill the constraints $\xi^*_{ij} \in \{0, 1\}$ as well as the equations (2) and (3) for the unknown decision variables x_{ij}. Thus for any stable points ξ^*_{ij} the condition of a permutation matrix is fulfilled. Therefore, in the last step, the stable point in which the dynamic process ends can be assigned to the decision variables $x_{ij} \in \{0, 1\}$. The nonvanishing elements ξ^*_{ij} with $\xi^*_{ij} = 1$ correspond to the final choice of positive decisions. The nonexistence of further stable points can be proved. Under consideration of the constraints, the dynamic competing processes prefer larger ξ_{ij} (i. e. smaller costs c_{ij}) to change them to '1' and the others to '0'. This guarantees that the decisions x_{ij} are made by using a selforganizing process, selecting the lowest costs with respect to the constraints. The procedure of the COCP can be illustrated as

$$c_{ij} \longrightarrow \xi_{ij}(t_0) \longrightarrow \xi_{ij}(t) \longrightarrow \xi^*_{ij} \longrightarrow x_{ij}\,.$$

These competing processes are realized with a system of nonlinear coupled differential equations. Similar to the equation of motion of the *synergetic computer* (Haken, 1990), the dynamical system can be derived in a vivid manner by using a potential function

$$V = -\frac{1}{2}\sum_k \sum_l \xi^2_{kl} + \frac{1}{4}\sum_k \sum_l \xi^4_{kl}$$

$$+\frac{1}{2}\sum_{k}\sum_{l}\sum_{\substack{k' \\ k'\neq k}}\xi_{kl}^2\xi_{k'l}^2+\frac{1}{2}\sum_{k}\sum_{l}\sum_{\substack{l' \\ l'\neq l}}\xi_{kl}^2\xi_{kl'}^2 \quad \text{with} \quad k,k',l,l'=1,...,N \tag{5}$$

and a gradient descent method

$$\dot{\xi}_{ij} \;=\; -\frac{\partial V}{\partial \xi_{ij}}\,. \tag{6}$$

The potential function V has minima, i.e. stable points, if and only if the constraints of a permutation matrix are fulfilled. A proof by examination of the Hesse matrix $\frac{\partial^2 V}{\partial \xi_{ij} \partial \xi_{kl}}$ can be found in (Starke, 1994). The negative quadratic terms of the potential function (5) and the following terms of order 4 cause a stable point for $\xi_{ij} = 1 \quad \forall i,j$ only. The last two terms prevent the case of more than one nonvanishing element of the matrix (ξ_{ij}) in one row or column by putting 'hills' in these areas. Hereby the former stable point is replaced by stable points next to the 'hill'. Using (6) and the potential function (5) the equation of motion can be derived to

$$\dot{\xi}_{ij} \;=\; \xi_{ij}+3\xi_{ij}^3-2\xi_{ij}\sum_{k}\xi_{kj}^2-2\xi_{ij}\sum_{l}\xi_{il}^2. \tag{7}$$

The potential function (5) has some saddlepoints. To avoid the persistence at a saddelpoint one has to add some noise to the gradient descent method (6).

A simulation example of a linear assignment problem using the COCP is shown in figure 1. The figure shows 4 time steps of the dynamic evolution of the matrix (ξ_{ij}). The size

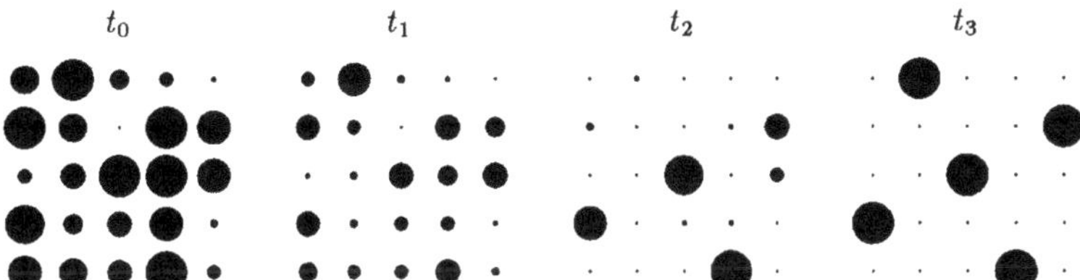

Figure 1 4 time steps of a simulation of the COCP handling a linear assignment problem.

of the circles in figure 1 is proportional to the size of the dynamic variables ξ_{ij}. In the following the cost matrix (c_{ij}) and the rounded initial values $(\xi_{ij}(t_0))$ of the simulation example are shown:

$$(c_{ij}) := \begin{pmatrix} 8 & 3 & 11 & 13 & 16 \\ 2 & 8 & 17 & 2 & 7 \\ 12 & 9 & 4 & 4 & 6 \\ 5 & 11 & 9 & 7 & 14 \\ 6 & 8 & 9 & 3 & 13 \end{pmatrix} \quad \text{and} \quad (\xi_{ij}(t_0)) = \begin{pmatrix} 0.6 & 0.9 & 0.4 & 0.3 & 0.1 \\ 1.0 & 0.6 & 0.0 & 1.0 & 0.7 \\ 0.3 & 0.5 & 0.9 & 0.9 & 0.7 \\ 0.8 & 0.4 & 0.5 & 0.7 & 0.2 \\ 0.7 & 0.6 & 0.5 & 0.9 & 0.3 \end{pmatrix}.$$

The result of the simulation of the COCP is identical with the result in (Zimmermann, 1990) who applied the hungarian method.

Additional constraints

Additional constraints can easily be considered by adding further potential terms to V which remove the unwanted stable points of the differential equations. The case of two nonvanishing elements ξ_{ij} and ξ_{kl} can be prevented by an additional potential term

$$V' = \alpha \cdot \xi_{ij}^2 \cdot \xi_{kl}^2. \tag{8}$$

The constant $\alpha > 0$ has to be chosen in a suitable way so that the 'hill' fills up the unwanted minima sufficiently. This has to be checked by a stability analysis. By using the additional potential term

$$V'' = \beta \cdot \xi_{k_1 k_2}^2 \cdot \xi_{k_3 k_4}^2 \cdot \ldots \cdot \xi_{k_{2m-1} k_{2m}}^2 \tag{9}$$

with suitable $\beta > 0$, it is even possible to prevent the specific combination of the decisions

$$x_{k_1 k_2} = 1, \ldots, x_{k_{2m-1} k_{2m}} = 1.$$

The resulting potential function which has to be used for the equation of motion (6) is

$$\tilde{V} = V + V' \quad \text{or respectively} \quad \tilde{V} = V + V''.$$

An example of preventing specific combinations of decisions is given in (Starke, 1994) for the TSP.

Selection processes of biological macromolecules

The evolutionary processes of biological macromolecules (Eigen, 1971), (Eigen and Schuster, 1977), (Eigen and Schuster, 1978) are special cases of these competing processes. In contrast to the differential equations in the papers cited above, each variable ξ_{ij} of the *Cost Oriented Competing Processes* competes with different partners.

1.2 Competing process between two elements

The preference of the largest elements ξ_{ij} can easily be shown in the following example. The competing process between the two elements ξ_{ij} and ξ_{ir} in row i of matrix ξ is given by the following two differential equations:

$$\dot{\xi}_{ij} = \xi_{ij} \cdot \left(1 + 3\xi_{ij}^2 - 2\sum_k \xi_{kj}^2 - 2\sum_l \xi_{il}^2\right) \tag{10}$$

$$\dot{\xi}_{ir} = \xi_{ir} \cdot \left(1 + 3\xi_{ir}^2 - 2\sum_k \xi_{kr}^2 - 2\sum_l \xi_{il}^2\right) \tag{11}$$

In the special case of $\sum_k \xi_{kj}^2 = \sum_k \xi_{kr}^2$ from $\xi_{ij} > \xi_{ir}$, one gets the result $\dot{\xi}_{ij} > \dot{\xi}_{ir}$ because of the strictly monotonouse right hand side of the differential equations (10) and (11).

That means that the larger variable ξ_{ij} grows faster ($\dot{\xi}_{ij} > 0$) or decays slower ($\dot{\xi}_{ij} < 0$). Because of this fact and because of the special position of the stable points, the largest element ξ_{ij} of row i will grow to '1' and all other decay to '0'.

1.3 Possibilities of intervention for the human decision-maker

The COCP provides two possibilities of intervention for the decision-maker. First, general rules or experiences of the human decision-maker can be considered by using further potential terms in the form of (8) or (9) to prevent specific decisions. With this it is even possible to prevent a specific combination of several decisions. Second, single changes of the problem configuration, like missing material or broken machines, can be considered by initializing the corresponding elements of the matrix (ξ_{ij}) with '0' to prevent or '1' to enforce a particular assignment.

1.4 Neural network interpretation

Similar to (Haken, 1990), (Baird, 1990) and (Hirsch and Baird, 1995) the differential equations of the *Cost Oriented Competing Processes* can be treated as a neural network of higher order. The equation of motion of the COCP can be written in the form

$$\dot{\xi}_{ij} = \sum_{k,l} T_{ijkl} \cdot \xi_{kl} + \sum_{k,l,m,n} J_{ijklmn} \cdot \xi_{ij}\xi_{kl}\xi_{mn}.$$

The constants J_{ijklmn} of the higher order terms in the equation of motion can be interpreted as higher order couplings. In the COCP these higher order couplings and the couplings T_{ijkl} are chosen in a way so that the final output ξ^*_{ij}, i. e. the stable points, correspond to decision variables $x_{ij} \in \{0, 1\}$ with respect to the constraints of the assignment problem.

2 ASSIGNMENT PROBLEMS OF HIGHER ORDER

The principles of the COCP explained in the section on linear assignment problems can be transferred to assignment problems of higher order by using an appropriate number of dimensions for the dynamic variables.

2.1 Quadratic assignment problems (QAP)

Handling *quadratic assignment problems (QAP)*, one has to minimize

$$\sum_{i,j,k,l} c_{ijkl} \cdot x_{ij} \cdot x_{kl}$$

with respect to the constraints $x_{ij} \in \{0, 1\}$,

$$\sum_{i} x_{ij} = 1 \quad \forall j \in \{1, ..., N\} \quad \text{and} \quad \sum_{j} x_{ij} = 1 \quad \forall i \in \{1, ..., N\}.$$

The costs c_{ijkl} have to be incurred for the total costs if and only if $x_{ij} = 1$ and $x_{kl} = 1$, i.e. if and only if both decisions are chosen positive. In this case, the *Cost Oriented Competing Processes* are applied to the 4-dimensional array (c_{ijkl}). For this reason the dynamic variable ξ_{ijkl} is 4-dimensional, too. In order to calculate the initial values $\xi_{ijkl}(t_0)$ the linear transformation (4) is used again. Then the equation of motion is

$$\begin{aligned} \dot{\xi}_{ijkl} &= -\frac{\partial V}{\partial \xi_{ijkl}} \\ &= \xi_{ijkl} + 7\xi_{ijkl}^3 - 2\xi_{ijkl}\left(\sum_{i',j',k'} \xi_{i'j'k'l}^2 + \sum_{i',j',l'} \xi_{i'j'kl'}^2 + \sum_{i',k',l'} \xi_{i'jk'l'}^2 + \sum_{j',k',l'} \xi_{ij'k'l'}^2\right) . \end{aligned}$$

The positions of the stable points of this dynamics ensure that $\xi_{ijkl} \in \{0, 1\}$ and

$$\sum_{i,j,k} \xi_{ijkl}^* = 1 \quad \forall l, \ldots, \sum_{j,k,l} \xi_{ijkl}^* = 1 \quad \forall i .$$

The decision variables x_{ij} are obtained by projecting the array ξ_{ijkl}^* onto those two dimensions belonging to the decision variables x_{ij} with

$$x_{ij} = \sum_{k,l} \xi_{ijkl}^* .$$

In time dependent optimization problems, the other dimensions can belong to previous decisions. The constraints of the assignment problem are fulfilled because of

$$\begin{aligned} \sum_i x_{ij} &= \sum_i \sum_{k,l} \xi_{ijkl}^* = 1 \quad \forall j \\ \sum_j x_{ij} &= \sum_j \sum_{k,l} \xi_{ijkl}^* = 1 \quad \forall i . \end{aligned}$$

2.2 Assignment problems of order n

In *assignment problems of order n* one has to minimize

$$\sum_{k_1, k_2, \ldots, k_{2n}} c_{k_1, k_2, \ldots, k_{2n}} \cdot x_{k_1 k_2} \cdot x_{k_3 k_4} \cdot \ldots \cdot x_{k_{2n-1} k_{2n}}$$

with respect to the constraints $x_{k_1 k_2} \in \{0, 1\}$,

$$\sum_{k_1} x_{k_1 k_2} = 1 \quad \forall k_2 \in \{1, \ldots, N\} \quad \text{and} \quad \sum_{k_2} x_{k_1 k_2} = 1 \quad \forall k_1 \in \{1, \ldots, N\}.$$

The costs $c_{k_1, \ldots, k_{2n}}$ have to be considered for the total costs if and only if all of the decision variables $x_{k_1 k_2}$, $x_{k_3 k_4}$, ..., $x_{k_{2n-1} k_{2n}}$ are equal 1, i.e. if and only if all decisions are chosen positive. In this case, the *Cost Oriented Competing Processes* are applied to the 2n-dimensional array $(c_{k_1,\dots,k_{2n}})$. As in the quadratic assignment problem, the linear transformation (4) is used here again. The result of the equation of motion is

$$\begin{aligned}
\dot{\xi}_{k_1,\dots,k_{2n}} &= -\frac{\partial V}{\partial \xi_{k_1,\dots,k_{2n}}} \\
&= \xi_{k_1,\dots,k_{2n}} + (4n-1)\cdot \xi^3_{k_1,\dots,k_{2n}} \\
&\quad -2\xi_{k_1,\dots,k_{2n}} \left(\sum_{k'_1,\dots,k'_{2n-1}} \xi^2_{k'_1,\dots,k'_{2n-1},k_{2n}} + \dots + \sum_{k'_2,\dots,k'_{2n}} \xi^2_{k_1,k'_2,\dots,k'_{2n}} \right) .
\end{aligned}$$

The decision variables $x_{k_1k_2}$ can be derived as in the case of the quadratic assignment problem by projecting the n dimensional array $\xi^*_{k_1,k_2,\dots,k_{2n}}$ onto the two dimensions which belong to the decision variables $x_{k_1k_2}$.

The considerations of competing processes between two elements, the possibilities of intervention for the decision maker and the neural network interpretation in the section of linear assignment problems can be directly transferred.

3 EXAMPLE OF INDUSTRIAL ASSIGNMENT PROBLEMS OF HIGHER ORDER

Assignment problems of quadratic or higher order often occur in manufacturing planning. The job scheduling of a weaving mill is a good example. In this problem, many (10 - 200) similar machines and about 4 times as many jobs with many possible assignments exist. The costs to be considered are setup costs, tardiness costs (for late jobs) and prohibition costs because of limited machine capacities and missing material. If one considers only one preceding job for the calculation of the costs, the method of the *Cost Oriented Competing Processes* can be adapted to this problem by calculating a 3-dimensional cost array to initialize the competing processes and using a linkage condition between the time steps (Starke and Berkemer, 1995). Additional dimensions allow the consideration of further preceding jobs in the optimization of the scheduling.

4 CONCLUSIONS AND OUTLOOK

For industrial applications it is remarkable that the algorithm of the *Cost Oriented Competing Processes* can easily be adapted to higher order assignment problems with specific constraints. It is possible to prohibit or enforce particular decisions. Therefore the experience of human decision-makers can be used additionally and one can pay attention to sudden changes of the constraints of the problem configuration.

Moreover, the COCP can be computed in a highly parallel way. Because of the close relationship of the equations of motion proposed here and of these of some physical systems like *e.g.* the laser (Haken, 1983), a hardware realization of the COCP should be possible. First theoretical steps with hardware realizations of some simpler selection processes are made with semiconductors and lasers in (Schindel, 1993) and (Beckert, 1994).

REFERENCES

Baird, B. (1990) A Learning Rule for CAM Storage of Continuous Periodic Sequences, in *Proceedings of the International Conference on Neural Networks*, San Diego, June 1990, III.

Beckert, S. (1994) *Modenselektion beim Laser zur Realisierung des Synergetischen Computers.* Dissertation, University of Stuttgart.

Eigen, M. (1971) Selforganization of Matter and the Evolution of Biological Macromolecules. *Die Naturwissenschaften*, **58**, 465–523.

Eigen, M. and Schuster, P. (1977) The Hypercycle - Part A: Emergence of the Hypercycle. *Die Naturwissenschaften*, **64**, 541–65.

Eigen, M. and Schuster, P. (1978) The Hypercycle - Part B: The Abstract Hypercycle. *Die Naturwissenschaften*, **65**, 7–41.

Eiselt, H. A., Pederzoli, G. and Sandblom, C.-L. (1987) *Operations Research: Continuous Optimization Models.* Walter de Gruyter, Berlin New York.

Haken, H. (1983) *Synergetics, An Introduction.* Springer-Verlag, Berlin Heidelberg.

Haken, H. (1990) *Synergetic Computers and Cognition - A Top-Down Approach to Neural Nets.* Springer Verlag, Berlin Heidelberg.

Hirsch, M. W. and Baird, B. (1995) Computing with dynamic attractors in neural networks. *BioSystems*, **34**, 173–95.

Schindel, M. (1993) *Theorie eines Halbleitersystems zur Realisierung der Ordnungsparameterdynamik eines Synergetischen Computers.* Dissertation, University of Stuttgart.

Starke, J. (1994) *Vergleichende Untersuchung lernfähiger Systeme.* Diploma thesis, University of Stuttgart.

Starke, J. and Berkemer, R. (1995) Manufacturing Planning with Cost Oriented Competing Processes - Example of a Weaving Mill, will be published.

Zimmermann, W. (1990) *Operations Research.* Oldenbourg Verlag, München Wien.

Modelling and solving of the allocation problem of non-convex polygons with rotations

Yu.G. Stoyan, M.V. Novozhilova
National Ukrainian Academy of Sciences
Institute for Problems in Machinery
2/10, Pozharsky Str., Kharkov-46, 310046, Ukraine.
Tel: (0572) 95 95 36. Fax: (0572) 94 29 14.
e-mail: stoyan@ipmach.kharkov.ua

Abstract

The work deals with the problem of optimal allocation of objects of the so-called irregular form. The objects are allocated on a strip of given width. The rotation of objects are allowed.

Keywords

Allocation, nonlinear optimization

1 INTRODUCTION

The problem is insufficiently studied, but it is of great applied importance and also of interest for the developing the theory of solving cutting and packing problems, as the problem in question belongs just that very set of problems. Problems of cutting and packing are frequently found in different spheres of human activity (cutting glass, metal, textile, leather etc., allocation of equipment in complex shape spaces, design of complex engineering systems).

A broad range of bibliography (see Sweeney and Ridenour (1989)) in certain directions of research exists. It is necessary to mention the well-known works of Dowsland (1992), Beasley (1985), Terno (1993), Li and Milenkovic (1993), Schwartz and Sharir (1983). However, very few works can practically be found in which the allocation problems are investigated taking into account the rotation possibility of allocation objects or the allocation region. Apparently it is caused by their complexity and non-triviality of mathematical models, necessary for their adequate description.

A wide class of problems connected with allocation of 2D and 3D complex shape geometric objects in domains of arbitrary shape has been studied for a number of years at the Institute for problems in Machinery (Ukraine, Kharkov).

In the work a mathematical model of the allocation problem being considered is constructed based on the original theory of Φ−functions and structures of nonlinear inequalities. The exact method searching for a problem local minimum from any feasible initial point uses the active set strategy ideas.

2 PROBLEM MATHEMATICAL MODEL CONSTRUCTION

Let there be in an arithmetic Euclidean space R^2 a connected region T_0 with a linear - piecewise frontier: $\{(x,y) \mid (x,y) \in R^2, 0 \leq y \leq A, 0 \leq x \leq z; A$ - *const*, z- *var*$\}$, given in a fixed orthogonal coordinate system XOY. There also are objects $T_i \subset R^2$, $i = 1,2,\ldots,n$ $(n > 1)$, which are canonically closed bounded point sets with an arcwise frontier and a connected interior (*int*). It is required to allocate objects T_i without overlapping in region T_0 so that value z reaches its minimum. In other words, it is necessary to determine

$$\min z \tag{1}$$

$$s.t.: \qquad T_i \subset T_0 \tag{2}$$

$$intT_i \cap intT_j = \emptyset, i,j = 1,2,\ldots,n, i \neq j, \tag{3}$$

where $\emptyset$ is empty set.

In general the allocation objects are non-convex polygons T_i with allocation parameters $u_i = (v_i, \varphi_i)$, where $v_i = (x_i, y_i)$, $v_i \in intT_i$ are coordinates of the origin of eigen coordinate system OX_iY_i , φ_i is angular parameter, which equals a rotation angle of the system OX_iY_i with respect to the system OXY.

Allocation parameters (v_i, φ_i) and variable z induce space R^{3n+1} of allocation parameters, in which the problem (1-3) ought to be considered, and the conditions (2-3) describe the feasible region $D \subset R^{3n+1}$ of the problem.

The initial information about allocation objects is an ordered set of vertices $\{t_i^k(x_i^k, y_i^k)\}$, $k = 1,..,N_i$ of object T_i given in the system OX_iY_i.

2.1 $\Phi_{ij}(u_i, u_j)$−functions

The constructive tool of formalizing the conditions of mutual non-overlapping of objects (condition (3)) is $\Phi_{ij}(u_i, u_j)$−functions introduced for each pair (i,j) of objects being allocated (Stoyan (1980)). Moreover, the $\Phi_{0j}(u_i, u_j)$−function can be also applied to describe the condition (2) of belonging of object T_i to region T_0.

Definition 1 *Any, everywhere defined and continuous function in R^6, possessing the following characteristic property:*

$$\Phi_{ij}(u_i, u_j) > 0, \ if \ clT_i(u_i) \bigcap clT_j(u_j) = \emptyset, \tag{4}$$

$$\Phi_{ij}(u_i, u_j) = 0, \ if \ \begin{array}{c} clT_i(u_i) \bigcap clT_j(u_j) \neq \emptyset, \\ intT_i(u_i) \bigcap intT_j(u_j) = \emptyset, \end{array} \tag{5}$$

$$\Phi_{ij}(u_i, u_j) < 0, \ if \ intT_i(u_i) \bigcap intT_j(u_j) \neq \emptyset, \tag{6}$$

where clT is topological closure of set T,
is called Φ-function.

Hence, the characteristic property of the Φ_{ij}-function allows to speak about non-intersection, tangency and intersection of corresponding point sets.

Later on we shall be interested in following properties of $\Phi_{0j}(u_i, u_j)$. We assume that all the $\varphi_i = const$.

Property 1 *For objects T_i and T_j the surface of the $\Phi_{ij}(v_i, v_j)$-function defined by equation $\Phi_{ij}(v_i, v_j) = 0$, is linear-piecewise.*

Property 2 *The region of non-negative values of the $\Phi_{ij}(v_i, v_j)$-function is a polyhedral closed set $D_{ij} \subset R^4$.*

Property 3 *The region $D_{ij} \subset R^4$ may be presented in the form $D_{ij} = \bigcup_{g=1}^{G_{ij}} Tr^g, G_{ij} \leq m_{ij} \leq O(K_i + K_j)$, where Tr^g is truncation, which is described by intersection of a finite number P_g closed half-spaces $H^l, l \in \{1, 2, \ldots, m_{ij}\}$, defined by generating planes E^l., i.e. $D_{ij} = \bigcup_{g=1}^{G_{ij}} \bigcap_{p=1}^{P_g} H^l$, where $l = f(p, g)$.*

The generating planes E^l are defined by equations $f^l(v_j, v_i) = 0$. Functions $f^l(v_j, v_i)$ have the form

$$f^l(v_j, v_i) = a^l \cdot (x_j - x_i) + b^l \cdot (y_j - y_i) + c^l, l = 1, 2, \ldots, m_{ij}, \tag{7}$$

where coefficients a^l, b^l, c^l are functions of **certain** vertices of objects T_i and T_j.

The position of point (v_j, v_i) in half-space H^l is determined by inequality $f^l(v_j, v_i) \leq 0$.

Let us consider two-valued predicate $S(f^l(v_i, v_j))$, as Rvachev (1967), such that

$$S(f^l(v_i, v_j)) = \begin{cases} 1, \ if & f^l(v_i, v_j) \leq 0, \\ 0, \ if & f^l(v_i, v_j) > 0. \end{cases} \tag{8}$$

Then predicate $\Omega_{D_{ij}}(v_i, v_j)$, characterizing region D_{ij}, may be presented in the following form:

$$\Omega_{D_{ij}}(v_i, v_j) = \bigvee_{g=1}^{G_{ij}} \left(\bigwedge_{p=1}^{P_g} S(f_l(v_i, v_j)) \right), l \in \{1, 2, \ldots, m_{ij}\}, \tag{9}$$

i.e. $\Omega_{D_{ij}}(v_i, v_j) = 1$, if $(v_i, v_j) \in D_{ij}$, and $\Omega(v_i, v_j) = 0$, if $(v_i, v_j) \notin D_{ij}$.

Structures of inequalities

To describe region D_{ij}, defined by predicate $\Omega_{D_{ij}}$, we shall use the structure of linear inequalities, following by Magas (1984). To simplify the exposition, we shall say that the l-th and r-th inequalities of the form (7) are connected by the conjunction operation, if $S(f^l)$ and $S(f^r)$ enter together at least into the one conjunct in expression (9). Otherwise, we shall say that the l-th and the r-th inequalities are connected by the disjunction operation.

Definition 2 *The structure of linear inequalities $S(F(x),\Delta,m)$ is an ordered set $F(x)$ of linear inequalities of the form $f_t(x)\geq 0\,(x\in R^k, t=1,2,\ldots,m)$ with defined operations of conjunction or disjunction for each pair of inequalities; the operations are determined by the relation matrix $\Delta=\|\ \delta_{lr}\ \|_{m\times m}$ such that $\delta_{lr}=1$ corresponds to the conjunction operation between the l-th and r-th inequalities ($l,r\in\{1,2,\ldots,m\}$) and $\delta_{lr}=0$ corresponds to the disjunction operation.*

As is obvious from the definition, the notion of a structure of inequalities generalizes the notion of a system of inequalities. That is, if all the elements of matrix Δ of the inequalities structure $S(F(x),\Delta,m)$ are equal to unit, then the inequalities structure shall be identical to a system of inequalities of set $F(x)$. We denote the matrix of operations with unit elements as Δ^1. Otherwise, if none pair of inequalities of structure $S(F(x),\Delta,m)$ is connected by conjunction operation, e.g. elements $\delta_{ll}=1$ only, then we denote such an operation matrix as Δ^0.

Therefore to describe region D_{ij} we shall use the structure $S_g(F_{ij}(v_i,v_j),\Delta,m_{ij})$,

Then under assumption that all the φ_i = *const* the conditions of mutual pairwise non-overlapping of n objects being allocated are described by the structure

$$\bigcap_{g=1}^{C_2^n} S_{g=y(i,j)}(F_{ij}(v_i,v_j),\Delta,m_{ij}), \tag{10}$$

Using such a methodology one can define the analytical description of the allocation conditions of n objects T_i into region T_0 as the system

$$\bigcap_{i=1}^{n} S_{i0}(F_{i0}(v_i,z),\Delta^1,m_{i0}), \tag{11}$$

where each system S_{i0} is a system of a four linear inequalities of the form:

$$\begin{array}{ll} x_i - z\leq -p_i^l, p_i^l=\max\limits_k x_i^k; & -x_i\leq p_i^r, p_i^r=\min\limits_k x_i^k \\ y_i\leq -p_i^d, p_i^d=\min\limits_k y_i^k; & y_i+A\leq -p_i^u, p_i^u=\max\limits_k x_i^k, \\ \multicolumn{2}{c}{k=1,\ldots,N_i,} \end{array} \tag{12}$$

In the case if φ_i are not *const* an ordered set $F(x)$ of non-linear inequalities with defined operations by analogy with above operations for each pair of inequalities is a structure of **non-linear** inequalities.

The feasible region D of the problem (1-3) is determined by the structure of non-linear inequalities.

2.2 Analytical model of the basic problem

Let set C_i=*conv* T_i be a convex hull of object T_i and m_i^c is the number of vertices of C_i. Then the structure Σ_0 of inequalities describing the object T_i position in T_0 must include $4 \times m_i^c$ systems consisting of both support function $f_c(v_i, \varphi_i)$ of kind

$$x_i - z \leq -x_i^c(\varphi_i), \quad y_i \leq -y_i^c(\varphi_i), \quad y_i + A \leq -y_i^c(\varphi_i), \quad -x_i \leq x_i^c(\varphi_i), \tag{13}$$

where $x_i^c(\varphi_i) = x_i^c \cos\varphi_i - y_i^c \sin\varphi_i$, $x_i^c(\varphi_i) = y_i^c \cos\varphi_i + x_i^c \sin\varphi_i$, and pair of inequalities which determine the range of parameter φ_i. Then structure Σ_0 has a form:

$$\begin{array}{l} S_1(F_1(x_i, 0, \varphi_i), \Delta^0, m_i^c) \bigcap S_2(F_2(x_i, z, \varphi_i), \Delta^0, m_i^c) \bigcap \\ \bigcap S_3(F_3(y_i, 0, \varphi_i), \Delta^0, m_i^c) \bigcap S_4(F_4(y_i, A, \varphi_i), \Delta^0, m_i^c) \end{array} \tag{14}$$

It is necessary to note that after introducing into consideration the angular parameters the analytical defining the condition (3) is getting more complex one.

1. The preimage of each generating plane E^l in space R^6 is the surface γ_l which has the analytical description in the form of system $\{f_1, f_2, f_3\}$ depending on the geometric characteristics of the surface γ_l like

$$\left\{ \begin{array}{ll} & (a^l \cos\varphi_i - b^l \sin\varphi_i)(x_j - x_i) + (b^l \cos\varphi_i + a^l \sin\varphi_i)(y_j - y_i) + \\ f_1 : & (a^l x_j^k + B^l y_j^k)\cos(\varphi_j - \varphi_i) + (b^l x_j^k - a^l y_j^k)\sin(\varphi_j - \varphi_i) - \\ & (a^l x_i^h + b^l y_i^h) \leq 0 \\ f_2 : & 0 \leq \varphi_i \leq 2\pi \\ f_3 : & \varphi_i - \hat{\varphi}_{jk}^- \leq \varphi_j \leq \varphi_i + \hat{\varphi}_{jk}^+ \end{array} \right. , \tag{15}$$

where $t_j^k = \{x_j^k, y_j^k\}, k \in \{1, ...N_j\}$ and $t_i^h = \{x_i^h y_i^h\}, h \in \{1, ...N_j\}$ are vertices of objects T_i and T_j respectively.

2. Using the representation of each of objects T_i in the form $T_i = \bigcup_{g_i=1}^{G_i} T_{g_i}$, where T_{g_i} is convex and $\bigcap_{g_i=1}^{G_i} T_{g_i} = \emptyset$ and taking into account that under consideration of the condition (3) for each pair of objects T_{g_i} and T_{g_j} coefficients a^l, b^l, c^l are functions of **arbitrary** vertices of objects T_{g_i} and T_{g_j}, we can write general structure S_{ij} describing condition (3) for objects T_i and T_j in the following way:

$$S_{ij} = \bigcap_{g_i}^{G_i} \bigcap_{g_j}^{L_j} S(F(v_{g_i}, v_{g_j}, \varphi_i, \varphi_j), m(\varphi_i, \varphi_j), \Delta^0). \tag{16}$$

Remark 1 *Number $m(\varphi_i, \varphi_j)$ defining the amount of inequalities in set $F(u_{g_i}, u_{g_j})$ is a variable value and it depends on values φ_i and φ_j.*

It is necessary to note that:

1. Support functions $f_c(v_i, \varphi_i)$ are concave at each point of region D.
2. Functions like f_1 from (11) in general (depending on the geometric characteristics of objects) have indefinite Hessian.
3. In general feasible region D is a non-connected set with a finite number η of connected components D_r, each of which in turn admits a representation in the form of a union of a finite number ω of such a non-convex subsets D_{rk}, that

$$D = \bigcup_{r}^{\eta} D_r = \bigcup_{r}^{\eta} (\bigcup_{k}^{\omega} D_{rk}) = \bigcup_{w=1}^{W} D_w, \tag{17}$$

4. It is follows from above properties the problem (1-3) is multiextremal.

3 METHOD OF SOLVING BASIC PROBLEM

The following strategy for solving the local optimization problem is proposed:

1. Having fixed all the parameters φ_i (e.g. $\varphi_i = 0$) so that all the problem constraints became linear ones we solve the problem of local minimization objective function z on a some non-convex connected component D_δ of region $D_{linear} \subset R^{2n+1}$ ($D_\delta = \bigcup_{q=1}^{Q} D_{\delta q}$, $D_{\delta q}$ is convex subregion). In other words we solve the problem searching for $z(v^*) = \min_{D_\delta \subset R^{2n+1}} z$ on a structure $S_\delta(F(v,z), \Delta, m)$ of linear inequalities (see Stoyan, Yu.G., Novozhilova, M.V., and Kartashov, A.V., (1994)) and structure S_δ may be represented by formula:

$$S_\delta(F(v,z), \Delta, m) = \bigcup_{q=1}^{Q} S_q(F^q(v,z), \Delta^1, (m^q = 4n + n(n-1)/2)), \tag{18}$$

where each $S_q(F^q(v,z), \Delta^1, m^q)$ is a system of linear inequalities.

The point v^* in general is described by several systems of equations $Fx = B$, which may belong to different systems S_q.

2. We choose one of the systems $S_q^*(F^q(v,z), \Delta^1, m^q)$ and construct the mapping of the set of linear systems $\{S_q(F^q(v,z), \Delta^1, m^q)\}$ onto a set of nonlinear systems $\{S_q^n(F^q(v,z), \Delta^1, m^q)\}$ by the following rule: each constraint of type (7) is substituted by constraint of type (15) and each constraint of type (12) is substituted by constraint of type (13). Then the system $S_q^{*n}(F^q(v,z), \Delta^1, m^q)$ will define some region $D \subset R^{3n+1}$.

3. Considering the problem of local optimization on the system $S_q^{*n}(F^q(v,z), \Delta^1, m^q)$ as a minimax problem in the space R^{3n} with the objective function $\Re = \max_i(x_i + x_i^c(\varphi_i))$, where function $\Re$ is convex, we solve the problem

$$\begin{aligned} & \min \Re \\ s.t. \quad & S_q^*(F^q(v,z), \Delta^1, m^q) \end{aligned} \tag{19}$$

The problem belongs to a well-known problems of non-linear programming, the methodology of projected Lagrangian was selected by the authors for solving the problem. More exactly, the method of sequential quadratic programming is utilized, where the approximation of Lagrange function of problem (19) is present in the objective function of auxiliary problem. The initial active set is chosen from set of constraints, which corresponds to linear ones active at the point v^*.

It is essential that

1. Hessian of problem (19) Lagrange function may be easily calculated at each point of the region D by means of obvious transformations, however it is undetermined at each point of the region D that is why in finding a descent direction of a objective function the modified LDL^T decomposition of Hessian is performed.
2. The step value is chosen not only from method convergence condition but with regard for geometric characteristics of objects being allocated, namely range of parameters φ_i, as well. It is necessary to note, that in defining descent direction the constraints determining the range of φ_i are not taken into consideration.

REFERENCES

Beasley, J.E. (1985) An exact two-dimensional non-guillotine cutting tree search procedure. *Operational Research* **33**, 49–65.

Dowsland, K.A. and Dowsland, W.B. (1992) Packing problems. *European Journal of Operational Research*, **56**, 2–14.

Gil, P.E., Murray, W. and Wright, M.H. (1981) *Practical Optimization.* Academic Press, London, New York.

Li, Zh. and Milenkovic, V. (1993) A compaction algorithm for non-convex polygons and its application, in *9th Annual ACM Symposium on Computational Geometry,* May 19-21.

Magas, S.L. (1984) *Methods of Solving Extremal Problems in Allocation of Polygonal Geometric Objects on a Strip.* Ph.D.Thesis, Moscow. (Russ.)

Schwartz, T.J. and Sharir, M. (1983) On the "Piano Movers" Problem. I. The case of Two-Dimensional Rigid Polygonal Body Moving Admist Polygonal Barriers. *Communications on Pure and Applied Mathematics,* **XXXVI**, 345–98.

Stoyan, Yu.G. (1980) On one Generalization of the Dense Allocation Function. *Reports Ukrainian SSR Academy of Science,* Ser.A. **8**, 70–4. (Russ.).

Stoyan, Yu.G., Novozhilova, M.V. and Kartashov, A.V. (1994) Mathematical Model and Method of Searching for a Local Extremum for the Non-convex Oriented Polygons Allocation Problem. The manuscript has been refereed and recommended for publication in *European Journal of Operational Research.*

Sweeney, P.E. and Ridenour, E.L. (1989) Cutting and Packing Problems: a Categorized, Application Orientated Research Bibliography. Working Paper 610, School of Business Administration, University of Michigan.

Terno, J.R. and Scheithauer, G. (1993) *Modelling of packing problems.* Working Paper.

Stochastic Problems

Parameters identification of a time-varying stochastic dynamic systems using Viterbi algorithm

Tarik Al-Ani, Yskander Hamam
Ecole Supérieure d'Ingénieurs en Electrotechnique et Electronique (E.S.I.E.E.), Département Automatique
Cité Descartes 2 bd Blaise Pascal - B.P. 99, 93162 Noisy-le-Grand Cedex. e-mail: alanit@esiee.fr, hamamy@esiee.fr

Abstract
An on-line non-linear parameter identification algorithm is introduced. The system parameters are considered to be time varying. This algorithm is based on Viterbi algorithm applied to finite state models. It allows the identification based on the dynamic continuous output of the system. The algorithm is first applied to first order dynamic systems, then it is extended to a higer order systems. In order to reduce the complexity of the parameters discretisation, these parameters are supposed to be decoupled with respect to noise. Preliminary Simulation results are given.

Keywords
Stochastic process, hidden Markov chain, Stochastic dynamic programming, Parameter identification.

1 INTRODUCTION

In adaptive control of Time varying dynamic systems, on-line identification is a fundemental step. The problem of on-line identification of stochastic time varying dynamic systems has been extensively studied in the literature, Goodwin and Sin (1984), Isermann et al. (1992). The identification algorithms, in general, are based on either a state space models or the input-output models. In this paper, we adopt the input-output stationnary linear or non-linear models such as the Linear Autoregressiv model LARX and the Nonlinear Autoregressive model NARX. This choise is justified by the fact that these models are more convenient in practice and that the number of parameters to be estimated is resonable. Our approach is based on the Viterbi algorithm developped for decoding convolutional codes, Viterbi (1979), Forney (1972). This recursive algorithm is optimal in the sense that it is the maximum-likelihood estimation of the entire received observation sequence, Forney (1973). It applies the dynamic programming principle to the detection of discrete-time finite state Markov processes with noisy observations. This algorithm, which

may be obtained by learning, uses a Hidden Markov Model (HMM), Rabiner (1989), to incorporate nonstationarity into input-output model. Thus our aim is to estimate time varying parameters in the LARX or NARX models. However, in this paper, the HMM is calculated on-line without learning.
The Viterbi algorithm is largely applied in communication theory, e.g. in convolutional coding, intersymbole interference, Viterbi and Omura (1979) and other domains such as signal processing, Streit and Barrett (1990). Furthermore, its computational complexity is linear with respect to the length of the measurment sequence.
As mentioned by Rabiner (1989), in a similar manner to the Kalman filter, the Viterbi algorithm tracks the states of stochastic processes with a recursive method that is optimum in a certain sense, and that lends itself readily to implementation and analysis. However, the underlying hidden process is assumed to be finite-state Markovian rather than Gaussian, which leads to marked difference in structure.
In order to apply the Viterbi algorithm, we proceed by descretizing the parameters spaces into N subspaces. Each subspace is then assigned its center to represent the hidden parameter value at time k. The aim of this paper is to estimate the parameter sequence using the available measurment sequence.
The system models are defined in section (2). Section (3) describes the Viterbi algorithm for a single and multiple parameter dynamic systems. Simulation results are given in section 4. The conclusions of this paper are presented in section (5).

2 SYSTEM MODELS

Let us generally define the following discrete Single Input Single Output NARX model

$$y(k) = \phi(\Theta(k);\ y(k-1),\ ...,\ y(k-p);\ u(k-1),\ ...,\ u(k-q)) + v(k),\ k = 0,\ 1,\ 2,\ ...,\ K, \quad (1)$$

where

$$\Theta(k) = [\theta_1(k),\ ...,\ \theta_{p+q}(k)]'. \quad (2)$$

In LARX model, ϕ, is defined by

$$\begin{aligned} \phi(.) &= \sum_{l=1}^{p} \alpha_l(k) y(k-l) + \sum_{l=1}^{q} \beta_l(k) u(k-l) + v(k), \\ \Theta(k) &= [\alpha_1(k)\ ...\ \alpha_p(k)\ \beta_1(k)\ ...\ \beta_q(k)]', \end{aligned}$$

where $y(k)$ is the output, $u(k)$ the input, $\alpha_.(k)$, $\beta_.(k)$ are the model parameters which may be time dependent and subject to zero mean white stationnary disturbance

$$\begin{aligned} \mathbf{w(k)} &= [w_1\ w_2\ ...\ w_{p+q}]', \\ \mathbf{E}(\mathbf{w}(k)\mathbf{w}'(k)) &= \mathbf{\Sigma_w},\ k = 0,\ 1,\ 2,\ ...,\ K. \end{aligned}$$

In this paper, we assume that the the components of the disturbance vector in equation (3) are decoupled (i. e. $\mathbf{\Sigma_w}$ is diagonal). The measurement noise $v(k)$ is considered to be a zero mean stationnary Gaussian random sequence, i.e $E(v(k)) = 0$ and $E(v(k)^2) = \sigma_v^2$, $k =$ 0, 1, 2, ..., K. These errors are also considered to be uncorrelated with disturbance input. The initial parameters $\theta_{.}(0)$ are considered to be a priori known. They may be constants or Gaussian random variables with mean values $m_{\theta(0)}$ and variance $\sigma^2_{\theta(0)}$.

3 THE VITERBI ALGORITHM

Based on dynammic programming methodes, this algorithm searches for a single best state (parameter) sequence (hidden Markov chain states). As mentioned in section (3.1), the underlying hidden process is assumed to be finite-state Markov. Thus we proceed by describing a Markov finite-state model for single parameter and then extend the results to Multiple parameter.

3.1 Finite-space model for single parameter

Let us denote the parameter value at time k by $\theta(k)$ and suppose that the allowed bounds of θ are known a priori, i.e. $S : \theta_{min} \leq \theta(k) \leq \theta_{max}$. This range is quantized into a finite number N of state cells $s_i \trianglerighteq [\theta_i, \theta_{i+1}]$, i.e. $S = \bigcup_{i=1}^{N} s_i$ and each cell is associated with a state of the Markov chain. Each state s_i of the Markov chain is represented by the cell center point $\theta_{ci} = \frac{(\theta_i + \theta_{i+1})}{2}$, $i \in \{1, 2, ..., N\}$, where $\theta_1 = \theta_{min}$ and $\theta_{N+1} = \theta_{max}$. Let the state of the Markov chain at time k be q_k, if $q_k \in s_i$ then the estimated parameter at time k is $\hat{\theta}(k) = \theta_{ci}$, $i \in \{1, 2, ..., N\}$.

Single state Viterbi algorithm

To find the single best Markov chain state sequence $\mathcal{F}_k^s \trianglerighteq \{q_1, q_2, ..., q_K\}$, $q_k \in s_i \in \mathbf{R}^1$, $i \in \{1, 2, ..., N\}$ for the given observation sequence $\mathcal{F}_k^y \trianglerighteq \{y(1), y(2), ..., y(k)\}$ we define the best high score (highest probability) along a single path, at time k, which accounts for the first k observations and ends in state s_i

$$\delta_k(i) = [\max_{\mathcal{F}_{k-1}^S} \Pr(q_k = s_i, \mathcal{F}_k^y | \lambda)]. \tag{3}$$

The parameter λ is called in the literature "Hidden Markov Model (HMM)", Rabiner (1989). This model is defined by the triplet $\lambda \trianglerighteq \{\mathbf{\Pi}, \mathbf{A}, \mathbf{B}\}$, where $\mathbf{\Pi}$, $\mathbf{A}$ and $\mathbf{B}$ are the Initial state distribution, state transition probabilities and measurement probability distribution, respectively.

$$\begin{aligned}
\mathbf{\Pi} &= [\pi_1, \pi_2, ..., \pi_N], \quad \pi_i = \Pr(q_0 \in s_i), \\
\mathbf{A} &= [a_{ij}], \qquad a_{ij} = \Pr(q_{k+1} \in s_j | q_k \in s_i), \\
\mathbf{B} &= [b_j(y(k))], \quad b_j(y(k)) = \Pr(y(k) | q_k \in s_j),
\end{aligned}$$

and $i, j \in \{1, 2, ..., N\}, k \in \{0, 1, , ..., K\}$. $y(k)$ may be, in general, discrete $y(k) = V(k)$, $V(k) \in \{1, 2, ..., M\}$ or continuous $y(k) \in \mathbf{R}^r$ which is the case in this paper. The model λ may be obtained off-line by training, Rabiner (1989). However, in our approach, training is not used. The parameters are directly calculated (off and on-line) using the observation model, equation (1). By induction on equation (3), the complete Viterbi algorithm may now be stated (using the natural logarithm to avoid data underflow) as follows

1. Initialization, $k = 0$

$$\begin{aligned} \delta_1(i) &= \ln \pi_i + \ln b_i(y(1)), \; 1 \leq i \leq N, \\ \Psi_1(i) &= \mathbf{0}, \end{aligned}$$

2. Recursion $1 \leq k \leq K$

$$\begin{aligned} \delta_k(j) &= \max_{1 \leq i \leq N} [\delta_{k-1}(i) + \ln a_{ij}] + ln b_j(y(k)), \; 1 \leq j \leq N, \\ \Psi_k(j) &= \arg \max_{1 \leq i \leq N} [\delta_{k-1}(i) + \ln a_{ij}], \qquad 1 \leq j \leq N, \end{aligned}$$

3. Termination

$$\begin{aligned} lnP^* &= \max_{1 \leq i \leq N} [\delta_K(i)], \qquad \text{(Viterbi score)}, \\ q_K^* &= \arg \max_{1 \leq i \leq N} [\delta_K(i)], \qquad \text{(Viterbi track)}, \end{aligned}$$

4. Path (optimal state sequence) backtracking $k = K-1, K-2, ..., 0$

$$q_k^* = \Psi_{k+1}(q_{k+1}^*),$$

The state sequence may be restrained by keeping track of the argument which maximizes equation, for each k and j. This is done using the array $\Psi_k(j)$. Thus the Viterbi algorithm may be used as a filtering as well as a smoothing algorithm. It should be noted that a treillis structure, efficiently implements the computation of the Viterbi algorithm, Forney (1973).

Deriviation of the parameters of hidden Markov chains

The two parameters $\mathbf{\Pi}$ and $\mathbf{A}$ may be calculated off-line. However the parameter $\mathbf{B}$ must be calculated on-line.

1. **Initial parameter distribution, $\mathbf{\Pi}$.** The choice of π depends, in general, on the application. In control applications, the initial parameter of the dynamic system or its expected value is, in general, known a priori. Thus, without loss of generality, we take $\pi_i = 1, \; q_0 \in s_i, \; i \in \{1, 2, ..., N\}, \; \pi_j = 0, \; j \in \{1, 2, ..., N\} \neq i$.
2. **State transition probabilities, $\mathbf{A}$.** The definition of the state transition probability a_{ij}, $i, j \in \{1, 2, ..., N\}$, is similar to that proposed by Streit and Barrett (1990). Based on the finite-space model discussed in section 3.1, we calculate, off or on-line, the state transition probabilities a_{ij}. If the state track lies in the ith cell at the current time step, the location of the track at the next time step is assumed to be characterized by a Gaussian distribution with mean x_{ci} and standard deviation σ_w. Hence, the transition probability that the state shifts from the ith cell to jth cell at the next time step is

$$a_{ij} = \frac{g_{ij}(k)}{\sum_{j=1}^{N} g_{ij}(k)}, \tag{4}$$

$$g_{ij}(k) = (2\pi\sigma_w)^{-\frac{1}{2}} \int_{\theta_j}^{\theta_{j+1}} \exp\{-[\frac{(\theta(k+1) - \theta_{ci}(k))^2}{2\sigma_w^2}]\}d\theta, \quad (5)$$

$1 \leq i \leq N$, $1 \leq j \leq N$. σ_w represents the standard deviation of the pertubation $w(k)$.

3. **The measurement probabilities, B**. The observation $y(k)$ in equation (1) is a linear function of the measurement Gaussian noise $v(k)$. Hence, the conditional probability density function of $y(k)$, given that $\theta(k) = \theta_{cj}$, is a Gaussian density function

$$b_j(y(k)) = (2\pi\sigma_v)^{-\frac{1}{2}} \exp\{-[\frac{(y(k) - \phi(\theta_{cj};\ y_.;\ u_.))^2}{2\sigma_v^2}]\},\ 1 \leq j \leq N. \quad (6)$$

Equation (6) must be evaluated for all states (parameter levels) θ_{cj}, $j = 1, 2, ..., N$.

3.2 Finite-space model for Multiple parameter dynamic system

In order to apply the Viterbi algorithm to multiparameter system identification, we modify the single state Viterbi algorithm. The idea is similar to that proposed by Xie and Evans (1991) for multiple frequency tracking. We consider the system described by equations (1)-(2). As mentioned in section 2, the parameters $\theta_l(k)$, $l = 1, 2, ..., p+q$, in equation (2) are decoupled with respect to perturbation noise (i. e. $\mathbf{\Sigma_w}$ is diagonal). In this case, each parameter $\theta_l(k)$, $l = 1, 2, ..., p+q$ is represented by a finite space model similar to that defined in section 3.1. The parameters $\theta_l(k)$, $l = 1, 2, ..., p+q$, in equation (1) are decoupled with respect to perturbation noise. Thus, the hidden Markov parameters for each state may be derived in the same way as in section 3.1.

4 COMPUTER IMPLEMENTATION AND SIMULATION

The Viterbi algorithm, conditioned on the whole past history of the possible parameters sequence, gives an optimum result. However, in practics this requires a large memory and long computation. Hence, the Viterbi algorithm is implemented in a suboptimum scheme in order to use a constant memory and achieve computation reduction. This scheme is based on a finite delay (defined by the memory). In order to keep a good performance, simulation or transfer function based approaches are used to evaluate this delay. Preliminary simulation has been carried out on PC using Matlab. The simulated ARX model used for the tests is

$$y(k) = a_1 y(k-1) + a_2 y(k-2) + b\ u(k) + v(k),$$

where, $a_1 \in \{1.6,\ 1.7,\ 1.8\}$, $a_2 \in \{0.7,\ 0.8,\ 0.9\}$ and $b \in \{0.9,\ 1.0,\ 1.1\}$. For each simulation, the observation noise v(k) and disturbance noise w(k) were taken to be stationnary white Gaussian with 0.1 standard deviation. The input signal u was generated at random. Figure 1 clearly shows that the Viterbi identifier works well. The accuracy may be increased by increasing the number of the discretization levels.

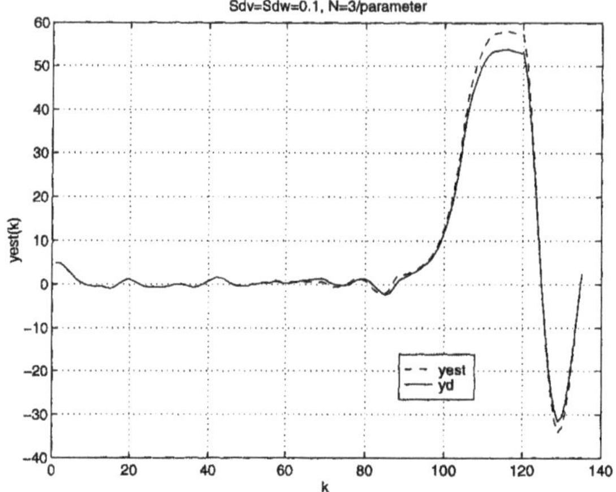

Figure 1 Computer simulation.

5 CONCLUSION AND PERSPECTIVES

In this paper, the application of the Viterbi algorithm to parameter identification of a general (linear or non-linear) dynamic systems has been described. The following conclusions may be drawn.

- Simulation has shown that the Viterbi algorithm may be used to identify time varying dynamic systems.
- The identification accuracy may be improved by increasing the discretization levels.
- The Viterbi algorithm allows discrete parameter identification depending on the cell resolution. To overcome this difficulty, continuous parameter identification may be obtained using the conditional average of the states, Streit and Barrett (1990).
- An on-line learning scheme of the Hidden Markov chain may considerably increase the accuracy of the identification.

The authors are presently working on the extensions of the above work to cover both continuous parameter representation and an on-line recusive extensions.

REFERENCES

Goodwin, G.C. and Sin, K.S. (1984) *Adaptive Filtering, Prediction and Control.* Prentice-Hall.

Forney, G.D. (1973) Maximum-Likelihood sequence estimation of digital sequences in the presence of intersymbol inference. *IEEE Trans. Inform. Theory,* **18**, 363–78.

Forney, G.D. (1973) The Vitrbi algorithm. Proc. IEEE, Mars, 268–78.

Isemann, R., Lachmann, K.-H. and Matko, D. (1992) *Adaptive Control Systems.* Prentice-Hall.

Rabiner, L.R. (1989) A tutorial on hidden Markov Models and selected applications in speech recognition. *Proc. IEEE,* **77**, 257–86.

Streit, R.L. and Barrett, R.F. (1990) Frequency line tracking using Hidden Markov Models. *IEEE Trans. Acoust., Speech, Signal Processing,* **38**, 586–98.

Viterbi, A.J. (1967) Error bounds for convolutional codes and asymptotically optimum decoding algorithm. *IEEE Trans. on Information Theory,* **IT-13**, 260–9.

Viterbi, A.J. and Omura, J.K. (1979) *Principles of Digital Communication and Coding.* McGraw-Hill, New York.

Xie Xianya and Evans Robins J. (1991) Multiple Target Tracking and Multiple Frequency Line tracking Using Hidden Markov Models. *IEEE Trans. on Signal Processing,* **39**, 2659–76.

Management of bond portfolios via stochastic programming – postoptimality and sensitivity analysis

Jitka Dupačová
Department of Probability and Mathematical Statistics
Charles University
Sokolovská 83, CZ–186 00 Prague 8, Czech Republic.
Tel: +42-2-21913280. Fax: +42-2-2323316.
e-mail: dupacova@karlin.mff.cuni.cz

Marida Bertocchi
Department of Mathematics, Statistics, Informatics and Applications
University of Bergamo
Piazza Rosate 2, I–24129 Bergamo, Italy. Tel: +39-35-277111.
Fax: +39-35-234693. e-mail: marida@ibguniv.bitnet

Abstract

Management of bond portfolio is formulated as a multiperiod scenario based stochastic program with random recourse. Sensitivity analysis of its optimal value with respect to the strategy applied in selection of input scenarios is detailed and applied on a real life problem from Italian bond market.

Keywords

Stochastic program, scenarios, sensitivity, postoptimality, application in finance

1 PROBLEM FORMULATION

We shall describe a stochastic programming model for management of portfolio of fixed income securities, called bonds for brevity. The main purpose of the portfolio management is to to maximize utility of the final wealth and, depending on the specific field of investment activities, to secure the prescribed or uncertain future payments. Similar problems arise in the context of management of one purpose investment funds, mutual funds, pension funds, investments of insurance companies, *etc.* Accordingly, the nature of liabilities can rank from fixed prescribed or planned external outflows (or inflows) to liabilities whose value depends on various external random factors such as mortality rates. One can formulate also problems related to portfolio protection where no explicit liabilities are considered. In contrast to the problem of a dedicated portfolio selection (see *e.g.* Hiller and Eckstein (1993), Shapiro (1988)), we allow for an *active trading strategy*. Accordingly, we formulate constraints on conservation of holdings for each asset at each time period and, similarly as in Golub et al. (1995), we extend the constraints on cashflows for each time period to include the possibility of *rebalancing* the portfolio.

From the point of view of stochastic programming, it is a multiperiod two-stage model with random relatively complete recourse and with additional nonlinearities due to the choice of the utility function. The prices of bonds and sometimes also the coupon cashflows f_t are driven by the assumed evolution of the interest rates: Given a sequence of equilibrium future forward short term interest rates r_t valid for the time interval $(t, t+1], t = 0, \dots, T$ the fair price of the j-th bond at time t equals the total cashflow generated by this bond in subsequent time instances discounted to t:

$$P_{jt}(\mathbf{r}) = \sum_{\tau=t+1}^{T} f_{j\tau} \prod_{h=t}^{\tau-1} (1 + r_h)^{-1} \tag{1}$$

where T is greater or equal to the time to maturity.

In reality, however, the sequence of the future short term forward rates that determines the prices (1) is not known, the sequences of interest rates are prescribed ad hoc or modeled in a probabilistic way. We shall consider a discrete distribution, say P, of S possible vectors $\mathbf{r}$ of interest rates concentrated with probabilities $p_s > 0 \;\; \forall s, \;\; \sum_s p_s = 1$ at points $\mathbf{r}^s \in R^T, s = 1, \dots, S$ called *scenarios*; this is the input information which is used to build the discussed model and which influences the results.

We shall mostly use the notation introduced in Golub et al. (1995):

$j = 1, \dots, J$ are indices of the considered bonds and T_j the dates of their maturities;

$t = 0, \dots, T_0$ is the considered discretization of the planning horizon;

b_j denote the initial holdings (in face value) of bond j;

b_0 is the initial holding in riskless asset;

f_{jt}^s is cashflow generated from bond j at time t under scenario s expressed as a fraction of its face value;

ξ_{jt}^s and ζ_{jt}^s are the selling and purchasing prices of bond j at time t for scenario s obtained from the corresponding fair prices (1) by subtracting or adding fixed transaction costs and spread; the initial prices ξ_{j0} and ζ_{j0} are known, i. e., scenario independent;

L_t is liability due at time t;

x_j/y_j are face values of bond j purchased / sold at the beginning of the planning period, i.e., at $t = 0$;

z_{j0} is the face value of bond j held in portfolio after the initial decisions x_j, y_j have been made.

The first stage decision variables x_j, y_j, z_{j0} are nonnegative,

$$y_j + z_{j0} = b_j + x_j \quad \forall j \tag{2}$$

and

$$y_0^+ + \sum_j \zeta_{j0} x_j = b_0 + \sum_j \xi_{j0} y_j \tag{3}$$

where the auxiliary nonnegative variable y_0^+ denotes the surplus.

The second-stage decisions on rebalancing the portfolio, borrowing or reinvestment of the surplus depend on individual scenarios and have to fulfil constraints on conservation of holdings in each bond at each time period and for each of scenarios

$$z_{jt}^s + y_{jt}^s = z_{j,t-1}^s + x_{jt}^s \quad \forall j, s, 1 \le t \le T_0 \tag{4}$$

where $x_{jt}^s, y_{jt}^s, z_{jt}^s$ denote the face value of bond j purchased, sold, held in the portfolio at time $t, t = 1, \ldots, T_0$ under scenario s and constraints on rebalancing the portfolio at each time period $1 \le t \le T_0$

$$\sum_j \xi_{jt}^s y_{jt}^s + \sum_j f_{jt}^s z_{j,t-1}^s + (1 + r_{t-1}^s) y_{t-1}^{+s} + y_t^{-s} =$$
$$L_t + \sum_j \zeta_{jt}^s x_{jt}^s + (1 + \delta + r_{t-1}^s) y_{t-1}^{-s} + y_t^{+s} \quad \forall s, t \tag{5}$$

The optimization problem consists in maximization of the expected utility of the final wealth

$$\sum_s p_s U(W_{T_0}^s) \tag{6}$$

subject to constraints (2)–(5) and nonnegativity constraints on all variables, with

$$W_{T_0}^s = \sum_j \xi_{jT_0}^s z_{jT_0}^s + y_{T_0}^{+s} - \alpha y_{T_0}^{-s} \quad \forall s \tag{7}$$

The multiplier α should be fixed according to the problem area. For instance, a pension plan assumes repeated application of the model with rolling horizon and values $\alpha > 1$ take into account the debt service in the future.

Thanks to the possibility of reinvestments and of unlimited borrowing, the problem has always a feasible solution. The existence of optimal solutions is quaranteed for a large class of utility functions that are *increasing and concave* what will be assumed henceforth.

In this way, we obtain a large scale deterministic program with a concave objective function and numerous linear constraints. The main outcome is the optimal value of the objective function (the maximal expected utility of the final wealth) and the optimal values of the first-stage variables x_j, y_j (and z_{j0}) for all j. We can view it as an output of our model that depends on the input given by the set of considered bonds and their characteristics (initial prices and cashflows), by their initial holdings, by the scheduled liabilities and by the used scenarios of evolution of interest rates and their probabilities.

We shall analyze the sensitivity of the optimal value of (6) on the selected scenarios of interest rates. This is an important task because there is an arbitrariness in constructing the probability distribution of the interest rates; there are scenarios designed only by experts or required by local authorities and those based on the binomial lattice techniques (*e.g.* , Black, Derman, Toy (1990)) or on continuous stochastic models. Moreover, due to the size of the resulting problem, not all scenarios can be used and more or less sofisticated sampling procedures have been suggested (cf. Zenios and Shtilman (1993)). A natural question is *the impact of the chosen sampling strategy* and *the influence of including additional or out-of-sample scenarios on the output* based on an inital manageable sample of scenarios. The first question will be treated in the subsequent Section and we refer to Dupačová (1995 a, b) for a postoptimality technique with respect to additional scenarios.

2 INFLUENCE OF SAMPLING STRATEGY

We shall treat now in detail sensitivity of the outcome on the nonrandom sampling s-trategy of Zenios and Shtilman (1993) using results known from sensitivity analysis for parametric nonlinear programs, *e.g.* , Gol'shtein (1970). We shall derive a formula for directional derivative of the optimal value in a form separable with respect to scenarios and time periods that allows for quick sensitivity analysis with respect to changes of the sampling strategy. We shall assume that the liabilities and cashflows are fixed and that the interest rate scenarios have been generated according to Black, Derman, Toy (1990); for the given horizon T it means that there are 2^T scenarios. They can be represented by random equiprobable binary fractions with T 0-1 digits, say

$$\omega^s = 0.\omega_1^s\omega_2^s \dots \omega_T^s$$

with $\omega_t^s = 0$ or 1 and their probabilities $p_s = 2^{-T} \forall s$. The digit 1 at the t-th position corresponds to the "up" move, the digit 0 corresponds to the "down" move of the forward short term interest rate in the period t. This binomial lattice is calibrated by the existing term structure to get the base rates r_{0t} and the volatility factors k_t for all t. The corresponding short term forward rates for scenario s and period t are then given as

$$r_t^s = r_{tl_t(s)} \tag{8}$$

where

$$r_{tl} = r_{t0}k_t^l,\ l_t(s) = \sum_{\tau=1}^{t} \omega_\tau^s \tag{9}$$

That is, $l_t(s)$ equals the number of the "up" moves for the given scenario s until time t. We denote $\mathbf{r}^s$ the vector of t components r_t^s.

Our sensitivity analysis will be related to a simplified version of the deterministic sampling strategy by Zenios and Shtilman (1993): We fix $L, 1 < L < T$ and assign the index $s, s = 1, \dots, 2^L$ to each possible binary fraction of length L. The sample point ω^s from (0,1) is determined by one of these binary fractions and by an arbitrary continuation up to binary fraction of length T. According to (8), (9) we build then $S = 2^L$ scenarios $\mathbf{r}^s$ and we denote

$$l^s = \sum_{t=1}^{L} \omega_t^s.$$

The lower and upper bounds for r_t^s with $t \geq L$ are evident:

$$r_t^{s-} = r_{t0} k_t^{l^s} \leq r_t^s \leq r_{t0} k_t^{t-l^s} = r_t^{s+},\ t = L+1, \ldots, T\, \forall s \tag{10}$$

and for $t \leq L$, r_t^s are fully determined by the described choice of the path ω^s. The input of our problem (2)–(7) consists thus of S T–dimensional scenarios $\mathbf{r}^s$ whose first L components are fixed whereas the subsequent $T - L$ components are subject to perturbations Δ^s such that

$$\begin{aligned} \Delta_t^s &= 0,\ t = 1, \ldots, L \\ \Delta_{L+\tau}^s &= r_{L+\tau} - r_{L+\tau}^s,\ t = L+1, \ldots, T \end{aligned} \tag{11}$$

where $r_{L+\tau}$ satisfies (10).

In our setting of problem, the objective function (6) does not contain any coefficients depending on scenarios; such coefficients enter equations (5) and (7) and they are differentiable in $\mathbf{r}$:

The derivatives of the purchasing and selling prices $\zeta_{jt}^s = \zeta_{jt}(\mathbf{r}^s)$ and $\xi_{jt}^s = \xi_{jt}(\mathbf{r}^s)$ equal to those of the fair prices (1); the fixed spread and transaction costs evidently do not enter the formulas. The directional derivatives of $P_{jt}(\mathbf{r}^s)$ in the direction of Δ^s equal (see (1))

$$P_{jt}^{s\prime}(0^+) = -\sum_{\tau=t+1}^{T} f_{j\tau} D_t^\tau \sum_{l=t}^{\tau-1} \frac{\Delta_l^s}{(1+r_l^s)} = -\sum_{l=t}^{T} \frac{\Delta_l^s}{(1+r_l^s)} \sum_{\tau=l+1}^{T} f_{j\tau} D_t^\tau \tag{12}$$

for all t; we have used notation $D_t^\tau = \prod_{h=t}^{\tau-1}(1+r_h)^{-1}$. Notice that these directional derivatives are *linear* in Δ^s $\forall s$.

We denote further $\varphi(\mathbf{r}^1, \ldots, \mathbf{r}^S)$ the optimal value of (2)–(7) for the initial "input" $\mathbf{r}^1, \ldots, \mathbf{r}^S$ and we indicate by asterisk the components of the corresponding optimal solution and of Lagrange multipliers.

Besides of the fixed number of scenarios, the basic assumptions that simplify the sensitivity analysis are existence of the unique optimal solution of (2)–(7), and unique Lagrange multipliers for the initial choice of scenarios, and a fixed rank of the matrix of the system (5), (7) for all considered perturbations. If these assumptions are fulfilled, the linearly perturbed problem that corresponds to the input $\mathbf{r}^s + \mu\Delta^s, s = 1, \ldots, S$ has an optimal solution for μ small enough and for arbitrary feasible perturbances Δ^s, there exists the directional derivative of the optimal value function at the given input $\mathbf{r}^1, \ldots, \mathbf{r}^S$ in any feasible direction $\Delta^s, s = 1, \ldots, S$ and equals the derivative at $\mu = 0^+$ of the Lagrange function of the corresponding linearly perturbed problem evaluated at the initial optimal solution and multipliers

$$\varphi'(0^+) = \frac{\partial}{\partial \mu} L(\mathbf{x}^*, \mathbf{y}^*, \mathbf{z}^*, \mathbf{W}^*; \lambda^*; \mathbf{r}^s + \mu\Delta^s, s = 1, \ldots, S)|_{\mu=0^+} \tag{13}$$

(cf. Gol'shtein (1970)). The perturbed coefficients appear only in equations (5) and (7); we denote the corresponding Lagrange multipliers by subscripts 5 and 7. Using the form of (5), (7) we get

$$\begin{aligned} \varphi'(0^+) &= \sum_{s=1}^{S} \left\{ \sum_{t=1}^{T_0} \lambda_{5t}^{s*} \sum_j P_{jt}^{s\prime}(0^+)(y_{jt}^{s*} - x_{jt}^{s*}) + \right. \\ & \left. \sum_{t=L+2}^{T_0} \lambda_{5t}^{s*}(y_{t-1}^{+s*} - y_{t-1}^{-s*})\Delta_{t-1}^s + \lambda_{7T}^{s*} \sum_j P_{jT}^{s\prime}(0^+) z_{jT_0}^{s*} \right\} \end{aligned} \tag{14}$$

The obtained expression is separable with respect to scenarios. Substituting (12) for $P_{jt}^{s\prime}$ and rearranging a bit we obtain separability with respect scenarios *and* time periods:

$$\varphi'(0^+) = \sum_{s=1}^{S} \sum_{l=L+1}^{T} \frac{\Delta_l^s}{1 + r_l^s} H_l^s \tag{15}$$

where

$$\begin{aligned} H_l^s &= - \sum_{t=1}^{\min(l,T_0)} \lambda_{5t}^{s*} \sum_j \sum_{\tau=l+1}^{T} f_{j\tau} D_t^\tau(\mathbf{r}^s)(y_{jt}^{s*} - x_{jt}^{s*}) + \lambda_{5,l+1}^{s*}(1 + r_l^s)(y_l^{+s*} - y_l^{-s*}) \\ & \quad L + 1 \le l \le T_0 - 1 \\ H_{T_0}^s &= - \sum_{t=1}^{T_0} \lambda_{5t}^{s*} \sum_j \sum_{\tau=T_0+1}^{T} f_{j\tau} D_t^\tau(\mathbf{r}^s)(y_{jt}^{s*} - x_{jt}^{s*}) \\ H_l^s &= - \sum_{t=1}^{T_0} \lambda_{5t}^{s*} \sum_j \sum_{\tau=l+1}^{T} f_{j\tau} D_t^\tau(\mathbf{r}^s)(y_{jt}^{s*} - x_{jt}^{s*}) + \lambda_7^{s*} \sum_j \sum_{\tau=l+1}^{T} f_{j\tau} D_{T_0}^\tau(\mathbf{r}^s) z_{jT_0}^{s*} \\ & \quad T_0 < l \le T \end{aligned}$$

The desired directions of changes in r_l^s for $l > L$ that result in decrease and/or in increase of the optimal value function can be thus obtained by inspection of the *signs* of H_l^s only. The magnitude of these changes is limited by (10) (and also by the fact that this result is of a *local* character).

The introduced model can be generalized to a multistage stochastic program with random complete recourse and the suggested sensitivity analysis can be extended to the multistage case as well.

Further problems of interest are sensitivity with respect to the input information used for calibration and fitting the binomial lattice, i.e., with respect to the yield curve and volatilities of the considered bond market, criteria for deleting "noninfluential" scenarios, inclusion of other random factors or additional scenarios, *etc.*

3 APPLICATION TO ITALIAN BOND MARKET

A simplified model has been developed and used to simulate the behaviour of an investment portfolio of fixed income securities on the Italian bond market within the time horizon of one year ($T_0 = 12$). The sample of bonds and all the informations come from a local bank; the spread was fixed at 2.5% and the transaction costs at 1%. Details are given in Bertocchi et al. (1995).

In this type of application, no liabilities are considered, liquidity can be obtained from the interbank market at the corresponding rate and the surplus can be always reinvested. For the numerical illustration, scenarios based on real life data from Italian bond market were generated. The initial portofolio we considered is composed of five typical governmental bonds, paying semi-annual coupon and covering two year forward till 29 years maturities:

Bonds	Qt	coupon (%)	redemption (%)	payment	dates	maturity
BTP36658	10	3.9375	100.1875	01/04	01/10	01/10/1996
BTP36631	20	5.03125	99.5313	01/03	01/09	01/03/1998
BTP12687	15	5.25	99.2312	01/01	01/07	01/01/2002
BTP36693	10	3.71875	99.3875	01/08	01/02	01/10/2004
BTP36665	5	3.9375	99.2188	01/05	01/11	01/11/2023

and of cash (500 mil. Liras), so that its (nominal) value is 6500 mil. Liras. The portfolio and the initial term structure are related to October 1994 (after the coupon was paid), the coupon yields and the redemption prices are after tax.

In this first study we worked with linear utility function. Over different numerical examples we report here results based on 32 scenarios (L=5). The most demanding part of the numerical study (done on DEC 5000/240 under ULTRIX 4.3 and using GAMS 2.25.062) was computation of the fair prices along the branches of various considered trees of the interest rate scenarios for $T = 349$. It took more than 68 minutes for all cases based on 32 scenarios.

The optimal strategy of rebalancing is to sell all bonds and to build a new portfolio of one or two bonds depending on scenario; purchasing starts only in the 5th month and no borrowing is used. The maximal expected wealth after 12 months is 7047.287 mil. Liras. Application of (15) for sensitivity of this result on the choice of scenarios for $5 < t \leq 12$ provides values $H_t^s = 0 \quad \forall s$ and positive values for $t > 12$; the reason is that there are non-zero multipliers only for $t = 1, \ldots, 4$, i.e. for periods where no trading takes place. This holds true even if some lower bounds on holdings of bonds are prescribed for the to reach portfolio diversification. Because of evident degeneracy of the dual problem we recomputed the problem with perturbed interest rates; in agreement with the theoretical value of derivative (15), the optimal value of the objective function did not change, neither did the character of optimal first-stage and second-stage solutions or the values of multipliers that appear in (15). Another outcome of the reported numerical experiment and of the sensitivity analysis is an evidence that the number of scenarios can be reduced from 32 to 16 or even to 8 without any change in the optimal value and in the trading strategy; it means that the output is rather robust with respect to the choice of interest rate scenarios from the full binomial lattice as described in Zenios and Shtilman (1993).

Acknowledgement. Joint research was supported from the visiting professors program by GNIM of CNR. Partly supported also by grants from Charles University (GAUK-357), from the Grant Agency of the Czech Republic (402/93/0631) and CNR grant n.94.00538.ct11.

REFERENCES

Bertocchi, M., Dupačová, J. and Moriggia, V. (1995) *Sensitivity analysis of a bond portfolio model for the Italian market.* Technical report, University of Bergamo.

Black, F., Derman, E. and Toy, W. (1990) A one-factor model of interest rates and its application to treasury bond options. *Financial Analysts J.*, Jan./Feb., 33–9.

Dupačová, J. (1995 a) Postoptimality for multistage stochastic linear programs. *Annals of Oper. Res.*, **57**.

Dupačová, J. (1995 b) Scenario based stochastic programs: Resistance with respect to sample. To appear in *Annals of Oper. Res.*, (Lillehammer 1994).

Gol'shtein, E.G. (1970) *Vypukloje Programmirovanie. Elementy Teoriji.* Nauka, Moscow. [*Theory of Convex Programming*, Translations of Mathematical Monographs **36**, American Mathematical Society, Providence RI (1972)].

Golub, B. et al. (1995) Stochastic programming models for portfolio optimization with mortgage-backed securities. *EJOR* .

Hiller, R.S. and Eckstein, J. (1993) Stochastic dedication: Designing fixed income portfolios using massively parallel Benders decomposition. *Manag. Sci.*, **39**, 1422–38.

Shapiro, J.F. (1988) Stochastic programming models for dedicated portfolio selection, in *Math. Models for Decision Support* (ed. B. Mitra), NATO ASI Series, **F48**. Springer, Berlin.

Zenios, S.A. and Shtilman, M.S. (1993) Constructing optimal samples from a binomial lattice. *Journal of Information & Optimization Sciences*, **14**, 125–47.

A note on objective functions in multistage stochastic nonlinear programming problems

Vlasta Kaňková
Academy of Sciences of the Czech Republic,
Institute of Information Theory and Automation
P. O. Box 18, CZ–182 08 Prague 8, Czech Republic. Tel: 42-2-66052501.
Fax: 42-2-66414903. e-mail: kankova@utia.cas.cz

Abstract

Multistage stochastic programming problems well correspond to many practical situations in which a random element exists and moreover it is reasonable to treat them with respect to some discrete time interval. In particular, this type of problems correspond to practical situations that can be considered with respect to some time interval and simultanously decomposed with respect to the individual time points. The aim of this paper is to investigate the objective functions corresponding to the individual problems belonging to the one multistage stochastic programming problem. A special attention is paid to the Lipschitz property.

Keywords

multistage stochastic programming problem, discrete time interval, individual time points, objective functions, Lipschitz property

1 INTRODUCTION

Let (Ω, S, P) be a probability space,
$\xi^j = \xi^j(\omega) = [\xi_1^j(\omega), \xi_2^j(\omega), \ldots, \xi_s^j(\omega)]$, $j = 0, 1, \ldots, M$, $M \geq 1$ be an s–dimensional random vector defined on (Ω, S, P),
$\bar{\xi}^k = \bar{\xi}^k(\omega) = [\xi^0(\omega), \xi^1(\omega), \ldots, \xi^k(\omega)]$, $k = 0, 1, \ldots, M$,
$F^{\xi^j}(z^j)$, $F^{\bar{\xi}^k}(\bar{z}^k)$, $F^{\xi^k|\bar{\xi}^{k-1}}(z^k|\bar{z}^{k-1})$, $z^j \in E_s$, $j = 0, 1, \ldots, M$, $\bar{z}^k \in E_{(k+1)s}$, $k = 0, 1, \ldots, M$ denote the distribution functions of the ξ^j, $\bar{\xi}^k(\omega)$ and the conditional distribution function ($\xi^k(\omega)$ conditioned by $\bar{\xi}^{k-1}(\omega)$),
$\bar{z}^k = [z^0, z^1, \ldots, z^k]$, $z^j \in E_s$, $z^j = [z_1^j, \ldots, z_s^j]$, $j = 0, 1, \ldots, k$, $k = 0, 1, \ldots, M$,
$Z^k = Z^{\xi^k} \subset E_s$ and $Z^{\bar{\xi}^k} \subset E_{(k+1)s}$, $k = 0, 1, \ldots, M$ denote the supports corresponding to the distribution functions $F^{\xi^k}(z^k)$ and $F^{\bar{\xi}^k}(\bar{z}^k)$, $z^k \in E_s$, $\bar{z}^k \in E_{(k+1)s}$.

Let, moreover, $f_i^k(x^k)$ and $h_i^k(\bar{x}^{k-1}, \bar{z}^{k-1})$, $i = 1, 2, \ldots, l_k$, $k = 1, 2, \ldots, M$ be real–

valued, continuous functions defined on E_n and $E_{kn} \times E_{ks}$,
$g_0^M(\bar{x}^M, \bar{z}^M)$ be a real-valued function defined on $E_{(M+1)n} \times E_{(M+1)s}$,
$\bar{x}^k = [x^0, x^1, \ldots, x^k]$, $x^j \in E_n$, $x^j = [x_1^j, \ldots, x_n^j]$, $j = 0, 1, \ldots, k$, $k = 0, 1, \ldots, M$,
$X^j \subset E_n$, $j = 0, 1, \ldots, M$ be nonempty sets,
$\bar{X}^k = X^0 \times X^1 \times \ldots \times X^k$, $\bar{Z}^k = Z^{\xi^0} \times Z^{\xi^1} \times \ldots, Z^{\xi^k}$, $k = 0, 1, \ldots, M$.
(E_n, $n \geq 1$ denotes an n-dimensional Euclidean space.)

A general $(M+1)$-stage stochastic programming problem ($M \geq 1$) can be introduced as an optimization problem considered with respect to some rather general space, see e.g. Dupačová (1995), Hartley (1980) or Rockafellar and Wets (1978). It is well-known that such defined optimization problem is very complicated and mostly suitable for some theoretical investigation. However, in the literature there also exists another (perhaps for a practical treatment more suitable) definition of this type of the stochastic optimization problems, see e.g. Birge (1988), Dupačová (1995) or King (1988). We can introduce it in the form:

Find

$$\inf \{ \mathsf{E}_{F^{\xi^0}} g_0^0(x^0, \xi^0) | \ x^0 \in X^0 \}, \tag{1}$$

where the function $g_0^0(x^0, z^0)$ is given by the recurrent system of the equations

$$\begin{aligned} g_0^k(\bar{x}^k, \bar{z}^k) = \ & \inf \{ \mathsf{E}_{F^{\xi^{k+1}} | \bar{\xi}^k = \bar{z}^k} \, g_0^{k+1}(\bar{x}^{k+1}, \bar{\xi}^{k+1}) \, | \, x^{k+1} \in X^{k+1} : \\ & f_i^{k+1}(x^{k+1}) \leq h_i^{k+1}(\bar{x}^k, \bar{z}^k), \ i = 1, 2, \ldots, l_{k+1} \}, \quad k = 0, 1, \ldots, M-1. \end{aligned} \tag{2}$$

(The symbols $\mathsf{E}_{F^{\bar{\xi}^k}}$, $\mathsf{E}_{F^{\xi^k | \bar{\xi}^{k-1}}}$ denote the operators of the mathematical expectation considered with respect to the distribution functions $F^{\bar{\xi}^k}$ and $F^{\xi^k | \bar{\xi}^{k-1}}$, $k = 0, 1, \ldots, M$, $\mathsf{E}_{F^{\xi^0 | \xi^{-1}}} = \mathsf{E}_{F^{\xi^0}}$.)

The problems given by the relation (2) are parametric problems and they can be considered as inner optimization problems connected with the stage $k+1$, $k = 0, 1, \ldots, M-1$. Moreover, the problems given by (2) can be, under some assumptions, considered as a decomposition of the corresponding problem considered with respect to a general space.

In this paper, we shall follow the definition given by (1) and (2). In particular, we shall deal with the behaviour of the objective functions $g_0^k(\bar{x}^k, \bar{z}^k)$, $k = 0, \ldots, M-1$. We focus mostly our investigation to determine assumptions under which these functions are Lipschitz. Namely, this property give the possibility to generalize the results achieved for one and two-stage stochastic programming problems in Kaňková (1980b, 1994a, 1994b) to the multistage case. In detail, the new results enable to investigate successively stability of the multistage problems, to construct the approximative problems with approximation error not greater than some prescribed value as well as to plan a random sample size to be the empirical estimates of the optimal value and the optimal solution enough "accurate" with the high probability. Moreover, we expect that the new results will be useful for the choice of scenarios in the corresponding approach.

Remark 1 *In general, it may happen that some symbols in (1) and (2) are not well defined. However, this situation cannot appear under the assumptions considered in this paper.*

2 SOME DEFINITIONS AND AUXILIARY ASSERTIONS

Evidently, to investigate the behaviour of the functions $g_0^k(\bar{x}^k, \bar{z}^k)$, $k = 0, \ldots, M-1$ it is necessary to investigate mostly the behaviour of the multifunctions $\mathcal{K}^{k+1}(\bar{x}^k, \bar{z}^k)$, $\bar{x}^k \in E_{(k+1)n}$, $\bar{z}^k \in E_{(k+1)s}$ $k = 0, \ldots, M-1$ given by

$$\mathcal{K}^{k+1}(\bar{x}^k, \bar{z}^k) = \{x^{k+1} \in X^{k+1} : f_i^{k+1}(x^{k+1}) \leq h_i^{k+1}(\bar{x}^k, \bar{z}^k), \, i = 1, 2, \ldots, l_{k+1}\}. \tag{3}$$

To this end, we substitute

$$y_i^k = h_i^{k+1}(\bar{x}^k, \bar{z}^k), \, i = 1, 2, \ldots, l_{k+1}, \, k = 0, 1, \ldots, M-1, \, \bar{x}^k \in E_{(k+1)n}, \, \bar{z}^k \in E_{(k+1)s} \tag{4}$$

and define for $k = 0, 1, \ldots, M-1$, $y^k \in E_{l_{k+1}}$ the multifunctions $\bar{\mathcal{K}}^{k+1}(y^k)$ by the relation

$$\bar{\mathcal{K}}^{k+1}(y^k) = \{x^{k+1} \in X^{k+1} : f_i^{k+1}(x^{k+1}) \leq y_i^k, \, i = 1, 2, \ldots, l_{k+1}\}. \tag{5}$$

Evidently

$$\bar{\mathcal{K}}^{k+1}(y^k) = \mathcal{K}^{k+1}(\bar{x}^k, \bar{z}^k) \quad \text{for } y^k = (y_1^k, \ldots, y_{l_{k+1}}^k) \quad \text{fulfilling relation (4)}. \tag{6}$$

The assumptions under which $\bar{\mathcal{K}}^{k+1}(y^k)$, for every given $k = 0, 1, \ldots, M-1$, is a uniformly continuous (with respect to the Hausdorff distance – for definition see e.g. Kaňková (1980a)) multifunction are introduced, for example in Kaňková (1980a). By the technique of that paper we obtain also sufficient assumptions under which the functions $g_0^k(\bar{x}^k, \bar{z}^k)$, $k = 0, 1, \ldots, M-1$ are uniformly continuous. To investigate the assumptions, under which the multifunctions $\bar{\mathcal{K}}^{k+1}(y^k)$, $k = 1, 2, \ldots, M-1$ are Lipschitz (with respect to the Hausdorff distance), we denote by Y_0^{k+1} the convex hull of the set Y^{k+1}

$$\begin{aligned} Y^{k+1} = \ & \{y^k \in E_{l_{k+1}}, y^k = (y_1^k, \ldots, y_{l_{k+1}}^k) : \text{there exists } \hat{x}^k \in \hat{X}^k, \bar{z}^k \in \bar{Z}^k \text{ such that} \\ & y_i = h_i^{k+1}(\bar{x}^k, \bar{z}^k) \quad \text{for every } i \in \{1, 2, \ldots, l_{k+1}\}\}. \end{aligned}$$

Furthermore, for $\varepsilon' > 0$ we denote by $Y(\varepsilon')$ the ε' neigbourhood of the nonempty set $Y \subset E_l$, $l \geq 1$.

We shall introduce the following system of the assumptions.

i.1 For every $k \in \{0, 1, \ldots, M-1\}$

a. $f_i^{k+1}(x^{k+1})$, $i = 1, 2, \ldots, l_{k+1}$ are linear functions, $X^{k+1} = E_n$, without loss of generality, we can consider in this case the constraints in (2) (consequently in (3)) to be in the form of equalities,

b. for every $y^k \in Y^{k+1}$, $\bar{\mathcal{K}}^{k+1}(y^k)$ is a nonempty, compact set,

c. the matrix A^{k+1} of the type $(l_{k+1} \times n)$, $l_{k+1} \leq n$ fulfils the relation

$$\bar{\mathcal{K}}^{k+1}(y^k) = \{x^{k+1} \in X^{k+1} : A^{k+1}x^{k+1} = y^k\}, \, y^k \in Y^{k+1}$$

and, moreover, all its submatrices of the types $(l_{k+1} \times l_{k+1})$, $A^{k+1}(1)$, $A^{k+1}(2)$, $\ldots$, $A^{k+1}(m)$ are nonsingular,

d. $C(k+1) = l_{k+1} \max_{i\,r\,s} a_{ir}^{k+1}(s)$, where $a_{ir}^{k+1}(s)$ for $s \in \{1, 2, \ldots, m\}$ denote the elements of the inverse matrix to $A^{k+1}(s)$,

i.2 For every $k \in \{0, 1, \ldots, M-1\}$

a. X^{k+1} is a convex, compact set, $M_1^{k+1} = \sup_{x^1, x^2 \in X^{k+1}} \|x^1 - x^2\|$,

b. $f_i^{k+1}(x^{k+1})$, $i = 1, 2, \ldots, l_{k+1}$ are convex functions on X^{k+1},

c. $\bar{\mathcal{K}}^{k+1}(y^k)$ is a nonempty set for every $y^k \in Y^{k+1}(\varepsilon)$ and some $\varepsilon > 0$,

d. $C(k+1) = \frac{M_1^{k+1}}{\varepsilon_0}$ for an $\varepsilon_0 \in (0, \varepsilon)$,

i.3 For every $k \in \{0, 1, \ldots, M-1\}$ there exists real-valued constants d_1^{k+1}, $\gamma_2^{k+1} > 0$ such that

a. if $x^{k+1} \in X^{k+1}$, $y^k \in Y_0^{k+1}(d_1^{k+1}\gamma_2^{k+1})$ fulfil the relations $f_j^{k+1}(x^{k+1}) \leq y_j^k$, $j = 1, \ldots, l_{k+1}$ and simultanously $f_i^{k+1}(x^{k+1}) = y_i^k$ for at least one $i \in \{1, \ldots, l_{k+1}\}$, then there exists a vector $x^{k+1}(0) = x^{k+1}(0, x^{k+1}) \in E_n$ fulfilling for every $d \in (0, d_1^{k+1})$ and every $i = 1, 2, \ldots, l_{k+1}$ the relations

$$\|x^{k+1}(0)\| = 1, \quad x^{k+1} + dx^{k+1}(0) \in X^{k+1},$$
$$f_i^{k+1}(x^{k+1}) - f_i^{k+1}(x^{k+1} + dx^{k+1}(0)) \geq \gamma_2^{k+1} d,$$

b. for every $y^k \in Y_0^{k+1}(\gamma_2^{k+1} d_1)$, $\bar{\mathcal{K}}^{k+1}(y^k)$ is a nonempty set,

c. $C(k+1) = \frac{1}{\gamma_2^{k+1}}$.

($\|\cdot\|$ denotes the Euclidean norm in E_n.)

We introduce the next auxiliary assertion.

Lemma 1 *If at least one of the assumptions (i.1), (i.2), (i.3) is fulfilled, then*

$$\Delta[\bar{\mathcal{K}}^{k+1}(y^k(1)), \bar{\mathcal{K}}^{k+1}(y^k(2))] \leq C(k+1)\,\|y^k(1) - y^k(2)\|$$

for every $y^k(1), y^k(2) \in Y^{k+1}$, $\quad k = 0, 1, \ldots, M-1$.
($\Delta[\cdot, \cdot]$ *denotes the Hausdorff distance of the nonempty, closed subsets of* E_n. *)*

Proof. If the assumptions (i.1) or (i.2) are fulfilled, then we refer the reader to the papers Kaňková (1980b, 1994a). Consequently, it remains to consider only the assumptions (i.3). To this end, let $k \in \{0, ,1, \ldots, M-1\}$, $y^k(1), y^k(2) \in Y_0^{k+1}$, $y^k(1) = (y_1^k(1), \ldots, y_{l_{k+1}}^k(1))$, $y^k(2) = (y_1^k(2), \ldots, y_{l_{k+1}}^k(2))$ be arbitrary given. We distinguish two cases

a. $\|y^k(1) - y^k(2)\| < \gamma_2^{k+1} d_1$, $\qquad$ b. $\|y^k(1) - y^k(2)\| \geq \gamma_2^{k+1} d_1$.

First, let us consider the case a. If $\underline{y}, \bar{y} \in E_{l_{k+1}}$, $\underline{y} = (\underline{y}_1, \underline{y}_2, \dots, \underline{y}_{l_{k+1}})$, $\bar{y} = (\bar{y}_1, \bar{y}_2, \dots, \bar{y}_{l_{k+1}})$ are defined by the relations

$$\underline{y}_i = \min(y_i^k(1), y_i^k(2)), \quad \bar{y}_i = \max(y_i^k(1), y_i^k(2)), \ i = 1, 2, \dots, l_{k+1},$$

then

$$\|y^k(1) - y^k(2)\| = \|\underline{y} - \bar{y}\|, \quad \underline{y}, \bar{y} \in Y_0^{k+1}(\gamma_2^{k+1} d_1).$$

If, furthermore, we define the set X_0^{k+1} by the relation

$$\begin{aligned} X_0^{k+1} = \ & \{x^{k+1} \in X^{k+1} : \text{there exists } y^k \in Y_0^{k+1}(\gamma_2^{k+1} d_1) \text{ such that} f_j(x^{k+1}) \leq y_j^k, \\ & \text{for every } j \in \{1, 2, \dots, l_{k+1}\} \text{ and simultanously } f_i(x^{k+1}) = y_i^k \text{ for} \\ & \text{at least one } i \in \{1, 2, \dots, l_{k+1}\}, y^k = (y_1^k, \dots, y_{l_{k+1}}^k)\}, \end{aligned}$$

then we can formulate the next implication

$$\begin{aligned} x \in \bar{\mathcal{K}}^{k+1}(\bar{y}), \ x \in X_0^{k+1} \implies & \text{there exists } x' \in \bar{\mathcal{K}}^{k+1}(\underline{y}) \text{ such that} \\ & \|x - x'\| \leq \tfrac{1}{\gamma_2^{k+1}} \|\underline{y} - \bar{y}\|. \end{aligned} \tag{7}$$

If we define the point x' by the relation

$$x' = x + \frac{\|\underline{y} - \bar{y}\|}{\gamma_2^{k+1}} x^{k+1}(0),$$

then it follows from the assumptions of Lemma 1 that

$$\|x - x'\| = \frac{\|\underline{y} - \bar{y}\|}{\gamma_2^{k+1}} < d_1 \tag{8}$$

and simultanously

$$f_i^{k+1}(x) - f_i^{k+1}(x') \geq \|\underline{y} - \bar{y}\| \geq \bar{y}_i - \underline{y}_i, \ i = 1, 2, \dots l_{k+1}.$$

However, it follows immediately from the last inequality and from the implication assumption that $x' \in \mathcal{K}^{k+1}(\underline{y})$. We have proven the validity of the implication (7). Since evidently $\underline{y} \leq y^k(1) \leq \bar{y}$, $\underline{y} \leq y^k(2) \leq \bar{y}$ we get

$$\bar{\mathcal{K}}^{k+1}(\underline{y}) \subset \bar{\mathcal{K}}^{k+1}(y^k(1)) \subset \bar{\mathcal{K}}^{k+1}(\bar{y}), \quad \bar{\mathcal{K}}^{k+1}(\underline{y}) \subset \bar{\mathcal{K}}^{k+1}(y^k(2)) \subset \bar{\mathcal{K}}^{k+1}(\bar{y}),$$

It follows from the last relation, the relation (8), the definition of the multifunction $\bar{\mathcal{K}}^{k+1}(y^k)$ and from the properties of the Hausdorff distance that also

$$\Delta[\bar{\mathcal{K}}^{k+1}(y^k(1)), \bar{\mathcal{K}}^{k+1}(y^k(2))] \leq C(k+1) \|y^k(1) - y^k(2)\|.$$

We have proven the validity of the assertion of Lemma 1 for $y^k(1)$, $y^k(2)$ fulfilling the condition a. It remains to consider the case b. However, the assertion of Lemma 1 follows, in this case, from the results already proven in the case a., from the convex behaviour of the set Y_0^{k+1} and from the triangular inequality due to the Hausdorff distance (Salinetti and Wets 1979) in the space of the closed subsets of E_n.

Since $k \in \{0, 1, \dots, M-1\}$ was arbitrary we have finished the proof of Lemma 1. □

3 THE MAIN RESULTS

Combining the achieved auxiliary result and the appropriate behaviour of the functions $h_i^k(\bar{x}^{k-1}, \bar{z}^{k-1})$, $g_0^M(\bar{x}^M, \bar{z}^M)$, $i = 1, 2, \dots, l_{k+1}$, $k = 1, 2, \dots, M$ we can obtain the behaviour of the functions $g_0^k(\bar{x}^k, \bar{z}^k)$, $k = 0, 1, \dots, M-1$. We shall focus our investigation mostly to the Lipschitz behaviour. To this end let us introduce the following system of assumptions.

L.2 a. For every $k = 0, 1, \dots, M-1$, $i = 1, 2, \dots, l_{k+1}$, $h_i^{k+1}(\bar{x}^k, \bar{z}^k)$ are Lipschitz functions on $\bar{X}^k \times \bar{Z}^k$ with the Lipschitz constants $\bar{L}_i(k)$,

b. $g_0^M(\bar{x}^M, \bar{z}^M)$ is a Lipschitz function on $\bar{X}^M \times \bar{Z}^M$ with the Lipschitz constant $\bar{L}_0(M)$.

We shall prove the next theorem.

Theorem 1 *If*

1. *at least one of the assumptions (i.1), (i.2), (i.3) is fulfilled,*
2. *the assumptions (L.2) are fulfilled,*
3. *for every $k \in \{0, 1, \dots, M-1\}$ at least one of the following assumptions is fulfilled*
   a. *$\mathcal{K}^{k+1}(\bar{x}^k, \bar{z}^k)$ is a compact set,*
   b. *$g_0^M(\bar{x}^M, \bar{z}^M)$ is a bounded function,*
4. *the vectors $\xi^j(\omega)$, $j = 0, 1, \dots, M$ are stochastically independent and moreover a finite $\mathsf{E}_{F^{\xi^M}} \xi^M$ exists,*

then for every $k = 0, 1, \dots, M$

1. *there exists a finite*

$$\mathsf{E}_{F^{\xi^k}} g_0^k(\bar{x}^k, \xi^k), \ \bar{x}^k \in \bar{X}^k,$$

2. *$g_0^k(\bar{x}^k, \bar{z}^k)$ is a Lipschitz function on $\bar{X}^k \times \bar{Z}^k$ with the Lipschitz constant*

$$\bar{L}_0(k) = \bar{L}_0(M) \prod_{r=k+1}^{M} [1 + C(r) \sum_{i=1}^{l_{r+1}} \bar{L}_i(r-1)].$$

Proof. Since it follows from the assumptions of Theorem 1 that

$$|g_0^M(\bar{x}^M, \bar{z}^M(1)) - g_0^M(\bar{x}^M, \bar{z}^M(2))| \leq \bar{L}_0(M) \|\bar{z}^M(1)) - \bar{z}^M(2))\|,$$

for every $\bar{z}^M(1)), \bar{z}^M(2)) \in \bar{Z}^M$, $\bar{x}^M \in \bar{X}^M$, we can see that under the assumption 3 of Theorem 1 a finite $\mathsf{E}_{F^{\bar{\xi}^M}} g_0^M(\bar{x}^M, \bar{\xi}^M)$ exists. Moreover, since it follows from the assumptions and from the assertions of Lemma 1 that

$$\Delta[\bar{\mathcal{K}}^M(y^{M-1}(1)), \bar{\mathcal{K}}^M(y^{M-1}(2))] \leq C(M) \|y^{M-1}(1) - y^{M-1}(2)\|,$$

for every $y^{M-1}(1), y^{M-1}(2) \in Y^M$, we can obtain by the assumptions L.2.a and the relation (6) that

$$\begin{aligned} \Delta \quad & [\mathcal{K}^M(\bar{x}^{M-1}(1), \bar{z}^{M-1}(1)), \mathcal{K}^M(\bar{x}^{M-1}(2), \bar{z}^{M-1}(2))] \leq \\ & C(M) \sum_{i=1}^{l_M} \bar{L}_i(M-1) \|(\bar{x}^{M-1}(1), \bar{z}^{M-1}(1)) - (\bar{x}^{M-1}(2), \bar{z}^{M-1}(2))\| \end{aligned}$$

for every $\bar{x}^{M-1}(1), \bar{x}^{M-1}(2) \in X^{M-1}$, $\bar{z}^{M-1}(1), \bar{z}^{M-1}(2) \in \bar{Z}^{M-1}$.

Employing the technique of the proof of Lemma 2 in Kaňková (1978) we see that $g_0^{M-1}(\bar{x}^{M-1}, \bar{z}^{M-1})$ is a Lipschitz function on $\bar{X}^{M-1} \times Z^{M-1}$ with the Lipschitz constant

$$\bar{L}_0(M-1) = L_0(M)[1 + C(M) \sum_{i=1}^{l_M} \bar{L}_i(M-1)].$$

Evidently, if we substitute successively $M =: k$, $k = M-1, M-2, \ldots, 1$ and repeat the proof for $M = k$ we can see that the assertion of Theorem 1 holds. □

Remark 2 *The assumption on the independence of the vectors $\xi^i(\omega)$, $i = 1, 2, \ldots, M$ can be replaced by another assumptions. For example, if the assumption 3.b is fulfilled and a bounded, conditioned probability density (with respect to the Lebesgue measure) corresponding to $F_{\xi^k|\bar{\xi}^{k-1}}(z^k|\bar{z}^{k-1})$, $k = 1, 2, \ldots, M$ exists such that it is a Lipschitz function on $\bar{Z}^k$. Of course, the value of the corresponding Lipschitz constants to $\bar{L}_0(k)$, $k = 0, 1, \ldots, M-1$ depend then also on the characteristics of the probability densities.*

Theorem 1 introduces the assumptions under which the objective functions defined by (2) are Lipschitz. Some special results for the two–stage stochastic programming problems have been already presented in Kaňková (1994a). To obtain sufficient assumptions under which the objective functions in (2) are convex or strongly convex (for definition see e.g. Kaňková (1994a)) we can employ usual techniques of the convex analysis, cf. Rockafellar (1970) and Pšeničnyi and Danilin (1975) (see also Kaňková (1978), (1994)). Moreover, employing the results of the convex analysis we can obtain also other assumptions under which the objective functions in (2) are Lipschitz.

4 CONCLUSION

The aim of the paper is to investigate the behaviour of the objective functions in the problems defined by the relation (1) and (2). We focused our consideration mostly on the Lipschitz property of the objective functions in the inner problems (2). Namely it is known from the literature that just this property give the possibility to investigate the convergence rate in the case of the empirical estimates as well as to investigate the stability or to construct the approximative problems more suitable for the numerical treatment. Moreover, sufficient assumptions of differentiability can be obtained utilizing the result of this paper and the proof technique similar to that used in Kaňková (1978). However, the corresponding investigation of these problems cannot be included in the framework of this paper.

REFERENCES

Birge, J.R. (1988) An L-shaped method computer code for multistage stochastic linear programs, in *Numerical Techniques for Stochastic Optimization Problems* (eds. Yu. Ermoliev and R.J.-B. Wets), Springer–Verlag, Berlin.

Dupačová, J. (1995) Multistage stochastic programs: the state–of–the–art and selected bibliography. *Kybernetika*, **31**, 151–74.

Hartley, R. (1980) Inequalities in complete convex stochastic programming. *J. Math. Anal. Appl.*, **75**, 373–84.

Kaňková, V. (1978) Differentiability of the optimalized function in a two stages stochastic nonlinear programming problem (in Czech). *Ekonomicko–matematický obzor*, **14**, 322–30.

Kaňková, V. (1980a) Optimization problem with parameter and its application to the problems of two–stage stochastic nonlinear programming. *Kybernetika*, **16**, 413–24.

Kaňková, V. (1980b) Approximative solution of problems of two–stage stochastic nonlinear programming (in Czech). *Ekonomicko–matematický obzor*, **16**, 64–76.

Kaňková, V. (1994a) On stability in two–stage stochastic nonlinear programming, in *Proceedings of the Fifth Prague Symposium* (eds. P. Mandl and M. Hušková), Springer–Verlag, Berlin.

Kaňková, V. (1994b) *A Note on Distribution Function Sensitivity in Stochastic Programming*. Research Report ÚTIA AV ČR No. 1826, Prague.

Kaňková, V. (1994c) A note on estimates in stochastic programming. *J. Comput. Appl. Math.*, **56**, 97–112.

King, A.J. (1988) Stochastic programming problems: Examples from the literature, in *Numerical Techniques for Stochastic Optimization Problems* (eds. Yu. Ermoliev and R.J.-B. Wets), Springer–Verlag, Berlin.

Pšeničnyi, B.N. and Danilin, Yu.M. (1970) *Numerical Methods in Extremal Problems* (in Russian). Nauka, Moscow 1975.

Rockafellar, R. T. (1970) *Convex Analysis*. Princeton Univ. Press, Princeton, New Jersey.

Rockafellar, R. T. and Wets, R. J.-B. (1978) The optimal recourse problem in discrete time: L^1-multipliers for inequality constraints. *SIAM Control Optim.*, **16**, 16–36.

Salinetti, G. and Wets, R.J.-B. (1979) On the convergence of sequences of convex sets in finite dimension. *SIAM Review*, **21**, 18–33.

Transportation Systems

72

Dynamic search for shortest multimodal paths in a transportation network

A. Di Febbraro, S. Sacone
Department of Communication, Computer, and System Sciences
University of Genova, Via Opera Pia 13, I-16145 Genova, Italy.
Tel: +39-10-3532983. Fax: +39-10-3532948.
e-mail: angela@dist.unige.it, simona@dist.unige.it

Abstract

Many transportation systems can be suitably represented as Discrete Event Systems. This is the case of the multimodal urban transportation network considered in this paper. In the simulation tool designed to reproduce its functioning, a module is included which allows to find the best multimodal paths between any pair of nodes in the network. The basic algorithm is an ad-hoc modified version of the Dijkstra algorithm to find the shortest paths in an oriented graph. Such an algorithm is complicated to include some choices left to users, regarding the number of paths required, the maximum time to travel, and the modes of transport to use. Then, a heuristic algorithm to find the best multimodal paths in a network integrating public and private modes of transport is proposed.

Keywords

Transportation, urban systems, Discrete Event Dynamic Systems.

1 INTRODUCTION

The integration between public and private transport services, with the aim of increasing the 'attractiveness' of public transportation means with respect to private ones, is a research topic of growing interest. Such an integration should lead to realize intermodality in transportation (Gedeon et al., 1993; Fernandez et al., 1994). With reference to the problem of integrating passenger transportation services in a urban area, the realization of efficient intermodal transportation systems should make people give up using their cars to move in the cities, which would result in both an optimization of the travelling times and a decrease in air pollution.

In this paper, a model of an intermodal transportation network which realizes the integration among urban public transportation services and private traffic is proposed. The system is represented as an oriented graph, in which the nodes are stations where it is possible to switch from a mode of transport to another one. In order to suit its peculiar characteristics, the considered transportation network is modelled as a Discrete Event Dynamic System (see, for instance, Cassandras, 1993). Due to the stochastic characteristics of such a system, some disturbances are also included. The behaviour of the discrete event system modelling the transportation network is studied by means of a special purpose designed simulation tool, which is the kernel of a urban traffic simulation program, designed to perform two major functions, almost independent of each other. The first objective is the validation of integrated timetables for the different transportation modes, so as to consider the various transportation services to be parts of a whole intermodal transportation system. In this sense, the traffic simulation program can evaluate the performance of the system, supporting the user in modifying adaptively a given timetable (Di Febbraro et al., 1994, 1995).

The second objective pursued is to give the users of the intermodal transportation network some real-time updated information about the state of the network itself. The optimal situation to realize is that in which a user can know, when approaching the transportation network, which path he had better follow to reach his destination, according to his requirements. Each user could enter the information system in the stations of the network, or even from his car. In particular, if a user approaches the network by car, he can evaluate where it is possible (and convenient) to leave his car and carry on with his journey using public modes of trasport. To this end, some intermodal stations have a parking area associated with.

This paper centres round the algorithms on which such an information system is founded. Two peculiar algorithms, resulting from the well-known Dijkstra algorithm (Christofides, 1975) and the method proposed by Yen (1971), are proposed to find the best multimodal paths, considering, respectively, only public modes of transport and both private and public modes of transport. Information made available to users regard essentially : i) the travelling times along all the possible paths, intermodal or not, between any pair of nodes in the network; ii) the availability of places in the parking areas associated with some macronodes; iii) the occurrence of irregular traffic conditions in some part of the network. Some examples of application of the algorithms proposed are presented.

2 THE DISCRETE EVENT MODEL OF THE TRANSPORTATION NETWORK

The multimodal transportation network is modelled as an oriented graph, whose fundamental elements are nodes, macronodes, and links. A node is a station for a single mode of transport, and it can only exist as a part of a macronode. A macronode is an intermodal station, that is, a place where people can change mode of transport, or simply enter the transportation system. Therefore, a macronode is composed of one or more nodes. A link is a unidirectional path which connects two macronodes and is devoted to a single transportation mode. There are special links, named inner links, which connect two nodes in a macronode and can be traversed only on foot.

It is assumed that users can approach the urban integrated transportation system using any means of transport. To model the arrivals of passengers by private means of transport, parking areas are considered to be located near some macronodes of the network, where passengers can leave their means of transport, if room is available. The existence of a parking area at a macronode and the availability of places in it play an important role in modelling the integration between public and private transport means.

For the sake of simplicity, it is assumed that the considered transportation system consists of three different kinds of public transportation services : underground, railway, and bus. However, there is no difficulty in adding and/or changing modes of transport. In this framework, the links which connect the N macronodes in the network are divided into three classes: a) bus links (*B-links*); b) railway links (*R-links*); c) underground links (*U-links*). As shown in Figure 1, each macronode can have at most six links in common with another macronode.

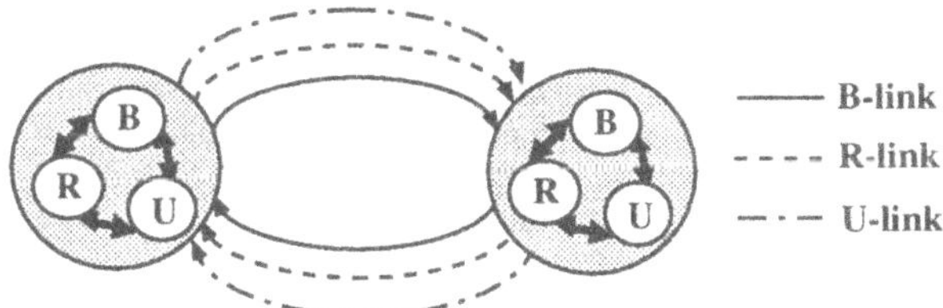

Figure 1 A generic pair of macronodes.

Macronodes n, n=1,...,N-1, represent the real intermodal stations of the transportation system under study, while the last macronode, macronode N, is a special macronode, i.e., a 'virtual' intermodal station which represents the outer world with respect to the network. In other words, each macronode which connects the urban area with its surroundings has a link towards macronode N. The presence of such a virtual macronode allows to model the considered transportation system as a closed network. This yields a major advantage consisting in the possibility of applying particular storage methods for closed graphs, which reduce sensibly the computational load. Moreover, the presence of macronode N also allows to model the movement of those means of transport which go, or come from outside the considered urban transportation network.

The entities of the model, i.e., its components requiring explicit representation (Banks and Carson, 1984), are : macronodes, nodes, links, and transportation means. With each class of entities, static quantities (parameters) and dynamic quantities (state variables) are associated. The discrete events that describe the evolution of the considered urban transportation network, and then rule the functioning of the traffic simulation tool, are divided into two classes (Di Febbraro et al., 1995):

1. events that describe the nominal traffic conditions;
2. events that represent unpredictable conditions affecting the system.

The evolution of the system is reproduced by means of discrete event simulation, which can be viewed as the modelling over time of a system in which changes in the state only occur at discrete time instants, i.e., those instants when events occur. Each type of event is associated with an operational procedure which, once invoked, modifies the state variables of the model in dependence of the event occurred. Such procedure also schedules the events which will occur as a consequence of the current event, and cancels those events which the current event prevents from occurring.

3 HEURISTIC SEARCH FOR THE BEST INTERMODAL PATHS

To find the best intermodal paths between a pair of nodes, two different algorithms are proposed. The very basic algorithm is an ad-hoc modified version of the well-known Dijkstra algorithm to find the shortest paths in an oriented graph (Christofides, 1975). Before describing in detail the proposed algorithm, it is worth introducing some notation. The transportation network is modelled as an oriented graph $G = \{S,L\}$, where S is the set of the macronodes of the network, $card(S) = N_s$, and L is the set of links of the network, $card(L) = N_l$. The origin macronode of a link is identified by the function $O : L \rightarrow S$, such that $O(l) = s$, $l \in L$, $s \in S$, means that macronode s is the origin macronode of link l. In an analogous way, function $D : L \rightarrow S$ identifies the destination macronode of a link, i.e., $D(l)$ represents the destination macronode of link l. As described above, the links are classified in dependence of the mode of transport to which they are devoted; let M be the set of the modes of transport present in the considered network and $card(M) = N_m$. In the considered case, $N_m = 4$ and $M = \{B,R,U,C\}$, where B, R, U, and C stand for bus, railway, underground, and cars, respectively. Moreover, let $M : L \rightarrow M$ be the function such that $M(l) = m$, $l \in L$, $m \in M$ means that link l is devoted to mode of transport m.

Using the variables introduced so far, the topology of graph G is defined by means of an adjacency matrix **A**, $\dim(\mathbf{A}) = N_s \times N_s \times N_m$, whose elements are:

$$a(i,j,m) = \begin{cases} 1 & \text{if } \exists\, l \in L : O(l)=i,\ D(l)=j,\ M(l)=m \\ 0 & \text{otherwise} \end{cases} \qquad \forall\, i,j \in S,\ m \in M$$

Other matrices represent the travel times on the links and the waiting times in the nodes. Let **T**, $\dim(\mathbf{T}) = N_s \times N_s \times N_m$, be the matrix containing the time-varying travel times on the links; if $\tau_{i,j}$, $i,j \in S$, denotes the present travel time on link l, $l \in L$, $O(l) = i$, $D(l) = j$, the generic element of **T** can be written as:

$$t(i,j,m) = \begin{cases} \tau_{i,j} & \text{if } a(i,j,m) = 1 \\ +\infty & \text{otherwise} \end{cases} \qquad \forall\, i,j \in S,\ m \in M$$

Then, let **IL** denote a matrix, $\dim(\mathbf{IL})=N_s\times N_m\times N_m$, whose generic element $il(i,m,n)$, $i\in S$, $m,n \in M$, represents the travel time on the inner link within macronode i, and connecting the node (station) of mode m with the node (station) of mode n. Of course, $il(i,m,m) = 0$. The generic element of matrix **WT**, $\dim(\mathbf{WT}) = N_s \times N_m$, namely $wt(i,m)$, $i \in S$, $m \in M$ represents the time to wait presently in macronode i before a means of transport of mode m arrives. Last, to model the time-varying availability of places in the parking areas associated with the nodes, a vector **p**, $\dim(\mathbf{p}) = N_s$, is introduced, and its generic element is defined as

$$p(i) = \begin{cases} 0 & \text{if there is no place in the parking area at node i} \\ 1 & \text{if there is at least a place in the parking area at node i} \\ +\infty & \text{if there is no parking area associated with node i} \end{cases} \qquad \forall\, i \in S$$

If macronode i is associated with a parking area, the value of p(i) is randomly determined by taking into account the probability of finding free places in it. Such a probability varies in dependence of macronode i and of the time of the day. It is worth noting that only matrices **A** and **IL** have constant elements. Matrices **T** and **WT** are updated in real time based on the data provided by the simulation kernel, whereas the components of vector **p** are stochastic variables.

As said above, the algorithm which solves the problem of finding the shortest path from an origin macronode o to a destination macronode d, o,d ∈ S, considering only public means of transport, is derived from the Dijkstra algorithm. The main difference between the classical version of Dijkstra algorithm and the proposed one stands in the cost function associated with each link. Here, such a cost not only depends on the origin and destination nodes of the link, but also on the mode of transport currently used. Then, the cost associated with link l, l ∈ L, can be expressed by a function c(i,j,m), where i = O(l), j = D(l), and m ∈ M \ {C} is the mode of transport used to reach node i, taking on the form:

$$c(i,j,m) = \min_{n \in M \setminus \{C\}} \{t(i,j,n) + il(i,m,n) + wt(i,n)\}$$

The basic version of the algorithm has been complicated to include some further possibilities offered to users, who can choose about:

1. the number k, k≥1, of best paths to find between the indicated pair of nodes;
2. the maximum travel time t_{max} of the path from node o to node d, i.e., a user can fix the maximum time he is willing to spend to reach his destination;
3. the modes of transport utilized, i.e., a user can choose which modes he prefers to use to reach his destination.

It is apparent that the introduction of the above possibilities for the user results in additional constraints to impose on the solution of the optimization problem. If only point 1 is specified, the problem to solve is that of finding the k best paths to move in the network from node o to node d. We solve such a problem by applying the method proposed by Yen (1971), based on the modified version of Dijkstra algorithm that we have proposed above. Almost the same holds if also the maximum travel time is specified. In fact, the structure of the problem to solve is left essentially untouched with respect to the previous case. Only an additional constraint on the travel time has to be taken into account, so that only those paths with overall travel times less than t_{max}, if any, should appear in the list of paths given as an output.

As for decisional aspect 3, the possibility for the users of neglecting some mode of transport does not lead to a trivial complication of the problem. Actually, if a user chooses to eliminate one or more modes of transport from his path, the resulting graph can become not connected. If so, we would not be able to solve the problem using the algorithm proposed so far, as the Dijkstra algorithm only applies to connected graphs. To deal with this aspect, a special procedure has been implemented to check the type of structure of the graph and act consequently. If the graph does not turn out to be connected, such a procedure removes from it the isolated macronodes. Once a connected topology has been reached again, the algorithm to apply is the same as above, the only difference standing in the fact that the set of possible modes of transport to consider is now only a subset of M.

The algorithms described so far only work on the public transportation network, without taking specifically into account the presence of private traffic. Being the primary aim of our work the integration of the public transportation services with the private traffic, we state now a heuristic algorithm which serves this purpose. It is assumed that this algorithm is run to answer a question posed by a user approaching a macronode of the network travelling by car. In fact, the algorithm searches for the best multimodal paths in which the first part (possibly the whole path) has to be covered by car, if any such a path exists, according to the following steps.

Algorithm

Given an origin macronode o and a destination macronode d :

1. Define the set of nodes $S' = \{s \in S, s \neq o \text{ and } p(s) = 1\}$;
2. $\forall$ s $\in$ S', find, by applying the Dijkstra algorithm on graph $G' = \{S, L'\}$, where L' is the set of the links relevant to mode C, the shortest path from macronode o to macronode s and, then, define vector $\mathbf{t}_c$, $\dim(\mathbf{t}_c) = \text{card}(S')$, such that $\mathbf{t}_c(s)$, s $\in$ S', is the travel time on the path determined;
3. $\forall$ s $\in$ S', find, by applying the modified version of the Dijkstra algorithm described above, the shortest multimodal path from s to d; then, define vector $\mathbf{t}_{mm}$, $\dim(\mathbf{t}_{mm}) = \text{card}(S')$, such that $\mathbf{t}_{mm}(s)$, s $\in$ S', is the travel time of such a path;
4. Determine the best path to move from o to d, by choosing macronode i $\in$ S', such that:
   $$\mathbf{t}_c(i) + \mathbf{t}_{mm}(i) = \min_{s \in S'} \{\mathbf{t}_c(s) + \mathbf{t}_{mm}(s)\}$$

The above algorithm can be easily combined with the three additional possibilities offered to users described above.

4 SIMULATION RESULTS

In this section, some experimental results showing the effectiveness of the heuristic algorithm proposed in the previous section are presented and discussed. The considered case study is relevant to the transportation network. Such a network is composed of 80 links and 21 macronodes, each of which consists of at least two nodes. In particular, in the network there are 21 bus stations, 13 railway stations, and 11 underground stations. In order to evaluate its behaviour with reference to the case study, the algorithm, which has been implemented in C programming language, has been run with many different input requests. Two significant examples are reported hereafter. The first example consists in a request to find the four best paths to move from macronode 6 to macronode 19 using private and/or public means of transport. The output of the algorithm is shown in Table 1, in which, for each path found, the following data are reported:

1. the total travel time (*TTT*) in seconds;
2. the sequence of the macronodes which compose the path (*Paths*); the nodes where the mode of transport changes are followed by the indication of the new mode adopted;
3. the number of macronodes (*NM*) in the path;
4. the number of mode changes (*NMC*) in the path.

Table 1

	TTT(sec)	*Paths*	*NM*	*NMC*
1	900	6-R-3-5-8-12-U-15-R-19	7	2
2	960	6-C-9-U-8-12-15-R-19	6	2
3	960	6-B-9-U-8-12-15-R-19	6	2
4	1020	6-R-3-5-8-U-12-15-B-19	7	2

As a second example, we report the case in which the three best paths to move from macronode 2 to macronode 20, with a maximum travel time of 800 seconds and by using only public transportation services, have been asked to the program. The solution is shown in the following table.

Table 2

	TTT(sec)	*Paths*	*NM*	*NMC*
1	720	2-B-5-R-8-U-12-15-20	6	2
2	780	2-B-4-U-8-12-15-20	6	1
3	780	2-B-5-R-8-12-U-15-20	6	2

It is worth noting the importance of the information about the number of changes of modes of transport to be made along a path. In this simple case, the user would probably choose the second path found, because it presents a total travel time very close to the optimum one, and a number of changes of modes less than the one requested by the best path determined.

5 CONCLUSIONS

The problems related to the integration of public and private means of transport in urban areas have been dealt with in this paper. The resulting intermodal transportation network has been modelled as a discrete event system, whose behaviour has been studied by means of a special-purpose discrete event simulator. Such a simulator includes two major modules: a traffic simulation kernel and a passenger information service, which perform its main functions.

In particular, the paper was focused on the functioning of the on-line information service included in the simulator, which is able to provide updated information about the best intermodal paths to reach any destination in the network. Such best paths are time-varying as they follow the system dynamics, i.e., the traffic conditions. The heuristic algorithm used to find the best paths, including not only public modes of transport but also private ones, is based on the Dijkstra algorithm and the method proposed by Yen. Examples of application of this special algorithm to find the best multimodal paths have been presented and commented on.

The usefulness of the proposed approach can be improved. In fact, whenever the passenger information system receives a request for a best multimodal path from a user, it provides this path based only on the present state of the transportation network. Actually, this yields good results only if the network is not too big, and the traffic has slow dynamics. Work is in progress to extend the proper applicability of the proposed heuristic algorithm by finding a good technique to estimate the future values of the variables under concern.

REFERENCES

Banks, J., and J.S. Carson (1984) *Discrete-Event System Simulation*. Prentice-Hall, Englewoods Cliffs, NJ.

Cassandras, C. G. (1993) *Discrete Event Systems*. Irwin and Aksen Ass., Boston, MA.

Christofides, N. (1975) *Graph Theory*. Academic Press, London, UK.

Di Febbraro, A., V. Recagno, and S. Sacone (1994) Discrete Event Simulation of an Intermodal Transportation System. *Proc. 1994 European Simulation Multiconference*, Barcelona, Spain 1100-1104.

Di Febbraro, A., V. Recagno, and S. Sacone (1995) INTRANET : a new Simulation Tool for Intermodal Transportation Systems, to appear in *Simulation Practice and Theory*.

Fernandez, E., J. De Cea, M. Florian, and E. Cabrera (1994) Network equilibrium models with combined modes, *Transportation Science* **28** , 182-192.

Gedeon, C., M. Florian, and T.G. Crainic (1993) Determining origin-destination matrices and optimal multiproduct flows for freight transportation over multimodal networks, *Transportation Research, Part B* **27B**, 351-368.

Yen, J.Y. (1971) Finding the k-shortest, loopless paths in a network. *Man. Sci.*, **17**, 712.

73

Arc routing for rural Irish networks

P. Keenan
University College Dublin
Department of Management Information Systems, Belfield, Dublin 4, Ireland. Tel : +353 1 706-8130.
e-mail: **pkeenan@irlearn.ucd.ie**

M. Naughton
University Software Systems
3 Terminus Mills, Clonskeagh, Dublin 14, Ireland.

Abstract

This paper discusses the capacitated arc routing problem in the context of large sparse networks drawn from sections of the Irish rural road network. The particular features of this form of arc routing are discussed. Various arc routing techniques are reviewed in this context. Two heuristics, a route first/cluster second heuristic and a clustering heuristic based on shortest path trees are proposed. These heuristics are shown to provide efficient routes and the clustering heuristic to provide routes which could form the basis of routes for practical use.

Keywords

Arc routing, heuristic, Chinese Postman

1 ARC ROUTING PROBLEMS

Arc Routing Problems are an important class of problems with many real world applications. This class of problems is defined on a graph $G=(V,A)$ where V is the vertex set and A is the arc set. Associated with the graph G is a non negative cost matrix C_{ij} giving the weight of the arc from vertex i to vertex j. The simplest form of arc routing problem is the Chinese Postman Problem(CPP). The CPP involves finding a minimum cost cycle that traverses each arc in A at least once. If in a graph G a tour exists which traverses all edges exactly once then the graph is Eulerian. A single tour which traverses such a graph is an Euler tour. The fundamental condition for an Eulerian graph is that every vertex be of even degree. If this is not the case a minimum cost perfect matching algorithm (MCPM) can be used to add additional arcs between vertices of odd degree. For undirected graphs, the CPP can be solved in polynomial time. A number of variations on the basic CPP exist, a recent review of work in this area can be found in Eiselt, Gendreau and Laporte (1995).

The rural postman problem (RPP) is an extension of the CPP to a case where it is only necessary to service a subset of the arcs in a network. The capacitated arc routing problem (CARP) is an extension of the CPP to a situation where a number of vehicles of limited capacity are used to service the arc network. Each arc has a non-negative weight q_{ij} and each vehicle a capacity of W. The restriction on vehicle capacity may be distance travelled, duration of route or quantity of goods carried. Arc routing problems may be considered for cases where the arcs are directed or undirected, or a mixture of both. The RPP and CARP are reviewed in Eiselt, Gendreau and Laporte (1994).

As capacitated arc problems are NP hard a variety of heuristics have been developed. While the objective of a CARP optimisation is usually to minimise distance there are a number of other objectives that may be present in a practical problem. These include balancing the workload on different routes, not only in terms of time but also in terms of number of customers serviced or distance travelled. The shape of the routes may also be an important factor in determining customer or driver acceptance of proposed routes. Time windows may be required, these may be relatively fixed or flexible. The presence of these additional constraints has influenced the design of heuristics and has meant that, as Eiselt et al point out, research in arc routing problems is relatively fragmented. The diverse nature of arc based problems, from postal delivery to refuse collection, means that heuristics suitable in one application are often unsuitable in other contexts.

2 IRISH RURAL ROUTING

The problem presented in this paper is routing on Irish rural road networks, for example the delivery of letters. However the principles could also be applied to electricity meter reading, election canvassing, or other applications where all residents had to be visited.

By European standards Ireland has a substantial rural population which is dispersed over an extensive network of rural roads. The vast majority of these road segments have one or more houses. Therefore for problems such as postal delivery where every house must be visited almost all roads will have to be serviced. Since the roads to be serviced will generally form a connected network, the problem of visiting all houses in a given area will be a capacitated Chinese postman problem (CCPP). However in some sparsely populated districts the arcs to be visited might not be initially connected and a capacitated rural postman problem (CRPP) is encountered.

The nature of Irish rural road networks is such that for many problems the service time for each arc is a small proportion of the capacity of a vehicle. A postman working a seven hour day in a van could service more than 200 road segments in some cases. A typical rural area serviced by a number of vehicles from the nearest town might have more than 1000 arcs to be serviced. The rural road network gives rise to a graph that is extremely sparse compared to the networks frequently encountered in routing literature. Typically the arc/vertex ratio is between 1.1 and 1.4.

However each road segment need only be serviced once as it is possible to service customers on both sides of the road. The rural nature of the problem means that many difficulties found in urban areas do not arise. These include problems with one way streets, no right or left turns, traffic restrictions etc. Many urban applications are based on pedestrian routes, these introduce problems such as delays in crossing the road (Roy and Rousseau 1989). In most urban applications both sides of the road are visited separately, effectively generating two arcs for each road in the network. Other urban applications use vehicles, but the actual service is based on walking tours around the parked vehicle. The problem then becomes a two phase one as the locations where the vehicle is to be parked must also be identified (Bodin and Levy 1991). Other arc routing problems use vehicles in applications such as refuse collection or road gritting where vehicle capacity is an important constraint (Eglese and Li 1992). In comparison to these situations, this paper address a problem which is relatively free of constraints in comparison to other real world problems. However the size of the problems encountered is larger and the networks are very sparse when compared to most examples in the literature.

3 ALGORITHMIC APPROACHES TO THE PROBLEM

The solution of arc routing problems has received less attention in the past than the more common node routing problems. However as for other types of NP hard vehicle routing problem, various heuristic based solution procedures have been proposed. These techniques can be grouped into those where clustering and routing takes place in a single step and those where two separate steps are used for establishing routes and the clusters to be serviced by each vehicle.

Lower Bounds for CARP

A number of lower bounds for the CARP have been proposed. These bounds require the solution of a matching algorithm on a modified graph H derived from G. Early bounds were developed by Golden and Wong(1981), Assad, Pearn and Golden(1987) and Pearn(1988). The more recent bounds are those of Surawatari, Hirabayashi and Nishida (1992) and Benavent, Campos, Corberán and Mota (1992). The latter authors review all previous bounds and propose a number of improvements to them. Hirabayashi, Suruwatari, and Nishida (1992) propose a branch and bound algorithm based on their bound that provided an optimal solution for the CARP. While this algorithm allowed optimal solutions it was tested only on networks with less than 50 arcs, owing to its slow execution time. As we will demonstrate later in this paper this algorithm in its basic form is too slow for use on large Irish rural networks.

Single Pass Heuristics

A variety of single pass heuristic algorithms have been proposed. These have been tested on relatively small networks where the service time for each arc is a large proportion of the vehicle capacity. Pearn (1989 and 1991) provides a review of these algorithms. For dense graphs Pearn proposes a modification of the construct strike algorithm proposed by Christofides in 1973. This procedure is based on the repeated identification and removal of cycles from the graph with the addition of additional artificial arcs where required. Pearn found that this procedure performed well on dense graphs, but not on sparse graphs. The Irish rural networks are very much more sparse than the networks used by Pearn. In 1991 Pearn proposed augment-insert algorithms for sparse networks with high demand ratios per arc. These algorithms aimed to identify cycles where the sum of demands on the arcs in the cycle exceeded the capacity of the vehicle. This does not arise in the Irish rural situation where the demand on each arc is a small proportion of the capacity of the vehicle.

While these and other single pass algorithms could be modified to cater for larger and sparser networks, these modifications would largely mean the design of a completely new algorithm. We would argue that development of these existing techniques is not likely to provide effective routes in an Irish rural context.

An alternative heuristic approach is to separate the clustering and routing of the clusters. This two phase approach has provided good heuristic solutions to the related point based vehicle routing problem. The separation of routing and clustering in CARP may be even more attractive given that optimal CPP routes can be generated within each cluster. Furthermore a separate clustering procedure may allow consideration of constraints other than distance minimisation. These include the generation of 'logical' routes, where the routes use by services such as postal delivery can be justified to the public. Postal delivery problems also require the sorting of letters and this can be more easily done if adjacent addresses are on the same route.

Route first, cluster second heuristics.

These heuristics generate a large route covering the entire network, which can then be broken up into clusters. We test a heuristic based the generation of a large Euler tour on a matched graph. Within a matched graph a number of possible Euler tours exist, we generate a large tour which passes close to the depot at the point where a vehicle's capacity would be exhausted.

Cluster First Route Second Heuristics

In point based vehicle routing problems a variety of strategies exist to cluster points before routing the clusters. In the arc routing domain a clustering algorithm has to identify a group of arcs that is likely to form a good route when the routing phase is complete. In a practical problem the clusters generated should also be acceptable, this generally entails adjacent arcs being visited on the same route. In the point based vehicle routing problem many successful procedures exist to generate routes which radiate from the depot in some type of cone shape. These group neighbouring point together and then use a travelling salesman algorithm to route within that cluster of points. A similar approach, grouping arcs, would give rise to reasonable CARP clusters. In the context of Irish rural networks we have experimented with the generation of arc based routes using shortest path trees rooted at the depot. This strategy provides radial routes which ensure that neighbouring arcs are on the same route. These clusters can be generated using the original network and within each cluster a matching algorithm can be used to generate an Euler tour.

4 COMPUTATIONAL RESULTS

Test networks

The networks used in this paper are derived from rural Irish road networks. Each section of the network has a travel time and a service time, which is representative of typical routing applications. In order to facilitate the operation of the algorithms a network simplification routine was employed to derive networks small enough to be solved optimally. This procedure eliminated unnecessary detail, including the removal of dead end roads and the inclusion of the time taken to service that arc in the section of road to which the dead end arc connects. This network simplification does not greatly affect the operation of the algorithms, but it reduces the size of the problem by up to two-thirds.

Optimal solutions and lower bounds

We attempted to find an optimal solution to the problem using the branch and bound algorithm by Hirabayashi, Suruwatari, and Nishida (1992). This algorithm proves to be unsuitable for large problems owing to the large number of sub-problems generated. We made a number of modifications to the branch and bound algorithm. We needed to modify it to take account of the fact that we used time, rather than distance, as our vehicle constraint. The most significant alteration to the operation of the algorithm was the use of an alternative branching strategy. In the initial algorithm a number of matchings are performed and the sub-problem with the highest bound selected. While this is an excellent strategy in terms of selecting a good branching arc it is computationally inefficient owing to the complexity of the MCPM algorithm, which might be solved *n/2* times.

An alternative strategy is to select the arc with the smallest weight as the branching arc. This alteration greatly speeds up the selection of the branching arc, although this may mean that more subproblems are generated. The original branching strategy lead to very long solution times in some cases. The use of the smallest arc as a branching strategy lead to solutions whose solution time varied much less. Therefore in some cases where the original strategy performed poorly it required solution times ten times those using the selection of the smallest arc strategy The solution time for the new branching procedure was less than half of the alternative in most cases. In some cases where the original strategy performed well the solution times were similar. A more complete discussion of these alterations may be found in Montwill and Naughton (1994). A disadvantage of the generation of an increased number of subproblems, in a practical procedure, is that the memory of the computer used is often an important limitation on the operation of this branch and bound algorithm.

In practice the branch and bound algorithm, including our amendments, remains unsuitable for large problems. A Pentium 100 MHz microcomputer with 16MB of memory was used for these computations, for problems larger than 70 arcs the number of subproblems generated meant that the algorithm ran out of memory. However in all cases a lower bound was available, which provided a basis for comparison for the heuristics used. This bound, derived by Surawatari, Hirabayashi and Nishida (1992) is not the best available, but is close to it Benavent, Campos, Corberán and Mota (1992).

Route First/ Cluster Second Heuristic

This approach generates a large Euler tour, which is then split up into routes with the maximum vehicle time. Initially a matching algorithm is used to make the network Eulerian. Within this matched network a large number of possible Euler tour exists. However the large tour generated needs to be suitable for splitting into smaller tours. Therefore a number of heuristics are used when building the large tour to ensure that suitable routes result from the splitting of this large tour.

Phase one of the building the Euler tour adds arcs, adding demand arcs where possible, generally moving in a direction away from the depot. This phase continues until approximately half of the route time of a vehicle has been allocated. In the second phase of the algorithm we attempt to keep the algorithm in the region of the most distant point routed in phase one. This attempts to ensure that routes are reasonably compact. A third phase begins when the route has completed more than 75% of the route time for a single vehicle. In this phase the large route makes it way back towards the depot. When the route time for a single vehicle is exhausted, the algorithm goes back to phase one. At the end of the routing process, when the large route has returned to the depot, checks are made to ensure that all sections of the network have been routed. If the large route returns prematurely to the depot, the addition of these unrouted arcs may affect quality of the routes generated.

This strategy should ensure that the routes are fairly compact and that the split points on the route occur close to the depot. By and large this approach provided reasonably efficient routes, but despite our fine tuning of the arc selection rules used, it proved difficult to ensure that the route stayed in the same region of the graph.

Cluster First Heuristic

This approach generated shortest path trees rooted at the depot. Sub trees whose traversal time is less than half the capacity of a vehicle are identified. Neighbouring subgroups are then combined to derive routes whose estimated travel time was close to the capacity of the vehicles. An Euler tour was then generated for each of these clusters using an MCPM algorithm. The test results present the initial clustering, further improvements are possible using an improvement algorithm to refine the boundaries of routes.

Table 1 50 arc network, total arc distance 15.32 km

Vehicle Capacity (minutes)	Number of vehicles	Lower Bound (km)	Optimal Solution	Route First/ Cluster second	Cluster First
300	2	17.95	17.95	17.95	18.19
240	2	17.95	17.95	19.43	18.19
150	3	18.39	*	21.17	21.86
120	3	18.39	*	20.03	20.98
100	4	19.01	*	23.15	22.57

Table 2 65 arc network, total arc distance 17.54 km

Vehicle Capacity (minutes)	Number of vehicles	Lower Bound (km)	Optimal Solution	Route First/ Cluster second	Cluster First
360	2	22.50	22.50	24.76	23.28
210	2	22.50	22.50	23.38	23.35
180	3	22.50	*	23.38	23.32
150	3	22.50	*	27.32	25.20

* owing to machine limitations these problems could not be solved optimally

Table 3 Large areas, e.g. complete postal delivery regions

Area	Number of arcs	Number of Vehicles	Lower bound (Km)	Route first (Km)	Cluster first (Km)
1	649	4	269.7	288.4	292.8
2	600	3	307.1	330.2	322.5
3	483	3	168.7	183.7	184.2
4	362	3	203.8	224.8	213.8

5 CONCLUSIONS

Analysis of test results

While the cluster first and route first heuristics have a similar performance, the cluster first heuristic provides superior routes when considerations such as geographical compactness are considered. The use of the shortest path based clustering approach provides routes which are practically useful without any great penalty in terms of distance travelled. These clusters can be derived with a negligible solution time on a modern PC, indicating their suitability for use in practical software for this problem.

This paper presents an initial examination of techniques for solving the capacitated arc routing problem on large sparse networks. These initial results indicate that the use of shortest path trees rooted at the depot can provide a good basis for clustering of routes. These clusters are not only efficient in terms of the distance of the routes provided, but also substantially satisfy the need for practical routes to be geographically compact. While other approaches may provide marginally superior solutions in terms of distance travelled, these may not ensure that routes meet the other requirements of practical routing.

Further work

The heuristic cluster could be further improved by using an Eulerian graph and explicitly taking account of the additional arc traversals needed. A variety of improvement routines could be used to swap arcs at the boundary between routes, both to improve the overall travel time and to make the routes more geographically compact. These would provide routes that were close to optimal and practically feasible, without requiring excessive program execution time.

While the branch and bound techniques tested proved too slow for practical use the possibility exists of improving these. The algorithm could make use of the superior lower bounds proposed by Benevant et al (1992). The heuristic techniques could be used to provide good initial upper bounds. Further examination of the branching strategies should lead to faster solution times. A combination of all of these might allow larger problems be solved optimally in a reasonable period of time. Even if optimality cannot be achieved, a procedure such as branch and bound could move much closer to the optimal solution in a shorter solution time. It seems likely, however, that heuristic techniques will continue to have an important role to play in solving problems on large networks.

REFERENCES

Assad, Pearn and Golden (1987) The Capacitated Chinese Postman Problem: Lower bounds and solvable cases. *American Journal of Mathematical and Management Science*, **7**, 63-88.

Benavent, E., Campos, V., Corberán, A. and Mota, E. (1992) The Capacitated Arc Routing Problem. Lower Bounds. *Networks*, **22**, 669-690.

Bodin, L.D. and Levy, L. (1991) The Arc Partitioning Problem. *European Journal of Operational Research*, **53**, 393-401.

Eglese, R.W. and Li L.Y.O. (1992) Efficient Routeing for Winter Gritting. *Journal of the Operational Research Society*, **43**, 1031-1043.

Eiselt, Gendreau and Laporte (1995) Arc Routing Problems : part I: The Chinese Postman Problem, *Operations Research*, **43**, 2, 231-242.

Eiselt, Gendreau and Laporte (1994), Arc Routing Problems. Part II : The Rural Postman Problem. Working Paper CRT-961, Centre de Recherche sur les Transports, Université de Montréal. (also forthcoming in *Operations Research*)

Golden and Wong (1981) Capacitated arc routing problems, *Networks*, **11**, 305-315.

Hirabayashi, R., Suruwatari, Y., and Nishida, N. (1992) Tour Construction Algorithm for the Capacitated Arc Routing Problems. *Asia-Pacific Journal of Operational Research*, **9**, 155-175.

Montwill, P. and Naughton, M. (1994) The Capacitated Chinese Chinese Postman Problem, MMangtSc disseration, University College Dublin.

Pearn, W.-L. (1991) Augment-Insert Algorithms for the Capacitated Arc Routing Problem. *Computers and Operations Research*, **18**, 189-198.

Pearn, W.-L. (1989) Approximate solutions for the Capacitated Arc Routing Problem.. *Computers and Operations Research*, **16**, 598-600.

Pearn, W.-L. (1988) New lower bounds for the capacitated arc routing problem. *Networks*, **18**, 181-191.

Roy, S. and Rousseau, J.-M. (1989) The Capacitated Canadian Postman Problem. *INFOR*, **27**, 58-73.

Surawatari, Y., Hirabayashi, R. and Nishida, N. (1992) Node Duplication Lower Bounds for the Capacitated Arc Routing Problem. *Journal of the Operations Research Society of Japan*, **35**, 2, 119-133.

Arc routing vehicle routing problems with vehicle/site dependencies

J. Sniezek and L. Bodin
College of Business and Management
University of Maryland
College Park, MD 20742, U.S.A.
Tel: (301) 405-2210, (301) 596-3100.
Fax: (410) 964-2688.

Abstract

In this paper, an algorithm is discussed for solving the Vehicle Arc Routing Problem when there exists the constraints of Vehicle/Site Dependencies. The traditional Capacitated Arc Routing Problem assumes a homogeneous fleet mix of vehicles available to service the arcs in the network. In essence, any vehicle in the vehicle fleet mix can travel or service any of the arcs in the network and all vehicles in the vehicle fleet mix have the same capacity. In actuality, most real world applications of the Capacitated Arc Routing Problems do not use a homogeneous fleet mix. Solid waste collection tends to utilize a heterogeneous rather than a homogeneous fleet mix. A city may have many different sized trash vehicles which pick up the garbage on the streets. As a result of the vehicles being of different sizes, the capacities of the vehicles will be different. More important is that the different size vehicles may not be able to travel on the same streets. For example, larger vehicles may not be able to travel on small alleys in the middle of the city. This second issue, where certain vehicles can not travel on certain streets is what is referred to as Vehicle/Site Dependencies. This paper analyzes and presents a practical solution to the Capacitated Arc Routing Problem while satisfying Vehicle/Site Dependencies with a heterogeneous fleet mix.

1 INTRODUCTION

Residential sanitation problems can be considered a special case of arc routing and partitioning problems. These problems are named neighborhood (or saturated) routing and scheduling problems. In a neighborhood routing and scheduling problem, most of the street segments in a geographic region require service. There is much literature on solving a neighborhood routing and scheduling problem over a street network. The **RouteSmart™** routing and scheduling system for scheduling sanitation vehicles was developed by Bowne Distinct LTD of Columbia, MD, USA to solve neighborhood routing and scheduling problems over a street network. Several sanitation companies in the United States are using **RouteSmart™** for scheduling their sanitation vehicles. Traditionally, the solution of a neighborhood routing and scheduling problem assumes a homogeneous set of vehicles and a homogeneous street network used for travel. All vehicles are assumed to have the same capacity and all vehicles can travel on all streets in the network.

An additional complication many sanitation companies encounter is the problem of Vehicle/Site Dependencies. In this complication there exists a fleet of vehicles where each vehicles may be different in terms of size, capacity, etc. In addition, many of the vehicles in the fleet are restricted, due to size or weight constraints, from traveling on or servicing certain streets in the network. In the neighborhood routing problem for sanitation vehicles, a Vehicle/Site Dependency is when a particular street segment can only be handled by a particular subset of the available vehicles. We have developed an algorithm to handle Vehicle/Site Dependencies and this algorithm has been implemented within **RouteSmart™**.

The traditional neighborhood routing and scheduling problem is generally solved in three steps. In Step 1, a workload estimation procedure is used to determine either the number of vehicles required to service the area under consideration, given a target time for each vehicle, or a vehicle target time is determined, given a specified number of vehicles. In Step 2, the area over which the routing is to be performed is broken down in partitions where each partition represents the region to be serviced by a particular vehicle. In Step 3, an Euler circuit problem is solved over each of the partitions to give a minimum deadhead travel time path that services all of the required streets in the partition.

For sanitation scheduling, this problem can have the following complications:

1. The vehicle can become filled so that trips to the recycling facility, transshipment point or landfill could be required.
2. Some of the streets in a partition can be directed or 1-way; other streets in the partition can be bi-directional so that two traversals of the street segment become necessary; other streets in the partition can be meandered so that only one traversal of the street segment is required but this traversal can begin at either one of the intersections and end at the other intersection. As such, the determination of the travel path is significantly more complicated than the traditional directed or undirected Euler circuit problem.
3. In both Steps 2 and 3, the service network, which is the network consisting of the set of street segments to be serviced, can be disconnected. Therefore, the partitioning and travel path algorithms must be implemented considering the fact that the underlying service network can be disconnected.
4. Turn restrictions can exist at specific intersections. For example, large garbage trucks do not make U-turns unless absolutely necessary. The need to restrict U-turns and other undesirable travel patterns complicates the minimum deadhead Euler circuit problem of step 3.

In addition to the above complications, an additional complication now being addressed by **RouteSmart™** is the constraint of Vehicle/Site Dependencies. The vehicles being used may differ in terms of size and capacity. Also, each particular street segment can only be serviced by a particular subset of the vehicles in the fleet mix. These constraints can occur because certain streets may be too narrow for certain vehicles to drive on or there may be a constraint on the weight that the street segment can handle. Thus, only vehicles in a vehicle type that meets certain weight/size restrictions can be used to service a street segment with a Vehicle/Site Dependency constraint.

A particular case of the Vehicle/Site Dependency problem is the arc routing problem with multiple vehicle types. In this problem, there is a non-homogeneous fleet mix such that the vehicles may have different capacities, but there are no Vehicle/Site Dependencies on any of the street segments. All street segments can be serviced by all of the vehicle types. The analysis being presented here applies to both the multiple vehicle type arc routing problem with Vehicle/Site Dependencies and the arc routing problem without Vehicle/Site Dependencies.

To solve this neighborhood routing problem with Vehicle/Site Dependencies adds significant complications to Steps 1 and 2 above. We have developed an algorithm, and have implemented this algorithm within **RouteSmart**™.

2 DESCRIPTION OF THE VEHICLE/SITE DEPENDENCY ROUTING PROBLEM

The Vehicle/Site Dependency routing problem being studied starts with a street network composed of individual street segments. A subset of the street segments in the network must be serviced by a vehicle from a vehicle fleet mix. Each vehicle in the vehicle fleet mix belongs to a specific vehicle class. Each street segment in the network can have restrictions as to which vehicle classes can service the street segment and which vehicle classes can drive the street segment without servicing it. Each vehicle in a given vehicle class will have a certain vehicle capacity. If this capacity is reached, a trip to a dump site is required.

The street network is represented as a set of arcs. The objective is to create subsets of the arcs, called partitions, where each partition is assigned to a vehicle in the vehicle fleet mix. An Euler circuit is then formed for each of the partitions. Every street segment which requires service must be in one of the partitions. Each partition and corresponding circuit represents the set of streets each vehicle is responsible for servicing and the line of travel the vehicle will follow when servicing those streets.

2.1 Attributes of the Arcs representing the Street Network

Each street segment in the street network is represented as an arc. The following attributes must be associated with each arc in the set of arcs to be partitioned.

1. Whether or not the arc requires service. Not all streets in a city have trash to pick up and thus not all arcs in the set of arcs to be partitioned require service. Those arcs which do not require service may still be used to travel between arcs which do require service.

2. The volume or weight associated with the arc, assuming the arc requires service. This volume or weight will represent the expected amount of trash to be picked up on that street segment. This metric is used to keep track of the total volume or weight a vehicle has picked up. This is required to account for the time associated with traveling to a dump or transfer site when the vehicle has filled itself to capacity.
3. The time it takes to service the arc assuming service is required.
4. The time it takes to travel the arc when not servicing the arc.
5. Every arc, whether it requires service or not, must contain a list of all vehicle classes that can travel on the arc when not servicing the arc. Some streets are too small for certain vehicle classes to travel on and thus the corresponding arcs can not be traveled on by any vehicles in those vehicle classes.
6. Every arc requiring service must contain a list of all vehicle classes that can service the arc. It is assumed that for every arc, this list is a subset of the list of vehicle classes that can travel the arc without servicing it.
7. The type of service associated with each service arc. The service required may be only on the right side of the street, only on the left side or on both sides of the street.
8. Is the arc directed or undirected. If a street can have the trash picked up on both sides with a single traversal of the street (meander) then the corresponding service arc will be undirected. If the street can not be meandered, then the corresponding service arcs (one for each side of the street) will be directed to reflect traffic orientation.

2.2 Attributes of the Vehicle Class

The following attributes must be associated with each vehicle class.

1. The total number of vehicles available in the vehicle class. This serves as an upper bound for the number of vehicles from each vehicle class that can be utilized.
2. The maximum weight or volume that a vehicle in the class can hold. Once a vehicle in the vehicle class has collected this maximum weight or volume, it is required to make a trip to a dump site before continuing service.
3. The target length of workday for that vehicle class plus a soft upper and lower bound for the target time.
4. All overhead time associated with the vehicle class.
5. The dumptime associated with the vehicle class. This time is the time it takes to empty the vehicle once its capacity has been reached.

2.3 Other Constraints

Other constraints associated with the Vehicle/Site Dependency Routing Problem are as follows:

1. A priority or preference list of vehicle classes is required. This will list in descending order of priority, a vehicle class, and the minimum and maximum number of vehicles to use from the vehicle class at that priority level. For example, the smallest

vehicles may be undesirable because they require several trips to the dump site. The smaller vehicles could be placed at the bottom of the preference list and thus would only be used if required due to Vehicle/Site Dependent constraints on the streets. A vehicle class may appear more than once in the preference list.
2. A single depot location is used as the starting and ending point for all vehicles.
3. A single landfill site is used where all vehicles must go when their capacity has been reached in order to empty the vehicle.
4. All vehicles must return to the landfill at the end of the route to empty the vehicle.

3 THE ALGORITHM

An algorithm has been developed which provides a practical solution to the Vehicle Arc Routing Problem while satisfying the complications of Vehicle/Site Dependencies with a non-homogeneous fleet mix. This algorithm has been implemented using **RouteSmart**™ and is currently being used by a large sanitation company in the United States. The steps of the algorithm have been broken into 3 sections. Each section corresponds to the three steps generally used to solve the Vehicle Arc Routing Problem without Vehicle/Site Dependencies, as discussed in the introduction of the paper. The steps of the algorithm used to solve the Vehicle Arc Routing Problem with Vehicle/Site Dependencies are as follows:

1a) Determine an initial fleet mix estimate. The fleet mix is the number of vehicles from each vehicle class which will be used to solve the problem. This fleet mix is determined based upon the given fleet mix preference list, the total service time associated with all street segments requiring service, the total volume associated with all street segments requiring service, the relative connectivity of the street segments requiring service, and the time constraints on route length and dump time. This estimate will become the actual fleet mix for the first iteration of the algorithm.

1b) Order all vehicles in the fleet mix to be used based upon their vehicle class. The ordering is done from the least restrictive to the most restrictive vehicle class.

2a) At this point, only consider the street segments which have yet to be assigned to a partition and which must be serviced by a vehicle class no larger than the i'th least restrictive vehicle class in the fleet mix. (i = the number of times we have reached step 2a in this iteration) Thus the first time step 2a is reached, only consider the streets which must be serviced by the smallest (least restrictive) vehicle class in the fleet mix. Based upon these street segments, seed a subset of the feasible vehicle classes' remaining vehicles.

2b) Assign each of the unassigned street segments considered in 2a to one of the vehicles seeded in 2a. Partition balance and geographic compactness is considered during the assignment of street segments to partitions.

2c) Now considering all remaining unassigned street segments continue assigning street segments to the partitions formed in 2b until the partitions reach a certain percentage of maximum route time. Again, partition balance and geographic compactness is considered during the assignment of street segments to partitions.

2d) If there are still unassigned street segments which can not be serviced by each and every unassigned vehicle classes in the fleet mix, then return to step 2a.
2e) Seed all remaining vehicles in the fleet mix and then assign all street segments still not assigned to a partition to any feasible partition while still taking route balance and geographic compactness into consideration.
2f) Perform swapping of service streets between the grown partitions in an attempt to balance the total route time for each partition. At this point, the total route times are only estimates, as the actual travel path has yet to be solved for each partition.

3a) Solve a Chinese Postman Problem for each partition to determine actual route times for each partition. If all partitions are balanced within an upper and lower bound, then output the travel paths and statistics for each partition. The problem is finished.
3b) If the partitions are out of balance, perform a revision of workload estimation based upon the out of balance partitions that currently exist. It is possible to change the fleet mix at this point. For example, it may be determined that due to Vehicle/Site Dependencies, an additional vehicle from the smallest vehicle class may be required.
3c) Return to step 2a for a new iteration using the revised fleet mix determined in 3b.

4 CONCLUSIONS

The algorithm presented in this paper is a modification of the traditional methods for solving Vehicle Arc Routing Problems in order to handle the real-world constraints of Vehicle/Site Dependencies. This algorithm has been implemented in RouteSmart and is currently being used by a large sanitation company in the United States. The problems being solved using this algorithm generally contain 3 different vehicle classes and require anywhere from 10-20 vehicles per solution. The problems are being solved over a street network of approximately 20,000 street segments.

There is still much work that needs to be done in the area of Vehicle/Site Dependencies. The solution approach discussed in this paper is a first attempt at handling Vehicle/Site Dependencies in an Arc Routing Vehicle Routing problem.

REFERENCES

There are numerous papers written on Vehicle Arc Routing and Scheduling without the consideration of Vehicle/Site Dependencies. Three excellent references are the following:

1. Bodin, L., B. L. Golden, A. Assad, and M. Ball (1983), "Routing and Scheduling of Vehicles and Crews: The State of the Art". Computers and Operations Research, 10(2), 63-211.

2. Golden, B. L. and A. Assad, eds. (1986). Special Issue on Time Windows. American Journal of Mathematical and Management Analysis, 6 (3 and 4), 251-399.

3. Golden, B. L. and A. Assad (1988). Vehicle Routing: Methods and Studies. North-Holland, Amsterdam.

To the best of our knowledge, there are no known papers addressing Vehicle Arc Routing and Scheduling which consider Vehicle/Site Dependencies. There is the following reference on Vehicle Node Routing with Vehicle/Site Dependencies:

4. Nag, B., B. L. Golden, and A. Assad 1988, "Vehicle Routing With Site Dependencies", Vehicle Routing: Methods and Studies, Ed. by B. Golden and A. Assad, 149-159.

INDEX OF CONTRIBUTORS

Abramovich, K. 497
Al Ani, T. 567
Albertini, F. 155

Bertocchi, M 574
Bertazzi, L. 535
Bianco, L. 543
Bodin, L. 607
Boggs, P.T. 3
Brännlund, U. 387
Bronisz, P. 300

Chernousko, F.L. 13
Casas, E. 187
Castro, J. 451

D'Alessandro, D. 155
Dai Pra, P. 163
Dell'Olmo, P. 543
Di Febbraro, A. 591
Di Marco, S.C. 285
Doležal, J. 325
Dolgui, A. 519
Döring, U. 279
Dourado Correia, A. 227
Duarte-Ramos, H. 227
Düchting, W. 309
Dupačová, J. 574
Dzemyda, G. 317

Ehtamo, H. 435

Gil, P. 227
Ginsberg, T. 309
Giordani, S. 543
Glück, R. 137
Goh, C.J. 442
Gonzáles, R.L.V. 285
Granat, J. 363

Hämäläinen, R.P. 435
Hamam, Y. 567
Han, Z. 107
Henry, J. 195
Hofer, E.P. 115
Hraba, T. 325

Iri, M. 24
Ishii, H. 273
Itoh, T. 273

Juhás, G. 147

Kabziński, J. 91
Kang, H.S. 292
Kaňková, V. 582
Karbuz, S. 259
Kearsley, A.J. 3
Keenan, P. 599
Kirjner-Neto, C. 235
Kiwiel, K.C. 387,459, 466
Kocian, M. 147
Kohlas, J. 37
Kręglewski, T. 363
Kruś, L. 300
Kučera, V. 54

Leblond, J. 99
Leitmann, G. 64
Lemaréchal, C. 395
Lepp, R. 411
Lindberg, P.O. 459
Locatelli, M. 473
Lopes, F.B. 481
Lopuch, B. 466
Lorena, L.A.N. 481

Marti, K. 75
Mäkelä, M.M. 379
Mäkinen, R.A.E. 379
Makowski, M. 371
Malafeev, O.A. 243
Malanowski, K. 419
Maublanc, J. 489
Miettinen, K. 379
Minciardi, R. 511
Mulholland, M. 251
Myslinski, A. 218

Nabona, N. 451
Nakashige, R. 137
Narotam, N.K. 251
Naughton, M. 599
Neck, R. 259
Nitka-Styczeń, K. 427
Nõu, A. 459
Novozhilova, M.V. 559

Olivi, M. 99
Outrata, J.V. 203

Paczyński, J. 363
Palucka, K. 331
Pellegrino, F. 395
Piekarski, J. 218
Polak, E. 235
Porwit, A. 331
Portmann, M.C. 519

Profumo, M. 511
Proth, J.M. 519

Quilliot, A. 489

Raivio, T. 435
Raymond, J.P. 211
Redfern, D.A. 442
Renaud, A. 395
Ross, F. 279
Rousselet, B. 218
Rudari, C. 163

Sacone, S. 511, 591
Sadeh, A. 267
Sagastizábal, C. 395
Sakamoto, T. 527
Šaltenis, V. 317
Schoen, F. 473
Semenkin, E. 497
Skulimowski, A.M.J. 171
Sladký, K. 179
Smeers, Y. 339
Sniezek, J. 607
Souissi, M. 339
Speranza, M.G. 527
Stachurski, A. 363, 403
Starke, J. 551
Stoyan, Yu.G. 559
Szczerbicki, E. 347

Tibken, B. 115
Tiešis, V. 317
Tishkovskaya, S.V. 355
Tolle, J.W. 3
Tong, S. 107
Troeva, M.S. 243
Tsoi, Ye.B. 355

Ukovich, W. 535
Ulmer, W 309
Urbaniak, A. 120

Waniewski, J. 331
Wiechert, W. 128
Wierzbicki, A.P. 363

Yvon, J.P. 195

Zhang, Z. 107
Zöchling, R. 137
Zidani, H. 211
Zhadan, V.G. 502

KEYWORD INDEX

Algebraic system theory 54
Allocation 559
Application in finance 574
Applications 24
Arc routing 599
Armijo rule 115
Aspiration-reservation-led decision support 371
Assembly manufacturing systems 519
Assignment
  problems 551
  problems of higher order 551
Automatic differentiation 115
Autonomous manufacturing agents 347

Bayesian characteristics 355
Best-first search 481
Bicriteria 275
Binary costs 543
Binding-time analysis 137
Biological system 120
Biomathematical modeling 115
Bounded measurable controls 411
Branch-and-bound 535
Bregman functions 459

Cancer 309
$CD4^{+}$ lymphocyte depletion 325
$CD8^{+}$ lymphocyte dynamics 325
Chinese Postman 599
Clustering 473
Combinational optimization 551
Complex systems 355
Computational
  benchmarks 451
  complexity 527
Computer
  algebra 128
  simulation 309
Concentrator 218, 227
Conjugate formation 331
Consistent approximations 235
Control system synthesis 54
Control-state and pure state constants 419
Controllability 13
Convex feasibility problems 466
Cutting stock 481
Cutting-stock algorithms 527

Decision
  making 292
  making support 371
  support systems 300
  trees 347
Decomposition 13, 497
Descent method 427
Deterministic control 54
Differential
  game 243
  games 13
Direct methods 435
Directional derivative 203
Discrete
  convergence 411
  event dynamic systems 591
  event system 147
  time interval 582
  -time 147
  -time control systems 171
Discretization 435
Discretization theory 235
Distributed system 251
Disturbance attentuation problem (LSDA) 107
Dominance rules 535
Dual problem 502
Duality 24
Dynamic
  games 163
  programming 171, 243, 543

Economics 259
Elliptic equations 187
Energy development 317
Enzyme kinetics 331
Evaporator 218, 227
Exact penalty function 403
Expert system 120

Feedback control 54, 442
Finite
  difference method 243
  element method 187
First order kinetics 331
Flow cytofluorometry 331
Forecasting 259
Fractionation 309
Frequency domain identification 99
Future 24
Fuzzy capacity 275
Fuzzy-controller 279

Game theory 107
Generalized Jacobian 203
Global convergence 3
Global optimization 473
Gradient projection method 502
Grouped data 355

Hardy spaces 99
Hereditary
  autonomous process 427
  shift operator 427
Hermite normal form 489
Heuristic 543, 599
Heuristics 481
Hidden Markov chain 567
Hierarchical control 218, 227
Hierarchical structure 292
History 24
HIV infection 325
HIV/AIDS infection 317

Identifiability 128
Individual time points 582
Inference 279
Information flow 347
Integer flow 275
Integer linear programming 489
Interior point methods 451
Interval matrices 179
Invariant imbedding 195
Inventory 535
Inventory control 519
Isotope labelling 128

Keywords
Knowledge acquisition 347

Lagrange–Newton method 419
Large deviations 163
Level methods 387
Linear dynamic matrix control 251
Linear
  extensions 543
  programming 451, 502,519
Linear systems 54
Linearization 403
Lipschitz property 582
Liquid–liquid extraction 251
Local optimization 527

Manufacturing
  planning 551
  systems 511
Mathematical
  model 325
  system 427
Medical diagnosis 120
Medium term planning 519
Membership function 279
Merge operation 527
Merit functions 3
Metabolic fluxes 128
Minimax problems 285
Model predictive control 251
Modelling 218, 227, 309
Multi Level Single Linkage 473
Multi-scale optimization 497
Multicommodity network flows 451
Multicriteria
  decision making 300
  decision support 317
  optimization 171, 371, 379
  optimization methods 300
Multistage stochastic programming problem 582

Natural Family Planning (NFP) 120
Network flow 24
Neural networks 347, 442, 551
Newton's method 502
Nondifferentiable
  convex optimization 403
  optimization 387
Nondominated flow pattern 275
Nonhomogeneous matrix products 179
Nonlinear
  boundary controls 211
  control 91
  control equation 411
  dynamical systems 13
  optimization 559
  ordinary differential equations 419
  programming 339, 435
  systems 155
Nonsmooth optimization 379
Numerical
  algorithms 137
  methods 218, 243
  solution 285

Objective functions 582
Observability 155
One-to-one mapping 527
Optimal
  control 13, 91, 259, 419, 442
  control of distributed parameter systems 195
  design 235
Optimality condition 218
Optimization 115, 259, 339, 511

Parameter
  estimation 115
  identification 567
Partial evaluation 137
Periodic control problem 427
Petri nets 147, 511
Piecewise-programmed strategies 243
Pontryagin's minimum principle 211
Portfolio theory 171
Posets 543
Postoptimality 574
Primal direction search problem 403
Primal-Dual algorithm 451

Probability functions 74
Production rules 347
Proximal
  bundle methods 387
  point methods 459

Quadratic assignment problems (QAP) 551
Quasi-variational
  inequalities 285
  inequality 203

Radiation therapy 309
Random supply 519
Real patient data 115
Redundancy 128
Relation 147
Relaxation methods 466
Relaxed controls 187
Reliability 339
Reliability characteristics 355
Resource management 54
Risk-sensitive control 163
Robust
  control 54, 91
  controller 107

Scenarios 574
Scheduling 551
Scientific computing 137
Self organization 551
Semilinear parabolic equations 211
Semismoothness 203
Sensitivity 128, 259, 574
Sequential quadratic programming 3, 419
Set covering problems 459
Shape optimization 218
Sharing problem 273
Side constraints 451
Side effects 309
Simulation 115, 218, 227, 317, 519
Singular controls 435
Software system 497
Space
  technology 292
  transformer 502
Spectral model 251
Stability of linear systems 179
State constraints 187, 442
Steepest descent 502
Stochastic
  algorithms 473
  dynamic programming 567
  games 292
  optimization 74
  process 567
  program 574
Structural
  analysis 74
  design 74
  optimization 379
Successive projections 466
Surrogate inequalities 466
System theory 115

T-invariants 147
Tabu search 481
Theory of cooperative games 300
Trajectory optimization 435
Transfer functions 54
Transportation 535, 591
Trust region 3
TSP 543
Tumor growth 309

Uncertain nonlinear system 107
Uncertain systems 64
Unification 497
Urban systems 591

Weighted rational approximation 99

www.ingramcontent.com/pod-product-compliance
Ingram Content Group UK Ltd.
Pitfield, Milton Keynes, MK11 3LW, UK
UKHW021900260726
13966UKWH00006B/67

* 9 7 8 1 4 7 5 7 6 6 7 0 7 *